T0406230

Organizational Structure of Science of Synthesis*

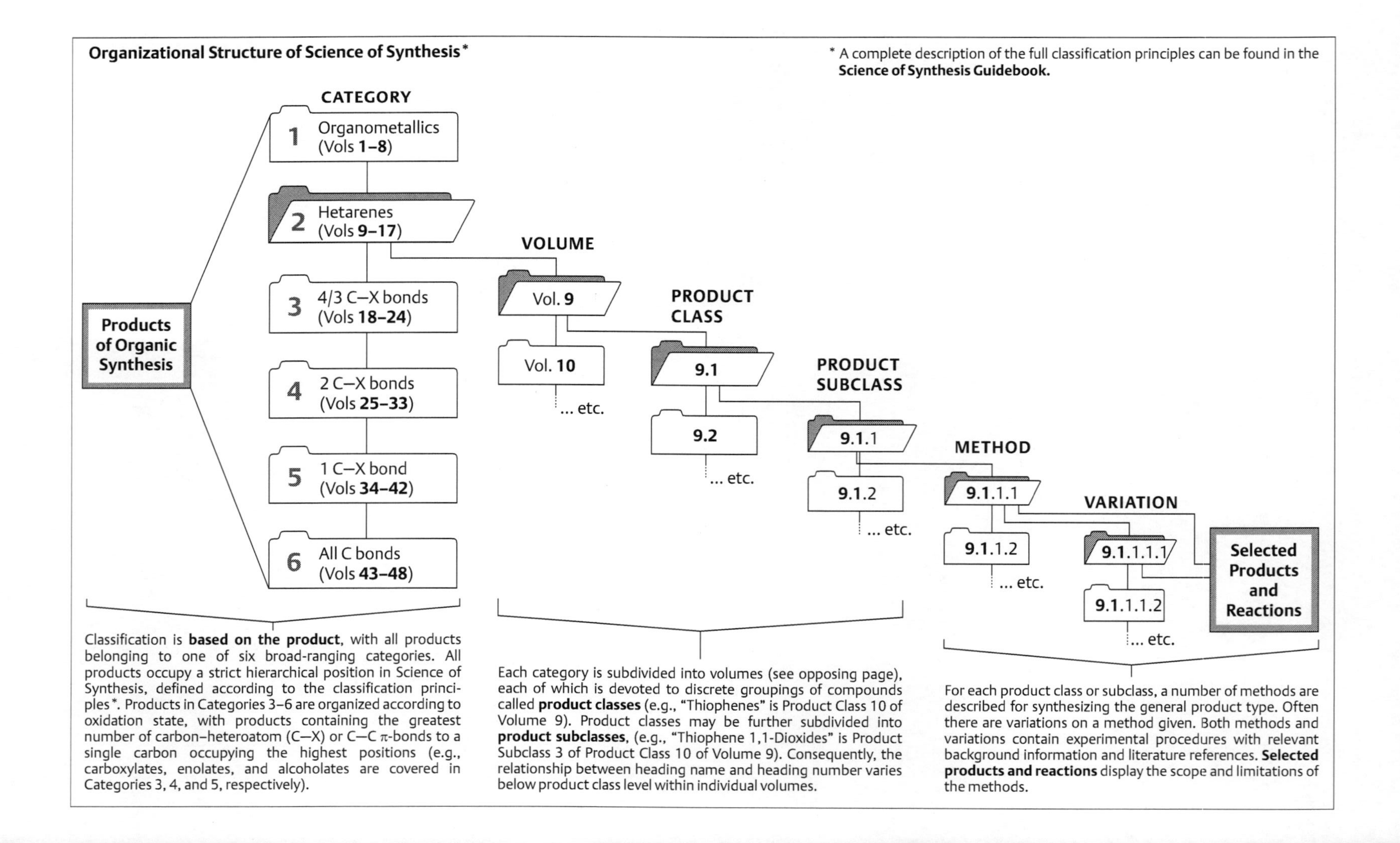

Classification is **based on the product**, with all products belonging to one of six broad-ranging categories. All products occupy a strict hierarchical position in Science of Synthesis, defined according to the classification principles*. Products in Categories 3–6 are organized according to oxidation state, with products containing the greatest number of carbon–heteroatom (C–X) or C–C π-bonds to a single carbon occupying the highest positions (e.g., carboxylates, enolates, and alcoholates are covered in Categories 3, 4, and 5, respectively).

Each category is subdivided into volumes (see opposing page), each of which is devoted to discrete groupings of compounds called **product classes** (e.g., "Thiophenes" is Product Class 10 of Volume 9). Product classes may be further subdivided into **product subclasses**, (e.g., "Thiophene 1,1-Dioxides" is Product Subclass 3 of Product Class 10 of Volume 9). Consequently, the relationship between heading name and heading number varies below product class level within individual volumes.

For each product class or subclass, a number of methods are described for synthesizing the general product type. Often there are variations on a method given. Both methods and variations contain experimental procedures with relevant background information and literature references. **Selected products and reactions** display the scope and limitations of the methods.

* A complete description of the full classification principles can be found in the **Science of Synthesis Guidebook.**

Science of Synthesis Reference Library

The **Science of Synthesis Reference Library** comprises volumes covering special topics of organic chemistry in a modular fashion, with six main classifications: (1) Classical, (2) Advances, (3) Transformations, (4) Applications, (5) Structures, and (6) Techniques. Volumes in the **Science of Synthesis Reference Library** focus on subjects of particular current interest with content that is evaluated by experts in their field. **Science of Synthesis**, including the **Knowledge Updates** and the **Reference Library**, is the complete information source for the modern synthetic chemist.

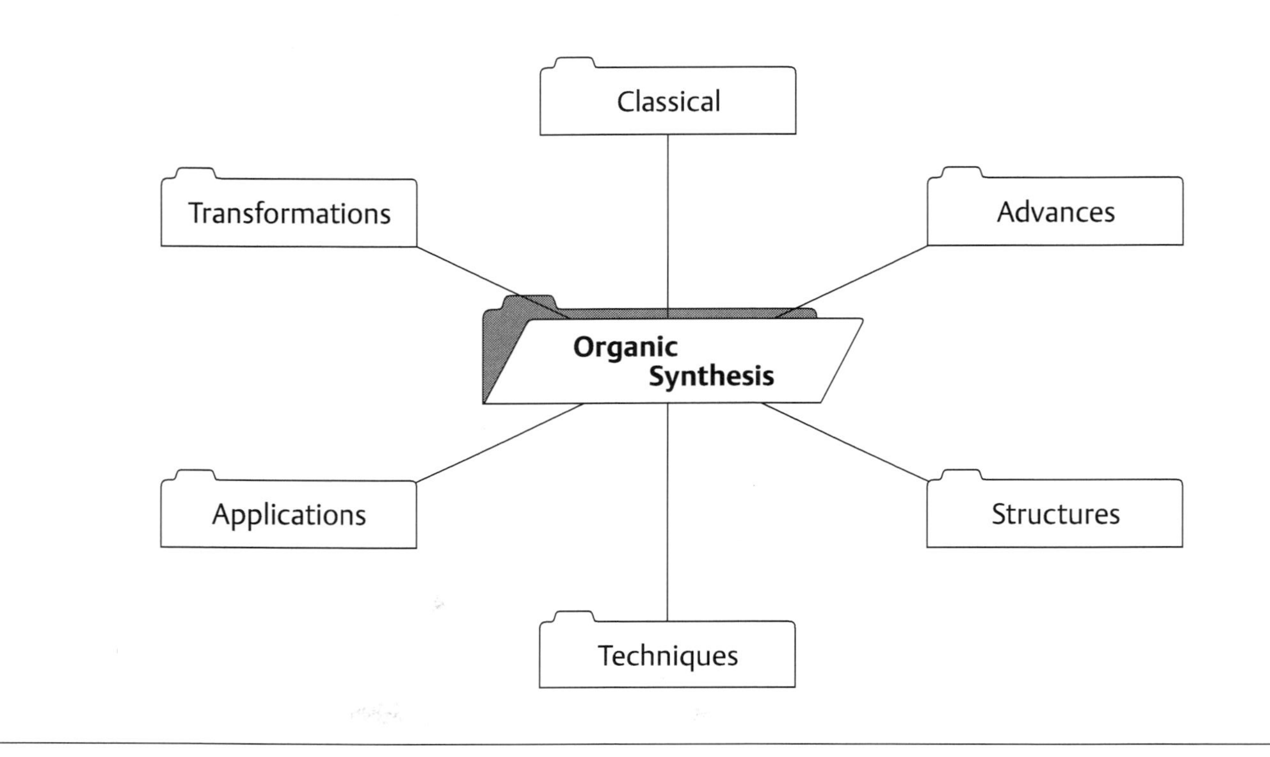

Science of Synthesis

Science of Synthesis is the authoritative and comprehensive reference work for the entire field of organic and organometallic synthesis.

Science of Synthesis presents the important synthetic methods for all classes of compounds and includes:
- Methods critically evaluated by leading scientists
- Background information and detailed experimental procedures
- Schemes and tables which illustrate the reaction scope

Science of Synthesis

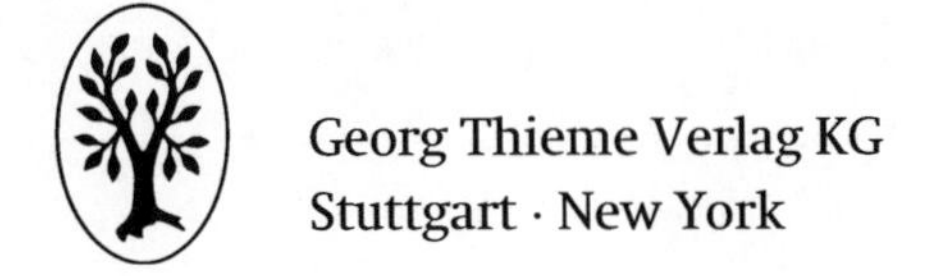

Georg Thieme Verlag KG
Stuttgart · New York

Science of Synthesis

Catalytic Reduction in Organic Synthesis 1

Volume Editor **J. G. de Vries**

Authors

W. Bonrath
C. S. J. Cazin
Z.-P. Chen
X. Dai
J. G. de Vries
K. Ding
B. Ghosh
R. Hudson
K. Kaneda
Y. Li
H. Lv
R. E. Maleczka, Jr.
J. A. Medlock
T. Mitsudome
A. Moores
M.-A. Müller
F. Nahra
Y. Nakagawa
P. Poechlauer
N. Ravasio
F. Shi
M. Tamura
X. Tan
S. Tin
K. Tomishige
F. Zaccheria
X. Zhang
Y.-G. Zhou
A. Zimmermann

2018
Georg Thieme Verlag KG
Stuttgart · New York

Rüdigerstrasse 14
D-70469 Stuttgart

Printed in Germany

Typesetting: Ziegler + Müller, Kirchentellinsfurt
Printing and Binding: AZ Druck und Datentechnik GmbH, Kempten

Bibliographic Information published by Die Deutsche Bibliothek

Die Deutsche Bibliothek lists this publication in the Deutsche Nationalbibliografie; detailed bibliographic data is available on the internet at <http://dnb.ddb.de>

Library of Congress Card No.: applied for

British Library Cataloguing in Publication Data

A catalogue record for this book is available from the British Library

ISSN (print) 2510-5469
ISSN (online) 2566-7297

ISBN (print) 978-3-13-240622-3
ISBN (PDF) 978-3-13-240623-0
ISBN (EPUB) 978-3-13-240624-7
DOI 10.1055/b-005-145236

Structure searchable version available at:
sos.thieme.com

ISBN 978-3-13-240622-3

Date of publication: June 13, 2018

This reference work mentions numerous commercial and proprietary trade names, registered trademarks and the like (not necessarily marked as such), patents, production and manufacturing procedures, registered designs, and designations. The editors and publishers wish to point out very clearly that the present legal situation in respect of these names or designations or trademarks must be carefully examined before making any commercial use of the same. Industrially produced apparatus and equipment are included to a necessarily restricted extent only and any exclusion of products not mentioned in this reference work does not imply that any such selection of exclusion has been based on quality criteria or quality considerations.

Warning! Read carefully the following: Although this reference work has been written by experts, the user must be advised that the handling of chemicals, microorganisms, and chemical apparatus carries potentially life-threatening risks. For example, serious dangers could occur through quantities being incorrectly given. The authors took the utmost care that the quantities and experimental details described herein reflected the current state of the art of science when the work was published. However, the authors, editors, and publishers take no responsibility as to the correctness of the content. Further, scientific knowledge is constantly changing. As new information becomes available, the user must consult it. Although the authors, publishers, and editors took great care in publishing this work, it is possible that typographical errors exist, including errors in the formulas given herein. Therefore, **it is imperative that and the responsibility of every user to carefully check whether quantities, experimental details, or other information given herein are correct based on the user's own understanding as a scientist.** Scale-up of experimental procedures published in **Science of Synthesis** carries additional risks. In cases of doubt, the user is strongly advised to seek the opinion of an expert in the field, the publishers, the editors, or the authors. When using the information described herein, the user is ultimately responsible for his or her own actions, as well as the actions of subordinates and assistants, and the consequences arising therefrom.

Preface

As the pace and breadth of research intensifies, organic synthesis is playing an increasingly central role in the discovery process within all imaginable areas of science: from pharmaceuticals, agrochemicals, and materials science to areas of biology and physics, the most impactful investigations are becoming more and more molecular. As an enabling science, synthetic organic chemistry is uniquely poised to provide access to compounds with exciting and valuable new properties. Organic molecules of extreme complexity can, given expert knowledge, be prepared with exquisite efficiency and selectivity, allowing virtually any phenomenon to be probed at levels never before imagined. With ready access to materials of remarkable structural diversity, critical studies can be conducted that reveal the intimate workings of chemical, biological, or physical processes with stunning detail.

The sheer variety of chemical structural space required for these investigations and the design elements necessary to assemble molecular targets of increasing intricacy place extraordinary demands on the individual synthetic methods used. They must be robust and provide reliably high yields on both small and large scales, have broad applicability, and exhibit high selectivity. Increasingly, synthetic approaches to organic molecules must take into account environmental sustainability. Thus, atom economy and the overall environmental impact of the transformations are taking on increased importance.

The need to provide a dependable source of information on evaluated synthetic methods in organic chemistry embracing these characteristics was first acknowledged over 100 years ago, when the highly regarded reference source **Houben–Weyl Methoden der Organischen Chemie** was first introduced. Recognizing the necessity to provide a modernized, comprehensive, and critical assessment of synthetic organic chemistry, in 2000 Thieme launched **Science of Synthesis, Houben–Weyl Methods of Molecular Transformations**. This effort, assembled by almost 1000 leading experts from both industry and academia, provides a balanced and critical analysis of the entire literature from the early 1800s until the year of publication. The accompanying online version of **Science of Synthesis** provides text, structure, substructure, and reaction searching capabilities by a powerful, yet easy-to-use, intuitive interface.

From 2010 onward, **Science of Synthesis** is being updated quarterly with high-quality content via **Science of Synthesis Knowledge Updates**. The goal of the **Science of Synthesis Knowledge Updates** is to provide a continuous review of the field of synthetic organic chemistry, with an eye toward evaluating and analyzing significant new developments in synthetic methods. A list of stringent criteria for inclusion of each synthetic transformation ensures that only the best and most reliable synthetic methods are incorporated. These efforts guarantee that **Science of Synthesis** will continue to be the most up-to-date electronic database available for the documentation of validated synthetic methods.

Also from 2010, **Science of Synthesis** includes the **Science of Synthesis Reference Library**, comprising volumes covering special topics of organic chemistry in a modular fashion, with six main classifications: (1) Classical, (2) Advances, (3) Transformations, (4) Applications, (5) Structures, and (6) Techniques. Titles will include *Stereoselective Synthesis*, *Water in Organic Synthesis*, and *Asymmetric Organocatalysis*, among others. With expert-evaluated content focusing on subjects of particular current interest, the **Science of Synthesis Reference Library** complements the **Science of Synthesis Knowledge Updates**, to make **Science of Synthesis** the complete information source for the modern synthetic chemist.

The overarching goal of the **Science of Synthesis** Editorial Board is to make the suite of **Science of Synthesis** resources the first and foremost focal point for critically evaluated information on chemical transformations for those individuals involved in the design and construction of organic molecules.

Throughout the years, the chemical community has benefited tremendously from the outstanding contribution of hundreds of highly dedicated expert authors who have devoted their energies and intellectual capital to these projects. We thank all of these individuals for the heroic efforts they have made throughout the entire publication process to make **Science of Synthesis** a reference work of the highest integrity and quality.

The Editorial Board July 2010

Science of Synthesis Reference Library

Catalytic Reduction in Organic Synthesis (2 Vols.)
Catalytic Oxidation in Organic Synthesis
N-Heterocyclic Carbenes in Catalytic Organic Synthesis (2 Vols.)
Metal-Catalyzed Cyclization Reactions (2 Vols.)
Applications of Domino Transformations in Organic Synthesis (2 Vols.)
Catalytic Transformations via C—H Activation (2 Vols.)
Biocatalysis in Organic Synthesis (3 Vols.)
C-1 Building Blocks in Organic Synthesis (2 Vols.)
Multicomponent Reactions (2 Vols.)
Cross Coupling and Heck-Type Reactions (3 Vols.)
Water in Organic Synthesis
Asymmetric Organocatalysis (2 Vols.)
Stereoselective Synthesis (3 Vols.)

Abstracts

p 7

1.1.1 Homogeneous Reduction of Alkenes

X. Tan, H. Lv, and X. Zhang

This chapter is focused on recent progress in the asymmetric hydrogenation of substituted alkenes, and the application of this methodology in the construction of a variety of chiral centers. The asymmetric hydrogenation of nonfunctionalized alkenes, α,β-unsaturated carbonyl compounds, enamides, enols, and other heteroatom-substituted alkenes is covered.

H_2
metal precursor
chiral ligand

Keywords: alkenes · alkanes · olefins · transition-metal catalysis · asymmetric hydrogenation · phosphorus ligands · homogeneous catalysis · reduction · rhodium · iridium · ruthenium

p 67

1.1.2 Reduction of Alkenes Using Nanoparticle Catalysis

R. Hudson and A. Moores

The transformation of alkenes to alkanes via hydrogenation represents a cornerstone of synthetic chemistry. Herein are outlined methods for alkene hydrogenations and transfer hydrogenations catalyzed by supported or unsupported palladium-, nickel-, iridium-, and iron-based nanoparticles.

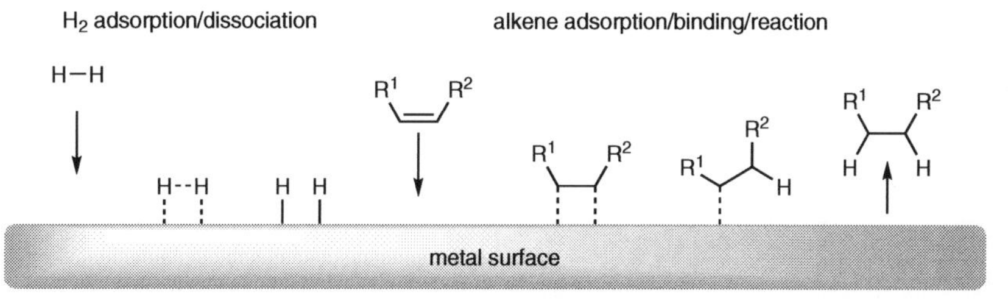

Keywords: alkenes · alkanes · olefins · nanostructures · carbon–carbon double bonds · catalysis · hydrogenation · reduction · palladium catalysts · nickel catalysts · iridium catalysts · iron catalysts · heterogeneous catalysis

p 91

1.2 Partial Reduction of Polyenes

F. Zaccheria and N. Ravasio

The selective hydrogenation of polyenes represents an important transformation in organic synthesis and requires a proper design and choice of the catalyst used for C=C bond hydrogenation, as well as careful tuning of the reaction conditions. This chapter illustrates some selected examples of partial reduction of polyenes via both homogeneous and heterogeneous catalysis, including hydrogenation of terpenes, cyclic dienes, and vegetable oils, to obtain products and intermediates useful for the chemical industry.

R1 R2 —catalyst→ R1 R2

Keywords: cyclooctadienes • dienes • fatty acid esters • geraniol • limonenes • polyenes • selective hydrogenation • reduction • palladium catalysts • platinum catalysts • ruthenium catalysts • rhodium catalysts

p 127

1.3 Reduction of Arenes

X. Dai and F. Shi

The group VIII metals, boranes, and Lewis pairs can catalyze the reduction of arenes to afford cycloalkanes. Cycloalkenes, as the intermediate product in the reduction of arenes, can also be generated by the partial reduction of arenes in the presence of ruthenium- and rhodium-based catalysts, but the selective partial reduction of polycyclic arenes to cycloalkenes still remains a challenge.

Ru, Rh, Ir, Pd

Ru, Rh

Ru, Rh, Pt, Pd
borane
Lewis pairs

Keywords: reduction • arenes • cycloalkanes • cycloalkenes • partial reduction • total reduction • hydrogenation

p 153

1.4 Reduction of Hetarenes

Z.-P. Chen and Y.-G. Zhou

The reduction of hetarenes provides a practical and efficient route to the corresponding saturated or partially saturated heterocycles, which are important structural motifs in many biologically active reagents and synthetic drugs. In the past decades, this approach has been extensively developed and it now represents a very powerful tool in organic synthesis. This chapter provides an overview of the reduction of hetarenes. Both heterogeneous and homogeneous catalysis approaches involving hydrogenations and transfer hydrogenations with transition-metal catalysts are discussed. Moreover, enantioselective approaches are also covered.

hydrogen source
transition-metal catalyst

Keywords: hetarenes • heterocycles • transition-metal catalysts • heterogeneous catalysis • homogeneous catalysis • asymmetric catalysis • hydrogenation • transfer hydrogenation • reduction

p 195

1.5 Catalytic Reduction of Alkynes and Allenes

W. Bonrath, J. A. Medlock, and M.-A. Müller

Catalytic reductions are one of the most important transformations in the chemical industry. In the field of alkyne and allene reduction, the most widely used method is hydrogenation. Numerous processes have been developed and implemented in the fine chemical and pharmaceutical industries for the production of a wide variety of alkenes and alkanes. This review provides an overview of the best (selective) reduction methods, from the use of the classic supported transition metal catalysts (e.g., the Lindlar catalyst) to more recently developed homogeneous catalysts which show alternative reactivity and selectivity, including preferential formation of *E*-alkenes.

Keywords: alkynes • alkenes • allenes • alkanes • hydrogenation • semi-hydrogenation • transfer hydrogenation • reduction • metal catalysts • Lindlar catalyst

p 229

1.6 Catalytic Reduction of Phenols, Alcohols, and Diols

S. Tin and J. G. de Vries

The catalytic deoxygenation of organic molecules has attracted a lot of attention in recent years because of interest in the use of biomass-derived fuels and chemicals. The raw materials used may contain up to 50 wt% of oxygen. In this chapter, some practical methods for the selective catalytic hydrodeoxygenation of phenols and alcohols to give arenes and alkanes, respectively, and the deoxydehydration of diols using hydrogen gas or transfer-hydrogenation methods are described.

$R^1OH \xrightarrow{H_2,\ catalyst} R^1H$

$R^1CH(OH)CH(OH)R^2 \xrightarrow{H_2,\ catalyst} R^1CH{=}CHR^2$

Keywords: hydrogenolysis • hydrogenation • reduction • alcohols • diols • alkanes • alkenes

p 243

1.7 Hydrogenolysis of Ethers

Y. Nakagawa, M. Tamura, and K. Tomishige

Selective hydrogenolysis of ethers to alcohols and hydrocarbons is becoming possible with appropriate metal catalysts. Total removal of oxygen atoms from functionalized ethers to give alkanes, especially from furan derivatives toward biofuels, is catalyzed by a combination of metal and acid.

$PhOR^1 \xrightarrow{H_2,\ catalyst} PhOH + R^1H$

$R^1OCH(R^2)CH_2OH \xrightarrow{H_2,\ catalyst} R^1OH + R^2CH_2CH_2OH$

2-R^1-5-R^2-furan $\xrightarrow{H_2,\ catalyst}$ hydrocarbons

Keywords: hydrogenolysis • hydrogenation • reduction • aryl ethers • unsymmetrical ethers • cyclic ethers • furan • biomass resources • biofuels

p 269

1.8 Catalytic Reduction of Carbonates

Y. Li and K. Ding

Carbonates are basic chemicals that are widely used in both industry and academia. Their reduction under either homogeneous or heterogeneous catalytic conditions generates formates, methanol, or methane. Carbonates can also act as a C_1 building block for the reductive methylation of amines.

$$\mathrm{MeOH} \xleftarrow[-\,R^1OH]{H_2} R^1O\text{–}C(=O)\text{–}OR^1 \xrightarrow[-\,R^1OH,\ -\,H_2O]{H_2,\ R^2_2NH} R^2\text{–}N(Me)\text{–}R^2$$

$$\mathrm{MHCO_3} \xrightarrow{H_2} \mathrm{HCO_2M} \text{ or } \mathrm{CH_4}$$

Keywords: reduction • hydrogenation • homogeneous catalysis • heterogeneous catalysis • carbonates • alcohols • hydrolysis • formates • methylation

p 289

1.9 Hydrogenation of Carbon Dioxide

F. Nahra and C. S. J. Cazin

Carbon dioxide is an economical, safe, and renewable C_1 source. This attractive C_1 building block is mainly used in the synthesis of organic chemicals, materials, and carbohydrates. As a feedstock to produce chemicals and fuel derivatives, carbon dioxide utilization will most certainly become an important tool in the quest for more sustainable chemistry. The atom-economical hydrogenation of carbon dioxide using dihydrogen offers a unique opportunity to achieve that goal. The main products of carbon dioxide hydrogenation or reduction fall into two categories: fuels and chemicals. The main topics discussed in this chapter are the hydrogenation of carbon dioxide to formic acid, methanol, and methane, as well as the reductive methylation of amines and C—H bonds. Both homogeneous and heterogeneous catalytic metal systems are reviewed herein.

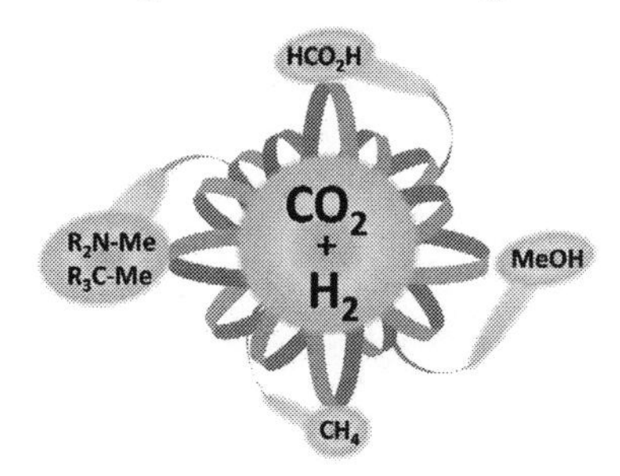

Keywords: carbon dioxide • formic acid • formates • methanol • hydrogenation • homogeneous catalysis • heterogeneous catalysis • reductive methylation • reduction

p 315

1.10 Reduction of Peroxo Compounds, Ozonides, and Molozonides

P. Poechlauer and A. Zimmermann

Research in the field of heterogeneous catalytic hydrogenation of peroxo compounds, ozonides, and molozonides has delivered results in a diverse range of fields of organic chemistry. It has revealed details of the reaction mechanisms that take place at the catalyst surface, and has enabled further understanding of the factors governing the chemoselectivity of various catalysts. Apart from enabling yield increases in single transformations, this has also opened up opportunities to apply the introduction and subsequent hydrogenation of peroxo moieties in multistep syntheses of complex molecules. The facile introduction of the peroxide moiety via very different reactions at positions otherwise difficult to access, and its subsequent transformation in the presence of other functional groups, has allowed the design of concise synthetic pathways to complex molecules, including pharmaceuticals and natural products.

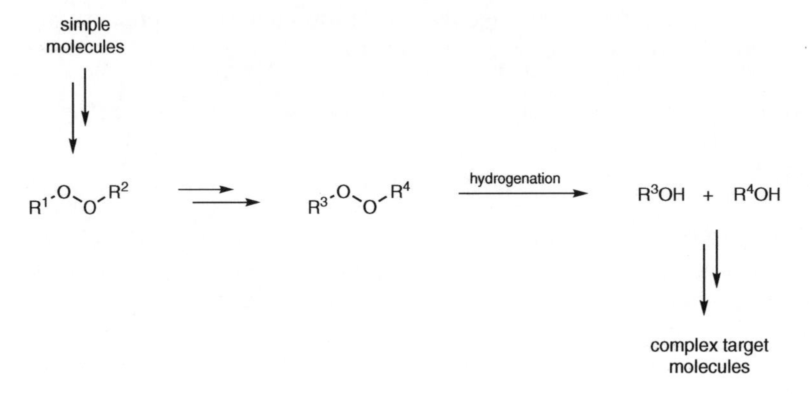

Keywords: heterogeneous catalysis • reduction • oxidation • hydrogenation • peroxides • endoperoxides • ozonolysis • alcohols

p 339

1.11 Reduction of Sulfur Compounds Using Metal Catalysts

K. Kaneda and T. Mitsudome

Homogeneous and heterogeneous metal catalysts based on molybdenum, rhenium, ruthenium, and platinum promote the reduction of sulfur compounds such as sulfoxides, sulfones, and disulfides by using alcohols or molecular hydrogen as reductants.

H_2 or alcohol
metal catalyst

H_2
metal catalyst
R^1SH

Keywords: reduction • hydrogenation • transfer hydrogenation • sulfides • thioethers • disulfides • sulfoxides • sulfones • thiols • deoxygenation • sulfur compounds • homogeneous catalysts • heterogeneous catalysts • green chemistry

p 355

1.12 Catalytic Hydrodehalogenation Reactions

B. Ghosh and R. E. Maleczka, Jr.

Hydrodehalogenation, or reductive dehalogenation, is an important organic transformation that is often used as a detoxification process in industry. A number of methods have been employed to effect this transformation in organic synthesis. Metal-catalyzed hydrodehalogenation is among the popular methods and is typically performed with molecular hydrogen or via transfer hydrogenation from other reagents. The current review highlights development in metal-catalyzed hydrodehalogenation reactions in the last 15 years, where protocols to afford spectroscopically characterized reaction products have been established.

Pr^i Pr^i Pr^i Pr^i Cl Cl Pd Pd Cl Cl N N N N Pr^i Pr^i Pr^i Pr^i (1 mol%)

Cl Cl Cl Cl Cl Cl Cl Cl Cl Cl

NaOH, iPrOH, 80 °C, 24 h

84%

Keywords: dehalogenation • heterogeneous catalysis • homogeneous catalysis • reduction • haloalkanes • alkanes • haloarenes • arenes • transition-metal catalysis • photoredox catalysis

Catalytic Reduction in Organic Synthesis 1

Preface V

Abstracts IX

Table of Contents XIX

Introduction
J. G. de Vries 1

1.1 **Reduction of Alkenes** 7

1.1.1 **Homogeneous Reduction of Alkenes**
X. Tan, H. Lv, and X. Zhang 7

1.1.2 **Reduction of Alkenes Using Nanoparticle Catalysis**
R. Hudson and A. Moores 67

1.2 **Partial Reduction of Polyenes**
F. Zaccheria and N. Ravasio 91

1.3 **Reduction of Arenes**
X. Dai and F. Shi 127

1.4 **Reduction of Hetarenes**
Z.-P. Chen and Y.-G. Zhou 153

1.5 **Catalytic Reduction of Alkynes and Allenes**
W. Bonrath, J. A. Medlock, and M.-A. Müller 195

1.6 **Catalytic Reduction of Phenols, Alcohols, and Diols**
S. Tin and J. G. de Vries 229

1.7 **Hydrogenolysis of Ethers**
Y. Nakagawa, M. Tamura, and K. Tomishige 243

1.8 **Catalytic Reduction of Carbonates**
Y. Li and K. Ding 269

1.9 **Hydrogenation of Carbon Dioxide**
F. Nahra and C. S. J. Cazin 289

1.10 **Reduction of Peroxo Compounds, Ozonides, and Molozonides**
P. Poechlauer and A. Zimmermann 315

1.11 **Reduction of Sulfur Compounds Using Metal Catalysts**
K. Kaneda and T. Mitsudome 339

1.12 **Catalytic Hydrodehalogenation Reactions**
B. Ghosh and R. E. Maleczka, Jr. 355

Keyword Index 375

Author Index 391

Abbreviations 411

Table of Contents

Introduction
J. G. de Vries

Introduction 1

1.1 **Reduction of Alkenes**

1.1.1 **Homogeneous Reduction of Alkenes**
X. Tan, H. Lv, and X. Zhang

1.1.1 **Homogeneous Reduction of Alkenes** 7
1.1.1.1 Reduction of Nonfunctionalized Alkenes 8
1.1.1.2 Reduction of Nitrogen-Substituted Alkenes 14
1.1.1.2.1 Reduction of Enamides 20
1.1.1.2.2 Reduction of α,β-Dehydro-α-amino Acids and Their Esters 27
1.1.1.2.3 Reduction of α,β-Dehydro-β-amino Acids and Their Esters 32
1.1.1.3 Reduction of Oxygen-Substituted Alkenes 34
1.1.1.3.1 Reduction of Enol Esters 35
1.1.1.3.2 Reduction of Enol Phosphinates 38
1.1.1.3.3 Reduction of Enol Ethers 39
1.1.1.3.4 Reduction of Silyl Enol Ethers 41
1.1.1.4 Reduction of Carbonyl-Substituted Alkenes 42
1.1.1.4.1 Reduction of α,β-Unsaturated Carboxylic Acids 43
1.1.1.4.2 Reduction of α,β-Unsaturated Esters 47
1.1.1.4.3 Reduction of α,β-Unsaturated Amides 48
1.1.1.4.4 Reduction of α,β-Unsaturated Ketones 52
1.1.1.5 Reduction of Alkenes Bearing Other Heteroatoms 54
1.1.1.5.1 Reduction of Phosphorus-Substituted Alkenes 55
1.1.1.5.2 Reduction of Vinylboronates 57
1.1.1.5.3 Reduction of Vinyl Fluorides 58
1.1.1.5.4 Reduction of Sulfur-Substituted Alkenes 59

1.1.2 Reduction of Alkenes Using Nanoparticle Catalysis

R. Hudson and A. Moores

1.1.2 Reduction of Alkenes Using Nanoparticle Catalysis 67

1.1.2.1 Reduction of Alkenes by Hydrogenation Using Palladium Nanoparticles 68

1.1.2.1.1 Hydrogenation with Commercial Palladium Nanoparticles 68

1.1.2.1.2 Hydrogenation with Palladium Nanoparticles in Poly(ethylene glycol) 69

1.1.2.1.3 Hydrogenation with Palladium Nanoparticles Stabilized in Mesocellular Foam 69

1.1.2.1.4 Hydrogenation with Phenanthroline-Stabilized Palladium Nanoparticles in Poly(ethylene glycol) 70

1.1.2.1.5 Hydrogenation with Palladium Nanoparticles Embedded in Polystyrene 71

1.1.2.1.6 Hydrogenation with Palladium Nanoparticles on Amphiphilic Supports 72

1.1.2.1.7 Hydrogenation with Palladium Nanoparticles in Biphasic Media 73

1.1.2.1.8 Asymmetric Hydrogenation with Palladium Nanoparticles 75

1.1.2.1.9 Hydrogenation with Palladium on Ferrite Nanoparticles 76

1.1.2.1.10 Hydrogenation with Palladium Nanoparticles Supported on Magnetic Carbon-Coated Cobalt Nanobeads 77

1.1.2.1.11 Hydrogenation with Magnetic Carbon-Supported Palladium Nanoparticles ... 78

1.1.2.2 Reduction of Alkenes by Hydrogenation Using Iron Nanoparticles 79

1.1.2.2.1 Hydrogenation with Unsupported Iron Nanoparticles 79

1.1.2.2.2 Hydrogenation with Graphene-Supported Iron Nanoparticles 81

1.1.2.2.3 Hydrogenation with Core-Shell Iron/Iron Oxide Nanoparticles 82

1.1.2.2.4 Hydrogenation with Amphiphilic Polymer-Supported Iron Nanoparticles 82

1.1.2.3 Reduction of Alkenes Using Ionic-Liquid-Stabilized Nanoparticles 83

1.1.2.3.1 Reduction of Alkenes with Iridium Nanoparticles in Ionic Liquids 84

1.1.2.4 Reduction of Alkenes by Transfer Hydrogenation 84

1.1.2.4.1 Selective Reduction of Alkenes with Hydrazine Using Nickel Nanoparticles Supported on Clay 85

1.1.2.4.2 Selective Reduction of Alkenes with Isopropanol Using Nickel Nanoparticles 86

1.1.2.4.3 Selective Reduction of Alkenes with Isopropanol Catalyzed by Nickel/Ruthenium/Platinum/Gold Heteroquatermetallic Nanoparticles 87

1.1.2.5 Conclusions and Future Perspectives 88

1.2 Partial Reduction of Polyenes

F. Zaccheria and N. Ravasio

1.2	**Partial Reduction of Polyenes**	91
1.2.1	Selective Hydrogenation of Polyunsaturated Terpenes	91
1.2.1.1	Isoprene Reduction	92
1.2.1.2	Limonene Reduction	95
1.2.1.3	Myrcene Reduction	97
1.2.1.4	Geraniol Reduction	99
1.2.1.4.1	Homogeneous Catalysts for Geraniol Reduction	99
1.2.1.4.2	Heterogeneous Catalysts for Geraniol Reduction	101
1.2.2	Partial Hydrogenation of Cyclic Polyenes	102
1.2.2.1	Cyclooctadiene Reduction	103
1.2.2.2	Cyclododecatriene Reduction	106
1.2.3	Selective Hydrogenation of Vegetable Oils and Related Compounds	109
1.2.3.1	Two-Phase Systems	111
1.2.3.2	Homogeneous Systems	113
1.2.3.3	Noble Metal Based Heterogeneous Systems	114
1.2.3.4	Non-Noble Metal Based Heterogeneous Systems	116
1.2.4	Conclusions	122

1.3 Reduction of Arenes

X. Dai and F. Shi

1.3	**Reduction of Arenes**	127
1.3.1	Reduction of Monocyclic Arenes to Cycloalkanes	127
1.3.1.1	Ruthenium-Catalyzed Reduction of Monocyclic Arenes to Cycloalkanes	128
1.3.1.2	Rhodium-Catalyzed Reduction of Monocyclic Arenes to Cycloalkanes	131
1.3.1.3	Iridium-Catalyzed Reduction of Monocyclic Arenes to Cycloalkanes	133
1.3.1.4	Palladium-Catalyzed Reduction of Monocyclic Arenes to Cycloalkanes	134
1.3.2	Reduction of Polycyclic Arenes	135
1.3.2.1	Ruthenium-Catalyzed Reduction of Polycyclic Arenes	135
1.3.2.2	Rhodium-Catalyzed Reduction of Polycyclic Arenes	139
1.3.2.3	Platinum-Catalyzed Reduction of Polycyclic Arenes	140
1.3.2.4	Palladium-Catalyzed Reduction of Polycyclic Arenes	141
1.3.2.5	Borane-Catalyzed Reduction of Polycyclic Arenes	143
1.3.2.6	Reduction of Polycyclic Arenes Catalyzed by Lewis Pairs	145

1.3.3 Reduction of Monocyclic Arenes to Cycloalkenes 145
1.3.3.1 Ruthenium-Catalyzed Reduction of Monocyclic Arenes to Cycloalkenes 147
1.3.3.2 Rhodium-Catalyzed Reduction of Monocyclic Arenes to Cycloalkenes 148

1.4 **Reduction of Hetarenes**
Z.-P. Chen and Y.-G. Zhou

1.4 **Reduction of Hetarenes** 153
1.4.1 Heterogeneous Catalysis 153
1.4.1.1 Reduction of Quinoline Derivatives 153
1.4.1.2 Reduction of Isoquinoline Derivatives 156
1.4.1.3 Reduction of Pyridine Derivatives 157
1.4.2 Homogeneous Catalysis 160
1.4.2.1 Racemic Reductions 160
1.4.2.1.1 Reduction of Quinoline Derivatives 160
1.4.2.1.2 Miscellaneous Substrates 165
1.4.2.2 Enantioselective Reductions 168
1.4.2.2.1 Neutral Iridium Complexes as Catalysts 170
1.4.2.2.2 Cationic Iridium Complexes as Catalysts 174
1.4.2.2.3 Ruthenium Complexes as Catalysts 177
1.4.2.2.4 Rhodium Complexes as Catalysts 184
1.4.2.2.5 Palladium Complexes as Catalysts 186
1.4.3 Conclusions and Future Perspectives 189

1.5 **Catalytic Reduction of Alkynes and Allenes**
W. Bonrath, J. A. Medlock, and M.-A. Müller

1.5 **Catalytic Reduction of Alkynes and Allenes** 195
1.5.1 Reduction of Alkynes 195
1.5.1.1 Total Hydrogenation of Alkynes to Alkanes 196
1.5.1.2 Selective Hydrogenation of Alkynes to Alkenes 196
1.5.1.2.1 Semi-hydrogenations of Terminal Alkynes and *Z*-Selective Reduction of Internal Alkynes 199
1.5.1.2.1.1 Heterogeneous Hydrogenations 199
1.5.1.2.1.2 Heterogeneous Transfer Hydrogenations 202
1.5.1.2.1.3 Homogeneous Hydrogenations 204
1.5.1.2.1.4 Homogeneous Transfer Hydrogenations 212

1.5.1.2.2 *E*-Selective Semi-hydrogenations of Internal Alkynes 214
1.5.1.2.2.1 Homogeneous Hydrogenations 214
1.5.1.2.2.2 Homogeneous Transfer Hydrogenations 219
1.5.2 Reduction of Allenes 220
1.5.2.1 Hydrogenation of Functionalized Allenes 221
1.5.2.2 Hydrogenation of Allene Ketones 223
1.5.3 Conclusions 224

1.6 **Catalytic Reduction of Phenols, Alcohols, and Diols**
S. Tin and J. G. de Vries

1.6 **Catalytic Reduction of Phenols, Alcohols, and Diols** 229
1.6.1 Hydrodeoxygenation of Phenols To Give Arenes 229
1.6.2 Hydrodeoxygenation of Aliphatic Alcohols To Give Alkanes 231
1.6.3 Reduction of Diols To Give Alkenes 233
1.6.3.1 Reduction of Diols by Metal-Free Catalysis 233
1.6.3.2 Reduction of Diols Using Rhenium Catalysts 236
1.6.4 Conclusions 239

1.7 **Hydrogenolysis of Ethers**
Y. Nakagawa, M. Tamura, and K. Tomishige

1.7 **Hydrogenolysis of Ethers** 243
1.7.1 Hydrogenolysis of Aryl Ethers 243
1.7.1.1 Homogeneous Nickel Catalysts for the Hydrogenolysis of Diaryl Ethers and Alkyl Aryl Ethers 244
1.7.1.2 Solid Catalysts for the Hydrogenolysis of Aryl Ethers 247
1.7.1.3 Hydrogenolysis of the β-O-4 Linkage in Lignin Model Compounds 249
1.7.1.4 Hydrogenolysis of Methoxy Groups in Lignin Model Monomers 251
1.7.2 Hydrogenolysis of Saturated Ethers 253
1.7.2.1 Ethers Bearing Directing Hydroxy Groups 253
1.7.2.2 Ethers without Directing Hydroxy Groups 256
1.7.2.3 Total Hydrodeoxygenation of Cyclic Ethers to Alkanes 257
1.7.3 Hydrogenolysis of Furans 258
1.7.3.1 Hydrogenolysis via Tetrahydrofuran Derivatives 258
1.7.3.2 Ring Opening of Furan Derivatives not Involving a Tetrahydrofuran Intermediate 260
1.7.4 Conclusions 264

1.8 **Catalytic Reduction of Carbonates**
Y. Li and K. Ding

1.8 **Catalytic Reduction of Carbonates** 269
1.8.1 Catalytic Reduction of Carbonates to Formates 269
1.8.1.1 Homogeneous Reductions 270
1.8.1.2 Heterogeneous Reductions 275
1.8.2 Catalytic Reduction of Dialkyl Carbonates to Methanol 275
1.8.2.1 Homogeneous Reductions 275
1.8.2.2 Heterogeneous Reductions 279
1.8.3 Catalytic Reduction of Carbonates to Methane 282
1.8.4 Other Reductive Transformations 282
1.8.4.1 Catalytic Reductive Methylation 282
1.8.4.2 Photo-/Electro- and Enzymatic Reduction of Carbonates 283
1.8.5 Conclusions 286

1.9 **Hydrogenation of Carbon Dioxide**
F. Nahra and C. S. J. Cazin

1.9 **Hydrogenation of Carbon Dioxide** 289
1.9.1 Hydrogenation of Carbon Dioxide to Formic Acid or Formate Salts 290
1.9.1.1 Homogeneous Systems 290
1.9.1.1.1 Rhodium-Based Catalysts 291
1.9.1.1.2 Ruthenium-Based Catalysts 292
1.9.1.1.3 Iridium-Based Catalysts 294
1.9.1.1.4 Other Metal-Based Catalysts 295
1.9.1.2 Heterogeneous Systems 296
1.9.2 Hydrogenation of Carbon Dioxide to Methanol 299
1.9.2.1 Homogeneous Systems 299
1.9.2.2 Heterogeneous Systems 301
1.9.3 Methanation of Carbon Dioxide 302
1.9.3.1 Heterogeneous Systems 302
1.9.4 Reductive Methylation Using Carbon Dioxide 304
1.9.4.1 Methylation of Amines 305
1.9.4.1.1 Homogeneous Systems 305
1.9.4.1.1.1 Methylation of Amines Using Hydrogen as Reductant 305
1.9.4.1.2 Heterogeneous Systems 307
1.9.4.2 Methylation of C—H Bonds 310

1.10 **Reduction of Peroxo Compounds, Ozonides, and Molozonides**

P. Poechlauer and A. Zimmermann

1.10 **Reduction of Peroxo Compounds, Ozonides, and Molozonides** 315

1.10.1 Reduction of Peroxo Compounds 316

1.10.1.1 Mechanistic Aspects and New Types of Catalysts 316

1.10.1.2 Reduction of Peroxo Compounds in the Presence of Other Reducible Functional Groups 318

1.10.1.2.1 Reduction of Peroxo Compounds in the Presence of Alkenes, Esters, Lactones, and Lactams 318

1.10.1.2.2 Reduction of Peroxo Compounds in the Presence of Oxo Groups 321

1.10.1.2.2.1 Hydrogenation of α-Peroxy Carbonyl Compounds 321

1.10.1.2.2.2 Hydrogenation of β-Peroxy Carbonyl Compounds 322

1.10.1.2.2.3 Hydrogenation of γ-Peroxy Carbonyl Compounds 323

1.10.1.2.3 Reduction of Peroxo Compounds in the Presence of Nitro Groups 324

1.10.1.3 Reduction of Peroxides as Part of Multistep Syntheses 324

1.10.1.3.1 Hydrogenation of Hydroperoxides Formed by an Ene Reaction of Singlet Oxygen with a C=C Bond 325

1.10.1.3.2 Hydrogenation of Dialkyl Peroxides Formed by Addition of an Alkyl Hydroperoxide to a C=C Bond To Effect a Net Hydration of this Double Bond 326

1.10.1.3.3 Hydrogenation of 1,2-Dioxin Derivatives Formed by [2+4] Cycloaddition of Singlet Oxygen to 1,3-Dienes To Effect 1,4-Dihydroxylation 327

1.10.1.3.3.1 Dihydroxylation and Hydrogenation of 4-Substituted 1,2-Dioxins 327

1.10.1.3.4 Hydrogenation of Endoperoxides Formed by [2+4] Cycloaddition of Singlet Oxygen to Cyclic 1,3-Dienes To Effect *cis*-1,4-Dihydroxylation 329

1.10.1.3.4.1 Cyclohexanetetraols by Hydrogenation of 2,3-Dioxabicyclo[2.2.2]octane-5,6-diols 329

1.10.1.3.4.2 4-Hydroxycyclohexenones by Hydrogenation of 1-Alkoxy-2,3-dioxabicyclo[2.2.2]oct-5-enes 329

1.10.1.3.4.3 2,7-Dioxabicyclo[2.2.1]heptanes by Hydrogenation of 2,3,5-Trioxabicyclo[2.2.2]oct-7-enes 330

1.10.2 Reduction of Molozonides, Ozonides, and Hydroxy and Alkoxy Hydroperoxides 331

1.11 **Reduction of Sulfur Compounds Using Metal Catalysts**
K. Kaneda and T. Mitsudome

1.11 **Reduction of Sulfur Compounds Using Metal Catalysts** 339

1.11.1 Deoxygenation of Sulfoxides to Thioethers Using Alcohols 339

1.11.2 Deoxygenation of Sulfoxides and Sulfones to Thioethers Using Molecular Hydrogen 344

1.11.3 Reduction of Disulfides to Thiols Using Molecular Hydrogen 351

1.11.4 Conclusions and Future Perspectives 352

1.12 **Catalytic Hydrodehalogenation Reactions**
B. Ghosh and R. E. Maleczka, Jr.

1.12 **Catalytic Hydrodehalogenation Reactions** 355

1.12.1 Metal-Catalyzed Hydrodehalogenation 355

1.12.1.1 Heterogeneous Catalysis 355

1.12.1.2 Homogeneous Catalysis 362

1.12.2 Photoinduced Hydrodehalogenation 369

1.12.3 Conclusions 372

Keyword Index 375

Author Index 391

Abbreviations 411

Introduction

J. G. de Vries

I have been somehow engaged with reduction reactions my entire working life. My doctoral thesis was on chiral NADH models that were used for asymmetric reduction of activated ketones, and, in passing, I also invented the reduction of aldehydes and ketones using sodium dithionite. None of this was catalytic, of course. From my six years as a medicinal chemist in the pharma industry, I particularly remember the hydrogenolysis of *para*-nitrobenzyl esters as the last step in a lengthy total synthesis of carbapenems, and later the hydrogenolysis of the benzyloxycarbonyl protecting groups of peptides that were going to be the ultimate painkillers. When I moved to DSM I was encouraged to work on asymmetric hydrogenation. Initially, we developed a water-soluble rhodium–bisphosphine catalyst for asymmetric imine hydrogenation that was highly enantioselective, but too expensive. Thus, after I became part-time professor in Groningen, together with Adri Minnaard and Ben Feringa, we focused on the development of cheap and simple ligands, an endeavor that culminated in the invention of the monodentate phosphoramidite ligands for the rhodium-, iridium-, and ruthenium-catalyzed homogeneous hydrogenation of a range of different substrates. At DSM, together with Laurent Lefort, we managed to robotize the ligand synthesis, allowing us to make and screen 96 ligands in two days. This eventually also led to a large-scale asymmetric hydrogenation process. In my current research, here at the Leibniz-Institut für Katalyse in Rostock, I also use hydrogenation and hydrogenolysis extensively for the deoxygenation of biomass and biomass-derived platform chemicals.

In 2007, I edited the three-volume *Handbook of Homogeneous Hydrogenation* together with Kees Elsevier (University of Amsterdam), so when Thieme approached me with a proposal for a new addition to the *Science of Synthesis* Reference Library on catalytic reduction, I thought it would be a great opportunity to check where homogeneous hydrogenation stands now, 10 years later, and at the same time make a comparison with heterogeneous hydrogenation. It is clear that heterogeneous hydrogenation is used frequently and on large scale in the bulk and fine chemical industries. However, when it comes to published work, I, and most of the authors who have participated in this work, discovered that there is a very noticeable difference between homogeneous and heterogeneous catalysis: Whereas in the former the products are always isolated and are extensively characterized, the focus in the latter is generally on the preparation and characterization of the heterogeneous catalyst, and the catalysis itself is usually a minor part. Worse still, in none of these publications are the products actually isolated; all conversions and yields reported are based on the use of GC. Thus, the reader must be forewarned that, in practice, the actual yields may be lower than the values given in the tables.

In comparison to 10 years ago, a number of trends in catalytic reduction can be seen: First and foremost, there has, of course, been a renaissance in the use of inexpensive metals (iron, cobalt, and recently also manganese) in homogeneous hydrogenation. The ligands that are used are still a bit too expensive, but this situation may change very soon. In connection with this theme, there has also been a resurgence in the use of pincer ligands.

A second trend is the use of metal nanoparticles. In heterogeneous catalysis, the synthesis of well-defined nanoparticles on a solid support can now be achieved in a highly controlled manner. There is also more information available that allows us to predict whether smaller or larger nanoparticles are required to optimize the selectivity of a particular reaction. In addition, the homogeneous catalysis community has also embraced the use of metal nanoparticles that are not on a support, but are instead suspended in solution. Here, the art is in the use of stabilizers as well as ligands, and the first asymmetric conversions using nanoparticles have even been reported.

A third major trend is the application of catalytic reduction in the conversion of biomass and biomass-derived platform chemicals. In contrast to the conversions of fossil-derived raw materials, here the game is to remove oxygen in a highly selective manner. This topic surfaces in many of the chapters, and was at the basis of many new developments in the field of ether and alcohol hydrogenolysis.

Finally, the return of photocatalysis in full force has been rather overwhelming. Only weak echoes of this movement can be found in these volumes, as hydrogenation reactions are exothermic and usually can proceed very fast catalytically without any light. Nevertheless, a few examples can be found in some chapters.

Luckily, I have been able to find excellent authors for all of the chapters. We decided to drop the chapter on epoxide hydrogenation, as not much progress has been made in the past 15 years. By the way, most authors have adhered nicely to the planned restriction of only covering the period from 2000 to 2017.

Homogeneous hydrogenation of alkenes is, of course, 99% enantioselective hydrogenation. This can be seen in Section 1.1.1 on homogeneous alkene reduction, which was produced by Xuefeng Tan, Hui Lv, and Xumu Zhang. This is a large chapter, reflecting the enormous amount of work that has been performed in this area in the past 17 years, with many new ligands being developed. Interestingly, the metals used are still the old favorites: rhodium, ruthenium, and iridium. Thus, there is an obvious gap in this area of research that one hopes will be filled soon, namely asymmetric hydrogenation using catalysts based on inexpensive metals. Not accidentally, the authors are Chinese; a tremendous number of publications on this topic have come from China, particularly in the past 10 years.

Not much progress has been made in the heterogeneous reduction of alkenes during this period, except in the area of nanoparticle catalysis. Audrey Moores and Reuben Hudson have nicely summarized the latest developments in this lively field in Section 1.1.2. Although palladium is obviously still everyone's friend, there are some nice examples using nickel and iron nanoparticles as well.

Section 1.2 on the partial reduction of polyenes has been expertly put together by Nicoletta Ravasio and Federica Zaccheria, who have worked in this area for a long time. Interestingly, the driver in this field has been, and continues to be, the conversion of polyene natural products such as terpenes and polyunsaturated fatty acids. Indeed, the number of publications in this area is rising as a result of the interest in renewable chemicals. Selectivity remains an issue.

Section 1.3 is on the reduction of arenes to cycloalkenes and cycloalkanes, a field of enormous importance in the production of bulk chemicals. The reduction reactions of benzene to cyclohexane or cyclohexene are important steps in the large scale production of nylon intermediates such as caprolactam and adipic acid. The progress in this field is nicely summarized by Feng Shi and Xingchao Dai. Although in the past many homogeneous ruthenium species were reported to be active catalysts, it is now generally accepted that these compounds form ruthenium nanoparticles that are the true catalysts. Thus, this area largely belongs to the heterogeneous catalysis domain.

Section 1.4, the reduction of hetarenes, is of particular importance for the pharma industry and hence, unsurprisingly, there is also a strong emphasis on enantioselective reduction in this area. The recent work in this field has been expertly reviewed by Yong-Gui Zhou and Zhang-Pei Chen. The Zhou group has indeed made many contributions to this field.

In the reduction of alkynes and allenes (Section 1.5), or more specifically the semireduction of these substrates to the monoalkenes, selectivity is the major issue. This is, of course, a field that has emanated from the vitamins industry, where this type of conversion is frequently required. Indeed, the Lindlar catalyst originates from one of the companies that produce vitamins. I am happy to say that I managed to persuade veteran vitamin chemists Werner Bonrath, Jonathan Medlock, and Marc-André Müller to write this chapter, and their experience shows in the composition of this interesting chapter, which also contains a wealth of information from an industrial viewpoint.

Section 1.6, the catalytic reduction of alcohols, phenols, and diols, is on a topic that was a bit of an oddity in the past, but has now assumed an entirely new meaning in the context of the conversion of renewables to chemicals and fuels. There is a lot of glycerol around since the production of biodiesel has begun on large scale. And, of course, the removal of hydroxy groups from sugars and phenols to transform them into useful chemicals is also desired. However, this is not an easy task and this field is only at the beginning of a long period of development. Because there are not many people working in this field, we had a hard time finding an author and therefore this chapter is authored by my colleague Sergey Tin and myself. We both think these are amazing reactions and we plan to investigate this topic ourselves in the future.

Section 1.7, on the hydrogenolysis of ethers, is very similar in this respect. Who would want to destroy an ether, if the ether was not obtained from a sugar [as in furfural and 5-(hydroxymethyl)furfural]? The expert in the field Keiichi Tomishige, aided by his colleagues Yoshinao Nakagawa and Masazumi Tamura, has performed an excellent piece of work in summarizing this emerging area.

Something that was also not on the horizon 10 years ago was the reduction of carbonates. In Section 1.8, Kuiling Ding and Yuehui Li describe this interesting new area that is closely connected to the use of carbon dioxide as a renewable building block. Not only is the reduction of organic carbonates described, but also the reduction of carbonate salts.

As mentioned above, the hydrogenation of carbon dioxide has taken on enormous importance in recent times. This is mainly related to the fact that the conversion of carbon dioxide into methanol (or methane) using hydrogen that is obtained by water electrolysis using renewable energy (windmills, photovoltaic cells) is seen as a feasible way to store surplus renewable energy. There already exists a genuine problem in the use of wind turbines in countries such as Germany and Denmark; if the wind blows hard, these countries produce so much electricity that they need to sell it at a negative price. Thus, storage would represent a much better option. Catherine Cazin and Fady Nahra have done an excellent job of summarizing the recent work in this area in Section 1.9. Not only that, they have also summarized the work on using carbon dioxide/hydrogen as methylation reagents.

We are all used to the existence of sluggish substrates that need special catalysts, and high temperatures and pressure, to be reduced. One could almost forget that there are also classes of chemicals that are too easily reduced as they are strong oxidants. Here, safety becomes the most important issue. In Section 1.10, Peter Poechlauer and Axel Zimmermann describe the reduction of peroxo compounds, ozonides, and molozonides, based for a large part on their own industrial experience in this field. This is very interesting chemistry, but please heed the safety warnings if you want to try these reactions yourself.

Section 1.11, on the reduction of sulfur compounds, is relatively small in view of the limited substrate scope. Nevertheless, Kiyotomi Kaneda and Takato Mitsudome have done an excellent job in summarizing the catalytic reduction of sulfoxides, sulfones, and disulfides. Yes, there are catalysts that keep working in the presence of sulfur compounds!

In Section 1.12, Banibrata Ghosh and Robert Maleczka, Jr. expertly describe catalytic dehalogenation reactions using metal catalysts. This field is of course strongly associated with remediation of polluting organic halides, but the chemistry occasionally also comes in handy for a total synthesis. Not surprisingly, this is an area where photocatalysis can bring some interesting results.

Volume 2 opens with Section 2.1 on the reduction of aldehydes, another area that has gained recent importance because of the renewables angle. Think, for instance, of the reduction of furfural to 2-(hydroxymethyl)furan, an important and large-scale industrial process, or the reduction of 5-(hydroxymethyl)furfural to 2,5-bis(hydroxymethyl)furan. My colleagues from Rostock, Kathrin Junge and Norbert Steinfeldt, have done an excellent job in summarizing the methodologies based on homogeneous catalysis as well as nanoparticle catalysis for these transformations, with a particular emphasis on the selective reduction of an aldehyde in the presence of other functional groups.

Who could be better suited to write Section 2.2, on transfer hydrogenation of ketones to alcohols, than Takao Ikariya, one of the pioneers of asymmetric transfer hydrogenation, together with his colleagues Yoshito Kayaki and Asuka Matsunami? Unfortunately, during the preparation of this book we learned of the passing of Professor Ikariya. He was an excellent scientist and a good friend and colleague, and he will be sorely missed by the chemistry community. Nevertheless, we are very grateful to his colleagues for finishing this excellent chapter.

Virginie Ratovelomanana-Vidal and her colleagues Quentin Llopis, Phannarath Phansavath, and Tahar Ayad had the daunting task of reviewing the catalytic hydrogenation of ketones (Section 2.3). Although initially the plan was to cover both enantioselective and non-chiral reductions, this was simply too much and hence it was decided to concentrate solely on the enantioselective hydrogenations. Even then, it was still an immense task that has been expertly performed. Ruthenium, iridium, and iron seem to be the metals of choice in this area. The diversity of ligands used is enormous.

Section 2.4, the hydrogenolysis of aryl ketones and aldehydes, benzylic alcohols and amines, and their derivatives, describes technology that is used quite frequently in synthesis. Think, for example, of the benzyloxycarbonyl (Z- or Cbz-group) deprotection via hydrogenolysis. Here, Ivana Fleischer and Benjamin Ciszek have done an excellent job of summarizing the recent developments in this field. Most reactions are catalyzed by heterogeneous catalysts, but the reduction of aromatic aldehydes and ketones can, of course, also be performed with homogeneous catalysts.

The hydrogenation of carboxylic acids and derivatives is an area where there has been an explosion of interest recently, and hence it was decided that this should be divided into separate parts covering heterogeneous catalysis and homogeneous catalysis. Section 2.5.1, on the homogeneous catalytic hydrogenation of carboxylic acids, anhydrides, esters, amides, and acid chlorides has been written by Yiping Shi, David Cole-Hamilton, and Paul Kamer. They have put together a splendid piece of work, even though David could not resist the temptation to narrate the entire history of homogeneous ester hydrogenation, which goes back to well before 2000; it is indeed interesting to be able to put everything in the correct context. An enormous amount of work was involved, but this has led to a fascinating chapter. In particular, the earlier work in the field of ester hydrogenation and also acid hydrogenation was performed at rather high temperatures. This is probably the main reason that the more stable metal pincer complexes seem to dominate this field.

To date, all industrial hydrogenations of carboxylic acid derivatives are performed at high temperatures using heterogeneous catalysts. Michèle Besson and Catherine Pinel have done an excellent job in Section 2.5.2, where they give an overview of the recent developments in the heterogeneous catalytic hydrogenation of carboxylic acids and derivatives. In this chapter, one can find discussion of the classical fatty acid hydrogenations, but also little gems such as the gas-phase hydrogenation of carboxylic acids to aldehydes. That is a great reaction!

Section 2.6 covers the reduction of imines and the reductive amination of aldehydes and ketones, an area where again much work has been done in the past two decades. This chapter is a very nice contribution from a French/Spanish consortium consisting of Carmen Claver, Philippe Kalck, Martine Urrutigoïty, and Itziar Peñafiel. Although the emphasis is on enantioselective reductions of imines and direct reductive amination, the heterogeneous methods are also summarized, making this a pretty complete overview.

And then there are all these other nitrogen-containing compounds that can also be reduced. Sandra Hinze, Pim Puylaert, and Arianna Savini, co-workers of mine in Rostock, have done a splendid job in Section 2.7, on the reduction of nitro compounds to amines, azo compounds, hydroxylamines, and oximes, and also of *N*-oxides to amines. Again, the emphasis here is very much on the selective reduction of these functional groups in the presence of other reducible entities. Heterogeneous catalysis is of course very important for these transformations, but some homogeneous catalysis, particularly based on iron catalysts is also included.

Section 2.8, on the reduction of azides, was written by Hironao Sajiki and Yasunari Monguchi. This transformation is, of course, very important to the synthetic chemist, as it is often used to transform an alcohol into an amine. The authors have done an excellent job in charting the relative selectivities of often-used catalysts and reaction conditions to effect only the reduction of azides in the presence of other functional groups. As a bonus, they also describe the selective reduction of alkynes in the presence of azides that remain untouched. Many synthetic examples are given in this chapter.

Section 2.9, the catalytic reduction of nitriles, was written by Bhalchandra Bhanage and Dattatraya Bagal. This excellent chapter is an interesting mix that includes coverage of old, well-established chemistry based on heterogeneous catalysis, but is mostly concerned with the new developments using metal nanoparticles and homogeneous catalysts, including systems based on iron, cobalt, and manganese. Of course, selectivity remains a very important issue in this chemistry, as the formation of secondary or even tertiary amines needs to be suppressed.

After having read all the chapters, I am happy to say that much has happened in the last 17 years. There have been many interesting new developments, some of which are only at an early stage. I think someone may need to do a new book on this subject 10 years from now.

Many thanks to all of the authors, who have dedicated so much of their precious time to writing the excellent contributions.

1.1 Reduction of Alkenes

1.1.1 Homogeneous Reduction of Alkenes

X. Tan, H. Lv, and X. Zhang

General Introduction

Transition-metal-catalyzed asymmetric hydrogenation of C=C bonds has become one of the most effective methods to synthesize chiral compounds. The wide variety of available alkene substrates, including α,β-unsaturated carbonyl compounds, enamides, enols, and other heteroatom-substituted alkenes, means that the asymmetric hydrogenation of C=C bonds can be used to construct a variety of chiral centers. Asymmetric hydrogenation has been extensively used in industry, and is one of the most successful homogeneous catalytic reactions employed. In 2001, William Knowles and Ryoji Noyori received the Nobel Prize for Chemistry for their great contributions to the development of asymmetric hydrogenation.[1]

Alkene substrates can be divided into two types, based on the mode of catalysis: simple alkenes without an auxiliary that can interact with the catalyst, and alkenes that do bear an auxiliary group that can interact with the catalyst. For simple alkenes, the predominant catalysts are iridium complexes, with the early prototype in this area developed by Crabtree;[2] asymmetric versions have been extensively studied over the last 20 years (Section 1.1.1.1).[3,4] With regard to alkenes containing coordinating functional groups, for the purpose of this review they have been divided into four categories: nitrogen-substituted alkenes (Section 1.1.1.2), oxygen-substituted alkenes (Section 1.1.1.3), carbonyl-substituted alkenes (Section 1.1.1.4), and alkenes substituted by other heteroatoms (Section 1.1.1.5).

There have been hundreds of reports of catalytic systems for the asymmetric hydrogenation of alkenes since the seminal discoveries by Knowles[5] and Kagan and Dang,[6] and a large number of prochiral unsaturated compounds have been successfully hydrogenated with excellent enantioselectivities. Ligands have played a key role in the development of the transition-metal-catalyzed hydrogenation of alkenes; representative examples have been selected for inclusion in this chapter. Phosphorus ligands, both mono- and di-, have been demonstrated to be the most important, while other types of ligand, such as those based on nitrogen and sulfur, have also gained importance, mostly as part of mixed heteroatom–phosphorus ligands.

On account of the large number of publications in this area, only the most important and representative methods are summarized here. The intent is to present the most significant recent results to the reader; thus, the selected examples were mainly published since 2000. It is hoped that together with other review articles[3,4,7] and books,[8–10] including a previous *Science of Synthesis* review on this topic,[11] this chapter will serve as an excellent reference source for chemists of all levels.

1.1.1.1 Reduction of Nonfunctionalized Alkenes

The asymmetric hydrogenation of nonfunctionalized alkenes has thus far been achieved with high enantioselectivity only when iridium catalysts have been used. Since Crabtree and co-workers reported the use of the mixed ligand complex $[Ir(cod)(py)(PCy_3)]PF_6$ in the 1970s,[2] great progress has been made in this area. The aforementioned compound, commonly referred to as Crabtree's catalyst, exhibits high catalytic reduction efficiency for both tri- and tetrasubstituted nonfunctionalized alkenes. Although Crabtree's catalyst is unusually air stable, both as a solid and in solution, in cases where the coordination of the alkene to the metal is poor, the active catalyst decomposes into an inactive trinuclear iridium hydride cluster. In 1997, Pfaltz and co-workers reported the first chiral mimic of Crabtree's catalyst $\{[Ir(cod)(PHOX)]PF_6\}$, using a phosphinodihydrooxazole (PHOX) ligand as a chiral N,P-chelating species.[12] On the basis of the conclusion by Crabtree et al. that the catalyst is deactivated in cases of poor alkene coordination, the extremely weakly coordinating counterion BARF {tetrakis[3,5-bis(trifluoromethyl)phenyl]borate} was tested as a replacement for PF_6, forming complexes of the type [Ir(cod)(PHOX)]BARF.[13] This change in the counterion led to catalysts that achieve higher turnover frequencies as well as good stability.

Following the pioneering work on catalytic asymmetric hydrogenation of nonfunctional alkenes reported by Pfaltz, other groups focused on the modification of the PHOX-type ligands by replacing the phosphine moiety with other phosphorus donor groups (i.e., phosphinite, carbene, and phosphite), or the dihydrooxazole moiety with different N-donor groups (such as pyridine, thiazole, and oxazole). Because of the unique and robust catalytic activities of the iridium–P,N-ligand systems, this section focuses on the iridium-catalyzed hydrogenation of nonfunctionalized alkenes employing a variety of ligands. In Scheme 1, the most distinguished ligands **1–29** for the asymmetric hydrogenation of nonfunctionalized alkenes **30** to give alkane products **31** are shown, and their catalytic performance is summarized in Scheme 2.[14–37] The criteria for the selection of examples in Scheme 2 include high enantioselectivity, good yield, and wide substrate scope.

Nonfunctionalized alkene substrates suitable for asymmetric hydrogenation can be divided into three classes: 1,1-disubstituted alkenes, trisubstituted alkenes, and tetrasubstituted alkenes. Compared to the more widely investigated trisubstituted alkenes, the asymmetric hydrogenation of 1,1-disubstituted and tetrasubstituted alkenes is more challenging. For 1,1-disubstituted alkenes, due to the absence of a third substituent on the double bond, the catalyst has to distinguish solely between the two alkyl/aryl substituents for enantiodiscrimination. This is a more demanding requirement compared to distinguishing between hydrogen and an alkyl or aryl group as in the case of the trisubstituted alkenes. Furthermore, alkenes with a 1,1-disubstituted double bond can often isomerize to form a more stable internal alkene, which usually leads to the predominant formation of the other enantiomer of the hydrogenated product.[3]

Scheme 1 Selected Ligands for the Reduction of Nonfunctionalized Alkenes

(2-Tol)$_2$P N Bu^t O — **1** (PHOX)

PPh$_2$ N O Bu^t — **2**

PCy$_2$ N O Bu^t — **3**

S PCy$_2$ N O Bu^t — **4**

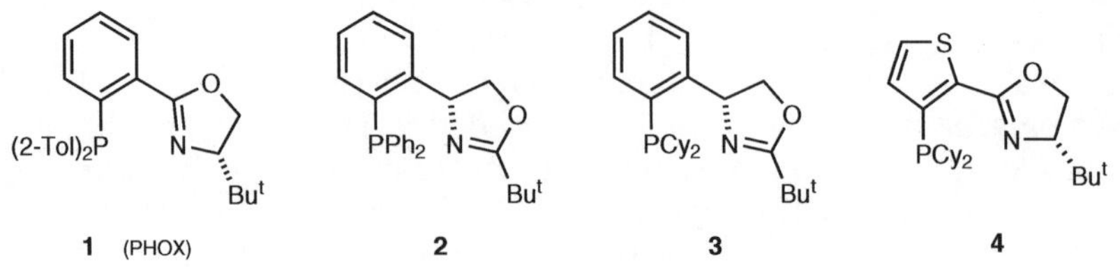

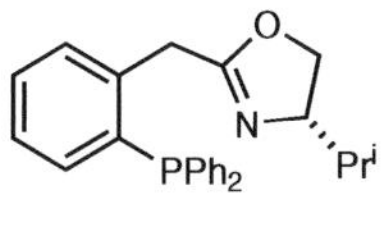

PPh2 Pr^i N O

5

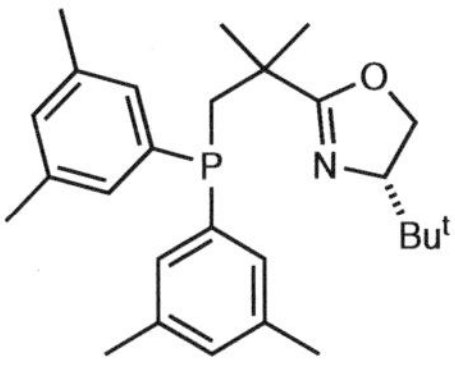

P N O Bu^t

6

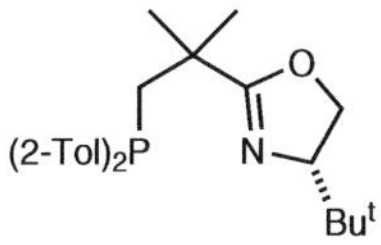

(2-Tol)2P N O Bu^t

7

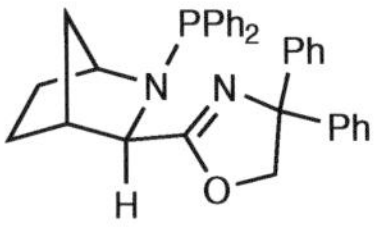

PPh2 N Ph Ph N O H

8

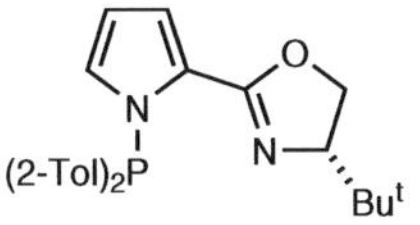

N (2-Tol)2P N O Bu^t

9

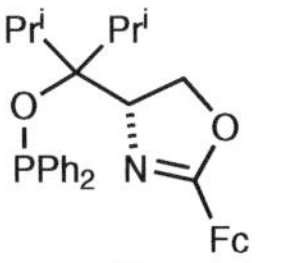

Pr^i Pr^i O PPh2 N O Fc

10

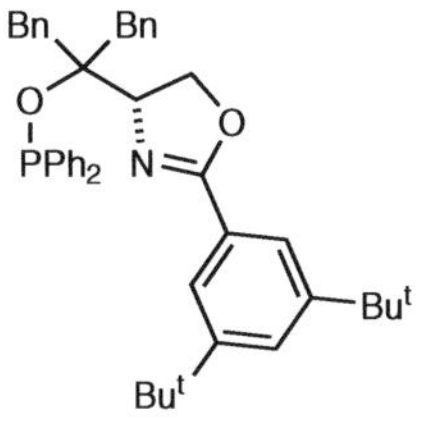

Bn Bn O PPh2 N O Bu^t Bu^t

11

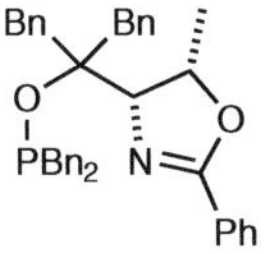

Bn Bn O PBn2 N O Ph

12

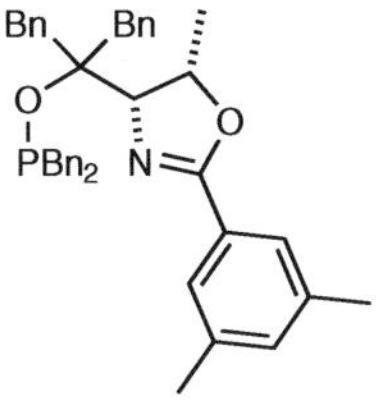

Bn Bn O PBn2 N O

13

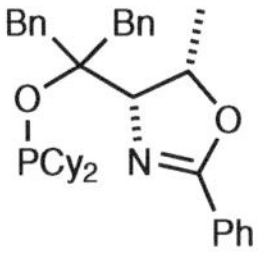

Bn Bn O PCy2 N O Ph

14

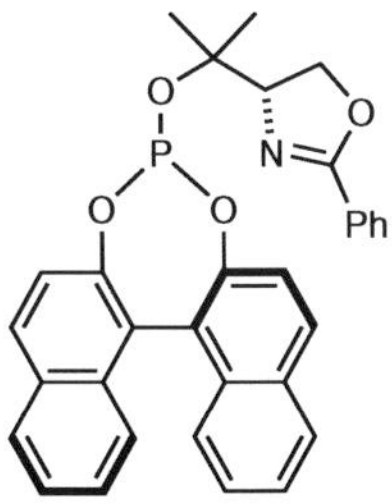

O P O O N O Ph

15

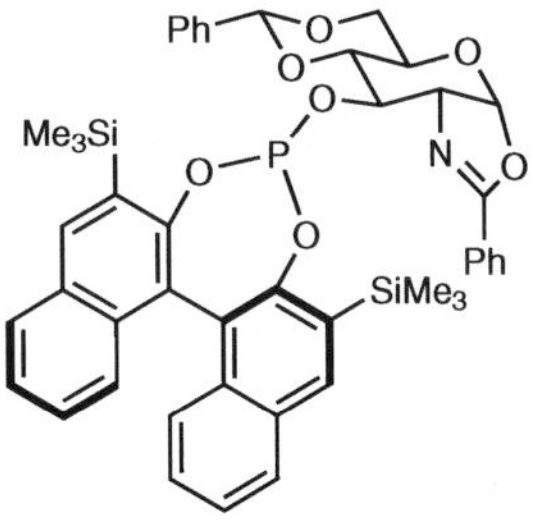

Ph O O O O Me3Si O P O N O Ph SiMe3

16

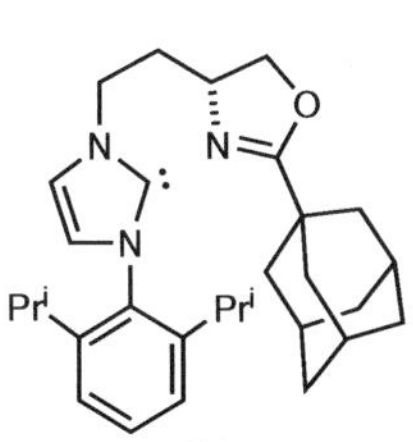

N N N O Pr^i Pr^i

17

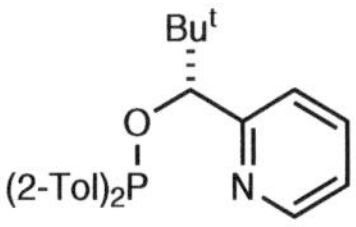

Bu^t O (2-Tol)2P N

18

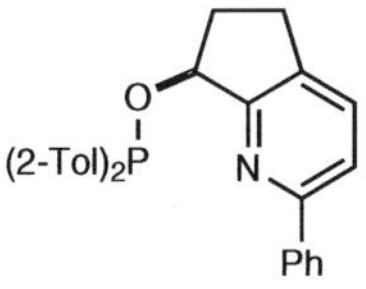

O (2-Tol)2P N Ph

19

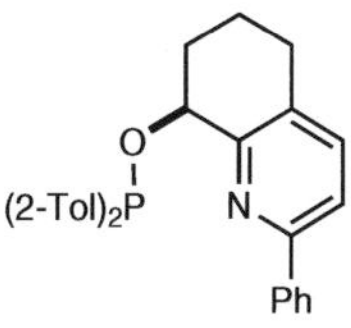

O (2-Tol)2P N Ph

20

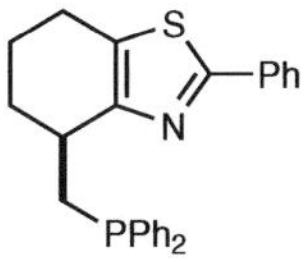

S Ph N PPh2

21

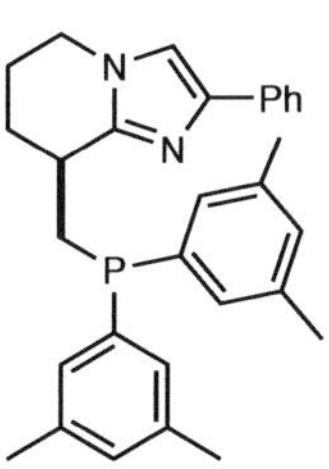

N Ph N P

22

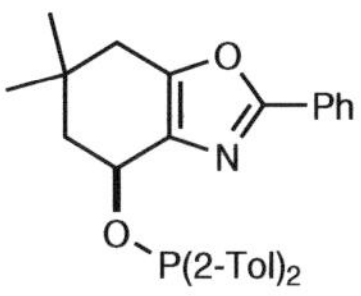

O Ph N O P(2-Tol)2

23

for references see p 63

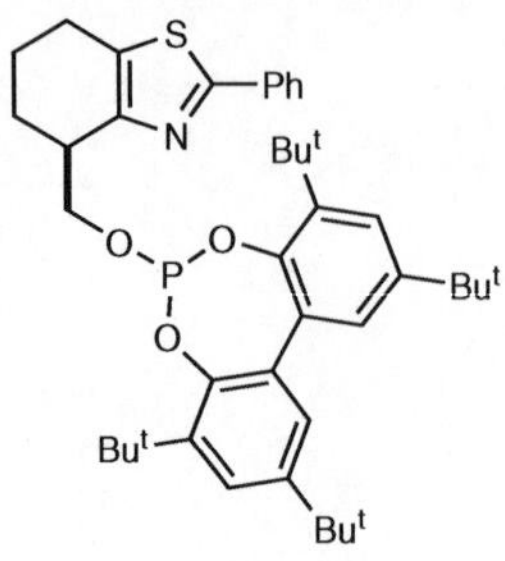

Scheme 2 Iridium-Catalyzed Hydrogenation of Nonfunctionalized Alkenes[14–37]

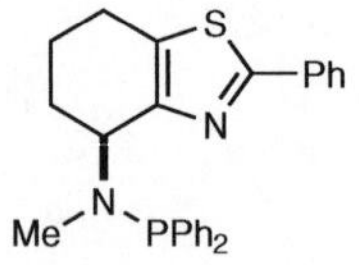

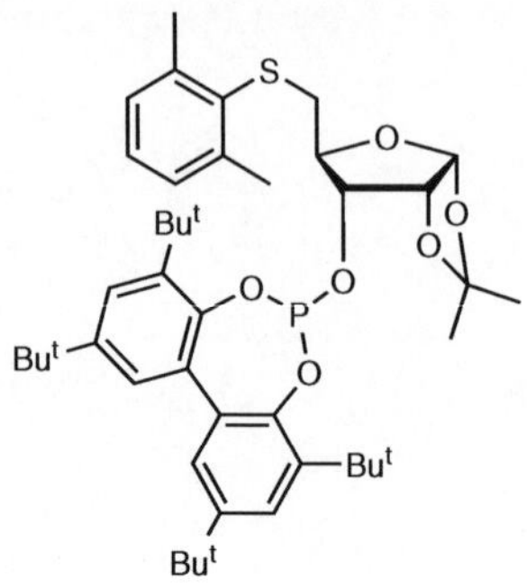

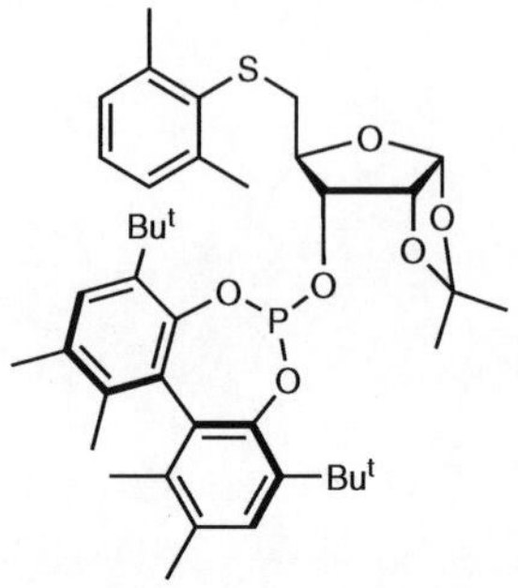

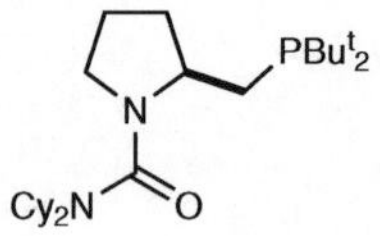

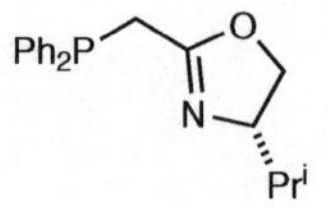

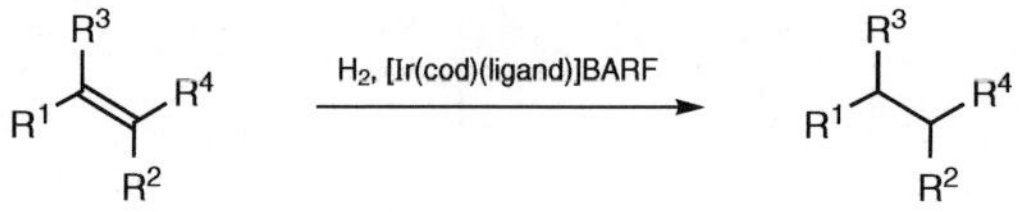

H_2, [Ir(cod)(ligand)]BARF

BARF = $[3,5\text{-}(F_3C)_2C_6H_3]_4B^-$

R^1	R^2	R^3	R^4	Ligand	Conversion or Yield (%)	ee (%)	Ref
Ph	H	Me	Ph	**1**	>99	97	[14]
Ph	H	Me	Ph	**3**	>99	99	[15]
Ph	H	Me	Ph	**4**	>99	99	[16]
Ph	H	Me	Ph	**5**	>99	97	[17]
Ph	H	Me	Ph	**6**	>99	98	[18]
Ph	H	Me	Ph	**8**	99	98	[19]
Ph	H	Me	Ph	**9**	100	99	[20]
Ph	H	Me	Ph	**10**	100	97	[21]
Ph	H	Me	Ph	**11**	100	98	[21]
Ph	H	Me	Ph	**12**	100	99	[22]
Ph	H	Me	Ph	**13**	100	99	[22]
Ph	H	Me	Ph	**15**	100	99	[23]
Ph	H	Me	Ph	**16**	100	>99	[24]

R^1	R^2	R^3	R^4	Ligand	Conversion or Yield (%)	ee (%)	Ref
Ph	H	Me	Ph	**17**	99	98	[25]
Ph	H	Me	Ph	**20**	>99	>99	[26]
Ph	H	Me	Ph	**23**	>99	>99	[27]
Ph	H	Me	Ph	**21**	>99	>99	[28]
Ph	H	Me	Ph	**24**	100	98	[29]
Ph	H	Me	Ph	**27**	100	99	[30]
Ph	H	Me	Ph	**28**	>99	>99	[31]
4-ClC_6H_4	H	Me	Ph	**1**	98	95	[14]
4-ClC_6H_4	H	Me	Ph	**5**	98	95	[17]
4-$MeOC_6H_4$	H	Me	Ph	**1**	>99	98	[14]
4-$MeOC_6H_4$	H	Me	Ph	**5**	>99	98	[17]
4-$MeOC_6H_4$	H	Et	Ph	**1**	97	95	[14]
4-$MeOC_6H_4$	H	Et	Ph	**2**	>99	97	[15]
4-$MeOC_6H_4$	H	Et	Ph	**5**	97	95	[17]
4-Tol	H	Ph	Ph	**21**	>99	>99	[32]
4-$MeOC_6H_4$	H	Ph	Ph	**21**	>99	95	[32]
3,5-$Me_2C_6H_3$	H	Ph	Ph	**21**	>99	>99	[32]
4-$MeOC_6H_4$	Me	Me	H	**6**	>99	96	[18]
4-$MeOC_6H_4$	Me	Me	H	**13**	100	92	[22]
4-$MeOC_6H_4$	Me	Me	H	**16**	100	95	[24]
4-$MeOC_6H_4$	Me	Me	H	**19**	>99	98	[26]
4-$MeOC_6H_4$	Me	Me	H	**27**	100	94	[30]
4-Tol	$(CH_2)_4Me$	Ph	H	**22**	>99	>99	[32]
Ph	H	Me	Me	**15**	100	99	[23]
Ph	H	Me	Me	**16**	100	>99	[24]
Ph	H	Me	Me	**21**	>99	99	[28]
Ph	H	Me	Me	**24**	100	99	[29]
Ph	H	Me	Me	**26**	100	99	[33]
Ph	H	Me	Me	**27**	100	99	[30]
4-$MeOC_6H_4$	H	Me	Me	**8**	99	99	[19]
4-$MeOC_6H_4$	H	Me	Me	**11**	100	95	[21]
4-$MeOC_6H_4$	H	Me	Me	**12**	100	99	[22]
4-$MeOC_6H_4$	H	Me	Me	**13**	100	99	[22]
4-$MeOC_6H_4$	H	Me	Me	**15**	100	99	[23]
4-$MeOC_6H_4$	H	Me	Me	**16**	100	99	[24]
4-$MeOC_6H_4$	H	Me	Me	**19**	>99	>99	[26]
4-$MeOC_6H_4$	H	Me	Me	**22**	>99	98	[34]
4-$MeOC_6H_4$	H	Me	Me	**23**	>99	96	[27]
4-$MeOC_6H_4$	H	Me	Me	**21**	>99	99	[28]
4-$MeOC_6H_4$	H	Me	Me	**24**	100	99	[29]
4-$MeOC_6H_4$	H	Me	Me	**26**	100	99	[33]
4-$MeOC_6H_4$	H	Me	Me	**27**	100	98	[30]
4-$MeOC_6H_4$	H	Me	iPr	**17**	100	97	[25]

for references see p 63

R^1	R^2	R^3	R^4	Ligand	Conversion or Yield (%)	ee (%)	Ref
4-BrC$_6$H$_4$	H	Ph	Me	**22**	>99	95	[32]
4-BrC$_6$H$_4$	H	Ph	Me	**25**	>99	95	[32]
4-PhC$_6$H$_4$	H	Ph	(CH$_2$)$_4$Me	**22**	99	99	[32]
4-PhC$_6$H$_4$	H	Ph	(CH$_2$)$_4$Me	**25**	99	99	[32]
4-Tol	H	Ph	(CH$_2$)$_4$Me	**25**	>99	97	[32]
4-MeOC$_6$H$_4$	H	Et	H	**14**	>99	94	[35]
4-MeOC$_6$H$_4$	H	Et	H	**15**	100	97	[23]
4-MeOC$_6$H$_4$	H	Et	H	**16**	100	>99	[24]
4-MeOC$_6$H$_4$	H	Et	H	**23**	>99	97	[27]
4-MeOC$_6$H$_4$	H	Et	H	**24**	100	97	[29]
4-BrC$_6$H$_4$	H	Et	H	**14**	>99	94	[35]
3-BrC$_6$H$_4$	H	Et	H	**14**	>99	94	[35]
Ph	H	Et	H	**15**	100	95	[23]
Ph	H	Et	H	**16**	100	99	[24]
Ph	H	*t*-Bu	H	**15**	100	>99	[23]
Ph	H	*t*-Bu	H	**16**	100	97	[24]
4-F$_3$CC$_6$H$_4$	H	*t*-Bu	H	**16**	100	97	[36]
Ph	H	Bu	H	**24**	100	95	[29]
Ph	H	iPr	H	**24**	100	95	[29]
Ph	H	Cy	H	**24**	100	94	[29]
MeO		Me	H	**7**	>99	96	[18]
MeO		Me	H	**8**	99	95	[19]
MeO		Me	H	**9**	100	92	[20]
MeO		Me	H	**15**	100	96	[23]
MeO		Me	H	**24**	100	99	[29]
		H	Me	**21**	>99	98	[28]

R^1	R^2	R^3	R^4	Ligand	Conversion or Yield (%)	ee (%)	Ref
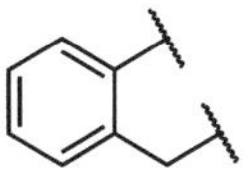		Me	Me	**29**	>99	94	[37]
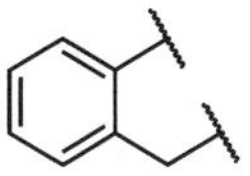		Bu	Me	**29**	>99	94	[37]
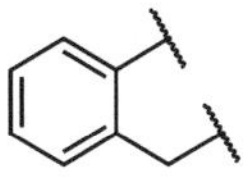		Ph	Me	**29**	99	95	[37]

Thus, a number of distinguished ligands for the iridium-catalyzed asymmetric hydrogenation of nonfunctionalized alkenes are shown in Scheme 1. Most of these are phosphine–dihydrooxazole, *N*-phosphine–dihydrooxazole, and phosphinite–dihydrooxazole ligands, and to a lesser extent phosphite/phosphoramidite–dihydrooxazole and carbene-based ligands.

Along with PHOX (**1**), the phosphine–dihydrooxazole ligands **2–7** show good performance in the iridium-catalyzed asymmetric hydrogenation of a number of simple alkenes, giving the target alkanes with high enantioselectivities and yields.[14–18] The most commonly used counterion is BARF. Catalyst loadings range from 0.1 to 1 mol%. Heteroatom-containing *N*-phosphine–dihydrooxazole ligands **8** and **9**[19,20] and phosphinite/phosphite–dihydrooxazole ligands **10–16**,[21–24] which have more flexible backbones, can also provide high enantioselectivities for a wide substrate scope. Carbene containing ligand **17** has also been successfully used for the asymmetric hydrogenation of a range of nonfunctionalized alkenes.[25]

Although most of the ligands that have been developed for the iridium-catalyzed asymmetric hydrogenation of minimally functionalized alkenes contain a dihydrooxazole unit, other nitrogen donor groups have also been successfully used. The most successful cases are ligands containing a pyridine (**18–20**),[26,38] a thiazole (**21**, **24**, and **25**),[28,29,32] an imidazole **22**,[32] and an oxazole **23**.[27]

Most of the research in the design of new chiral ligands has been focused on developing chiral mimics of Crabtree's catalyst; thus, the possibility of developing ligand motifs where the nitrogen donor atom is exchanged for a different heteroatom has not been investigated to such a large extent. Representative chiral mimics are phosphite–sulfur ligands **26**[33] and **27**,[30] and proline-based P,O-ligand **28**.[31]

(*R*)-2-(4-Methoxyphenyl)-1-phenylpropane (31, **R^1 = 4-$MeOC_6H_4$; R^2 = H; R^3 = Me; R^4 = Ph); Typical Procedure Using [Ir(cod)(1)]BARF as Catalyst:**[14]

To a 35-mL autoclave with a magnetic stirrer was added [Ir(cod)(**1**)]BARF (1.6 mg, 1 μmol, 0.3 mol%), alkene **30** (R^1 = 4-$MeOC_6H_4$; R^2 = H; R^3 = Me; R^2 = Ph; 74 mg, 0.33 mmol), and CH_2Cl_2 (0.3 mL). The autoclave was sealed and pressurized with H_2 to 50 atm, and the mixture was stirred for 2 h. After depressurizing the autoclave, the contents were removed and the solvent was removed and replaced with heptane (3 mL). The soln was passed through a short plug of silica gel (0.5 cm) to remove the metal salts. Analysis by GC indicated 99.8% conversion into (*R*)-2-(4-methoxyphenyl)-1-phenylpropane (**31**, R^1 = 4-$MeOC_6H_4$; R^2 = H; R^3 = Me; R^2 = Ph). Kugelrohr distillation (130–140 °C/0.04 mbar) afforded a clear oil; yield: 75 mg (99%); 98% ee [determined by HPLC (Chiralcel OJ, iPrOH/heptane

5:95, flow rate: 0.5 mL·min^{-1}, 20 °C, 254 nm, t_R (*R*) 15.6 min, t_R (*S*) 20.3 min]. A reaction using 0.3 mol% of [Ir(cod)(**1**)]BARF and 4 mmol of **30** (R^1 = 4-$MeOC_6H_4$; R^2 = H; R^3 = Me; R^2 = Ph) gave the same results.

(*S*)-1,2-Diphenylpropane (31, R^1 = R^4 = Ph; R^2 = H; R^3 = Me); Typical Procedure Using [Ir(cod)(17)]BARF as Catalyst:[25]

(*E*)-1,2-Diphenylprop-1-ene (**30**, R^1 = R^4 = Ph; R^2 = H; R^3 = Me; 0.2 mmol), iridium complex [Ir(cod)**17**]BARF (1.2 μmol, 0.6 mol%), and CH_2Cl_2 (100 μL) were added to a test tube containing a small stirrer bar. The tube was placed in a bomb, which was pressurized to 50 atm with H_2. The mixture was stirred at 300 rpm for 2 h. The bomb was then vented, and the mixture was passed through a short silica gel plug using EtOAc/hexanes (3:7) as the eluent. The product soln was collected in a vial containing a known amount of dodecane. The yield (99%) and ee (98%) of the product were determined by GC analysis using a chiral column.

1-(Butan-2-yl)-4-methoxybenzene (31, R^1 = 4-$MeOC_6H_4$; R^2 = H; R^3 = R^4 = Me); Typical Procedure Using [Ir(cod)(21)]BARF as Catalyst:[28]

A vial was charged with (*E*)-2-(4-methoxyphenyl)but-2-ene (**30**, R^1 = 4-$MeOC_6H_4$; R^2 = H; R^3 = R^4 = Me; 0.5 mmol) and [Ir(cod)(**21**)]BARF (0.5–1 mol%), and anhyd CH_2Cl_2 (2.0 mL) was added. The vial was placed in a high-pressure reactor, which was purged with argon (3 ×) before it was pressurized to 50 atm and held at this pressure for 2 h. The pressure was released and the solvent was removed by evaporation. Et_2O/pentane (1:1; 1.5 mL) was added and the soln was filtered through a short plug of silica gel, and then the solvent was evaporated. The ee (99%) was determined by GC or HPLC, and conversion (>99%) was determined by ^1H NMR spectroscopy.

(*R*)-1,2-Diphenylpropane (31, R^1 = R^4 = Ph; R^2 = H; R^3 = Me); Typical Procedure Using [Ir(cod)(28)]BARF as Catalyst:[31]

In a glovebox, a stock soln was prepared of the ligand and $[Ir(cod)_2]$BARF (1:1), or of the precatalyst in CH_2Cl_2 (5 mM soln). The alkene **30** (R^1 = R^4 = Ph; R^2 = H; R^3 = Me; 250 μmol) was weighed into a separate vial, and a 0.5-mL portion of the stock soln {1 mol% of [Ir(cod)(**28**)]BARF} was added. A stirrer bar was added and four vials (1.5 mL) were placed into a 60-mL autoclave. The autoclave was purged and then pressurized with H_2 (99.995%) to 50 atm. The mixtures were stirred at rt for 2 h. After pressure release, the soln was concentrated in a stream of N_2 and the residue was taken up in heptane (3 mL) and filtered through a short plug of silica gel (0.5 × 6 cm), eluting with heptane/propan-2-ol (7:3). The filtrates were analyzed by GC and HPLC.

1.1.1.2 Reduction of Nitrogen-Substituted Alkenes

Chiral amines are important organic compounds and chemical intermediates, and are frequently found in natural and nonnatural products. They can be used as resolution reagents and chiral auxiliaries, or as intermediates for the synthesis of a variety of biologically active molecules. The catalytic enantioselective hydrogenation of enamides, which have an acyl group at the nitrogen atom, using transition-metal complexes bearing chiral ligands has been found to be one of the most efficient and convenient methods for the preparation of chiral nonracemic amines and their derivatives.

In 1972, Kagan and Dang reported the first example of the enantioselective hydrogenation of *N*-acetyl enamides using a rhodium catalyst containing the ligand Diop [2,3-*O*-isopropylidene-2,3-dihydroxy-1,4-bis(diphenylphosphino)butane] for the synthesis of protected chiral amines.[6] Since then, many examples of the catalytic enantioselective hydrogenation of enamides to prepare optically pure amines and their derivatives have been reported. A large number of efficient transition-metal catalysts with chiral ligands have been developed for the hydrogenation of a wide range of enamides, leading to protected chiral primary amines with good to excellent enantioselectivities.[7]

In this section, the intent is to provide an overview of transition-metal-catalyzed methods for the enantioselective hydrogenation of enamides that have been reported since 2000. The enamide substrates are classified into three types: simple enamides, dehydro-α-amino acid derivatives, and dehydro-β-amino acid derivatives. Some representative diphosphorus and monophosphorus ligands **32–96** that have displayed excellent performance in the catalytic hydrogenation of enamides are shown in Schemes 3 and 4.[39–79]

Scheme 3 Representative Diphosphorus Ligands Applied in the Reduction of N-Substituted Alkenes

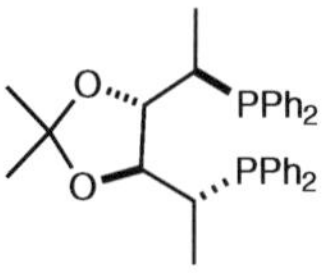

32 (Me-Diop) 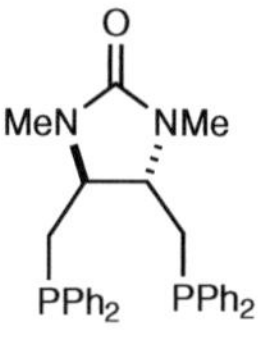 **33** (BDPMI) 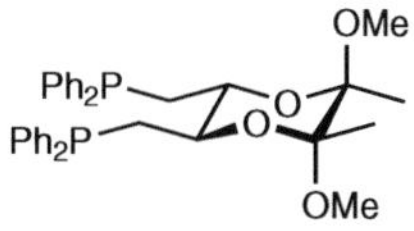 **34** 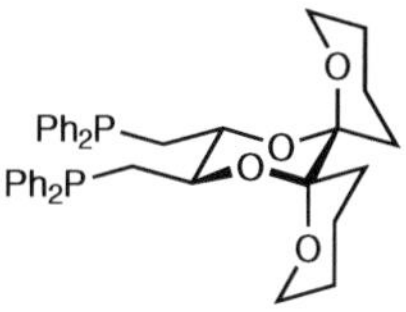 **35** (SK-Phos)

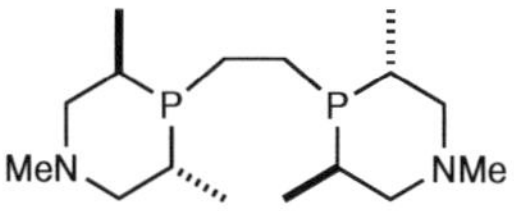

36 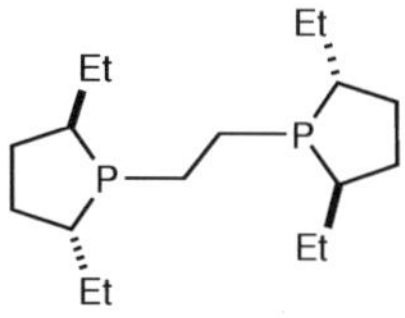 **37** (Et-BPE) 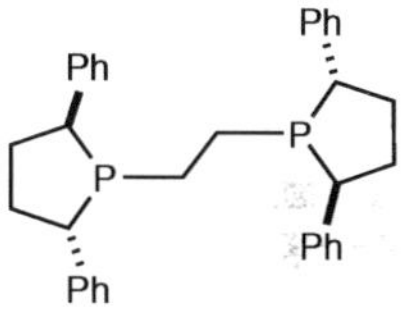 **38**

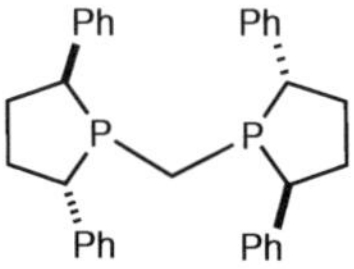

39 (Ph-BPM) 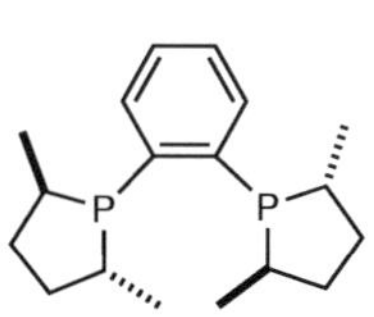 **40** (Me-DuPhos) 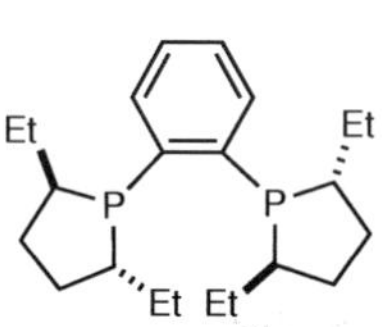 **41** (Et-DuPhos)

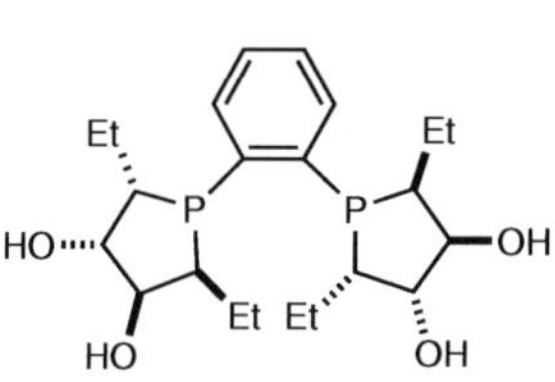

42 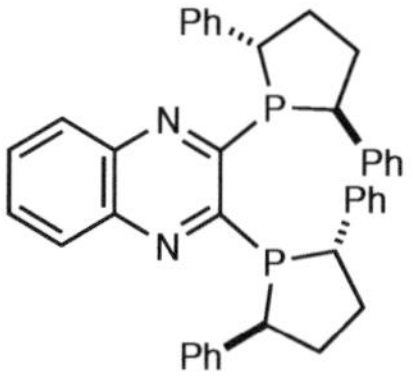 **43** (Ph-Quinox) 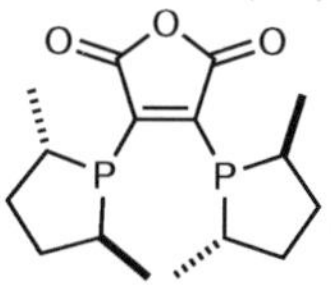 **44** (MalPhos)

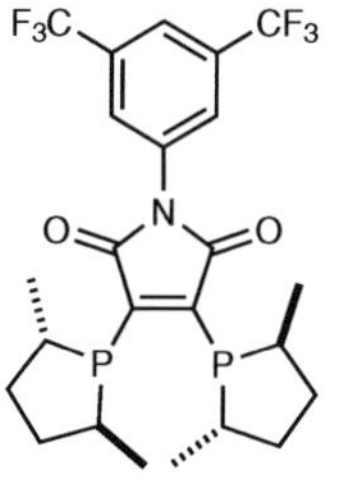

45 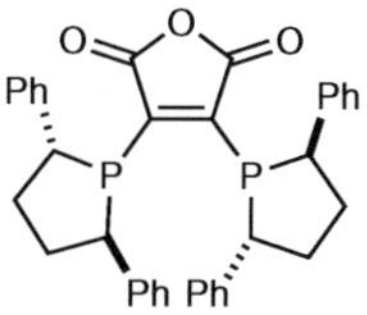 **46** (Ph-MalPhos) 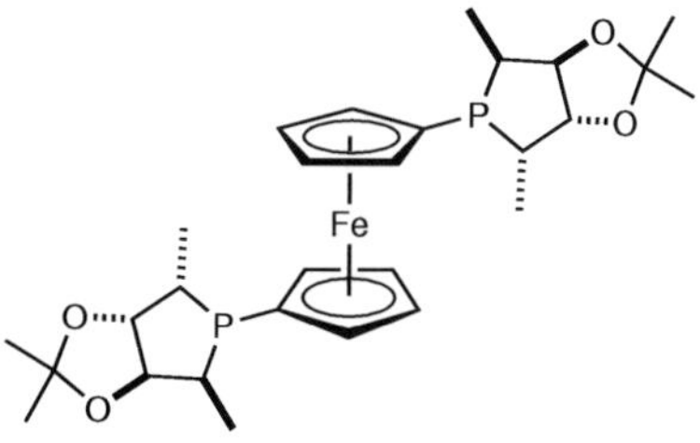 **47**

for references see p 63

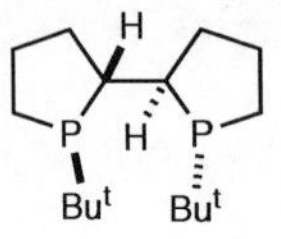

48 (TangPhos) 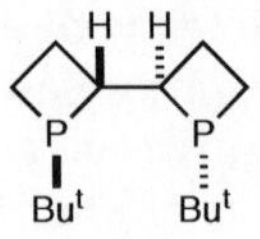 **49** (DisquareP*) 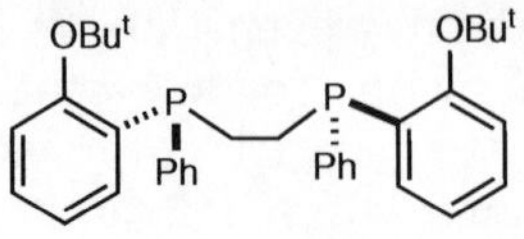 **50** (*t*-Bu-SMS-Phos) 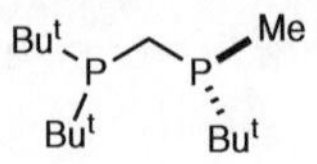 **51** (TrichickenfootPhos)

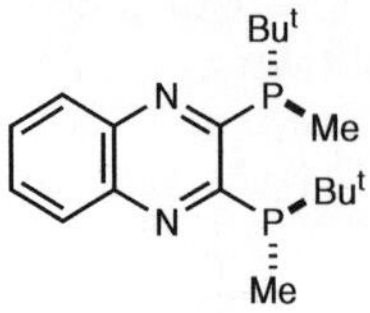

52 (QuinoxP*) 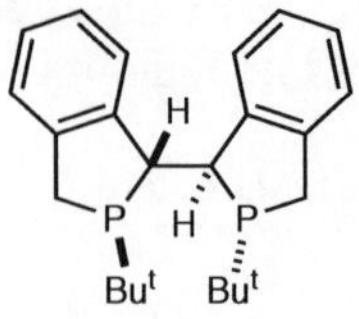 **53** (DuanPhos) 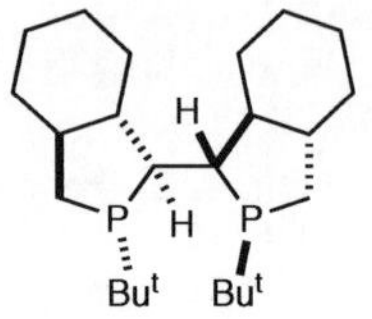 **54** (ZhangPhos)

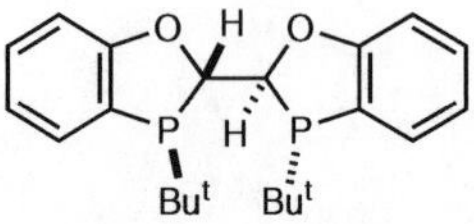

55 (BIBOPs) 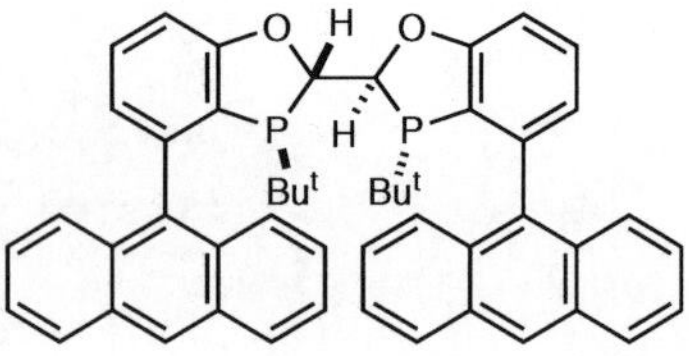 **56** (WingPhos)

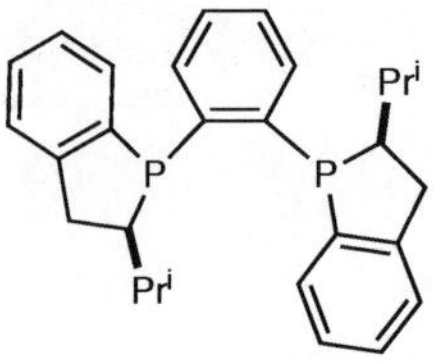

57 (iPr-BeePhos) 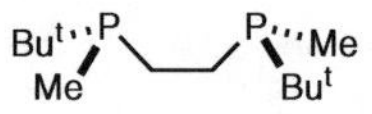 **58** (*t*-Bu-BisP*) 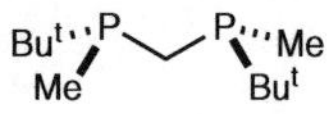 **59** (*t*-Bu-MiniPhos)

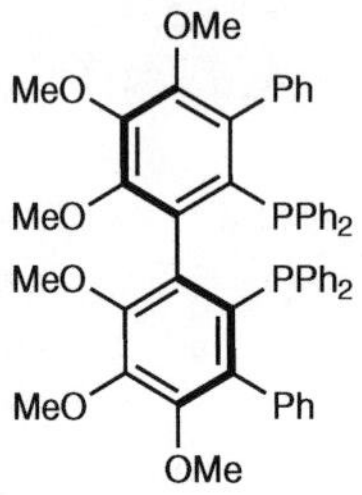

60 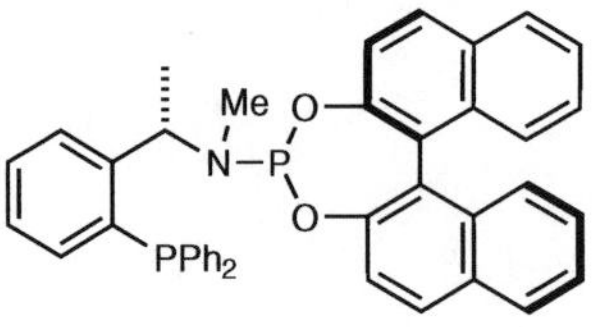 **61** 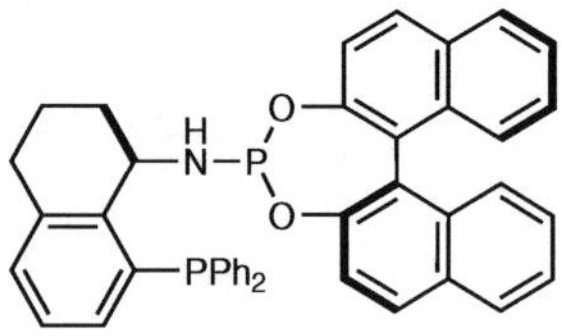 **62** (THNAPhos)

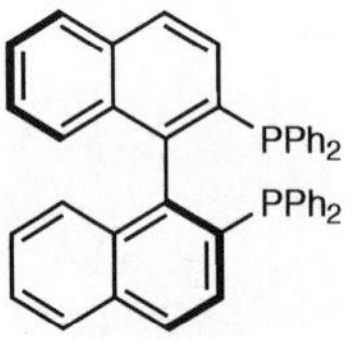

63 (BINAP) 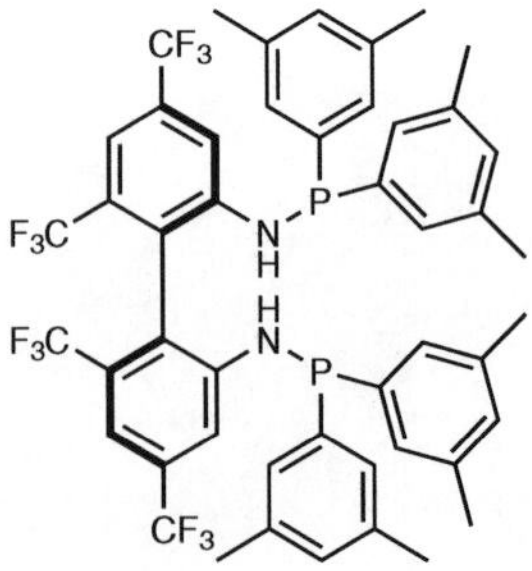 **64** 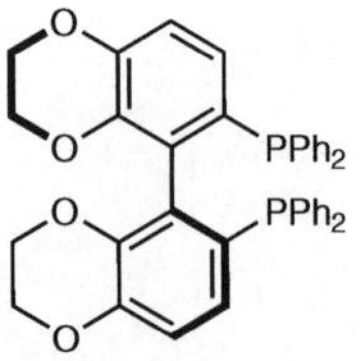 **65** (SynPhos)

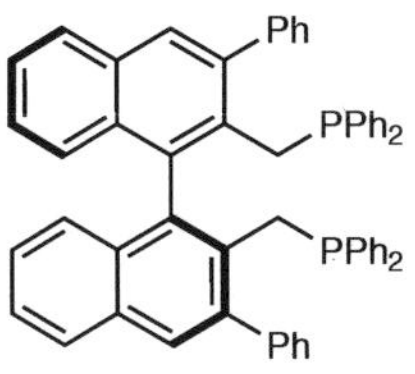

66 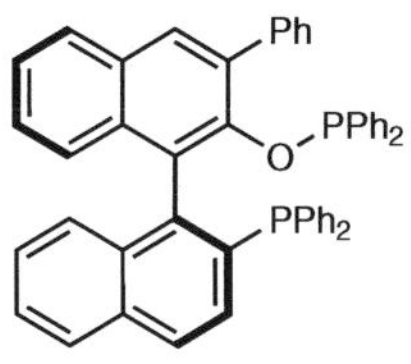 **67** 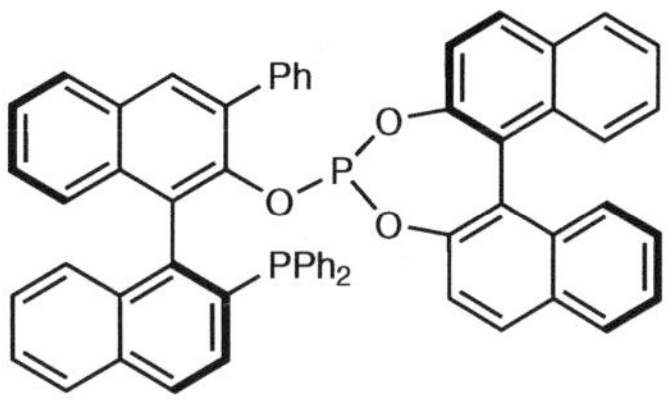 **68**

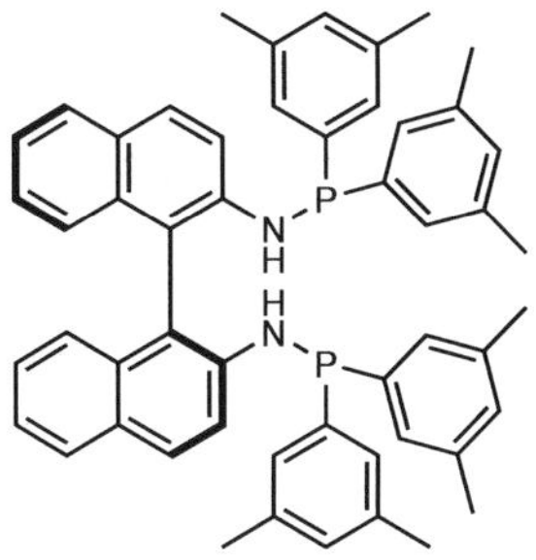

69 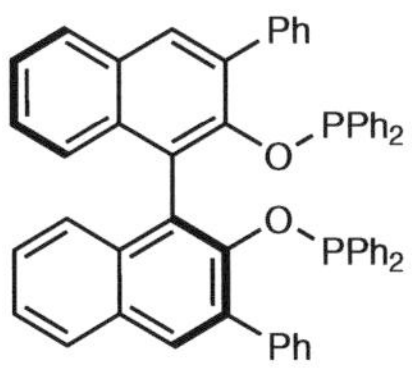 **70**

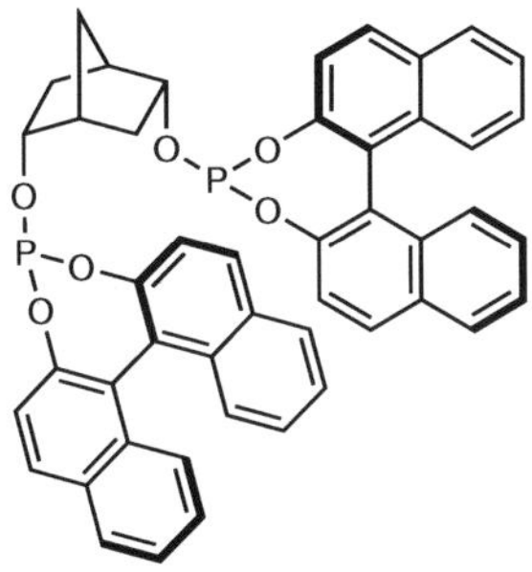

71 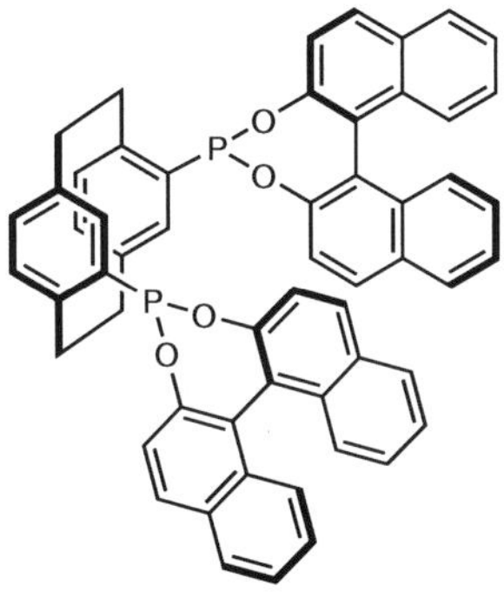 **72**

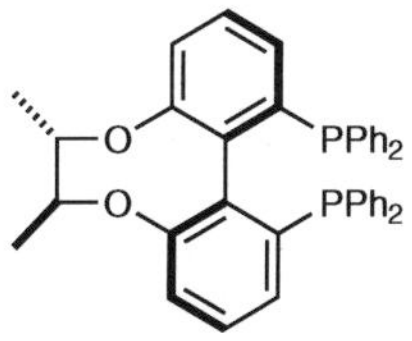

73 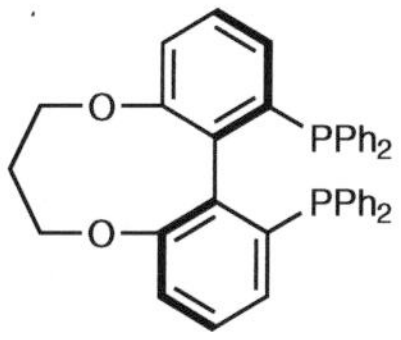 **74** (C$_3$-TunaPhos)

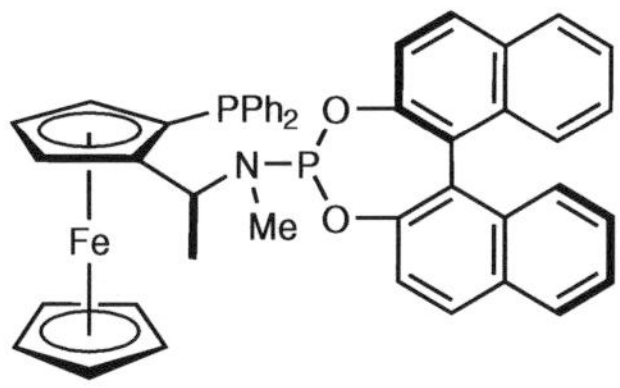

75 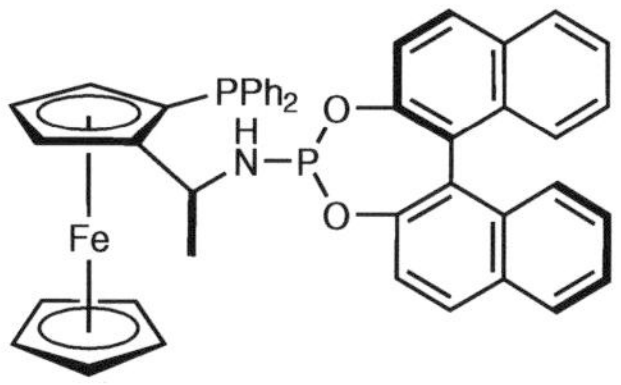 **76**

for references see p 63

F_3C CF_3 N Me P CF_3 PPh_2 Fe CF_3 PBu^t_2 PPh_2 Fe PBu^t_2 P CF_3 Fe CF_3

77 **78** **79**

Ph PPh_2 Ph_2P N N N Fe Me N PPh_2 PPh_2 Fe

80 (ClickFerroPhos) **81** (BoPhoz)

Scheme 4 Representative Monophosphorus Ligands Applied in the Reduction of N-Substituted Alkenes

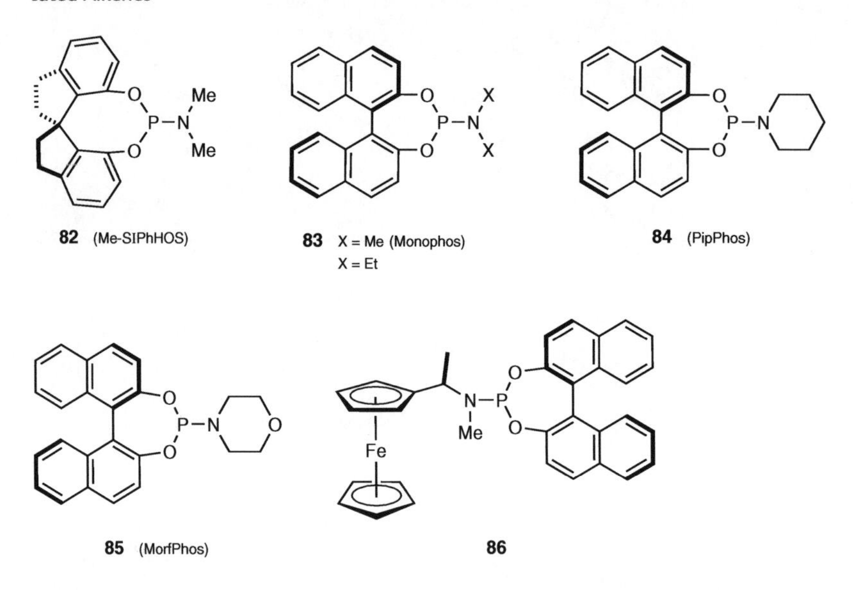

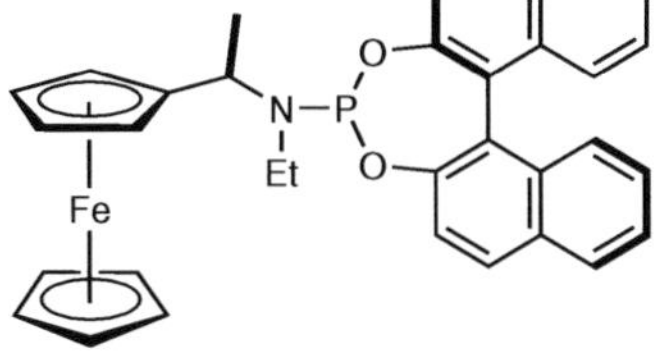

87

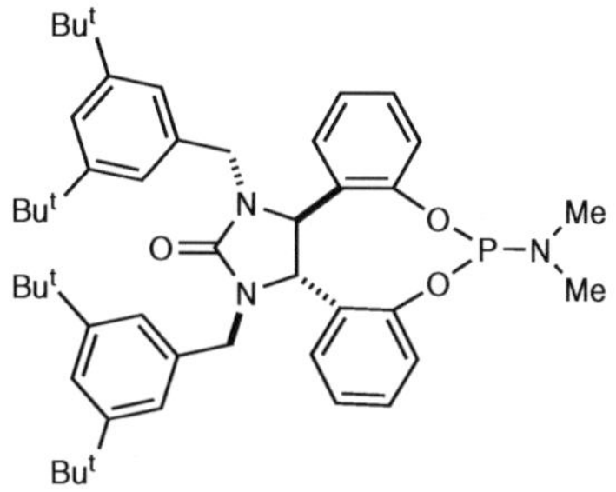

88 (3,5-t-Bu_2-Bn-DpenPhos)

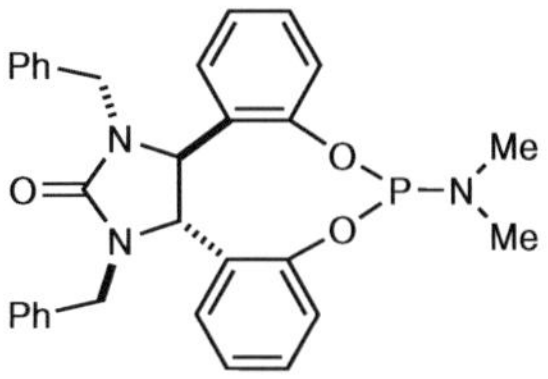

89 (Bn-DpenPhos)

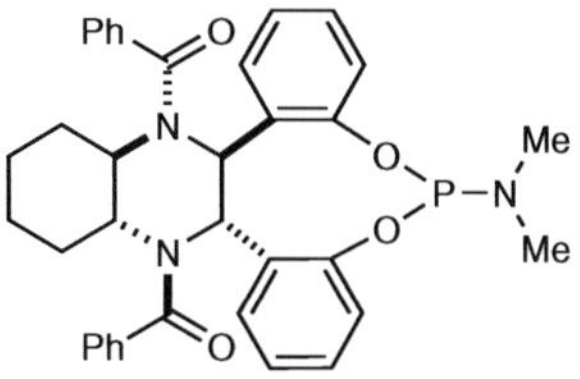

90 (CydamPhos)

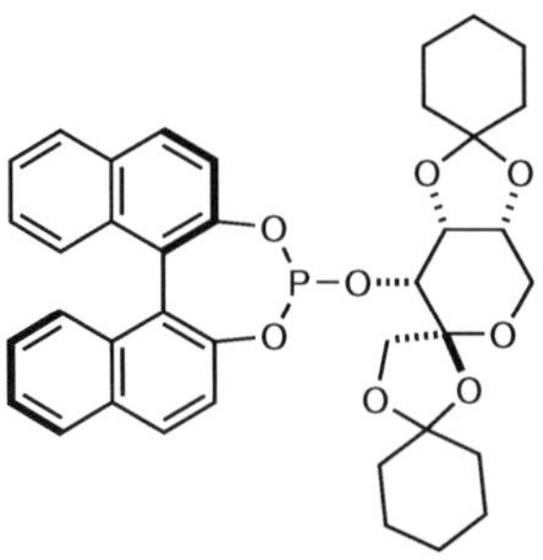

91

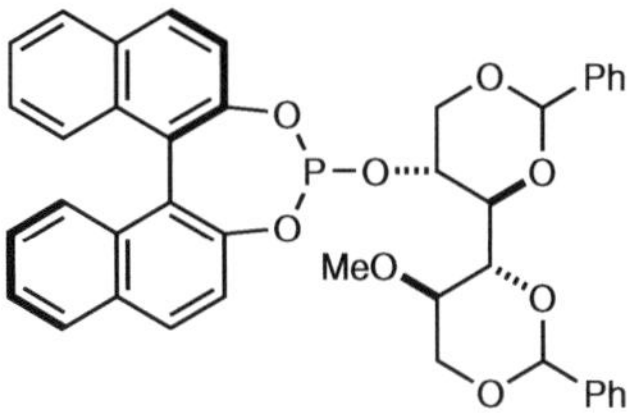

92 (ManniPhos)

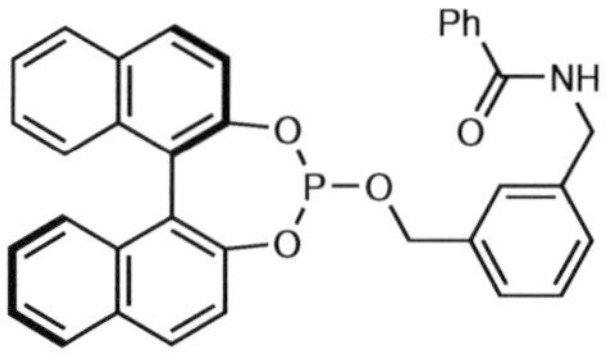

93

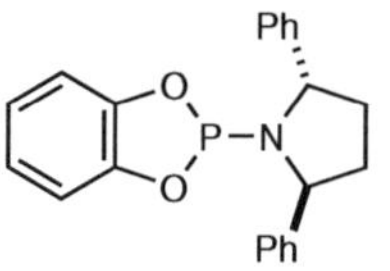

94

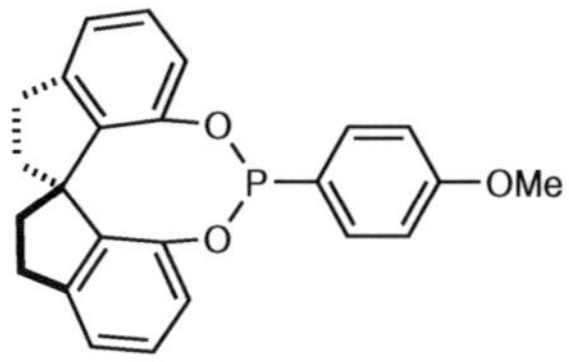

95

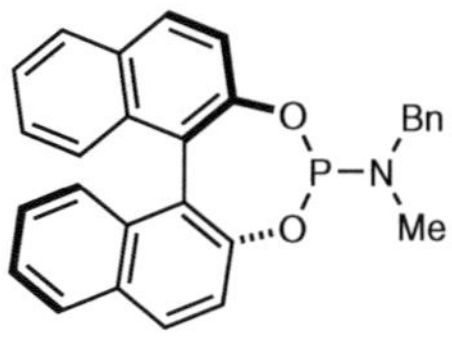

96

for references see p 63

1.1.1.2.1 Reduction of Enamides

Since Kagan and Dang reported that a rhodium catalyst containing the chiral bisphosphine ligand Diop is efficient for the enantioselective hydrogenation of *N*-acetyl enamides,[6] a number of rhodium catalysts bearing chiral ligands have been developed. These rhodium catalysts, as well as several ruthenium and iridium catalysts, show high reactivity for the hydrogenation of a wide range of acyclic and cyclic enamides to chiral acylamines with high enantioselectivities.

In general, the most efficient catalytic asymmetric hydrogenation systems for enamides are based on cationic rhodium complexes with a chiral phosphorus ligand, a diene (cycloocta-1,5-diene or norbornadiene), and a non-binding counterion such as tetrafluoroborate, perchlorate, trifluoromethanesulfonate, hexafluoroantimonate, or hexafluorophosphate. The reductions are typically carried out with hydrogen gas in an alcohol or halogenated solvent. The catalysts can be generated in situ or by can be pre-prepared. The diene is hydrogenated off to generate the active catalyst. Examples of the hydrogenations of enamides to give *N*-alkylamide products **97** and **98** are shown in Schemes 5 and 6.[39–79]

Acyclic β-unsubstituted enamides, such as *N*-(1-arylvinyl)amides, have been the most investigated substrates in the asymmetric hydrogenation of enamides (Scheme 5). Many chiral diphosphorus and monophosphorus ligands have been used in systems for the asymmetric hydrogenation of these substrates, inducing high yields and enantioselectivities. The DuPhos analogue **42**,[40] Diop analogue **32**,[39] the related ligand BDPMI (**33**),[41] and BPE analogue **36**[48] have designs that are based on carbon chirality. Their use results in excellent enantioselectivity in the asymmetric hydrogenation of *N*-(1-arylvinyl)amides. The P-chiral ligands **48**,[42] **49**,[43] **50**,[50] **55**,[51] **52**,[52] **53**,[53,68] **54**,[54] and **51**[55] all induce high enantioselectivities, especially the chiral ligands TangPhos (**48**), DuanPhos (**53**), and ZhangPhos (**54**) developed by Zhang et al. Chiral ligands based on axial chirality or a combination of axial chirality and carbon chirality (e.g., **61**,[46] **62**,[47] and **64**[49]) are also highly effective in the asymmetric hydrogenation of *N*-(1-arylvinyl)amides. Usually, examples of ligands with planar chirality have a structure based on a ferrocene backbone; thus, ferrocene-based ligands have played an important role in such asymmetric hydrogenations. Ligands of particular note include **75**[44] and **77**.[45]

The development of chiral monodentate phosphorus ligands was an important breakthrough in the history of asymmetric hydrogenation. Because these ligands are quite easily prepared, in particular those based on 1,1′-binaphthalen-2,2′-diol (BINOL) and its analogues, they are much less expensive than the bisphosphine ligands. In 2000, monodentate phosphonites were introduced by Pringle and co-workers,[80] monodentate phosphites by Reetz and co-workers,[81] and monodentate phosphoramidites by Minnaard, Feringa, and de Vries.[82] Monodentate phosphites and phosphoramidites have both been used for the asymmetric hydrogenation of enamides.[58,83] Subsequently, Zhou et al. developed the chiral spiro monophosphoramidite ligand SIPHOS (**82**),[56] which is also highly efficient for the rhodium-catalyzed asymmetric hydrogenation of *N*-(1-arylvinyl)amides. Monophosphorus ligands **83** (X = Me, Et),[57] **84**,[58] **85**,[58] **86**,[59] **91**,[62] **92**,[63] and **93**,[64] based on axial chirality, are excellent ligands for the rhodium-catalyzed *N*-(1-arylvinyl)-amide reduction. Other ligands **88**[60] and **90**[61] based on carbon-centered chirality are also effective for the hydrogenation of these substrates.

In addition to *N*-(1-arylvinyl)amides, *N*-(1-alkylvinyl)amides such as *N*-(3,3-dimethylbut-1-en-2-yl)amides and *N*-[1-(1-adamantyl)vinyl]amides can also be hydrogenated to the corresponding alkylamides using ligands **58**,[67] **52**,[52] **51**,[55] and **44**.[66]

Scheme 5 Rhodium-Catalyzed Asymmetric Hydrogenation of Enamides with a 1,1-Disubstituted Alkene[39–68,72]

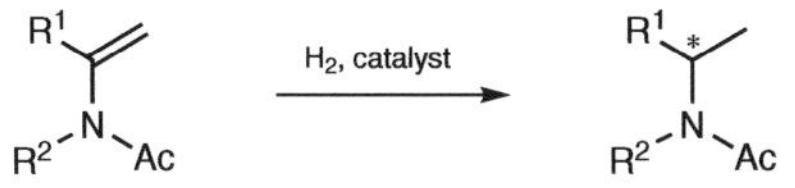

R^1	R^2	Catalyst[a]	Mol% of Catalyst	Temp (°C)	Pressure (atm)	ee (%)	Ref
Ph	H	$[Rh(cod)_2]SbF_6$/**32**	2	25	10	98.3	[39]
Ph	H	$[Rh(cod)_2]PF_6$/**42**	1	25	10	96	[40]
Ph	H	$[Rh(cod)_2]BF_4$/**33**	1	25	1	98.5	[41]
Ph	H	$[Rh(nbd)_2]SbF_6$/**48**	1	25	1.4	>99	[42]
Ph	H	$[Rh(nbd)_2]PF_6$/**49**	0.1	25	1	>99	[43]
Ph	H	$[Rh(cod)_2]BF_4$/**75**	0.1	25	10	99.6	[44]
Ph	H	$[Rh(cod)_2]BF_4$/**77**	0.1	25	20	95.8	[45]
Ph	H	$[Rh(cod)_2]BF_4$/**61**	1	25	10	99.5	[46]
Ph	H	$[Rh(cod)_2]BF_4$/**62**	1	25	10	99.7	[47]
Ph	H	$[Rh(nbd)_2]SbF_6$/**36**	1	25	1	96	[48]
Ph	H	$[Rh(cod)_2]BF_4$/**64**	0.5	5	10	99.5	[49]
Ph	H	$[Rh(nbd)_2]BF_4$/**50**	0.2	25	1	99.3	[50]
Ph	H	$[Rh(nbd)_2]BF_4$/**55**	1	0	7	99	[51]
Ph	H	$[Rh(cod)_2]SbF_6$/**52**	0.1	25	3	99.4	[52]
Ph	H	$[Rh(nbd)_2]SbF_6$/**53**	1	25	1.4	>99	[53]
Ph	H	$[Rh(nbd)_2]SbF_6$/**54**	1	25	1.4	>99	[54]
Ph	H	$[Rh(cod)_2]BF_4$/**51**	1	25	3	99	[55]
Ph	H	$[Rh(cod)_2]BF_4$/**82**	1	25	10	98.7	[56]
Ph	H	$[Rh(cod)_2]BF_4$/**83** (X = Et)	0.25	5	20.4	96	[57]
Ph	H	$[Rh(cod)_2]BF_4$/**84**	2	25	25	99	[58]
Ph	H	$[Rh(cod)_2]BF_4$/**85**	2	25	25	99	[58]
Ph	H	$[Rh(cod)_2]BF_4$/**86**	1	25	10	99	[59]
Ph	H	$[Rh(cod)_2]BF_4$/**88**	1	25	40	97.6	[60]
Ph	H	$[Rh(cod)_2]BF_4$/**90**	1	25	10	98.1	[61]
Ph	H	$[Rh(cod)_2]BF_4$/**91**	1	25	10	95	[62]
Ph	H	$[Rh(cod)_2]BF_4$/**92**	0.1	20	10	99.5	[63]
Ph	H	$[Rh(cod)_2]BF_4$/**93**	1	25	1	>99	[64]
Ph	H	$[Rh(cod)_2]BF_4$/**94**	1	25	1	96.5	[65]
4-Tol	H	$[Rh(cod)_2]BF_4$/**33**	1	25	1	98.6	[41]
4-Tol	H	$[Rh(cod)_2]BF_4$/**77**	0.2	5	20	99.4	[45]
4-Tol	H	$[Rh(cod)_2]BF_4$/**61**	1	25	10	99.1	[46]
4-Tol	H	$[Rh(cod)_2]BF_4$/**62**	1	25	10	99.8	[47]
4-Tol	H	$[Rh(cod)_2]BF_4$/**64**	0.5	5	10	99.7	[49]
4-Tol	H	$[Rh(nbd)_2]SbF_6$/**54**	1	25	1.4	>99	[54]
4-Tol	H	$[Rh(cod)_2]BF_4$/**82**	1	5	50	99.7	[56]

R^1	R^2	Catalyst[a]	Mol% of Catalyst	Temp (°C)	Pressure (atm)	ee (%)	Ref
4-Tol	H	$[Rh(cod)_2]BF_4$/**83** (X = Et)	1	5	20.4	98	[57]
4-Tol	H	$[Rh(cod)_2]BF_4$/**86**	1	25	10	>99	[59]
4-Tol	H	$[Rh(cod)_2]BF_4$/**88**	1	25	40	99.3	[60]
4-Tol	H	$[Rh(cod)_2]BF_4$/**90**	1	25	10	98.4	[61]
4-Tol	H	$[Rh(cod)_2]BF_4$/**94**	1	25	1	97	[65]
$4-MeOC_6H_4$	H	$[Rh(cod)_2]PF_6$/**42**	1	25	10	95	[40]
$4-MeOC_6H_4$	H	$[Rh(cod)_2]BF_4$/**33**	1	25	1	99	[41]
$4-MeOC_6H_4$	H	$[Rh(nbd)_2]PF_6$/**49**	0.1	25	1	>99	[43]
$4-MeOC_6H_4$	H	$[Rh(cod)_2]BF_4$/**77**	0.2	5	20	99.3	[45]
$4-MeOC_6H_4$	H	$[Rh(cod)_2]BF_4$/**61**	1	25	10	99.8	[46]
$4-MeOC_6H_4$	H	$[Rh(cod)_2]BF_4$/**64**	0.5	5	10	99.5	[49]
$4-MeOC_6H_4$	H	$[Rh(nbd)_2]BF_4$/**55**	1	0	7	99	[51]
$4-MeOC_6H_4$	H	$[Rh(nbd)_2]SbF_6$/**54**	1	25	1.4	>99	[54]
$4-MeOC_6H_4$	H	$[Rh(cod)_2]BF_4$/**83** (X = Et)	1	5	20.4	98	[57]
$4-MeOC_6H_4$	H	$[Rh(cod)_2]BF_4$/**84**	2	25	25	99	[58]
$4-MeOC_6H_4$	H	$[Rh(cod)_2]BF_4$/**85**	2	25	25	99	[58]
$4-MeOC_6H_4$	H	$[Rh(cod)_2]BF_4$/**86**	1	25	10	98	[59]
$4-MeOC_6H_4$	H	$[Rh(cod)_2]BF_4$/**88**	1	25	40	97.4	[60]
$4-MeOC_6H_4$	H	$[Rh(cod)_2]BF_4$/**90**	1	25	10	95.7	[61]
$4-MeOC_6H_4$	H	$[Rh(cod)_2]BF_4$/**91**	1	25	10	95.9	[62]
$4-MeOC_6H_4$	H	$[Rh(cod)_2]BF_4$/**92**	1	20	10	99.5	[63]
$4-MeOC_6H_4$	H	$[Rh(cod)_2]BF_4$/**94**	1	25	1	97	[65]
$4-ClC_6H_4$	H	$[Rh(cod)_2]BF_4$/**33**	1	25	1	97.8	[41]
$4-ClC_6H_4$	H	$[Rh(cod)_2]BF_4$/**75**	0.1	25	10	98.8	[44]
$4-ClC_6H_4$	H	$[Rh(cod)_2]BF_4$/**61**	1	25	10	98.5	[46]
$4-ClC_6H_4$	H	$[Rh(cod)_2]BF_4$/**62**	1	25	10	99.9	[47]
$4-ClC_6H_4$	H	$[Rh(cod)_2]BF_4$/**64**	0.5	5	10	98.2	[49]
$4-ClC_6H_4$	H	$[Rh(nbd)_2]SbF_6$/**54**	1	25	1.4	>99	[54]
$4-ClC_6H_4$	H	$[Rh(cod)_2]BF_4$/**82**	1	5	50	99.3	[56]
$4-ClC_6H_4$	H	$[Rh(cod)_2]BF_4$/**84**	2	25	25	99	[58]
$4-ClC_6H_4$	H	$[Rh(cod)_2]BF_4$/**85**	2	25	25	99	[58]
$4-ClC_6H_4$	H	$[Rh(cod)_2]BF_4$/**86**	1	25	10	99	[59]
$4-ClC_6H_4$	H	$[Rh(cod)_2]BF_4$/**88**	1	25	40	99.8	[60]
$4-ClC_6H_4$	H	$[Rh(cod)_2]BF_4$/**90**	1	25	10	95	[61]
$4-ClC_6H_4$	H	$[Rh(cod)_2]BF_4$/**91**	1	25	10	98.5	[62]
$4-ClC_6H_4$	H	$[Rh(cod)_2]BF_4$/**92**	1	20	10	99.7	[63]
2-naphthyl	H	$[Rh(cod)_2]PF_6$/**42**	1	25	10	99	[40]
2-naphthyl	H	$[Rh(cod)_2]BF_4$/**33**	1	25	1	>99	[41]
2-naphthyl	H	$[Rh(nbd)_2]SbF_6$/**48**	1	25	1.4	>99	[42]
2-naphthyl	H	$[Rh(nbd)_2]SbF_6$/**36**	1	25	1	>99	[48]
2-naphthyl	H	$[Rh(cod)_2]BF_4$/**64**	0.5	5	10	99.3	[49]

R^1	R^2	Catalyst[a]	Mol% of Catalyst	Temp (°C)	Pressure (atm)	ee (%)	Ref
2-naphthyl	H	$[Rh(nbd)_2]BF_4$/**55**	1	0	7	98	[51]
2-naphthyl	H	$[Rh(cod)_2]SbF_6$/**52**	0.1	25	3	99.7	[52]
2-naphthyl	H	$[Rh(nbd)_2]SbF_6$/**54**	1	25	1.4	>99	[54]
2-naphthyl	H	$[Rh(cod)_2]BF_4$/**88**	1	25	40	98.4	[60]
2-naphthyl	H	$[Rh(cod)_2]BF_4$/**90**	1	25	10	96.8	[61]
2-naphthyl	H	$[Rh(cod)_2]BF_4$/**91**	1	25	10	96.9	[62]
2-naphthyl	H	$[Rh(cod)_2]BF_4$/**92**	1	20	10	99.5	[63]
2-furyl	H	$[Rh(cod)_2]SbF_6$/**32**	2	25	10	99	[39]
2-furyl	H	$[Rh(cod)_2]BF_4$/**82**	1	5	50	98.7	[56]
t-Bu	H	$[Rh(cod)_2]SbF_6$/**52**	0.1	25	5	99	[52]
t-Bu	H	$[Rh(cod)_2]BF_4$/**51**	1	25	3	99	[55]
t-Bu	H	$[Rh(cod)_2]BF_4$/**44**	1	25	2.5	96	[66]
1-adamantyl	H	[Rh(nbd)(**58**)][b]	1	25	3	99	[67]
(*E*)-CH=CHPh	H	$[Rh(cod)_2]BF_4$/**53**	1	25	1	99.6[c]	[68]
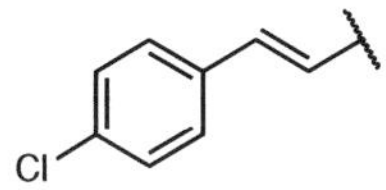 Cl	H	$[Rh(cod)_2]BF_4$/**53**	1	25	1	99.4[c]	[68]
C≡CTMS	H	$[Rh(cod)_2]BF_4$/**53**	1	25	1	99.2[c]	[68]
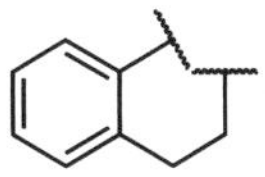		$[Rh(cod)_2]BF_4$/**45**	1	25	2.5	96	[66]
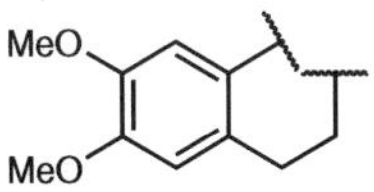 MeO, MeO		$[Rh(nbd)_2]SbF_6$/**48**	1	25	1.4	97	[42]
		$[Rh(nbd)_2]SbF_6$/**34**	1	25	3	98	[72]

[a] For ligand structures see Schemes 3 and 4, Section 1.1.1.2.
[b] The counterion of the rhodium(I) catalyst was not specified.
[c] Only the acetylamino-substituted C=C bond was reduced.

Acyclic β-substituted enamides are generally prepared as *E*/*Z* mixtures, and the separation of the isomers is problematic; however, a number of rhodium catalysts containing chiral bisphosphine or monophosphorus ligands are efficient for the hydrogenation of such *E*/*Z* mixtures (Scheme 6). For example, the use of Diop (**32**)[39] or its analogue BDPMI (**33**)[41] gives excellent results in the asymmetric hydrogenation of *E*/*Z* mixtures of acyclic β-alkyl-substituted enamides. Catalysts based on sterically encumbered and electron-rich chiral phosphine ligands [TangPhos (**48**)[42] and ZhangPhos (**54**)[54]] exhibit extremely high reactivity and selectivity for *E*/*Z* mixtures of acyclic β-alkyl-substituted enamides. The 1,4-bisphosphine ligand **34**[72] and SK-Phos (**35**)[72] also induce high enantioselectivities for this type of substrate. Monophosphorus ligands **91**[62] and **92**[63] can also be used for the asymmetric hydrogenation of *E*/*Z* mixtures of acyclic β-alkyl-substituted enamides. Acyclic β-(methoxymethyl)oxy-substituted enamides can be hydrogenated to the corresponding alkylamides using $[Rh(cod)_2]PF_6$/**34**.[72]

Whereas several different catalysts are capable of hydrogenating acyclic β-alkyl-substituted α-arylenamides with high enantioselectivities, acyclic β-substituted α-alkylenamides remain a challenge. To date, only a few chiral catalysts have been reported for

for references see p 63

the efficient hydrogenation of such substrates. Zhang et al. disclosed that the rhodium catalyst based on TangPhos (**48**) is highly enantioselective for the hydrogenation of the *Z*-isomers of β-aryl-substituted α-alkylenamides.[69] WingPhos (**56**)[70] can be used for the highly enantioselective hydrogenation of the *E*-isomers of β-aryl-substituted α-alkylenamides. Zhang et al. reported that DuanPhos (**53**)[71] is an efficient ligand for the hydrogenation of α-trifluoromethyl-substituted β-aryl-substituted enamides.

The asymmetric hydrogenation of cyclic enamides has been extensively studied due to the importance of cyclic amines in pharmaceuticals (Scheme 6). Bisphosphine ligand **60**,[75] has been used for the reduction of a series of cyclic enamides, with high enantioselectivities in the alkylamide products obtained. Other bisphosphine ligands **48**,[42] **56**,[70] **53**,[74] **34**,[72] **63**,[76] **78**,[78,79] **45**,[66] and **40**[73] also induce high enantioselectivity for a variety of substrates; however, the substrate scope is limited for most of these ligand systems.

Scheme 6 Rhodium-Catalyzed Asymmetric Hydrogenation of Enamides with a Tri- or Tetrasubstituted Alkene[39,41,42,47,48,53,54,59,62–64,69–79]

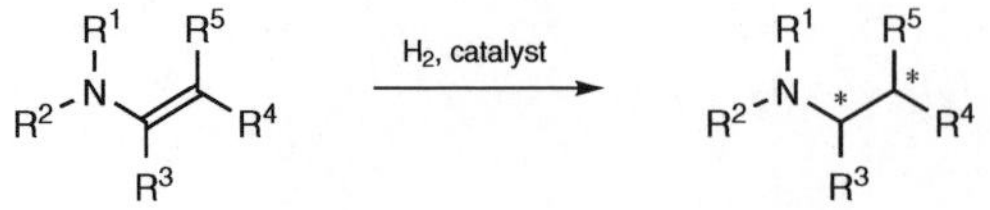

H_2, catalyst

98

R^1	R^2	R^3	R^4	R^5	Catalyst[a]	Mol% of Catalyst	T[b]	P[b]	ee (%)	Ref
Ac	H	Me	H	Ph	$[Rh(cod)_2]BF_4$/**48**	1	25	30	99.3	[69]
Ac	H	Me	Ph	H	$[Rh(cod)_2]BF_4$/**56**	0.5	50	51	97	[70]
Ac	H	Me	$4-MeOC_6H_4$	H	$[Rh(cod)_2]BF_4$/**56**	0.5	50	51	>99	[70]
Ac	H	Me	$4-FC_6H_4$	H	$[Rh(cod)_2]BF_4$/**56**	0.5	50	51	99	[70]
Ac	H	Me	H	$2-MeOC_6H_4$	$[Rh(cod)_2]BF_4$/**48**	1	25	30	99	[69]
Ac	H	Me	H	$3-MeOC_6H_4$	$[Rh(cod)_2]BF_4$/**48**	1	25	30	99.1	[69]
Ac	H	Me	H	$4-MeOC_6H_4$	$[Rh(cod)_2]BF_4$/**48**	1	25	30	96.6	[69]
Ac	H	CF_3	H	Ph	$[Rh(cod)_2]BF_4$/**53**	1	25	5	99	[71]
Ac	H	CF_3	H	$2-ClC_6H_4$	$[Rh(cod)_2]BF_4$/**53**	1	25	5	97	[71]
Ac	H	CF_3	H	$3-ClC_6H_4$	$[Rh(cod)_2]BF_4$/**53**	1	25	5	99	[71]
Ac	H	CF_3	H	$4-ClC_6H_4$	$[Rh(cod)_2]BF_4$/**53**	1	25	5	99	[71]
Ac	H	CF_3	H	$4-MeOC_6H_4$	$[Rh(cod)_2]BF_4$/**53**	1	25	5	98	[71]
Ac	H	Ph	Me	H	$[Rh(cod)_2]SbF_6$/**32**	2	25	10	97.3[c]	[39]
Ac	H	Ph	Me	H	$[Rh(cod)_2]BF_4$/**33**	1	25	1	>99[c]	[41]
Ac	H	Ph	Me	H	$[Rh(nbd)_2]SbF_6$/**48**	1	25	1.4	98[c]	[42]
Ac	H	Ph	Me	H	$[Rh(nbd)_2]BF_4$/**54**	1	25	1.4	>99[c]	[54]
Ac	H	Ph	Me	H	$[Rh(nbd)_2]SbF_6$/**34**	1	25	3	98[c]	[72]
Ac	H	Ph	Me	H	$[Rh(nbd)_2]SbF_6$/**35**	1	25	3	97[c]	[72]
Ac	H	Ph	Me	H	$[Rh(cod)_2]BF_4$/**91**	1	25	10	96.7[c]	[62]
Ac	H	Ph	Me	H	$[Rh(cod)_2]BF_4$/**92**	1	20	10	99.2[c]	[63]
Ac	H	$4-MeOC_6H_4$	Me	H	$[Rh(cod)_2]SbF_6$/**32**	2	25	10	98[c]	[39]
Ac	H	$4-MeOC_6H_4$	Me	H	$[Rh(cod)_2]BF_4$/**33**	1	25	1	>99[c]	[41]
Ac	H	$4-MeOC_6H_4$	Me	H	$[Rh(nbd)_2]SbF_6$/**48**	1	25	1.4	98[c]	[42]
Ac	H	$4-MeOC_6H_4$	Me	H	$[Rh(nbd)_2]SbF_6$/**36**	1	25	1	97[c]	[48]
Ac	H	$4-MeOC_6H_4$	Me	H	$[Rh(nbd)_2]SbF_6$/**53**	1	25	1.4	>99[c]	[53]

R^1	R^2	R^3	R^4	R^5	Catalyst[a]	Mol% of Catalyst	T[b]	P[b]	ee (%)	Ref
Ac	H	4-$MeOC_6H_4$	Me	H	$[Rh(nbd)_2]SbF_6$/**35**	1	25	3	98[c]	[72]
Ac	H	4-ClC_6H_4	Me	H	$[Rh(cod)_2]BF_4$/**33**	1	25	1	>99[c]	[41]
Ac	H	Ph	Et	H	$[Rh(cod)_2]BF_4$/**33**	1	25	1	>99[c]	[41]
Ac	H	Ph	Et	H	$[Rh(cod)_2]BF_4$/**62**	1	25	10	99.1[c]	[47]
Ac	H	Ph	Et	H	$[Rh(cod)_2]BF_4$/**86**	1	25	10	96[c]	[59]
Ac	H	2-naphthyl	Me	H	$[Rh(cod)_2]SbF_6$/**32**	2	25	10	>99[c]	[39]
Ac	H	2-naphthyl	Me	H	$[Rh(nbd)_2]SbF_6$/**48**	1	25	1.4	99[c]	[42]
Ac	H	2-naphthyl	Me	H	$[Rh(nbd)_2]SbF_6$/**36**	1	25	1	>99[c]	[48]
Ac	H	2-naphthyl	Me	H	$[Rh(nbd)_2]SbF_6$/**35**	1	25	3	97[c]	[72]
Ac	H	Ph	OMOM	H	$[Rh(cod)_2]PF_6$/**34**	1	25	15	98[c]	[72]
Ac		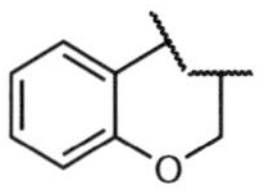	H	Ph	$[Rh(cod)_2]BF_4$/**40**	1	25	4	97.2	[73]
Ac	H	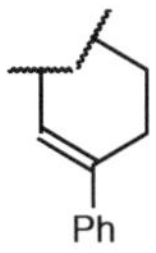		H	$[Rh(cod)_2]BF_4$/**53**	1	25	1	98[d]	[74]
Ac	H	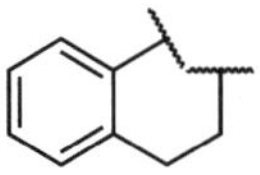		H	$[Rh(nbd)_2]SbF_6$/**60**	0.5	−20	1.7	98	[75]
Ac	H	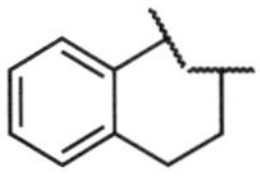		H	$[Rh(cod)_2]BF_4$/**93**	1	25	20	98	[64]
Ac	H	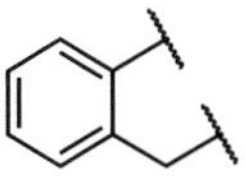		H	$[Rh(nbd)_2]SbF_6$/**60**	0.5	−20	1.7	96	[75]
Ac	H	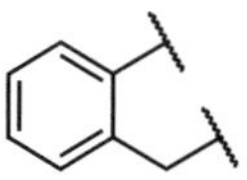		H	$[Rh(cod)_2]BF_4$/**92**	1	20	10	96	[63]
Ac	H	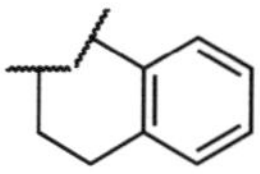		H	$[Rh(cod)_2]BF_4$/**56**	0.5	50	51	96	[70]
Ac	H	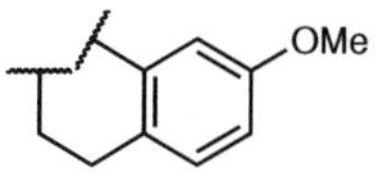		H	$[Rh(cod)_2]BF_4$/**56**	0.5	50	51	96	[70]
Bz	H	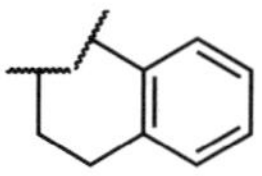		H	Ru(**63**)$(O_2CCF_3)_2$	0.5	25	10	96	[76]

for references see p 63

R^1	R^2	R^3	R^4	R^5	Catalyst[a]	Mol% of Catalyst	T[b]	P[b]	ee (%)	Ref
Bz	H	Br O		H	[RuCl(*p*-cymene)(**65**)]Cl	1	50	50	96	[77]
Ac	H			Me	$[Rh(nbd)_2]SbF_6$/**60**	0.5	−20	1.7	96	[75]
Ac	H	NEt		F	$Ru(cod)(O_2CCF_3)_2$/**78**	0.5	20	20	99.1	[78]
Ac	H	4-ClC_6H_4	4-ClC_6H_4	OMOM	$[Rh(cod)_2]BF_4$/**78**	1	40	40	98	[79]
Ac	H	4-Tol	4-Tol	OMOM	$[Rh(cod)_2]BF_4$/**78**	1	40	40	96	[79]

[a] For ligand structures see Schemes 3 and 4, Section 1.1.1.2.
[b] T = temperature (°C); P = pressure (atm).
[c] The starting alkene was a mixture of *E*/*Z* isomers.
[d] Only the acetylamino-substituted C=C bond was reduced.

(*R*)-*N*-(1-Phenylethyl)acetamide (97, **R^1 = Ph; R^2 = H); Typical Procedure Using $[Rh(nbd)_2]SbF_6$/48 as Catalyst:**[42]

A 0.05 M soln of (1*S*,1*S*′,2*R*,2*R*′)-TangPhos (**48**) in MeOH (0.10 mL, 5 μmol) was added to a soln of $[Rh(nbd)_2]SbF_6$ (2.3 mg, 4.5 μmol) in MeOH (3 mL) in a glovebox. After the mixture was stirred for 10 min, *N*-(1-phenylvinyl)acetamide (0.5 mmol) was added. The hydrogenation was performed at rt under H_2 (1.4 atm) for 12 h. After carefully releasing the H_2, the mixture was passed through a short silica gel plug to remove the catalyst. The resulting soln was analyzed directly by chiral GC or HPLC to determine the ee (>99%).

(*R*)-*N*-(1-Phenylethyl)acetamide (97, **R^1 = Ph; R^2 = H); Typical Procedure Using $[Rh(cod)_2]BF_4$/** 82 **as Catalyst:**[56]

N-(1-Phenylvinyl)acetamide (0.5 mmol), $[Rh(cod)_2]BF_4$ (2.0 mg, 5 μmol), and ligand **82** (3.6 mg, 11 μmol) were mixed in a 25-mL autoclave, and anhyd toluene (5 mL) was introduced under N_2. After three vacuum/H_2 refill cycles, the hydrogenation was performed at rt under H_2 (10 atm) for 12 h. The excess H_2 was released and the mixture was passed through a short silica gel column (EtOAc/petroleum ether 1:1) to remove the catalyst. The ee of the product (98.7%) was determined by chiral capillary GC using a Varian Chirasil-L-Val column (25 m).

(*R*)-*N*-(1-Phenylpropyl)acetamide (98, **R^1 = Ac; R^2 = R^5 = H; R^3 = Ph; R^4 = Me); Typical Procedure Using [Rh(nbd)(**54**)]BF_4 as Catalyst:**[54]

In a glovebox filled with N_2, [Rh(nbd)(**54**)]BF_4 (6.7 mg, 0.01 mmol) was dissolved in MeOH (10 mL). To 1 mL of this soln, the substrate (*E*/*Z*)-*N*-(1-phenylprop-1-enyl)acetamide (0.1 mmol) was added. The resulting soln was then transferred into an autoclave and charged with H_2 (1.4 atm). The hydrogenation was performed at rt for 12 h. After carefully releasing the pressure in the hood, the mixture was passed through a short silica gel plug to remove the catalyst. The ee was determined from the resulting soln by chiral GC or HPLC.

(R)-N-(1-Phenylpropan-2-yl)acetamide (98, R^1 = Ac; R^2 = R^5 = H; R^3 = Me; R^4 = Ph); Typical Procedure Using $[Rh(nbd)_2]BF_4$/56 as Catalyst:[70]

A 4-mL vial equipped with a magnetic stirrer bar was charged with (E)-N-(1-phenylprop-1-en-2-yl)acetamide (0.1 mmol), $[Rh(nbd)_2]BF_4$/**56** (0.5 mg, 0.5 μmol), and CH_2Cl_2 (0.5 mL) under N_2. After stirring for 2 min at rt, the mixture was transferred to an autoclave. The autoclave was purged with H_2 (3 ×) and charged to 51 atm. The mixture was stirred at 50 °C for 12 h, cooled to rt, and depressurized carefully in a well-ventilated hood. A crude reaction sample was passed through Celite to remove the metal precipitate, and the filtrate was directly analyzed by chiral HPLC to determine the conversion and ee; yield: quant; 97% ee.

(S)-N-(1,2,3,4-Tetrahydronaphthalen-1-yl)acetamide [98, R^1 = Ac; R^2 = R^5 = H; R^3,R^4 = 2-$(CH_2)_2C_6H_4$]; Typical Procedure Using $[Rh(nbd)_2]SbF_6$/60 as Catalyst:[75]

To a soln of $[Rh(nbd)_2]SbF_6$ (2.3 mg, 4.5 μmol) in CH_2Cl_2 (2 mL) in a glovebox was added a 0.05 M soln of **60** in MeOH (0.10 mL, 5 μmol). After the mixture was stirred for 10 min, N-(3,4-dihydronaphthalen-1-yl)acetamide (1 mmol) was added. The hydrogenation was performed at −20 °C under H_2 (1.7 atm) for 12–48 h. After carefully releasing the H_2 pressure, the mixture was passed through a short silica gel plug to remove the catalyst. The ee was determined directly by chiral GC or HPLC, without any further purification. The absolute configuration of the product was determined by comparing the optical rotation with the reported value, and was further confirmed by the GC retention time.

1.1.1.2.2 Reduction of α,β-Dehydro-α-amino Acids and Their Esters

The great importance of chiral α-amino acids and their derivatives in pharmaceutical, agricultural, and biological chemistry has made their synthesis a central subject in organic chemistry. Asymmetric hydrogenation of α,β-dehydro-α-amino acids and their derivatives has been demonstrated to be one of the most successful synthetic approaches. The asymmetric hydrogenation of dehydroamino acids has now become a typical reaction to evaluate the efficiency of new chiral phosphine ligands. A large number of catalysts have been examined and considerable success has been achieved. The most representative results reported since 2000 are listed in Scheme 7.[40,42,53,57–61,63,84–99]

2-Acetamidoacrylic acid (**99**, R^1 = R^2 = R^3 = H) and its esters are the simplest dehydro-α-amino acids, and these often serve as the standard substrates for the evaluation of new catalysts. Many rhodium catalysts can hydrogenate this substrate with excellent enantioselectivity and reactivity.

β-Substituted dehydro-α-amino acids are the most studied substrates in the hydrogenation of dehydro-α-amino acids. β-Alkyl-substituted dehydro-α-amino acids are often synthesized as mixtures of E/Z isomers. In contrast to the high turnover numbers (TONs) and selectivities achieved with the Z-isomer, asymmetric hydrogenation of the E-isomer is often much slower and less selective. Ligands with chirality based on a carbon backbone, such as **42**,[40] **39**,[84] **38**,[92] **43**,[87] **46**,[87] **47**,[88] and **57**,[95] can be used for the highly selective hydrogenation of the Z-isomer of β-substituted dehydro-α-amino acids. The P-chiral phosphine ligands TangPhos (**48**)[42] and TrichickenfootPhos (**51**)[85] are also efficient chiral ligands for this kind of substrate. Ligands with chirality based on atropisomerism, such as **76**,[44] **66**,[93] **67**,[91] **68**,[91] **71**,[94] and **69**,[96] induce excellent enantioselectivities but have moderate catalytic activity. Another important class of chiral phosphine ligands possess planar chirality; examples are **80**[86] and **81**,[90] which are based on ferrocene backbones, and **72**,[89] which is based on a paracyclophane backbone. The monophosphorus ligands **83**,[57,58] **87**,[59] **89**,[60] **90**,[61] **92**,[63] **93**,[64] and **95**[97] are also effective ligands for the hydrogenation of β-substituted dehydro-α-amino acids. Dehydro-α-amino acid substrates possess-

for references see p 63

ing two unsaturated double bonds can be hydrogenated smoothly using $[Rh(cod)_2]BF_4$/**48** with low catalyst loading, giving γ,δ-unsaturated α-amino acid products with both good enantio- and chemoselectivities.[98]

The transition-metal-catalyzed hydrogenation of tetrasubstituted dehydro-α-amino acids is a challenging task due to the low reactivity of the hindered substrates. In the past few years, only a few successful examples have been reported. Hoge et al. reported the use of three-hindered-quadrant bisphosphine ligand **51** for the rhodium-catalyzed hydrogenation of cyclic dehydro-α-amino acids.[85] Molinaro et al. have systematically studied the asymmetric hydrogenation of β,β-diaryl-α-amino acids;[99] excellent results are obtained with a range of substrates using the $[Rh(nbd)_2]BF_4$/**79** catalyst.

Scheme 7 Rhodium-Catalyzed Asymmetric Hydrogenation of Dehydro-α-amino Acids and Their Esters[40,42,44,53,57–61,63,64,84–99]

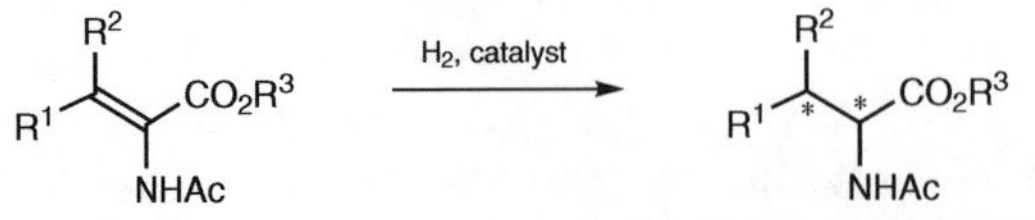

R^1–C(=C(NHAc)CO_2R^3)–R^2 (**99**) → (H_2, catalyst) → R^1–C*H(R^2)–C*H(NHAc)CO_2R^3 (**100**)

R^1	R^2	R^3	Catalyst[a]	Mol% of Catalyst	Temp (°C)	Pressure (atm)	ee (%)	Ref
H	H	H	$[Rh(cod)_2]PF_6$/**42**	1	25	3	>99	[40]
H	H	H	$[Rh(cod)_2]BF_4$/**39**	0.02	30	6	>99	[84]
H	H	H	$[Rh(cod)_2]BF_4$/**51**	1	25	3.3	>99	[85]
H	H	H	$[Rh(nbd)_2]BF_4$/**80**	1	25	1	98	[86]
H	H	H	$[Rh(cod)_2]BF_4$/**43**	0.1	25	10	98.7	[87]
H	H	H	$[Rh(cod)_2]PF_6$/**47**	1	25	1	99.4	[88]
H	H	H	$[Rh(cod)_2]BF_4$/**72**	0.1	25	3.5	97	[89]
H	H	H	$[Rh(cod)_2]OTf$/**81**	1	25	1	96.1	[90]
Ph	H	H	$[Rh(cod)_2]PF_6$/**42**	1	25	3	>99	[40]
Ph	H	H	$[Rh(nbd)_2]SbF_6$/**48**	1	25	1.4	>99	[42]
Ph	H	H	$[Rh(cod)_2]BF_4$/**51**	1	25	3.3	>99	[85]
Ph	H	H	$[Rh(nbd)_2]BF_4$/**80**	1	25	1	98	[86]
Ph	H	H	$[Rh(cod)_2]BF_4$/**43**	0.1	25	10	99.5	[87]
Ph	H	H	$[Rh(cod)_2]PF_6$/**67**	1	25	1	>99	[91]
Ph	H	H	$[Rh(cod)_2]PF_6$/**68**	1	25	1	99	[91]
Ph	H	H	$[Rh(cod)_2]BF_4$/**72**	0.1	25	3.5	99	[89]
Ph	H	H	$[Rh(cod)_2]OTf$/**81**	1	25	1	99.4	[90]
2-naphthyl	H	H	$[Rh(cod)_2]PF_6$/**42**	1	25	3	>99	[40]
2-naphthyl	H	H	$[Rh(nbd)_2]SbF_6$/**48**	1	25	1.4	>99	[42]
2-naphthyl	H	H	$[Rh(cod)_2]PF_6$/**47**	1	25	1	>99.9	[88]
2-naphthyl	H	H	$[Rh(cod)_2]PF_6$/**67**	1	25	1	>99	[91]
2-naphthyl	H	H	$[Rh(cod)_2]PF_6$/**68**	1	25	1	99	[91]
H	H	Me	$[Rh(cod)_2]PF_6$/**42**	1	25	3	>99	[40]
H	H	Me	$[Rh(cod)_2]BF_4$/**39**	0.02	30	6	>99	[84]
H	H	Me	$[Rh(cod)_2]BF_4$/**51**	1	25	3.3	>99	[85]
H	H	Me	$[Rh(nbd)_2]BF_4$/**80**	1	25	1	>99	[86]

R[1]	R[2]	R[3]	Catalyst[a]	Mol% of Catalyst	Temp (°C)	Pressure (atm)	ee (%)	Ref
H	H	Me	$[Rh(cod)_2]BF_4$/**43**	0.1	25	10	99.9	[87]
H	H	Me	$[Rh(cod)_2]BF_4$/**46**	0.1	25	10	99.7	[87]
H	H	Me	$[Rh(cod)_2]PF_6$/**47**	1	25	1	99.0	[88]
H	H	Me	$[Rh(nbd)_2]SbF_6$/**53**	0.01	25	1.3	>99	[53]
H	H	Me	$[Rh(cod)_2]BF_4$/**72**	0.1	25	3.5	99	[89]
H	H	Me	$[Rh(cod)_2]BF_4$/**83** (X = Me)	2	25	5	97	[58]
H	H	Me	$[Rh(cod)_2]BF_4$/**83** (X = Et)	1	25	20.4	96.5	[57]
H	H	Me	$[Rh(cod)_2]BF_4$/**89**	0.1	25	20	98.8	[60]
H	H	Me	$[Rh(cod)_2]BF_4$/**90**	1	25	20	98.8	[61]
H	H	Me	$[Rh(cod)_2]BF_4$/**92**	1	20	1.2	97.7	[63]
H	H	Me	$[Rh(cod)_2]BF_4$/**93**	1	25	1	99	[64]
H	H	Me	$[Rh(cod)_2]OTf$/**81**	1	25	1	98.5	[90]
H	H	Me	$[Rh(cod)_2]BF_4$/**84**	2	25	5	99	[58]
H	H	Me	$[Rh(cod)_2]BF_4$/**85**	2	25	5	99	[58]
Me	H	Me	$[Rh(cod)_2]BF_4$/**89**	1	25	20	98.4	[60]
Me	H	Me	$[Rh(cod)_2]BF_4$/**90**	1	25	20	98.7	[61]
Ph	H	Me	$[Rh(cod)_2]PF_6$/**42**	1	25	3	>99	[40]
Ph	H	Me	$[Rh(nbd)_2]SbF_6$/**48**	1	25	1.4	99	[42]
Ph	H	Me	$[Rh(cod)_2]BF_4$/**39**	0.03	30	10	>99	[84]
Ph	H	Me	$[Rh(cod)_2]BF_4$/**51**	1	25	3.3	>99	[85]
Ph	H	Me	$[Rh(nbd)_2]BF_4$/**80**	1	25	1	>99	[86]
Ph	H	Me	$[Rh(cod)_2]BF_4$/**38**	0.03	28	10	99	[92]
Ph	H	Me	$[Rh(cod)_2]BF_4$/**43**	0.1	25	10	>99.5	[87]
Ph	H	Me	$[Rh(cod)_2]BF_4$/**46**	0.1	25	10	98.8	[87]
Ph	H	Me	$[Rh(cod)_2]PF_6$/**47**	1	25	1	99.5	[88]
Ph	H	Me	$[Rh(nbd)_2]SbF_6$/**53**	1	25	1.3	>99	[53]
Ph	H	Me	$[Rh(cod)_2]PF_6$/**66**	1	25	3	97.8	[93]
Ph	H	Me	$[Rh(cod)_2]PF_6$/**67**	1	25	1	>99	[91]
Ph	H	Me	$[Rh(cod)_2]PF_6$/**68**	1	25	1	>99	[91]
Ph	H	Me	$[Rh(cod)_2]BF_4$/**72**	0.1	25	3.5	99	[89]
Ph	H	Me	$[Rh(cod)_2]BF_4$/**83** (X = Et)	1	25	20.4	98	[57]
Ph	H	Me	$[Rh(cod)_2]BF_4$/**87**	1	25	10	98	[59]
Ph	H	Me	$[Rh(cod)_2]BF_4$/**89**	1	25	20	99.6	[60]
Ph	H	Me	$[Rh(cod)_2]BF_4$/**90**	0.1	25	20	98.2	[61]
Ph	H	Me	$[Rh(cod)_2]BF_4$/**92**	1	20	1.2	96.8	[63]
Ph	H	Me	$[Rh(cod)_2]BF_4$/**93**	1	25	1	96	[64]
Ph	H	Me	[Rh][b]/**71**	1	25	1	>99.9	[94]
Ph	H	Me	$[Rh(cod)_2]OTf$/**57**	0.5	30	4	98	[95]
Ph	H	Me	$[Rh(cod)_2]BF_4$/**69**	0.2	25	3.4	98	[96]
Ph	H	Me	$[Rh(cod)_2]OTf$/**81**	1	25	1	99.1	[90]

for references see p 63

R^1	R^2	R^3	Catalyst[a]	Mol% of Catalyst	Temp (°C)	Pressure (atm)	ee (%)	Ref
Ph	H	Me	$[Rh(cod)_2]BF_4$/**75**	0.01	25	10	99	[44]
Ph	H	Me	$[Rh(cod)_2]$ BF_4/**95**	1	25	10	99	[97]
Ph	H	Me	$[Rh(cod)_2]BF_4$/**83** (X = Me)	2	25	5	95	[58]
Ph	H	Me	$[Rh(cod)_2]BF_4$/**84**	2	25	5	99	[58]
Ph	H	Me	$[Rh(cod)_2]BF_4$/**85**	2	25	5	99	[58]
2-naphthyl	H	Me	$[Rh(cod)_2]PF_6$/**42**	1	25	3	>99	[40]
2-naphthyl	H	Me	$[Rh(nbd)_2]SbF_6$/**48**	1	25	1.4	>99	[42]
2-naphthyl	H	Me	$[Rh(cod)_2]PF_6$/**47**	1	25	1	99	[88]
2-naphthyl	H	Me	$[Rh(cod)_2]PF_6$/**66**	1	25	3	97.8	[93]
2-naphthyl	H	Me	$[Rh(cod)_2]PF_6$/**67**	1	25	1	>99	[91]
2-naphthyl	H	Me	$[Rh(cod)_2]PF_6$/**68**	1	25	1	>99	[91]
2-naphthyl	H	Me	$[Rh(cod)_2]BF_4$/**89**	1	25	20	96.9	[60]
2-naphthyl	H	Me	$[Rh(cod)_2]BF_4$/**90**	1	25	20	98.4	[61]
$4-MeOC_6H_4$	H	Me	$[Rh(cod)_2]PF_6$/**42**	1	25	3	>99	[40]
$4-MeOC_6H_4$	H	Me	$[Rh(nbd)_2]SbF_6$/**48**	1	25	1.4	>99	[42]
$4-MeOC_6H_4$	H	Me	$[Rh(cod)_2]PF_6$/**47**	1	25	1	97.8	[88]
$4-MeOC_6H_4$	H	Me	$[Rh(cod)_2]PF_6$/**66**	1	25	3	95.8	[93]
$4-MeOC_6H_4$	H	Me	$[Rh(cod)_2]BF_4$/**83** (X = Et)	1	25	20.4	99	[57]
$4-MeOC_6H_4$	H	Me	$[Rh(cod)_2]BF_4$/**87**	1	25	10	>99	[59]
$4-MeOC_6H_4$	H	Me	$[Rh(cod)_2]BF_4$/**89**	1	25	20	97.2	[60]
$4-MeOC_6H_4$	H	Me	$[Rh(cod)_2]BF_4$/**90**	1	25	20	97.4	[61]
$4-MeOC_6H_4$	H	Me	$[Rh(cod)_2]BF_4$/**92**	1	20	1.2	98	[63]
$4-MeOC_6H_4$	H	Me	[Rh][b]/**71**	1	25	1	>99.9	[94]
$4-MeOC_6H_4$	H	Me	$[Rh(cod)_2]BF_4$/**69**	0.2	25	3.4	98	[96]
$4-MeOC_6H_4$	H	Me	$[Rh(cod)_2]$ BF_4/**95**	1	25	10	98	[97]
$4-FC_6H_4$	H	Me	$[Rh(cod)_2]PF_6$/**42**	1	25	3	>99	[40]
$4-FC_6H_4$	H	Me	$[Rh(nbd)_2]SbF_6$/**48**	1	25	1.4	>99	[42]
$4-FC_6H_4$	H	Me	$[Rh(cod)_2]PF_6$/**66**	1	25	3	96.6	[93]
$4-FC_6H_4$	H	Me	$[Rh(cod)_2]PF_6$/**67**	1	25	1	>99	[91]
$4-FC_6H_4$	H	Me	$[Rh(cod)_2]PF_6$/**68**	1	25	1	99	[91]
$4-FC_6H_4$	H	Me	$[Rh(cod)_2]BF_4$/**83** (X = Et)	1	25	20.4	99.8	[57]
$4-FC_6H_4$	H	Me	$[Rh(cod)_2]BF_4$/**69**	0.2	25	3.4	98	[96]
2-thienyl	H	Me	$[Rh(cod)_2]PF_6$/**42**	1	25	3	>99	[40]
$4-FC_6H_4$	H	Me	$[Rh(nbd)_2]SbF_6$/**48**	1	25	1.4	>99	[42]
$4-FC_6H_4$	H	Me	$[Rh(cod)_2]PF_6$/**67**	1	25	1	95	[91]
$4-FC_6H_4$	H	Me	$[Rh(cod)_2]PF_6$/**68**	1	25	1	95	[91]
(*E*)-CH=CHPh	H	Me	$[Rh(cod)_2]BF_4$/**48**	0.033	25	2	>99[c]	[98]
	H	Me	$[Rh(cod)_2]BF_4$/**48**	0.033	25	2	99[c]	[98]

R[1]	R[2]	R[3]	Catalyst[a]	Mol% of Catalyst	Temp (°C)	Pressure (atm)	ee (%)	Ref
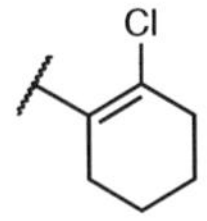 Cl	H	Me	$[Rh(cod)_2]BF_4$/**48**	0.033	25	2	99[c]	[98]
$(CH_2)_5$		Me	$[Rh(cod)_2]BF_4$/**51**	1	25	3	>99	[85]
Ph	Ph	Me	$[Rh(nbd)_2]BF_4$/**79**	5	25	27	97	[99]
Ph	4-ClC_6H_4	Me	$[Rh(nbd)_2]BF_4$/**79**	5	25	27	96	[99]
Ph	4-$MeOC_6H_4$	Me	$[Rh(nbd)_2]BF_4$/**79**	5	25	27	97	[99]
Ph	4-$F_3CC_6H_4$	Me	$[Rh(nbd)_2]BF_4$/**79**	5	25	27	95	[99]
Ph	4-pyridyl	Me	$[Rh(nbd)_2]BF_4$/**79**	5	25	27	96	[99]

[a] For ligand structures see Schemes 3 and 4, Section 1.1.1.2.
[b] The rhodium source was not specified.
[c] The only product is the γ,δ-unsaturated α-amido ester.

(*S*)-*N*-Acetylalanine (100, **R¹ = R² = R³ = H); Typical Procedure Using $[Rh(cod)_2]BF_4$/**39 **as Catalyst:**[84]
The reaction was carried out in an Argonaut Endeavor hydrogenation vessel. The glass liner was charged with 2-acetamidoacrylic acid (1.29 g, 10.0 mmol) and Rh(cod)]BF_4/**39** (0.16 mg, 2 μmol, 0.02 mol%). The vessel was charged with N_2 to 10 atm and then vented (5 ×). Degassed MeOH (4 mL) was added, and then the vessel was charged with N_2 to 10 atm and vented twice. The reaction was stirred at 1000 rpm and heated to 30 °C. The vessel was charged with H_2 (6 atm). H_2 uptake was complete after 15 min. The mixture was cooled to rt, vented, and concentrated to give (*S*)-*N*-acetylalanine; conversion: 100%; >99% ee.

Methyl (*R*)-*N*-Acetylphenylalaninate (100, **R¹ = Ph; R² = H; R³ = Me); Typical Procedure Using $[Rh(cod)_2]BF_4$/**46 **as Catalyst:**[87]
The glass liner of a vessel was charged with methyl (*Z*)-2-acetamido-3-phenylacrylate (**99**, R¹ = Ph; R² = H; R³ = Me; 2.0 mmol) and $[Rh(cod)_2]BF_4$/**46** (1.7 mg, 2 μmol, 0.1 mol%). The vessel was charged with N_2 to 10 atm and vented (5 ×). Degassed MeOH (4 mL) was added, and then the vessel was charged with N_2 to 10 atm and vented (2 ×). The reaction was stirred at 1000 rpm and heated to 25 °C. The vessel was charged with H_2 (10 atm). H_2 uptake was complete after 16 h. The mixture was vented and concentrated to give methyl (*R*)-*N*-acetylphenylalaninate; conversion: 100%; 98.8% ee.

Methyl (*R*,*E*)-2-Acetamido-5-phenylpent-4-enoate [100, **R¹ = (*E*)-CH=CHPh; R² = H; R³ = Me]; Typical Procedure Using $[Rh(cod)_2]BF_4$/**48 **as Catalyst:**[98]
In a N_2-filled glovebox, the catalyst was prepared in situ by mixing $[Rh(cod)_2]BF_4$ with (1*S*,1*S*′,2*R*,2*R*′)-TangPhos (**48**) (1:1.1 molar ratio) in CH_2Cl_2 (1 mL) at rt for 30 min. An aliquot of the catalyst soln (0.1 mL, 1 μmol) was transferred by syringe into a vial charged with methyl (2*Z*,4*E*)-2-acetamido-5-phenylpenta-2,4-dienoate [**99**, (*E*)-CH=CHPh; R² = H; R³ = Me; 0.1 mmol] in anhyd CH_2Cl_2 (2 mL). The vial was subsequently transferred into an autoclave into which H_2 was finally charged. The reaction was then stirred under H_2 (1–2 atm) at rt for 1 h. The H_2 was released slowly and carefully. The soln was passed through a short column of silica gel and concentrated to remove the metal complex. The ee (>99%) was determined by HPLC analysis on a chiral stationary phase. Reaction using 0.033 mol% of catalyst afford the same conversion and ee value.

for references see p 63

Methyl (*S*)-2-Acetamido-3,3-diphenylpropanoate (100, $R^1 = R^2 = Ph$; $R^3 = Me$); Typical Procedure Using $[Rh(nbd)_2]BF_4$/79 as Catalyst:[99]

In a glovebox, $[Rh(nbd)_2]BF_4$/**79** (5 mol%) was added to methyl 2-acetamido-3,3-diphenylacrylate (**99**, $R^1 = R^2 = Ph$; $R^3 = Me$; 1 equiv) in MeOH ($18 mL \cdot g^{-1}$) and the mixture was placed in a bomb. The bomb was charged with H_2 (27 atm) and the reaction was stirred overnight. The solvent was removed under reduced pressure and the crude mixture was purified by column chromatography (silica gel); yield: 94%; 97% ee.

1.1.1.2.3 Reduction of α,β-Dehydro-β-amino Acids and Their Esters

In addition to the successes achieved in the preparation of α-amino acids, there has also been rapid recent development in the asymmetric hydrogenation of dehydro-β-amino acid derivatives. The simplicity and effectiveness have made this approach one of the most efficient methods to synthesize novel chiral β-amino acids and their derivatives. Frequently, dehydro-β-amino acids are synthesized as a mixture of *Z*- and *E*-isomers; the *Z*-isomers are less reactive toward asymmetric hydrogenation, and afford the products with lower selectivity.

A large number of catalytic systems, most notably based on rhodium complexes of chiral bisphosphine ligands, have been employed in the hydrogenation of dehydro-β-amino acid esters **101** (Scheme 8).[48,53,100–110] In most cases, the *E*-isomer substrates can be hydrogenated smoothly with high enantioselectivities. Successful ligands employed include **76**,[105] **36**,[48] **51**,[103] **44**,[101] **40**,[100] **37**,[100] **58**,[102] **59**,[102] **73**,[104] and **96**.[106] In a few cases, the *Z*-dehydro-β-amino acid derivatives can also be reduced with high enantioselectivities, such as when using systems involving ligands **76**,[105] **36**,[48] and **53**.[53] The ability to selectively hydrogenate mixtures of *Z*/*E* isomers is very important for industrial application, as the isomeric substrates cannot be easily separated on large scale. In this context, the bisphosphine ligands **48**[107] and **51**[103] with a rhodium catalytic system and **70**[108] with a ruthenium catalytic system have been successfully used for the asymmetric hydrogenation of *Z*/*E* mixtures, resulting in high enantioselectivities in the β-amino acid products.

There have been very few reports of the successful asymmetric hydrogenation of tetrasubstituted dehydro-β-amino acid derivatives. The Zhang group reported the hydrogenation of α-acetoxy-β-enamido esters using a rhodium catalyst with DuanPhos (**53**) as ligand;[109] high enantioselectivities are obtained. The same group also revealed the hydrogenation of cyclic β-(acylamino)acrylates using a ruthenium catalyst based on **74** (C_3-TunaPhos) as ligand.[110]

Scheme 8 Rhodium- and Ruthenium-Catalyzed Asymmetric Hydrogenation of Dehydro-β-amino Acid Esters[48,53,100–110]

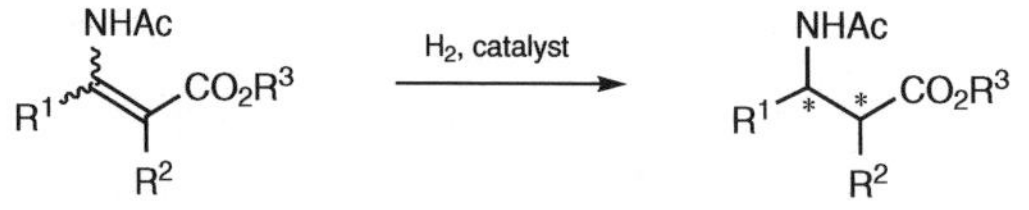

NHAc, R^1, CO_2R^3, R^2 **101** —(H_2, catalyst)→ NHAc, R^1, *, *, CO_2R^3, R^2 **102**

R^1	R^2	R^3	Config	Catalyst^a	Mol% of Catalyst	Temp (°C)	Pressure (atm)	ee (%)	Ref
Me	H	Me	*E*	$[Rh(cod)_2]BF_4$/**41**	1	25	2	95	[100]
Me	H	Me	*E*	$[Rh(cod)_2]BF_4$/**37**	1	40	2	98	[100]
Me	H	Me	*E*	$[Rh(cod)_2]BF_4$/**44**	1	25	1	97.9	[101]
Me	H	Me	*E*	$[Rh(nbd)_2]SbF_6$/**36**	1	25	1	>99	[48]
Me	H	Me	*E*	$[Rh(nbd)_2]BF_4$/**58**	1	25	3	98.7	[102]

R^1	R^2	R^3	Config	Catalyst[a]	Mol% of Catalyst	Temp (°C)	Pressure (atm)	ee (%)	Ref
Me	H	Me	*E*	$[Rh(nbd)_2]BF_4$/**59**	1	25	3	96.4	[102]
Me	H	Me	*E*	$[Rh(cod)_2]BF_4$/**51**	1	25	1.4	99	[103]
Me	H	Me	*E*	$[Ru(\eta^6\text{-}C_6H_6)$(**73**)Cl]Cl	1	25	17	96.4	[104]
Me	H	Me	*E*	$[Rh(cod)_2]BF_4$/**76**	1	5	10	98	[105]
Me	H	Me	*E*	$[Rh(cod)_2]BF_4$/**96**	2	25	10	98	[106]
Me	H	Me	*Z*	$[Rh(nbd)_2]SbF_6$/**36**	1	25	1	96	[48]
Me	H	Me	*E*/*Z*	$[Rh(nbd)_2]SbF_6$/**48**	1	25	1.4	99.5	[107]
Me	H	Me	*E*/*Z*	$[Rh(cod)_2]BF_4$/**51**	1	25	1.4	98	[103]
Me	H	Et	*E*	$[Rh(nbd)_2]SbF_6$/**53**	1	25	1.3	>99	[53]
Me	H	Et	*E*	$[Rh(cod)_2]BF_4$/**41**	1	25	2	99	[100]
Me	H	Et	*E*	$[Rh(cod)_2]BF_4$/**37**	1	40	2	98	[100]
Me	H	Et	*E*	$[Rh(nbd)_2]SbF_6$/**36**	1	25	1	>99	[48]
Me	H	Et	*E*	$[Rh(nbd)_2]BF_4$/**58**	1	25	3	99.7	[102]
Me	H	Et	*E*	$[Rh(nbd)_2]BF_4$/**59**	1	25	3	99.3	[102]
Me	H	Et	*E*	$[Ru(\eta^6\text{-}C_6H_6)$(**73**)Cl]Cl	1	25	17	95.7	[104]
Me	H	Et	*E*	$[Rh(cod)_2]BF_4$/**76**	1	5	10	98	[105]
Me	H	Et	*E*	$[Rh(cod)_2]BF_4$/**96**	0.5	25	25	98	[106]
Me	H	Et	*Z*	$[Rh(nbd)_2]SbF_6$/**53**	1	25	1.4	97	[53]
Ph	H	Me	*E*	$[Rh(cod)_2]BF_4$/**44**	1	25	1	97.9	[101]
Ph	H	Me	*E*/*Z*	[Ru(*p*-cymene)(**70**)Cl]Cl	2	50	13.6	99	[108]
Ph	H	Et	*Z*	$[Rh(cod)_2]BF_4$/**76**	1	5	10	>99	[105]
4-$MeOC_6H_4$	H	Me	*Z*	$[Rh(cod)_2]BF_4$/**76**	1	5	10	98	[105]
4-$MeOC_6H_4$	H	Me	*E*/*Z*	$[Rh(nbd)_2]SbF_6$/**48**	1	25	1.4	98.5	[107]
4-$MeOC_6H_4$	H	Me	*E*/*Z*	[Ru(*p*-cymene)(**70**)Cl]Cl	2	50	13.6	99	[108]
4-FC_6H_4	H	Me	*Z*	$[Rh(cod)_2]BF_4$/**76**	1	5	10	98	[105]
4-FC_6H_4	H	Me	*E*/*Z*	$[Rh(nbd)_2]SbF_6$/**48**	1	25	1.4	95	[107]
4-FC_6H_4	H	Me	*E*/*Z*	[Ru(*p*-cymene)(**70**)Cl]Cl	2	50	13.6	99	[108]
Ph	OAc	Me	*E*	$[Rh(nbd)_2]BF_4$/**53**	5	25	30	96	[109]
4-ClC_6H_4	OAc	Me	*E*	$[Rh(nbd)_2]BF_4$/**53**	5	25	30	97	[109]
4-Tol	OAc	Me	*E*	$[Rh(nbd)_2]BF_4$/**53**	5	25	30	95	[109]
2-naphthyl	OAc	Me	*E*	$[Rh(nbd)_2]BF_4$/**53**	5	25	30	95	[109]
$(CH_2)_3$		Me	–	[Ru(cod)(**74**)]$(BF_4)_2$	5	25	50	5	[110]

[a] For ligand structures see Schemes 3 and 4, Section 1.1.1.2.

Methyl (*R*)-3-Acetamidobutanoate (102**, $R^1 = R^3 = Me$; $R^2 = H$); Typical Procedure Using $[Rh(cod)_2]BF_4$/96 as Catalyst:**[106]

In a Schlenk tube equipped with a septum and stirrer bar, a mixture of $Rh(cod)_2BF_4$ (5.1 mg, 12.5 μmol) and the ligand (*S*)-**96** (10.9 mg, 25 μmol) were dissolved in CH_2Cl_2 (1.25 mL). An aliquot of this soln (1 mL) was added to a glass tube containing methyl (*E*)-3-acetamidobut-2-enoate (0.5 mmol) in CH_2Cl_2 (4 mL). The glass tube was placed in an autoclave (Endeavor reactor), and the autoclave was purged twice with N_2 and once with H_2. The autoclave was pressurized to 10 atm H_2, and the reaction was stirred at rt. The mixture was filtered through a short column of silica gel. The conversion was determined by ^{1}H NMR spectroscopy, and the ee was determined by chiral GC (CP-Chirasil-Dex-CB column, 160 °C).

1.1.1.3 Reduction of Oxygen-Substituted Alkenes

The asymmetric hydrogenation of oxygen-substituted alkenes has been widely investigated since the 1970s. It not only serves as an alternative method to the asymmetric hydrogenation of ketones in the preparation of chiral alcohols, but also gives chiral products, such as chiral cyclic ethers, that are difficult to access via ketone reduction. Generally, oxygen-substituted alkenes with a coordinating group, such as enol esters or enol phosphinates, can be hydrogenated smoothly and give the desired products with high yields and high enantioselectivities. Noncoordinating alkylated enol ethers and silyl enol ethers are more slowly hydrogenated, and investigations into the asymmetric hydrogenation of enol ethers are rare. In addition, due to steric effects, there are only a few successful examples of hydrogenation of enol esters or enol ethers with a tetrasubstituted C=C bond, and relatively high catalyst loadings are needed to obtain high conversions and high enantioselectivities. Rhodium complexes based on bisphosphine ligands dominate this area (Scheme 9).

Scheme 9 Representative Ligands for the Reduction of Oxygen-Substituted Alkenes

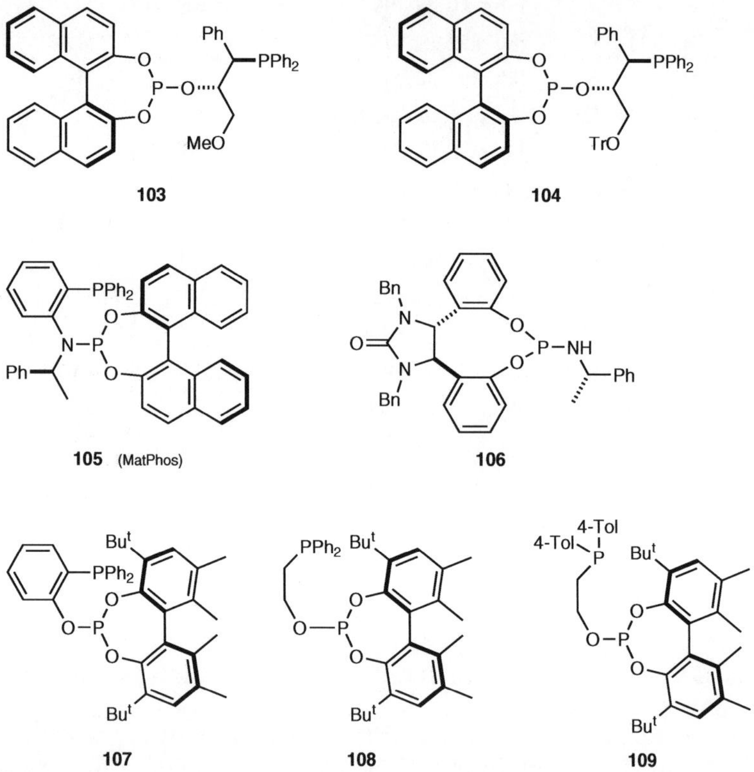

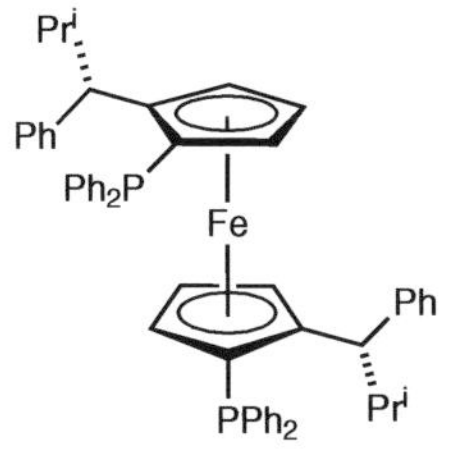

110 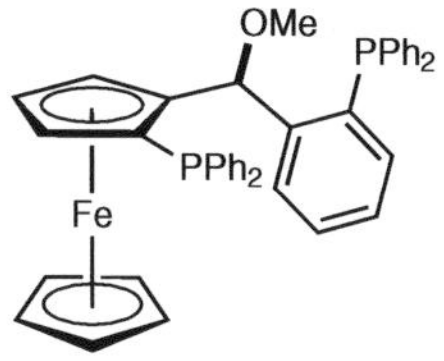 111 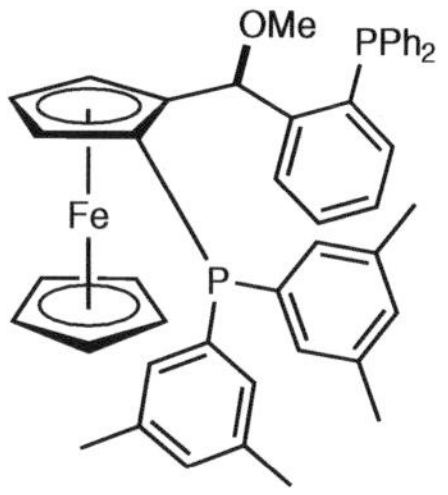 112

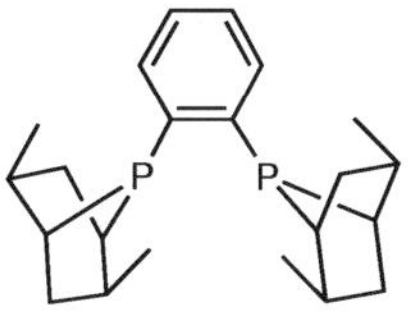

113 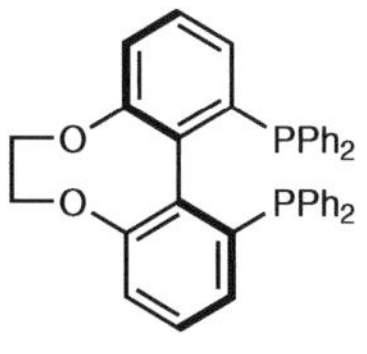 114 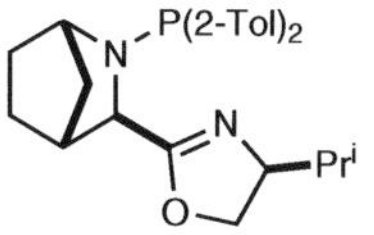 115

1.1.1.3.1 Reduction of Enol Esters

The asymmetric hydrogenation of enol esters readily gives chiral alcohols after hydrolysis of the ester groups. Therefore, enol esters have been the most frequently used oxygen-substituted alkenes in asymmetric hydrogenation. Rhodium complexes with monodentate or bidentate phosphorus ligands have been the predominant catalysts for the asymmetric hydrogenation of enol esters. Acyclic enol esters can be hydrogenated effectively with high enantiomeric excess values, but the reduction of cyclic enol esters is less effective due to the increased rigidity of the substrate. Only two reports exist in which the successful hydrogenation of such substrates is noted, but with a very limited substrate scope.[111,112]

α-Substituted enol esters **116** are easily reduced to the chiral esters **117** with excellent conversion and high enantiomeric excess using rhodium catalysts under mild conditions (Scheme 10).[50,53,54,113–121] Both alkyl- and aryl-substituted enol esters are well tolerated in this reaction, but only a few catalytic systems can achieve hydrogenation of alkyl-substituted enol esters with high enantioselectivity.

for references see p 63

Scheme 10 Rhodium-Catalyzed Asymmetric Hydrogenation of α-Substituted Enol Esters[50,53,54,113–121]

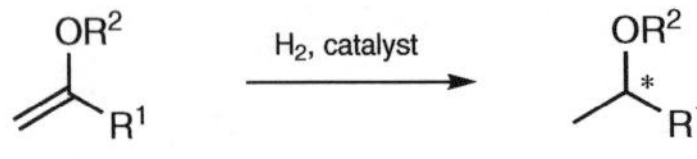

H_2, catalyst

116 → 117

R^1	R^2	Conditions[a]	Conversion (%)	ee (%)	Ref
Ph	Ac	[Rh(nbd)(**48**)]SbF_6 (1 mol%), EtOAc, 25 °C, 1.4 atm	100	96	[113]
Ph	Ac	[Rh(nbd)(**53**)]SbF_6 (1 mol%), THF, 25 °C, 1.4 atm	100	97	[53]
Ph	Ac	[Rh(nbd)(**54**)]BF_4 (1 mol%), EtOAc, 25 °C, 1.4 atm, 12 h	100	97	[54]
Ph	Ac	[Rh(nbd)$_2$]BF_4/**103** (1 mol%), THF, 25 °C, 20 atm, 18 h	>99	96	[114]
Ph	Ac	[Rh(nbd)$_2$]BF_4/**104** (1 mol%), THF, 25 °C, 20 atm, 18 h	>99	97	[114]
Ph	Ac	[Rh(nbd)(**105**)]BF_4 (0.5 mol%), THF, 25 °C, 20 atm, 1 h	>99	99	[115]
Ph	Ac	Rh(cod)$_2$$BF_4$/**106** (1 mol%), 1,2-dichloroethane, –20 °C, 5 atm, 10 h	>99	93	[116]
Ph	Ac	[Rh(nbd)$_2$]BF_4/**50** (0.1 mol%), MeOH, 25 °C, 1 atm, 5 h	100	99	[50]
Ph	Bz	[Rh(nbd)$_2$]BF_4/**107** (0.5 mol%), CH_2Cl_2, 40 °C, 4 atm, 20 h	100	98	[117]
Ph	Bz	[Rh(nbd)$_2$]BF_4/**108** (0.5 mol%), CH_2Cl_2, 40 °C, 4 atm, 20 h	100	98	[117]
Ph	Bz	[Rh(nbd)$_2$]BF_4/**109** (0.2 mol%), CH_2Cl_2, 40 °C, 4 atm, 20 h	100	98	[117]
Pr	Bz	[Rh(nbd)(**105**)]BF_4 (0.5 mol%), THF, 25 °C, 20 atm, 1 h	>99	99	[115]
Pr	Bz	[Rh(nbd)$_2$]BF_4/**108** (0.2 mol%), CH_2Cl_2, 40 °C, 4 atm, 24 h	100	96	[117]
CO_2Me	Ac	[Rh(nbd)$_2$]BF_4/**110** (1 mol%), acetone, 25 °C, 1 atm, 12 h	95	95	[118]
CO_2Me	Ac	[Rh(nbd)$_2$]BF_4/**111** (1 mol%), MeOH, 25 °C, 1 atm, 20 h	100	98	[119]
CO_2Me	Ac	[Rh(nbd)$_2$]BF_4/**112** (1 mol%), MeOH, 25 °C, 1 atm, 20 h	100	99	[119]
CO_2Et	Ac	[Rh(cod)(**41**)]OTf (0.1 mol%), MeOH, 25 °C, 2 atm, 2 h	100	99	[120]
Et	Ac	[Rh(cod)(**105**)]BF_4 (0.5 mol%), THF, 25 °C, 20 atm, 1 h	>99	99	[115]
Pr	Ac	[Rh(nbd)(**105**)]BF_4 (0.5 mol%), THF, 25 °C, 20 atm, 1 h	>99	98	[115]
$(CH_2)_5Me$	Ac	[Rh(nbd)(**105**)]BF_4 (0.1 mol%), THF, 25 °C, 20 atm, 1 h	>99	99	[115]
cyclopropyl	Ac	[Rh(nbd)(**105**)]BF_4 (0.5 mol%), THF, 25 °C, 20 atm, 1 h	>99	86	[115]
Cy	Ac	[Rh(nbd)(**105**)]BF_4 (0.5 mol%), THF, 25 °C, 20 atm, 1 h	>99	80	[115]
t-Bu	Ac	[Rh(cod)(**105**)]BF_4 (2 mol%), THF, 25 °C, 20 atm, 18 h	>99	96	[115]
CN	Ac	[Rh(nbd)(**105**)]BF_4 (1 mol%), THF, 25 °C, 20 atm, 1 h	>99	97	[115]
cyclopropyl	Bz	[Rh(nbd)$_2$]BF_4/**108** (0.2 mol%), CH_2Cl_2, 40 °C, 4 atm, 24 h	>99	96	[121]
cyclopentyl	Bz	[Rh(nbd)$_2$]BF_4/**109** (0.2 mol%), CH_2Cl_2, 40 °C, 4 atm, 24 h	>99	90	[121]
Cy	Bz	[Rh(nbd)$_2$]BF_4/**109** (0.2 mol%), CH_2Cl_2, 40 °C, 4 atm, 24 h	>99	90	[121]

[a] For ligand structures see Scheme 3, Section 1.1.1.2, and Scheme 9, Section 1.1.1.3.

Rhodium- or ruthenium-catalyzed reduction of α,β-disubstituted enol esters **118** also allows the highly enantioselective formation of chiral esters **119** (Scheme 11).[50,122–124] However, the substrate scope is generally limited to enol esters substituted with electron-withdrawing groups; the reduction of dialkyl- or diaryl-substituted enol esters is rare.

Scheme 11 Rhodium- and Ruthenium-Catalyzed Reduction of Trisubstituted Enol Esters[50,122–124]

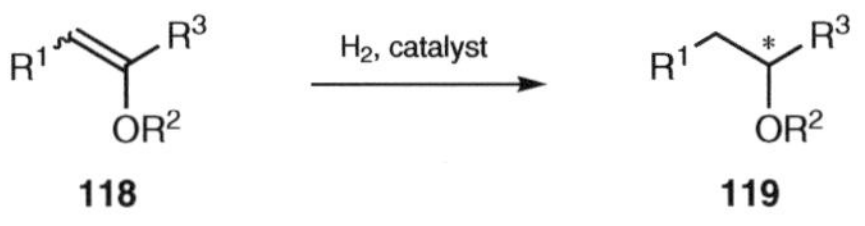

R^1	R^2	R^3	Conditions[a]	Conversion (%)	ee (%)	Ref
Ph	Ac	CO_2Et	$[Rh(nbd)_2]BF_4$/**50** (1 mol%), MeOH, 25 °C, 1 atm, 2 h	100	>99	[50]
Ph	Ac	CO_2Et	[Rh(cod)(**41**)]OTf (0.2 mol%), MeOH, 25 °C, 4 atm, 48 h	100	96	[122]
Ph	Ac	CO_2Me	$[Rh(cod)_2]BF_4]$/**106** (1 mol%), 1,2-dichloroethane, 25 °C, 60 atm, 10 h	100	>99	[123]
Pr	Ac	CO_2Me	$[Rh(cod)_2BF_4]$/**106** (1 mol%), 1,2-dichloroethane, 25 °C, 60 atm, 10 h	100	99	[123]
Pr	Bz	CO_2Me	[Rh(cod)(**41**)]OTf (0.2 mol%), MeOH, 25 °C, 4.1 atm, 48 h	100	98	[122]
iPr	Ac	CO_2Me	$[Rh(cod)_2]BF_4$/**106** (1 mol%), 1,2-dichloroethane, 25 °C, 60 atm, 10 h	100	99	[123]
iPr	Ac	CO_2Et	[Rh(cod)(**41**)]OTf (0.2 mol%), MeOH, 25 °C, 6.1 atm, 48 h	100	96	[122]
4-FC_6H_4	Ac	CO_2Me	$[Rh(cod)_2]BF_4$/**106** (1 mol%), 1,2-dichloroethane, 25 °C, 40 atm, 10 h	100	99	[123]
3-ClC_6H_4	Ac	CO_2Me	$[Rh(cod)_2]BF_4$/**106** (1 mol%), 1,2-dichloroethane, 25 °C, 40 atm, 10 h	100	96	[123]
2-BrC_6H_4	Ac	CO_2Me	$[Rh(cod)_2]BF_4$/**106** (1 mol%), 1,2-dichloroethane, 25 °C, 40 atm, 10 h	100	96	[123]
$(CH_2)_8Me$	Ac	CF_3	[RuCl(*p*-cymene)(**63**)]Cl (1 mol%), MeOH, 50 °C, 4 atm, 42 h	99	99	[124]

[a] For ligand structures see Scheme 3, Section 1.1.1.2, and Scheme 9, Section 1.1.1.3.

The reduction of cyclic enol esters has rarely been investigated, and the enantioselectivity is usually poor. An exception to this is when rhodium/Me-Pennphos (**113**) or ruthenium/C_2-TunaPhos (**114**) are used as the catalyst (Scheme 12).[111,112]

Scheme 12 Reduction of Cyclic Enol Acetates[111,112]

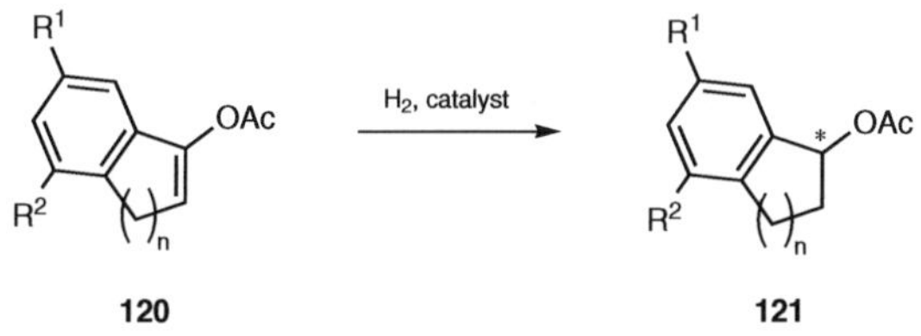

R^1	R^2	n	Conditions[a]	Conversion (%)	ee (%)	Ref
H	H	1	$[Rh(cod)_2]BF_4$/**113** (1 mol%), MeOH, 25 °C, 1.7 atm, 24 h	100	99	[111]
H	H	1	$[Ru(cod)_2]BF_4$/**114** (1 mol%), EtOH/CH_2Cl_2 (4:1), 50 °C, 3 atm, 4 h	98	99	[112]
H	H	2	$[Rh(cod)_2]BF_4$/**113** (1 mol%), MeOH, 25 °C, 1.7 atm, 24 h	100	99	[111]
Me	Me	2	$[Rh(cod)_2]BF_4$/**113** (1 mol%), MeOH, 25 °C, 1.7 atm, 24 h	100	99	[111]

[a] For ligand structures see Scheme 9, Section 1.1.1.3.

(*R*)-1-Phenylethyl Acetate (117, $R^1 = Ph$; $R^2 = Ac$); Typical Procedure Using $[Rh(nbd)_2]BF_4$/105 as Catalyst:[115]

A 10-mL high-pressure autoclave equipped with a glass inlet and stirrer bar was charged under argon with alkene **116** ($R^1 = Ph$; $R^2 = Ac$; 126.9 mg, 0.780 mmol) and 1 mL of a 0.0039 M catalyst stock soln of $[Rh(nbd)_2]BF_4$ and ligand **105** in THF. The autoclave was then pressurized to 20 atm H_2 and stirred at 1000 rpm for 1 h at rt. After releasing the pressure, a small sample of the mixture was immediately analyzed by ^1H NMR spectroscopy to determine the conversion (>99%). The rest of the sample was purified by removing the THF, followed by addition of CH_2Cl_2 (3 mL) to the crude mixture and filtration through a small silica gel pad. The filtrate was concentrated under reduced pressure to afford a colorless liquid; yield: 88 mg (69%); 99% ee (*R*).

Methyl 2-Acetoxy-3-phenylpropanoate (119, $R^1 = Ph$; $R^2 = Ac$; $R^3 = CO_2Me$); Typical Procedure Using $[Rh(cod)_2]BF_4$/106 as Catalyst:[123]

$[Rh(cod)_2]BF_4$ (2.0 mg, 5 μmol) and ligand **106** (11 μmol) were dissolved in CH_2Cl_2 (1 mL) under N_2 and the soln was stirred at rt for 10 min. Methyl 2-acetoxy-3-phenylacrylate (**118**, $R^1 = Ph$; $R^2 = Ac$; $R^3 = CO_2Me$; 0.5 mmol) in CH_2Cl_2 (1.5 mL) was added to the catalyst soln. The mixture was then transferred to a stainless-steel autoclave under N_2, and then the autoclave was sealed. After purging with H_2 (3 ×), the final H_2 pressure was adjusted to 60 atm. The mixture was stirred at rt for 10 h, and then H_2 was released. After removal of the solvent under reduced pressure, the residue was passed through a pad of Celite and then analyzed by ^1H NMR spectroscopy to determine the conversion.

2,3-Dihydro-1*H*-inden-1-yl Acetate (121, $R^1 = R^2 = H$; n = 1); Typical Procedure Using $[Rh(cod)_2]BF_4$/113 as Catalyst:[111]

To a soln of $[Rh(cod)_2]BF_4$ (5.0 mg, 12 μmol) in MeOH (10 mL) in a glovebox was added a 0.1 M soln of (*R*,*S*,*R*,*S*)-Me-PennPhos (**113**) in MeOH (0.15 mL, 0.015 mmol). After the mixture was stirred for 30 min, the enol acetate **120** (1.2 mmol) was added. The hydrogenation was performed at rt under 1.7 atm of H_2 for 24 h. After the H_2 was released, the mixture was passed through a short silica gel column to remove the catalyst. The ee was measured by capillary GC without any further purification.

1.1.1.3.2 Reduction of Enol Phosphinates

In contrast to enol esters, the rhodium-catalyzed asymmetric hydrogenation of enol phosphinates is generally less effective and only moderate enantioselectivities are achieved.[125] However, an iridium complex based on *N*-phosphine–dihydrooxazole ligand **115** is effective for the hydrogenation of α-substituted and α,β-disubstituted enol phosphinates. Using this system, a range of enol phosphinates **122** with both aromatic and aliphatic substituents are hydrogenated in excellent yields and enantioselectivities (Scheme 13).[126,127]

Scheme 13 Iridium-Catalyzed Reduction of Enol Phosphinates[126,127]

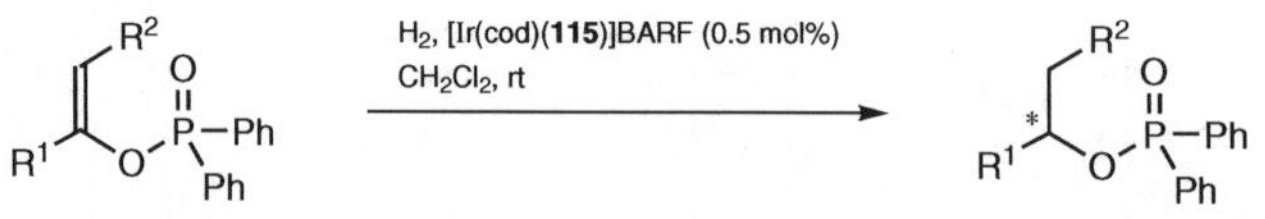

H_2, [Ir(cod)(**115**)]BARF (0.5 mol%), CH_2Cl_2, rt

122 → **123**

BARF = $[3,5-(F_3C)_2C_6H_3]_4B^-$

R^1	R^2	Pressure (atm)	Conversion (%)	ee (%)	Ref
Ph	H	30	>99	95	[126]
4-Tol	H	30	97	96	[126]
4-$F_3CC_6H_4$	H	30	>99	>99	[126]
4-BrC_6H_4	H	30	>99	>99	[126]
t-Bu	H	30	>99	>99	[126]
Ph	Me	30	>99	96	[127]
Ph	CO_2Me	50	>99	>99	[127]
4-Tol	CO_2Me	50	>99	>99	[127]
4-$F_3CC_6H_4$	CO_2Me	50	>99	>99	[127]
Me	CO_2Me	50	>99	>99	[127]
iPr	CO_2Me	50	>99	>99	[127]

Ethyl Diphenylphosphinates 123; **General Procedure:**[126]
A vial was charged with the enol phosphinate **122** (0.5 mol) and [Ir(cod)(**115**)]BARF (0.5 mol%). Anhyd CH_2Cl_2 (2 mL) was added, and the vial was placed into a high-pressure hydrogenation apparatus fitted with a steel bomb. After repeated purging (3 ×) with H_2, the pressure of H_2 was adjusted to 30–50 atm and the mixture was stirred for 1–3 h at rt. After releasing the pressure, the solvent was removed under reduced pressure. The crude product was chromatographed (pentane/iPrOH 70:30). The conversion was determined by ^1H NMR spectroscopy and the ee was determined by chiral HPLC.

1.1.1.3.3 Reduction of Enol Ethers

In this section, only alkylated enol ethers will be discussed. Because of the lack of a coordinating group, there are limited reports on asymmetric hydrogenations of alkylated enol ethers, and only a few cyclic enol ethers and alkylated enol ethers can be hydrogenated with high enantioselectivities. The best results have been obtained with analogues of Crabtree's iridium catalyst.

Very few examples of the asymmetric reduction of acyclic alkyl-substituted enol ethers have been reported. Good results are obtained only with enol ethers that are also allylic alcohols or their derivatives. As shown in Scheme 14, the enol ethers **124** can be reduced to chiral ethers **125** under basic conditions using an iridium–N-heterocyclic carbene **17** catalyst.[128]

Scheme 14 Iridium-Catalyzed Reduction of Acyclic Enol Ethers[128]

H_2 (50 atm)
[Ir(cod)(**17**)]BARF (1 mol%)
K_2CO_3, CH_2Cl_2, rt

124 → **125**

BARF = $[3,5\text{-}(F_3C)_2C_6H_3]_4B^-$

R^1	R^2	R^3	Conversion (%)	ee (%)	Ref
CH_2OH	Me	Me	>99	96	[128]
CH_2OH	Me	Et	>99	98	[128]
CH_2OH	Me	iPr	>99	91	[128]
CH_2OH	Me	Bu	>99	93	[128]
CH_2OAc	Me	Me	>99	92	[128]

In terms of the hydrogenation of cyclic enol ethers, 2-substituted 4*H*-1-benzopyrans **126** can be hydrogenated at room temperature to the 2-substituted 3,4-dihydro-2*H*-1-benzopyrans (chromanes) **127** in excellent yields and with excellent enantioselectivities using an iridium complex based on ligand **14** (Scheme 15).[129]

Scheme 15 Iridium-Catalyzed Reduction of 2-Substituted 4*H*-1-Benzopyrans[129]

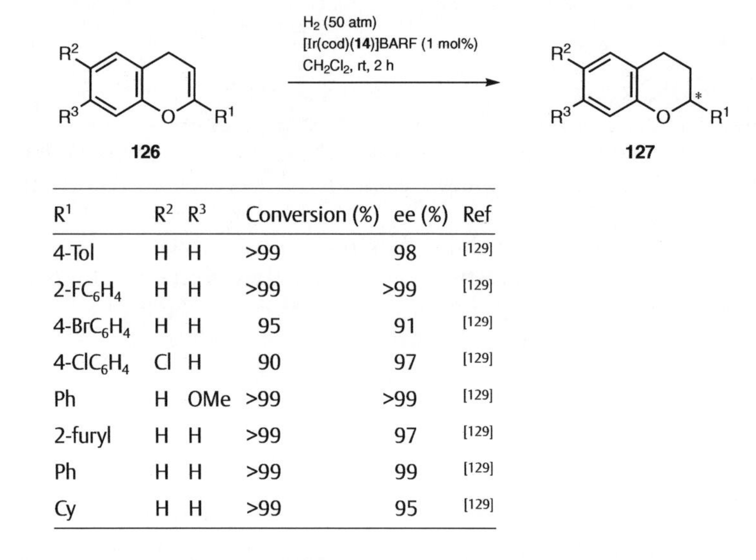

R^1	R^2	R^3	Conversion (%)	ee (%)	Ref
4-Tol	H	H	>99	98	[129]
$2\text{-}FC_6H_4$	H	H	>99	>99	[129]
$4\text{-}BrC_6H_4$	H	H	95	91	[129]
$4\text{-}ClC_6H_4$	Cl	H	90	97	[129]
Ph	H	OMe	>99	>99	[129]
2-furyl	H	H	>99	97	[129]
Ph	H	H	>99	99	[129]
Cy	H	H	>99	95	[129]

Alkyl Ethers 125; **General Procedure:**[128]

The alkene **124** was dissolved in CH_2Cl_2 (0.5 M) and the catalyst [Ir(cod)**17**]BARF (1 mol%) was then added. The resulting soln was degassed by three freeze–pump–thaw cycles, and then the soln was transferred to a Parr bomb. The bomb was flushed with H_2 for 1 min without stirring. The mixture was stirred at 700 rpm under 50 atm of H_2. After 12 h, the bomb was vented and the solvent was evaporated. The conversion was determined by ^1H NMR analysis. The crude product was passed through a silica gel plug (EtOAc/hexanes 3:7) to obtain the purified product.

3,4-Dihydro-2*H*-1-benzopyrans 127; **General Procedure:**[129]
A soln of 4*H*-1-benzopyran **126** (0.5 mmol) and [Ir(cod)(**14**)]BARF (8.7 mg, 5 mmol, 1 mol%), in anhyd CH_2Cl_2 (2.5 mL), under an argon atmosphere, was placed in an autoclave. The autoclave was sealed and pressurized with H_2 (50 atm), and the reaction was stirred at rt for 2 h. After releasing the pressure, the solvent was evaporated and the catalyst was removed by filtration through a short silica gel column (3 × 1 cm), using pentane/EtOAc (1:1) as eluent. Evaporation of the solvent from the filtrate afforded the product.

1.1.1.3.4 Reduction of Silyl Enol Ethers

Silyl enol ethers are noncoordinative substrates and are sensitive to acid. Although the hydrogenation of silyl enol ethers using transition-metal catalysts has been reported, the enantioselectivities are extremely low.[130] Recently, the advent of frustrated Lewis pair chemistry has provided a new approach for metal-free homogeneous hydrogenation. Silyl enol ethers **129** can be efficiently hydrogenated by a chiral frustrated Lewis pair catalyst, which is generated in situ by the hydroboration of chiral diene **128** with bis(pentafluorophenyl)borane [$HB(C_6F_5)_2$], giving a variety of optically active secondary alcohols **130** in high yields and with excellent enantiomeric excesses (Scheme 16).[131]

Scheme 16 Reduction of Silyl Enol Ethers Using a Frustrated Lewis Pair Catalyst[131]

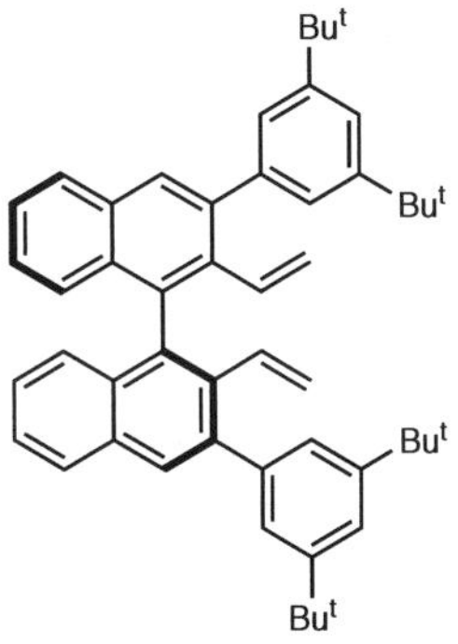

for references see p 63

1. $HB(C_6F_5)_2$ (10 mol%)
128 (5 mol%), t-Bu_3P (10 mol%)
toluene, H_2 (40 atm), 50 °C, 24 h
2. TBAF, toluene, rt, 0.5 h

129 → **130**

R^1	R^2	ee (%)	Yield (%)	Ref
Ph	H	98	98	[131]
4-ClC_6H_4	H	>99	97	[131]
4-EtC_6H_4	H	97	97	[131]
3-$MeOC_6H_4$	H	>99	97	[131]
2-$MeOC_6H_4$	H	>99	97	[131]
2-naphthyl	H	99	99	[131]
1-naphthyl	H	99	98	[131]
1-thienyl	H	97	97	[131]
cyclohexen-1-yl	H	99	98	[131]
(indane)		99	98	[131]

1-Phenylethanol (130, R^1 = Ph; R^2 = H); Typical Procedure:[131]
In a N_2 atmosphere glovebox, $HB(C_6F_5)_2$ (10.4 mg, 0.03 mmol), chiral diene **128** (10.2 mg, 0.015 mmol), and dry toluene (0.15 mL) were added to a 10-mL glass test tube. The resulting mixture was stirred at rt for 5 min, and then t-Bu_3P (10 wt% in pentane; 60.7 mg, 0.03 mmol) was added. The mixture was stirred for another 5 min at rt. To the resulting soln was added trimethyl[(1-phenylvinyl)oxy]silane (**129**, R^1 = Ph; R^2 = H; 57.7 mg, 0.3 mmol). The tube was then placed in a stainless-steel autoclave. After being sealed, the autoclave was purged with H_2 (3 ×) and the final pressure of H_2 was adjusted to 40 atm. The mixture was stirred at 50 °C for 24 h. The mixture was cooled to rt, a 1.0 M soln of TBAF in THF (0.3 mL, 0.3 mmol) was added, and the mixture was stirred at rt for 30 min and then washed with H_2O. The solvent was removed under reduced pressure, and the crude residue was purified by flash chromatography (silica gel, pentane/CH_2Cl_2 10:1) to give a colorless oil; yield: 35.9 mg (98%); 98% ee.

1.1.1.4 Reduction of Carbonyl-Substituted Alkenes

Chiral carbonyl compounds can be easily prepared by asymmetric reduction of the C=C bond in α,β-unsaturated carbonyls. Thus, the asymmetric hydrogenation of α,β-unsaturated carbonyl compounds, including α,β-unsaturated carboxylic acids, α,β-unsaturated esters, α,β-unsaturated amides, and α,β-unsaturated ketones, has been widely investigated. In general, asymmetric hydrogenation of acyclic α,β-unsaturated carbonyl compounds is well developed, giving the corresponding chiral carbonyl compounds with high yields and excellent enantioselectivities. By contrast, only a few methodologies have been reported for the asymmetric reduction of cyclic α,β-unsaturated carbonyl compounds. In addition, the asymmetric reduction of α,β-unsaturated carbonyl compounds with a tetrasubstituted C=C bond with high enantioselectivities has proven to be challenging. Representative ligands used in systems for the asymmetric reduction of carbonyl-group-substituted alkenes are collected in Scheme 17.

Scheme 17 Representative Ligands for the Reduction of Carbonyl-Substituted Alkenes

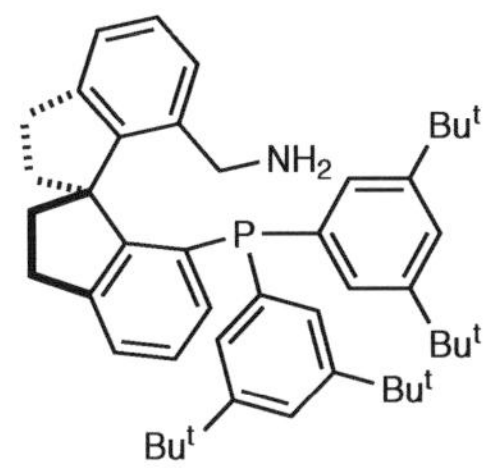

131

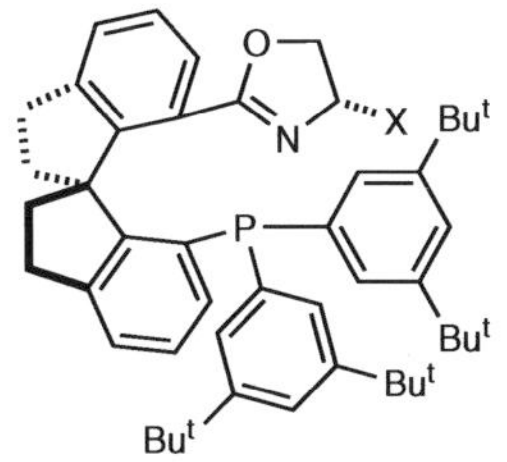

132 X = H, Bn, Me, iPr

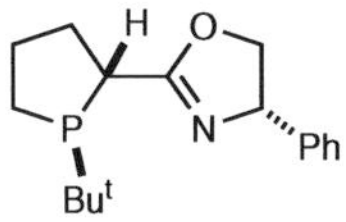

133

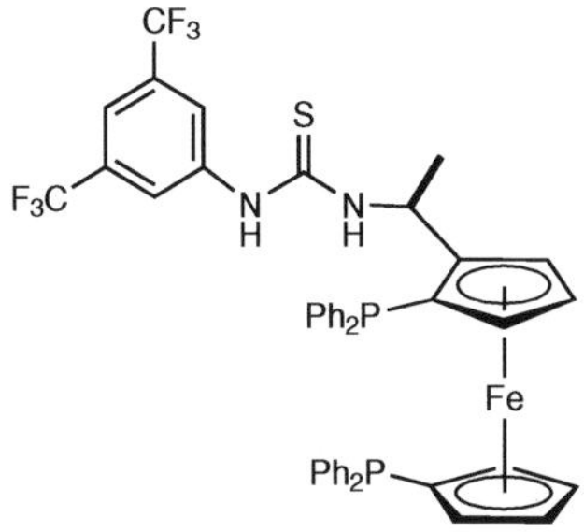

134

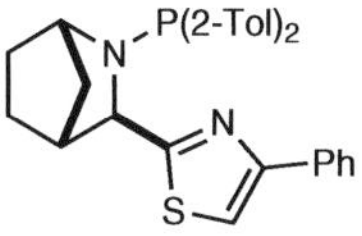

135

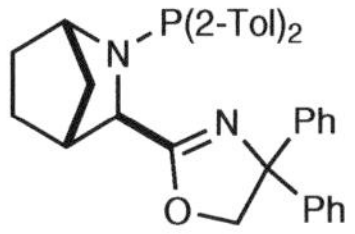

136

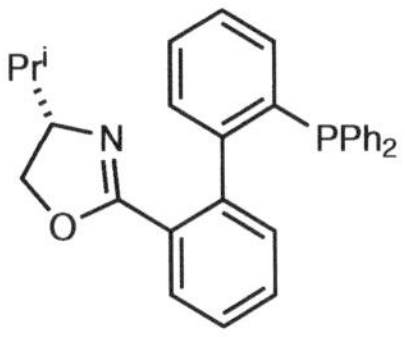

137

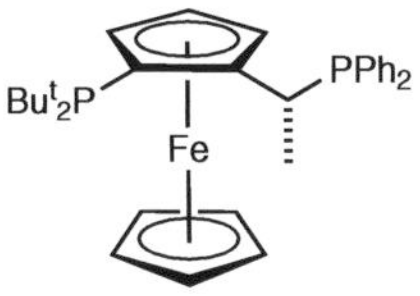

138

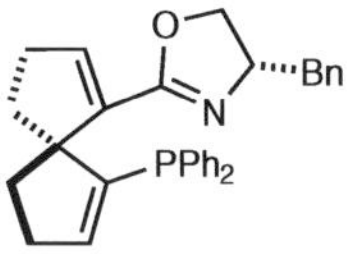

139

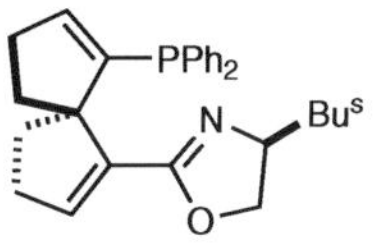

140

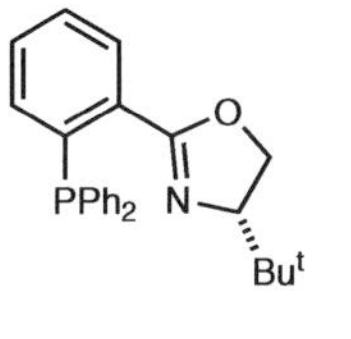

141

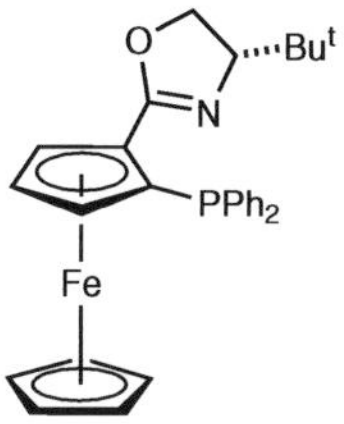

142

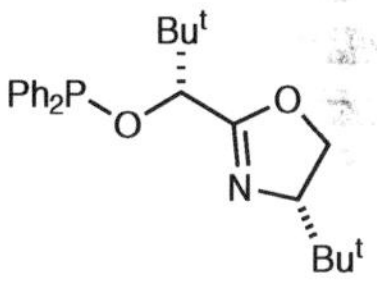

143

1.1.1.4.1 **Reduction of α,β-Unsaturated Carboxylic Acids**

Transition-metal-catalyzed enantioselective hydrogenation of unsaturated carboxylic acids is the most straightforward method to synthesize chiral acids. Traditionally, the asymmetric hydrogenation of unsaturated acids is dominated by rhodium and ruthenium catalysts modified with chiral phosphorus ligands, and excellent yields and enantiomeric excesses are obtained with these catalytic systems. Progress in this area was well-summarized in *Science of Synthesis: Stereoselective Synthesis*, Vol. 1, Sections 1.5.5.1 and 1.5.5.2.[11] Recently, iridium complexes with chiral spiro-phosphine–dihydrooxazole ligands and chiral spiro-phosphine–benzylamine ligands developed by the Zhou group have been shown to exhibit excellent activity and enantioselectivity in the asymmetric hydrogenation of α,β-unsaturated carboxylic acids.[132–135] In addition, a base-free catalytic system based on secondary interactions between catalyst and substrate has also been developed by the Zhang group; turnover numbers (TON) of up to 20 000 were achieved, but the substrate scope of this catalytic system is limited to aryl-substituted terminal vinyl carboxylic acids.[136]

The asymmetric reduction of α-substituted acrylic acids **144** using an iridium complex of chiral spiro-aminophosphine ligand **131** as catalyst, in the presence of cesium carbonate, gives α-substituted chiral acids **145** in excellent yields and with excellent enantioselectivities (Scheme 18).[132] A variety of α-substituted acrylic acids give excellent results under mild conditions at low catalyst loadings.

Scheme 18 Iridium-Catalyzed Reduction of α-Substituted Acrylic Acids[132]

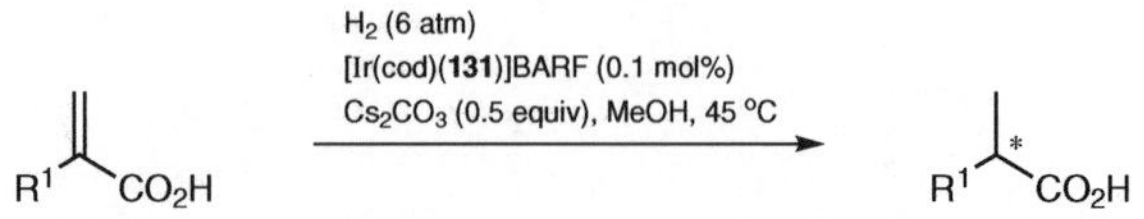

R^1 CO_2H **144** → H_2 (6 atm), [Ir(cod)(**131**)]BARF (0.1 mol%), Cs_2CO_3 (0.5 equiv), MeOH, 45 °C → R^1 * CO_2H **145**

BARF = $[3,5\text{-}(F_3C)_2C_6H_3]_4B^-$

R^1	Time (min)	ee (%)	Yield (%)	Ref
Ph	15	98	97	[132]
4-$MeOC_6H_4$	15	98	98	[132]
4-ClC_6H_4	30	97	98	[132]
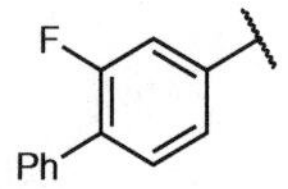 F, Ph	15	96	99	[132]
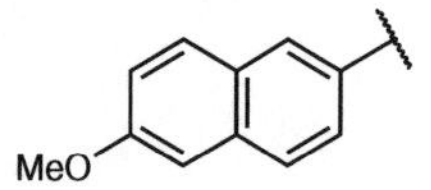 MeO	15	97	99	[132]
4-$iBuC_6H_4$	15	98	98	[132]
Bn	15	98	99	[132]
iPr	25	96	97	[132]
$(CH_2)_5Me$	30	94	97	[132]
CH_2CO_2Me	15	95	99	[132]
CH_2OEt	15	95	98	[132]

α,β-Disubstituted α,β-unsaturated carboxylic acids **146** can also be reduced to the corresponding chiral acids **147** with high yields and excellent enantioselectivities by employing chiral spiro-phosphine–dihydrooxazole iridium complexes (Scheme 19).[133,134] A variety of substituents on the α,β-unsaturated carboxylic acids are well tolerated, and the enantiopure carboxylic acids are efficiently prepared.

Scheme 19 Iridium-Catalyzed Reduction of α,β-Disubstituted α,β-Unsaturated Carboxylic Acids[133,134]

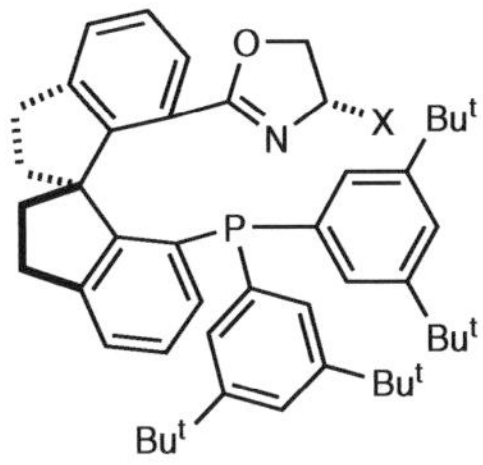

132 X = H, Bn, Me, iPr

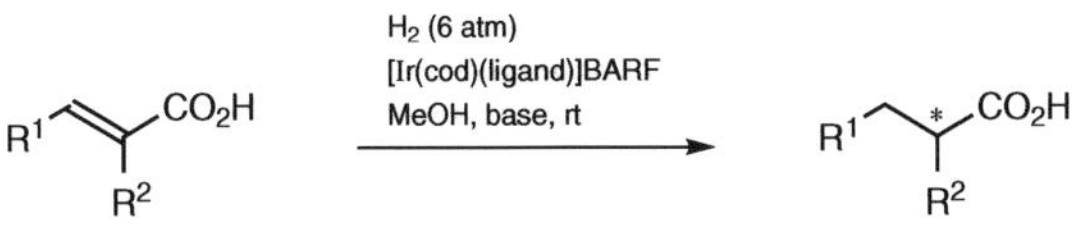

H_2 (6 atm), [Ir(cod)(ligand)]BARF, MeOH, base, rt

146 → **147**

R^1	R^2	Ligand	Mol% of Catalyst	Base	Time (h)	Yield (%)	ee (%)	Ref
Ph	Me	**132** (X = Bn)	0.25	Et_3N	0.5	99	99	[133]
Ph	iPr	**132** (X = Bn)	0.25	Et_3N	3	97	99	[133]
Ph	Ph	**132** (X = Bn)	1	Et_3N	5	95	94	[133]
4-$MeOC_6H_4$	Me	**132** (X = Bn)	0.25	Et_3N	0.58	97	99	[133]
4-$F_3CC_6H_4$	Me	**132** (X = Bn)	0.25	Et_3N	1	98	97	[133]
Me	Me	**132** (X = iPr)	0.25	Cs_2CO_3	0.5	92	99	[133]
Me	Pr	**132** (X = iPr)	0.5	Cs_2CO_3	18	92	98	[133]
Pr	Me	**132** (X = iPr)	0.5	Cs_2CO_3	18	89	99	[133]
iBu	Me	**132** (X = iPr)	1	Cs_2CO_3	18	97	90	[133]
2-naphthyl	Me	**132** (X = Bn)	0.25	Et_3N	8	96	99	[133]
2-furyl	Me	**132** (X = Bn)	0.25	Et_3N	18	98	98	[133]
Me	OPh	**132** (X = Bn)	0.5	Cs_2CO_3	10	95	>99	[134]
Ph	OPh	**132** (X = Bn)	0.5	Cs_2CO_3	8	95	>99	[134]
Ph	OMe	**132** (X = Me)	0.25	Cs_2CO_3	6	95	>99	[134]
Ph	OBn	**132** (X = Me)	0.25	Cs_2CO_3	4	96	>99	[134]
4-$MeOC_6H_4$	OBn	**132** (X = Me)	0.25	Cs_2CO_3	5	95	>99	[134]
4-$F_3CC_6H_4$	OBn	**132** (X = Me)	0.25	Cs_2CO_3	2	93	>99	[134]
Me	OMe	**132** (X = Me)	0.25	Cs_2CO_3	6	91	>99	[134]

The enantioselective hydrogenation of unsaturated heterocyclic acids **148** to furnish the corresponding chiral acids **149** can be achieved by using chiral spiro-phosphine–dihydrooxazole iridium complexes (Scheme 20).[135] This method is suitable for the hydrogenation of five- to seven-membered-ring unsaturated heterocyclic acids, leading to the corresponding chiral heterocyclic acids in high yields and with high enantioselectivities.

for references see p 63

Scheme 20 Iridium-Catalyzed Reduction of Unsaturated Heterocyclic Acids[135]

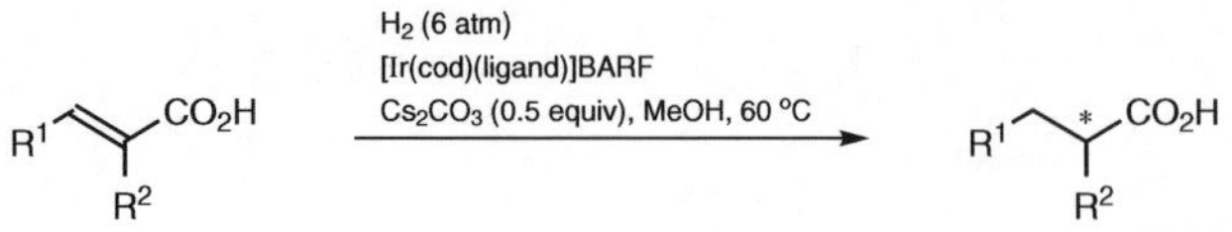

H_2 (6 atm), [Ir(cod)(ligand)]BARF, Cs_2CO_3 (0.5 equiv), MeOH, 60 °C

148 → **149**

R^1	R^2	Catalyst	Mol% of Catalyst	ee (%)	Yield (%)	Ref
$CH_2N(Boc)CH_2$		**132** (X = H)	2	90	94	[135]
$(CH_2)_2N(Boc)CH_2$		**132** (X = H)	0.5	97	94	[135]
$(CH_2)_2N(Boc)(CH_2)_2$		**132** (X = H)	0.2	97	95	[135]
$(CH_2)_2N(H)CH_2$		**132** (X = Bn)	1	96	99	[135]
$(CH_2)_2N(Me)CH_2$		**132** (X = Bn)	1	99	98	[135]
CH_2OCH_2		**132** (X = H)	2	88	94	[135]
$(CH_2)_2OCH_2$		**132** (X = H)	1	97	95	[135]
$(CH_2)_2O(CH_2)_2$		**132** (X = H)	0.1	97	94	[135]
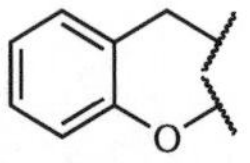 (chroman structure)		**132** (X = Bn)	0.5	92	97	[135]

α-Substituted Acetic Acids 145**; General Procedure:**[132]
In an argon-filled glovebox, a hydrogenation tube was charged with a stirrer bar, the α-substituted acrylic acid **144** (0.5 mmol), [Ir(cod)(**131**)]BARF (0.9 mg, 0.5 μmol), and Cs_2CO_3 (82 mg, 0.25 mmol). MeOH (2 mL) was injected into the hydrogenation tube using a syringe while the mixture was stirring. The hydrogenation tube was then placed in an autoclave, and the autoclave was purged with H_2 (3×). The autoclave was then charged with H_2 to 6 atm, and the mixture was stirred at 45 °C for the specified time before releasing the H_2. The mixture was acidified with 3 M HCl, and extracted with Et_2O. Evaporation of the solvent afforded the crude product. The conversion of the substrate was determined by ^{1}H NMR spectroscopy. The product was converted into the corresponding ester or amide and the ee value was determined by GC, HPLC, or supercritical fluid chromatography (SFC) with a chiral column.

2-Methyl-3-phenylpropanoic Acid (147, **R^1 = Ph; R^2 = Me); Typical Procedure:**[133]
A hydrogenation tube was charged with a stirrer bar, (*E*)-2-methyl-3-phenylacrylic acid (**146**, R^1 = Ph; R^2 = Me; 81 mg, 0.5 mmol), [Ir(cod){**132** (X = Bn)}]BARF (2.4 mg, 1.25 μmol), Et_3N (25 mg, 0.25 mmol), and MeOH (2 mL) under air. The hydrogenation tube was then put into an autoclave. The air in the autoclave was replaced with H_2 (three cycles). The autoclave was then charged with H_2 to 6 atm, and the mixture was stirred at rt for 30 min. After releasing H_2, the mixture was concentrated on a rotary evaporator. The conversion of the substrate was determined by ^{1}H NMR analysis. The crude product was purified by flash chromatography (silica gel) to give the pure product as a colorless liquid; yield: 99%; 99% ee.

Saturated Heterocyclic Acids 149**; General Procedure:**[135]
A hydrogenation tube was charged with a stirrer bar, the unsaturated heterocyclic acid **148** (0.5 mmol), [Ir(cod)(**132**)]BARF, and Cs_2CO_3 (81 mg, 0.25 mmol) in an argon-filled glovebox. MeOH (2 mL) was injected into the hydrogenation tube using a syringe. The

tube was placed in an autoclave and purged with H_2 (5 ×). The autoclave was charged with H_2 to 6 atm, and the mixture was stirred at 60 °C for 12 h before the release of the H_2 pressure. The mixture was acidified with 3 M HCl (pH 1) and extracted with Et_2O (3 × 10 mL). The combined extracts were washed with sat. aq NaCl (15 mL), dried ($MgSO_4$), and concentrated under reduced pressure. The crude product was purified by flash chromatography (silica gel) to give the pure chiral acid.

1.1.1.4.2 Reduction of α,β-Unsaturated Esters

α,β-Unsaturated esters have been widely investigated as substrates for asymmetric hydrogenation reactions due to their applications in the preparation of functionalized chiral esters or acids. The reduction of nitrogen- and oxygen-substituted α,β-unsaturated esters is discussed in Sections 1.1.1.2 and 1.1.1.3; therefore, the examples in this category are focused on the hydrogenation of nonfunctionalized α,β-unsaturated esters. As these substrates lack a strongly coordinating functional group, fewer examples of systems for their asymmetric hydrogenation can be found in the literature. Iridium complexes bearing phosphine–dihydrooxazole ligands exhibit good activity and enantioselectivity for this transformation. To date, the asymmetric hydrogenation of α-substituted or α,β-disubstituted α,β-unsaturated esters has been well documented; however, examples of the corresponding transformations of α,β-unsaturated esters with a tetrasubstituted C=C bond are still rare. A selection of results for the asymmetric hydrogenation of α,β-unsaturated esters **150** using rhodium-, iridium-, and ruthenium-based catalysts is given in Scheme 21.[137–142]

Scheme 21 Asymmetric Reductions of α,β-Unsaturated Esters[137–142]

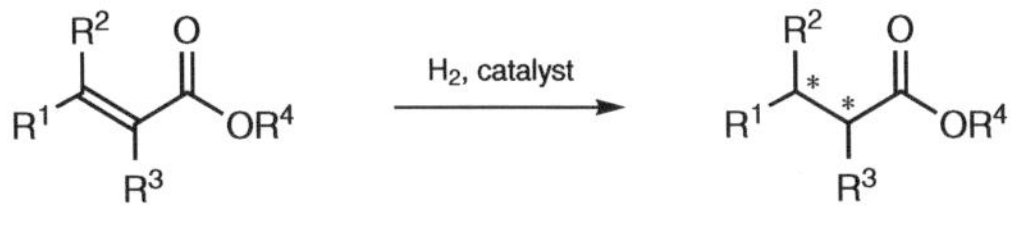

R^1–C(R²)=C(R³)–C(O)OR⁴ **150** → (H_2, catalyst) → R^1–C*H(R²)–C*H(R³)–C(O)OR⁴ **151**

R^1	R^2	R^3	R^4	Catalyst[a]	Mol% of Catalyst	Pressure (atm)	Conversion (%)	ee (%)	Ref
Ph	Me	H	Me	[Ir(cod)(**133**)]BARF	1	50	–	98	[137]
Ph	Me	H	Me	$\{Rh(cod)Cl\}_2$/**134**	0.5	50	99	96	[138]
Ph	Me	H	Et	[Ir(cod)(**135**)]BARF	0.5	50	>99	98	[139]
4-$MeOC_6H_4$	Me	H	Et	[Ir(cod)(**135**)]BARF	0.5	50	>99	98	[139]
4-$O_2NC_6H_4$	Me	H	Et	[Ir(cod)(**135**)]BARF	0.5	50	>99	97	[139]
Ph	Et	H	Et	[Ir(cod)(**135**)]BARF	0.5	50	>99	>99	[139]
Ph	iPr	H	Et	[Ir(cod)(**135**)]BARF	0.5	50	>99	>99	[139]
Ph	Cy	H	Et	[Ir(cod)(**135**)]BARF	0.5	50	>99	>99	[139]
Ph	CO_2Et	H	Et	[Ir(cod)(**135**)]BARF	0.5	50	>99	95	[139]
Bn	Me	H	Et	[Ir(cod)(**136**)]BARF	0.5	50	>99	93	[139]
Me	Ph	H	Et	[Ir(cod)(**115**)]BARF	1	50	>99	98	[139]
iPr	Ph	H	Et	[Ir(cod)(**115**)]BARF	1	50	>99	99	[139]
Me	Cy	H	Et	[Ir(cod)(**115**)]BARF	1	50	>99	93	[139]
Ph	H	Me	Et	[Ir(cod)(**136**)]BARF	0.5	50	>99	95	[139]
Ph	H	Me	Bn	[Ir(cod)(**136**)]BARF	0.5	50	>99	97	[139]
Ph	H	Me	CHMePh	[Ir(cod)(**136**)]BARF	0.5	50	>99	99	[139]

for references see p 63

R^1	R^2	R^3	R^4	Catalyst[a]	Mol% of Catalyst	Pressure (atm)	Conversion (%)	ee (%)	Ref
Ph	H	iBu	Et	[Ir(cod)(**136**)]BARF	0.5	50	>99	90	[139]
Me	H	Me	Et	[Ir(cod)(**136**)]BARF	0.5	20	>99	97	[139]
Me	H	Me	Bn	[Ir(cod)(**136**)]BARF	0.5	20	>99	99	[139]
Et	H	Me	Et	[Ir(cod)(**136**)]BARF	0.5	20	>99	96	[139]
Me	H	$(CH_2)_3$		[Ir(cod)(**136**)]BARF	0.5	50	>99	96	[139]
Ph	H	$(CH_2)_3$		[Ir(cod)(**136**)]BARF	0.5	20	>99	99	[139]
Ph	H	$(CH_2)_3$		[Ir(cod)(**140**)]BARF	1	5	>99	92	[140]
Ph	H	$(CH_2)_2$		[Ir(cod)(**137**)]BARF	1	20	100	95	[141]
cyclopropyl	Ph	Me	Me	Ru(cod)(2-methylallyl)$_2$/ **138**	20	34	100	98	[142]

[a] For ligand structures see Scheme 9, Section 1.1.1.3, and Scheme 17, Section 1.1.1.4; BARF = $[3,5\text{-}(F_3C)_2C_6H_3]_4B^-$.

Ethyl 3-Phenylbutanoate (151, R^1 = Ph; R^2 = Me; R^3 = H; R^4 = Et); Typical Procedure Using [Ir(cod)(135)]BARF as Catalyst:[139]
A vial was charged with ethyl (*E*)-3-phenylbut-2-enoate (**150**, R^1 = Ph; R^2 = Me; R^3 = H; R^4 = Et; 0.25 M) and [Ir(cod)(**135**)]BARF (0.5 mol%). Anhyd CH_2Cl_2 was added (2 mL), and the vial was placed in a high-pressure hydrogenation apparatus. The reactor was purged with argon (3 ×), and then filled with H_2 to a pressure of 50 atm. The reaction was stirred at rt for 15 h before the H_2 pressure was released and the solvent was removed under reduced pressure. The crude product was filtered through a short plug of silica gel. Conversion was determined by ^{1}H NMR spectroscopy, and the ee was determined by chiral GC or HPLC.

1.1.1.4.3 Reduction of α,β-Unsaturated Amides

In contrast to the reduction of α,β-unsaturated esters, there are only a few examples of asymmetric hydrogenation where α,β-unsaturated amides have been used as substrates. Iridium complexes with phosphine–dihydrooxazole ligands are the most efficient catalysts for this transformation. The asymmetric hydrogenation of α,β-unsaturated amides with a tetrasubstituted C=C bond, as well as that of endocyclic α,β-unsaturated amides, is still problematic.

The catalytic asymmetric hydrogenation of acyclic α,β-unsaturated amides **152** and **154** using iridium complexes based on chiral ligands **139** or **142** gives chiral amides **153** and **155**, respectively, with excellent enantiomeric excess values (Schemes 22 and 23).[143,144] In addition, the combination of rhodium and ZhaoPhos ligand (**134**) also shows good performance in the asymmetric reduction of acyclic α,β-unsaturated amides **156** (Scheme 24).[138,145]

Scheme 22 Iridium-Catalyzed Asymmetric Reduction of β,β-Disubstituted α,β-Unsaturated Weinreb Amides[143]

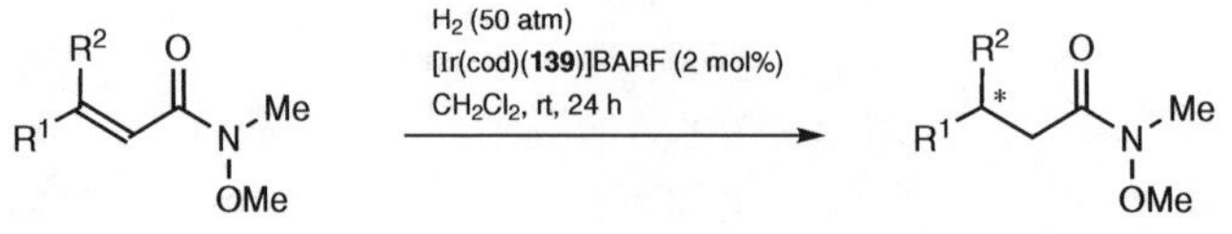

H_2 (50 atm)
[Ir(cod)(**139**)]BARF (2 mol%)
CH_2Cl_2, rt, 24 h

152 → **153**

BARF = $[3,5\text{-}(F_3C)_2C_6H_3]_4B^-$

R^1	R^2	Conversion (%)	ee (%)	Ref
Ph	Me	>99	95	[143]
4-$MeOC_6H_4$	Me	>99	96	[143]
4-FC_6H_4	Me	>99	96	[143]
4-$F_3CC_6H_4$	Me	>99	92	[143]
3-ClC_6H_4	Me	>99	95	[143]
2-Tol	Me	>99	82	[143]
2-naphthyl	Me	>99	95	[143]
Bn	Me	>99	91	[143]
Ph	Et	>99	96	[143]
Ph	iPr	94	90	[143]

Scheme 23 Iridium-Catalyzed Asymmetric Reduction of Acyclic α,β-Disubstituted α,β-Unsaturated Amides[144]

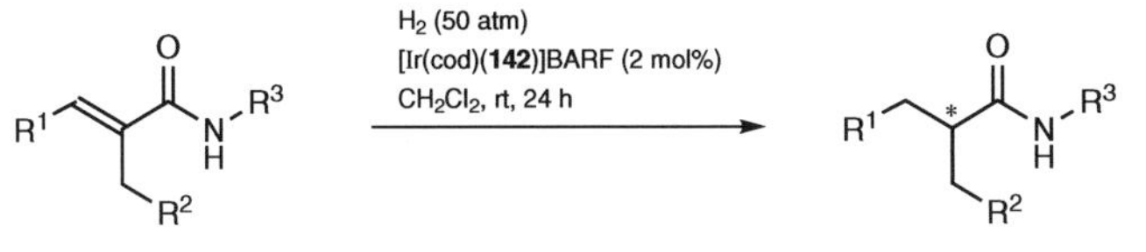

H_2 (50 atm)
[Ir(cod)(**142**)]BARF (2 mol%)
CH_2Cl_2, rt, 24 h

154 → **155**

R^1	R^2	R^3	Conversion (%)	ee (%)	Ref
Ph	H	Ph	100	96	[144]
Ph	H	iBu	100	93	[144]
Ph	H	Bn	100	95	[144]
Ph	*t*-Bu	iBu	100	96	[144]
Ph	Bn	iBu	100	97	[144]
Ph	OMe	iBu	100	95	[144]
3-ClC_6H_4	Et	iBu	100	98	[144]
4-$MeOC_6H_4$	Et	iBu	100	96	[144]
Pr	H	Bn	100	84	[144]
t-Bu	H	Bn	100	87	[144]
Me	H	Ph	100	95	[144]

Scheme 24 Rhodium-Catalyzed Asymmetric Reduction of Acyclic α,β-Unsaturated Amides[138,145]

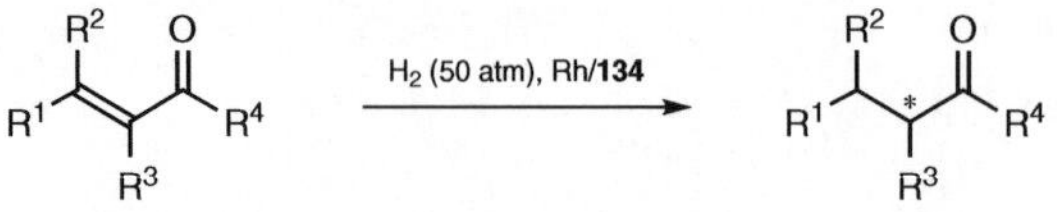

156 → 157 (H_2 (50 atm), Rh/**134**)

R^1	R^2	R^3	R^4	Solvent	ee (%)	Yield (%)	Ref
Ph	Me	H	NH_2	CH_2Cl_2/iPrOH (3:1)	95	97	[138]
4-$MeOC_6H_4$	Me	H	NH_2	CH_2Cl_2/iPrOH (3:1)	95	97	[138]
4-ClC_6H_4	Me	H	NH_2	CH_2Cl_2/iPrOH (3:1)	95	95	[138]
2-naphthyl	Me	H	NH_2	CH_2Cl_2/iPrOH (3:1)	93	98	[138]
Ph	Me	H	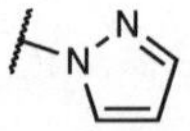 pyrazol-1-yl	CH_2Cl_2/EtOH (1:5)	96	94	[145]
4-$MeOC_6H_4$	Me	H	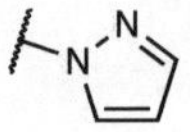 pyrazol-1-yl	CH_2Cl_2/EtOH (1:5)	93	94	[145]
4-ClC_6H_4	Me	H	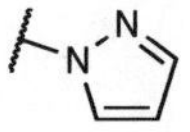 pyrazol-1-yl	CH_2Cl_2/EtOH (1:5)	90	96	[145]
3-Tol	Me	H	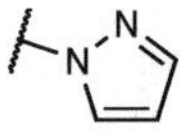 pyrazol-1-yl	CH_2Cl_2/EtOH (1:5)	93	97	[145]
Ph	H	Me	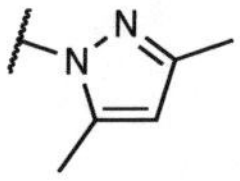 3,5-dimethylpyrazol-1-yl	CH_2Cl_2	97	94	[145]
4-$MeOC_6H_4$	H	Me	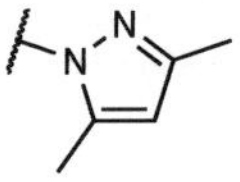 3,5-dimethylpyrazol-1-yl	CH_2Cl_2	91	92	[145]
4-ClC_6H_4	H	Me	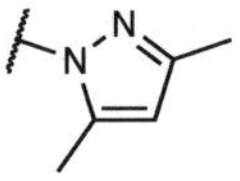 3,5-dimethylpyrazol-1-yl	CH_2Cl_2	98	97	[145]
2-furyl	H	Me	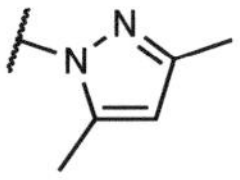 3,5-dimethylpyrazol-1-yl	CH_2Cl_2	99	95	[145]

Cyclic α,β-unsaturated amides **158** with an exocyclic C=C bond can be reduced by using iridium complexes based on phosphine–dihydrooxazole ligands **137**[141] or **140**,[140] affording α-substituted chiral lactams **159** in high yields and with excellent enantioselectivities (Scheme 25).

Scheme 25 Iridium-Catalyzed Asymmetric Reduction of Cyclic α,β-Unsaturated Amides[140,141]

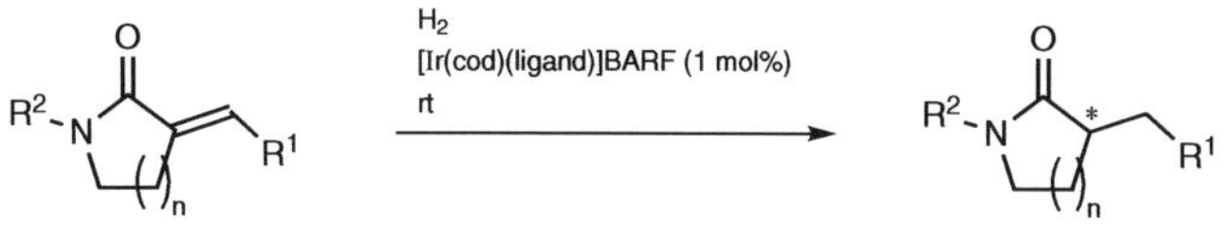

158 159

R^1	R^2	n	Ligand[a]	Solvent	Pressure (atm)	Conversion (%)	ee (%)	Ref
Ph	Ac	1	**137**	toluene	20	100	97	[141]
Ph	Bn	1	**137**	toluene	20	100	98	[141]
4-$MeOC_6H_4$	Bn	1	**137**	toluene	20	100	98	[141]
4-FC_6H_4	Bn	1	**137**	toluene	20	100	98	[141]
Ph	Boc	2	**140**	CH_2Cl_2	1	>99[b]	93	[140]
4-Tol	Boc	2	**140**	CH_2Cl_2	1	>99[b]	91	[140]
4-$F_3CC_6H_4$	Boc	2	**140**	CH_2Cl_2	1	>99[b]	95	[140]
3-$F_3CC_6H_4$	Boc	2	**140**	CH_2Cl_2	1	>99[b]	95	[140]
2-furyl	Boc	2	**140**	CH_2Cl_2	1	>99[b]	96	[140]
Ph	Boc	3	**140**	CH_2Cl_2	5	>99	97	[140]
4-$MeOC_6H_4$	Boc	3	**140**	CH_2Cl_2	5	>99	98	[140]
4-$F_3CC_6H_4$	Boc	3	**140**	CH_2Cl_2	5	>99	97	[140]
2-furyl	Boc	3	**140**	CH_2Cl_2	5	>99	97	[140]

[a] For ligand structures see Scheme 17, Section 1.1.1.4.
[b] Ligand with opposite configuration at the spirocyclic center was used.

β,β-Disubstituted Weinreb Amides 153; **General Procedure:**[143]
The catalyst [Ir(cod)(**139**)]BARF (2 μmol) and the α,β-unsaturated Weinreb amide **152** (0.1 mmol) were dissolved in CH_2Cl_2 (1.0 mL) in a vial under an argon atmosphere. In a glovebox, the vial was transferred to a Parr steel autoclave, which was purged with H_2 (3 ×) and finally pressurized to 50 atm. The mixture was stirred at rt for 24 h, before the H_2 was released in a fume hood. The conversion was determined by ^1H NMR analysis of an aliquot of the crude mixture. The mixture was filtered through a short pad of silica gel, eluting with petroleum ether/EtOAc (5:1).

α-Substituted Amides 155; **General Procedure:**[144]
A mixture of [Ir(cod)(**142**)]BARF (2 μmol), the α,β-unsaturated amide **154** (0.1 mmol), and CH_2Cl_2 (2 mL) in a tube containing a magnetic stirrer bar was placed in an autoclave under air, and the autoclave was sealed. The autoclave was pressurized to 50 atm with H_2, and the soln was stirred at rt for 24 h. The pressure was then carefully released and the mixture was passed through a short column of silica gel, eluting with (EtOAc/petroleum ether). The soln was concentrated and the resulting oil or solid was analyzed.

Amides 157; **General Procedure:**[138]
In a N_2-filled glovebox, a soln of $\{RhCl(cod)\}_2$ (4.9 mg, 0.01 mmol) and ligand **134** (2.1 equiv) in anhyd solvent (5.0 mL) was stirred at rt for 30 min. A specified volume of the resulting soln (0.5 mL, 1 mol% Rh catalyst) was transferred by syringe to a Score-Break ampule charged with a soln of the substrate **156** (0.2 mmol) in 0.5 mL of the solvent. The ampule was placed in an autoclave, which was then charged with H_2 (50 atm). The mixture was stirred at the desired temperature for 48 h. After release of H_2, the resulting

mixture was concentrated under reduced pressure. The residue was passed through a silica gel plug to remove the metal complex, and the filtrate was concentrated under reduced pressure.

3-Benzyl-1-(*tert*-butoxycarbonyl)azepan-2-one (159, R^1 = Ph; R^2 = Boc; n = 3); Typical Procedure Using [Ir(cod)(140)]BARF as Catalyst:[140]
To a vial containing CH_2Cl_2 (1.5 mL) and a magnetic stirrer bar was added the precatalyst [Ir(cod)(**140**)]BARF (1.5 μmol) and (*E*)-3-benzylidene-1-(*tert*-butoxycarbonyl)azepan-2-one (**158**, R^1 = Ph; R^2 = Boc; n = 3; 0.15 mmol) under an argon atmosphere. The vial was transferred in a glovebox to a Parr steel autoclave, which was purged with H_2 (3 ×) and finally pressurized to 5 atm. The mixture was stirred at rt for 5–10 h. The H_2 was released in a hood, and the conversion was determined by ^1H NMR analysis of an aliquot of the crude mixture. The mixture was filtered through a short pad of silica gel, and the solvent of the filtrate was removed under reduced pressure to afford the product; conversion: >99%; 97% ee (determined by chiral HPLC).

1.1.1.4.4 Reduction of α,β-Unsaturated Ketones

Only a few examples of alkene-selective asymmetric reduction of α,β-unsaturated ketones with good yields and good regio- and enantioselectivities have been reported. An iridium complex based on P,N-ligand **141** shows good performance in the asymmetric hydrogenation of acyclic α,β-unsaturated ketones and cyclic α,β-unsaturated ketones with an exocyclic C=C bond.[146] With regard to the transition-metal-catalyzed hydrogenation of endocyclic α,β-unsaturated ketones, it remains a challenge to achieve this transformation with high chemoselectivity and high enantioselectivity. However, chiral counteranion directed asymmetric transfer hydrogenation seems to be advantageous in the reduction of such alkenes, giving the target products with excellent enantiomeric excesses.[147]

The chemoselective asymmetric hydrogenation of α,β-unsaturated ketones **160** proceeds very smoothly in the presence of iridium complexes with phosphorus–dihydrooxazole ligands **141** and **143**, giving chiral ketones **161** in high yields and with excellent enantioselectivity (Scheme 26).[146,148] This method has a broad substrate scope and works well for the reduction of α-substituted or α,β-disubstituted α,β-unsaturated ketones.

Scheme 26 Iridium-Catalyzed Asymmetric Reduction of Acyclic α,β-Unsaturated Ketones[146,148]

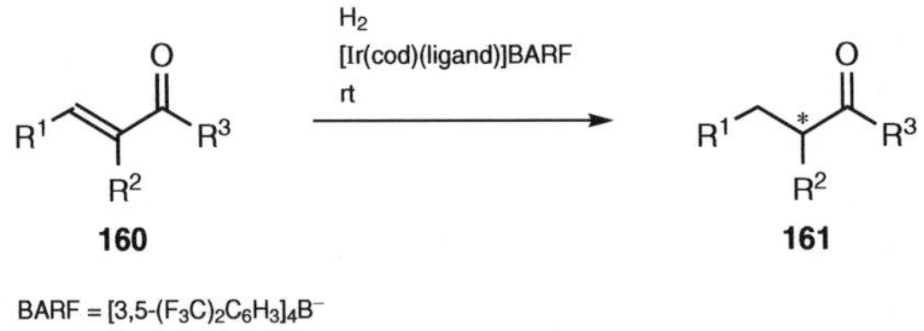

R^1	R^2	R^3	Ligand	Solvent	Pressure (atm)	Time (h)	ee (%)	Yield (%)	Ref
Ph	Me	Me	**141**	toluene	2	3	98	91	[146]
Ph	Me	Me	**143**	CH_2Cl_2	50	24	>99	>99	[148]
Ph	Me	Ph	**141**	toluene	2	3	99	96	[146]
Ph	Pr	Me	**141**	toluene	2	3	99	88	[148]
Ph	Ph	Me	**141**	toluene	2	3	98	94	[146]
Ph	Ph	Ph	**141**	toluene	2	3	99	93	[146]
2-MeOC$_6$H$_4$	Me	Me	**141**	toluene	2	3	99	90	[146]

R^1	R^2	R^3	Ligand	Solvent	Pressure (atm)	Time (h)	ee (%)	Yield (%)	Ref
3-$MeOC_6H_4$	Me	Me	**141**	toluene	2	3	99	93	[146]
4-$MeOC_6H_4$	Me	Me	**141**	toluene	2	3	98	97	[146]
4-$MeOC_6H_4$	Me	Me	**143**	CH_2Cl_2	50	24	99	>99	[148]
3-$O_2NC_6H_4$	Me	Me	**141**	toluene	2	3	98	91	[146]
4-$O_2NC_6H_4$	Me	Me	**141**	toluene	2	3	99	88	[146]

Reports of the hydrogenation of the alkene unit in cyclic α,β-unsaturated ketones are rare, and the successful examples are mainly focused on the hydrogenation of cyclic unsaturated ketones where the C=C bond is exocyclic. For example, the iridium-catalyzed reduction of α,β-unsaturated ketones **162** gives chiral cyclic ketones **163** in good yields and with excellent enantiomeric excesses (Scheme 27).[140,141,146,148]

Scheme 27 Iridium-Catalyzed Asymmetric Reduction of Exocyclic α,β-Unsaturated Ketones[140,141,146,148]

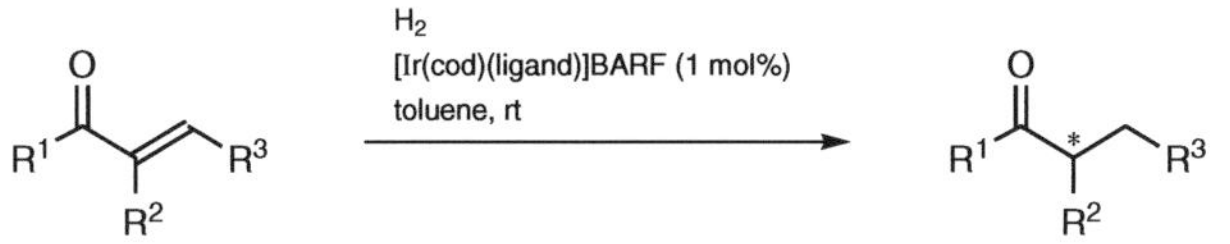

H_2
[Ir(cod)(ligand)]BARF (1 mol%)
toluene, rt

162 → **163**

R^1	R^2	R^3	Ligand	Solvent	Pressure (atm)	Time (h)	ee (%)	Yield (%)	Ref
$(CH_2)_3$		Ph	**141**	toluene	2	3	92	91	[146]
$(CH_2)_3$		Ph	**137**	CH_2Cl_2	2	24	96	100	[141]
$(CH_2)_4$		Ph	**141**	toluene	2	3	94	93	[146]
$(CH_2)_4$		Ph	**140**	CH_2Cl_2	5	10	97	>99	[140]
$(CH_2)_4$		Ph	**143**	CH_2Cl_2	50	24	99	>99	[148]
$(CH_2)_5$		Ph	**141**	toluene	2	3	99	91	[146]
$(CH_2)_5$		Ph	**140**	CH_2Cl_2	5	10	96	>99	[140]
$(CH_2)_6$		Ph	**141**	toluene	2	3	97	94	[146]
$(CH_2)_4$		4-ClC_6H_4	**141**	toluene	2	3	97	94	[146]
$(CH_2)_4$		4-$MeOC_6H_4$	**141**	toluene	2	3	98	97	[146]
$(CH_2)_4$		3-$O_2NC_6H_4$	**141**	toluene	25	3	99	96	[146]
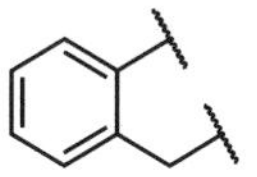		Ph	**137**	CH_2Cl_2	2	24	95	100	[141]

R^1	R^2	R^3	Ligand	Solvent	Pressure (atm)	Time (h)	ee (%)	Yield (%)	Ref
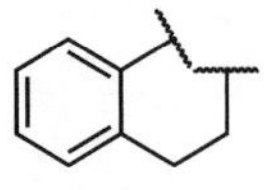		Ph	**141**	toluene	2	3	99	96	[146]
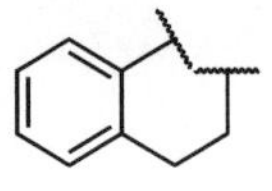		Ph	**140**	CH_2Cl_2	5	10	96	>99	[140]
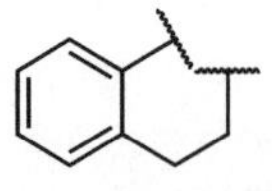		Pr	**141**	toluene	5	24	91	89	[146]

α-Substituted Ketones 161 **or Cyclic Ketones** 163**; General Procedure Using [Ir(cod)(**141**)]BARF as Catalyst:**[146]
Complex [Ir(cod)(**141**)]BARF (3.9 mg, 2.5 μmol) and the α,β-unsaturated ketone **160** or **162** (0.25 mmol) were placed in a 5-mL vial equipped with a stirrer bar. The vial was then placed in an argon-filled steel autoclave. Toluene (1.0 mL) was added to the mixture under an argon atmosphere. The autoclave was then closed, purged with H_2 (3 ×) (at lower pressure than that used for the reaction), and finally pressurized to the required value. The mixture was stirred for the indicated period of time, and then the H_2 was slowly released. The conversion of the substrate was determined by ^{1}H NMR analysis of the crude mixture, and the product was purified by chromatography (pentane/EtOAc 10:1).

1.1.1.5 Reduction of Alkenes Bearing Other Heteroatoms

Remarkable progress has been reported in the enantioselective hydrogenation of heteroatom-substituted alkenes that are not enamides or enol esters. Most of these hydrogenations have been performed using rhodium and iridium catalyst precursors, while the substrates are mainly phosphorus-, boron-, fluorine-, and sulfur-substituted alkenes. Some representative ligand structures **164–171** used in such asymmetric hydrogenations that are additional to those ligands already mentioned in this chapter, are shown in Scheme 28.

Scheme 28 Representative Ligands for the Asymmetric Reduction of Alkenes Bearing Heteroatoms Other than Nitrogen and Oxygen

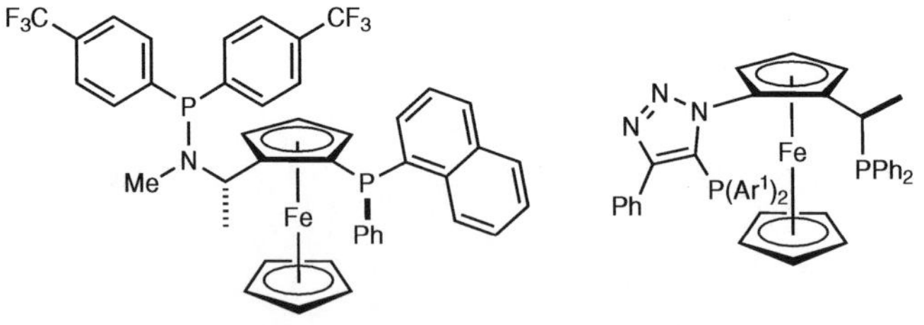

164

165 Ar^1 = 3,5-Me$_2$C$_6$H$_3$

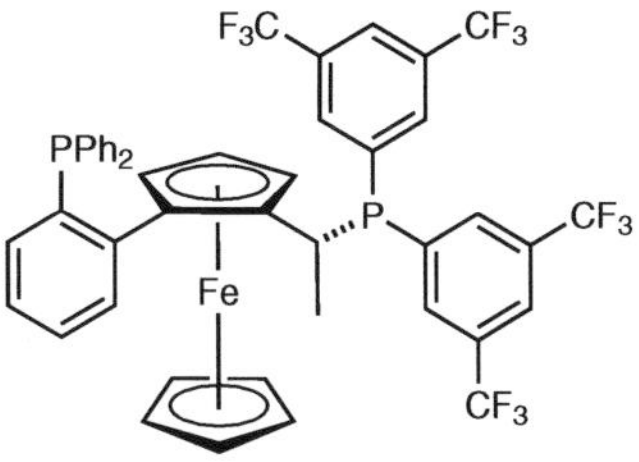

166 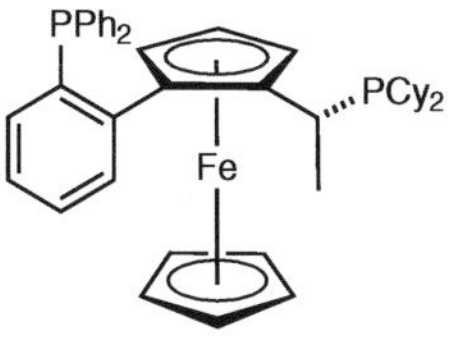 167 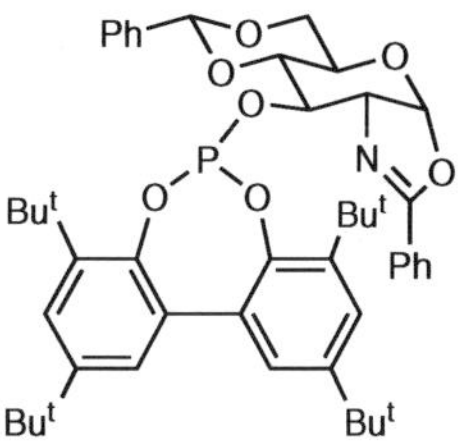 168

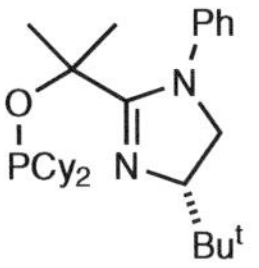

169 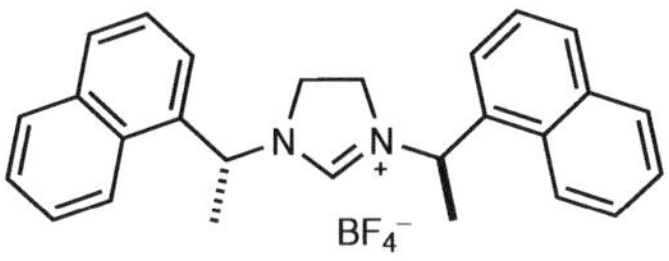 170 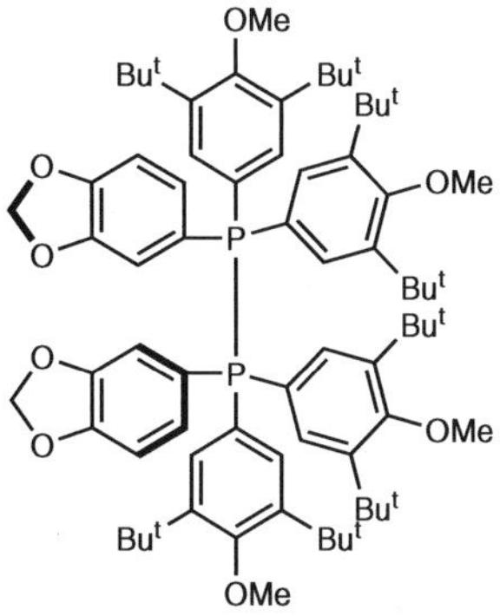 171 [(*S*)-(+)-DTBM-SEGPHOS]

1.1.1.5.1 Reduction of Phosphorus-Substituted Alkenes

Rhodium-based catalyst systems have frequently been reported for the asymmetric hydrogenation of vinylphosphonates, while there are also a few examples involving an iridium catalyst. Some representative results are given in Scheme 29.[149–151] In 2009, Zheng and coworkers reported the hydrogenation of α,β-unsaturated phosphonates **172** with diverse catalysts, obtaining remarkable results with a catalyst based on BoPhoz-type ligand **164**.[149] Thus, high enantioselectivities are obtained with a range of (1-arylvinyl)phosphonates **172** (R^1 = aryl; 93–97% ee). It is noteworthy that an alkyl-substituted analogue (R^1 = iPr) could also be hydrogenated with outstanding enantioselectivity (98% ee). Fukuzawa et al. have reported a rhodium catalyst based on ClickFerrophos ligand **165** for the reduction of relatively hindered (2-arylprop-1-enyl)phosphonates.[150] This catalyst provides acceptable catalyst activity and high enantioselectivity (92–96% ee). Andersson has reported the use of an analogue of the Crabtree catalyst {[Ir(cod)(**25**)]BARF} for the highly selective catalytic hydrogenation of diphenyl(vinyl)phosphine oxides.[151]

Scheme 29 Rhodium- and Iridium-Catalyzed Asymmetric Hydrogenation of Vinylphosphonates and Vinylphosphine Oxides[149–151]

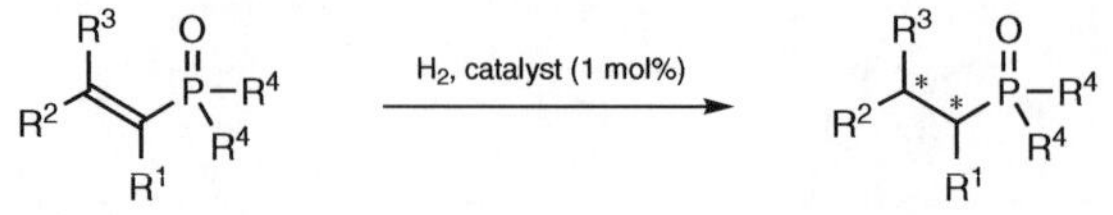

R^1	R^2	R^3	R^4	Catalyst[a]	Conversion or Yield (%)	ee (%)	Ref
Ph	H	H	OEt	$[Rh(cod)]BF_4$/**164**	95	95	[149]
Ph	H	H	OiPr	$[Rh(cod)]BF_4$/**164**	96	95	[149]
3-$MeOC_6H_4$	H	H	OEt	$[Rh(cod)]BF_4$/**164**	96	96	[149]
4-$MeOC_6H_4$	H	H	OEt	$[Rh(cod)]BF_4$/**164**	92	96	[149]
4-FC_6H_4	H	H	OEt	$[Rh(cod)]BF_4$/**164**	96	95	[149]
iPr	H	H	OEt	$[Rh(cod)]BF_4$/**164**	95	98	[149]
H	Ph	Me	OEt	[Rh(cod)(**165**)]BF_4	95	95	[150]
H	4-$O_2NC_6H_4$	Me	OEt	[Rh(cod)(**165**)]BF_4	92	95	[150]
H	4-BrC_6H_4	Me	OEt	[Rh(cod)(**165**)]BF_4	98	96	[150]
H	2-thienyl	Me	OEt	[Rh(cod)(**165**)]BF_4	99	95	[150]
H	4-Tol	OBz	OMe	[Rh(cod)(**165**)]BF_4	91	96	[150]
Ph	H	H	Ph	[Ir(cod)(**25**)]BARF	>99	>99	[151]
4-Tol	H	H	Ph	[Ir(cod)(**25**)]BARF	>99	>99	[151]
4-$MeOC_6H_4$	H	H	Ph	[Ir(cod)(**25**)]BARF	>99	>99	[151]
4-FC_6H_4	H	H	Ph	[Ir(cod)(**25**)]BARF	>99	>99	[151]
2-Tol	H	H	Ph	[Ir(cod)(**25**)]BARF	>99	>99	[151]
Cy	H	H	Ph	[Ir(cod)(**25**)]BARF	>99	99	[151]
$(CH_2)_2OAc$	H	H	Ph	[Ir(cod)(**25**)]BARF	>99	>99	[151]
$(CH_2)_2Ph$	H	H	Ph	[Ir(cod)(**25**)]BARF	>99	>99	[151]

[a] For ligand structures see Scheme 1, Section 1.1.1.1, and Scheme 28, Section 1.1.1.5; BARF = $[3,5-(F_3C)_2C_6H_3]_4B^-$.

Diethyl (*S*)-(1-Phenylethyl)phosphonate (173, **R^1 = Ph; R^2 = R^3 = H; R^4 = OEt); Typical Procedure Using [Rh(cod)/(164)]BF_4 as Catalyst:**[149]

In a N_2-filled glovebox, ligand **164** (2.75 μmol) was added to a soln of $[Rh(cod)_2]BF_4$ (1.0 mg, 2.5 μmol) in CH_2Cl_2 (1 mL). The mixture was stirred at rt for 30 min, and then a soln of diethyl (1-phenylvinyl)phosphonate (**172**, R^1 = Ph; R^2 = R^3 = H; R^4 = OEt; 0.25 mmol) in CH_2Cl_2 (1 mL) was added. The mixture was transferred to a Parr stainless-steel autoclave. The autoclave was purged with H_2 (3 ×), and then the H_2 pressure was maintained at 10 atm. The hydrogenation was performed at rt for 24 h. After carefully releasing the H_2, the solvent was removed. The residue was filtered through a short silica gel column to remove the catalyst. The filtrate was concentrated under reduced pressure; yield: 95%; 95% ee (determined by HPLC on a chiral column).

1.1.1.5.2 Reduction of Vinylboronates

The controlled construction of secondary organoboron derivatives is a problematic issue in organic synthesis, and yet, because the boron atom may be replaced with a variety of functional groups, solutions to this problem would have important ramifications for the preparation of chiral compounds. The asymmetric hydrogenation of vinylboronates provides an alternative route to the synthesis of chiral secondary organoboron compounds. Some results that are excellent in terms of the yields and enantiomeric excess obtained for the asymmetric hydrogenation of vinylboronates **174** are listed in Scheme 30.[36,152–154]

In 2006, Morken et al. reported a rhodium catalyst with Walphos ligand **166** for the reduction of (1-alkylvinyl)boronates.[152] Moderate to high enantioselectivities are obtained, but high catalyst loading is needed (5 mol%). Andersson applied a series of P,N-ligands for the iridium-catalyzed hydrogenation of vinylboronates, but only a few ligands (**115**,[153] **135**,[153] and **168**[36]) are effective and the substrate scope is limited. In 2012, Pfaltz et al. systematically studied the use of P,N-ligands for the hydrogenation of vinylboronates.[154] Ligand **169** gives the best results in the asymmetric hydrogenation of α-alkyl-substituted vinylboronates. Performing the reactions at low temperatures (−35 °C) leads to higher enantioselectivities. In the hydrogenation of α,β-disubstituted *E*-isomer substrates, high enantioselectivities are obtained at room temperature and the substrate scope is wide.

Scheme 30 Rhodium- and Iridium-Catalyzed Asymmetric Hydrogenation of Vinylboronates[36,152–154]

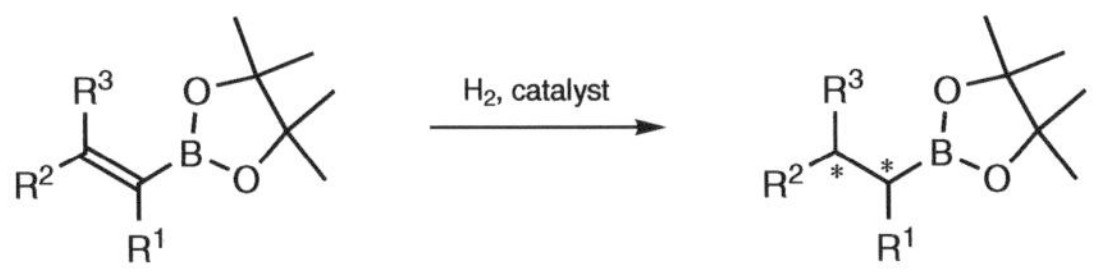

H_2, catalyst

174 → **175**

R^1	R^2	R^3	Catalyst[a]	Mol% of Catalyst	Conversion or Yield (%)	ee (%)	Ref
Cy	H	H	$[Rh(cod)]BF_4$/**166**	5	>95	97	[152]
Ph	Ph	H	[Ir(cod)(**115**)]BARF	0.5	>99	98	[153]
Ph	H	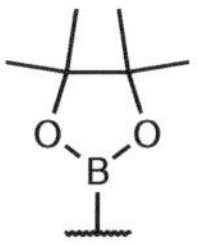	[Ir(cod)(**135**)]BARF	0.5	>99	96	[153]
Ph	H	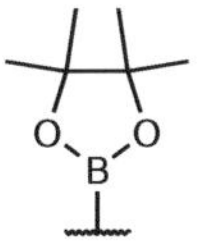	[Ir(cod)(**168**)]BARF	1	100	>99	[36]
Bu	H	H	[Ir(cod)(**169**)]BARF	0.5	>99	95	[154]
Cy	H	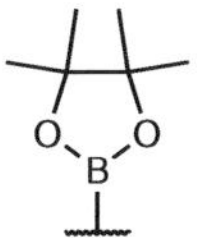	[Ir(cod)(**169**)]BARF	1	>99	95	[154]

for references see p 63

R^1	R^2	R^3	Catalyst[a]	Mol% of Catalyst	Conversion or Yield (%)	ee (%)	Ref
Ph	H	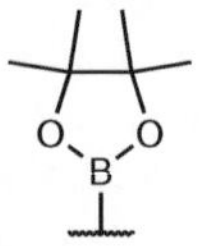	[Ir(cod)(**169**)]BARF	1	>99	98	[154]
Cy	H	4-Tol	[Ir(cod)(**169**)]BARF	1	>99	97	[154]
$(CH_2)_5Me$	H	H	[Ir(cod)(**169**)]BARF	0.1	>99	96	[154]
Ph	H	Ph	[Ir(cod)(**169**)]BARF	1	>99	>99	[154]
Cy	H	Ph	[Ir(cod)(**169**)]BARF	1	>99	95	[154]
Cy	H	$4-MeOC_6H_4$	[Ir(cod)(**169**)]BARF	1	>99	97	[154]
Cy	H	$4-F_3CC_6H_4$	[Ir(cod)(**169**)]BARF	1	>99	98	[154]
Cy	H	$3-FC_6H_4$	[Ir(cod)(**169**)]BARF	1	>99	99	[154]
Cy	H	Bn	[Ir(cod)(**169**)]BARF	1	>99	96	[154]
Cy	H	Et	[Ir(cod)(**169**)]BARF	1	>99	97	[154]

[a] BARF = $[3,5-(F_3C)_2C_6H_3]_4B^-$.

(*R*)-4,4,5,5-Tetramethyl-2-(octan-2-yl)-1,3,2-dioxaborolane [175, **$R^1 = (CH_2)_5Me$; $R^2 = R^3 = H$**]; **Typical Procedure Using [Ir(cod)(**169)**]BARF as Catalyst:**[154]

A soln of (*E*)-4,4,5,5-tetramethyl-2-(oct-1-en-2-yl)-1,3,2-dioxaborolane [**174**, $R^1 = (CH_2)_5Me$; $R^2 = R^3 = H$; 303 mg, 1.27 mmol] and [Ir(cod)(**169**)]BARF (2.07 mg, 1.28 μmol, 0.1 mol%) in CH_2Cl_2 (6 mL) was placed in an autoclave. The autoclave was pressurized with N_2 (1 atm) and cooled to −20 °C for 1 h. The autoclave was then pressurized with H_2 (5 ×; up to 10 atm) and the pressure was released. The reaction was performed under 2 atm H_2 pressure over 4 h at rt. After releasing the H_2 pressure, the solvent was removed under reduced pressure. The crude product was taken up in heptane (3 mL) and purified through a plug of silica gel (0.5 cm × 2 cm, heptane/*t*-BuOMe 10:1) to give the analytically pure hydrogenation product.

1.1.1.5.3 Reduction of Vinyl Fluorides

Fluorinated compounds are ubiquitous in our daily life and have significantly influenced progress in science and key technologies. Chiral fluorinated compounds are widely used in the pharmaceutical industry. The asymmetric hydrogenation of fluorine-substituted C=C bonds represents a direct method to obtain chiral fluorinated compounds. Only a few examples of asymmetric hydrogenation of fluoroalkene compounds are known, with two examples provided in Scheme 31.[155,156] Anderson reported the use of an iridium complex based on P,N-ligand **25** in the asymmetric hydrogenation of vinyl fluorides that also contain an allylic alcohol.[155] Only a few substrates could be hydrogenated with high enantioselectivities. Krska et al. disclosed the hydrogenation of a cyclic vinyl fluoride that also contained an amine group; thus, the $[Rh(cod)]BF_4$/**167** complex is efficient in this asymmetric hydrogenation, using the hydrochloric acid salt of the amine as the substrate.[156]

Scheme 31 Rhodium- and Iridium-Catalyzed Asymmetric Hydrogenation of Vinyl Fluorides[155,156]

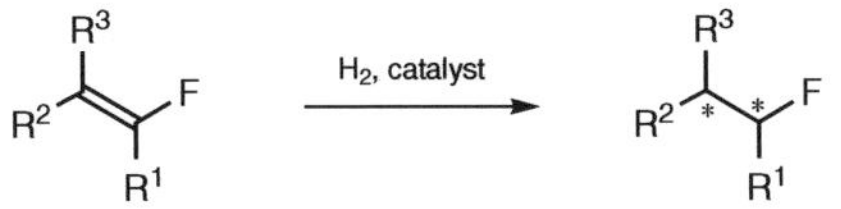

R^1	R^2	R^3	Catalyst[a]	Mol% of Catalyst	Conversion or Yield (%)	ee (%)	Ref
CH_2OAc	H	Ph	[Ir(cod)(**25**)]BARF	1	82	99	[155]
CH_2OH	H	Ph	[Ir(cod)(**25**)]BARF	1	97	99	[155]
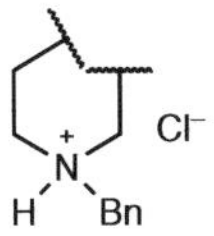 (N-benzylpiperidinium chloride, H, Bn, Cl^-)		CH_2OH	$[Rh(cod)_2]BF_4$/**167**	0.1	99	>99	[156]

[a] BARF = $[3,5-(F_3C)_2C_6H_3]_4B^-$.

(*R*)-2-Fluoro-3-phenylpropan-1-ol (177, $R^1 = CH_2OH$; $R^2 = H$; $R^3 = Ph$)**; Typical Procedure Using [Ir(cod)(**25**)]BARF as Catalyst:**[155]

A vial was charged with (*Z*)-2-fluoro-3-phenylprop-2-en-1-ol (**176**, $R^1 = CH_2OH$; $R^2 = H$; $R^3 =$ Ph; 0.5 mmol) and [Ir(cod)(**25**)]BARF (1 mol%), and CH_2Cl_2 (2 mL) was added. The vessel was purged with argon (3×; 10 atm), and then flushed and pressurized with H_2 (20 atm) and stirred at 700 rpm for 24 h. Conversion was determined by ^{1}H NMR analysis after evaporation of solvent. Et_2O/pentane (1:1; 1.5 mL) was added, and the soln was filtered through a short plug of silica gel, rinsing with Et_2O/pentane (1:1; 3 mL). The solvent was evaporated and the ee (99%) was determined by HPLC [Chiralcel OB-H, iPrOH/hexane 3:97, flow rate: 0.5 mL•min^{-1}, 220 nm; t_R 36.0 min (major), t_R 39.5 min (minor)].

1.1.1.5.4 Reduction of Sulfur-Substituted Alkenes

Because of the strong coordination ability of sulfur atoms, the use of sulfur-containing substrates often results in deactivation of catalysts. Thus, the asymmetric reduction of alkenes bearing a sulfur atom is a challenge, and very few successful examples are known. In 2016, the Glorius group achieved the asymmetric hydrogenation of vinyl sulfides **178** to furnish dihydrobenzo[*b*][1,4]thiazepin-4(5*H*)-one derivatives **179** (Scheme 32).[157] Very recently, Lv and Zhang achieved the asymmetric hydrogenation of 2-(acetylamino)vinyl sulfides.[158] The asymmetric hydrogenation of alkenes **180** bearing sulfonyl groups has also been reported, leading to alkyl sulfones **181** (Scheme 33).[159–161]

Scheme 32 Rhodium- and Ruthenium-Catalyzed Asymmetric Reduction of Vinyl Sulfides[157,158]

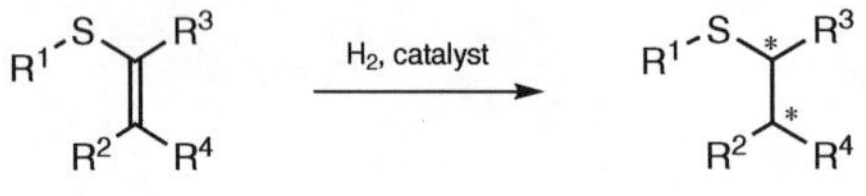

178 → (H_2, catalyst) → 179

R^1	R^2	R^3	R^4	Catalyst	Conditions	ee (%)	Yield (%)	Ref
Me	NHAc	H	Ph	$[Rh(nbd)_2]BF_4$/**53**	H_2 (80 atm), iPrOH, rt, 24 h	98	97	[158]
Me	NHAc	H	4-$MeOC_6H_4$	$[Rh(nbd)_2]BF_4$/**53**	H_2 (80 atm), iPrOH, rt, 24 h	94	96	[158]
Me	NHAc	H	4-BrC_6H_4	$[Rh(nbd)_2]BF_4$/**53**	H_2 (80 atm), iPrOH, rt, 24 h	99	98	[158]
Me	NHAc	H	3-Tol	$[Rh(nbd)_2]BF_4$/**53**	H_2 (80 atm), iPrOH, rt, 24 h	94	94	[158]
Me	NHAc	H	2-FC_6H_4	$[Rh(nbd)_2]BF_4$/**53**	H_2 (80 atm), iPrOH, rt, 24 h	96	94	[158]
Me	NHAc	H	2-furyl	$[Rh(nbd)_2]BF_4$/**53**	H_2 (80 atm), iPrOH, rt, 24 h	99	99	[158]
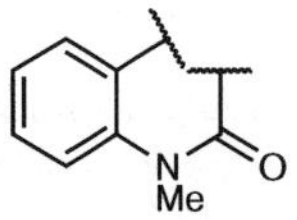		Ph	H	$[Ru(cod)(2\text{-methylallyl})_2]$/**170**	*t*-BuOK, H_2 (100 bar), hexane, 25 °C, 24–48 h	93	99	[157]
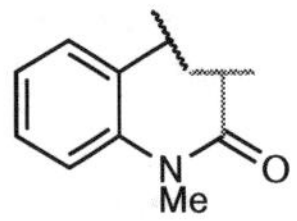		3-$MeOC_6H_4$	H	$[Ru(cod)(2\text{-methylallyl})_2]$/**170**	*t*-BuOK, H_2 (100 bar), hexane, 25 °C, 24–48 h	94	93	[157]
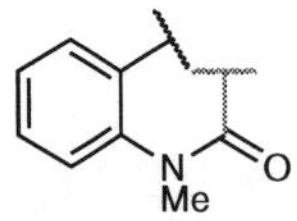		4-$F_3CC_6H_4$	H	$[Ru(cod)(2\text{-methylallyl})_2]$/**170**	*t*-BuOK, H_2 (100 bar), hexane, 25 °C, 24–48 h	90	70	[157]
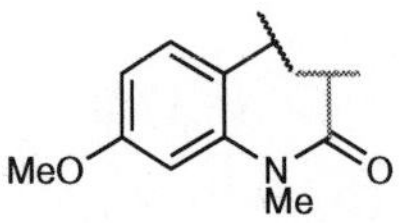		Ph	H	$[Ru(cod)(2\text{-methylallyl})_2]$/**170**	*t*-BuOK, H_2 (100 bar), hexane, 25 °C, 24–48 h	85	99	[157]

Scheme 33 Rhodium- and Iridium-Catalyzed Asymmetric Reduction of Vinyl Sulfones[159,160]

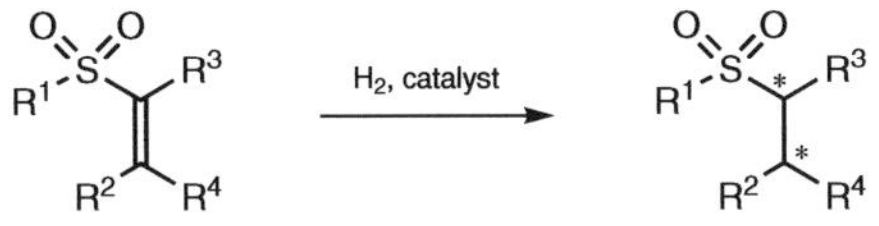

R^1	R^2	R^3	R^4	Catalyst[a]	Conditions	ee (%)	Yield (%)	Ref
4-Tol	H	Ph	H	$[Rh(nbd)_2]BF_4$/**171**	H_2 (10 atm), CH_2Cl_2, rt, 2–5 h	97	99	[159]
4-Tol	H	$4-MeOC_6H_4$	H	$[Rh(nbd)_2]BF_4$/**171**	H_2 (10 atm), CH_2Cl_2, rt, 2–5 h	99	99	[159]
4-Tol	H	$4-BrC_6H_4$	H	$[Rh(nbd)_2]BF_4$/**171**	H_2 (10 atm), CH_2Cl_2, rt, 2–5 h	98	98	[159]
4-Tol	H	cyclopropyl	H	$[Rh(nbd)_2]BF_4$/**171**	H_2 (10 atm), CH_2Cl_2, rt, 2–5 h	88	95	[159]
OMe	H	Ph	H	$[Rh(nbd)_2]BF_4$/**171**	H_2 (10 atm), CH_2Cl_2, rt, 2–5 h	94	96	[159]
Bn	Me	H	Ph	[Ir(cod)(**135**)]BARF	H_2 (50 atm), CH_2Cl_2, rt, 17 h	96	>99	[160]
Ph	Me	H	Ph	[Ir(cod)(**135**)]BARF	H_2 (50 atm), CH_2Cl_2, rt, 17 h	94	>99	[160]
Bn	Bu	H	Ph	[Ir(cod)(**135**)]BARF	H_2 (50 atm), CH_2Cl_2, rt, 17 h	91	>99	[160]
Bn	Ph	H	Me	[Ir(cod)(**135**)]BARF	H_2 (50 atm), CH_2Cl_2, rt, 17 h	96	61	[160]
Bn	Me	H	Bu	[Ir(cod)(**135**)]BARF	H_2 (50 atm), CH_2Cl_2, rt, 17 h	93	>99	[160]
Bn	Me	H	$4-MeOC_6H_4$	[Ir(cod)(**135**)]BARF	H_2 (50 atm), CH_2Cl_2, rt, 17 h	92	>99	[160]
Bn	Me	H	$4-F_3CC_6H_4$	[Ir(cod)(**135**)]BARF	H_2 (50 atm), CH_2Cl_2, rt, 17 h	95	>99	[160]
4-Tol	H	Me	Ph	[Ir(cod)(**135**)]BARF	H_2 (50 atm), CH_2Cl_2, rt, 17 h	97	>94	[160]
$(CH_2)_4$		H	Ph	[Ir(cod)(**135**)]BARF	H_2 (50 atm), CH_2Cl_2, rt, 17 h	93	>99	[160]

[a] BARF = $[3,5-(F_3C)_2C_6H_3]_4B^-$.

N-[1-Aryl-2-(methylsulfanyl)ethyl]acetamides 179 **(R^1 = Me; R^2 = NHAc; R^3 = H; R^4 = Aryl); General Procedure Using Ligand** 53:[158]

In a N_2-filled glovebox, a stock soln was prepared by mixing $[Rh(nbd)_2]BF_4$ with (S_C,R_P)-DuanPhos (**53**) (1:1.1 molar ratio) in CH_2Cl_2 at rt for 30 min. Aliquots of the catalyst soln (0.5 mL, 0.5 μmol) were transferred by syringe into vials charged with the vinyl sulfides **178** (R^1 = Me; R^2 = NHAc; R^3 = H; R^4 = aryl; 0.05 mmol) in anhyd iPrOH (3 mL). The vials were subsequently transferred into an autoclave, which was then charged with H_2. The reaction was stirred under H_2 (80 atm) at rt for 24 h, and then the H_2 was released slowly and carefully. The soln was concentrated and passed through a short column of silica gel (eluent: EtOAc) to yield the desired products. The ee was determined by HPLC analysis on a chiral stationary phase.

2-Aryl-5-methyl-2,3-dihydrobenzo[*b*][1,4]thiazepin-4(5*H*)-ones 179; **General Procedure Using Ligand** 170:[157]

In a glovebox, to a flame-dried screw-capped tube equipped with a magnetic stirrer bar was added $Ru(cod)(2\text{-methylallyl})_2$ (0.02 mmol), ligand **170** (0.04 mmol), and dry *t*-BuOK (0.05 mmol). The mixture was suspended in hexane (1 mL) and stirred at 70 °C for 16 h. The mixture was then transferred under argon to a glass vial containing the benzo[*b*]-[1,4]thiazepin-4(5*H*)-one (0.20 mmol) in hexane (6 mL). The glass vial was placed in a 150-mL stainless-steel autoclave. The autoclave was pressurized and depressurized with H_2 (3 ×), before the indicated pressure (100 bar) was set. The mixture was stirred at 25 °C for 24–48 h. After the autoclave was carefully depressurized, the mixture was concentrated under reduced pressure and purified by flash column chromatography (silica gel, pentane/EtOAc 4:1).

1-Arylethyl 4-Tolyl Sulfones 181 **(R^1 = 4-Tol; R^2 = R^4 = H; R^3 = Aryl); General Procedure Using $[Rh(nbd)_2]BF_4$/**171 **as Catalyst:**[159]

In a N_2-filled glovebox, a stock soln was prepared by mixing $[Rh(nbd)_2]BF_4$ with (*S*)-(+)-DTBM-SEGPHOS (**171**; 1:1.1 molar ratio) in CH_2Cl_2 at rt for 1 h. Aliquots of the catalyst soln (0.1 mL, 1 μmol) were transferred by syringe into vials charged with the different substrates **180** (R^1 = 4-Tol; R^2 = R^4 = H; R^3 = aryl; 0.1 mmol) in anhyd CH_2Cl_2 (1.4 mL). The vials were subsequently transferred into an autoclave, which was then charged with H_2. The reaction was stirred under H_2 (10 atm) at rt for 2–5 h. The H_2 was released slowly and carefully, and then the products were purified by column chromatography. The ee was determined by HPLC analysis on a chiral stationary phase.

β-Substituted Sulfones 181; **General Procedure Using [Ir(cod)(**135**)]BARF as Catalyst:**[160]

A vial was charged with a vinyl sulfone **180** (0.25 mmol) and [Ir(cod)(**135**)]BARF (0.5 mol%). Dry, distilled CH_2Cl_2 (2 mL) was added and the vial was placed in a high-pressure hydrogenation apparatus. The reactor was purged with H_2 (3 ×), and then filled to 50 atm pressure with H_2. The mixture was stirred at rt for 17 h before the H_2 pressure was released and the solvent was removed under reduced pressure. The crude product was filtered through a short plug of silica gel. The conversion was determined by ^{1}H NMR spectroscopy, and ee values were determined by chiral HPLC or GC.

References

[1] Knowles, W. S., *Angew. Chem. Int. Ed.*, (2002) **41**, 1998.
[2] Crabtree, R. H.; Felkin, H.; Morris, G. E., *J. Organomet. Chem.*, (1977) **141**, 205.
[3] Verendel, J. J.; Pàmies, O.; Diéguez, M.; Andersson, P. G., *Chem. Rev.*, (2014) **114**, 2130.
[4] Ager, D. J.; de Vries, A. H. M.; de Vries, J. G., *Chem. Soc. Rev.*, (2012) **41**, 3340.
[5] Vineyard, B. D.; Knowles, W. S.; Sabacky, M. J.; Bachman, G. L.; Weinkauff, D. J., *J. Am. Chem. Soc.*, (1977) **99**, 5946.
[6] Kagan, H. B.; Dang, T.-P., *J. Am. Chem. Soc.*, (1972) **94**, 6429.
[7] Tang, W.; Zhang, X., *Chem. Rev.*, (2003) **103**, 3029.
[8] Shang, G.; Li, W.; Zhang, X., In *Catalytic Asymmetric Synthesis*, 3rd ed., Ojima, I., Ed.; Wiley: Hoboken, NJ, (2010); p 343.
[9] Zhou, Q.-L.; Xie, J.-H., In *Stereoselective Formation of Amines*, Li, W.; Zhang, X., Eds.; Springer: Berlin, (2014); p 75.
[10] *The Handbook of Homogeneous Hydrogenation*, de Vries, J. G.; Elsevier, C. J., Eds.; Wiley-VCH: Weinheim, Germany, (2007); Vols. 1–3.
[11] Ager, D., In *Science of Synthesis: Stereoselective Synthesis*, de Vries, J. G., Ed.; Thieme: Stuttgart, (2010); Vol. 1, p 185.
[12] Schnider, P.; Koch, G.; Prétôt, R.; Wang, G.; Bohnen, F. M.; Krüger, C.; Pfaltz, A., *Chem.–Eur. J.*, (1997) **3**, 887.
[13] Blackmond, D. G.; Lightfoot, A.; Pfaltz, A.; Rosner, T.; Schnider, P.; Zimmermann, N., *Chirality*, (2000) **12**, 442.
[14] Lightfoot, A.; Schnider, P.; Pfaltz, A., *Angew. Chem. Int. Ed.*, (1998) **37**, 2897.
[15] Liu, D.; Tang, W.; Zhang, X., *Org. Lett.*, (2004) **6**, 513.
[16] Cozzi, P. G.; Menges, F.; Kaiser, S., *Synlett*, (2003), 833.
[17] Lu, W.-J.; Chen, Y.-W.; Hou, X.-L., *Adv. Synth. Catal.*, (2010) **352**, 103.
[18] Schrems, M. G.; Pfaltz, A., *Chem. Commun. (Cambridge)*, (2009), 6210.
[19] Trifonova, A.; Diesen, J. S.; Andersson, P. G., *Chem.–Eur. J.*, (2006) **12**, 2318.
[20] Cozzi, P. G.; Zimmermann, N.; Hilgraf, R.; Schaffner, S.; Pfaltz, A., *Adv. Synth. Catal.*, (2001) **343**, 450.
[21] Blankenstein, J.; Pfaltz, A., *Angew. Chem. Int. Ed.*, (2001) **40**, 4445.
[22] Menges, F.; Pfaltz, A., *Adv. Synth. Catal.*, (2002) **344**, 40.
[23] Diéguez, M.; Mazuela, J.; Pàmies, O.; Verendel, J. J.; Andersson, P. G., *Chem. Commun. (Cambridge)*, (2008), 3888.
[24] Diéguez, M.; Mazuela, J.; Pàmies, O.; Verendel, J. J.; Andersson, P. G., *J. Am. Chem. Soc.*, (2008) **130**, 7208.
[25] Perry, M. C.; Cui, X.; Powell, M. T.; Hou, D.-R.; Reibenspies, J. H.; Burgess, K., *J. Am. Chem. Soc.*, (2003) **125**, 113.
[26] Kaiser, S.; Smidt, S. P.; Pfaltz, A., *Angew. Chem. Int. Ed.*, (2006) **45**, 5194.
[27] Källström, K.; Hedberg, C.; Brandt, P.; Bayer, A.; Andersson, P. G., *J. Am. Chem. Soc.*, (2004) **126**, 14 308.
[28] Hedberg, C.; Källström, K.; Brandt, P.; Hansen, L. K.; Andersson, P. G., *J. Am. Chem. Soc.*, (2006) **128**, 2995.
[29] Mazuela, J.; Paptchikhine, A.; Pàmies, O.; Andersson, P. G.; Diéguez, M., *Chem.–Eur. J.*, (2010) **16**, 4567.
[30] Coll, M.; Pàmies, O.; Diéguez, M., *Adv. Synth. Catal.*, (2013) **355**, 143.
[31] Rageot, D.; Woodmansee, D. H.; Pugin, B.; Pfaltz, A., *Angew. Chem. Int. Ed.*, (2011) **50**, 9598.
[32] Tolstoy, P.; Engman, M.; Paptchikhine, A.; Bergquist, J.; Church, T. L.; Leung, A. W.-M.; Andersson, P. G., *J. Am. Chem. Soc.*, (2009) **131**, 8855.
[33] Coll, M.; Pàmies, O.; Diéguez, M., *Chem. Commun. (Cambridge)*, (2011) **47**, 9215.
[34] Kaukoranta, P.; Engman, M.; Hedberg, C.; Bergquist, J.; Andersson, P. G., *Adv. Synth. Catal.*, (2008) **350**, 1168.
[35] McIntyre, S.; Hörmann, E.; Menges, F.; Smidt, S. P.; Pfaltz, A., *Adv. Synth. Catal.*, (2005) **347**, 282.
[36] Mazuela, J.; Norrby, P.-O.; Andersson, P. G.; Pàmies, O.; Diéguez, M., *J. Am. Chem. Soc.*, (2011) **133**, 13 634.
[37] Schrems, M. G.; Neumann, E.; Pfaltz, A., *Angew. Chem. Int. Ed.*, (2007) **46**, 8274.
[38] Drury, W. J., III; Zimmermann, N.; Keenan, M.; Hayashi, M.; Kaiser, S.; Goddard, R.; Pfaltz, A., *Angew. Chem. Int. Ed.*, (2004) **43**, 70.

[39] Li, W.; Zhang, X., *J. Org. Chem.*, (2000) **65**, 5871.
[40] Li, W.; Zhang, Z.; Xiao, D.; Zhang, X., *J. Org. Chem.*, (2000) **65**, 3489.
[41] Lee, S.-g.; Zhang, Y. J.; Song, C. E.; Lee, J. K.; Choi, J. H., *Angew. Chem. Int. Ed.*, (2002) **41**, 847.
[42] Tang, W.; Zhang, X., *Angew. Chem. Int. Ed.*, (2002) **41**, 1612.
[43] Imamoto, T.; Oohara, N.; Takahashi, H., *Synthesis*, (2004), 1353.
[44] Hu, X.-P.; Zheng, Z., *Org. Lett.*, (2004) **6**, 3585.
[45] Li, X.; Jia, X.; Xu, L.; Kok, S. H. L.; Yip, C. W.; Chan, A. S. C., *Adv. Synth. Catal.*, (2005) **347**, 1904.
[46] Huang, J.-D.; Hu, X.-P.; Duan, Z.-C.; Zeng, Q.-H.; Yu, S.-B.; Deng, J.; Wang, D.-Y.; Zheng, Z., *Org. Lett.*, (2006) **8**, 4367.
[47] Qiu, M.; Hu, X.-P.; Wang, D.-Y.; Deng, J.; Huang, J.-D.; Yu, S.-B.; Duan, Z.-C.; Zheng, Z., *Adv. Synth. Catal.*, (2008) **350**, 1413.
[48] Yan, Y.; Zhang, X., *Tetrahedron Lett.*, (2006) **47**, 1567.
[49] Wang, C.-J.; Gao, F.; Liang, G., *Org. Lett.*, (2008) **10**, 4711.
[50] Zupančič, B.; Mohar, B.; Stephan, M., *Org. Lett.*, (2010) **12**, 1296.
[51] Tang, W.; Qu, B.; Capacci, A. G.; Rodriguez, S.; Wei, X.; Haddad, N.; Narayanan, B.; Ma, S.; Grinberg, N.; Yee, N. K.; Krishnamurthy, D.; Senanayake, C. H., *Org. Lett.*, (2010) **12**, 176.
[52] Imamoto, T.; Tamura, K.; Zhang, Z.; Horiuchi, Y.; Sugiya, M.; Yoshida, K.; Yanagisawa, A.; Gridnev, I. D., *J. Am. Chem. Soc.*, (2012) **134**, 1754.
[53] Liu, D.; Zhang, X., *Eur. J. Org. Chem.*, (2005), 646.
[54] Zhang, X.; Huang, K.; Hou, G.; Cao, B.; Zhang, X., *Angew. Chem. Int. Ed.*, (2010) **49**, 6421.
[55] Gridnev, I. D.; Imamoto, T.; Hoge, G.; Kouchi, M.; Takahashi, H., *J. Am. Chem. Soc.*, (2008) **130**, 2560.
[56] Hu, A.-G.; Fu, Y.; Xie, J.-H.; Zhou, H.; Wang, L.-X.; Zhou, Q.-L., *Angew. Chem. Int. Ed.*, (2002) **41**, 2348.
[57] Jia, X.; Li, X.; Xu, L.; Shi, Q.; Yao, X.; Chan, A. S. C., *J. Org. Chem.*, (2003) **68**, 4539.
[58] Bernsmann, H.; van den Berg, M.; Hoen, R.; Minnaard, A. J.; Mehler, G.; Reetz, M. T.; de Vries, J. G.; Feringa, B. L., *J. Org. Chem.*, (2005) **70**, 943.
[59] Zeng, Q.-H.; Hu, X.-P.; Duan, Z.-C.; Liang, X.-M.; Zheng, Z., *J. Org. Chem.*, (2006) **71**, 393.
[60] Liu, Y.; Ding, K., *J. Am. Chem. Soc.*, (2005) **127**, 10488.
[61] Zhao, B.; Wang, Z.; Ding, K., *Adv. Synth. Catal.*, (2006) **348**, 1049.
[62] Huang, H.; Zheng, Z.; Luo, H.; Bai, C.; Hu, X.; Chen, H., *Org. Lett.*, (2003) **5**, 4137.
[63] Huang, H.; Zheng, Z.; Luo, H.; Bai, C.; Hu, X.; Chen, H., *J. Org. Chem.*, (2004) **69**, 2355.
[64] Pignataro, L.; Bovio, C.; Civera, M.; Piarulli, U.; Gennari, C., *Chem.–Eur. J.*, (2012) **18**, 10368.
[65] Hoen, R.; van den Berg, M.; Bernsmann, H.; Minnaard, A. J.; de Vries, J. G.; Feringa, B. L., *Org. Lett.*, (2004) **6**, 1433.
[66] Enthaler, S.; Erre, G.; Junge, K.; Addis, D.; Kadyrov, R.; Beller, M., *Chem.–Asian J.*, (2008) **3**, 1104.
[67] Gridnev, I. D.; Yasutake, M.; Higashi, N.; Imamoto, T., *J. Am. Chem. Soc.*, (2001) **123**, 5268.
[68] Liu, T.-L.; Wang, C.-J.; Zhang, X., *Angew. Chem. Int. Ed.*, (2013) **52**, 8416.
[69] Chen, J.; Zhang, W.; Geng, H.; Li, W.; Hou, G.; Lei, A.; Zhang, X., *Angew. Chem. Int. Ed.*, (2009) **48**, 800.
[70] Liu, G.; Liu, X.; Cai, Z.; Jiao, G.; Xu, G.; Tang, W., *Angew. Chem. Int. Ed.*, (2013) **52**, 4235.
[71] Jiang, J.; Lu, W.; Lv, H.; Zhang, X., *Org. Lett.*, (2015) **17**, 1154.
[72] Li, W.; Waldkirch, J. P.; Zhang, X., *J. Org. Chem.*, (2002) **67**, 7618.
[73] Zhou, Y.-G.; Yang, P.-Y.; Han, X.-W., *J. Org. Chem.*, (2005) **70**, 1679.
[74] Zhou, M.; Liu, T.-L.; Cao, M.; Xue, Z.; Lv, H.; Zhang, X., *Org. Lett.*, (2014) **16**, 3484.
[75] Tang, W.; Chi, Y.; Zhang, X., *Org. Lett.*, (2002) **4**, 1695.
[76] Renaud, J. L.; Dupau, P.; Hay, A.-E.; Guingouain, M.; Dixneuf, P. H.; Bruneau, C., *Adv. Synth. Catal.*, (2003) **345**, 230.
[77] Wu, Z.; Ayad, T.; Ratovelomanana-Vidal, V., *Org. Lett.*, (2011) **13**, 3782.
[78] Stumpf, A.; Reynolds, M.; Sutherlin, D.; Babu, S.; Bappert, E.; Spindler, F.; Welch, M.; Gaudino, J., *Adv. Synth. Catal.*, (2011) **353**, 3367.
[79] Meng, J.; Gao, M.; Lv, H.; Zhang, X., *Org. Lett.*, (2015) **17**, 1842.
[80] Claver, C.; Fernandez, E.; Gillon, A.; Heslop, K.; Hyett, D. J.; Martorell, A.; Orpen, A. G.; Pringle, P. G., *Chem. Commun. (Cambridge)*, (2000), 961.
[81] Reetz, M. T.; Mehler, G., *Angew. Chem. Int. Ed.*, (2000) **39**, 3889.
[82] van den Berg, M.; Minnaard, A. J.; Schudde, E. P.; van Esch, J.; de Vries, A. H. M.; de Vries, J. G.; Feringa, B. L., *J. Am. Chem. Soc.*, (2000) **122**, 11539.
[83] Reetz, M. T.; Mehler, G.; Meiswinkel, A.; Sell, T., *Tetrahedron Lett.*, (2002) **43**, 7941.
[84] Jackson, M.; Lennon, I. C., *Tetrahedron Lett.*, (2007) **48**, 1831.

[85] Hoge, G.; Wu, H.-P.; Kissel, W. S.; Pflum, D. A.; Greene, D. J.; Bao, J., *J. Am. Chem. Soc.*, (2004) **126**, 5966.
[86] Fukuzawa, S.-i.; Oki, H.; Hosaka, M.; Sugasawa, J.; Kikuchi, S., *Org. Lett.*, (2007) **9**, 5557.
[87] Fox, M. E.; Jackson, M.; Lennon, I. C.; Klosin, J.; Abboud, K. A., *J. Org. Chem.*, (2008) **73**, 775.
[88] Liu, D.; Li, W.; Zhang, X., *Org. Lett.*, (2002) **4**, 4471.
[89] Zanotti-Gerosa, A.; Malan, C.; Herzberg, D., *Org. Lett.*, (2001) **3**, 3687.
[90] Boaz, N. W.; Debenham, S. D.; Mackenzie, E. B.; Large, S. E., *Org. Lett.*, (2002) **4**, 2421.
[91] Yan, Y.; Chi, Y.; Zhang, X., *Tetrahedron: Asymmetry*, (2004) **15**, 2173.
[92] Pilkington, C. J.; Zanotti-Gerosa, A., *Org. Lett.*, (2003) **5**, 1273.
[93] Zhou, Y.-G.; Zhang, X., *Chem. Commun. (Cambridge)*, (2002), 1124.
[94] Cai, C.; Deng, F.; Sun, W.; Xia, C., *Synlett*, (2007), 3007.
[95] Shimizu, H.; Saito, T.; Kumobayashi, H., *Adv. Synth. Catal.*, (2003) **345**, 185.
[96] Guo, R.; Li, X.; Wu, J.; Kwok, W. H.; Chen, J.; Choi, M. C. K.; Chan, A. S. C., *Tetrahedron Lett.*, (2002) **43**, 6803.
[97] Fu, Y.; Hou, G.-H.; Xie, J.-H.; Xing, L.; Wang, L.-X.; Zhou, Q.-L., *J. Org. Chem.*, (2004) **69**, 8157.
[98] Gao, M.; Meng, J.-j.; Lv, H.; Zhang, X., *Angew. Chem. Int. Ed.*, (2015) **54**, 1885.
[99] Molinaro, C.; Scott, J. P.; Shevlin, M.; Wise, C.; Ménard, A.; Gibb, A.; Junker, E. M.; Lieberman, D., *J. Am. Chem. Soc.*, (2015) **137**, 999.
[100] Jerphagnon, T.; Renaud, J.-L.; Demonchaux, P.; Ferreira, A.; Bruneau, C., *Tetrahedron: Asymmetry*, (2003) **14**, 1973.
[101] Holz, J.; Monsees, A.; Jiao, H.; You, J.; Komarov, I. V.; Fischer, C.; Drauz, K.; Börner, A., *J. Org. Chem.*, (2003) **68**, 1701.
[102] Yasutake, M.; Gridnev, I. D.; Higashi, N.; Imamoto, T., *Org. Lett.*, (2001) **3**, 1701.
[103] Wu, H.-P.; Hoge, G., *Org. Lett.*, (2004) **6**, 3645.
[104] Qiu, L.; Wu, J.; Chan, S.; Au-Yeung, T. T.-L.; Ji, J.-X.; Guo, R.; Pai, C.-C.; Zhou, Z.; Li, X.; Fan, Q.-H.; Chan, A. S. C., *Proc. Natl. Acad. Sci. U. S. A.*, (2004) **101**, 5815.
[105] Hu, X.-P.; Zheng, Z., *Org. Lett.*, (2005) **7**, 419.
[106] Peña, D.; Minnaard, A. J.; de Vries, J. G.; Feringa, B. L., *J. Am. Chem. Soc.*, (2002) **124**, 14552.
[107] Tang, W.; Zhang, X., *Org. Lett.*, (2002) **4**, 4159.
[108] Zhou, Y.-G.; Tang, W.; Wang, W.-B.; Li, W.; Zhang, X., *J. Am. Chem. Soc.*, (2002) **124**, 4952.
[109] Wang, Q.; Huang, W.; Yuan, H.; Cai, Q.; Chen, L.; Lv, H.; Zhang, X., *J. Am. Chem. Soc.*, (2014) **136**, 16120.
[110] Tang, W.; Wu, S.; Zhang, X., *J. Am. Chem. Soc.*, (2003) **125**, 9570.
[111] Jiang, Q.; Xiao, D.; Zhang, Z.; Cao, P.; Zhang, X., *Angew. Chem. Int. Ed.*, (1999) **38**, 516.
[112] Wu, S.; Wang, W.; Tang, W.; Lin, M.; Zhang, X., *Org. Lett.*, (2002) **4**, 4495.
[113] Tang, W.; Liu, D.; Zhang, X., *Org. Lett.*, (2003) **5**, 205.
[114] Núñez-Rico, J. L.; Etayo, P.; Fernández-Pérez, H.; Vidal-Ferran, A., *Adv. Synth. Catal.*, (2012) **354**, 3025.
[115] Konrad, T. M.; Schmitz, P.; Leitner, W.; Franciò, G., *Chem.–Eur. J.*, (2013) **19**, 13299.
[116] Liu, Y.; Wang, Z.; Ding, K., *Tetrahedron*, (2012) **68**, 7581.
[117] Kleman, P.; González-Liste, P. J.; García-Garrido, S. E.; Cadierno, V.; Pizzano, A., *Chem.–Eur. J.*, (2013) **19**, 16209.
[118] Lotz, M.; Ireland, T.; Almena Perea, J.; Knochel, P., *Tetrahedron: Asymmetry*, (1999) **10** 1839.
[119] Lotz, M.; Polborn, K.; Knochel, P., *Angew. Chem. Int. Ed.*, (2002) **41**, 4708.
[120] Burk, M. J., *J. Am. Chem. Soc.*, (1991) **113**, 8518.
[121] Kleman, P.; González-Liste, P. J.; García-Garrido, S. E.; Cadierno, V.; Pizzano, A., *ACS Catal.*, (2014) **4**, 4398.
[122] Burk, M. J.; Kalberg, C. S.; Pizzano, A., *J. Am. Chem. Soc.*, (1998) **120**, 4345.
[123] Liu, Y.; Sandoval, C. A.; Yamaguchi, Y.; Zhang, X.; Wang, Z.; Kato, K.; Ding, K., *J. Am. Chem. Soc.*, (2006) **128**, 14212.
[124] Kuroki, Y.; Asada, D.; Sakamaki, Y.; Iseki, K., *Tetrahedron Lett.*, (2000) **41**, 4603.
[125] Hayashi, T.; Kanehira, K.; Kumada, M., *Tetrahedron Lett.*, (1981) **22**, 4417.
[126] Cheruku, P.; Gohil, S.; Andersson, P. G., *Org. Lett.*, (2007) **9**, 1659.
[127] Cheruku, P.; Diesen, J.; Andersson, P. G., *J. Am. Chem. Soc.*, (2008) **130**, 5595.
[128] Zhu, Y.; Burgess, K., *Adv. Synth. Catal.*, (2008) **350**, 979.
[129] Valla, C.; Baeza, A.; Menges, F.; Pfaltz, A., *Synlett*, (2008), 3167.
[130] Tanaka, M.; Watanabe, Y.; Mitsudo, T.-a.; Yasunori, Y.; Takegami, Y., *Chem. Lett.*, (1974) **3**, 137.
[131] Wei, S.; Du, H., *J. Am. Chem. Soc.*, (2014) **136**, 12261.

[132] Zhu, S.-F.; Yu, Y.-B.; Li, S.; Wang, L.-X.; Zhou, Q.-L., *Angew. Chem. Int. Ed.*, (2012) **51**, 8872.
[133] Li, S.; Zhu, S.-F.; Zhang, C.-M.; Song, S.; Zhou, Q.-L., *J. Am. Chem. Soc.*, (2008) **130**, 8584.
[134] Li, S.; Zhu, S.-F.; Xie, J.-H.; Song, S.; Zhang, C.-M.; Zhou, Q.-L., *J. Am. Chem. Soc.*, (2010) **132**, 1172.
[135] Song, S.; Zhu, S.-F.; Pu, L.-Y.; Zhou, Q.-L., *Angew. Chem. Int. Ed.*, (2013) **52**, 6072.
[136] Chen, C.; Wang, H.; Zhang, Z.; Jin, S.; Wen, S.; Ji, J.; Chung, L. W.; Dong, X.-Q.; Zhang, X., *Chem. Sci.* (2016) **7**, 6669.
[137] Tang, W.; Wang, W.; Zhang, X., *Angew. Chem. Int. Ed.*, (2003) **42**, 943.
[138] Wen, J.; Jiang, J.; Zhang, X., *Org. Lett.*, (2016) **18**, 4451.
[139] Li, J.-Q.; Quan, X.; Andersson, P. G., *Chem.–Eur. J.*, (2012) **18**, 10609.
[140] Liu, X.; Han, Z.; Wang, Z.; Ding, K., *Angew. Chem. Int. Ed.*, (2014) **53**, 1978.
[141] Tian, F.; Yao, D.; Liu, Y.; Xie, Y.; Zhang, W., *Adv. Synth. Catal.*, (2010) **352**, 1841.
[142] Christensen, M.; Nolting, A.; Shevlin, M.; Weisel, M.; Maligres, P. E.; Lee, J.; Orr, R. K.; Plummer, C. W.; Tudge, M. T.; Campeau, L.-C.; Ruck, R. T., *J. Org. Chem.*, (2016) **81**, 824.
[143] Shang, J.; Han, Z.; Li, Y.; Wang, Z.; Ding, K., *Chem. Commun. (Cambridge)*, (2012) **48**, 5172.
[144] Lu, W.-J.; Hou, X.-L., *Adv. Synth. Catal.*, (2009) **351**, 1224.
[145] Li, P.; Hu, X.; Dong, X.-Q.; Zhang, X., *Chem. Commun. (Cambridge)*, (2016) **52**, 11677.
[146] Lu, S.-M.; Bolm, C., *Angew. Chem. Int. Ed.*, (2008) **47**, 8920.
[147] Martin, N. J. A.; List, B., *J. Am. Chem. Soc.*, (2006) **128**, 13368.
[148] Maurer, F.; Huch, V.; Ullrich, A.; Kazmaier, U., *J. Org. Chem.*, (2012) **77**, 5139.
[149] Wang, D.-Y.; Hu, X.-P.; Deng, J.; Yu, S.-B.; Duan, Z.-C.; Zheng, Z., *J. Org. Chem.*, (2009) **74**, 4408.
[150] Konno, T.; Shimizu, K.; Ogata, K.; Fukuzawa, S.-i., *J. Org. Chem.*, (2012) **77**, 3318.
[151] Cheruku, P.; Paptchikhine, A.; Church, T. L.; Andersson, P. G., *J. Am. Chem. Soc.*, (2009) **131**, 8285.
[152] Moran, W. J.; Morken, J. P., *Org. Lett.*, (2006) **8**, 2413.
[153] Paptchikhine, A.; Cheruku, P.; Engman, M.; Andersson, P. G., *Chem. Commun. (Cambridge)*, (2009), 5996.
[154] Ganić, A.; Pfaltz, A., *Chem.–Eur. J.*, (2012) **18**, 6724.
[155] Engman, M.; Diesen, J. S.; Paptchikhine, A.; Andersson, P. G., *J. Am. Chem. Soc.*, (2007) **129**, 4536.
[156] Krska, S. W.; Mitten, J. V.; Dormer, P. G.; Mowrey, D.; Machrouhi, F.; Sun, Y.; Nelson, T. D., *Tetrahedron*, (2009) **65**, 8987.
[157] Li, W.; Schlepphorst, C.; Daniliuc, C.; Glorius, F., *Angew. Chem. Int. Ed.*, (2016) **55**, 3300.
[158] Gao, W.; Lv, H.; Zhang, X., *Org. Lett.*, (2017) **19**, 2877.
[159] Shi, L.; Wei, B.; Yin, X.; Xue, P.; Lv, H.; Zhang, X., *Org. Lett.*, (2017) **19**, 1024.
[160] Zhou, T.; Peters, B.; Maldonado, M. F.; Govender, T.; Andersson, P. G., *J. Am. Chem. Soc.*, (2012) **134**, 13592.
[161] Peters, B. K.; Zhou, T.; Rujirawanich, J.; Cadu, A.; Singh, T.; Rabten, W.; Kerdphon, S.; Andersson, P. G., *J. Am. Chem. Soc.*, (2014) **136**, 16557.

1.1.2 Reduction of Alkenes Using Nanoparticle Catalysis

R. Hudson and A. Moores

General Introduction

Nearly 200 years after the discovery of catalytic hydrogenation, the process remains an active area of research today due to its far-ranging industrial relevance in organic synthesis and the petrochemicals and food sectors. This reaction, which is often metal-catalyzed, first relied on bulk, heterogeneous precious metals such as platinum, palladium, and others.[1] Later, homogeneous systems emerged,[2] followed by demonstrations of nanoparticle-mediated hydrogenation[3–5] as a seminal example of the development of nanoparticle heterogeneous catalysis.[6] This chapter focuses specifically on methods for nanoparticle-catalyzed alkene hydrogenation.

Nanoparticle-mediated hydrogenation relies on the ability of the metal to both dissociatively adsorb molecular hydrogen and interact with the unsaturated reagent, while nanoscale typically allows an increased surface area to volume ratio, as well as the preponderance of low-coordinate edge, corner, and defect atoms. Catalytic hydrogenation also bears an additional, unique dependence on subsurface effects.[3–5,7] As molecular hydrogen dissociates, the hydrogen atoms may not all remain on the metal surface, but penetrate the metal lattice. This subsurface hydrogen plays a crucial role in the hydrogenation process. Considering palladium specifically, Shaikutdinov and co-workers demonstrated the importance of subsurface hydrogen by revealing situations where the subsurface hydrogen in bulk palladium simply diffused too deep to appreciably participate in subsequent reaction steps. In nanoparticles, on the other hand, subsurface hydrogen remains accessible, as the whole particle volume is, by construction, within a depth of a few nanometers. Using temperature-programmed desorption under low pressures, Shaikutdinov and co-workers elucidated overlap of molecular hydrogen and alkene desorption curves as the Goldilocks conditions for alkane production in nanoparticle systems.[5] Under similar conditions, the molecular hydrogen and alkene desorption curves did not overlap for palladium(111) bulk crystals, and no alkane was formed. Wilde, Schauermann, and co-workers provided evidence for sustained hydrogenation when the particles presented a carbonaceous deposit, positing that such a layer facilitated hydrogen diffusion from the surface into the subsurface, allowing a continuous and abundant supply of subsurface hydrogen.[7]

Horiuti, Ogden, and Polanyi proposed a mechanism relying on adsorption and dissociation of molecular hydrogen followed by addition to the alkene, and this is still generally accepted today (Scheme 1).[1] In the active pathway, the alkene most likely binds to the metal surface in a di-σ fashion. Addition of one hydrogen atom provides a mono-σ-bound alkyl species remaining on the surface, which represents a common intermediate in two pathways: incorporation of a second hydrogen atom for the formation of the alkane product, or potential isomerization via β-hydride elimination for reversion to the alkene.

for references see p 89

Scheme 1 Heterogeneous Hydrogenation Mechanism[1]

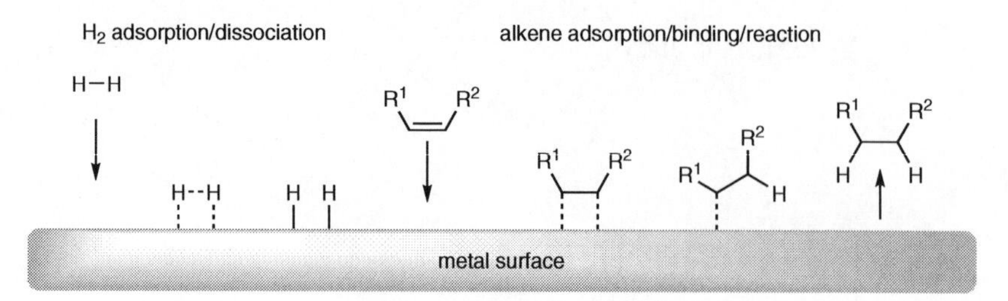

SAFETY: In a typical alkene hydrogenation reaction, the most important safety consideration is the potentially high pressure of the reactor. If the procedure requires a pressure of H_2 in access of 1 atm, be sure to select a reactor setup with an appropriate relief valve.

1.1.2.1 Reduction of Alkenes by Hydrogenation Using Palladium Nanoparticles

1.1.2.1.1 Hydrogenation with Commercial Palladium Nanoparticles

Nanoparticle-catalyzed hydrogenation, although no longer in its naissance, is not yet "out of the catalogue" or "off the shelf" chemistry today. Several commercial vendors sell prepared nanoparticles of the metals typically used in hydrogenation reactions (Pd, Pt, Ru, Ir, etc.) as powders, suspensions, or on supports, but available literature for their use in hydrogenation reactions is limited. Most methods highlighted in this chapter, therefore, require synthesis of the catalyst itself prior to use in the desired reduction step. The first example, however, uses commercially available palladium nanoparticles on aluminum hydroxide, albeit with a limited demonstrated scope: the single reaction of hydrogenation of cholesterol (**1**) to give cholestanol (**2**) (Scheme 2).[8]

Scheme 2 Hydrogenation of an Alkene Using Commercial Palladium Nanoparticles[8]

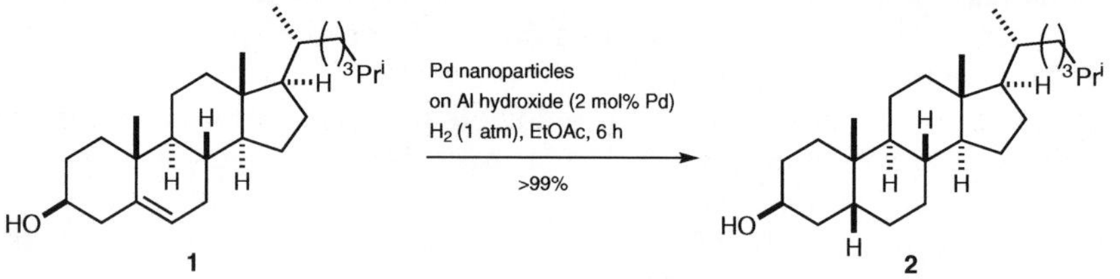

Pd nanoparticles on Al hydroxide (2 mol% Pd), H_2 (1 atm), EtOAc, 6 h; >99%

The catalyst is robust, and provides the same analytically pure product even after nine rounds of recycling. The catalyst could either be reused for the same transformation, or even for a subsequent alcohol oxidation step.

Cholestanol (2); Typical Procedure:[8]

Palladium nanoparticles on aluminum hydroxide (247 mg, 2 mol% Pd) were added to a soln of cholesterol (**1**; 387 mg, 1 mmol) in EtOAc (5 mL), and the mixture was stirred for 6 h under H_2 (1 atm). The catalyst was filtered off and dried for reuse. The filtrate was concentrated under reduced pressure; yield: quant.

1.1.2.1.2 Hydrogenation with Palladium Nanoparticles in Poly(ethylene glycol)

For use as an easily recyclable catalyst, palladium nanoparticles can be synthesized in poly(ethylene glycol) (PEG) and used for the hydrogenation of alkenes **3**, yielding alkanes **4** (Scheme 3).[9] The catalyst (Pd@PEG2000) is easy to synthesize, easy to recover, and highly active for reduction with molecular hydrogen. The very long shelf life of the catalyst allows for excellent reactivity, even after months of storage.[9]

Scheme 3 Hydrogenation of Alkenes Using Palladium Nanoparticles in Poly(ethylene glycol)[9]

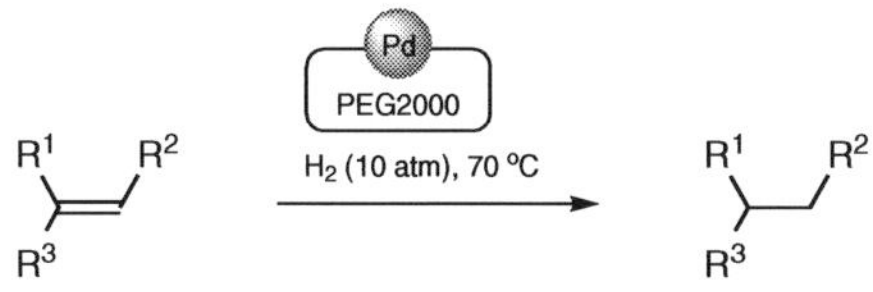

Pd PEG2000, H_2 (10 atm), 70 °C; **3** → **4**

R^1	R^2	R^3	Time (min)	Yield (%)	Ref
Bu	H	H	165	99	[9]
Ph	H	H	15	100	[9]
CN	H	H	80	>99.5	[9]
CHO	H	H	90	>99.5	[9]
CO_2H	H	H	25	>99.5	[9]
CO_2H	H	Me	90	>99.5	[9]
CO_2Me	H	H	25	>99.5	[9]
CO_2Me	H	Me	120	>99.5	[9]
$(CH_2)_4$		H	80	100	[9]

Palladium Nanoparticles in PEG (Pd@PEG2000):[9]
Under vigorous stirring, $Pd(OAc)_2$ (2 mg, 8.9 μmol) was added to PEG2000 (16 g) in a 50-mL, round-bottomed flask at 90 °C and the mixture was stirred for 2.3 h. The soln solidified upon cooling to rt and was used without further purification.

Alkanes 4; General Procedure:[9]
An alkene **3** (1.5 mmol) and Pd@PEG2000 (0.5 g, 0.278 μmol Pd) were added to a high-pressure reactor. The reactor was held for 20 min at 70 °C before H_2 (10 atm) was introduced under vigorous stirring. Upon completion of the reaction, the reactor was cooled with ice water. The product was isolated by decantation and analyzed by GC.

1.1.2.1.3 Hydrogenation with Palladium Nanoparticles Stabilized in Mesocellular Foam

As an excellent high-surface-area material, polyurea-modified siliceous mesocellular foam (MCF) can effectively support palladium nanoparticles for alkene reduction under molecular hydrogen. The catalyst (Pd@MCF) is easily recovered, and can be recycled up to 10 times with no appreciable decrease in yield, allowing for minimal waste generation. Alkenes such as **5** are hydrogenated to alkanes **6** in excellent yields (Scheme 4).[10]

Scheme 4 Hydrogenation of Alkenes Using Palladium Nanoparticles in Mesocellular Foam[10]

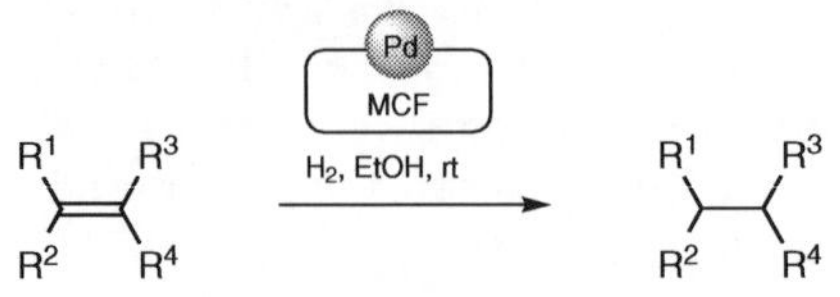

R^1	R^2	R^3	R^4	H_2 (psi)	Time (h)	Yield[a] (%)	Ref
CO_2Me	CH_2CO_2Me	H	H	40	6	99.9	[10]
CO_2Et	CH_2CO_2Et	H	H	40	6	99.9	[10]
2-naphthyl	H	NHAc	CO_2Me	40	6	99.9	[10]
3-MeO-4-AcOC_6H_3	H	NHAc	CO_2Me	40	6	99.9	[10]
Ph	H	H	Ac	40	8	99.9	[10]
Ph	H	H	CO_2Me	40	8	99.9	[10]
4-BrC_6H_4	H	H	H	100	18	99	[10]
CH_2OBn	H	H	H	100	18	99	[10]

[a] Determined by GC.

Palladium Nanoparticles in Mesocellular Foam (Pd@MCF):[10]
Spherical mesocellular foam microparticles (1 g) were dried for 24 h at 100 °C and cooled to rt under argon. Anhyd toluene (20 mL) was then added, followed by a soln of 1-[3-(trimethoxysilyl)propyl]urea (2.2 mmol) in toluene (2 mL). The mixture was stirred under argon for 10 min, and then heated to 80 °C for 24 h and cooled to rt. The solid was collected by filtration, and washed several times with toluene, EtOH, acetone, and CH_2Cl_2 to remove any unreacted precursor. The resulting material was suspended in EtOH, and heated to 60 °C for 18 h. The solid was again collected by filtration, washed, dried, and resuspended in anhyd toluene (20 mL). A soln of $Pd(OAc)_2$ (123 mg, 0.55 mmol) in CH_2Cl_2 (2 mL) was added dropwise. The mixture was heated under argon at 60 °C for 24 h. The soln was cooled to rt, and the solid material was collected by filtration, washed, and dried. Elemental analysis indicated a Pd loading of 5 wt%.

Alkanes 6; General Procedure:[10]
An alkene **5** (1 mmol), Pd@MCF (1 mol% Pd), and EtOH (5 mL) were added to an oven-dried vial under argon. The vial was pressurized with H_2 (40 or 100 psi) and the mixture was then stirred at rt for 6–18 h. The reaction progress was monitored by GC. Upon reaction completion, the catalyst was recovered by filtration, washed with MeOH (5 × 5 mL), and dried under reduced pressure. The filtrate was concentrated under reduced pressure.

1.1.2.1.4 Hydrogenation with Phenanthroline-Stabilized Palladium Nanoparticles in Poly(ethylene glycol)

To provide improved stabilization beyond what a simple polymer matrix may offer, additional ligands can be incorporated when the nanoparticles are synthesized. Phenanthroline-stabilized palladium nanoparticles in poly(ethylene glycol) (1,10-phenanthroline/Pd@PEG) can be generated this way to provide excellent alkene hydrogenation catalysts that are easy to recycle and reuse. Alkenes such as **7** are hydrogenated to alkanes **8** in excellent yields (Scheme 5).[11]

Scheme 5 Hydrogenation of Alkenes Using Phenanthroline-Stabilized Palladium Nanoparticles in Poly(ethylene glycol)[11]

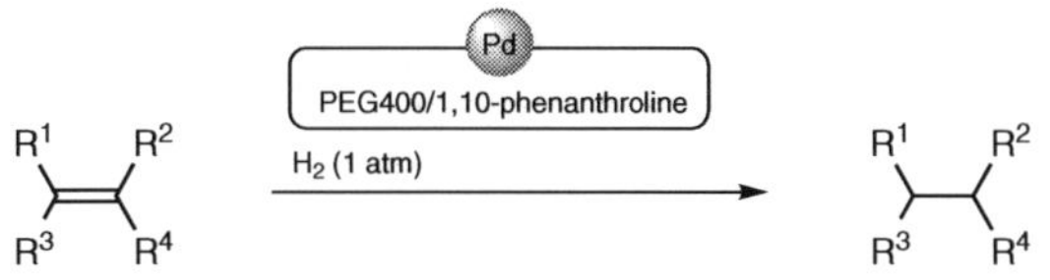

Pd, PEG400/1,10-phenanthroline, H_2 (1 atm); 7 → 8

R^1	R^2	R^3	R^4	Temp (°C)	Time (h)	Yield[a] (%)	Ref
Ph	H	H	H	50	4	100	[11]
Me	$(CH_2)_4Me$	H	H	50	20	100	[11]
$(CH_2)_5Me$	H	H	H	50	8	100	[11]
Bu	H	H	H	30	20	100	[11]
Ph	H	H	Ph	50	4	100	[11]
Ph	H	H	CH_2OH	40	20	100	[11]
Me	Me	H	CHO	40	8	92	[11]
Me	H	Me	Ac	40	8	94	[11]
$(CH_2)_4$		H	H	30	20	100	[11]
$(CH_2)_4$		Me	H	50	20	54	[11]
$(CH_2)_4$		Ph	H	50	20	100	[11]
$(CH_2)_6$		H	H	50	8	100	[11]
 [bicyclic structure]		H	H	40	4	100	[11]

[a] Determined by GC.

Phenanthroline-Stabilized Palladium Nanoparticles in Poly(ethylene glycol) (Phenanthroline/Pd@PEG):[11]
In a 25-mL, round-bottomed flask were mixed $Pd(OAc)_2$ (5 mg, 22 μmol), 1,10-phenanthroline (3 mg, 17 μmol), and PEG400 (4 g). The mixture was stirred under H_2 (1 atm) for 15 min. The catalyst was used without purification.

Alkanes 8; General Procedure:[11]
The phenanthroline/Pd@PEG catalyst and an alkene **7** (10 mmol) were mixed in a 25-mL, round-bottomed flask equipped with a reflux condenser and a magnetic stirrer bar. The mixture was purged with N_2, and then vigorously stirred under H_2 (1 atm) for the indicated time at the indicated temperature. Upon completion of the reaction, the mixture was extracted with Et_2O, and the extract was analyzed by GC/MS.

1.1.2.1.5 Hydrogenation with Palladium Nanoparticles Embedded in Polystyrene

Palladium nanoparticles can be generated within a polystyrene support as a durable, reusable catalyst (Pd@PS) for the hydrogenation of various alkenes **9** providing alkanes **10** in excellent yields. The catalyst can be recycled up to 16 times with no drop in activity (Scheme 6).[12]

Scheme 6 Hydrogenation of Alkenes Using Palladium Nanoparticles in Polystyrene[12]

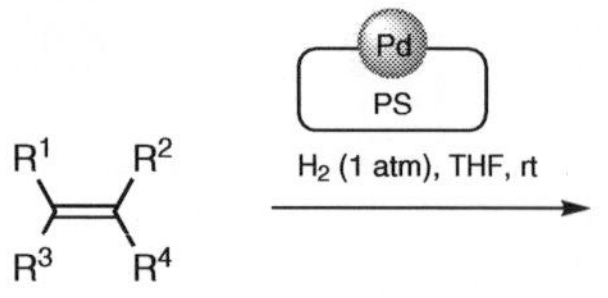

R^1	R^2	R^3	R^4	Time (h)	Yield (%)	Ref
Ph	H	H	Ph	1	100	[12]
Ph	H	H	Ac	1.5	97	[12]
Ph	H	H	CO_2H	1.5	>99	[12]
Ph	H	H	CO_2Me	1	>99	[12]
Ph	H	H	Bz	1	84	[12]
$(CH_2)_7Me$	$(CH_2)_7CO_2H$	H	H	12	100	[12]
Me	H	$OC(O)CH_2$		9	100	[12]
Me	H	$(CH_2)_3C(O)$		12	100	[12]

Palladium Nanoparticles in Polystyrene (Pd@PS):[12]
BuOH (64.0 mg, 0.86 mmol), styrene (809 mg, 7.8 mmol), and 1,4-divinylbenzene (102 mg, 0.78 mmol) were added to a soln of $Pd(PPh_3)_4$ (99.4 mg, 0.086 mmol) in THF (1 mL) in a 50-mL round-bottomed flask equipped with a condenser. The mixture was stirred at 90 °C for 6 h and cooled to rt. AIBN (2.8 mg, 17 μmol) was added and the mixture was then stirred for another 6 h at rt. The product was collected by filtration, and then suspended in THF (10 mL), collected again by filtration, and dried at rt. The solid was crushed into a grey powder.

Alkanes 10; General Procedure:[12]
To a soln of an alkene **9** (1.0 mmol) in THF (5 mL) was added Pd@PS (37.0 mg, 0.5 mol% Pd). The mixture was stirred under H_2 (balloon) for 1–12 h at rt. Upon completion of the reaction, hexane was added, and the catalyst was filtered off and washed with Et_2O. The product was obtained by concentration of the filtrate under reduced pressure.

1.1.2.1.6 Hydrogenation with Palladium Nanoparticles on Amphiphilic Supports

Palladium nanoparticles can be generated on amphiphilic polystyrene/poly(ethylene glycol) (PS-PEG) supports and used for the hydrogenation of alkenes **11** under aqueous conditions. The process generates the corresponding alkanes **12** in high yield, and the catalyst (Pd@PS-PEG) can be easily recovered and recycled (Scheme 7).[13]

Scheme 7 Hydrogenation of Alkenes Using Palladium Nanoparticles on Amphiphilic Supports[13]

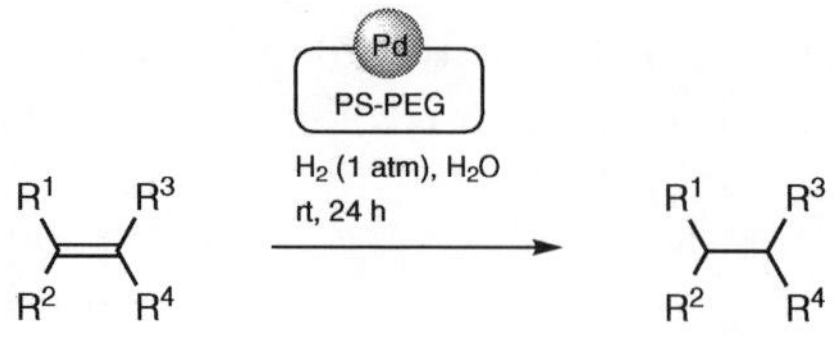

R^1	R^2	R^3	R^4	Yield (%)	Ref
Ph	H	H	H	>99	[13]
4-Tol	H	H	H	>99	[13]
$4\text{-}MeOC_6H_4$	H	H	H	>99	[13]
$4\text{-}F_3CC_6H_4$	H	H	H	>99	[13]
Ph	H	H	Me	>99	[13]
Ph	H	H	CH_2OH	81	[13]
Ph	H	H	CHO	87	[13]
Ph	H	H	Ac	97	[13]
Ph	H	H	CO_2Me	99	[13]
Ph	H	H	CO_2H	98	[13]
Ph	H	Me	CO_2H	99	[13]
Ph	H	Ph	CO_2H	99	[13]

Palladium Nanoparticles on Amphiphilic Polystyrene/Poly(ethylene glycol) (Pd@PS-PEG):[13]
To a suspension of PS-PEG terminated with a bispyridine ligand[14] (5.5 g) in toluene (60 mL) was added $Pd(OAc)_2$ (0.457 g, 2.04 mmol). The suspension was shaken at rt for 1 h, and the beads were collected by filtration, rinsed with CH_2Cl_2 (3 ×), and dried under reduced pressure. A mixture of the beads, benzyl alcohol (18 mL), and H_2O (56 mL) was heated at reflux for 12 h. The product was collected by filtration, rinsed with H_2O (3 ×) and acetone (3 ×), and dried under reduced pressure.

Alkanes 12**; General Procedure:**[13]
Pd@PS-PEG (65 mg, 26 μmol Pd) was added to a suspension of an alkene **11** (0.5 mmol) in H_2O (1.0 mL). The mixture was shaken under H_2 (1 atm) for 24 h at rt. Upon completion of the reaction, the catalyst was filtered off and rinsed with Et_2O (4 × 5 mL). The combined rinses were dried (Na_2SO_4), filtered, and subjected to chromatography (silica gel).

1.1.2.1.7 Hydrogenation with Palladium Nanoparticles in Biphasic Media

For use in aqueous/organic biphasic conditions, palladium nanoparticles can be embedded in the walls of polymeric microreactors (Pd@MR). The microreactors are composed of (a) an outer shell of hydrophilic polyacrylamide, which helps maintain the microreactor in the aqueous phase; (b) a hydrophobic wall of poly[styrene-*co*-(acetoacetoxy)ethyl methacrylate], which forms the structural integrity of the hollow sphere; and (c) palladium nanoparticles embedded in the microreactor walls. In this system, the microreactor catalyst prefers the aqueous phase, whereas the hydrogenation product prefers the organic phase, which makes the process well suited for continuous large-scale application, with no need for complicated reaction workup. The catalyst is robust, and can easily be recycled up to eight times with no decrease in catalytic activity. Alkenes such as **13** are hydrogenated to alkanes **14** in excellent yields (Scheme 8).[15]

for references see p 89

Scheme 8 Hydrogenation of Alkenes Using Palladium Nanoparticles in Biphasic Media[15]

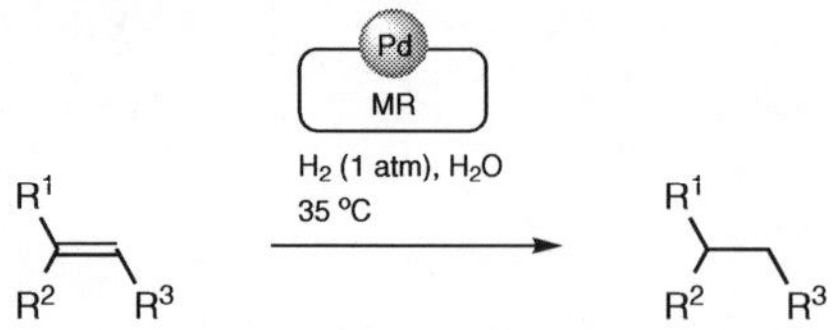

Pd MR

H_2 (1 atm), H_2O
35 °C

13 → 14

R^1	R^2	R^3	Yield[a] (%)	Ref
Ph	H	H	100	[15]
4-($ClCH_2$)C_6H_4	H	H	100	[15]
Ph	H	CH_2OH	99.5	[15]
CO_2Bu	H	H	100	[15]
Me	CO_2Bu	H	100	[15]
H	(CH_2)$_4$		100	[15]

[a] Yield determined by ^{1}H NMR spectroscopy or HPLC.

Palladium Nanoparticles Embedded in Polymeric Microreactors (Pd@MR):[15]
Styrene (5.21 g, 50.0 mmol) was added to a soln of methacrylic acid (0.43 g, 5.0 mmol) in H_2O (100 mL). The mixture was degassed at rt with N_2, and $K_2S_2O_8$ (0.297 g, 1.10 mmol) was added. After degassing again, the soln was vigorously stirred at 80 °C for 24 h under N_2. The product was isolated by centrifugation, washed with H_2O, and then dispersed in H_2O (150 mL).

H_2O (55 mL) and $K_2S_2O_8$ (0.135 g, 0.55 mmol) were added to this suspension (45 mL). The suspension was degassed with N_2 at rt under vigorous stirring for 30 min. Upon raising the temperature to 80 °C with continued magnetic stirring, a mixture of styrene (1.04 g, 10 mmol), 2-(acetoacetoxy)ethyl methacrylate (2.14 g, 10.0 mmol), acrylamide (0.355 g, 5.0 mmol), and 1,4-divinylbenzene (0.135 g, 1.0 mmol) was added dropwise. The mixture was stirred at 80 °C for 24 h, and the product was isolated by centrifugation and washed with H_2O.

The microspheres obtained were then dispersed in DMF (100 mL) for 12 h to remove the core. The hollow spheres were collected by centrifugation, washed with DMF (3 × 25 mL) and H_2O (2 × 25 mL), and dispersed in H_2O (200 mL).

The dispersion of hollow microspheres (20 mL) and a soln of $PdCl_2$ (18 mg, 0.10 mmol) in H_2O (80 mL) were added to a 250-mL conical flask. The mixture was stirred at rt for 4 h. Upon completion of the reaction, the pH was adjusted to 7 with aq NaOH. Upon cooling in an ice bath, cool 0.05 M aq $NaBH_4$ (10 mL) was slowly added. The resulting microreactors were dialyzed against H_2O for 7 d to remove impurities. The resultant product was dispersed in deionized H_2O (120 mL).

Alkanes 14; **General Procedure:**[15]
An alkene **13** (33 mmol) was added to a suspension of Pd@MR in H_2O (20 mL, 0.0167 μmol Pd) in a glass tube. The tube was fitted with a H_2 balloon. The mixture was first degassed with N_2 and stirred with a magnetic stirrer bar, and then the hydrogenation was started at 35 °C. The reaction was monitored by HPLC or ^{1}H NMR spectroscopy until completion. Upon completion of the reaction, the organic layer (product) was decanted.

1.1.2.1.8 Asymmetric Hydrogenation with Palladium Nanoparticles

Prochiral alkenes (e.g., **15**) can be asymmetrically hydrogenated under hydrogen gas with palladium nanoparticles supported on graphene with cinchonidine to impart stereochemical influence. The supported catalyst can be prepared by deposition of palladium onto graphene oxide, followed by simultaneous reduction of the palladium and the graphene oxide to afford the Pd@graphene catalyst. With cinchonidine as a chiral modifier, the catalyst affords the corresponding enantioenriched alkanes **16** in high yield (Scheme 9).[16] An optional benzylamine additive could be used to facilitate interactions between the support, the carboxylic acid, the chiral cinchonidine modifier, and the metal surface.

Scheme 9 Asymmetric Hydrogenation of Alkenes Using Palladium Nanoparticles/Cinchonidine[16]

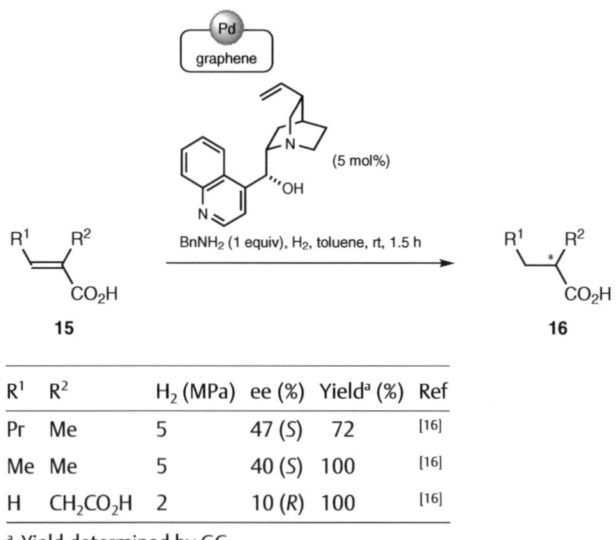

R^1	R^2	H_2 (MPa)	ee (%)	Yield[a] (%)	Ref
Pr	Me	5	47 (*S*)	72	[16]
Me	Me	5	40 (*S*)	100	[16]
H	CH_2CO_2H	2	10 (*R*)	100	[16]

[a] Yield determined by GC.

Palladium Nanoparticles on Graphene (Pd@graphene):[16]
Graphite oxide (1 g) was sonicated in H_2O (250 mL). After 2 h, a soln of $PdCl_2$ (88 mg, 0.5 mmol) in H_2O (50 mL) acidified with concd HCl (72 μL) was added and the mixture was then sonicated for another 30 min. Upon completion of sonication, the slurry was cooled in an ice bath, and a freshly prepared soln of $NaBH_4$ (724.4 mg) in H_2O (20 mL) was added. After 5 min, the mixture was brought to rt, and stirred for an additional 4 h. The resultant solid was isolated by centrifugation, washed with H_2O (3 × 20 mL) and MeOH (3 × 20 mL), and dried at 50 °C for 12 h under reduced pressure.

Alkanes 16; **General Procedure:**[16]
Pd@graphene (15 mg) and toluene (10 mL) were added to a stainless-steel autoclave equipped with a glass liner. The reactor was then flushed and pressurized with H_2 (5 MPa). After 30 min, the pressure was released and cinchonidine (0.05 mmol, 5 mol%), $BnNH_2$ (107 mg, 1 mmol, 1 equiv), and an alkene **15** (1 mmol) were added. The reactor was again flushed and pressurized with H_2 (2–5 MPa), and the mixture was magnetically stirred (1000 rpm) at rt for 90 min. H_2 was released, and the soln was treated with 10% aq HCl, dried (Na_2SO_4), and analyzed by GC.

for references see p 89

1.1.2.1.9 Hydrogenation with Palladium on Ferrite Nanoparticles

Magnetic nanoparticles, serving as either the catalyst itself (see Section 1.1.2.2)[17,18] or just as a catalyst support,[19–22] offer an easy and ecologically benign means for catalyst separation by the simple application of an external magnet or internal stirrer bar.[23–29] This strategy has been developed with palladium nanoparticles for alkene hydrogenation.

One of the most popular techniques for catalysis with magnetically retrievable particles is the use of the particle simply as a vehicle for recovery. By this method, the nature of the particle is irrelevant; the important property being that it can effectively bind via a linker to the active catalytic species. Manorama and co-workers functionalized the exterior of magnetic nickel ferrite ($NiFe_2O_4$) with dopamine, presenting the free amine for interaction with the active palladium species.[30] The resulting catalyst (Pd@ferrite) is robust, demonstrating the ability for 10 cycles of catalysis with no appreciable loss in activity, and is selective for the reduction of alkenes over aromatics or carbonyls. Thus, alkenes such as **17** are hydrogenated to alkanes **18** in excellent yields (Scheme 10).

Scheme 10 Hydrogenation of Alkenes Using Palladium on Amine-Terminated Ferrite Nanoparticles[30]

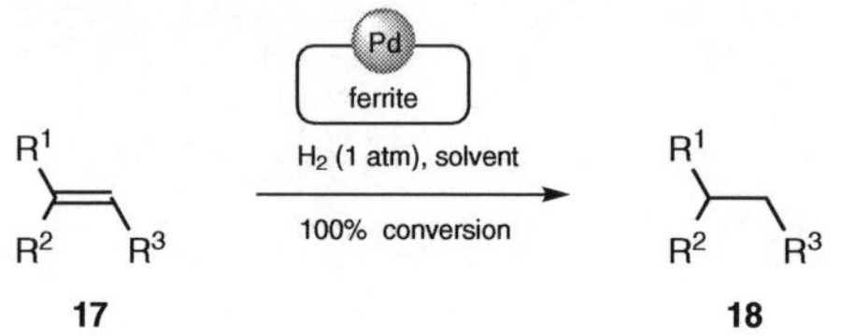

Pd ferrite
R1, R2, R3 **17** → H_2 (1 atm), solvent; 100% conversion → R1, R2, R3 **18**

R^1	R^2	R^3	Solvent[a]	Yield (%)	Ref
Ph	H	CO_2Me	EtOH (20 min) or EtOAc (45 min)	97	[30]
Ph	H	CHO	EtOH (20 min) or EtOAc (45 min)	75	[30]
Ph	H	CO_2H	EtOH (25 min) or EtOAc (55 min)	98	[30]
Ph	H	Ph	EtOH (15 min) or EtOAc (30 min)	98	[30]
Bz	H	Ph	EtOH (25 min) or EtOAc (55 min)	99	[30]
Me	$(CH_2)_3C(O)$		EtOH (25 min) or EtOAc (55 min)	95	[30]

[a] Time required for reaction completion in each solvent in parentheses; it was not explicitly stated in the original article which solvent the reported isolated yields correspond to.

Palladium on Amine-Terminated Ferrite Nanoparticles (Pd@ferrite):[30]

CAUTION: *Hydrazine hydrate is a severe skin and mucous membrane irritant and a possible human carcinogen.*

To a suspension of $NiFe_2O_4$ nanoparticles (1 g) in H_2O was added dopamine (2 g). The mixture was heated for 12 h at reflux, and the product was precipitated with acetone and centrifuged to obtain the amine-functionalized magnetic particle. The resultant particles (1 g) were then dispersed in H_2O, and a soln of $NaPdCl_4$ was added until 10 wt% Pd was achieved. The pH was adjusted to 9 by dropwise addition of dilute $H_2NNH_2 \cdot H_2O$. The mixture was stirred for 18 h at rt. After the product had settled, it was washed multiple times with H_2O, collected by centrifugation, and dried at rt.

Alkanes 18**; General Procedure:**[30]
To a soln of an alkene **17** (2 mmol) in EtOAc or EtOH (10 mL) was added Pd@ferrite (25 mg). The mixture was stirred under H_2 (1 atm) and the reaction was monitored by TLC. After

completion of the reaction, the catalyst was retrieved by application of an external magnet and washed with EtOAc in preparation for a subsequent round of catalysis.

1.1.2.1.10 Hydrogenation with Palladium Nanoparticles Supported on Magnetic Carbon-Coated Cobalt Nanobeads

Rather than through a pendent ligand on the surface of a magnetic particle, catalytically active particles can be adsorbed directly to a surface coating. Through microwave decomposition, palladium(0) can be deposited directly onto carbon-coated cobalt nanoparticles. The catalyst (Pd@CoNPs) is robust enough for use in subsequent recycling reactions, and maintains a higher conversion than commercial palladium on charcoal through six recycling reactions. Alkenes such as **19** are hydrogenated to alkanes **20** with quantitative conversion (Scheme 11).[31]

Scheme 11 Hydrogenation of Alkenes Using Palladium Nanoparticles Supported on Magnetic Carbon-Coated Cobalt Nanobeads[31]

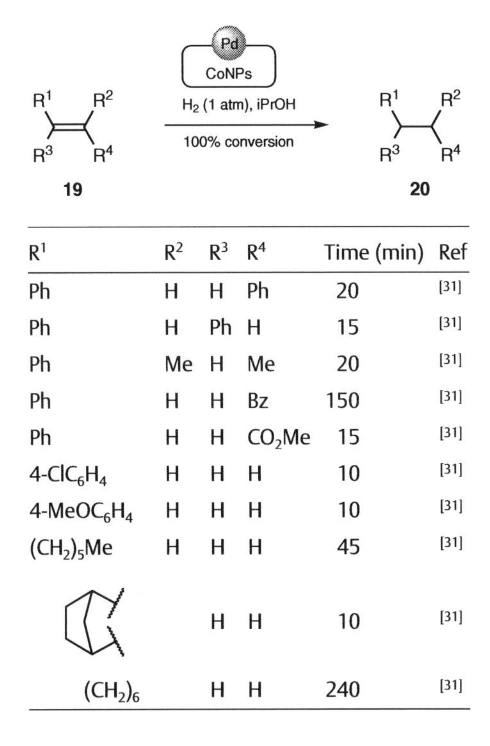

R^1	R^2	R^3	R^4	Time (min)	Ref
Ph	H	H	Ph	20	[31]
Ph	H	Ph	H	15	[31]
Ph	Me	H	Me	20	[31]
Ph	H	H	Bz	150	[31]
Ph	H	H	CO_2Me	15	[31]
4-ClC_6H_4	H	H	H	10	[31]
4-$MeOC_6H_4$	H	H	H	10	[31]
$(CH_2)_5Me$	H	H	H	45	[31]
		H	H	10	[31]
$(CH_2)_6$		H	H	240	[31]

Palladium Nanoparticles Supported on Magnetic Carbon-Coated Cobalt Nanobeads (Pd@CoNPs):[31]
$Pd_2(dba)_3$•$CHCl_3$ (7.8 mg, 7.5 μmol) was added to a soln of carbon-coated cobalt nanoparticles (1.0 g, 7 wt% C) in toluene (5 mL) in a microwave vial under N_2. The vial was sonicated for 10 min and then heated to 110 °C under microwave irradiation for 2 min. The resultant product was isolated by an external magnet, washed with CH_2Cl_2 (5 × 5 mL), and dried under reduced pressure.

Alkanes 20; General Procedure:[31]
Pd@CoNPs (0.5 μmol, 0.2 wt% Pd) and dodecane (as an internal standard) were added to a soln of an alkene **19** (0.5 mmol) in iPrOH (5 mL) in a Schlenk flask. The slurry was sonicat-

for references see p 89

ed for 10 min, and the flask was evacuated and flushed with H_2. The mixture was vigorously stirred under H_2 (1 atm). The reaction was monitored by GC, and upon reaction completion, the particles were magnetically recovered. The conversion was determined by GC.

1.1.2.1.11 Hydrogenation with Magnetic Carbon-Supported Palladium Nanoparticles

The grafting and calcination of cellulosic carbon onto ferrite nanoparticles and subsequent deposition of palladium leads to another class of particles (Pd@carbon/ferrite), where palladium serves as the active catalytic species and the magnetic support serves the role of facilitating catalyst retrieval. The catalyst, made from cellulose as an abundant biopolymer, is robust, easy to synthesize on gram scale in water, and functionalizable. Alkenes such as **21** are hydrogenated to alkanes **22** with quantitative conversion (Scheme 12).[32]

Scheme 12 Hydrogenation of Alkenes Using Magnetic Carbon-Supported Palladium Nanoparticles[32]

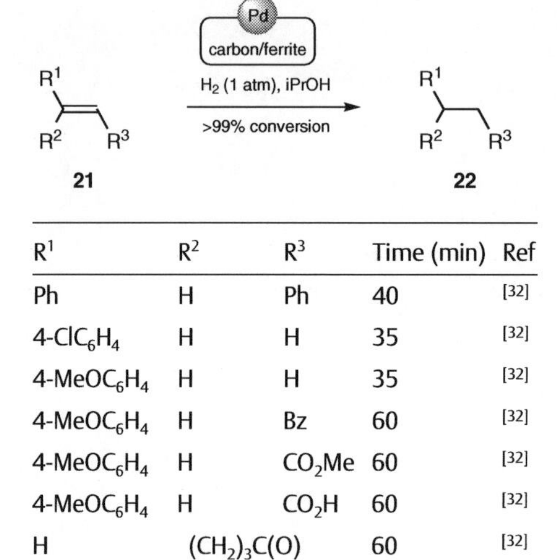

R^1	R^2	R^3	Time (min)	Ref
Ph	H	Ph	40	[32]
4-ClC_6H_4	H	H	35	[32]
4-$MeOC_6H_4$	H	H	35	[32]
4-$MeOC_6H_4$	H	Bz	60	[32]
4-$MeOC_6H_4$	H	CO_2Me	60	[32]
4-$MeOC_6H_4$	H	CO_2H	60	[32]
H	$(CH_2)_3C(O)$		60	[32]

Magnetic Carbon-Supported Palladium Nanoparticles (Pd@carbon/ferrite):[32]
To a soln of $FeSO_4$•$7H_2O$ (2.78 g, 10 mmol) and $Fe_2(SO_4)_3$ (4.0 g, 10 mmol) in H_2O (500 mL) was added 25% aq NH_4OH to adjust the pH to 10. The mixture was stirred for 1 h at 50 °C. Upon cooling to rt, cellulose (10 g) was added with vigorous stirring, which was continued for 6 h at rt. $PdCl_2$ was added and the mixture was stirred for 8 h. The resulting particles were separated with an external magnet, washed with H_2O, and calcinated at 450 °C for 3 h. The weight percent of Pd was determined to be 4.81% by ICP-AES.

Alkanes 22**; General Procedure:**[32]
To a soln of an alkene **21** (5 mmol) in iPrOH (25 mL) was added Pd@carbon/ferrite (100 mg) in a 100-mL, round-bottomed flask. The mixture was stirred with a magnetic stirrer bar under H_2 (1 atm) at rt for 35–60 min. The reaction progress was monitored by GC/MS and TLC. After completion of the reaction, the catalyst was magnetically recovered on the stirrer bar, and the products were isolated by removal of the solvent under reduced pressure.

1.1.2.2 Reduction of Alkenes by Hydrogenation Using Iron Nanoparticles

Rather than serve as simply a vehicle for recovery, magnetic particles themselves can serve as catalysts for hydrogenation directly. Under such a strategy, iron particles can effectively catalyze hydrogenation reactions, albeit under typically more forcing conditions than those for palladium-catalyzed hydrogenation. Despite generally requiring higher temperatures, pressures, or longer reaction times, methods for the direct use of iron particles as the active catalytic species avoid the use of the more precious, toxic, and less earth-abundant metals, such as the more commonly used platinum series metals.[33,34] The rate of hydrogenation with these catalysts is strongly dependent on steric factors. Tetra- and trisubstituted alkenes are not hydrogenated. *E*-Alkenes and some *Z*-alkenes require higher temperatures. Monosubstituted and 1,1-disubstituted alkenes are hydrogenated at room temperature.

1.1.2.2.1 Hydrogenation with Unsupported Iron Nanoparticles

Iron nanoparticles can be synthesized by reduction of iron(III) chloride with 3 equivalents of ethylmagnesium chloride in tetrahydrofuran. The resulting particles can be used for hydrogenation of alkene substrates such as **23** to give alkanes **24** (Scheme 13).[35,36] Tri- and tetrasubstituted alkenes are not hydrogenated under these conditions.

Scheme 13 Hydrogenation of Alkenes Using Unsupported Iron Nanoparticles[35,36]

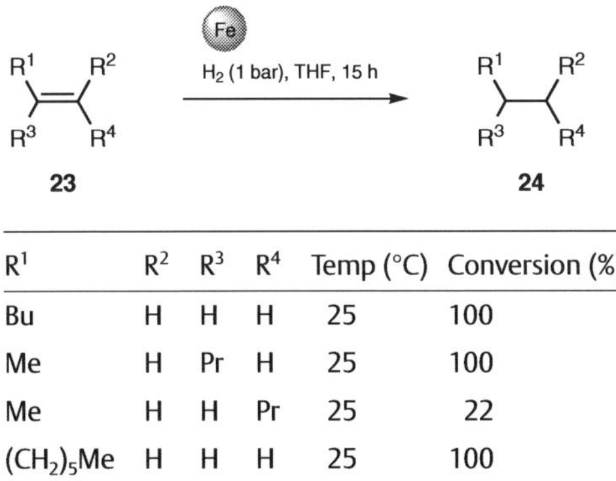

Fe
H_2 (1 bar), THF, 15 h
23 → **24**

R^1	R^2	R^3	R^4	Temp (°C)	Conversion (%)	Ref
Bu	H	H	H	25	100	[35,36]
Me	H	Pr	H	25	100	[35,36]
Me	H	H	Pr	25	22	[35,36]
$(CH_2)_5Me$	H	H	H	25	100	[35,36]
H	H	Me	Pr	25	100	[35,36]
Me	Me	Me	Me	25	0	[35,36]
Ph	H	H	Ph	25	6	[35,36]
Ph	H	H	Ph	100	100	[35,36]
$(CH_2)_4$		H	H	25	22	[35,36]
$(CH_2)_4$		H	H	100	100	[35,36]
$(CH_2)_6$		H	H	25	11	[35,36]
$(CH_2)_6$		H	H	100	100	[35,36]
$(CH_2)_4$		Me	H	25	0	[35,36]
$(CH_2)_4$		Me	H	100	0	[35,36]
$(CH_2)_6$		Me	H	25	0	[35,36]
$(CH_2)_6$		Me	H	100	0	[35,36]

Iron nanoparticles can also be generated by the decomposition of iron(II) bis(trimethylsilyl)amide ($[Fe\{N(TMS)_2\}_2]_2$) under hydrogen gas, and similarly used for the hydrogenation of alkenes **25** to alkanes **26** (Scheme 14).[37]

Scheme 14 Hydrogenation of Alkenes Using Iron Nanoparticles Generated from Iron(II) Bis(trimethylsilyl)amide[37]

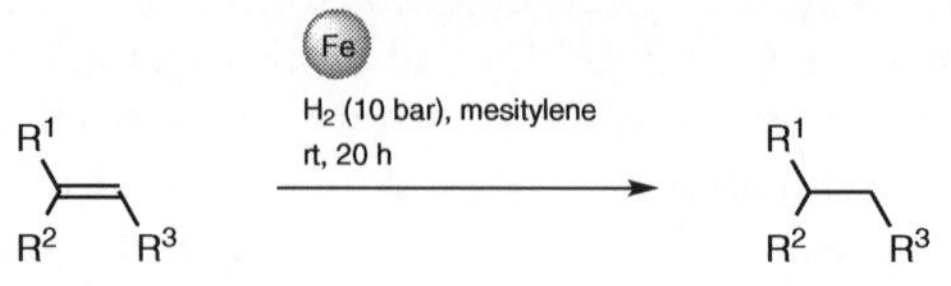

R^1	R^2	R^3	Yield[a] (%)	Ref
$(CH_2)_5Me$	H	H	>99	[37]
Pr	H	Pr	0	[37]
Me	$(CH_2)_4Me$	H	>99	[37]
Me	H	Et	95	[37]
Ph	H	H	87	[37]
3-ClC_6H_4	H	H	98	[37]
3-$F_3CC_6H_4$	H	Ph	97	[37]
Ph	H	Ph	>99	[37]
H	$(CH_2)_4$		>99	[37]
H	$(CH_2)_3$		>99	[37]
H	 [bicyclic structure]		>99	[37]

[a] Determined by GC.

Iron Nanoparticles from Iron(III) Chloride:[35,36]
In a Schlenk tube under argon, a 2.0 M soln of EtMgCl (2.5 mL, 5 mmol) was added via syringe to a soln of $FeCl_3$ (162 mg, 1.0 mmol) in THF (20.0 mL). The mixture was vigorously stirred for 30 min. The iron nanoparticles were used without further purification.

Alkanes 24; **General Procedure:**[35,36]
A soln of iron nanoparticles in THF (1 mL, [Fe] = 0.05 M), prepared as described above, was added to a 5-mL flask containing an alkene **23** (1 mmol) in THF (1 mL). The mixture was vigorously stirred under H_2 (1 bar) for 15 h at rt or 100 °C. The conversions were determined by GC.

Iron Nanoparticles from Iron(II) Bis(trimethylsilyl)amide:[37]
A soln of $[Fe\{N(TMS)_2\}_2]_2$ (376.5 mg, 0.5 mmol) in anhyd degassed mesitylene (20 mL) was magnetically stirred for 18 h at 150 °C under H_2 (3 bar). Over the course of the reaction, the soln turned from green to black. The resultant iron nanoparticles were isolated by removal of the solvent under reduced pressure. A black shiny powder (112 mg) was obtained and kept under argon in a glovebox.

Alkanes 26; **General Procedure:**[37]
The above obtained iron nanoparticles (1.3 mg, 2.4 mol%) were added to a soln of an alkene **25** (1 mmol) in mesitylene (1 mL) in an autoclave. The autoclave was pressurized

with H_2 (10 bar) and the mixture was stirred at rt for 20 h. Upon reaction completion, hexadecane was added as an external standard, and the yield was determined by GC.

1.1.2.2.2 Hydrogenation with Graphene-Supported Iron Nanoparticles

Iron nanoparticles can be synthesized with chemically derived graphene as a structured support. The resultant iron nanoparticles on graphene (Fe@graphene) are active for alkene hydrogenation (Scheme 15) and are magnetically recoverable.[38] Excess Grignard reagent is needed to ensure that the catalyst remains in the catalytically active zero-valent state.

Scheme 15 Hydrogenation of Alkenes Using Graphene-Supported Iron Nanoparticles[38]

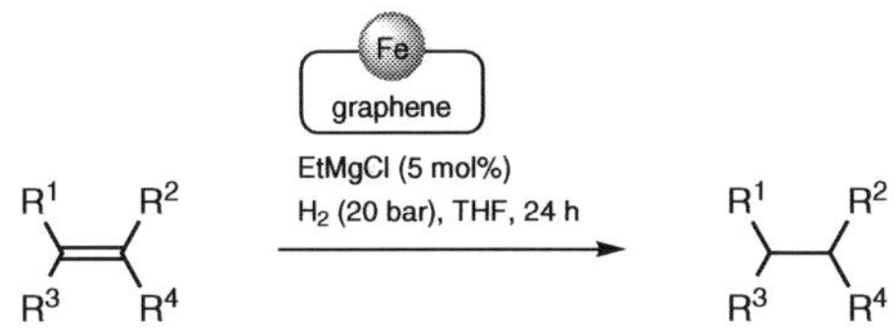

Fe graphene
EtMgCl (5 mol%)
H_2 (20 bar), THF, 24 h

27 (1 mmol) → **28**

R^1	R^2	R^3	R^4	Temp (°C)	Yield[a] (%)	Ref
Ph	H	H	H	100	99	[38]
Bu	H	H	H	100	100	[38]
Bu	H	H	H	25	100	[38]
$(CH_2)_5Me$	H	H	H	100	100	[38]
$(CH_2)_5Me$	H	H	H	25	11	[38]
Me	H	H	Pr	100	0	[38]
Me	Me	Me	H	100	0	[38]
Me	Me	Me	Me	100	0	[38]
Ph	H	H	Ph	100	7	[38]
$(CH_2)_4$		H	H	100	79	[38]
$(CH_2)_3$		H	H	100	73	[38]
$(CH_2)_6$		H	H	100	99	[38]
 (bicyclic structure)		H	H	100	100	[38]

[a] Determined by GC/MS and NMR spectroscopy.

Iron Nanoparticles on Graphene (Fe@graphene):[38]
To a soln of $Fe(CO)_5$ (447 mg, 2.28 mmol) in diphenylmethane (200 mL) was added chemically derived graphene (920 mg). The mixture was sonicated three times for 10 min at rt. The product was used for alkene hydrogenation with no further purification. The loading of Fe was 3.66 wt%.

Alkanes 28; General Procedure:[38]
To a soln of Fe@graphene (14 mg, 9.1 μmol, 0.9 mol% Fe) in THF (1 mL) was added a 2 M soln of EtMgCl in THF (25 μL, 5 mol%) and an alkene **27** (1 mmol). The Schlenk tube was purged with H_2 (5 ×) and the contents were transferred via syringe into an autoclave,

for references see p 89

which was purged with H_2 (3 ×). The mixture was stirred for 24 h at the specified temperature and pressure. Upon reaction completion, the catalyst was magnetically recovered, and the yield was determined by GC/MS and NMR spectroscopy.

1.1.2.2.3 Hydrogenation with Core-Shell Iron/Iron Oxide Nanoparticles

Instead of using particles that are comprised completely of zero-valent iron, nanoparticles with a zero-valent iron core and an iron oxide shell (core-shell nanoparticles; Fe@Fe_xO_y) can be used for alkene hydrogenation, without the need for strict inert conditions or for an excess of a Grignard reagent to maintain the zero-valent surface (Scheme 16).[39] The particles can be magnetically recycled for up to 10 rounds of catalysis, with very little decrease in yield.

Scheme 16 Hydrogenation of Alkenes Using Core-Shell Iron/Iron Oxide Nanoparticles[39]

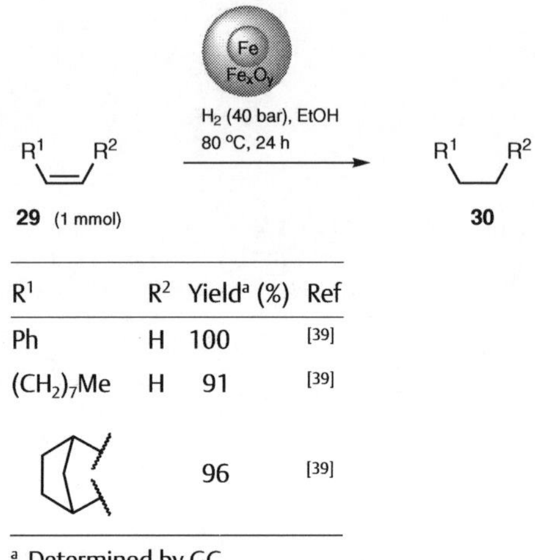

R1	R2	Yield^a (%)	Ref
Ph	H	100	[39]
$(CH_2)_7Me$	H	91	[39]
		96	[39]

^a Determined by GC.

Core-Shell Iron/Iron Oxide Nanoparticles (Fe@Fe_xO_y):[39]
To a soln of $FeSO_4$•$7H_2O$ (8.4 g, 30 mmol) in degassed MeOH/H_2O (3:7; 480 mL), a soln of $NaBH_4$ (2.4 g, 63.5 mmol) in H_2O (60 mL) was added using a syringe pump at 3 mL•min^{-1}, and the mixture was stirred for an additional 30 min upon complete addition of $NaBH_4$. The particles were magnetically recovered, and rinsed with degassed EtOH (3 × 200 mL).

Alkanes 30; **General Procedure:**[39]
An alkene **29** (1 mmol) and dodecane (1 mmol) as internal standard were added to a soln of Fe@Fe_xO_y (2.8 mg, 5 mol%) in EtOH (17 mL) in a high-pressure reactor. The mixture was stirred for 24 h at 80 °C under H_2. Upon reaction completion, the particles were magnetically recovered and used in a subsequent reaction. The yield was determined by GC.

1.1.2.2.4 Hydrogenation with Amphiphilic Polymer-Supported Iron Nanoparticles

In addition to methods that take advantage of their ability for magnetic recovery, iron nanoparticles can also be embedded in polystyrene(PS)/poly(ethylene glycol) (PEG) block copolymers with a terminal amine for use as an immobilized catalyst for the reduction of alkenes **31** under flow conditions. The catalyst (Fe@PS-PEG-NH_2) is resistant to oxidation and can be used for the production of alkanes **32** on multigram scale (Scheme 17).[33]

Scheme 17 Hydrogenation of Alkenes Using Amphiphilic Polymer-Supported Iron Nanoparticles under Flow Conditions[33]

Fe PS-PEG-NH$_2$, H$_2$ (40 bar), EtOH, 100 °C; **31** → **32**

R^1	R^2	R^3	R^4	Flow (mL•min^{-1})	Yield[a] (%)	Ref
Ph	H	H	H	1	100	[33]
$(CH_2)_7Me$	H	H	H	1	73	[33]
Ph	H	CH_2OAc	H	1	98	[33]
Ph	H	CH_2OH	H	1	100	[33]
4-ClC_6H_4	H	H	H	1	99	[33]
4-$O_2NC_6H_4$	H	H	H	1	84	[33]
$(CH_2)_3C(O)$		H	H	1	100	[33]
$(CH_2)_7Me$	H	H	H	0.5	90	[33]
Me	H	Bu	H	0.5	83	[33]
Me	Bu	H	H	0.5	87	[33]
H	$(CH_2)_5$		H	0.5	14	[33]
Me	H	$(CH_2)_4$		0.5	6	[33]

[a] Determined by GC.

Amphiphilic Polymer-Supported Iron Nanoparticles (Fe@PS-PEG-NH$_2$):[33]
A suspension of amine-terminated polystyrene/poly(ethylene glycol) beads (1 g) in octadec-1-ene (60 mL) in a 200-mL Schlenk flask was purged with N$_2$ at 120 °C for 30 min. While stirring, the temperature was raised to 180 °C, at which point Fe(CO)$_5$ (2.1 mL) was quickly added via syringe. The mixture was stirred for 30 min at 180 °C under N$_2$, and then allowed to cool to rt. The resulting polymer-supported iron nanoparticles were washed with hexanes (3 × 50 mL) and dried under reduced pressure.

Alkanes 32**; General Procedure:**[33]
A cartridge packed with Fe@PS-PEG-NH$_2$ (300 mg) was used in an H-Cube flow hydrogenation system. A 0.05 M soln of a substrate **31** in EtOH was forced through the system with the specified flow at 100 °C under H$_2$ (40 bar). Upon completion of the reaction, the yield was determined by GC.

1.1.2.3 Reduction of Alkenes Using Ionic-Liquid-Stabilized Nanoparticles

The range of functionality and physical properties of ionic liquids enable them to serve as a tunable, and often reusable, class of solvents for various organic transformations. Their interfacing with nanocatalysts has been intensely researched because of synergistic effects between this medium and the nanoparticles.[40] Nanoparticles of various metals (Pt, Pd, Ru, Ir) have been used for the hydrogenation of arenes and alkenes in this context,[41–45] although most reports typically use only cyclohexene and hex-1-ene as model alkene substrates.

1.1.2.3.1 Reduction of Alkenes with Iridium Nanoparticles in Ionic Liquids

Iridium nanoparticles can be effectively synthesized in, and stabilized by, imidazolium ionic liquids, specifically 1-butyl-3-methylimidazolium hexafluorophosphate ([bmim]PF_6), for use in alkene hydrogenation. The catalyst (Ir@[bmim]PF_6) is robust and can be recycled at least five times with no appreciable loss in activity. Alkenes such as **33** are hydrogenated to alkanes **34** with excellent conversion (Scheme 18).[46]

Scheme 18 Hydrogenation of Alkenes Using Iridium Nanoparticles in an Ionic Liquid[46]

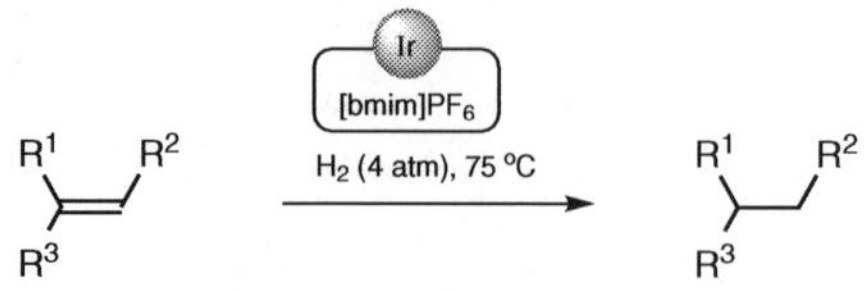

Ir, [bmim]PF_6, H_2 (4 atm), 75 °C

33 → **34**

R^1	R^2	R^3	Time (h)	Conversion[a] (%)	Ref
$(CH_2)_7Me$	H	H	0.5	100	[46]
Ph	H	H	1	63	[46]
Me	H	CO_2Me	17	100	[46]
$(CH_2)_4$		H	3.2	100	[46]
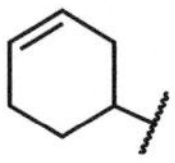	H	H	4	100[b]	[46]

[a] Determined by GC.
[b] Conversion into ethylcyclohexane.

Iridium Nanoparticles in 1-Butyl-3-methylimidazolium Hexafluorophosphate (Ir@[bmim]PF_6):[46]
To a soln of {IrCl(cod)}$_2$ (16 mg, 25 μmol) in CH_2Cl_2 (3 mL) was added 1-butyl-3-methylimidazolium hexafluorophosphate (3 mL) in a Fisher–Porter vessel. The soln was stirred at rt for 15 min. The volatiles were removed under reduced pressure (0.1 bar) at 75 °C. After 1 h, H_2 (4 bar) was introduced and the mixture was stirred for 10 min at 75 °C affording the product ready for use in catalysis.

Alkanes 34; General Procedure:[46]
An alkene **33** (60 mmol) and H_2 (4 atm) were added to the soln of iridium nanoparticles in 1-butyl-3-methylimidazolium hexafluorophosphate described above. The mixture was stirred at 75 °C for 0.5–17 h and the reaction was monitored by GC. Upon completion of the reaction the organic layer (top) was removed to provide the product.

1.1.2.4 Reduction of Alkenes by Transfer Hydrogenation

Transfer hydrogenation defines hydrogenation methods in which two hydrogen atoms are provided by a donor reagent other than molecular hydrogen. Typical hydrogen donors in this context are alcohols, which become oxidized in the process to aldehydes or ketones, or hydrazine. Despite being less atom-economical than direct hydrogenation, transfer hydrogenation has many advantages including mild conditions (notably ambient pressure), easy operation, and high selectivity. Many catalysts have been developed for this purpose, including both homogeneous and heterogeneous systems, and corresponding asymmetric versions. The following sections are focused on nanoparticulate catalysts developed for the transfer hydrogenation of alkenes to alkanes.[47] In most examples, nickel represents the metal of choice for nanoparticle-catalyzed transfer hydrogenation. Nickel has indeed shown its ability to perform this reaction in bulk form using isopropanol as the hydrogen source.[48]

SAFETY: Hydrazine is toxic and unstable, and should be handled as a solution.

1.1.2.4.1 Selective Reduction of Alkenes with Hydrazine Using Nickel Nanoparticles Supported on Clay

In 2008, Pitchumani et al. reported the first transfer hydrogenation method to reduce alkenes with nanoparticles. Nickel nanoparticles supported on clay (Ni@clay) and the moderately activated hydrogen source hydrazine are used to furnish the products **36** from mono- and disubstituted (both *E* and *Z*) alkenes **35** in good to excellent yields. Functional groups such as carbonyl, ether, and ester are tolerated (Scheme 19).[49]

Scheme 19 Transfer Hydrogenation of Alkenes Using Hydrazine as Hydrogen Donor and Nickel Nanoparticles Supported on Clay[49]

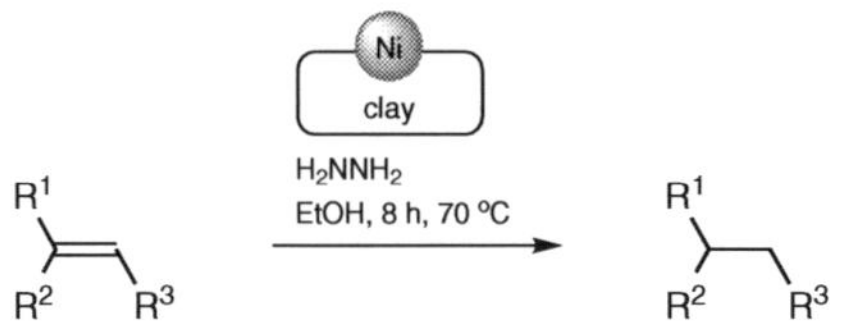

Ni / clay; H_2NNH_2; EtOH, 8 h, 70 °C

R^1, R^2, R^3 **35** → R^1, R^2, R^3 **36**

R^1	R^2	R^3	Yield (%)	Ref
Ph	H	H	82	[49]
Ph	Me	H	74	[49]
Ph	H	Me	68	[49]
4-$MeOC_6H_4$	H	H	88	[49]
4-$MeOC_6H_4$	H	Me	79	[49]
H	$(CH_2)_4$		52	[49]
H	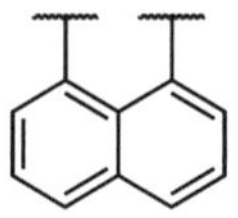 (naphthalene-1,8-diyl)		91	[49]
Bz	H	Ph	74	[49]
CO_2Me	H	Ph	81	[49]
H	$(CH_2)_3C(O)$		48	[49]
CH_2SPh	H	H	79	[49]

Nickel Nanoparticles Supported on Clay (Ni@clay):[49]

CAUTION: *Hydrazine hydrate is a severe skin and mucous membrane irritant and a possible human carcinogen.*

A mixture of montmorillonite K 10 clay (10 g) and 1 M aq $NiCl_2$ (250 mL) was stirred at rt for 72 h. The clay was collected by filtration, washed thoroughly with distilled H_2O, and dried. $H_2NNH_2 \cdot H_2O$ (2 mL) was added to a suspension of the Ni(II)/montmorillonite K 10

for references see p 89

clay (500 mg) in 25% aq NH_3 (1 mL). The suspension was stirred at 60 °C for 5 h, cooled, and centrifuged with continuous washing with distilled H_2O until the mother liquor was neutral.

Alkanes 36; General Procedure:[49]

CAUTION: *Hydrazine hydrate is a severe skin and mucous membrane irritant and a possible human carcinogen.*

An alkene **35** (0.9 mmol) and H_2NNH_2•H_2O (0.4 mL, added slowly) were added to a suspension of Ni@clay (100 mg) in EtOH (3 mL). The mixture was heated to 70 °C and stirred for 8 h. Upon completion of the reaction, the mixture was cooled, filtered, dried (Na_2SO_4), and concentrated to give the product, which was sometimes further purified by column chromatography.

1.1.2.4.2 Selective Reduction of Alkenes with Isopropanol Using Nickel Nanoparticles

Alonso, Yus, and co-workers used unsupported nickel nanoparticles produced in situ by reduction with lithium.[50] The resulting nickel nanoparticles were characterized by transmission electron microscopy and had a diameter of 1.75 ± 1.00 nm. These nanoparticles are active for carbonyl reduction,[50] and also for alkene reduction (Scheme 20).[51] The mild and selective reduction of 22 substrates **37** delivered the corresponding hydrogenated products **38** using isopropanol as hydrogen donor in a mixture of isopropanol and tetrahydrofuran at 76 °C. Isopropanol is a very popular hydrogen donor because of its nontoxicity, low cost, and stability. Under these conditions, the catalyst loading was 20 mol%. The final example shown in Scheme 20 is relevant as it leads to the natural product brittonin A [**38**, $R^1 = R^3 = 3,4,5\text{-}(MeO)_3C_6H_2$; $R^2 = H$] in excellent yield.

Scheme 20 Transfer Hydrogenation of Alkenes Catalyzed by Nickel Nanoparticles Using Isopropanol as Hydrogen Donor[51]

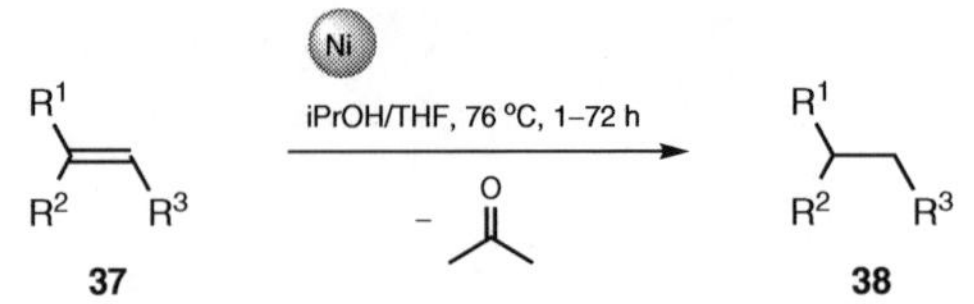

Ni
iPrOH/THF, 76 °C, 1–72 h
37 → **38**

R^1	R^2	R^3	Time (h)	Yield (%)	Ref
$(CH_2)_5Me$	H	H	3	>99	[51]
$(CH_2)_2Ph$	H	H	2	>99	[51]
Pr	H	Pr	2	>99	[51]
Bu	H	Bu	1	60	[51]
Ph	H	Ph	1	>99	[51]
Ph	Me	H	1	>99	[51]
Ph	Ph	H	4	>99	[51]
H	$(CH_2)_6$		2	>99	[51]
H	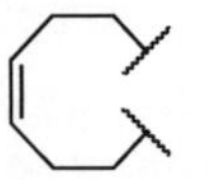		72	>99[a]	[51]

R^1	R^2	R^3	Time (h)	Yield (%)	Ref
H	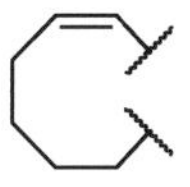		5	>99[a]	[51]
$(CH_2)_4CO_2Et$	H	H	2	89	[51]
$(CH_2)_4CO_2Et$	H	H	24	>99	[51]
CH_2CO_2Et	H	Et	2	>99	[51]
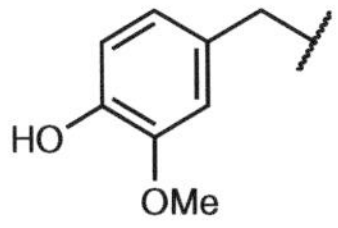 HO, OMe	H	H	48	>99	[51]
$3,4-(MeO)_2C_6H_3$	H	H	3	>99	[51]
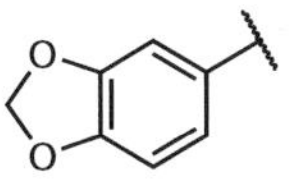 O, O	H	H	2	96	[51]
$CH(OH)(CH_2)_6Me$	H	H	4	>99	[51]
CH(OH)Ph	Me	H	3	90	[51]
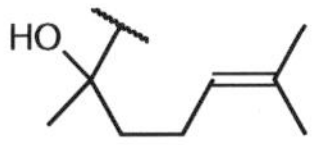 HO	H	H	72	51[a]	[51]
CH_2OBn	H	H	3	>99	[51]
CH_2NHCy	H	H	2	>99	[51]
$3,4,5-(MeO)_3C_6H_2$	H	$3,4,5-(MeO)_3C_6H_2$	2	95	[51]

[a] Both double bonds were hydrogenated.

Unsupported Active Nickel Nanoparticles:[51]
$NiCl_2$ (130 mg, 1 mmol) was added to a suspension of Li (14 mg, 2 mmol) and 4,4′-di-*tert*-butylbiphenyl (13 mg, 0.05 mmol) in anhyd THF (2 mL) at rt under argon. The dark blue soln changed to black indicating that nickel(0) nanoparticles were formed.

Alkanes 38**; General Procedure:**[51]
iPrOH (5 mL) and an alkene **37** (5 mmol) were added, in that order, to a freshly prepared soln of nickel(0) nanoparticles. The mixture was heated to 76 °C and the reaction was monitored by GLC/MS. Upon completion of the reaction, the suspension was diluted with Et_2O (20 mL), filtered through Celite, dried ($MgSO_4$), filtered, and concentrated under reduced pressure. The resulting residue was either analyzed directly, or purified further by column chromatography.

1.1.2.4.3 Selective Reduction of Alkenes with Isopropanol Catalyzed by Nickel/Ruthenium/Platinum/Gold Heteroquatermetallic Nanoparticles

In 2014, the Uozumi group reported an interesting system based on the synthesis of heteroquatermetallic nanoparticles composed of nickel, ruthenium, platinum, and gold. The nanoparticles were produced by reduction of a mixture of nickel(II) chloride, ruthenium(III) chloride, hexachloroplatinic acid, and potassium tetrachloroaurate(III) with excess lithium triethylborohydride in tetrahydrofuran at 24 °C in the presence of trioctylphosphine oxide as a capping agent. The resulting nanoparticles were found to have a diameter of about 3 nm (determined by transmission electron microscopy), and to have a

for references see p 89

ratio of metals of 1.1:1.0:1.1:1.0 for Ni/Ru/Pt/Au [determined by inductively coupled plasma atomic emission spectroscopy (ICP-AES) analysis]. This highly active catalyst works at 100 °C under low loading (0.5 mol%) for the transfer hydrogenation of alkenes **39** with isopropanol to furnish alkanes **40** (Scheme 21).[52]

Scheme 21 Transfer Hydrogenation of Alkenes with Nickel/Ruthenium/Platinum/Gold Heteroquatermetallic Nanoparticles Using Isopropanol as Hydrogen Donor[52]

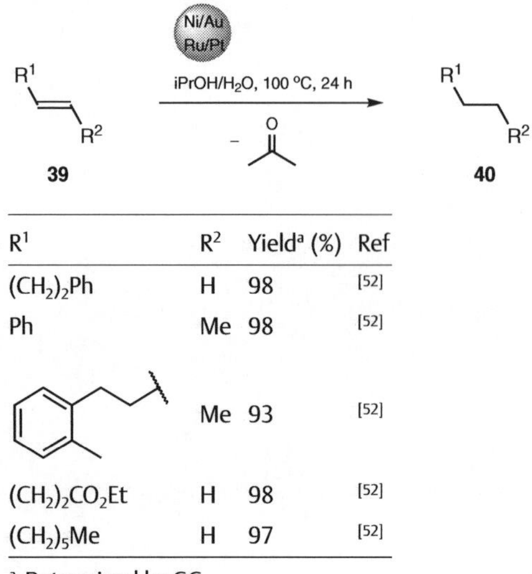

R^1	R^2	Yield[a] (%)	Ref
$(CH_2)_2Ph$	H	98	[52]
Ph	Me	98	[52]
(structure)	Me	93	[52]
$(CH_2)_2CO_2Et$	H	98	[52]
$(CH_2)_5Me$	H	97	[52]

[a] Determined by GC.

Ni/Ru/Pt/Au Heteroquatermetallic Nanoparticles:[52]
To freshly distilled THF (10 mL) under N_2 at 24 °C was added anhyd $NiCl_2$ (0.125 mmol), $RuCl_3$•H_2O (0.125 mmol), $KAuCl_4$ (0.125 mmol), H_2PtCl_6•$6H_2O$ (0.125 mmol), and trioctylphosphine oxide (0.50 mmol), followed by a 1 M soln of $LiBEt_3H$ (7.5 mL; added rapidly). The resulting suspension was stirred at 24 °C for 2 h under N_2. EtOH (50 mL) was added and the resulting agglomerates were purified and filtered with EtOH (3 × 30 mL), and ultimately stored in EtOH for further use.

Alkanes 40; **General Procedure:**[52]
A mixture of an alkene **39** (1.0 mmol) and Ni/Ru/Pt/Au heteroquatermetallic nanoparticles (0.5 mol% in total) in H_2O (0.75 mL) and iPrOH (2.5 mL) was stirred at 100 °C for 24 h under air. Upon completion of the reaction, the soln was cooled to rt, and *o*-xylene (0.25 mmol) and THF (6.0 mL) were added. The resulting soln was analyzed by GC and GC/MS and the yield was determined using *o*-xylene as an internal standard.

1.1.2.5 Conclusions and Future Perspectives

The merits of nanoparticle catalysts over their bulk counterparts in general include an increased surface area to volume ratio, as well as a preponderance of corner and edge atoms that are often vital for effective catalysis. Thanks to recent progress in the field, metal nanoparticles also compare well to classic molecular catalysts in term of both their activity and selectivity. Many of the hydrogenation reactions presented above are textbook examples of this point. Nanoparticles are also more easily synthesized than their molecular counterparts, which tend to rely more on custom-made designer ligands. Additionally, because of the possibilities for immobilization on a range of supports or through the simple use of magnetic particles, nanoparticles are amenable to separation from the reaction medium for reuse. Because nanoscale catalysts have important advantages over both molecular and bulk systems, and the difficulty of recovery can be mitigated through the use of supports, nanocatalysts are currently seen as very useful and cost effective by the synthesis community and industry.

References

[1] Horiuti, J.; Ogden, G.; Polanyi, M., *Trans. Faraday Soc.*, (1934) **30**, 663.
[2] Ramp, F. L.; DeWitt, E. J.; Trapasso, L. E., *J. Org. Chem.*, (1962) **27**, 4368.
[3] Doyle, A. M.; Shaikhutdinov, S. K.; Freund, H.-J., *J. Catal.*, (2004) **223**, 444.
[4] Doyle, A. M.; Shaikhutdinov, S. K.; Freund, H.-J., *Angew. Chem. Int. Ed.*, (2005) **44**, 629.
[5] Doyle, A. M.; Shaikhutdinov, S. K.; Jackson, S. D.; Freund, H.-J., *Angew. Chem. Int. Ed.*, (2003) **42**, 5240.
[6] Bell, A. T., *Science (Washington, D. C.)*, (2003) **299**, 1688.
[7] Wilde, M.; Fukutani, K.; Ludwig, W.; Brandt, B.; Fischer, J.-H.; Schauermann, S.; Freund, H.-J., *Angew. Chem. Int. Ed.*, (2008) **47**, 9289.
[8] Kwon, M. S.; Kim, N.; Park, C. M.; Lee, J. S.; Kang, K. Y.; Park, J., *Org. Lett.*, (2005) **7**, 1077.
[9] Ma, X.; Jiang, T.; Han, B.; Zhang, J.; Miao, S.; Ding, K.; An, G.; Xie, Y.; Zhou, Y.; Zhu, A., *Catal. Commun.*, (2008) **9**, 70.
[10] Erathodiyil, N.; Ooi, S.; Seayad, A. M.; Han, Y.; Lee, S. S.; Ying, J. Y., *Chem.–Eur. J.*, (2008) **14**, 3118.
[11] Pillai, U. R.; Sahle-Demessie, E., *J. Mol. Catal. A: Chem.*, (2004) **222**, 153.
[12] Park, C. M.; Kwon, M. S.; Park, J., *Synthesis*, (2006), 3790.
[13] Nakao, R.; Rhee, H.; Uozumi, Y., *Org. Lett.*, (2005) **7**, 163.
[14] Uozumi, Y.; Nakao, R., *Angew. Chem. Int. Ed.*, (2003) **42**, 194.
[15] Lan, Y.; Zhang, M.; Zhang, W.; Yang, L., *Chem.–Eur. J.*, (2009) **15**, 3670.
[16] Szőri, K.; Puskás, R.; Szőllősi, G.; Bertóti, I.; Szépvölgyi, J.; Bartók, M., *Catal. Lett.*, (2013) **143**, 539.
[17] Ishikawa, S.; Hudson, R.; Moores, A., *Heterocycles*, (2012) **86**, 1023.
[18] Shi, F.; Tse, M. K.; Pohl, M.-M.; Brückner, A.; Zhang, S.; Beller, M., *Angew. Chem. Int. Ed.*, (2007) **46**, 8866.
[19] Ishikawa, S.; Hudson, R.; Masnadi, M.; Bateman, M.; Castonguay, A.; Braidy, N.; Moores, A.; Li, C.-J., *Tetrahedron*, (2014) **70**, 8952.
[20] Hudson, R.; Chazelle, V.; Bateman, M.; Roy, R.; Li, C.-J.; Moores, A., *ACS Sustainable Chem. Eng.*, (2015) **3**, 814.
[21] Abu-Reziq, R.; Alper, H.; Wang, D.; Post, M. L., *J. Am. Chem. Soc.*, (2006) **128**, 5279.
[22] Polshettiwar, V.; Varma, R. S., *Org. Biomol. Chem.*, (2009) **7**, 37.
[23] Hudson, R.; Feng, Y.; Varma, R. S.; Moores, A., *Green Chem.*, (2014) **16**, 4493.
[24] Polshettiwar, V.; Varma, R. S., *Green Chem.*, (2010) **12**, 743.
[25] Hudson, R., *Synlett*, (2013) **24**, 1309.
[26] Hudson, R., *RSC Adv.*, (2016) **6**, 4262.
[27] Hudson, R.; Bishop, A.; Glaisher, S.; Katz, J. L., *J. Chem. Educ.*, (2015) **92**, 1892.
[28] Lu, A.-H.; Salabas, E. L.; Schüth, F., *Angew. Chem. Int. Ed.*, (2007) **46**, 1222.
[29] Shylesh, S.; Schünemann, V.; Thiel, W. R., *Angew. Chem. Int. Ed.*, (2010) **49**, 3428.
[30] Guin, D.; Baruwati, B.; Manorama, S. V., *Org. Lett.*, (2007) **9**, 1419.
[31] Kainz, Q. M.; Linhardt, R.; Grass, R. N.; Vilé, G.; Pérez-Ramírez, J.; Stark, W. J.; Reiser, O., *Adv. Funct. Mater.*, (2014) **24**, 2020.
[32] Nasir Baig, R.; Varma, R. S., *ACS Sustainable Chem. Eng.*, (2014) **2**, 2155.
[33] Hudson, R.; Hamasaka, G.; Osako, T.; Yamada, Y. M. A.; Li, C.-J.; Uozumi, Y.; Moores, A., *Green Chem.*, (2013) **15**, 2141.
[34] Welther, A.; Jacobi von Wangelin, A., *Curr. Org. Chem.*, (2013) **17**, 326.
[35] Phua, P.-H.; Lefort, L.; Boogers, J. A. F.; Tristany, M.; de Vries, J. G., *Chem. Commun. (Cambridge)*, (2009), 3747.
[36] Rangheard, C.; de Julián Fernández, C.; Phua, P.-H.; Hoorn, J.; Lefort, L.; de Vries, J. G., *Dalton Trans.*, (2010) **39**, 8464.
[37] Kelsen, V.; Wendt, B.; Werkmeister, S.; Junge, K.; Beller, M.; Chaudret, B., *Chem. Commun. (Cambridge)*, (2013) **49**, 3416.
[38] Stein, M.; Wieland, J.; Steurer, P.; Tölle, F.; Mülhaupt, R.; Breit, B., *Adv. Synth. Catal.*, (2011) **353**, 523.

[39] Hudson, R.; Riviere, A.; Cirtiu, C. M.; Luska, K. L.; Moores, A., *Chem. Commun. (Cambridge)*, (2012) **48**, 3360.
[40] Kaushik, M.; Feng, Y.; Boyce, N.; Moores, A., In *Nanocatalysis in Ionic Liquids*, Prechtl, M. H. G., Ed.; Wiley-VCH: Weinheim, Germany, (2016); p 3.
[41] Mu, X.-d.; Evans, D. G.; Kou, Y., *Catal. Lett.*, (2004) **97**, 151.
[42] Huang, J.; Jiang, T.; Han, B.; Gao, H.; Chang, Y.; Zhao, G.; Wu, W., *Chem. Commun. (Cambridge)*, (2003), 1654.
[43] Umpierre, A. P.; Machado, G.; Fecher, G. H.; Morais, J.; Dupont, J., *Adv. Synth. Catal.*, (2005) **347**, 1404.
[44] Scheeren, C. W.; Machado, G.; Dupont, J.; Fichtner, P. F. P.; Texeira, S. R., *Inorg. Chem.*, (2003) **42**, 4738.
[45] Huang, J.; Jiang, T.; Gao, H.; Han, B.; Liu, Z.; Wu, W.; Chang, Y.; Zhao, G., *Angew. Chem. Int. Ed.*, (2004) **43**, 1397.
[46] Dupont, J.; Fonseca, G. S.; Umpierre, A. P.; Fichtner, P. F. P.; Teixeira, S. R., *J. Am. Chem. Soc.*, (2002) **124**, 4228.
[47] Gladiali, S.; Alberico, E., *Chem. Soc. Rev.*, (2006) **35**, 226.
[48] Boldrini, G. P.; Savoia, D.; Tagliavini, E.; Trombini, C.; Umani-Ronchi, A., *J. Org. Chem.*, (1985) **50**, 3082.
[49] Dhakshinamoorthy, A.; Pitchumani, K., *Tetrahedron Lett.*, (2008) **49**, 1818.
[50] Alonso, F.; Riente, P.; Yus, M., *Tetrahedron*, (2008) **64**, 1847.
[51] Alonso, F.; Riente, P.; Yus, M., *Tetrahedron*, (2009) **65**, 10637.
[52] Ito, Y.; Ohta, H.; Yamada, Y. M. A.; Enoki, T.; Uozumi, Y., *Chem. Commun. (Cambridge)*, (2014) **50**, 12123.

1.2 Partial Reduction of Polyenes

F. Zaccheria and N. Ravasio

General Introduction

The partial hydrogenations, or semihydrogenations, of one or more C=C bonds in polyenes are important selective transformations used for a series of processes, ranging from petrochemistry via polymers to fine chemicals.

According to Augustine,[1] three main types of selectivity in the hydrogenation of carbon–carbon multiple bonds can be identified: Type I selectivity occurs when two simultaneous reactions take place starting from two reactants in the same mixture; this is the case, for example, in the reduction of propyne in the presence of >90% of propene in the C3-cut produced by fluid catalytic cracker (FCC) units in naphtha-processing plants.[2] Type II selectivity refers to the differentiation between two parallel reactions in which products derive from separate pathways starting from the same material (i.e., limonene to give menthene, or butadiene to give butane), and Type III selectivity is typically found in serial reactions (i.e., alkyne to alkene to alkane). The partial reduction of polyenes can belong either to the Type II or the Type III selectivity group.

Because the two or more alkene functionalities in the substrate are usually very similar in reactivity, saturation of each one has, in principle, the same kinetic order. In most cases, a regioselective transformation can be achieved by varying the reaction conditions, such as hydrogen availability, temperature, or catalyst environment, as well as by carefully choosing and designing the catalytic system in order to avoid oversaturation after the first C=C hydrogenation. Selectivity is often related to electronic interactions between catalyst and substrate, which can be influenced by varying the metal dispersion, particularly in the case of heterogeneous systems.

In many cases, the C=C moieties are not equally substituted and therefore the selective reduction is influenced by the size and the number of substituents on the C=C unit. The general ranking in hydrogenation activity is monosubstituted > 1,1-disubstituted ≈ *Z*-1,2-disubstituted > *E*-1,2-disubstituted > trisubstituted > tetrasubstituted. Strained double bonds are more easily hydrogenated than unstrained ones, and acyclic alkenes are more reactive than endocyclic ones.

In this chapter, selected examples of partial reduction of polyenes are presented, related to the selective hydrogenation of terpenes, cyclic polyenes, and vegetable oils and related materials. These classes of substrates have been especially chosen due to their importance in industrial processes.

1.2.1 Selective Hydrogenation of Polyunsaturated Terpenes

Terpenes represent an important class of molecule due to their availability from natural compounds, and they are widely used as intermediates for different sectors of the chemical industry, ranging from flavors and fragrances to polymers.[3] Some important industrial processes are based on the use of terpenes as starting materials and involve a hydrogenation step. An example of this procedure is the BASF synthesis of (–)-menthol starting from a geraniol/nerol mixture,[4] in turn derived from citral hydrogenation.[5]

The molecular structure of terpenes means that they are very versatile; the presence of C=C bonds and, in some cases, epoxy, hydroxy, or carbonyl groups allows one to plan a

for references see p 123

variety of synthetic routes to other products taking advantage of the natural carbon chain. However, to achieve selective transformations of terpenes is rather challenging, due to chemoselectivity issues as well as the easy isomerization of the double bonds. Several terpenes, cyclic as well as acyclic ones, contain a polyenic structure, and in some cases the partial reduction of C=C bonds is not trivial, particularly in the non-oxygenated hydrocarbon structures.

1.2.1.1 Isoprene Reduction

Starting with the roots of the terpene family, an interesting case is the partial reduction of isoprene (2-methylbuta-1,3-diene, **3**). Isoprene is the smallest possible terpene, occurring naturally in oak, poplar, and eucalyptus. However, because of its great importance in the industrial preparation of polymers, and in particular in the preparation of synthetic rubber, it is produced on a large scale by various condensation processes.[6]

The partial hydrogenation of isoprene, mainly carried out over heterogeneous catalysts, has considerations similar to those encountered in the selective hydrogenation of buta-1,3-diene. It is worth underlining that the latter diene has a significant role in petrochemistry and its hydrogenation has been widely studied using heterogeneous catalysts, both for applicative and mechanistic purposes.[7,8] For conjugated alkadiene hydrogenation reactions, the solid catalysts are mainly classified into type A systems, operating via 1,2-addition, and type B catalysts, promoting the 1,4-addition;[9] these processes occur, respectively, through a bis π-adsorbed species **1** and through a π-allyl intermediate **2** (Scheme 1).

Both noble and non-noble metals have been used in butadiene reduction, with palladium being the most studied. This metal catalyzes the selective reduction of butadiene to butene and falls into the Type B category.[1]

Scheme 1 1,2- and 1,4-Addition in the Hydrogenation of Buta-1,3-diene[9]

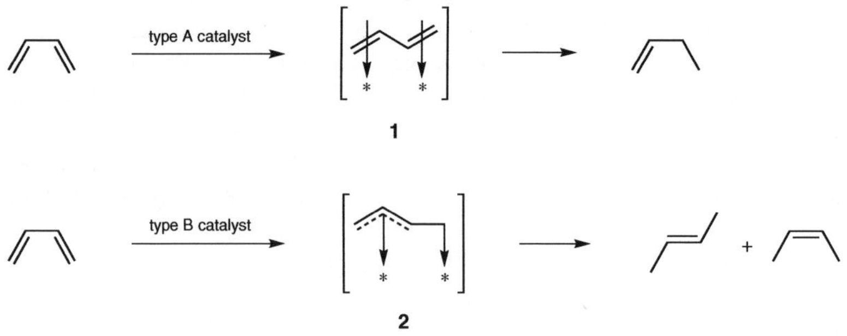

type A catalyst

1

type B catalyst

2

Isoprene partial reduction also represents an appropriate model for petrochemistry, as isopentenes are the desired species in the C5 gasoline fraction used as the feedstock for the *tert*-amyl methyl ether process (TAME),[10] in which it is important to increase the conversion of alkadienes into isopentenes, while limiting the further reaction to isopentane.

In the case of isoprene (**3**), in addition to electronic features affecting the 1,2- versus 1,4-addition, steric factors also play an important role. In this case, not only is the partial hydrogenation an issue, but also the regioselectivity of the reduction. Some catalytic systems have been proposed for this application, again mainly relying on the use of palladium, which generally gives a mixture of products **4**, **5**, and **6** in a ratio of 1:2:1, according to a simple mechanism proposed by Bond and depicted in Scheme 2.[11]

The isoprene molecule **3** is chemisorbed by both π-bonds and then the first H atom adds, with equal probability, to one of the terminal positions, thus forming two different adsorbed π-allylic radicals. The addition of the second H atom with equal probability to either of the terminal points of the delocalized allylic bond leads to the mixture of isopentenes **4–6**, which in turn can be further hydrogenated into isopentane.

Scheme 2 Partial Hydrogenation of Isoprene[11]

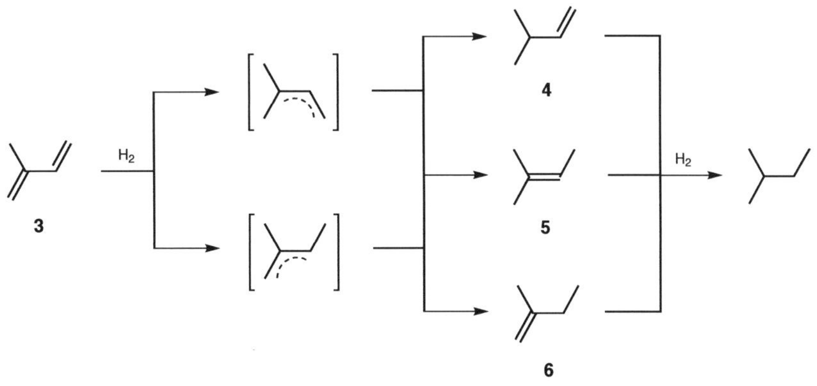

Different approaches and studies have been performed in order to elucidate the main parameters affecting both activity and selectivity obtained with palladium systems, examining catalyst deactivation,[12] metal dispersion,[13] and catalyst preparation procedure.[14]

As a general conclusion, the activity increases by increasing the metal dispersion, while the selectivity to isopentenes decreases. This effect can mainly be attributed on one hand to the increase in the number of active sites of the catalyst for isopentene to isopentane hydrogenation, and on the other to a higher adsorption strength between isopentenes and the catalyst active site. The electron-deficient character of smaller palladium particles actually leads to a decrease in the selectivity for isopentenes, as the desorption rate becomes slower than the hydrogenation rate.[15,16]

The most recent investigations have therefore been aimed at designing catalysts with optimized metal dispersion and/or electronic environment. This strategy has been pursued, for example, in the preparation of highly dispersed palladium particles on zeolitic imidazolate frameworks (ZIFs), which have been used to obtain selectivities of up to 97% at 99% conversion [flow reactor, H_2 (10 atm), 80 °C].[17] The good performance observed can be related to the high dispersion of palladium particles obtained thanks to the anchoring effect of the carboxy groups incorporated into the ligand of ZIF-8. The high activity is not achieved at the expense of selectivity, as the remarkable increase in the electron cloud density of palladium caused by the strong interaction with the carboxy groups allows the decrease in selectivity that is usually caused by the small metal particles to be overcome.

Similar considerations apply to palladium nanoparticles in third-generation poly(propylene imine) (PPI) dendrimers covalently grafted to a silica surface modified with poly(allylamine) (PAA), which are reported to reach selectivities of up to 99% for the monoene [H_2 (30 atm), 70 °C].[18] In this case, the electron-donor properties of the nitrogen-containing ligands not only displace chemisorbed alkene from the palladium surface but also reduce the adsorption activity of the latter, thus preventing coke formation and polymerization reactions, which are known to strongly affect the selectivity and activity of the catalysts.

In some reports, an increase in selectivity of the partial hydrogenation is ascribed mainly to the metal morphology. Supported palladium, platinum, and nickel catalysts obtained after deposition of well-faceted nanoparticles on alumina show high selectivity for the hydrogenation of isoprene due to the major exposition of (111) crystallographic facets.[19] The preparation method, based on the fast reduction of an aqueous solution of the

for references see p 123

metallic precursor in the presence of cetyltrimethylammonium bromide (CTAB) using sodium borohydride under hydrogen, provides metal particles on alumina with a proportion of (111) exposed crystallographic facets above 96%, considered to be responsible for the high monoene selectivity obtained. In particular, for platinum nanoparticles, exposing 96% of (111) facets increases the selectivity for monoene products from 70 to 90% in comparison with a reference catalyst.

The influence of nanoparticle structure and morphology on both activity and selectivity is a widely studied topic, particularly in the case of the selective hydrogenation of buta-1,3-diene.[20,21] Freund and co-workers showed that using the common procedure to normalize rates based on the total number of palladium surface atoms results in a particle-size dependence (see Figure 1, star line). In contrast, considering palladium atoms within incomplete (111) terraces as active sites (and using their number for rate normalization) leads to a reaction rate that is independent of particle size, at least for palladium particles ≥4 nm. Furthermore, the turnover frequencies of palladium particles ≥4 nm were in excellent agreement with the turnover frequency of a Pd(111) single crystal, demonstrating that large, well-faceted palladium particles exhibit the behavior of Pd(111) (Figure 1).

Figure 1 Turnover Frequencies for Buta-1,3-diene Hydrogenation as a Function of Catalytic Palladium Particle Size, Normalized by the Total Number of Palladium Surface Atoms (Star) and by the Total Number of Palladium Surface Atoms within Incomplete (111) Facets (Diamond)[20,21]a

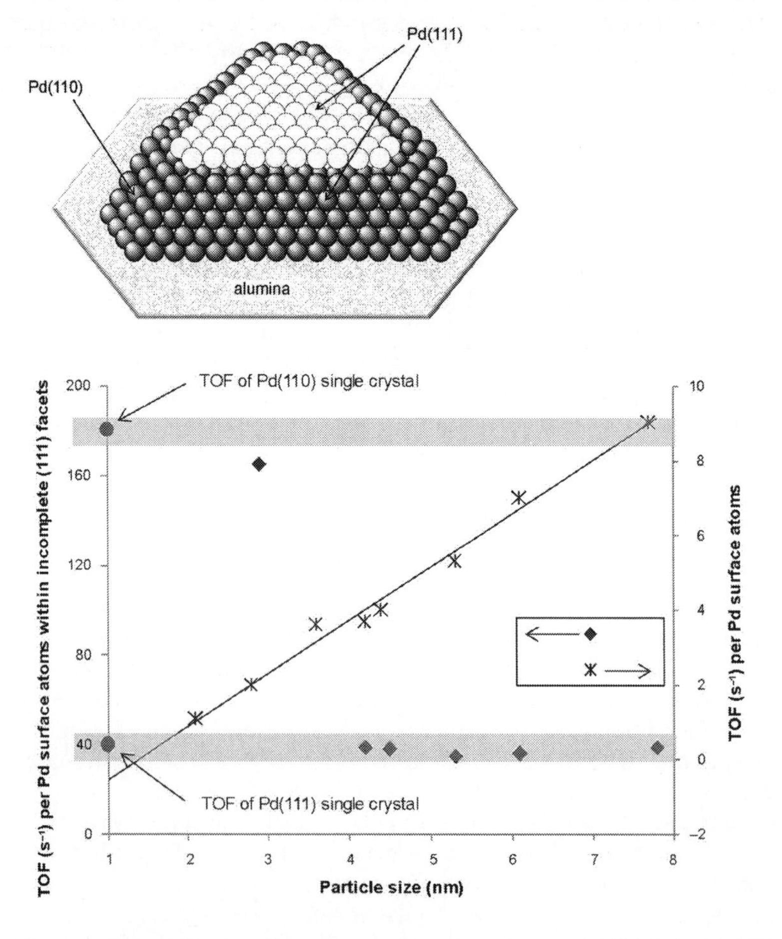

a Reprinted from (Silvestre-Albero; Rupprechter; Freund, *Journal of Catalysis*, Vol. 240), Copyright (2006), p 58 with permission from Elsevier.

Alumina-supported nickel catalysts have also been studied in the selective hydrogenation of isoprene (**3**) under flow conditions.[22] The authors underline the importance of the high dispersion and SMSI (strong metal support interaction) obtained over γ-alumina in order to gain higher activity and stability with respect to the lower dispersed systems obtained over κ-alumina. On the other hand, the weak interaction of nickel with κ-alumina has a positive effect on the formation of coke, thus leading to higher monoene-selectivity with increasing time on stream.

Gold catalysts have also been recently explored, with more attention to charge state issues rather than morphology. The study, carried out using anionic gold on zirconium(IV) oxide (ZrO_2), cationic gold on silica, metallic gold on amine-functionalized silica (N-SiO_2), and anionic gold with hydroxy groups on cerium(IV) oxide (CeO_2), revealed that negatively charged gold on irreducible monoclinic zirconium(IV) oxide favors the selective hydrogenation of isoprene, thus showing that this is a "charge-sensitive reaction". No isopentane formation was observed under the conditions tested.[23]

Isopentenes 4–6; General Procedure Using Pd Particles on Zeolitic Imidazolate Frameworks (ZIFs):[17]
Selective hydrogenation of isoprene (**3**) with Pd/ZIF-8 was carried out in a continuously flowing tubular fixed-bed microreactor of internal diameter 10.0 mm and length 500 mm, using 10.0 wt% isopentene [90% 2-methylbut-2-ene (**5**) and 10% 2-methylbut-1-ene (**6**)] and 5.0 wt% isoprene in heptane as a model compound (this mixture is considered a model for isoprene as it is obtained from the cracker, which is also contaminated with isopentene). 2.0 mL of the catalyst was diluted with quartz particles before being loaded into the reactor. Before the reaction, the catalyst was reduced in high purity H_2 at 423 K for 5 h. The selective hydrogenation reaction of isoprene was carried out with a weight hourly space velocity of 11 h^{-1}, at 353 K, 1.0 MPa and a H_2/oil volume ratio of 4. After a stabilization period of 24 h, the reaction products were collected and analyzed using an Agilent 1790 GC equipped with a flame ionization detector and a PONA capillary column (50 m × 0.2 mm).

1.2.1.2 Limonene Reduction

Two C=C bonds are also present in the structure of limonene (**7**), which has been used as a model compound for partial hydrogenation. In contrast to isoprene (**3**), limonene (**7**) possesses both an endocyclic and an acyclic double bond (Scheme 3).

Scheme 3 Partial Hydrogenation of Limonene

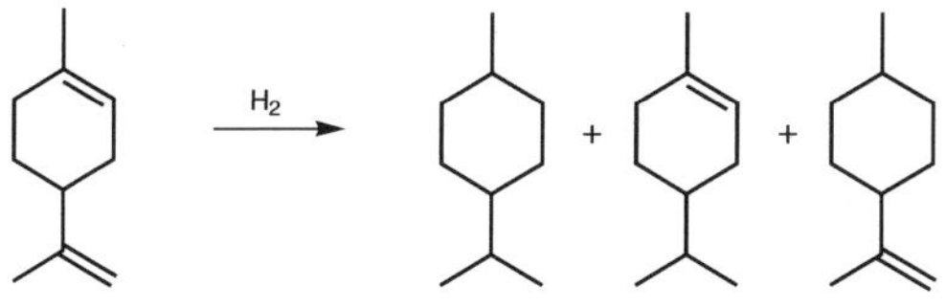

H_2

7 8 9 10

Partial reduction of limonene tends to be problematic due to overreduction, regioselectivity issues, and isomerizations, leading to *p*-menthane (**8**), *p*-menthene (**9**), 1,2-dihydrolimonene (**10**), or isomers of *p*-menthene (**9**). Hydrogenation of the acyclic double bond is favored and is usually achieved with 95% yield through the use of platinum(IV) oxide and hydrogen.[24,25] This reaction was used in the first step of the synthesis of analogues of linckoside B.[24]

Some other approaches to selectively obtain *p*-menthene (**9**) have been recently proposed and investigated using carbon dioxide close to, but below, the critical conditions,

for references see p 123

with platinum/synthetic carbon sorbent (SKN) as catalyst.[26] Under these conditions, the effect of the pressure on the two-phase region in carbon dioxide/hydrogen/liquid reagent mixtures proves to be the essential factor in determining hydrogenation rates and selectivity for *p*-menthene (**9**), which can vary between 50 and 70% by varying the pressure from 16 to 12.5 MPa. Selectivity of up to 70% was obtained by using a ruthenium/silica SBA-15 composite material prepared by reactive deposition of a ruthenium complex [Ru(tmhd)$_2$(cod); tmhd = 2,2,6,6-tetramethyl-3,5-heptanedionato] with hydrogen in supercritical carbon dioxide.[27]

Comparison with a ruthenium catalyst, prepared by impregnation of the same complex on silica SBA-15 in supercritical carbon dioxide, in which the nanoparticles have a smaller size, highlights the same trend already observed in the case of isoprene (Section 1.2.1.1): the higher the metal dispersion, the higher the activity and the lower the selectivity. The ruthenium catalyst obtained by impregnation contains particles with a smaller size, which leads to higher conversion, but almost no selectivity toward menthene.

In contrast, the selective hydrogenation of the endocyclic C=C bond in limonene (**7**) is rarely reported, and is an example of the difficult semihydrogenation of a more highly substituted double bond in the presence of a less-substituted one.

A recent detailed mechanistic study on limonene hydrogenation using Wilkinson's catalyst [RhCl(PPh$_3$)$_3$] and supported Wilkinson's catalysts confirmed that the hydrogenation of the endocyclic double bond is disfavored, with no formation of monoene **10** observed.[28] In addition, the study confirmed that hydrogenation of the acyclic double bond, or alternatively its isomerization, are the favorite pathways.

However, an interesting strategy to selectively obtain product **10**, based on temporary protection of the more-activated alkene, has been proposed (Scheme 4).[29] The strategy involves a three-step/one-pot procedure based on selective hydroboration of the less-substituted C=C bond with 9-borabicyclo[3.3.1]nonane, hydrogenation with palladium on carbon, and oxidation with 2-methyl-2-nitrosopropane. This approach provides an unprecedented yield of alkene **10** of 90%. Moreover, the synthetic scope of this protocol has been widely explored for many other polyenes besides limonene (**7**), proving to be successful and general. This strategy has the advantages of achieving the regioselective hydrogenation of the disfavored alkene using commercial reagents, under no harsh conditions, and without the need for the isolation of intermediates.

Scheme 4 Strategy for the Reduction of More-Substituted C=C Bonds, and Partial Reduction of Limonene[29]

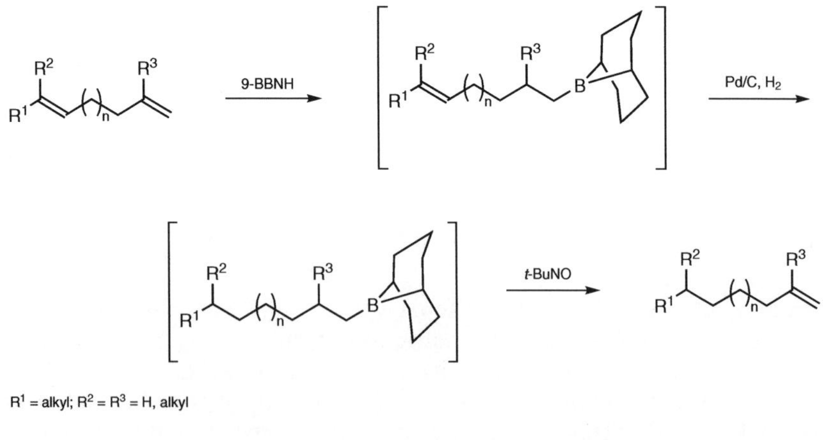

1. 9-BBNH, THF, 0–25 °C
2. Pd/C, H_2
3. *t*-BuNO, 45 min
90%

7 10

1,2-Dihydrolimonene (10):[29]
A 50-mL Schlenk flask attached to a Firestone valve (Ace Glass) was flame-dried and cooled under vacuum (0.01 Torr). The flask was subjected to three purge/evacuation cycles with dry argon and then charged with dry, degassed THF (7 mL). Diene **7** (3.66 mmol) was added, and the soln was cooled to 0 °C. After 10 min, a 0.5 M soln of 9-BBNH in THF (7.33 mL, 3.66 mmol) was added dropwise using a gastight syringe over 30 min. The mixture was stirred at 0 °C for 1 h, and then at 25 °C for 1 h. 10% Pd/C (150 mg, 4 mol% Pd) was then added in one portion. The atmosphere was replaced with H_2 (1 atm) using 7 evacuation/purge cycles, and the resulting black mixture was stirred vigorously. Reaction progress was monitored by ^{1}H NMR via disappearance of the alkenic protons: an NMR sample was prepared by removing an aliquot (100 μL), dilution with CH_2Cl_2, filtration through Celite, and concentration. Following consumption of the substrate alkene, the atmosphere was replaced with argon via seven evacuation/purge cycles. *t*-BuNO dimer (382 mg, 2.19 mmol) was added in one portion, and the resulting soln was stirred for 45 min. The mixture was diluted with Et_2O, filtered through Celite (rinsing with Et_2O), and concentrated. The residue was dissolved in Et_2O (10 mL), cooled to 0 °C, treated with ethanolamine (40 μL, 6.6 mmol), and stirred for 10 min. The resulting white mixture was diluted with pentane, filtered through a plug of silica (rinsing with pentane), and concentrated to give the crude product.

1.2.1.3 Myrcene Reduction

It is even more difficult to obtain high selectivity in the partial hydrogenation of myrcene (**11**) than in the partial hydrogenation of limonene. Myrcene is a monoterpene that occurs naturally in high concentration in hops and parsley, and is produced industrially by pyrolysis of β-pinene.[30] Its availability therefore makes it interesting as a substrate or an intermediate reagent, and it is an important starting material in manufacturing synthetic substitutes for natural aromas ranging from (−)-menthol to Nopol. Myrcene (**11**) contains three C=C bonds, two of them being conjugated, and its hydrogenation can in principle generate a plethora of partially reduced products **12–17**, as well as the fully saturated alkane **18** (Scheme 5).

Sol–gel palladium–silica composites have been proposed for the selective monohydrogenation of myrcene, and give 90–95% selectivity for the mixture of monohydrogenated products at near-complete conversion.[31] In this investigation, excellent results were obtained with catalysts treated at 1100 °C (selectivity up to 97%) before use in comparison with the same one calcined at 300 °C (0% selectivity for monohydrogenated compounds, but 100% for completely saturated myrcene). As a possible explanation for these results, it was suggested that the 300 °C sintered catalysts have a higher surface energy (more kinks or steps) compared to those treated at 1100 °C and they are active in the hydrogenation of any kind of double bond via both alkyl and more stable π-allyl intermediates. On the other hand, with the 1100 °C sintered catalysts, which have a lower surface energy, only the less energetic π-allyl pathway that operates in the hydrogenation of conjugated double bonds is favored. This leads to a higher selectivity for monohydrogenated products. The difference in selectivity is therefore mainly ascribed to morphological features of the metallic supported phase.

for references see p 123

Scheme 5 Partial Hydrogenation of Myrcene[31,32]

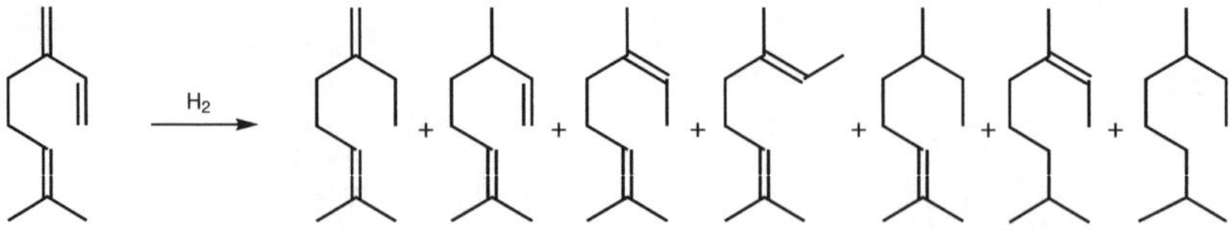

Promising results in terms of selectivity are also reported for a system based on palladium nanoparticles supported over a magnetic material (S_LPd^*).[32] Comparison with a traditional palladium-on-carbon catalyst in the hydrogenation of myrcene (**11**) clearly shows the higher selectivity of the supported nanoparticle catalyst in the partial hydrogenation into monohydrogenated products (86%), compared to the full reduction obtained with the conventional catalyst. This difference in selectivity can be related to the accessibility of the double bonds for surface palladium species. The presence of bulky ligands close to palladium on the surface of the S_LPd^* and $S_{NH2}Pd^*$ (Pd nanoparticles on an amino-functionalized support) materials makes the activation of the disubstituted and trisubstituted double bonds more difficult compared to the more available palladium nanoparticles on Pd/C. As a result, the selectivity for monohydrogenation drastically decreases with the reaction time using the Pd/C catalyst but not using the S_LPd^* and $S_{NH2}Pd^*$ catalysts. Moreover, the proposed catalyst is highly active, with a turnover frequency of 51 000 h^{-1}.

Following earlier research on the hydrogenation of limonene, Bogel-Łukasik and coworkers also reported the reduction of myrcene in high-density carbon dioxide.[33] In this investigation, the effect of the active metals used (palladium, rhodium, and ruthenium; supported on alumina), as well as the solvent properties of supercritical carbon dioxide on the selectivity, were explored. In the absence of steric or morphological effects, deriving from bulky ligands or high dispersion, as described for the two cases mentioned above (using high-temperature calcined palladium/silica[31] or S_LPd^*[32]), palladium on alumina in supercritical carbon dioxide did not offer any selectivity. However, the use of the rhodium and ruthenium catalysts favored the formation of the monounsaturated compound 2,6-dimethyloct-2-ene (**16**), i.e. a doubly hydrogenated product. The effect is mainly ascribed to the different hydrogenation mechanisms associated with palladium and rhodium. In the case of palladium, the hydrogenation mechanism is mainly based on the intermediacy of π-allyl-adsorbed species. In contrast, the mechanism of rhodium-catalyzed hydrogenation of alkenes is comparable to the platinum-catalyzed one;[34] this suggests that the hydrogenation of terpenes catalyzed by rhodium occurs via a 1,2-σ_2 adsorption, as described in the Horiuti–Polanyi mechanism.[35] Unfortunately, none of the catalysts are stable under the reaction conditions used and the metal leaches in noticeable amounts over the course of the reactions.

Homogeneous systems based on chromium, ruthenium, rhodium, and iridium have also been tested in the partial reduction of myrcene (**11**).[36] The catalytic activity strongly depends on the nature of the metal, showing the order Ru < Cr < Ir < Rh. Chemoselectivity toward the monohydrogenated products can be achieved through the appropriate choice of the metal and the reaction conditions. In particular, monohydrogenated products have been obtained with an excellent selectivity of 95–98% at a high conversion of myrcene (>80%) using carbonyl(hydrido)tris(triphenylphosphine)rhodium(I) [$RhH(CO)(PPh_3)_3$] as catalyst, at temperatures lower than 100 °C.

Partially Hydrogenated Myrcene Products 12–17; **General Procedure Using Pd Nanoparticles Supported over a Magnetic Material:**[32]

The catalytic reactions were carried out in a modified Fischer–Porter glass reactor connected to a pressurized H_2 tank. In a typical experiment, the magnetic nanocatalyst

(10 mg) [in some cases prereduced by keeping the powder catalyst under H_2 pressure (6 atm) for 30 min inside the reactor] and the substrate **11** were added under an inert atmosphere. The reactor was attached to the hydrogenation apparatus and purged with H_2, and the pressure was set to 6 atm. The catalyst was recovered magnetically by placing a magnet on the reactor wall. The organic phases were transferred into a glass vessel and analyzed by GC and GC/MS. The catalyst was reused by adding to the reactor a new amount of substrate.

1.2.1.4 Geraniol Reduction

When oxygenated groups are present in the polyene, as in the case of geraniol, other factors strongly influence the reaction and different hydrogenation conditions have been explored.

The selective reduction of geraniol has been studied with both homogeneous and heterogeneous catalysts, the former being mainly explored in the context of enantioselective transformations.

1.2.1.4.1 Homogeneous Catalysts for Geraniol Reduction

The use of rhodium–ChenPhos (**19**) complexes has been reported to give high enantioselectivities for the reduction of geraniol under traditional hydrogenation conditions (Scheme 6).[37] At room temperature and under 25 atm of H_2, 99% of (*S*)-citronellol [(*S*)-**20**] is obtained, albeit after a reaction time of 20 hours. The observed higher activity for hydrogenation of the C=C moiety closer to the hydroxy group is ascribed to the formation of a hydrogen-bonding secondary interaction between the substrate and the catalyst. Thus, the same complex is inactive for the hydrogenation of some similar substrates that lack a hydrogen-bonding donor. Moreover, the appropriate orientation of the dimethylamino group in the complex is crucial to achieve high enantioselectivity.

Scheme 6 Enantioselective Reduction of Geraniol into Citronellol[37]

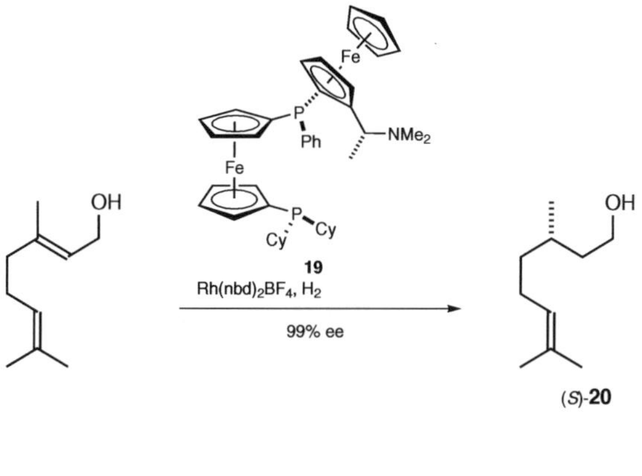

As mentioned in Section 1.2.1, the enantio- and regioselective partial reduction of geraniol into (*R*)- or (*S*)-citronellol is one of the key steps of a new BASF menthol synthesis.[4] The reaction, which produces one of the two citronellol enantiomers with 95–99% ee, is performed at 35–45 °C and 70–120 atm of hydrogen in an alcoholic solvent, preferentially methanol, using mainly homogeneous ruthenium complexes with atropisomeric biaryl bisphosphine ligands. The same process can also be applied to the enantioselective reduction of nerol.[4] The high efficiency of ruthenium complexes with phosphorus-containing biaryl ligands [Ru(II)–BINAP] in the reduction of allylic alcohols, and particularly of geraniol and nerol, was discovered earlier by Noyori.[38]

A 78% yield of (*R*)-citronellol [(*R*)-**26**, R^1 = Me] with 98% ee has also been obtained using dichloro(cycloocta-1,5-diene)ruthenium(II) {$[RuCl_2(cod)]_n$}/(*S*)-4-Tol-BINAP (**22**) as the catalyst in an asymmetric transfer hydrogenation with isopropanol.[39] The use of donor alcohols other than isopropanol, such as pentan-2-ol and cyclohexanol, results in good activity but lower stereoselectivity. The reaction protocol requires the presence of potassium hydroxide in order to promote the deprotonation of the alcohol as the first step. Deuterium incorporation studies suggest that the asymmetric-induction step occurs during the ruthenium-assisted 1,3-hydrogen shift through an enal intermediate **23** to an enolate **24** and enol **25** (Scheme 7). In this protocol, the presence of the hydroxy group is therefore exploited as both hydrogen source and as an auxiliary group for catalyst complexation, and as a consequence controls the regioselectivity.

Scheme 7 Asymmetric Transfer Hydrogenation of Geraniol and Farnesol[39]

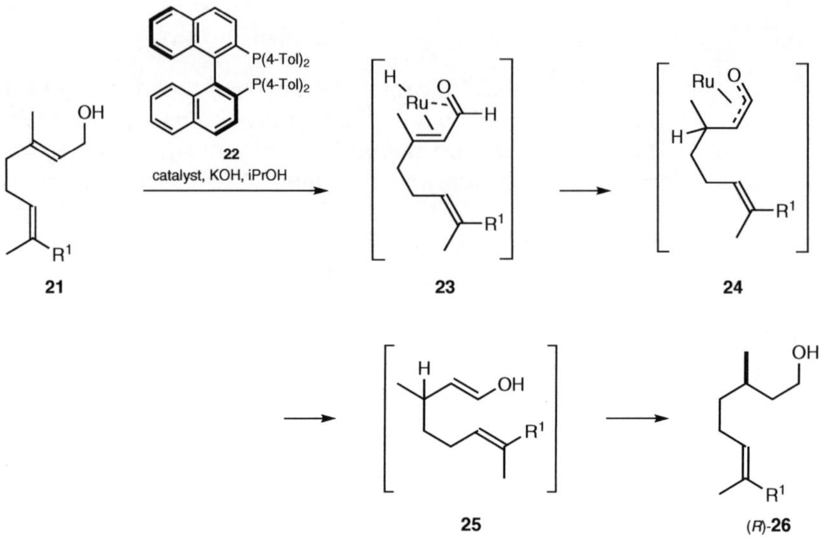

R^1	Catalyst	Temp (°C)	ee (%)	Yield (%)	Ref
Me	$RuCl_2$(cod)	100	98	78	[39]
$(CH_2)_2CH{=}CMe_2$	$RuCl_2$(*p*-cymene)	83	81	65	[39]

As shown in Scheme 7, a similar catalytic system has also been applied to the reduction of farnesol [**21**, $R^1 = (CH_2)_2CH{=}CMe_2$], in which the partial hydrogenation is complicated by the presence of a further double bond. In this case as well, the product is obtained with good regioselectivity, with 65% isolated yield (81% ee) of the product (*R*)-**26** [$R^1 = (CH_2)_2CH{=}CMe_2$] in which the double bond of the allylic alcohol is hydrogenated exclusively.

BASF have also reported a similar menthol process, that starts with the chemoselective asymmetric hydrogenation of neral, one of the isomers of citral, to citronellal. Here, a more expensive rhodium/ChiraPhos catalyst is used, which can be recycled.[40]

(*S*)-Citronellol [(*S*)-20]; Typical Procedure for Enantioselective Hydrogenation Using ChenPhos:[37]
A soln of (R_C,S_{Fc},S_P)-ChenPhos (**19**; 3.28 mg, 0.0021 mmol, as reported) and $Rh(nbd)_2BF_4$ (1.50 mg, 0.002 mmol, as reported) in CH_2Cl_2 (2 mL) was stirred under N_2. After 30 min, the clear yellow soln was transferred into an autoclave, and geraniol (0.2 mmol) was added. The air in the autoclave was replaced with H_2 (3 ×), and then the autoclave was charged with H_2 to 25 atm. The mixture was stirred at rt for 20 h and then the H_2 was carefully released. The mixture was concentrated under reduced pressure and the residue was purified by flash chromatography (silica gel, petroleum ether/EtOAc 4:1). The ee value of the product was determined by chiral HPLC or chiral GC.

(*R*)-Citronellol [(*R*)-26, R^1 = Me]; Typical Procedure for Enantioselective Transfer Hydrogenation:[39]
Under an atmosphere of argon or N_2, a degassed 0.01 M soln of geraniol in iPrOH (10 mL), $\{RuCl_2(cod)\}_n$ (0.001 M), KOH (0.002 M), and chiral ligand **22** (0.002 M) [alternatively, 0.001 M of **22**/$RuCl_2$(*p*-cymene)] was added to a Schlenk flask equipped with a magnetic stirrer bar and covered with a fresh rubber septum. Three freeze (liquid N_2)–pump–thaw cycles were performed and the flask was transferred to an oil bath at 100 °C and stirred in a closed system for 2 h. [With **22**/$RuCl_2$(*p*-cymene), degassing was performed by three pump–fill cycles and the reaction temperature was 83 °C.] The solvent was evaporated under reduced pressure and the product was purified by chromatography (silica gel, pentane/EtOAc).

1.2.1.4.2 Heterogeneous Catalysts for Geraniol Reduction

The use of isopropanol as reaction medium has been exploited as a strategy in order to tune the selectivity of heterogeneous catalysts. A low loading of copper on alumina, prepared by a nonconventional chemisorption–hydrolysis technique, has been used for the regioselective hydrogenation of geraniol into citronellol (**20**) in 98% yield, using isopropanol as the solvent under 1 atmosphere of hydrogen.[41] The main role of isopropanol is to suppress the acidic sites of the catalytic system and thus avoid dehydration reactions, which typically occur with other solid systems used for this application (and therefore require the use of basic additives).[42] The process can be also applied to a natural essential oil mixture such as palmarosa oil, comprised of 81% geraniol, thereby affording a 75% yield of citronellol.

The use of copper on alumina in combination with isopropanol under 1 atmosphere of hydrogen also provides 82% of the product **26** [$R^1 = (CH_2)_2CH{=}CMe_2$], monohydrogenated at the allylic alcohol C=C bond, when reducing farnesol [**21**, $R^1 = (CH_2)_2CH{=}CMe_2$].[41]

The use of copper on alumina for geraniol reduction is an interesting example of the versatility of terpenes in organic synthesis. Thus, this catalytic system allows the selectivity of the process to be tuned by varying the reaction conditions, namely the choice of hydrogen/nitrogen atmosphere, the solvent, and the catalyst activation temperature. As mentioned above, the use of an oxygenated solvent such as isopropanol under hydrogen provides citronellol (**20**) as the main product (Scheme 8). On the other hand, by using heptane as solvent, under nitrogen and with the same catalyst activated at higher temperature (270 °C), it is possible to obtain isopulegol (**27**) in 59% yield as the main product by means of a bifunctional process involving isomerization of the allylic alcohol to citronellal and a subsequent acid-catalyzed ene reaction, which is otherwise suppressed by using isopropanol as solvent. If the reaction is carried out under hydrogen, further hydrogenation of isopulegol (**27**) into menthol (**28**) takes place. The latter product is obtained in 41% yield from geraniol through this unprecedented direct triple cascade transformation.

for references see p 123

Scheme 8 Selective Transformations of Geraniol over Copper Catalysts[41]

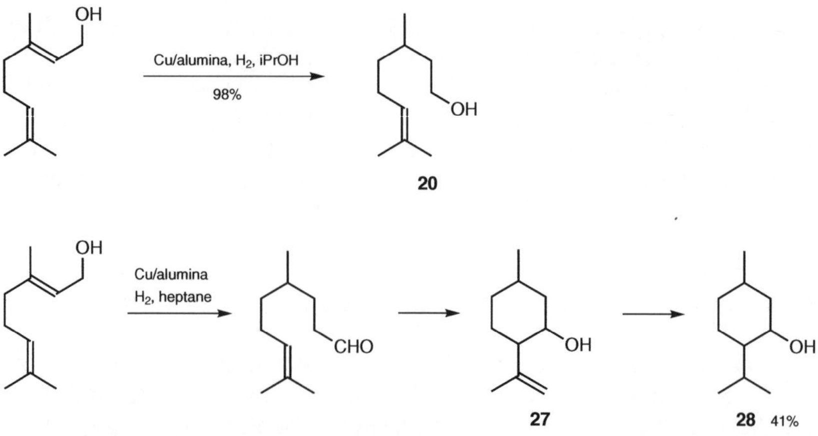

OH — Cu/alumina, H_2, iPrOH, 98% → OH **20**

OH — Cu/alumina, H_2, heptane → CHO → OH **27** → OH **28** 41%

The same auxiliary effect invoked in the hydrogenation with homogeneous complexes has been exploited and studied in the case of solid catalysts such as palladium(II) complexes supported on MCM-41.[43] In this study, a higher reaction rate obtained in the hydrogenation of hydroxy-substituted dienes such as geraniol or linalool compared to limonene is clearly evident when using palladium complexes supported over MCM-41. This support possesses surface hydroxy groups that can readily interact with the hydroxy groups of geraniol and linalool, thus resulting in higher activity and selectivity compared to similar catalysts prepared by partially capping the surface hydroxy groups with bulky, hydrophobic trimethylsilyl functionalities.

In the case of terpenes, the main strategies toward a partial hydrogenation are therefore based on the design of catalysts with controlled activity. This has been pursued by reducing the amount of the active species and the metal dispersion, or by increasing the steric hindrance of the metal ligands. Thus, as a general trend, activity increases with metal dispersion, thereby lowering chemoselectivity. A suitable trade-off can be obtained by increasing the metal particle size or by limiting the access to the catalytic site by means of bulky molecules in the active-site environment. When reducing substrates containing hydroxy groups (e.g., geraniol or farnesol), it is, on the other hand, possible to take advantage of this functionality by exploiting its interaction with the catalyst in order to control the regioselectivity of the process.

1.2.2 Partial Hydrogenation of Cyclic Polyenes

The selective hydrogenation of aliphatic polyenes is an important synthetic transformation for the preparation of intermediates for different sectors of the chemical industry, and, in particular, for the preparation of monomers.

Among these substrates, cyclododeca-1,5,9-triene and cycloocta-1,5-diene are of great importance. The corresponding monoenes are extensively used as intermediates in the synthesis of dicarboxylic aliphatic acids, ketones, cyclic alcohols, and lactones, as well as for other purposes.[44] Moreover, the selective hydrogenation of cyclododeca-1,5,9-triene to cyclododecene is industrially important in the synthesis of valuable organic and polymeric intermediates, such as 12-laurolactam and dodecanedioic acid, which are important monomers for the synthesis of nylon 12, nylon-6,12, co-polyamides, and polyesters and for applications in coatings.[45]

1.2.2.1 Cyclooctadiene Reduction

Both homogeneous and heterogeneous catalysts have been tested for this type of substrate, though the latter are much more studied. Mainly palladium-based catalysts have been explored for this application. In Schemes 9 and 10, the most recent systems proposed for the selective hydrogenation of cycloocta-1,5-diene into cyclooctene (**29**) using both molecular hydrogen and hydrogen-transfer conditions are summarized.

As far as homogeneous catalysts are concerned, good results in terms of yield have been obtained using palladium complexes as catalysts for hydrogenation[46] or hydrogen transfer.[47] In the latter case, reaction of a palladium(0)–N-heterocyclic carbene complex with formic acid allows the isolation of a unique formato(hydrido)palladium complex, which is involved in the dehydrogenation of formic acid. The hydrogen thus formed in situ can be used for the chemoselective hydrogenation of polyenes, providing an 82% yield of cyclooctene (**29**) from cycloocta-1,5-diene after 24 hours.[47] Similar complexes provide up to 93% yield in the partial reduction of cycloocta-1,5-diene under a hydrogen atmosphere (Scheme 9).[46]

Scheme 9 Partial Reduction of Cycloocta-1,5-diene Using Homogeneous Catalytic Systems[46,47]

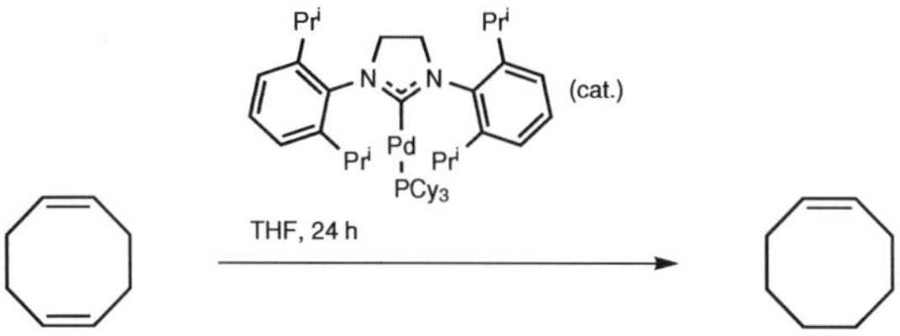

mol% of Catalyst	Conditions	Yield (%)	Turnover Frequency (h^{-1})	Ref
0.25	H_2 (1 bar), rt	93	16	[46]
2	HCO_2H, 60 °C	82	2	[47]

In spite of the good yields obtained, palladium-based homogeneous complexes suffer from quite low activity, as shown by the turnover frequency values. In this respect, heterogeneous systems perform much better (Scheme 10). High yields of up to 95% have been achieved using supported palladium nanoparticles under mild conditions [H_2 (2–10 atm), 40–70 °C].[48] A detailed kinetic study shows that activity increases with increasing reaction temperature and hydrogen pressure, and decreasing cyclooctadiene concentration. On the other hand, selectivity for cyclooctene (**29**) decreases slightly with increasing temperature and hydrogen pressure, but does not seem to be affected by cyclooctadiene concentration.

Palladium nanoparticles (PdNPs) generated on gel-type ion-exchange resins have also been used with good performance in terms of conversion (98.5%), selectivity (96.7%), and turnover frequency (418 h^{-1}).[49] These good results, obtained in batch mode, can be transferred to flow conditions, demonstrating the high durability and efficiency of the catalytic system. Flow conditions were also used by designing microreactors involving palladium nanoparticles supported on a pore-flow-through silica monolith with hierarchical porosity (Pd-MonoSil-3). Here, the catalyst robustness was remarkable, revealing stable activity for up to 70 hours.[50]

for references see p 123

Use of a noble metal catalyst based on platinum/carbon nanotubes (CNTs) was less satisfactory for this application, as it was impossible to avoid complete reduction to cyclooctane.[51] Supported rhodium nanoparticles (RhNPs) were also less efficient than the corresponding palladium catalyst.[52]

Some interesting examples of the use of non-noble metal based systems have also been reported. In situ generated nickel(0) nanoparticles (NiNPs), from the system nickel(II) chloride/lithium/4,4′-di-*tert*-butylbiphenyl/ethanol (which also generates molecular hydrogen in situ), have been employed in a new, mild, and simple methodology for the highly stereoselective *syn*-semihydrogenation of internal alkynes, semihydrogenation of terminal alkynes, reduction of dienes to alkenes, and reduction of alkynes and alkenes to alkanes.[53]

Copper catalysts (Cu/TiO_2), prepared by a nonconventional chemisorption–hydrolysis technique that results in a highly dispersed metallic phase, reveal outstanding activity in the partial hydrogenation of cycloocta-1,3-diene.[54] Comparison with the corresponding catalysts prepared by a traditional incipient-wetness technique emphasizes the better performance obtained on the basis of the high dispersion of the metallic phase and the well-formed metal crystallites, with a large fraction of step sites exposed that are expected to be very active in hydrogen dissociation. Moreover, among the catalysts prepared by the chemisorption–hydrolysis technique, a sample with a 4% copper loading leads to turnover frequency values higher than those obtained with an 8% copper loaded catalyst (530 h^{-1} vs 360 h^{-1}). FT-IR spectra of adsorbed carbon monoxide show that the 4% catalyst consists of small metal particles with mainly (111) microfacets and a high proportion of step and borderline sites, whereas the 8% catalyst in addition to step, borderline, and (111) sites, also contains (100) microfacets. Therefore, on the latter sample a lower fraction of (111) surface sites is present, thus resulting in a lower activity. The ability to obtain a highly active copper catalyst by virtue of the preparation method therefore allows one to combine this high activity with the peculiar selectivity of copper in reducing polyenes, which leaves isolated C=C bonds untouched. Thus, turnover frequencies comparable to those reported for noble metal based catalysts are obtained, as well as complete selectivity for the cyclooctene product.

Scheme 10 Partial Reduction of Cycloocta-1,5-diene Using Heterogeneous Catalytic Systems[45,48–55]

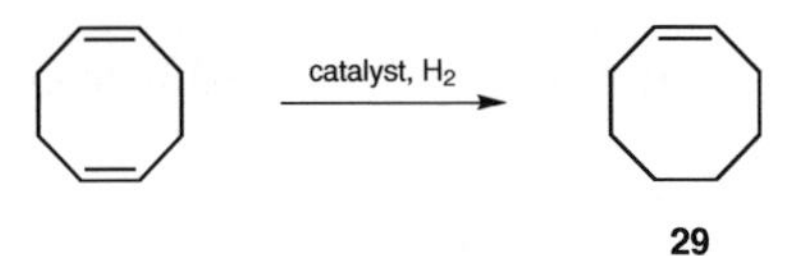

Catalyst	H_2 (atm)	Conversion[a] (%)	Selectivity[a] (%)	Yield[a] (%)	Turnover Frequency[a] (h^{-1})	Ref
Cu/TiO_2[b]	3	n.r.	100	n.r.	530	[54]
Pd/α-alumina	2–10	100	95	95	n.r.	[48]
PdNPs/polymer	1	98	97	95	418	[49]
PdNPs/polymer	2.5	87[c]	97[c]	84	395[c]	[49]
Pd-MonoSil-3[d]	2.6	95[c]	90[c]	86[c]	n.r.	[50]
Pt/CNTs	1	85	78	66	n.r.	[51]

Catalyst	H_2 (atm)	Conversion[a] (%)	Selectivity[a] (%)	Yield[a] (%)	Turnover Frequency[a] (h^{-1})	Ref
RhNPs/resin	1	91	64	58	259	[52]
NiNPs	–[e]	n.r.	n.r.	95	n.r.	[53]
Ru_6Sn/MCM-41	30	72	87	63	4090	[45,55]

[a] n.r. = not reported.
[b] Cycloocta-1,3-diene was used as the substrate.
[c] Under flow conditions.
[d] MonoSil = silica monoliths.
[e] No external H_2 was applied; H_2 is generated in situ from EtOH under the reaction conditions.

Bimetallic catalysts have also been designed that have both good activity and selectivity in this transformation. In particular, the Ru_6Sn/MCM-41 system leads to lower yields and needs a higher hydrogen pressure (30 atm), but operates under solvent-free conditions and results in high turnover frequencies (up to 4090 h^{-1} after 24 h).[45,55] The outstanding activity is ascribed to the single site heterogeneous catalysts (SSHC) obtained by anchoring homogeneous clusters on the silica matrix. The formation of isolated sites, which was convincingly demonstrated by extended X-ray absorption fine structure (EXAFS) studies, allows the optimization of the metal activity and additionally avoids complications arising from site interactions. Comparison with the corresponding monometallic systems, as well as with other bimetallic materials (i.e., Pd–Ru), underlines the good performance obtained with Ru–Sn in terms of both activity and selectivity. Fine tuning of temperature and contact times is critical in the optimization of selectivity for the partial hydrogenation product.

A study of a AuPd/CeO_2 bimetallic catalyst provides interesting insights into the effects the electronic and morphological features of the metal phase have on the selectivity of the hydrogenation.[56] The authors propose that the presence of easily reducible palladium particles is mandatory to obtain high activity. Indeed, by comparing Pd/CeO_2 with Pd/TiO_2, it is evident that the former does not show any catalytic activity if not previously reduced by molecular hydrogen. Thus, due to the high potential of cerium(IV) oxide to stabilize cationic species, no palladium(0) particles are formed under the reaction conditions. The addition of gold to these catalysts results in higher conversion and selectivity than observed with the corresponding monometallic catalysts, both Pd/TiO_2 as well as Pd/CeO_2, due to the promotional effect of gold in generating isolated palladium sites. This confirms the beneficial effect of well-dispersed sites, as was already observed with the Ru–Sn catalysts.

Cyclooctene (29); Typical Procedure Using Polymer-Supported Palladium Nanoparticles:[49]

The resin-supported Pd(II) catalyst (50 mg, ca. 1.25% Pd w/w, ca. 0.006 mmol of Pd) was added into a 100-mL flask containing a degassed soln of cycloocta-1,5-diene (2.08 mmol) in MeOH (12 mL). A flow of H_2 (15 mL•min^{-1}) at 1 bar was bubbled through the soln at rt, taking this as the start time of the reaction. The Pd(II)/resin became slowly black (ca. 20 min). After the desired time, the MeOH soln was completely removed under a stream of H_2 using a gastight syringe. A sample of this soln (0.5 μL) was used for GC (product yield), GC/MS (product identification), and ICP-OES analysis (Pd leaching). For recycling experiments, a fresh soln of the substrate (2.08 mmol) in MeOH (12 mL) was then transferred under H_2 via a gastight syringe into the flask containing the catalyst.

1.2.2.2 Cyclododecatriene Reduction

Although less studied than cycloocta-1,5-diene reduction, the selective hydrogenation of cyclododeca-1,5,9-triene leads to key substrates for the chemical industry. Cyclododecene (**30**), obtained from the partial reduction of cyclododeca-1,5,9-triene, is a chemical intermediate for the production of polymers, fragrances, and perfumes, as it is the most important starting material for cyclic and linear C_{12} compounds. In particular, its oxidation to cyclododecanone is the first step in the preparation of laurinlactam, which in turn is the precursor to nylon-12.

The partial hydrogenation of cyclododeca-1,5,9-triene is more complicated than that of cycloocta-1,5-diene because of the presence of three C=C bonds, and also because of the higher possibility of isomerization during the hydrogenation process. In addition, in this process three cyclododecatriene isomers (plus the reactant), seven cyclododecadiene isomers, two cyclododecene isomers (*E* and *Z*), and cyclododecane can be formed.[57] In order to obtain high selectivities in the hydrogenation of cyclododecatriene, two main approaches have been pursued. The first one is based on process design and optimization and the second one relies mainly on optimization of the catalyst.

The protocols proposed for cyclododeca-1,5,9-triene partial hydrogenation almost all rely on the use of solid palladium catalysts. After all, palladium is known for its activity and selectivity in semihydrogenation reactions.[7] In particular, the dependence of activity and selectivity on dispersion and metal structure has been extensively studied. Activity in polyene hydrogenation generally decreases with increasing metal dispersion. The strong structure sensitivity observed in the hydrogenation of conjugated dienes is mainly ascribed to the electronic structure of small palladium particles. The diene strongly chemisorbs on the small, electron-deficient clusters, thus leading to self-poisoning. A similar effect, although less pronounced, likely takes place also with unconjugated polyenes. On the other hand, the same strong interaction is the origin of the selectivity observed with palladium systems, as the desorption of the half-hydrogenated product is relatively fast, and takes place before further hydrogenation.

This behavior became apparent during a study of the use of palladium on carbon in the hydrogenation of (1*Z*,5*E*,9*E*)-cyclododeca-1,5,9-triene (Scheme 11).[57] A series of catalysts prepared over three carbon supports (L3SA, L3SB, and L3SC) with different surface areas were tested. The supported metal materials each had a different palladium dispersion and particle size as measured, respectively, by carbon monoxide chemisorptions and transmission electron microscopy (TEM). The influence of the different structures confirmed what was previously reported, namely that the initial rate is increased by increasing the dispersion, whereas the turnover frequency goes through a maximum for medium-sized metal particles (Figure 2). Moreover, an increase in selectivity for cyclododecene (**30**) is observed with larger palladium particles (from a maximum of 30% to a maximum of 45% of product versus cyclododecatriene conversion). These effects can be considered as an intermediate behavior between those observed in the hydrogenation of alkenes and conjugated alkadienes, maybe related to the stronger adsorption of cyclododecatriene and cyclododecadiene on palladium compared to cyclododecene.

Scheme 11 Partial Hydrogenation of (1*Z*,5*E*,9*E*)-Cyclododeca-1,5,9-triene[57]

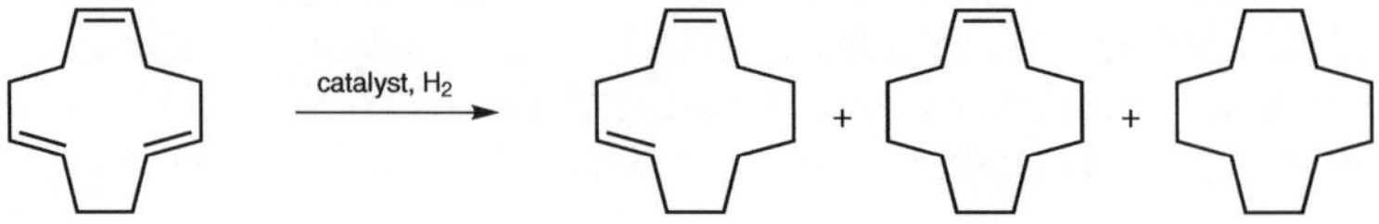

catalyst, H_2 → + +

30

Catalyst	CO/Pd (mol/mol)	Particle Size (nm)	Initial Rate ($\times 10^3$ mol•s^{-1}•g_{Pd}^{-1})	Turnover Frequency (s^{-1})	Ref
Pd/L3SA	0.34	5.3	21	6.5	[57]
Pd/L3SB	0.24	–	19	8.5	[57]
Pd/L3SC	0.13	10.1	9	7.5	[57]
Pd/L3SA-ox[a]	0.61	1.5	24	4.0	[57]

[a] L3SA modified by oxidizing treatment with NaOCl.

Figure 2 Influence of Pd/C Dispersion on Activity in (1*Z*,5*E*,9*E*)-Cyclododeca-1,5,9-triene Reduction[57]

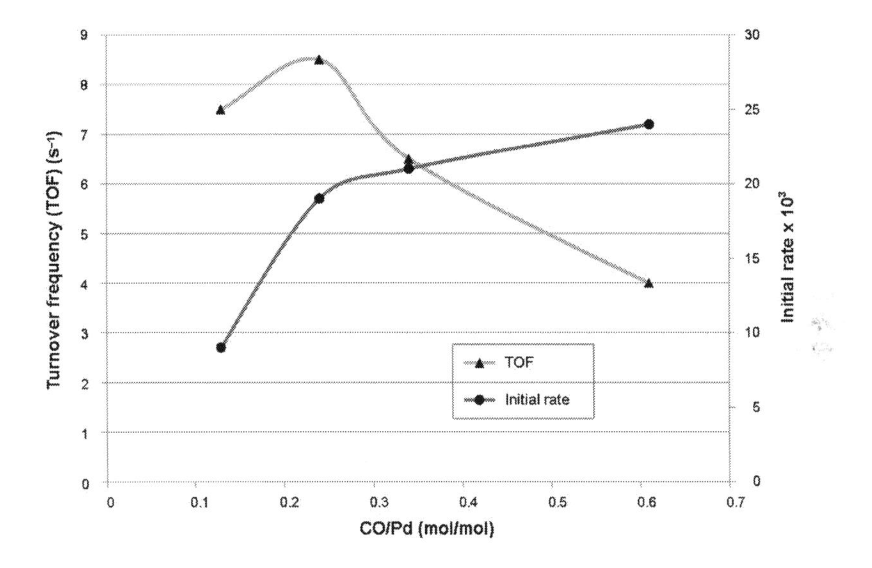

Another parameter strongly influencing both activity and selectivity is the hydrogen pressure: the lower the pressure, the higher the cyclododecene content. McAllister reported a cyclododecene yield of 90% by applying a protocol in batch liquid phase with a powdered catalyst of 5% palladium on charcoal at a reduced hydrogen pressure of between 2 and 0.25 atm.[58]

The liquid-phase hydrogenation of cyclododeca-1,5,9-triene using a palladium-on-alumina catalyst of the egg-shell type has been also studied in discontinuous operation at different hydrogen pressures and using a stepwise reduced hydrogen pressure. The yield of cyclododecene is considerably increased if the hydrogen concentration, and consequently the rate of the hydrogenation, is reduced. In this manner a cyclododecene yield of 93% at a (cyclododecatriene + cyclododecadiene) conversion of 98% could be reached.[59] The high yield obtained under reduced pressure was ascribed to the competitive rates of cyclododecadiene isomerization and hydrogenation.

A comparison of these results with those obtained in the hydrogenation of cyclooctadiene reveals the crucial importance of the adsorption of double bonds on the catalyst surface. Thus, calculation of the minimum-energy conformation of cyclooctadiene shows that the double bonds adopt a parallel orientation, and both can easily and simultaneously bind to the palladium surface. This affords a much stronger bonding of cyclooctadiene compared to cyclooctene, and consequently a relatively easy displacement of cyclooctene. On the other hand, cyclododecatriene has no parallel-oriented double bonds, and there-

for references see p 123

fore no preferential adsorption with respect to cyclododecene takes place (Figure 3). The same is true for the intermediate cyclododecadiene isomers except for (Z,Z)-cyclododeca-1,5-diene, which has a boat form and a nearly parallel orientation of both double bonds so that its strong adsorption, and consequently the displacement of cyclododecene, is possible.

Figure 3 Conformations of Cycloocta-1,5-diene and (1*Z*,5*E*,9*E*)-Cyclododeca-1,5,9-triene

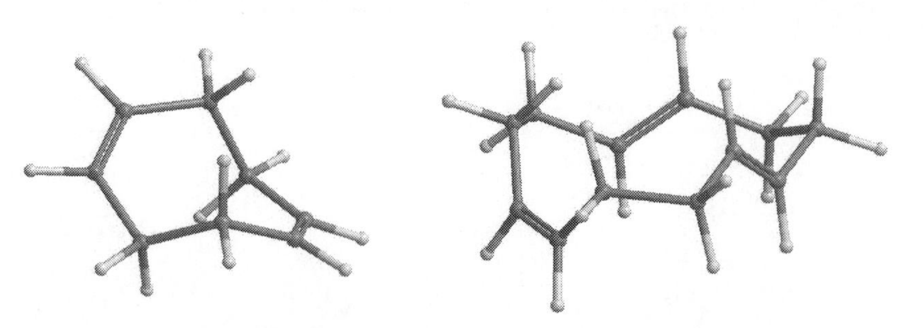

When working at reduced pressure, the ratio between isomerization to (Z,Z)-cyclododeca-1,5-diene and hydrogenation is high enough to reach a dense coverage of the catalyst surface with cyclododecadiene with consequent maximum displacement of cyclododecene. Comparison of experiments over a wide range of hydrogen pressures shows an increasing cyclododecene yield with decreasing pressure of hydrogen. At very low pressure of hydrogen, cyclododecene yields of >90% can be reached.

The implication of this study is that at high pressures the palladium surface is not completely covered with cyclododecatriene and cyclododecadiene isomers; cyclododecene can therefore access the surface and be hydrogenated to cyclododecane, thus lowering the cyclododecene yield. At low pressure, a regime of dense coverage of the surface by cyclododecatriene/cyclododecadiene takes place, so that readsorption and subsequent hydrogenation of formed cyclododecene is to a large extent prevented. The preference of cyclododecatriene/cyclododecadiene adsorption over cyclododecene is explained by adsorption of cyclododecatriene/cyclododecadiene via two double bonds.[60] On the basis of this strategy, several studies have been carried out with variations in the reaction parameters, and particularly which carry out the process under flow conditions.[61]

Bimetallic systems that are active and selective for cycloocta-1,5-diene have also been tested in the selective hydrogenation of cyclododeca-1,5,9-triene.[45,55] Turnover frequencies of 3280 h^{-1} were obtained after 8 hours with a conversion of 28.2% and a selectivity for cyclododecene of 87%; after 24 hours, selectivity is preserved with a conversion of 41.7% (with a turnover frequency value of 1785 h^{-1}).

Cyclododecene (30):[57]

The liquid-phase hydrogenation of (1*Z*,5*E*,9*E*)-cyclododeca-1,5,9-triene was carried out in a 100-mL sealed stainless-steel autoclave (Engineers Autoclave) equipped with a mechanical stirrer, a sample port, a gas inlet, and a vent. An aliquot of the prereduced catalyst (50 mg) was reactivated in situ in decane/dodecane (3:1; 40 mL) at 80 °C and 4 bar H_2 pressure for 2 h under stirring (1200 rpm). The reactant cyclododeca-1,5,9-triene (30 mL) was then pushed into the reactor by hydrogen flow. The H_2 pressure in the reactor was maintained at 4 bar by regulation of the H_2 feed. Samples were withdrawn periodically and analyzed by GC (Varian GC-3900) using an HP-FFAP capillary column (25 m × 0.32 mm, 0.52-mm film thickness). Dodecane was used as internal standard.

1.2.3 Selective Hydrogenation of Vegetable Oils and Related Compounds

The hydrogenation of vegetable oils is one of the oldest catalytic processes carried out on an industrial scale. The aim of this reaction in the past was the hardening of oil to produce margarines, or improving its oxidative stability to prevent the oil from becoming rancid. We are now moving toward a bio-based economy, and in this scenario vegetable oils are the only renewable source of long carbon-atom chains; therefore, they are valuable as raw materials for the synthesis of fuels, lubricants, surfactants, and many other chemicals. However, if an oil is to be used as a raw material for the chemical industry, it has to satisfy two prerequisites: long-term storage stability and constant composition. For this reason, partial hydrogenation of vegetable oils has become an important reaction and crucial for the development of a sustainable economy (Scheme 12). Oxidative stability, which determines the storage stability, relies on the absence of polyunsaturation in the oil.

Scheme 12 Partial Hydrogenation of Triglycerides or Methyl Esters of Polyunsaturated Fatty Acids into Triolein or Methyl Oleate

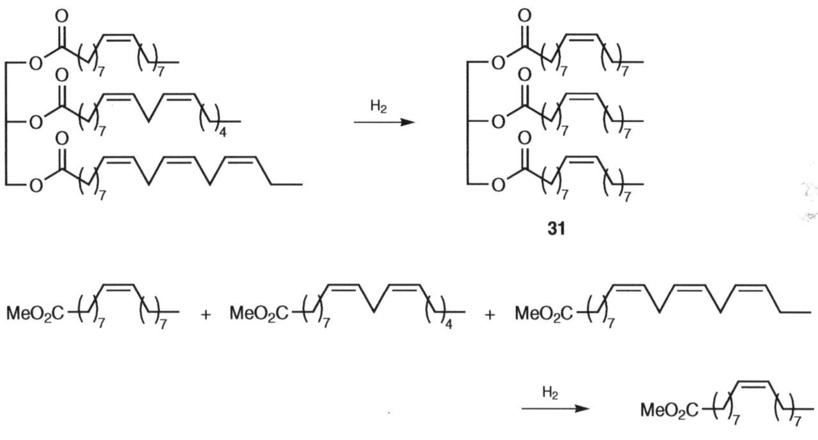

A vegetable oil is a triglyceride, a triester of glycerol with three fatty acids. Figure 4 shows the structure of triolein (**31**), a triglyceride in which all three acids are oleic acid. The most common fatty acids in vegetable oils are linolenic, linoleic, oleic, and stearic acids, which have a C_{18} chain and, respectively, 3, 2, 1, or 0 C=C bonds (i.e., C18:3, C18:2, C18:1, and C18:0 fatty acids). Their relative oxidation rates are reported in Table 1.[62]

Figure 4 Structure of Triolein

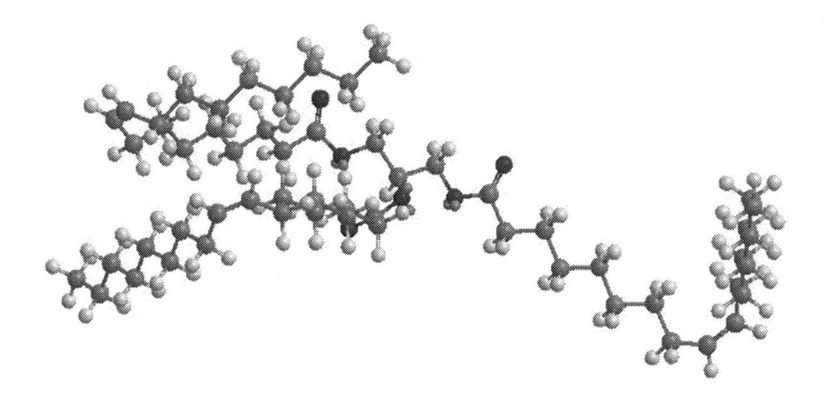

for references see p 123

Table 1 Relative Oxidation Rates and Melting Points of C_{18} Natural Fatty Acids[62]

Name	Structure	Relative Oxidation Rate	mp (°C)	Ref
stearic acid	$HO_2C(CH_2)_{16}CH_3$	0	70	[62]
oleic acid	$HO_2C(CH_2)_7CH{=}CH(CH_2)_7CH_3$	1	16	[62]
linoleic acid	$HO_2C(CH_2)_7CH{=}CHCH_2CH{=}CH(CH_2)_4CH_3$	40	−7	[62]
linolenic acid	$HO_2C(CH_2)_7CH{=}CHCH_2CH{=}CHCH_2CH{=}CHCH_2CH_3$	100	−13	[62]

From this data it is apparent that reducing the number of double bonds greatly improves the oxidative stability of the oil. However, overreduction to saturated compounds should be avoided because as unsaturation is removed, the melting point of the fatty acid increases and therefore the low-temperature fluidity of the oil, essential when dealing with the formulation of fuels or lubricants, decreases. In addition, *Z*/*E* and positional isomerization should be avoided, as both phenomena lead to an increase of the melting point, as shown in Table 2.[63] All these issues are hard to control in the hydrogenation with nickel-based catalysts, the most widely used in the fat and oils industry.

Table 2 Melting Points of Positional and Geometric Isomers of Octadecenoic Acids[63]

C18:1 Isomer	Config	mp (°C)	Ref
$HO_2C(CH_2)_4CH{=}CH(CH_2)_9CH_3$	*E*	53	[63]
	Z	28	[63]
$HO_2C(CH_2)_7CH{=}CH(CH_2)_6CH_3$	*E*	45	[63]
	Z	16	[63]
$HO_2C(CH_2)_{10}CH{=}CH(CH_2)_3CH_3$	*E*	52	[63]
	Z	27	[63]
$HO_2C(CH_2)_{13}CH{=}CHCH_3$	*E*	58	[63]
	Z	40	[63]

The formation of *E*-isomers is also an issue in the manufacture of margarines and spreadables. In this case, hardening of the oil is in fact the goal of the hydrogenation process, but the consumption of foods with a high content of trans fatty acids has been shown to raise low density lipoprotein cholesterol levels and to lower high density lipoprotein cholesterol levels, thus increasing the risk of developing coronary heart disease. For this reason, a specific legislative frame has been approved in both the European Union and the United States, although up to now a limit for trans fats has only been set for a specific category of food (infant and toddler formulas).[64]

Both these aspects, namely the search for low *trans* hydrogenated fats to be used in food formulations and the need to improve oxidation stability without threatening cold flow properties for oils that go into industrial applications, have fostered the search for innovative catalytic systems to be used in the partial hydrogenation of vegetable oils. The goal is therefore to maximize the content of C18:1 derivatives. Most of the catalytic systems studied rely on the use of noble metal based catalysts.

1.2.3.1 Two-Phase Systems

The group of Papadogianakis has developed a useful system for the partial reduction of vegetable oil that is based on aqueous-phase organometallic catalysis.[65] The use of aqueous media not only facilitates recovery and recycling of the catalyst while promoting unusual catalytic reactivity, but, in addition, the large heat capacity of water makes it an excellent medium to carry out exothermic reactions such as hydrogenation in a safer way.

In particular, hydrogenation of linseed oil methyl esters gives a remarkable 79.8% of C18:1 product in the presence of water-soluble rhodium/tris(3-sulfonatophenyl)phosphine trisodium salt (**32**, TPPTS) complexes in a hexane/water two-phase system at 80 °C, with a TPPTS/Rh ratio of 4:1, 10 bar of hydrogen, and dodecyltrimethylammonium chloride as surfactant. Unfortunately, a significant amount of C18:1 *E*-isomers was also obtained (42.5%; i.e. 53% of the total C18:1 product), while the amount of methyl stearate (C18:0) increased from ca. 4 to ca. 10%. In Scheme 13, some results achieved for the hydrogenation of sunflower oil methyl esters to give C18:1 product **34** using a water-soluble rhodium/TPPTS (**32**) catalyst are illustrated.[65]

Scheme 13 Ligands Used for Ruthenium- and Palladium-Catalyzed Hydrogenations of Polyunsaturated Fatty Acid Methyl Esters, and Partial Hydrogenation of Sunflower Oil Esters[65–67]

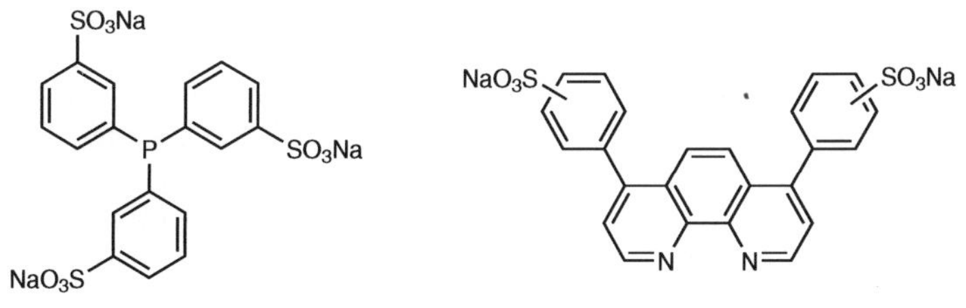

SO_3Na P SO_3Na NaO_3S

32

NaO_3S SO_3Na N N

33

for references see p 123

1.2% 63.3% 32%

3.5%

$RhCl_3 \cdot 3H_2O$, H_2 (50 bar)
32, H_2O, 120 °C

34

Molar Ratio (C=C/Rh)	Time (min)	Product Composition (%)				Ref
		C18:3	C18:2	C18:1	C18:0	
13500:1	5	0.7	32.2	58.1[a]	9.0	[65]
13500:1[b]	5	1.3	38.7	52.8[c]	7.2	[65]
1800:1	60	0.0	0.5	0.5[d]	99.0	[65]

[a] Ratio (*Z*/*E*) 38.6:19.5.
[b] Dodecyltrimethylammonium chloride was added.
[c] Ratio (*Z*/*E*) 39.5:13.3.
[d] *E*-Isomer only.

High catalytic activity, reaching a turnover frequency of 117000 h^{-1}, was obtained in the partial hydrogenation of sunflower oil methyl esters with the same catalyst at 120 °C under 50 bar of hydrogen without addition of surfactant and organic solvent. In this case, the final amount of C18:1 was 68.3% with 47.6% *E*-isomers, while methyl stearate increased dramatically from 3 to 25.7%.[66]

Cynara cardunculus (cardoon) oil methyl esters with a starting composition very similar to sunflower oil have been hydrogenated in the presence of a similar system based on ruthenium, but with a higher hydrogen pressure (80 bar), to give 54.9% C18:1 product and 13.8% methyl stearate.[65,68]

Palladium/tris(3-sulfonatophenyl)phosphine trisodium salt (**32**) complexes are also effective in aqueous/organic two-phase systems.[67] They were tested in the hydrogenation of soybean oil methyl esters and a remarkable 81.4% of C18:1 product was reached when using palladium(II) acetate as the catalyst precursor at 120 °C and 20 bar of hydrogen. No added surfactant was needed because of the presence of lecithin, a natural phospholipid present in several vegetable oils, which leads to micelle formation without the help of any external surfactant.

Micelles containing palladium(0) nanoparticles can also be obtained in the presence of various nitrogen-containing ligands in aqueous/organic two-phase systems. The highest catalytic activity in the reduction of soybean oil methyl esters, with a turnover frequency of 71300 h^{-1} and selectivity for C18:1 isomers of 78.4%, was observed using palladium(II) chloride as the catalyst precursor modified with water-soluble bathophenanthroline disulfonic acid disodium salt **33** (BPhDS) at 120 °C and 20 bar of hydrogen.[67] The heterogeneous nature of the catalyst was proven by poisoning in the presence of mercury(0). Dynamic light scattering (DLS) experiments showed the presence of micelles with an average hydrodynamic radius of 36 nm with in situ formed $[Pd(OAc)_2]_3$/BPhDS catalyst, and an average radius of 57 nm with preformed and recycled $PdCl_2$/BPhDS, showing that the micelle radius is influenced by the size of the counteranion in the precursor. The catalyst can be also recovered after the reaction by simple separation of the aqueous phase from the organic layer, and reused by adding a new portion of soybean oil methyl esters to the aqueous phase. The activity of the catalyst remains almost unchanged when recycled, although small amounts of palladium black separated from the aqueous phase both after the first reaction and after recycling. The stability of this system, confirmed by DLS measurement on the aqueous catalyst solution after reuse, is worth underlining, as the partial hydrogenation reaction is carried out at 120 °C, whereas transition metal(0) nanoparticles are commonly used at much lower temperatures.

Water-soluble rhodium and ruthenium/tris(3-sulfonatophenyl)phosphine trisodium salt (**32**) complexes can also be used for the full hydrogenation of vegetable oil methyl esters to methyl stearate.[65]

Hydrogenated Fatty Acid Methyl Esters, e.g. 34:[65]
The partial hydrogenation reaction of sunflower oil methyl esters (sunflower fatty acid ME) was carried out in the presence of water-soluble Rh/TPPTS (**32**) complexes in an aqueous/organic two-phase system. The water-soluble Rh/TPPTS catalyst precursor was prepared by dissolving $RhCl_3\cdot 3H_2O$ (1.32 mg, 0.005 mmol) and TPPTS (**32**; 9.24 mg, 0.015 mmol) in deaerated demineralized H_2O (15 mL) under argon, while stirring, within 1 min. This aqueous Rh/TPPTS catalyst precursor soln, having a Rh concentration of 34 ppm, was loaded into an autoclave previously evacuated and filled with argon together with sunflower oil methyl esters (1.94 g), which resulted in a two-phase system with a volume ratio (aqueous/organic phase) of 15:2.1. The autoclave was pressured with H_2 and the content was heated with stirring (850 rpm). The reaction temperature was 120 °C, the H_2 pressure was 50 bar, and the reaction time was 5 min. After the reaction, the autoclave was cooled to rt and H_2 was vented, and the biphasic mixture was removed. The upper organic layer was separated from the lower aqueous layer containing the catalyst using a separatory funnel, and then dried (Na_2SO_4). The organic layer containing the products was analyzed by GC after addition of methyl heptadecanoate as a standard.

1.2.3.2 Homogeneous Systems

A truly homogeneous system has been developed for the selective hydrogenation of cardanol,[69] a constituent of cashew nutshell liquid (CNSL). This is a very interesting raw material, particularly in scenarios where reuse of agro-food industry residues has become mandatory to foster sustainable development. CNSL is extracted from the shells of cashew nuts, the global production of which is around 3 million tons per year. Cardanol is a mixture of three phenols *meta*-substituted by a linear C_{15} chain with three, two, or one C=C bonds; the mixture comprises 38% of the triene, 17% of the diene, and 42% of the monoene (as well as 3% of material with the fully saturated side chain). The reduction of the polyunsaturated components to monounsaturated products **35** ($m + n = 12$) is useful in order to improve the oxidative stability of the carbon chain, and also to obtain a single compound for further use as a raw material (Scheme 14).

Ruthenium nanoparticles stabilized with poly(vinylpyrrolidinone) in butan-1-ol and rhodium/tris(3-sulfonatophenyl)phosphine trisodium salt in a two-phase water/hexane system give poor results in this hydrogenation. However, the use of ruthenium(III) chloride under hydrogen-transfer conditions, i.e. by simply using isopropanol as solvent at reflux temperature, gave 89% monounsaturated compound, which could even be increased to 97%. Unfortunately, positional isomerization along the chain takes place to some extent.[69] The structures of the proposed in situ formed cationic ruthenium complex catalysts are shown in Scheme 14.

Scheme 14 Partial Hydrogenation of Cardanol[69]

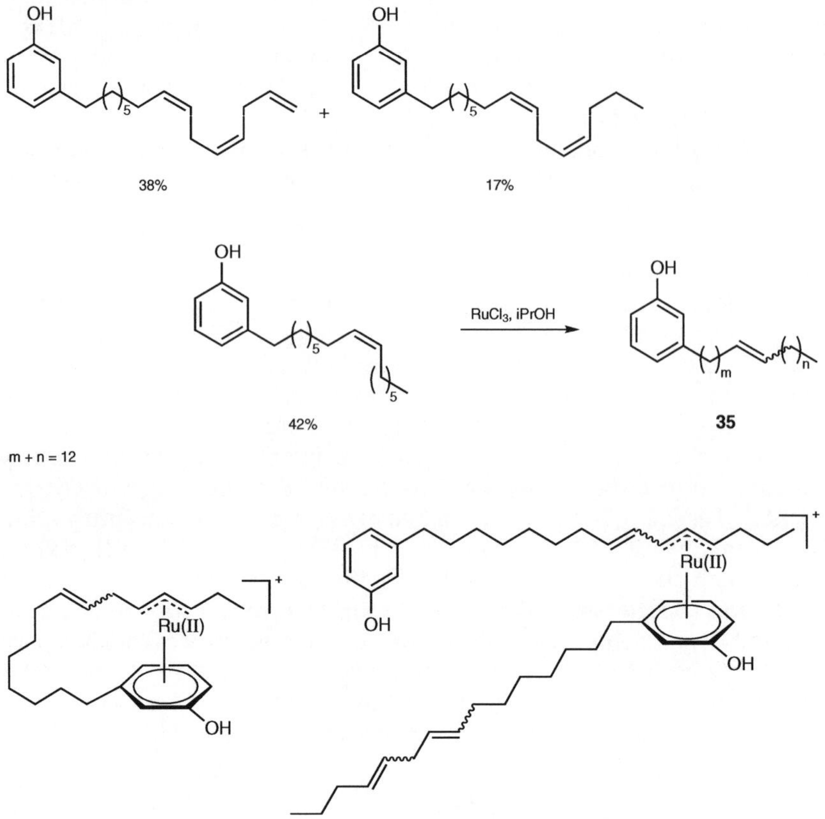

OH OH

38% 17%

OH OH

RuCl3, iPrOH

42% **35**

m + n = 12

Ru(II) Ru(II)

OH OH OH

This catalytic system has also been found to be effective for fatty acid polyunsaturated esters, but the presence of an aromatic ring is a prerequisite for this reaction. Thus, methyl linoleate gives mainly transesterification with isopropanol, whereas the benzyl ester of linoleic acid gives 95% of the monoene derivative as a mixture of positional isomers. The possible involvement of ruthenium nanoparticles, formed by reduction of ruthenium(III) chloride by the alcohol, was excluded by carrying out a test in the presence of an excess of mercury, which did not affect either the activity or the selectivity of the reaction.

3-Alkenylphenols 35; **Typical Procedure:**[69]

A 25-mL Schlenk flask equipped with a condenser was charged, under N_2, with cardanol (1.7 mmol), $RuCl_3$ (0.08 mmol), iPrOH (5 mL), and a magnetic stirrer bar. The mixture was heated at reflux for 18 h and the solvent was evaporated under reduced pressure. CH_2Cl_2 was added to the residue and the mixture was filtered through silica gel. The filtrate was concentrated, and the residue was analyzed by ^1H and ^{13}C NMR spectroscopy and HPLC to determine the ratio of poly-, mono-, and unsaturated products.

1.2.3.3 Noble Metal Based Heterogeneous Systems

Most vegetable oil hydrogenation processes are studied under classical gas–liquid–solid triphasic conditions; that is, the oil reacts with hydrogen in the presence of the solid catalyst. In such cases, selectivity is significantly dependent on the experimental conditions. In particular, high hydrogen availability at the catalyst surface, i.e. high hydrogen gas pressure, increases activity and lowers selectivity for the monounsaturated product. In contrast, low hydrogen pressure increases selectivity for oleic acid derivatives since polyunsaturated components are more strongly adsorbed on the catalyst surface than the corresponding monounsaturated ones, which are more easily desorbed; lower hydrogen pressure also increases selectivity for *Z*-isomers. On the other hand, high temperatures favor *E*-isomers and overreduction to fully saturated compounds.

Many studies have been devoted to the search for catalysts, involving both noble and non-noble metals, that produce low amounts of *E*-isomers to be used in the food sector; the best results have been obtained in the presence of platinum/zeolite systems.[70]

Significant results have been obtained for the selective hydrogenation of diene components to monoenes. Romanenko and co-workers reported 74% yield of the C18:1 product in the hydrogenation of sunflower oil, over palladium supported on Sibunit carbon, with an *E* content of 41%; however, 12% stearic acid is also formed, which increases during the following catalytic cycles. A commercial nickel catalyst (Nysosel) gave 80% C18:1 and 9% C18:0, although with a higher content of *E*-isomers.[71] Several noble and non-noble metals supported on silica or alumina have also been tested in the hydrogenation of sunflower oil. Palladium and platinum catalysts at 150 °C and 3.5 atm hydrogen gave the C18:1 product in 70–75% yield, with roughly 50% of *E*-isomers, and the C18:0 product in the range 16–22%. Among non-noble-metal based catalysts, cobalt on alumina at 180 °C and 4.5 atm hydrogen shows good catalytic activity with formation of 83% C18:1, and only 8% C18:0.[72] In general, palladium and platinum catalysts show very high activity and fairly good selectivity toward *cis*-isomers, but they suffer from poor selectivity in the reduction of polyenes to monenes, thus giving early formation of saturated C18:0.

An interesting system specifically developed to produce hydrogenated oils suitable as raw materials for the synthesis of chemical intermediates is based on palladium catalysts, with the reaction performed in organic solvent and in the presence of a small amount of water in a ratio of 5:1 to 100:1 with respect to the mass of palladium. Apparently, the presence of specific quantities of water improves both the activity and the selectivity of the catalytic system. For example, cardoon (*Cynara cardunculus*) oil in petroleum ether in the presence of palladium on alumina was hydrogenated at 5–6 °C by using hydrogen saturated with water, giving 77.6% of the C18:1 product, with a content of *E*-isomers <10% and of positional isomers <15%.[73,74]

The presence of positional isomers is very detrimental for oxidative cleavage reactions, in which inorganic and organic peroxides, peracids, nitric acid, permanganates and periodates, or oxygen and ozone, or gaseous mixtures thereof, are used as oxidizing agents. For example, the reaction shown in Scheme 15, which uses ozone to produce azelaic acid, has been carried out on an industrial scale since the 1950s; the presence of substrates containing C=C bonds in positions other than the natural 9,10-position of oleic acid would result in a plethora of products. It should be noted that the selectivity for diene versus monoene reduction is high for this system (>93%) with a limited increase in saturated C18:0 from 3 to 7%;[73,74] however, the saturated compound is not able to react in C=C cleavage reactions, therefore these few percent points translate to a loss of raw material.

Scheme 15 Oxidative Cleavage of Oleic Acid into Azelaic Acid and Pelargonic Acid[73,74]

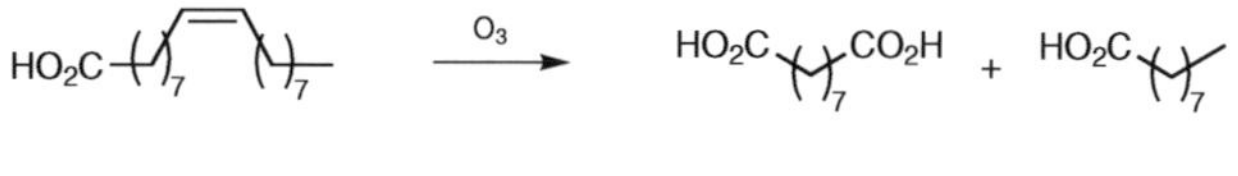

$HO_2C-(CH_2)_7-CH=CH-(CH_2)_7-CH_3 \xrightarrow{O_3} HO_2C-(CH_2)_7-CO_2H + HO_2C-(CH_2)_7-CH_3$

for references see p 123

1.2.3.4 Non-Noble Metal Based Heterogeneous Systems

Copper catalysts are known as low activity catalysts for the hydrogenation of polyenes. On the other hand, they are also known for their inactivity in the hydrogenation of monoenes to the fully saturated compounds. Therefore they show high selectivity for hydrogenation to the monounsaturated compounds. If we consider the simplified scheme of consecutive first-order reactions shown in Figure 5, we can define linolenic acid selectivity (S_{Ln}) as k_3/k_2 and linoleic acid selectivity (S_{Lo}) as k_2/k_1. High linolenic acid selectivity (S_{Ln}) indicates a much faster hydrogenation of triunsaturated to diunsaturated fatty acids compared to hydrogenation of diunsaturated to monounsaturated fatty acids.[75]

Figure 5 Stepwise Hydrogenation of Linolenic Acid Derivatives[75]

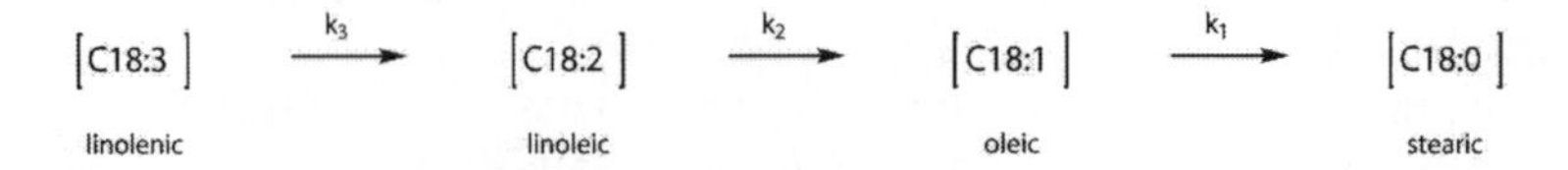

It was found in the 1960s that S_{Ln} for copper catalysts is between 7 and 14, whereas for nickel catalysts it is always <3. Moreover, no stearic acid is formed, thus showing that oleic acid selectivity (S_O) is infinite. Therefore copper chromites and copper/silica can be used to improve the quality of edible vegetable oils by eliminating linolenic acid, the main component responsible for the rancid flavor, while preserving nutritionally valuable linoleic acid and leaving oleic acid unaffected;[76,77] however, they have never been used on an industrial scale because of their much lower activity compared to the nickel catalysts. It should be noted that the catalysts were used without any reduction treatment and the generally accepted mechanism was based on the conjugation of the methylene-interrupted double bonds followed by hydrogenation of the conjugated polyene according to a Horiuti–Polanyi mechanism, resulting in both 1,2- and 1,4-addition. In turn, the conjugation reaction was considered to be due to addition of a hydrogen atom to one of the terminal double-bonded carbon atoms, i.e. to the terminal atom of a pentadienyl or octatrienyl system, followed by elimination of a hydrogen atom, giving the conjugated isomer (Scheme 16). However, this was not in agreement with the fact that increasing the hydrogen pressure favors the hydrogenation of conjugated polyenes but not the hydrogenation of methylene-interrupted ones, showing that a high concentration of hydrogen on the catalyst surface may favor hydrogen addition to a conjugated polyene but does not favor the conjugation step.

Scheme 16 Conjugation/Isomerization Mechanism via Hydrogen Addition

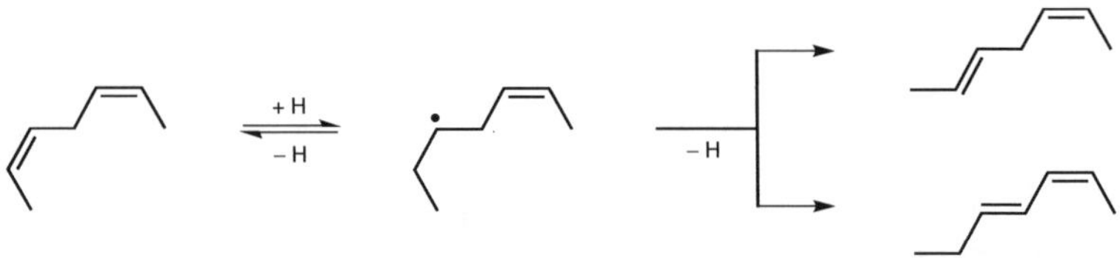

+ H

− H

− H

An alternative interpretation of the data suggests that the conjugation step could take place through abstraction of an allylic hydrogen atom and not through hydrogen addition.[78] This would explain why a high hydrogen concentration does not favor conjugation, and it is supported by the fact that allylic hydrogens are prone to dehydrogenation because of the stability of the allylic radicals that form.

This new hypothesis concerning the first step in the hydrogenation of a methylene-interrupted polyenes encouraged the investigation of the activity and selectivity of prereduced copper/silica and copper/alumina catalysts. On the surface of these systems, copper is present in its metallic state, which is much more active in dehydrogenation reactions[79] and therefore could be more effective in the hydrogen-abstraction step than conventional copper catalysts. Moreover, the high activity of these catalysts in the hydrogenation of conjugated dienes to monoenes had already been shown.[54]

In fact, prereduced copper/silica and copper/alumina catalysts turned out to be excellent candidates for this kind of application, i.e. the selective hydrogenation of both diene and triene components to the monoene, leaving stearic acid unchanged. By reducing the copper phase to the metallic state, S_{Lo} increases, indicating a much faster hydrogenation of diunsaturated to monounsaturated fatty acids compared to hydrogenation of monounsaturated to saturated ones. Therefore, high levels of monounsaturated fatty acids in the triglyceride can accumulate before significant levels of saturated (stearic) ones are formed.

The high diene-to-monoene selectivity (S_{Lo}) exhibited by supported copper catalysts can be exploited toward resolving two of the issues mentioned in Section 1.2.3. The first one is the heterogeneity in fatty acid composition of vegetable oils. The second one is the poor oxidative stability of vegetable oils, which often prevents their use as raw materials.

The availability of raw materials of constant composition is a prerequisite for the set-up of a chemical process, as well as its availability over time, independent of seasonal and climatic variations. The food versus feed issue, favoring the use of non-food oils, and the transition toward a circular economy strongly promoting the use of agro-industry residues, have also to be taken into account. Therefore, a method for the conversion of oils coming from very different origins, either from crushing of non-food crop seeds or from waste products of the food value chain, into a raw material with homogeneous composition to be transformed into biodiesel or biolubricants or other products would be hugely beneficial.

The use of prereduced copper/silica catalysts under very mild conditions allows the standardization of a very wide range of materials.[80–83] Thus, Table 3 displays the results obtained in the hydrogenation of polyunsaturated oils with very different composition (i.e., rich in C18:3 such as flax and camelina; rich in C18:2 such as tobacco seed and safflower; or with mixed composition), all of them with iodine values (see Table 3 footnote) of between 100 and 200. The table includes oils from non-food crops such as camelina (*Camelina sativa*), which is grown increasingly worldwide due to a good yield per hectare, high oil content, adaptability to marginal lands, and compatibility with existing farm equipment, and also oils stemming from residues from the food processing industry such as grapeseed oil and pumpkin seed oil.

Table 3 Hydrogenation of Various Methyl Esters over Heterogeneous Copper Catalysts[82]

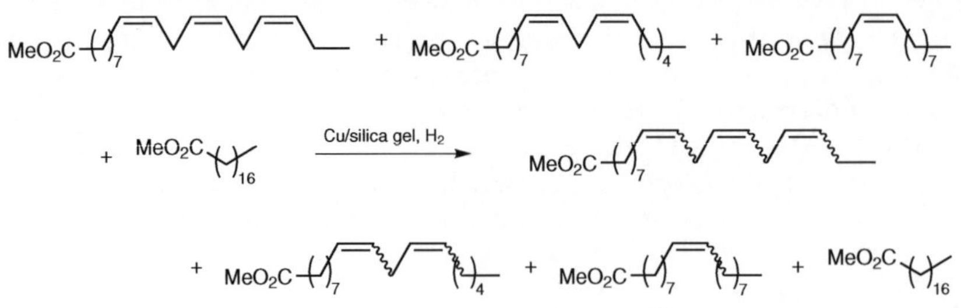

Methyl Ester[a]	Composition (%)				*E*-Isomers[b] (%)	Positional Isomers[c] (%)	Iodine Value[d]	Pour Point[e] (°C)	Ref
	C18:0	C18:1	C18:2	C18:3					
flax	3	16	17	58	–	–	186	−23	[82]
	4	63	27	–	20	23	112	−11	[82]
camelina	2	16	21	37	–	–	159	−24	[82]
	3	44	26	1	22	25	95	−10	[82]
Plukenetia volubilis	3	9	34	50	–	–	202	−25	[82]
	3	57	36	–	26	24	115	−13	[82]
hemp	3	14	57	17	–	–	164	−20	[82]
	3	49	39	–	35	24	118	−13	[82]
tobacco	2	12	79	–	–	–	145	−17	[82]
	3	57	33	–	45	12	111	−10	[82]
grapeseed	4	17	72	–	–	–	145	−18	[82]
	4	50	39	–	35	6	108	−12	[82]
safflower	3	16	73	–	–	–	140	−22	[82]
	3	56	30	–	36	4	95	−10	[82]
cardoon	3	20	63	–	–	–	128	n.d.	[82]
	3	68	16	–	39	19	87	n.d.	[82]
pumpkin	7	23	50	–	–	–	107	−8	[82]
	7	49	23	–	26	2	96	−2	[82]

[a] For each substance, the first row indicates the composition of starting material and the second row indicates the composition of hydrogenation product.
[b] Percentage of *E*-isomer in the C18:1 product.
[c] Percentage of C18:1 product with an isomerized C=C bond.
[d] The iodine value is the amount of I_2 in grams that is consumed by 100 g of a chemical substance, in this case a fat. It is a measure for the total unsaturation of the fat.
[e] n.d. = not determined.

All of the oils can be reduced to a similar composition, with the C18:1 product ranging from 43 to 68% and an iodine value in a narrow range (87–118), and in all cases below the 120 threshold. This limit has been set by the European EN14214 regulation for biodiesel as a parameter representative of the oil polyunsaturation. Polyunsaturation is directly linked with oxidative stability, and oils with iodine values >120 are considered unsuitable for the formulation of biodiesel because of the formation of tars and carbon residues in the engine, due to oxidation and polymerization of the double bonds.[84] In agreement with the lowering of the iodine value, the oxidative stability of flaxseed oil measured with a Rancimat increased from less than 1 hour to 5.3 hours (should be >6 h according to the European regulation and >3 h according to the US rules). Moreover, the reduction of polyunsaturation also has a positive impact on cetane number (CN). This is an indicator of the combustion speed of diesel fuel, and should be >51 according to the European regulation and >47 for the US rules. Cetane number increases with increasing chain length, i.e. going from C14:0 to C18:0, and decreases with the number of double bonds in the C_{18} chain. The oils mentioned in Table 3 (apart from cardoon and pumpkin) have a cetane number that is too low; however, after selective hydrogenation all of them exceed the 51 threshold (Figure 6).[81]

Figure 6 Cetane Number of Various Methyl Esters Before and After Selective Hydrogenation[81]

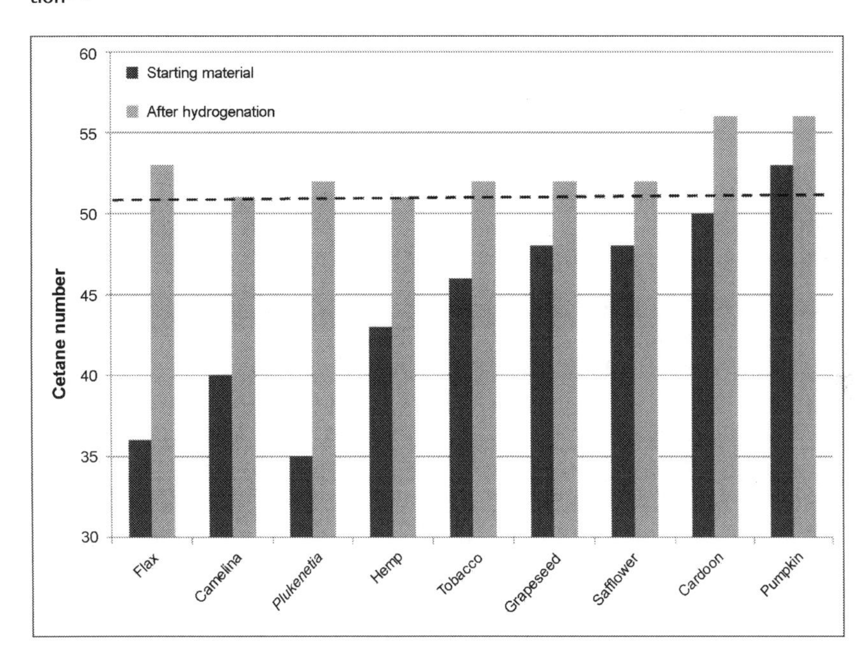

The amount of stearic acid remains almost unchanged, thus confirming the very high selectivity to oleic acid of these catalytic systems. Furthermore, *E*/*Z* and positional isomerization is limited, thus preserving acceptable cold properties, as shown by the pour point values reported in Table 3. This means that the selective hydrogenation treatment in the presence of copper catalysts is able to convert various polyunsaturated oils into a material with oxidative stability that allows it to be stored over time and with a composition that makes it suitable for the formulation of high-quality biodiesel.

The particular case of camelina oil[83] is worth underlining. The linolenic acid (C18:3) content of this oil could represent a serious issue for its use as a biofuel or biolubricant raw material. In the case of biodiesel, a maximum content of 12% of C18:3 is allowed by the European EN14 214 regulation (UNI EN 14 214:2013) because of the insufficient oxidative stability of polyunsaturated compounds. In fact, the presence of fatty acid esters containing three or more double bonds has dramatic effects on the oxidative stability because the bis-allylic position is particularly prone to oxidative attack. The predicted oxidizability, as reported in the literature, sharply drops after the hydrogenation step (Figure 7).[85]

for references see p 123

Figure 7 Oxidizability and C18:3 Content in Camelina Oil Methyl Esters Before and After Hydrogenation Compared with Rapeseed Oil Methyl Esters[85]

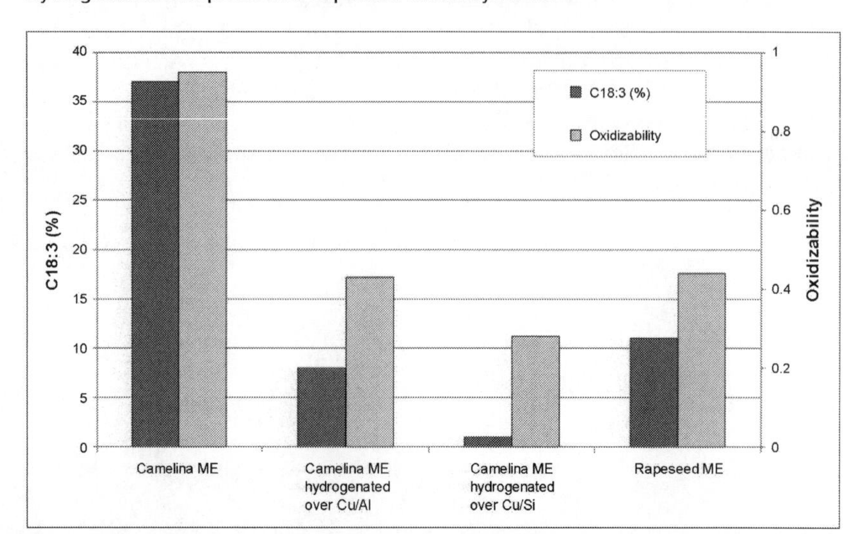

The starting oxidizability value of 0.95 for camelina oil methyl esters is quite high considering that rapeseed oil has a value commonly around 0.44. After 15 minutes of hydrogenation, this value reaches 0.43 in the case of copper/alumina and 0.28 in the case of copper/silica catalysts, following the same trend as the C18:3 content lowering. This shows that this simple treatment is able to convert a material highly unstable toward oxidation into an oil with high resistance to oxidation, even when compared to commonly used ones. It should also be kept in mind that cold properties are preserved thanks to the high linoleic acid selectivity (S_{Lo}) of copper catalysts (Table 3).

The catalytic system can also be successfully used to produce oils with a very high oleic acid content, which can be used in industrial applications. Oils with high oleic acid content are the ideal trade-off between oxidative stability and cold flow properties, which are fundamental features not only for biodiesel but also in other fields, typically in lubricants. This issue is also highly relevant for oxidative cleavage reactions to produce azelaic and pelargonic acids, as mentioned in Section 1.2.3.3.[73,74]

Supported-copper catalysts have been used for the hydrogenation of rapeseed oil methyl esters to produce base fluids for lubricant formulations. In particular, six different silicas with Brunauer–Emmett–Teller (BET) surface area ranging from 350 to 600 $m^2 \cdot g^{-1}$, both nonporous and mesoporous, have been used as support with excellent results.[80] All of them give complete removal of C18:3 and a very significant increase in C18:1 from 64% in the starting material up to 88%, without affecting the content in C18:0, both under 6 and under 20 bar of hydrogen. The amount of *E*-isomers remains <20%, thus allowing the preservation of good cold properties with a pour point around −15 °C. In contrast, the introduction of a very small amount of titanium(IV) oxide produced an increase in C18:0 from 2 to 5%, while a small amount of alumina brought the *E*-isomer content to 32%, resulting in an increase in pour point to −4 °C in the first case and to −9 °C in the second.[80]

This reaction is also effective in directly hydrogenating rapeseed oil (rapeseed triglycerides),[86] and both soybean oil (soybean triglycerides) and soybean oil methyl esters.[87] The results summarized in Figure 8 show that high amounts of oleic acid (C18:1) can be obtained starting from these two oils and their methyl esters. Although rapeseed oil hydrogenation can be carried out at 6 atm and 180 °C, which are the same conditions used for methyl esters, a higher content in C18:1 for soybean oil is obtained by using a higher hydrogen pressure (20 atm).

Figure 8 Composition of Rapeseed Oil, Rapeseed Oil Methyl Esters, Soybean Oil, and Soybean Oil Methyl Esters Before (Above) and After (Below) Selective Hydrogenation Using Copper/Silica[80]

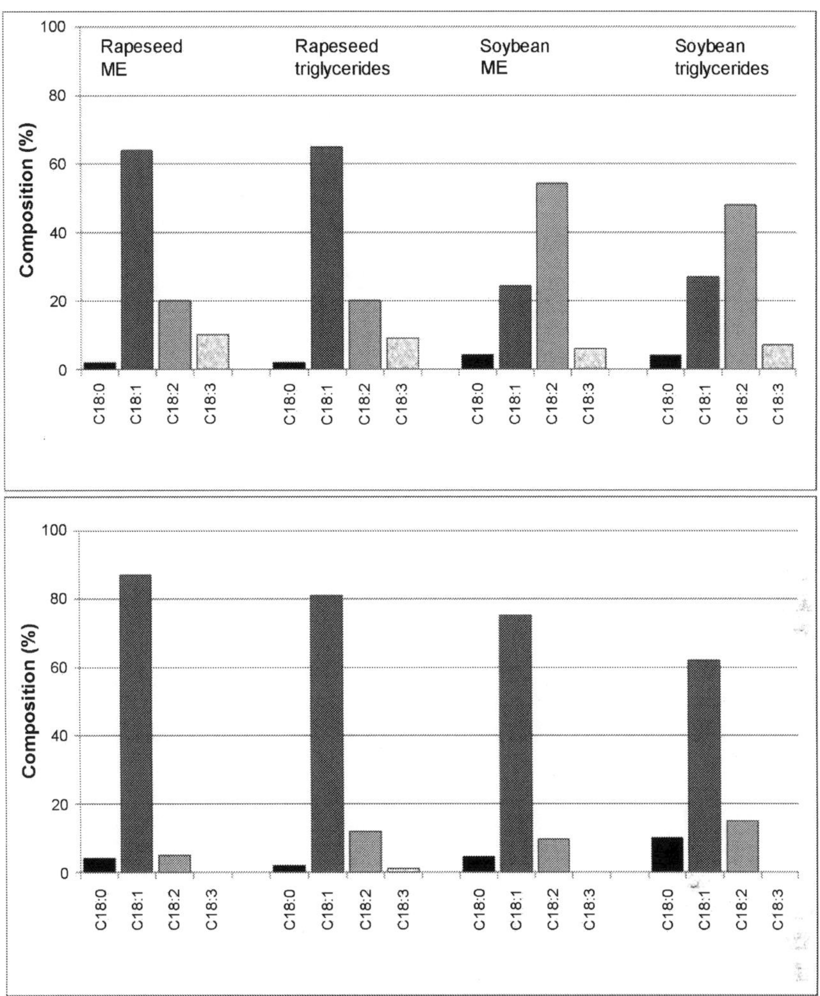

Hydrogenated Fatty Acid Methyl Esters, e.g. 34:[81,82]

The copper catalyst, with an 8% metal loading, was prepared as follows: silica gel (Davicat from Grace Davison; BET = 313 $m^2 \cdot g^{-1}$, PV = 1.79 $mL \cdot g^{-1}$) was added to a $[Cu(NH_3)_4]^{2+}$ soln prepared by adding aq NH_3 to a $Cu(NO_3)_2 \cdot 3H_2O$ soln until pH 9 had been reached. After 20 min under stirring, the slurry, held in an ice bath at 0 °C, was diluted with H_2O. The solid was separated by filtration, washed with H_2O, dried overnight at 110 °C, and calcined in air in a furnace heated at 350 °C. Before reaction, the catalyst was prereduced ex situ in H_2 (1 atm) at 270 °C, and transferred into the reactor under an inert atmosphere.

Hydrogenation reactions were carried out in a stainless steel autoclave (Parr series 4560) at 160 °C (180 °C in case of Tall Oil), under H_2 (4–6 atm), in the presence of 2% (by weight) powdered Cu/SiO_2 with an 8% copper loading. Mixtures, separated by simple filtration, were analyzed by GC (HP-6890) using a nonbonded, biscyanopropylpolysiloxane (100 m) capillary column at isotherm T = 190 °C.

for references see p 123

1.2.4 Conclusions

Although catalytic hydrogenation is one of the oldest reactions studied in the lab and carried out on an industrial scale, the issue of selectivity is still a challenge for both renewable and fossil-derived substrates. A deeper insight into reaction mechanisms operating in the presence of different metals or under different experimental conditions, e.g. hydrogen transfer or hydrogen addition, is therefore not redundant. Furthermore, investigation of relationships between the metal-phase morphology and the catalyst activity and selectivity is a powerful tool for the design of innovative catalysts.

References

[1] Augustine, R. L., *Heterogeneous Catalysis for the Synthetic Chemist*, Marcel Dekker: New York, (1996); p 94.
[2] Wu, W.; Li, Y.-L.; Chen, W.-S.; Lai, C.-C., *Ind. Eng. Chem. Res.*, (2011) **50**, 1264.
[3] Schwab, W.; Fuchs, C.; Huang, F.-C., *Eur. J. Lipid Sci. Technol.*, (2013) **115**, 3.
[4] Bergner, E. J.; Ebel, K.; Johann, T.; Löber, O., US 7 709 688, (2010).
[5] Jäkel, C.; Paciello, R., US 7 534 921, (2009).
[6] Weissermel, K.; Arpe, H.-J., *Industrial Organic Chemistry*, 4th ed., Wiley-VCH: Weinheim, Germany, (2003); p 117.
[7] Molnár, Á.; Sárkány, A.; Varga, M., *J. Mol. Catal. A: Chem.*, (2001) **173**,185.
[8] Gross, E.; Somorjai, G. A., *J. Catal.*, (2015) **328**, 91.
[9] Phillipson, J. J.; Wells, P. B.; Wilson, G. R., *J. Chem. Soc. A*, (1969), 1351.
[10] Solà, L.; Pericàs, M. A.; Cunill, F.; Izquierdo, J. F., *Ind. Eng. Chem. Res.*, (1997) **36**, 2012.
[11] Bond, G. C., *J. Mol. Catal. A: Chem.*, (1997) **118**, 333.
[12] Lin, T.-B.; Chou, T.-C., *Appl. Catal., A*, (1994) **108**, 7.
[13] Chang, J.-C.; Chou, T.-C., *Appl. Catal., A*, (1997) **156**, 193.
[14] Bachir, R.; Marecot, P.; Didillon, B.; Barbier, J., *Appl. Catal., A*, (1997) **164**, 313.
[15] Badalo Branco, J.; Pereira Gonçalves, A.; Pires de Mato, A., *J. Alloys Compd.*, (2008) **465**, 361.
[16] Su, W.-B.; Tang, M.-T.; Chang, J.-R., *Ind. Eng. Chem. Res.*, (2005) **44**, 1677.
[17] Jia, X.; Wang, S.; Fan, Y., *J. Catal.*, (2015) **327**, 54.
[18] Karakhanov, E.; Maximov, A.; Kardasheva, Y.; Semernina, V.; Zolotukhina, A.; Ivanov, A.; Abbott, G.; Rosenberg, E.; Vinokurov, V., *ACS Appl. Mater. Interfaces*, (2014) **6**, 8807.
[19] Thomazeau, C.; Cseri, T.; Bisson, L.; Aguilhon, J.; Pham Minh, D.; Boissière, C.; Durupthy, O.; Sanchez, C., *Top. Catal.*, (2012) **55**, 690.
[20] Silvestre-Albero, J.; Rupprechter, G.; Freund, H.-J.; *Chem. Commun. (Cambridge)*, (2006), 80.
[21] Silvestre-Albero, J.; Rupprechter, G.; Freund, H.-J., *J. Catal.*, (2006) **240**, 58.
[22] Wang, R.; Li, Y.; Shi, R.; Yang, M., *J. Mol. Catal. A: Chem.*, (2011) **344**, 122.
[23] Yang, K.; Liu, C., *J. Ind. Eng. Chem. (Amsterdam, Neth.)*, (2015) **28**, 161.
[24] Liu, Q.; Yu, Y.; Wang, P.; Li, Y., *New J. Chem.*, (2013) **37**, 3647.
[25] Izzo, I.; Pironti, V.; Della Monica, C.; Sodano, G.; De Riccardis, F., *Tetrahedron Lett.*, (2001) **42**, 8977.
[26] Bogel-Łukasik, E.; Fonseca, I.; Bogel-Łukasik, R.; Tarasenko, Y. A.; Nunes da Ponte, M.; Paiva, A.; Brunner, G., *Green Chem.*, (2007) **9**, 427.
[27] Morère, J.; Torralvo, M. J.; Pando, C.; Renuncio, J. A. R.; Cabañas, A., *RSC Adv.*, (2015) **5**, 38 880.
[28] Tanielyan, S. K.; Augustine, R. L.; Marin, N.; Alvez, G., *ACS Catal.*, (2011) **1**, 159.
[29] Graham, T. J. A.; Poole, T. H.; Reese, C. N.; Goess, B. C., *J. Org. Chem.*, (2011) **76**, 4132.
[30] Surburg, H.; Panten, J., *Common Fragrance and Flavor Materials: Preparation, Properties and Uses*, Wiley-VCH: Weinheim, Germany, (2006).
[31] Robles-Dutenhefner, P. A.; Speziali, M. G.; Sousa, E. M. B.; dos Santos, E. N.; Gusevskaya, E. V., *Appl. Catal., A*, (2005) **295**, 52.
[32] Guerrero, M.; Costa, N. J. S.; Vono, L. L. R.; Rossi, L. M.; Gusevskaya, E. V.; Philippot, K., *J. Mater. Chem. A*, (2013) **1**, 1441.
[33] Bogel-Łukasik, E.; Gomes da Silva, M.; Nogueira, I. D.; Bogel-Łukasik, R.; Nunes da Ponte, M., *Green Chem.*, (2009) **11**, 1847.
[34] Mitsui, S.; Shoinoya, M.; Gohke, K.; Watanabe, F.; Imaizumi, S.; Senda, Y., *J. Catal.*, (1975) **40**, 372.
[35] Horiuti, I.; Polanyi, M., *Trans. Faraday Soc.*, (1934) **30**, 1164.
[36] Speziali, M. G.; Moura, F. C. C.; Robles-Dutenhefner, P. A.; Araujo, M. H.; Gusevskaya, E. V.; dos Santos, E. N., *J. Mol. Catal. A: Chem.*, (2005) **239**, 10.
[37] Wang, Q.; Liu, X.; Liu, X.; Li, B.; Nie, H.; Zhang, S.; Chen, W., *Chem. Commun. (Cambridge)*, (2014) **50**, 978.
[38] Takaya, H.; Ohta, T.; Sayo, N.; Kumobayashi, H.; Akutagawa, S.; Inoue, S.; Kasahara, I.; Noyori, R., *J. Am. Chem. Soc.*, (1987) **109**, 1596.
[39] Wu, R.; Beauchamps, M. G.; Laquidara, J. M.; Sowa, J. R., Jr., *Angew. Chem. Int. Ed.*, (2012) **51**, 2106.
[40] Jäkel, C.; Paciello, R., In *Asymmetric Catalysis on Industrial Scale: Challenges, Approaches and Solutions*, 2nd ed., Blaser, H.-U.; Federsel, H.-J., Eds.; Wiley-VCH: Weinheim: Germany, (2010); pp 187–205.

[41] Zaccheria, F.; Ravasio, N.; Fusi, A.; Rodondi, M.; Psaro, R., *Adv. Synth. Catal.*, (2005) **347**, 1267.
[42] Tsuyuki, T.; Nishitani, M., JP 57 024 320, (1982); *Chem. Abstr.*, (1982) **96**, 181 460.
[43] Shimazu, S.; Baba, N.; Ichikuni, N.; Uematsu, T., *J. Mol. Catal. A: Chem.*, (2002) **182–183**, 343.
[44] Weissermel, K.; Arpe, H.-J., *Industrial Organic Chemistry*, 4th ed., Wiley-VCH: Weinheim, Germany, (2003); p 243.
[45] Thomas, J. M.; Johnson, B. F. G.; Raja, R.; Sankar, G.; Midgley, P. A., *Acc. Chem. Res.*, (2003) **36**, 20.
[46] Jurčík, V.; Nolan, S. P.; Cazin, C. S. J., *Chem.–Eur. J.*, (2009) **15**, 2509.
[47] Broggi, J.; Jurčík, V.; Songis, O.; Poater, A.; Cavallo, L.; Slawin, A. M. Z.; Cazin, C. S. J., *J. Am. Chem. Soc.*, (2013) **135**, 4588.
[48] Schmidt, A.; Schomäcker, R., *Ind. Eng. Chem. Res.*, (2007) **46**, 1677.
[49] Moreno-Marrodan, C.; Barbaro, P.; Catalano, M.; Taurino, A., *Dalton Trans.*, (2012) **41**, 12 666.
[50] Sachse, A.; Linares, N.; Barbaro, P.; Fajula, F.; Galarneau, A., *Dalton Trans.*, (2013) **42**, 1378.
[51] Chen, P.; Chew, L. M.; Xia, W., *J. Catal.*, (2013) **307**, 84.
[52] Moreno-Marrodan, C.; Liguori, F.; Mercadé, E.; Godard, C.; Claver, C.; Barbaro, P., *Catal. Sci. Technol.*, (2015) **5**, 3762.
[53] Alonso, F.; Osantey, I.; Yus, M., *Tetrahedron*, (2007) **63**, 93.
[54] Boccuzzi, F.; Chiorino, A.; Gargano, M.; Ravasio, N.; *J. Catal.*, (1997) **165**, 140.
[55] Hermans, S.; Raja, R.; Thomas, J. M.; Johnson, B. F. G.; Sankar, G.; Gleeson, D., *Angew. Chem. Int. Ed.*, (2001) **40**, 1211.
[56] Concepción, P.; García, S.; Hernández-Garrido, J. C.; Calvino, J. J.; Corma, A., *Catal. Today*, (2015) **259**, 213.
[57] Cabiac, A.; Delahay, G.; Durand, R.; Trens, P.; Plée, D.; Medevielle, A.; Coq, B., *Appl. Catal., A*, (2007) **318**, 17.
[58] McAllister, C. G., US 3 400 164, (1968).
[59] Gaube, J.; David, W.; Sanchayan, R.; Roy, S.; Müller-Plathe, F., *Appl. Catal., A*, (2008) **343**, 87.
[60] Gaube, J.; David, W.; Sanchayan, R.; Wuchter, N.; Klein, H.-F., *Appl. Catal., A*, (2011) **409–410**, 21.
[61] Stüber, F.; Delmas, H., *Ind. Eng. Chem. Res.*, (2003) **42**, 6.
[62] Frankel, E. N., *Lipid Oxidation*, 2nd ed., Oily Press: Bridgewater, UK, (2005).
[63] *Fatty Acids: Their Chemistry, Properties, Production and Uses*, Part 1, Markley, K. S., Ed.; Interscience: New York, (1960); pp 112, 113.
[64] *Inception Impact Assessment*; European Commission, (2016); available online at http://ec.europa.eu/food/safety/labelling_nutrition/labelling_legislation/trans-fats_en.
[65] Bouriazos, A.; Vasiliou, C.; Tsichla, A.; Papadogianakis, G., *Catal. Today*, (2015) **247**, 20.
[66] Bouriazos, A.; Mouratidis, K.; Psaroudakis, N.; Papadogianakis, G., *Catal. Lett.*, (2008) **121**, 158.
[67] Bouriazos, A.; Sotiriou, S.; Stathis, P.; Papadogianakis, G., *Appl. Catal., B*, (2014) **150–151**, 345.
[68] Bouriazos, A.; Ikonomakou, E.; Papadogianakis, G., *Ind. Crops Prod.*, (2014) **52**, 205.
[69] Perdriau, S.; Harder, S.; Heeres, H. J.; de Vries, J. G., *ChemSusChem*, (2012) **5**, 2427.
[70] Philippaerts, A.; Breesch, A.; Cremer, G.; Kayaert, P.; Hofkens, J.; Mooter, G.; Jacobs, P.; Sels, B., *J. Am. Oil Chem. Soc.*, (2011) **88**, 2023.
[71] Romanenko, A. V.; Voropaev, I. V.; Abdullina, R. M.; Chumachenko, V. A., *Solid Fuel Chem.*, (2014) **48**, 356.
[72] Cepeda, E. A.; Iriarte-Velasco, U.; Calvo, B.; Sierra, I., *J. Am. Oil Chem. Soc.*, (2016) **93**, 731.
[73] Borsotti, G.; Capuzzi, L.; Digioia, F., WO 2 016 102 509, (2016).
[74] Borsotti, G.; Capuzzi, L.; Digioia, F., WO 2 014 207 038, (2014).
[75] Philippaerts, A.; Jacobs, P.; Sels, B., In *Catalytic Hydrogenation for Biomass Valorization*, Rinaldi, R., Ed.; RSC: Cambridge, UK, (2015); p 223.
[76] Johansson, L. E.; *J. Am. Oil Chem. Soc.*, (1980) **57**, 16.
[77] Koritala, S.; Dutton, H. J., *J. Am. Oil Chem. Soc.*, (1969) **48**, 245.
[78] Dijkstra, A. J., *Eur. J. Lipid Sci. Technol.*, (2002) **104**, 29.
[79] Zaccheria, F.; Ravasio, N.; Psaro, R.; Fusi, A., *Chem. Commun. (Cambridge)*, (2005), 253.
[80] Ravasio, N.; Zaccheria, F.; Gargano, M.; Recchia, S.; Fusi, A.; Poli, N.; Psaro, R., *Appl. Catal., A*, (2002) **233**, 1.
[81] Zaccheria, F.; Psaro, R.; Ravasio, N., *Green Chem.*, (2009) **11**, 462.
[82] Zaccheria, F.; Psaro, R.; Ravasio, N.; Bondioli, P., *Eur. J. Lipid Sci. Technol.*, (2012) **114**, 24.

[83] Pecchia, P.; Galasso, I.; Mapelli, S.; Bondioli, P.; Zaccheria, F.; Ravasio, N., *Ind. Crops Prod.*, (2013) **51**, 306.
[84] Bondioli, P.; Della Bella, L.; Ravasio, N.; Zaccheria, F., In *Catalysis of Organic Reactions*, Prunier, M. L., Ed.; CRC: Boca Raton, FL, (2008); pp 271–278.
[85] McCormick, R. L.; Ratcliff, M.; Moens, L.; Lawrence, R., *Fuel Process. Technol.*, (2007) **88**, 651.
[86] Dworakowska, S.; Bogdał, D.; Zaccheria, F.; Ravasio, N., *Catal. Today*, (2014) **223**, 148.
[87] Trasarti, A. F.; Segobia, D. J.; Apesteguía, C. R.; Santoro, F.; Zaccheria, F.; Ravasio, N., *J. Am. Oil Chem. Soc.*, (2012) **89**, 2245.

1.3 Reduction of Arenes

X. Dai and F. Shi

General Introduction

Arene reduction can be considered as one of the most important organic transformations, finding applications in the synthesis of cyclohexane derivatives as relevant synthons for chemicals and pharmaceuticals,[1] and also in the removal of carcinogenic benzene content from gasoline,[2,3] which is required due to increasing government limitations. Traditionally, this reaction required severe temperature and/or pressure conditions because of the resonance stabilization of the aromatic ring.[1] Catalysts based on group VIII metals, such as ruthenium, rhodium, platinum, palladium, and iridium, have been used for the reduction of arenes. Of these, platinum-based catalysts have been extensively studied and are already used in commercial processes. Ruthenium-based catalysts are often used for partial reduction of arenes. Some new catalytic materials, such as amorphous metal alloys[4] and ionic-liquid-like copolymer-stabilized nanocatalysts in ionic liquids,[5] have also been used for arene reduction.

The reduction of arenes can be divided into of two kinds of reactions, depending on the resulting products: (i) total reduction to cycloalkanes (Section 1.3.1); and (ii) partial reduction to cycloalkenes (Section 1.3.3). Cycloalkanes, the products of total reduction of an arene, are important structural units for pharmaceuticals and functional materials. For example, cyclohexane, obtained by the total reduction of the simplest arene, benzene, is a valuable chemical[6,7] that is used in the manufacture of nylon-6 and nylon-66, which constitute about 90% of all polyamides. Moreover, cyclohexane is also an excellent and nontoxic solvent for cellulose ether, wax, asphalt, and rubber.[8,9] In addition, the depletion of benzene in transportation fuels by reduction to cyclohexane is an effective method to reduce the emission of particulate matters in exhaust gases, and to increase the cetane number of diesel fuel. Cycloalkenes, as intermediate products in the reduction of arenes, are thermodynamically less-favorable products. The selective reduction of benzene to cyclohexene is a very challenging task for two reasons: (i) thermodynamics favor the formation of cyclohexane over cyclohexene by about 75 $kJ \cdot mol^{-1}$; and (ii) kinetic studies have shown that the reactivity of the C=C bond in cyclohexene is much higher than the reactivity of benzene.[10] Therefore, kinetic control is key in manipulating the selectively reductive conversion of benzene to cyclohexane and cyclohexene.

1.3.1 Reduction of Monocyclic Arenes to Cycloalkanes

Reduction of monocyclic arenes is an active area of research due to the high industrial interest in the production of cycloalkanes from the corresponding arenes. To date, most monocyclic arenes may be transformed into the corresponding cycloalkanes by employing ruthenium, rhodium, iridium, or palladium catalysts. However, selective reduction of monocyclic arenes that bear reducible groups is still a challenge.

The first example of benzene reduction was reported by Sabatier and co-worker in 1901.[11] Following Sabatier's work, numerous groups have worked on developing new heterogeneous catalysts for the reduction of arenes (Scheme 1).[5,12–18] Historically significant was the work of Adams on finely divided "platinum-oxide, platinum black" (Adams' catalyst)[18] and the process of selectively digesting alloys to generate finely divided sam-

for references see p 150

ples of metals (e.g., Raney nickel) by the industrial chemist Murray Raney.[19] The development of transition-metal nanocluster catalysts allows the reduction of arenes under mild conditions due to the distinct catalytic activities. However, these catalysts generally suffer from aggregation problems, which lead to a momentous decrease in catalytic activity and lifetime.[5,16,17] The stability of the catalyst can be improved by depositing active metal on metal oxides or carbon supports to form supported catalysts, and adding another active metal to form alloy catalysts.[12–15] The development of homogeneous catalysts for the reduction of arenes has lagged well behind the heterogeneous systems. Since the original discovery of the so-called Wilkinson's catalyst,[20,21] there have been only a few reports of homogeneous catalysts for the reduction of arenes (Scheme 1).[22–26] Most of these catalysts are inactive as the metal complex; however, they convert into metal nanoparticles under the hydrogenation conditions,[27] and these metal nanoparticles are the true hydrogenation catalysts. A clear example is from the study by Finke,[28] where it is demonstrated that the so-called supramolecular benzene hydrogenation catalyst $[Ru_3(\mu_2\text{-}H)_3(\eta^6\text{-}C_6H_6)(\eta^6\text{-}C_6Me_6)_2(\mu_3\text{-}O)]^+$, a triruthenium cluster cation, is not the true hydrogenation catalyst as previously believed; instead, trace heterogeneous $Ru(0)_n$, derived from $[Ru_3(\mu_2\text{-}H)_3(\eta^6\text{-}C_6H_6)(\eta^6\text{-}C_6Me_6)_2(\mu_3\text{-}O)]^+$ under the reaction conditions, is responsible for at least ≥99.97% of the observed catalysis and behaves as the true, active catalyst.

Scheme 1 Evolution of Catalysts for the Reduction of Benzene to Cyclohexane[5,12–17,24–26]

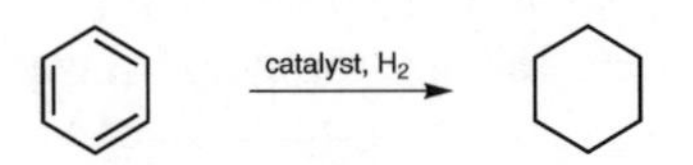

Catalyst	Temp (°C)	Pressure (MPa)	TOF[a] (h^{-1})	Ref
$[(\eta^6\text{-}C_6Me_6)Ru(\mu\text{-}H)_2(\mu\text{-}Cl)Ru(\eta^6\text{-}C_6Me_6)]Cl$	50	5	246	[24]
$[Ru_4H_4(\eta^6\text{-}C_6H_6)_4](BF_4)_2$	90	6	364	[26]
$[RuCl(\eta^6\text{-}p\text{-cymene})(\eta^2\text{-triphos})]PF_6$[b]	90	6	476	[25]
Rh nanoparticles	75	0.1	250	[5]
Rh nanoparticles embedded in poly(1-vinylpyrrolidin-2-one)	80	2	250	[17]
Ni/silica gel	80	8	3974.9	[13]
Co/silica gel	100	8	3352.2	[13]
Ru(0) nanoclusters	25	0.0689	5430	[16]
Ru nanoparticles on carbon nanotubes	80	4	6983.09	[15]
RuB nanoparticles on MOF MIL-53(AlCr)[c]	30	1	23 040	[14]
RuNi bimetallic nanoparticles on carbon black	60	5.3	18 210.3	[12]

[a] Turnover frequency.
[b] triphos = 1,1,1-tris(diphenylphosphinomethyl)ethane.
[c] MOF = metal–organic framework; MIL = materials of Institut Lavoisier.

1.3.1.1 Ruthenium-Catalyzed Reduction of Monocyclic Arenes to Cycloalkanes

Benzene and a large number of mono-, di-, and tri-substituted derivatives (e.g., **1** and **3**) can be reduced to give the corresponding cyclohexane derivatives (e.g., **2** and **4**) in good to excellent results using the ruthenium-based catalysts $Ru_2Cl_4(\eta^6\text{-}C_6H_6)_2$ (Scheme 2)[29] or ruthenium nanoparticles stabilized by hyperbranched polystyrene (HPS) bearing ammonium salts $[Ru@HPS\text{-}N^+Bu_3Cl^-]$ (Scheme 3).[30] Benzene and electron-rich analogues give excellent yields, whereas more electron-deficient arenes [e.g., benzoic acid (**1**, $R^1 = CO_2H$; $R^2 = R^3 = R^4 = R^5 = H$)] result in lower yields with catalyst $Ru_2Cl_4(\eta^6\text{-}C_6H_6)_2$. The hyperbranched-polystyrene-stabilized ruthenium catalyst has no activity in the case of starting materials with strong electron-withdrawing groups (e.g., **3**, $R^1 = NO_2$). Both ruthenium-based catalysts show poor tolerance of other reducible functional groups.

Scheme 2 Reduction of Monocyclic Arenes to Cycloalkanes Catalyzed by $Ru_2Cl_4(\eta^6\text{-}C_6H_6)_2$[29]

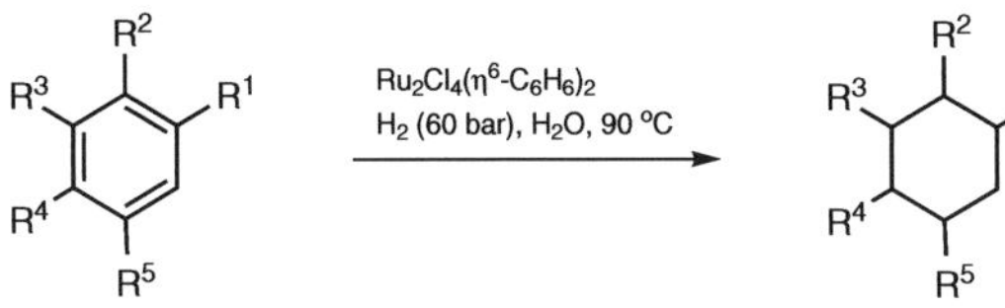

$Ru_2Cl_4(\eta^6\text{-}C_6H_6)_2$, H_2 (60 bar), H_2O, 90 °C: **1** → **2**

R^1	R^2	R^3	R^4	R^5	Time (h)	Conversion[a] (%)	TOF[b] (h^{-1})	Ref
H	H	H	H	H	0.50	100	1998	[29]
Me	H	H	H	H	0.10	99	900	[29]
Et	H	H	H	H	8.10	83	103	[29]
Pr	H	H	H	H	1.35	96	713	[29]
iPr	H	H	H	H	1.50	99	662	[29]
t-Bu	H	H	H	H	1.45	98	676	[29]
Cy	H	H	H	H	3.00	98	328	[29]
Ph	H	H	H	H	2.20	63[c]	396	[29]
OMe	H	H	H	H	4.40	95	217	[29]
Me	Me	H	H	H	1.50	96; (*cis*/*trans*) 98:2	640	[29]
Me	H	Me	H	H	2.00	98; (*cis*/*trans*) 85:15	490	[29]
Me	H	H	Me	H	2.30	93; (*cis*/*trans*) 73:27	404	[29]
Me	Me	Me	H	H	3.20	51	159	[29]
Me	Me	H	Me	Me	22.00	61[d]	14	[29]
$CMe{=}CH_2$	H	H	H	H	3.30	95[e]	288	[29]
OH	H	H	H	H	24.00	100[f]	21	[29]
Ac	H	H	H	H	24.00	75[g]	40	[29]
CO_2Me	H	H	H	H	21.00	88	42	[29]
CO_2Et	H	H	H	H	8.10	100	123	[29]
CO_2H	H	H	H	H	5.50	63[h]	91	[29]
NH_2	H	H	H	H	12.00	53[i]	44	[29]
NMe_2	H	H	H	H	9.30	70	75	[29]

[a] Measured by GC.
[b] Turnover frequency (mol substrate transformed per mol catalyst per hour).
[c] Substrate dissolved in cyclohexane (10 mL).
[d] Substrate dissolved in cyclohexane (10 mL); ratio (catalyst/substrate) 1:500.
[e] The product is isopropylcyclohexane.
[f] Ratio (catalyst/substrate; homogeneous phase) 1:500.
[g] The product is 1-cyclohexylethanol.
[h] Ratio (catalyst/substrate) 1:500; substrate dissolved in H_2O/EtOH.
[i] Homogeneous aqueous phase (pH = 2).

for references see p 150

Scheme 3 Reduction of Monocyclic Arenes to Cycloalkanes Catalyzed by Ruthenium Nanoparticles Stabilized by Hyperbranched Polystyrene Bearing Ammonium Salts[30]

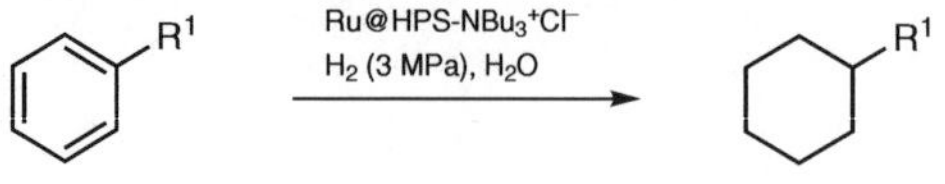

$Ru@HPS\text{-}NBu_3^+Cl^-$
H_2 (3 MPa), H_2O

3 → 4

R^1	Temp (°C)	Time (h)	Yield (%)	Ref
Me	30	1	>99[a]	[30]
OH	30	1	>99[a]	[30]
OMe	30	1	>99[a]	[30]
Ac	30	1	74[a,b]	[30]
CN	30	1	0	[30]
NO_2	30	1	0	[30]
Br	30	1	0	[30]
CO_2Et	30	12	95[c]	[30]
$CONMe_2$	30	18	92[c]	[30]
$(CH_2)_2CO_2Et$	30	12	96[c]	[30]
$(CH_2)_2CONMe_2$	30	24	93[c]	[30]
CH_2OAc	50	36	90[c]	[30]
$(CH_2)_3OAc$	50	12	95[c]	[30]
CH(OH)Me	30	12	92[c,d]	[30]
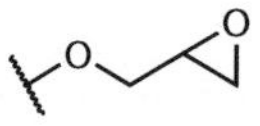	30	6	90[c]	[30]
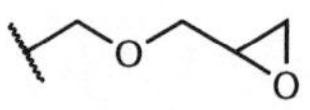	30	24	96[c]	[30]

[a] GC yield, with decane as an internal standard.
[b] Ratio (1-cyclohexylethanone/1-phenylethanol) 39:61 (determined by ^1H NMR spectroscopy).
[c] Isolated yield.
[d] >99% ee for both the starting material and the product (determined by chiral GC).

Cyclohexanes 2; Typical Procedure:[29]
$Ru_2Cl_4(\eta^6\text{-}C_6H_6)_2$ (20 mg, 0.04 mmol) was dissolved in H_2O (5 mL). The soln was placed in a 100-mL, stainless-steel autoclave and the arene **1** (4 mmol) was added. The autoclave was purged three times with H_2 and then pressurized with H_2 (60 bar). The vigorously stirred (900 rpm) mixture was heated to 90 °C. After the reaction had finished, the autoclave was cooled to rt and the pressure was released. The two-phase system was decanted. The aqueous phase was analyzed by NMR spectroscopy and the organic phase containing the products and substrate was filtered and analyzed by GC and NMR spectroscopy.

Cyclohexanes 4; General Procedure:[30]
A stirred mixture of a substituted benzene **3** (1 mmol) and catalyst $Ru@HPS\text{-}N^+Bu_3Cl^-$ [3 μmol Ru; ratio (substrate/catalyst) 333:1] in H_2O (1 mL) was treated with H_2 (3 MPa) at 30 or 50 °C in an autoclave with an inner glass tube. After the reaction had finished, the autoclave was cooled to rt and the pressure was released. The mixture was extracted with

EtOAc (5 × 3 mL) and the combined extracts were dried (Na_2SO_4), filtered, and concentrated under reduced pressure. The crude product was purified by chromatography using a short column (silica gel).

1.3.1.2 Rhodium-Catalyzed Reduction of Monocyclic Arenes to Cycloalkanes

Cycloalkanes (e.g., **7** and **9**) can be synthesized efficiently by the reduction of the corresponding arenes **6** and **8** in the presence of rhodium-based catalysts such as rhodium bis(imino)pyridine complex **5** (Scheme 4),[31] or solid-supported rhodium nanoparticles Rh@D400Li$^+$in (Scheme 5),[32] which are prepared in situ under the reaction conditions within the pores of the lithium salt of commercial DOWEX strong cation-exchange sulfonic gels [2% cross linkage, 400 mesh (75 μm) bead size]. For most arenes, yields of the desired products are high; however, poor selectivity is obtained in the case of substrates containing reducible groups (e.g., **8**, R^1 = Ac; R^2 = R^3 = R^4 = H). Moreover, with rhodium catalyst precursor **5**, a base such as potassium *tert*-butoxide or alumina is required.

Scheme 4 Reduction of Monocyclic Arenes to Cycloalkanes Catalyzed by Nanoparticles Generated from a Rhodium Bis(imino)pyridine Complex[31]

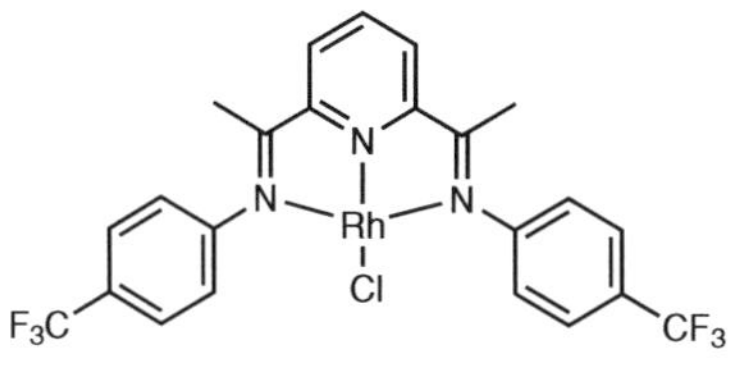

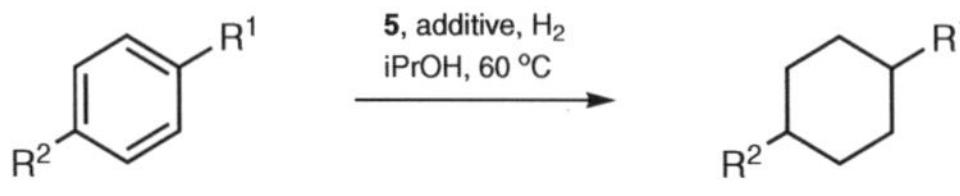

R^1	R^2	Time (h)	Additive	Yield[a] (%)	TOF[b] (h^{-1})	Ref
H	H	200	*t*-BuOK	100	45	[31]
Me	H	400	*t*-BuOK	93	21	[31]
Me	Me	400	*t*-BuOK	31	7	[31]
CH=CH_2	H	380	*t*-BuOK	67[c]	21	[31]
CMe=CH_2	H	455	*t*-BuOK	53[d]	14	[31]
NH_2	H	270	*t*-BuOK	100	33	[31]
OH	H	300	*t*-BuOK	100[e]	30	[31]
H	H	400	alumina	100	23	[31]
Me	H	420	alumina	51	11	[31]
Me	Me	420	alumina	13	3	[31]
CH=CH_2	H	260	alumina	14[c]	6	[31]

R^1	R^2	Time (h)	Additive	Yield[a] (%)	TOF[b] (h^{-1})	Ref
CMe=CH_2	H	300	alumina	20[d]	8	[31]
NH_2	H	350	alumina	100	26	[31]
OH	H	300	alumina	98[e]	29	[31]

[a] GC yield.
[b] Turnover frequency (mol of H_2 per mol of rhodium per hour).
[c] The product is ethylcyclohexane.
[d] The product is isopropylcyclohexane.
[e] The product is cyclohexanone.

Scheme 5 Reduction of Monocyclic Arenes to Cycloalkanes Catalyzed by In Situ Generated Solid-Supported Rhodium Nanoparticles[32]

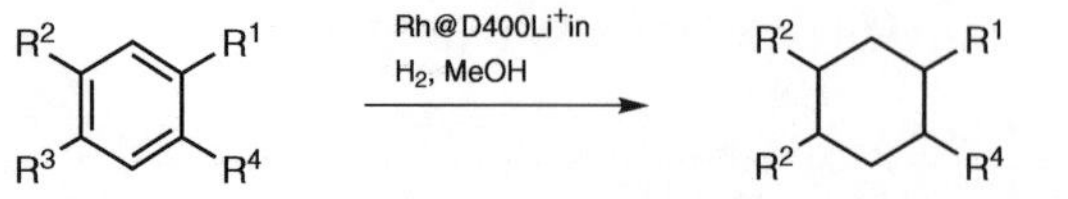

Rh@D400Li$^+$in, H_2, MeOH; **8** → **9**

R^1	R^2	R^3	R^4	Temp (°C)	H_2 (bar)	Time (min)	Conversion (%)	Selectivity[a] (%)	Ratio (*cis*/*trans*)	TOF[b] (h^{-1})	Ref
CH=CH_2	H	H	H	rt	1	380	100	90[c]	–	36	[32]
Me	H	H	H	rt	1	345	95	100	–	38	[32]
OMe	H	H	H	rt	10	240	98	64	–	56	[32]
OMe	H	H	H	60	15	240	100	72	–	58	[32]
CO_2Me	H	H	H	rt	10	240	74	97	–	n.r.	[32]
CO_2Me	H	H	H	60	10	240	100	100	–	58	[32]
Ac	H	H	H	rt	10	240	99	7	–	57	[32]
H	Bz	H	H	rt	10	240	8	17	–	n.r.	[32]
Me	H	Me	H	rt	10	240	100	100	76:24	58	[32]
Me	H	H	Me	rt	10	240	63	100	94:6	n.r.	[32]
Me	H	H	Me	60	15	240	100	100	92:8	58	[32]
OMe	H	H	Me	40	15	240	100	71	97:3	58	[32]
CO_2Me	H	H	Me	rt	10	240	32	85	100	n.r.	[32]
CO_2Me	H	H	NH_2	60	40	4320	76	37	91:9	n.r.	[32]

[a] Selectivity at the given conversion (mol product indicated/mol substrate converted).
[b] Turnover frequency (mol substrate transformed per mol catalyst per hour); calculated at the conversion indicated and based on bulk Rh content; n.r. = not reported.
[c] The product is ethylcyclohexane.

Cyclohexanes 7; General Procedure by Reduction in the Presence of *t*-BuOK:[31]
The Rh catalyst precursor **5** (14.1 mg, 0.024 mmol) and *t*-BuOK (135 mg, 0.12 mmol) were dissolved in iPrOH (5 mL) and the resulting soln was transferred under H_2 to a 25-mL flask. The flask was connected to the H_2 reservoir and then immersed in a 60 °C bath. The mixture was shaken (500 rpm) for 1 h. The arene **6** (1.2 mmol) was added via syringe, and H_2 uptake was recorded. The reactions were monitored by measuring the H_2 consumption and by periodic GC analysis of samples removed using a syringe.

Cyclohexanes 7; General Procedure by Reduction in the Presence of Alumina:[31]
The Rh catalyst precursor **5** (14.1 mg, 0.024 mmol) was dissolved in iPrOH (5 mL) and the soln was transferred under H_2 to a 25-mL flask containing basic alumina (122.4 mg). The flask was connected to the H_2 reservoir and then immersed in a 60 °C bath. The mixture was shaken (500 rpm) for 1 h. The arene **6** (1.2 mmol) was added via syringe and H_2 uptake was recorded. In the case of phenol (**6**, R^1 = OH), the substrate was dissolved in iPrOH (2 mL) and the soln was then added via syringe to a preactivated mixture containing the catalyst precursor and alumina in iPrOH (3 mL). The reaction was monitored by measuring the H_2 consumption and by periodic GC analysis of samples removed using a syringe.

Cyclohexanes 9; General Procedure:[32]
The Rh(I)-metalated resin (ca. 1.3% w/w Rh; ca. 6 μmol of Rh) was swollen in degassed MeOH (5 mL). After 5 min, a soln of an arene **8** (1.45 mmol) in MeOH (4.1 mL) was added under N_2. H_2 was then bubbled through (1 bar H_2; 10 mL•min^{-1}) at rt while stirring at 170 rpm with an orbital stirrer. After 10 min, Rh@D400Li+in had been generated in situ. Then, the mixture was exposed to the desired H_2 pressure and temperature. After the indicated time, the methanolic soln was completely removed under a stream of H_2 using a gastight syringe. A sample of this soln (0.5 μL) was used for GC and GC/MS analysis.

1.3.1.3 Iridium-Catalyzed Reduction of Monocyclic Arenes to Cycloalkanes

Iridium nanoparticles show good catalytic performance for the synthesis of cycloalkanes **11** via the reduction of the corresponding arenes **10** under solvent-free conditions, with maximum conversion up to 100% and turnover frequency (TOF) up to 79 h^{-1} (Scheme 6). Monoalkyl-substituted aromatic rings are more easily reduced in comparison to multisubstituted aromatic compounds. The presence of other reducible groups is not tolerated. The iridium nanoparticles were synthesized into the nanopores of modified montmorillonite clay.[33]

Scheme 6 Reduction of Monocyclic Arenes to Cycloalkanes Catalyzed by Iridium Nanoparticles[33]

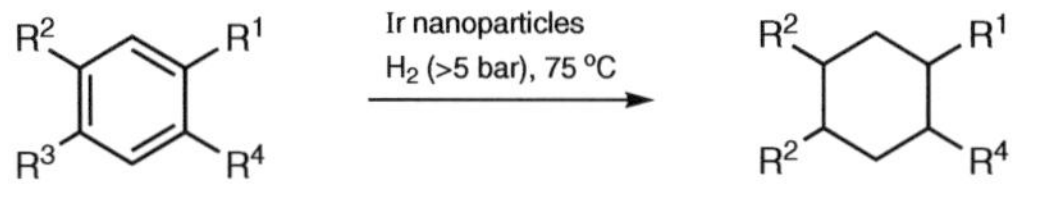

R2, R1, R3, R4 — 10 → Ir nanoparticles, H_2 (>5 bar), 75 °C → R2, R1, R2, R4 — 11

R^1	R^2	R^3	R^4	Conversion[a] (%)	TOF[b] (h^{-1})	Ref
Me	H	H	H	100	79	[33]
Me	H	H	Me	90	71	[33]
Me	Me	H	H	97	77	[33]
Me	H	Me	H	100	79	[33]
OH	H	H	Me	43	34	[33]
Me	OH	H	H	65	51	[33]
CH=CH_2	H	H	H	45[c]	35	[33]
Ac	H	H	H	73[d]	57	[33]

[a] Determined by GC analysis.
[b] Turnover frequency with respect to the product.
[c] The product is ethylcyclohexane.
[d] The product is 1-cyclohexylethanol.

Cyclohexanes 11; General Procedure:[33]

The reactions were carried out in a high-pressure 50-mL reactor (Parr Instrument Co.). An arene **10** (10 mmol) and Ir nanoparticles (50 mg) were placed into the reactor, which was then pressurized with H_2 to 5 bar at 25 °C. The reactions were performed at 75 °C under stirring at 500 rpm for 6 h. The products were analyzed by GC. The catalyst was recovered by simple filtration, and then washed and dried for recycling.

1.3.1.4 Palladium-Catalyzed Reduction of Monocyclic Arenes to Cycloalkanes

In the transformation of monocyclic arenes **12** into cycloalkanes **13**, a palladium-substituted polyoxometalate having a Keggin structure supported on active carbon ($K_5PPdW_{11}O_{39}$/C) is used as a catalyst precursor (Scheme 7).[34] The catalyst system enables fast reduction of arenes at 30 bar hydrogen and 200 °C. Most interesting is the finding that arenes can be selectively reduced in the presence of distal ketone groups under similar conditions. However, aromatic compounds with conjugated ketone moieties undergo complete reduction to saturated hydrocarbons.

Scheme 7 Reduction of Monocyclic Arenes to Cycloalkanes Catalyzed by a Palladium-Substituted Polyoxometalate Supported on Active Carbon[34]

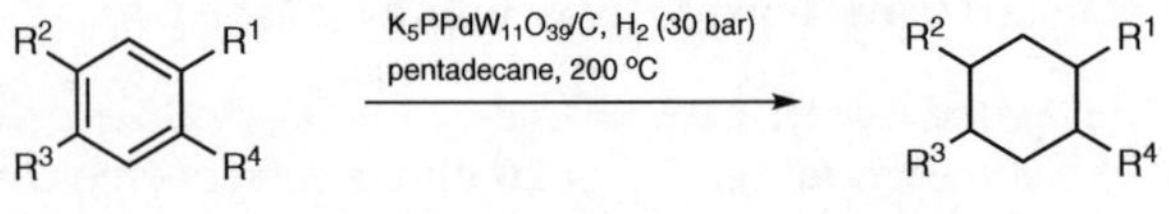

R2, R1, R3, R4 (**12**) → $K_5PPdW_{11}O_{39}$/C, H_2 (30 bar), pentadecane, 200 °C → R2, R1, R3, R4 (**13**)

R^1	R^2	R^3	R^4	Yield[a] (%)	Ref
Me	H	H	H	100[b]	[34]
Me	H	H	H	100[c]	[34]
Me	Me	Me	Me	100[b]	[34]
Bz	H	H	H	100[d]	[34]
Ac	H	H	H	100[e]	[34]
CH_2Ac	H	H	H	100	[34]
CH_2COEt	H	H	H	100	[34]
$(CH_2)_2Ac$	H	H	H	100	[34]
CMe_2Ac	H	H	H	5	[34]
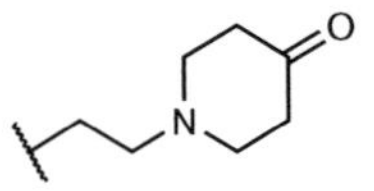 2-(4-oxopiperidin-1-yl)ethyl	H	H	H	68	[34]
CHO	H	H	H	100[f]	[34]

[a] Determined by GC.
[b] 230 °C, 1 h; yield determined by GC/MS.
[c] $K_5PPdW_{11}O_{39}$/alumina; yield determined by GC/MS.
[d] The product is dicyclohexylmethane.
[e] The product is ethylcyclohexane.
[f] The product is methylcyclohexane.

Cyclohexanes 13; **General Procedure:**[34]
The reactions were carried out in a 50-mL Parr autoclave. The arene **12** (1.75 mmol), 10 wt% $K_5PPdW_{11}O_{39}/C$ (0.3 g), and pentadecane (1 mL) were placed in the autoclave, which was then pressurized with H_2 to 30 bar. The mixture was stirred at 200 °C under H_2 for 4 h. The products were analyzed by GC and GC/MS, with comparison to reference samples.

1.3.2 Reduction of Polycyclic Arenes

Thermodynamic and kinetic studies have shown that the reduction of polyarenes is easier than that of benzene under moderate reaction conditions.[35] Polycyclic arenes can be reduced to the corresponding hydrogenated ring systems in the presence of metal catalysts such as ruthenium, rhodium, platinum, and palladium, as well as non-metal catalysts such as borane and Lewis pairs.

The reduction of polyarenes to afford cycloalkenes (e.g., 1,4-dihydronaphthalenes) with high selectivities remains a challenge. Some examples of the reduction of polycyclic arenes to cycloalkenes have been reported,[36] but an equivalent amount of manganese nanoparticles has to be used as reductant. The development of clean, efficient, and economic methods for the partial reduction of polycyclic arenes to alkenes with molecular hydrogen as reductant is still in demand.

1.3.2.1 Ruthenium-Catalyzed Reduction of Polycyclic Arenes

Ruthenium nanoparticles (Ru-NPs) stabilized by triphenylphosphine are competent reduction catalysts in the transformation of polycyclic arenes **14** containing two to four rings into hydrogenated products **15** under mild reaction conditions (Table 1).[37] Good to excellent selectivities can be obtained by optimizing the reaction conditions. Reduction of substrates bearing substituents at the 1-position is slower than that of substrates bearing substituents at the 2-position. In all cases, reduction takes place mainly at the less-substituted ring.

for references see p 150

Table 1 Reduction of Polycyclic Arenes Catalyzed by Ruthenium Nanoparticles[37]

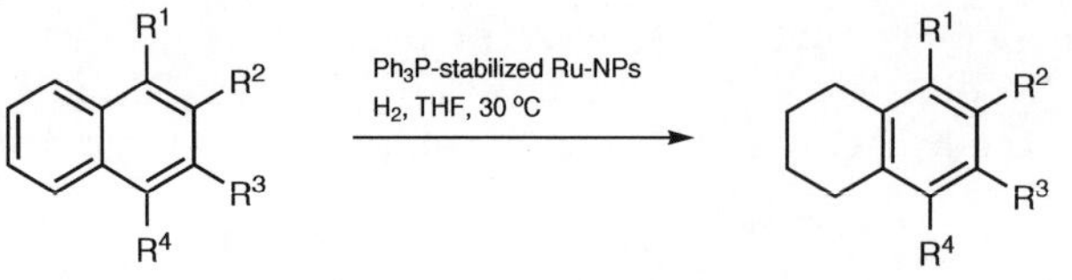

R^1, R^2, R^3, R^4 — **14** → (Ph_3P-stabilized Ru-NPs; H_2, THF, 30 °C) → R^1, R^2, R^3, R^4 — **15**

Substrate	Temp (°C)	Pressure (bar)	Time (h)	Product	Conversion[a] (%)	Selectivity (%)	TON[b]	Ref
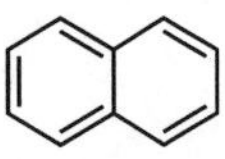	30	20	16	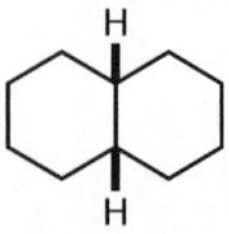 H H	100	84	39	[37]
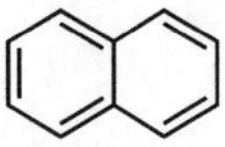	30	3	16	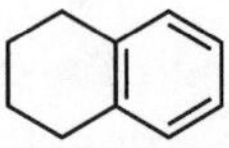	100	74	39	[37]
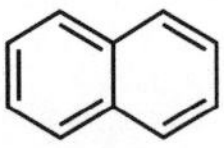	30	3	10	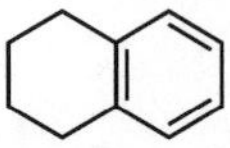	70	93	27	[37]
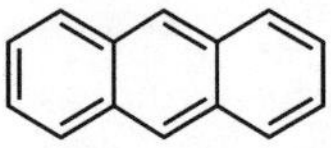	30	3	16	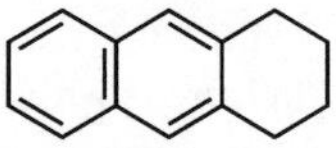	41	91	16	[37]
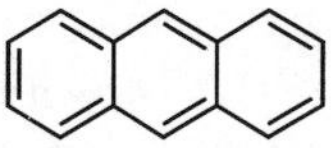	30	20	0.5	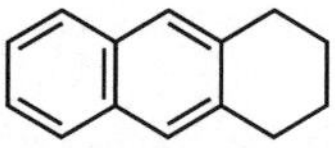	44	100	17	[37]
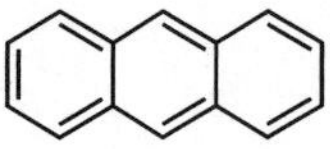	30	20	9	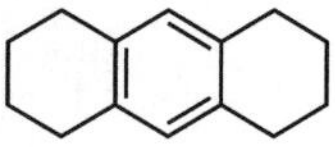	100	96	39	[37]
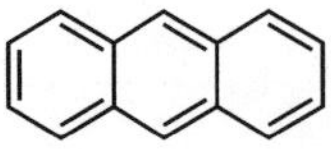	30	20	16	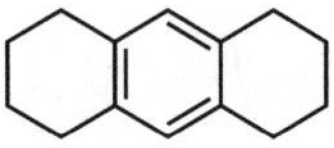	100	90	39	[37]
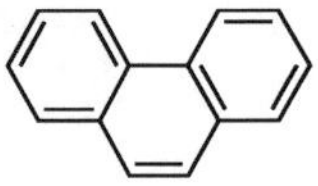	30	20	16	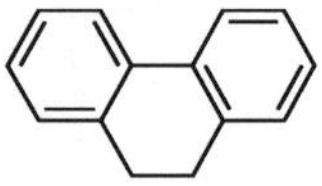	6	42	2	[37]
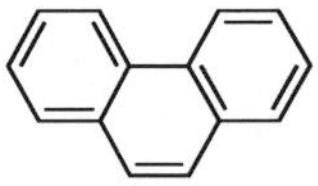	50	20	16	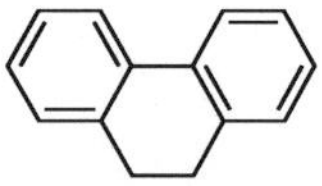	24	42	9	[37]
	30	20	16	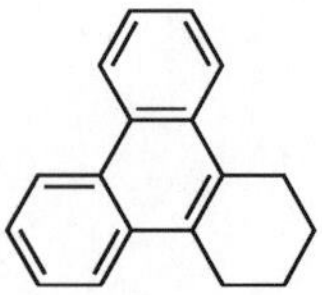	61	53	24	[37]

Table 1 (cont.)

Substrate	Temp (°C)	Pressure (bar)	Time (h)	Product	Conversion[a] (%)	Selectivity (%)	TON[b]	Ref
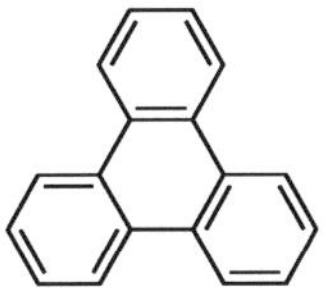	80	20	16	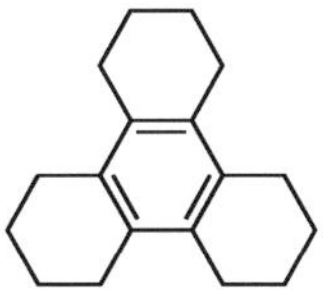	100	100	39	[37]
	80	20	60	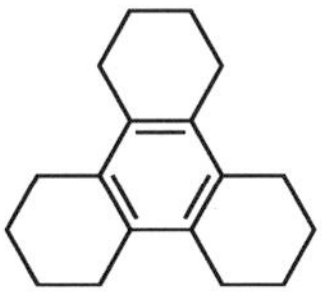	100	88	39	[37]
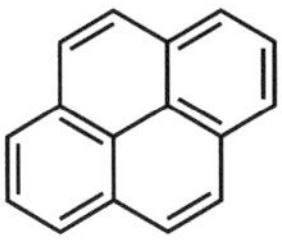	50	20	16	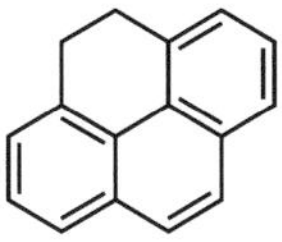	17	93	7	[37]
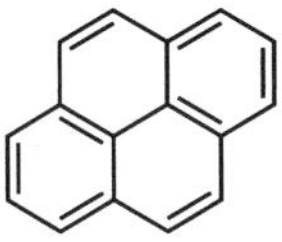	80	20	16	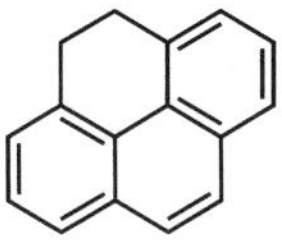	25	90	10	[37]
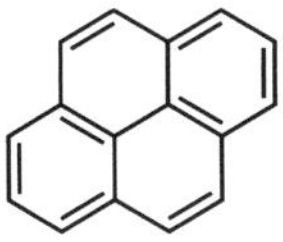	80	20	60	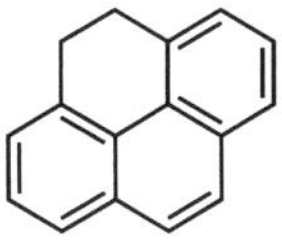	44	86	17	[37]
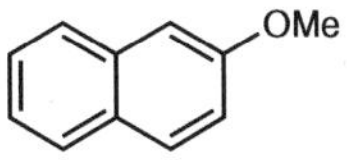 OMe	30	20	2.5	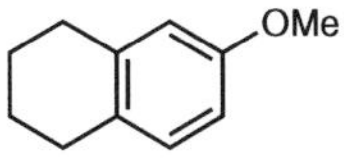 OMe	31[c]	83[c]	24[c]	[37]
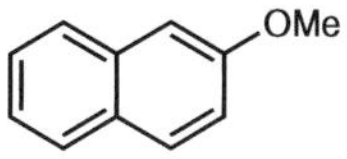 OMe	30	20	2.5	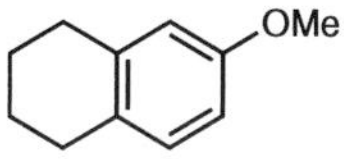 OMe	91	83	35	[37]
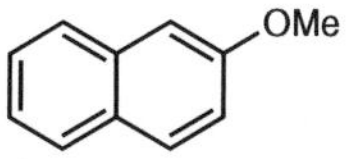 OMe	30	10	2.5	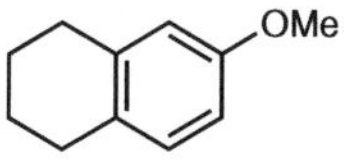 OMe	11	83	4	[37]
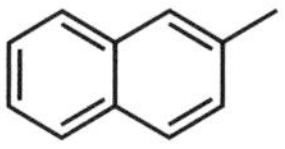	30	20	16	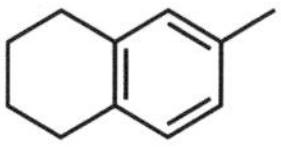	6	79	2	[37]
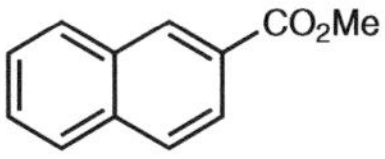 CO_2Me	30	20	16	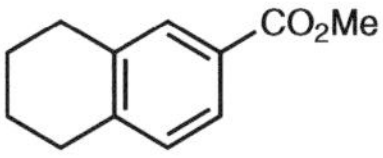 CO_2Me	0	–	0	[37]
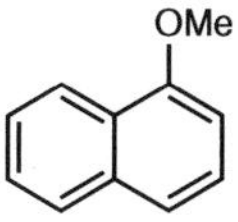 OMe	30	20	2.5	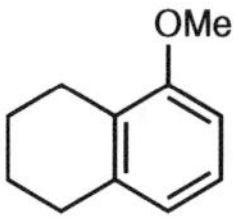 OMe	40	85	15	[37]

for references see p 150

Table 1 (cont.)

Substrate	Temp (°C)	Pressure (bar)	Time (h)	Product	Conversion[a] (%)	Selectivity (%)	TON[b]	Ref
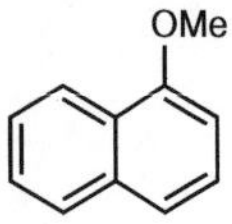 OMe	30	20	16	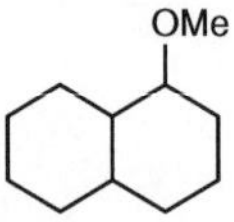 OMe	100	65	39	[37]
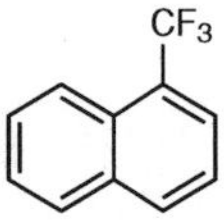 CF_3	30	20	2.5	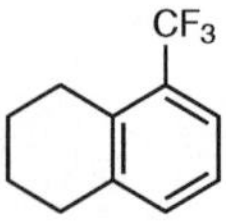 CF_3	45	63	17	[37]
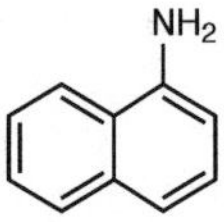 NH_2	30	20	16	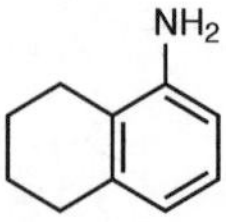 NH_2	16	100	6	[37]
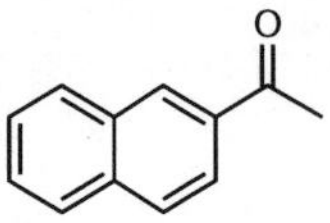 O	30	20	2.5	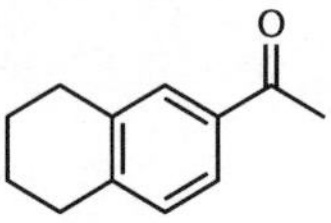 O	100	52	39	[37]
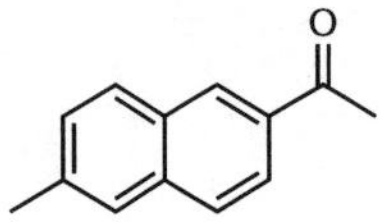 O	30	20	2.5	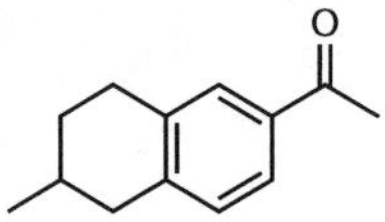 O	44	16	17	[37]
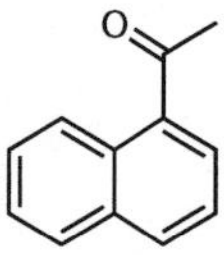 O	30	20	2.5	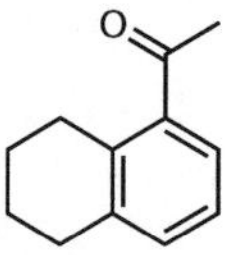 O	100	38	39	[37]
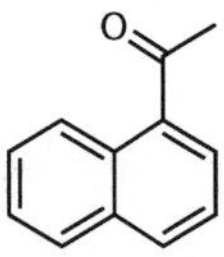 O	30	10	16	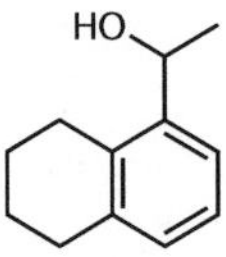 HO	100	36	39	[37]

[a] Determined by GC.
[b] Turnover number (mol substrate converted per mol of surface Ru).
[c] 1.24 mmol substrate was used; in all other cases, 0.62 mmol was used.

Hydrogenated Polyarenes 15**; General Procedure:**[37]
A five-entry autoclave or an autoclave Par 477 equipped with PID control temperature and a reservoir for kinetic measurements was charged in a glovebox with Ph_3P-stabilized Ru nanoparticles[37] (3 mg, 2 mol%; the catalyst concentration was calculated based on the total number of metallic atoms in the nanoparticles) and the arene **14** (0.62 mmol) in THF (10 mL). H_2 was then introduced until the desired pressure was reached. The mixture was stirred for the indicated time at the desired temperature. The autoclave was then depressurized and the soln was filtered through silica gel. The filtrate was analyzed by GC. The conversion and the selectivities of the product were determined using a Fisons instrument (GC9000 series) equipped with an HP-5MS column. The *cis*/*trans* selectivity was confirmed by NOE experiments where relevant.

1.3.2.2 Rhodium-Catalyzed Reduction of Polycyclic Arenes

Rhodium nanoparticles deposited on the surface of carboxylate-functionalized multi-walled carbon nanotubes using a simple one-step sonochemical method provide a promising catalyst for the reduction of polycyclic arenes. A variety of polycyclic arenes **16** can be reduced to afford the corresponding hydrogenated products **17**. Good to excellent selectivities can be obtained just by optimizing the reaction time (Table 2).[38]

Table 2 Partial Reduction of Polycyclic Arenes Catalyzed by Rhodium Nanoparticles Supported on Multi-walled Carbon Nanotubes[38]

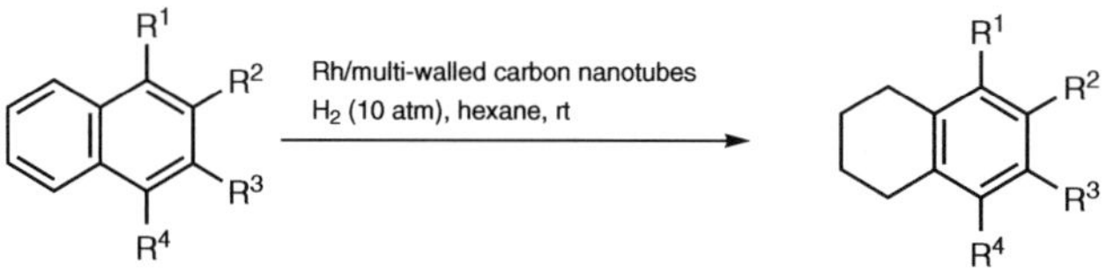

16 → **17**: Rh/multi-walled carbon nanotubes, H_2 (10 atm), hexane, rt

Substrate	Time (h)	Product	Conversion[a] (%)	Selectivity[a] (%)	Ref
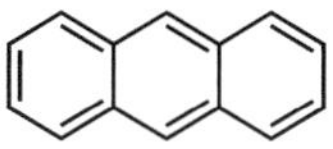	0.1	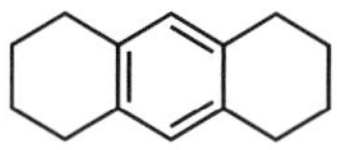	30.1	67.9	[38]
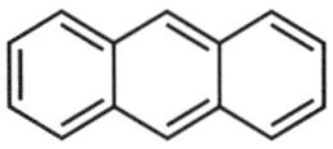	0.5	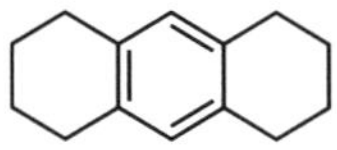	100	80.2	[38]
	0.5	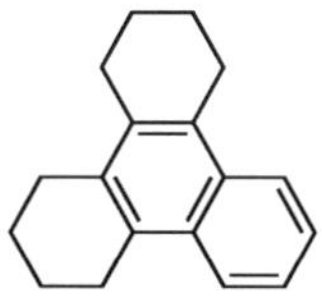	22.1	46.6	[38]
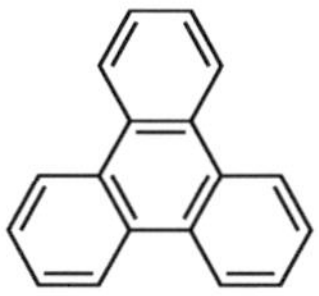	3	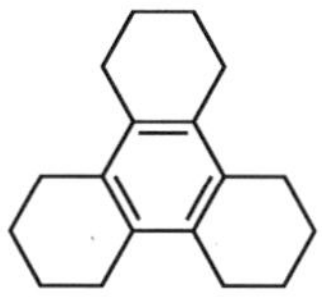	100	94.8	[38]
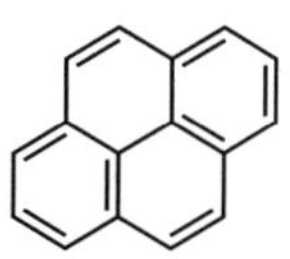	0.1	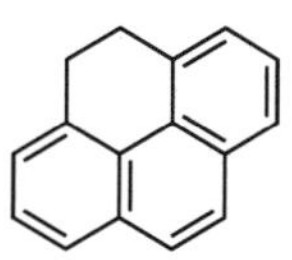	25.6	71.1	[38]
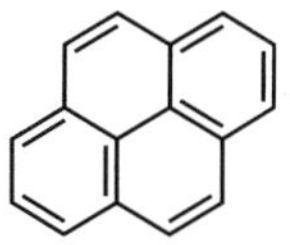	0.5	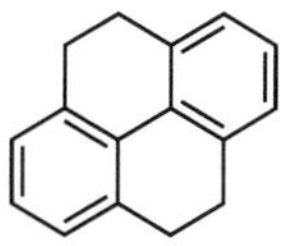	100	39.2	[38]
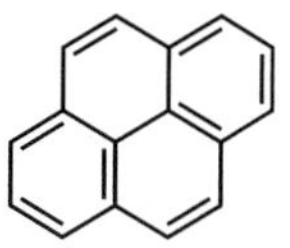	3	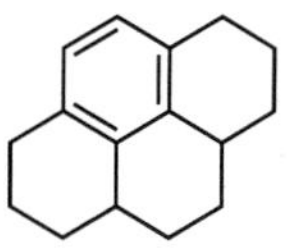	100	32.6	[38]

[a] Determined by GC/MS.

for references see p 150

Hydrogenated Polyarenes 17; Typical Procedure:[38]

The arene **16** (0.1 mmol), Rh nanoparticles supported on multi-walled carbon nanotubes (10 mg), and hexane (10 mL) were placed in a 20-mL cylindrical glass vial (2.5 cm in diameter and 6.5 cm in height) with a magnetic stirrer bar. The vial was then placed in a 130-mL homemade stainless-steel high-pressure reactor and the mixture was stirred (600 rpm) at rt (20 °C). The reaction cell was first flushed with H_2 for 1 min to replace the air. The outlet valve was then closed to maintain H_2 pressure (10 atm) in the system, controlled by a large reservoir of H_2 connected to the reaction cell. GC/MS (Shimadzu GCMS-QP2010) and the NIST mass spectral database 2005 were used for analysis of the products.

1.3.2.3 Platinum-Catalyzed Reduction of Polycyclic Arenes

The synthesis of partially reduced ring systems **19** from polycyclic arenes **18** can be achieved using platinum nanoparticles supported on carboxylate-functionalized multi-walled carbon nanotubes as the catalyst at room temperature (Table 3). Tetra-, octa-, or dodecahydro products can be produced just by changing the reaction time. However, the complete ring saturation of polycyclic arenes cannot be achieved.[38]

Table 3 Partial Reduction of Polycyclic Arenes Catalyzed by Platinum Nanoparticles Supported on Multi-walled Carbon Nanotubes[38]

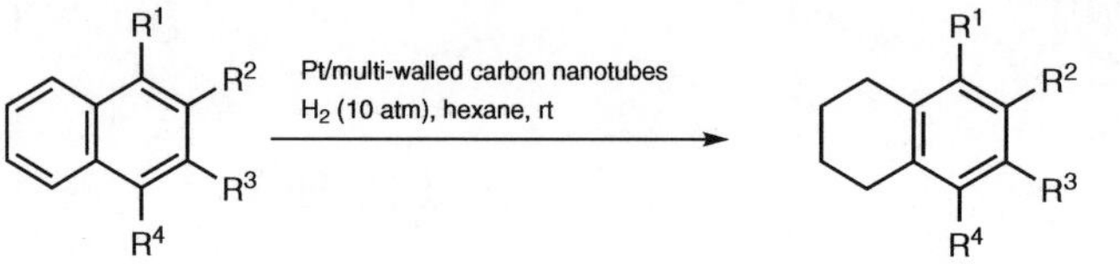

R1, R2, R3, R4 **18** → Pt/multi-walled carbon nanotubes, H_2 (10 atm), hexane, rt → R1, R2, R3, R4 **19**

Substrate	Time (h)	Product	Conversion[a] (%)	Selectivity[a] (%)	Ref
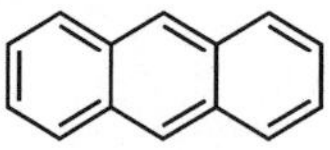	0.1	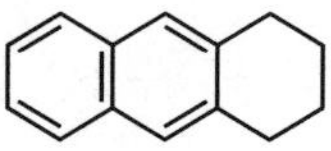	26.0	80.4	[38]
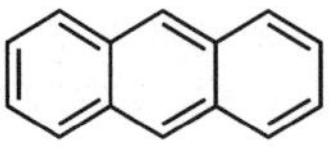	0.5	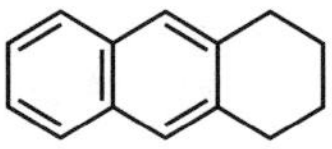	100	74.7	[38]
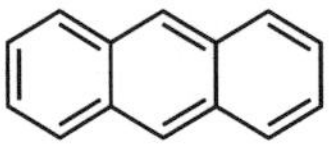	1	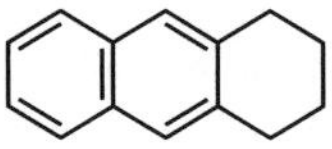	100	49.7	[38]
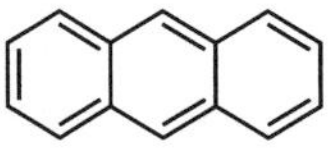	4	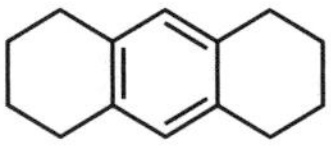	100	75.1	[38]
	0.5		21.2	45.5	[38]

Table 3 (cont.)

Substrate	Time (h)	Product	Conversion[a] (%)	Selectivity[a] (%)	Ref
	3		100	100	[38]
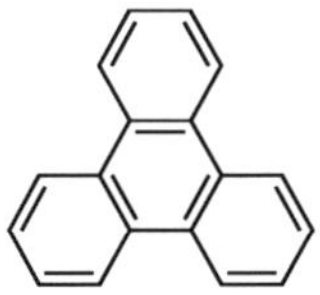	6	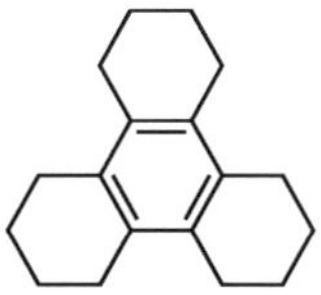	100	97.4	[38]
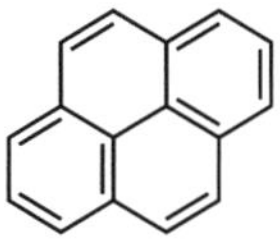	0.1	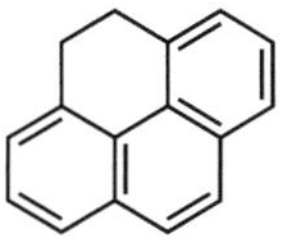	0	–	[38]
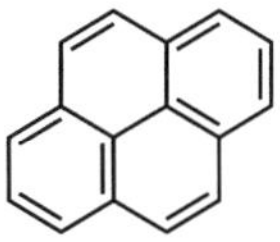	0.5	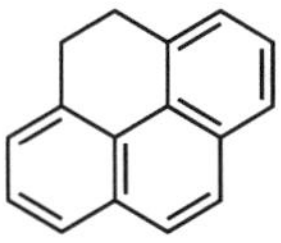	2.6	100	[38]
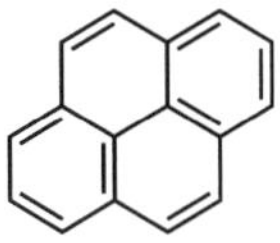	3	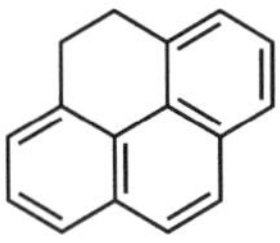	7.3	100	[38]

[a] Determined by GC/MS.

Hydrogenated Polyarenes 19**; General Procedure:**[38]
The arene **18** (0.1 mmol), Pt nanoparticles supported on multi-walled carbon nanotubes (10 mg), and hexane (10 mL) were placed in a 20-mL cylindrical glass vial (2.5 cm in diameter and 6.5 cm in height) with a magnetic stirrer bar. The vial was then placed in a 130-mL homemade stainless-steel high-pressure reactor and the mixture was stirred (600 rpm) at rt (20 °C) during the reaction. The reaction cell was first flushed with H_2 for 1 min to replace the air. The outlet valve was then closed to maintain H_2 pressure (10 atm) in the system controlled by a large reservoir of H_2 connected to the reaction cell. GC/MS (Shimadzu GCMS-QP2010) and the NIST mass spectral database 2005 were used for analysis of the products.

1.3.2.4 Palladium-Catalyzed Reduction of Polycyclic Arenes

A series of representative polycyclic arenes **20** can be transformed into the corresponding K-region hydroarenes **21** over palladium on charcoal at low pressure and ambient temperature (Table 4).[39] For most of the substrates, good to excellent yields are obtained by optimizing the reaction conditions.

Table 4 Partial Reduction of Polycyclic Arenes Catalyzed by Palladium on Charcoal[39]

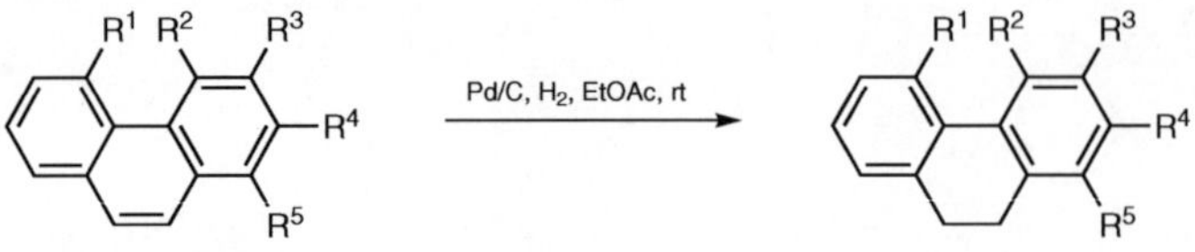

20 → 21 (Pd/C, H_2, EtOAc, rt); substituents R^1, R^2, R^3, R^4, R^5

Entry	Substrate	Pressure (psig)	Time (h)	Product	Yield (%)	Ref
1	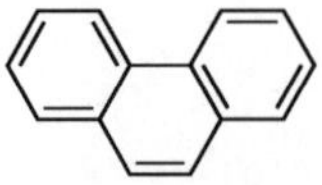	20	120	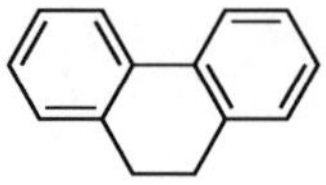	70	[39]
2	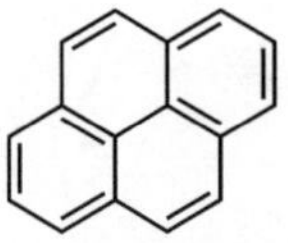	50	65	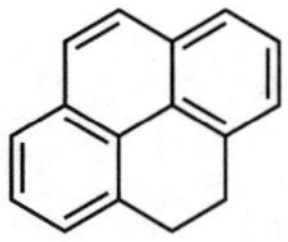	45	[39]
3	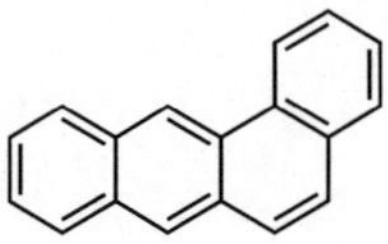	20	10	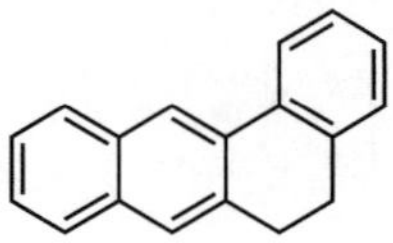	97	[39]
4	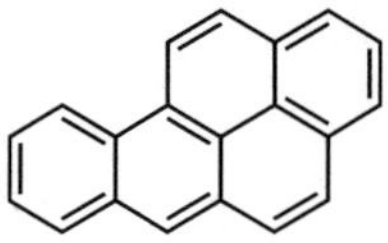	20	24	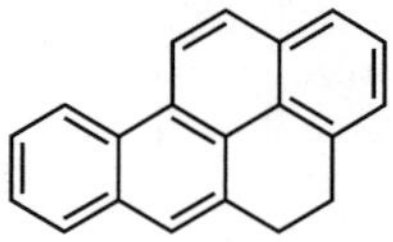	100	[39]
5	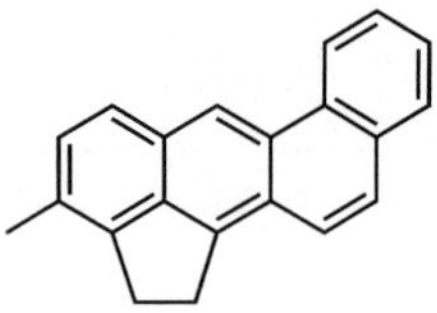	20	18	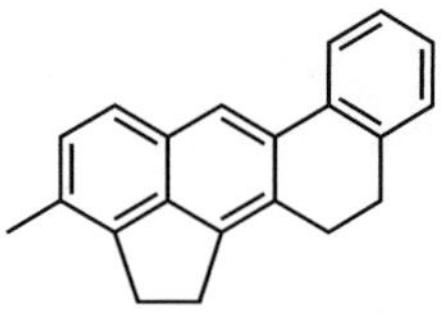	100	[39]
6	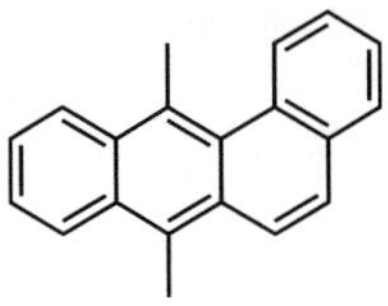	20	8	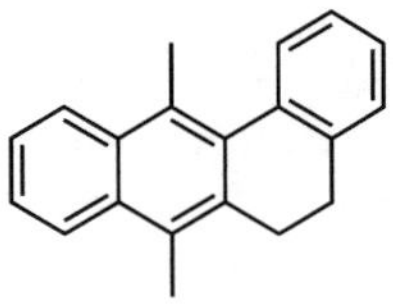	80	[39]
7	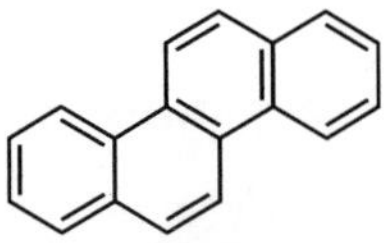	20	20	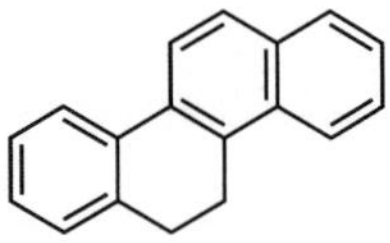	30	[39]
8	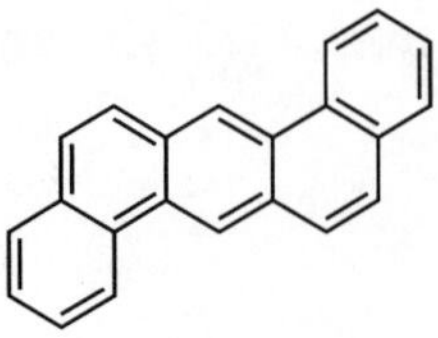	20	5	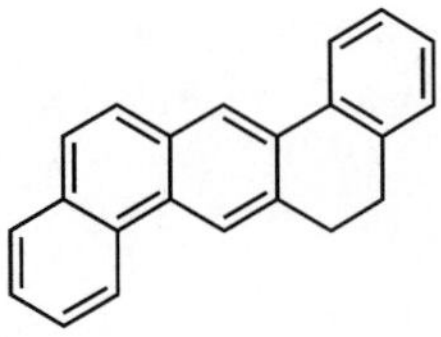	70	[39]

Table 4 (cont.)

Entry	Substrate	Pressure (psig)	Time (h)	Product	Yield (%)	Ref
9	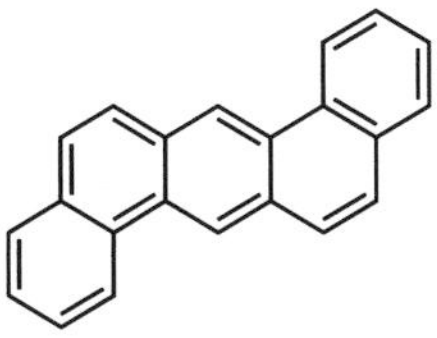	20	24	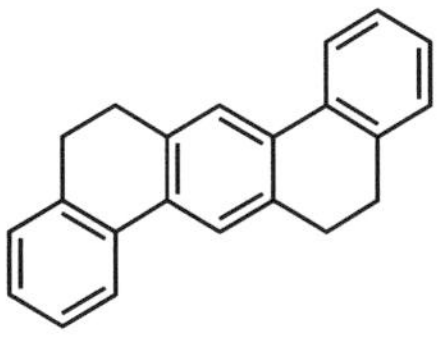	100	[39]
10	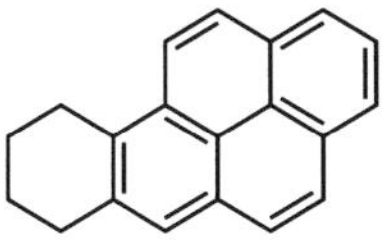	45	18	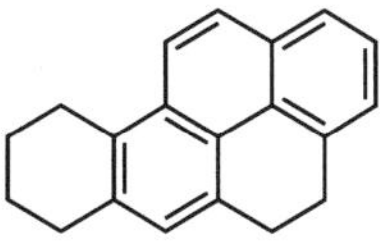	50	[39]
11	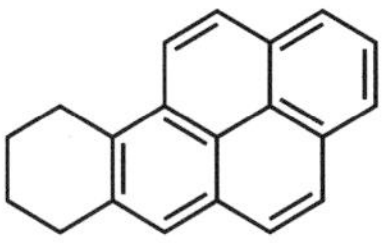	50	48	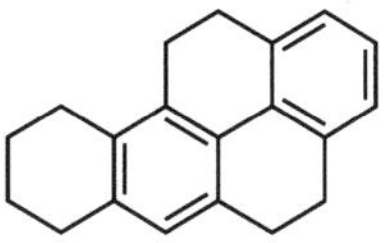	100	[39]
12	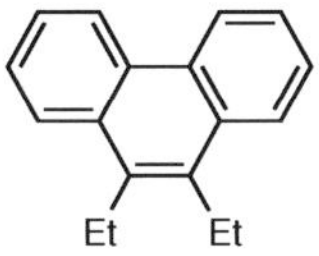 Et Et	50	65	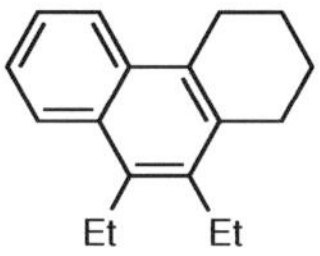 Et Et	64	[39]

5,6-Dihydrotetraphene (Table 4, Entry 3); Typical Procedure:[39]
The reaction was conducted in a Vortex low-pressure hydrogenator manufactured by J. B. Tetraphene (180 mg, 0.79 mmol), 10% Pd/C (40 mg), and EtOAc (15 mL) were introduced into a 500-mL Pyrex bottle which was flushed three times with H_2 before being pressurized to 20 psig. The mixture was stirred magnetically at moderate speed at ambient temperature for 10 h, and then filtered through Celite with several acetone washes. The filtrate was concentrated to dryness. The residue was taken up in hexane or another appropriate solvent, and purified by passage through a short column (Florisil); yield: 97%.

1.3.2.5 Borane-Catalyzed Reduction of Polycyclic Arenes

Tetrapropyldiborane can be used to catalyze the regioselective and partial reduction of naphthalene and its derivatives **22** at 200 °C under hydrogen pressures of 100 bar. Good to excellent yields of tetrahydroarenes **23** can be obtained by optimizing the amount of the catalyst. Hydrogenation of the naphthalene derivatives mainly occurs at the least substituted ring (Scheme 8).[40]

for references see p 150

Scheme 8 Partial Reduction of Polycyclic Arenes Catalyzed by Tetrapropyldiborane[40]

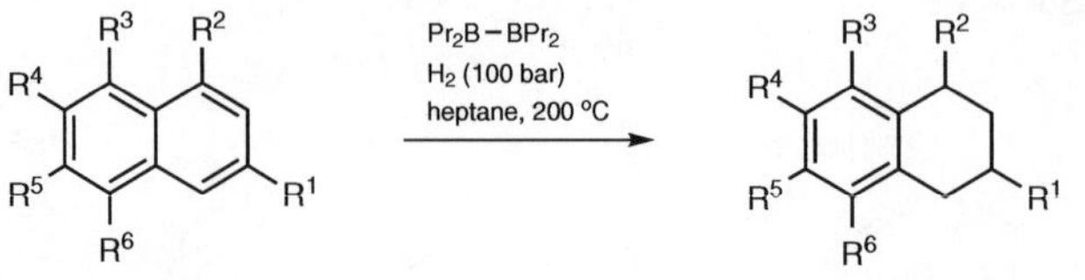

Pr_2B-BPr_2, H_2 (100 bar), heptane, 200 °C: **22** → **23**

R^1	R^2	R^3	R^4	R^5	R^6	Substrate (mmol)	Catalyst (mmol)	Yield[a] (%)	Ref
H	H	H	H	H	H	78	12	99	[40]
H	H	Me	H	H	H	14	18	18	[40]
H	H	Me	H	H	H	20	3	76	[40]
H	H	H	Me	H	H	14	12	85	[40]
H	H	Et	H	H	H	32	12	44	[40]
H	H	Et	H	H	H	20	3	70	[40]
H	H	H	Et	H	H	32	12	72	[40]
H	H	Me	H	Me	H	13	12	89	[40]
H	H	Me	H	H	Me	13	6	62	[40]
H	H	H	Me	Me	H	32	12	94	[40]
Me	H	H	Me	H	H	32	12	66	[40]
Me	H	H	Me	H	H	32	4	76	[40]
H	CH_2CH_2		H	H	H	33	12	14	[40]
H	CH=CH		H	H	H	32	12	7[b]	[40]
H	H	Ph	H	H	H	10	6	91	[40]
H	H	H	Ph	H	H	10	6	95	[40]
H	H	H	H	2-naphthyl	H	4	6	96[c]	[40]
H	H	OMe	H	H	H	32	12	39	[40]
H	H	H	OMe	H	H	32	12	31	[40]
H	H	$OBEt_2$	H	H	H	35	12	68[d]	[40]
H	H	Cl	H	H	H	31	12	38	[40]
H	H	H	Cl	H	H	25	6	75	[40]

[a] Determined by GC.
[b] The product is 1,2,2a,3,4,5-hexahydroacenaphthylene (**23**, $R^1 = R^4 = R^5 = R^6 = H$; $R^2,R^3 = CH_2CH_2$).
[c] The product is 5,5′,6,6′,7,7′,8,8′-octahydro-2,2′-binaphthalene (**23**, $R^1 = R^2 = R^3 = R^4 = R^6 = H$; R^5 = 5,6,7,8-tetrahydronaphthalen-2-yl).
[d] The product is 1,2,3,4-tetrahydronaphthalene (**23**, $R^1 = R^2 = R^3 = R^4 = R^5 = R^6 = H$).

1,2,3,4-Tetrahydronaphthalene (23, **$R^1 = R^2 = R^3 = R^4 = R^5 = R^6 = H$); Typical Procedure:**[40]
A soln of naphthalene (10.0 g, 78.1 mmol) and tetrapropyldiborane (2.44 g, 12.6 mmol) in heptane (100 mL) was placed in an autoclave, which was then pressurized to 100 bar with H_2. The mixture was heated at 200 °C for 4 h. After cooling to rt, the pressure decrease generally corresponded to an uptake of more than 2 equivalents of H_2 (dissolution of the propane formed from the catalyst). After venting the gas, the liquid suspension was filtered to separate the solid polyboranes formed. The filtrate was subjected to GC and GC/MS analysis; yield: 99%.

1.3.2.6 Reduction of Polycyclic Arenes Catalyzed by Lewis Pairs

The frustrated Lewis pair tris(pentafluorophenyl)borane/(pentafluorophenyl)diphenylphosphine is an efficient catalyst for the reduction of polycyclic arenes **24** at 80 °C and 8.2 MPa hydrogen gas pressure (Scheme 9).[41] The majority of the substrates can be reduced smoothly to afford the corresponding dihydroanthracenes and derivatives **25** in good yields. However, the reaction is less efficient in the case of tetraphene.

Scheme 9 Partial Reduction of Polycyclic Arenes Catalyzed by Tris(pentafluorophenyl)borane/(Pentafluorophenyl)diphenylphosphine[41]

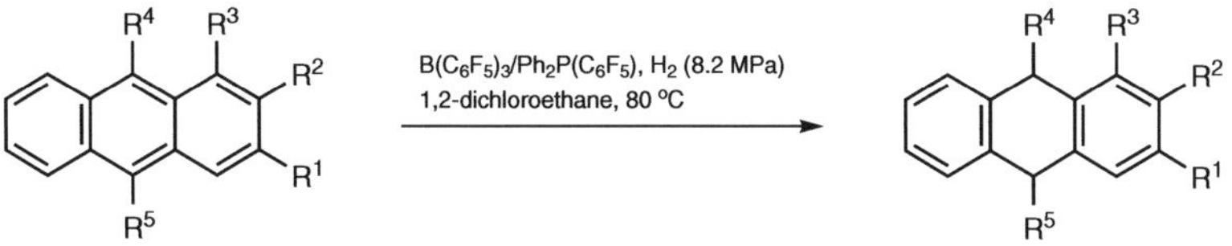

$B(C_6F_5)_3/Ph_2P(C_6F_5)$, H_2 (8.2 MPa)
1,2-dichloroethane, 80 °C

24 → **25**

R^1	R^2	R^3	R^4	R^5	Time (h)	Yield (%)	Ref
H	H	H	H	H	10	97	[41]
H	H	H	Me	H	10	95	[41]
H	H	H	Me	Me	48	80[a]	[41]
$(CH{=}CH)_2$		H	H	H	10	90	[41]
H	$(CH{=}CH)_2$		H	H	48	30	[41]

[a] Ratio (*cis*/*trans*) 3:1.

9-Methyl-9,10-dihydroanthracene (25, $R^1 = R^2 = R^3 = R^5 = H$; $R^4 = Me$); Typical Procedure:[41]
In a glovebox filled with N_2, a 6-mL glass vial equipped with a stirrer bar was charged with a yellow soln of $B(C_6F_5)_3$ (10 mg, 20 μmol), $Ph_2P(C_6F_5)$ (7 mg, 20 μmol), and 9-methylanthracene (**24**, $R^1 = R^2 = R^3 = R^5 = H$; $R^4 = Me$; 38 mg, 200 μmol) in 1,2-dichloroethane (2 mL). The vial was put into a pressure vessel reactor (Parr Instrument Company). The reactor was filled with H_2 (8.2 MPa), and the mixture was stirred at 80 °C for 10 h. The mixture was concentrated under reduced pressure and the crude product was purified by column chromatography (silica gel, hexane); yield: 95%.

1.3.3 Reduction of Monocyclic Arenes to Cycloalkenes

As the intermediate product of reduction of benzene to cyclohexane, cyclohexene and its derivatives can be synthesized by the partial reduction reaction of the corresponding monocyclic arenes. This process represents another important application of the reduction of arenes. However, the formation of cyclohexene through this route is thermodynamically unfavored. The standard free energy change for cyclohexene formation from benzene reduction (23 kJ·mol^{-1}) is much less negative than that for cyclohexane formation (98 kJ·mol^{-1}).[42] Hence, high-performance catalysts for benzene reduction aimed at kinetically manipulating the selectivity for cyclohexene have been intensively pursued. To date, ruthenium and rhodium are promising active metals for this selective reduction of monocyclic arenes to the corresponding cycloalkenes, although yields of the products are low.

As early as 1957, cyclohexene was detected among the products in the gas-phase reduction of benzene with nickel film as a catalyst.[43] The amount of cyclohexene formed was minor, but the selectivity was as high as 19%. Since then, several investigations have

for references see p 150

been reported for the partial reduction of benzene to cyclohexene in the gas phase.[44–46] However, the reported selectivities to cyclohexene in the gas-phase reduction of benzene are substantially lower than those attained in liquid-phase processes. Therefore, the liquid-phase reduction of benzene to cyclohexene has attracted more attention, and the use of various metal catalysts, such as ruthenium, nickel, platinum, and cobalt, has been investigated. To date, ruthenium is widely accepted as the most effective metal for selective reduction of benzene to cyclohexene. Various ruthenium-based catalysts, such as ruthenium nanoparticles,[47] ruthenium-alloy catalysts,[48–53] and supported ruthenium catalysts,[50,54–65] have been studied thoroughly (Scheme 10). The first industrially performed partial reduction of benzene in the liquid phase was carried out over a non-supported solid ruthenium catalyst, which was developed by Asahi Chemical Industry Co., Ltd.[66]

Scheme 10 Evolution of Catalysts for the Reduction of Benzene to Cyclohexene[48–54,56,61–66]

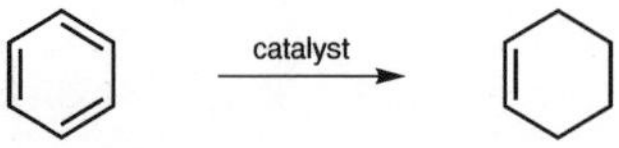

catalyst

Catalyst	Temp (°C)	Pressure (MPa)	Yield (%)	Ref
RuZn	150	0.5	60	[66]
Ru/ZrO_2	140	5	51	[56]
Ru/P25 TiO_2[a]	140	5	61	[54]
Ru/La_2O_3-ZnO	150	4.31	34	[65]
RuB/ZrO_2	140	4	47	[53]
RuB/silica gel	150	4	34	[48]
RuB/alumina•xH_2O	145	4.28	39.6	[51]
RuZn/*m*-ZrO_2[b]	145	4.28	43.4	[64]
RuZn/ZrO_2	150	5	48.5	[61]
RuZn/hydroxyapatite	150	5	33	[62]
RuCu/ZnO	150	4	49.4	[63]
RuCoB/γ-alumina	150	5	34.8	[52]
RuLaB/ZrO_2	140	4.5	53.7	[49]
RuBZn/ZrO•xH_2O	160	5	45.6	[50]

[a] P25 TiO_2 powder (51 m^2/g) purchased from Degussa.
[b] *m*-ZrO_2 = monoclinic zirconium(IV) oxide.

The technology is also applied in the commercial production of cyclohexanol from benzene through cyclohexene. The yield of cyclohexene can reach 60%. One of the most remarkable features of the reaction is that the system comprises four phases: vapor (hydrogen), oil, aqueous, and solid. The catalyst is maintained in the aqueous phase, and the reactants (benzene and hydrogen) are dissolved in the aqueous phase where the reaction proceeds. Thus, the reactants and products transfer between four phases through dissolution, diffusion, and extraction; achieving quick phase transfer is a very important factor in enhancing reaction selectivity.

1.3.3.1 Ruthenium-Catalyzed Reduction of Monocyclic Arenes to Cycloalkenes

A wide variety of aromatic hydrocarbons **26**, representative of components of petroleum-derived fuels, can be partially reduced to afford the corresponding cyclohexene derivatives **27** in the presence of ruthenium nanoparticles immobilized on poly(4-vinylpyridine), although yields of the cycloalkene products are low (Scheme 11).[67]

Scheme 11 Reduction of Monocyclic Arenes to Cycloalkenes Catalyzed by Ruthenium Nanoparticles Immobilized on Poly(4-vinylpyridine)[67]

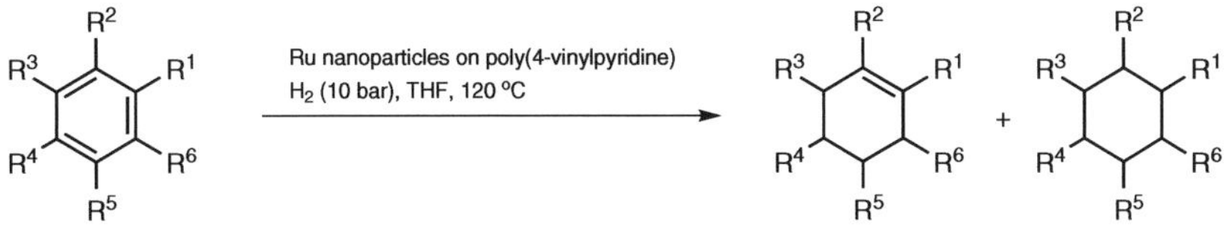

Ru nanoparticles on poly(4-vinylpyridine)
H_2 (10 bar), THF, 120 °C

26 → **27** + cycloalkane

R^1	R^2	R^3	R^4	R^5	R^6	Yield[a] (%)		TOF[b] (h^{-1})	Ref
						27	Cycloalkane		
H	H	H	H	H	H	–	100	–	[67]
Me	H	H	H	H	H	1	99	69	[67]
Me	Me	H	H	H	H	21	79	24	[67]
Me	H	Me	H	H	Me	19	81	23	[67]
Me	H	H	Me	H	H	29	71	16	[67]
Me	H	Me	H	Me	H	38	62	3	[67]
Me	H	Me	H	Me	H	10[c]	90	40	[67]
OMe	H	H	H	H	H	22	78	37	[67]
CF_3	H	H	H	H	H	2	98	56	[67]

[a] Determined by GC/MS.
[b] Turnover frequency (mol aromatic substrate converted per mol of Ru per hour).
[c] 150 °C, H_2 (50 bar).

1,3,5-Trimethylcyclohex-1-ene (27, $R^1 = R^3 = R^5 =$ Me; $R^2 = R^4 = R^6 =$ H); Typical Procedure:[67]
The reactor was loaded with the Ru nanoparticles immobilized on poly(4-vinylpyridine) [ratio (substrate/Ru) 100:1] and THF (20 mL) and then sealed. H_2 was introduced into the reactor through a high-pressure buret and released three times to deoxygenate the system, after which the reactor was repressurized to 6 bar and heated to 120 °C to reach 10 bar. After the catalyst had been incubated at 120 °C under 10 bar for 1 h, mesitylene (**26**, $R^1 = R^3 = R^5 =$ Me; $R^2 = R^4 = R^6 =$ H; 1 mL) was dissolved in THF (10 mL) and the soln was placed into the high-pressure buret, which was subsequently charged with H_2 (10 bar). The reactor was depressurized to about 2–3 bar and the soln of mesitylene (**26**, $R^1 = R^3 = R^5 =$ Me; $R^2 = R^4 = R^6 =$ H) in the buret was then quickly injected into the reactor. The pressure was adjusted to 10 bar (this was taken as the zero time of the reaction) and kept constant by feeding through an open connection to the H_2 tank. Samples of the mixture were periodically withdrawn from the reactor and analyzed immediately by use of a Varian 3900 gas chromatograph fitted with a FactorFour VF-5ms capillary column and a Saturn 2100T mass detector. The identity of each product was verified through comparison of its mass spectrum with the library of the instrument and the molar percentage of each product was calculated based on peak areas, previously calibrated by using a series of

for references see p 150

standard solns containing known amounts of each component. Each experiment was repeated at least twice to verify reproducibility; the variations in the calculated TOF values for repeat experiments were typically within 10%.

1.3.3.2 Rhodium-Catalyzed Reduction of Monocyclic Arenes to Cycloalkenes

Rhodium nanoparticles stabilized by the ionic-liquid-like copolymer poly[(1-vinylpyrrolidin-2-one)-*co*-(1-vinyl-3-butylimidazolium chloride)] [poly(PVP-*co*-VBIMCl] is a promising catalyst for the reduction of benzenes **28** in ionic liquids (Scheme 12). In addition to cyclohexanes **31**, the cycloalkene products **29** and **30** are obtained as well, although the selectivities are poor.[68]

Scheme 12 Reduction of Monocyclic Arenes to Cycloalkenes Catalyzed by Rhodium Nanoparticles[68]

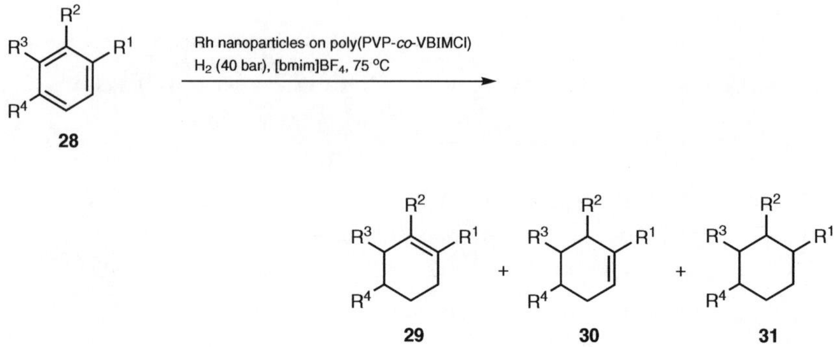

Rh nanoparticles on poly(PVP-*co*-VBIMCl)
H_2 (40 bar), [bmim]BF_4, 75 °C

28 → 29 + 30 + 31

R1	R2	R3	R4	Conversion[a] (%)	Selectivity[a] (%)			TOF[b] (h^{-1})	Ref
					29	**30**	**31**		
H	H	H	H	96	–	–	100	160	[68]
Me	H	H	H	95	<1	–	>99	158	[68]
Et	H	H	H	82	<1	–	>99	137	[68]
Pr	H	H	H	22	19	–	19	130	[68]
s-Bu	H	H	H	37	38	–	62	62	[68]
OH	H	H	H	37	29	–	71	247	[68]
OMe	H	H	H	84	–	–	>99	140	[68]
CH_2OH	H	H	H	15	26	–	74	100	[68]
Me	Me	H	H	33	16	16	68	55	[68]
Me	H	Me	H	42	6	6	88	70	[68]
Me	H	H	Me	41	10	–	90	68	[68]
Pr	H	H	OH	11	19	–	32	25	[68]
OMe	H	H	OH	21[c]	29	–	71	70[c]	[68]

[a] Determined by GC.
[b] Turnover frequency (mol substrate conversion per total mol Rh per hour) during the first 10 h.
[c] Ratio (substrate/Rh) 1000:1; in all other cases it was 2000:1.

1-*sec*-Butylcyclohex-1-ene (29, R1 = *s*-Bu; R2 = R3 = R4 = H); Typical Procedure:[68]
A mixture of *sec*-butylbenzene [**28**, R1 = *s*-Bu; R2 = R3 = R4 = H; 32 mmol; ratio (substrate/Rh) 2000:1] and Rh nanoparticles stabilized by poly(PVP-*co*-VBIMCl) [ratio (copolymer/Rh) 5:1; 16 μmol Rh; M_n (copolymer) = 50 400, M_w (copolymer) = 75 300] in 1-butyl-3-methylimida-

zolium tetrafluoroborate {[bmim]BF_4} (6 mL) was placed in an autoclave. The reactor was purged three times with H_2, and then the autoclave was pressurized to 40 bar with H_2. The mixture was stirred at 800 rpm at 75 °C for 12 h, which afforded a two-phase system. The lower ionic liquid phase contained the catalyst. The upper organic phase was decanted and analyzed by GC, GC/MS, IR, and ^1H NMR spectroscopy.

for references see p 150

References

[1] Nishimura, S., *Handbook of Heterogeneous Catalytic Hydrogenation for Organic Synthesis*; Wiley: New York, (2001).
[2] Harley, R. A.; Hooper, D. S.; Kean, A. J.; Kirchstetter, T. W.; Hesson, J. M.; Balberan, N. T.; Stevenson, E. D.; Kendall, G. R., *Environ. Sci. Technol.*, (2006) **40**, 5084.
[3] Gu, W.; Stalzer, M. M.; Nicholas, C. P.; Bhattacharyya, A.; Motta, A.; Gallagher, J. R.; Zhang, G.; Miller, J. T.; Kobayashi, T.; Pruski, M.; Delferro, M.; Marks, T. J., *J. Am. Chem. Soc.*, (2015) **137**, 6770.
[4] Qiao, M.-H.; Xie, S.-H.; Dai, W.-L.; Deng, J.-F., *Catal. Lett.*, (2001) **71**, 187.
[5] Mu, X.-d.; Meng, J.-q.; Li, Z.-C.; Kou, Y., *J. Am. Chem. Soc.*, (2005) **127**, 9694.
[6] Weissermel, K.; Arpe, H.-J., *Industrial Organic Chemistry*, 4th ed.; Wiley-VCH: Weinheim, Germany, (2008); p 313.
[7] Le Page, J.-F., *Applied Heterogeneous Catalysis: Design, Manufacture, Use of Solid Catalysts*; Éditions Technip: Paris, (1987).
[8] Struijk, J.; Scholten, J. J. F., *Appl. Catal., A*, (1992) **82**, 277.
[9] Saeys, M.; Reyniers, M.-F.; Neurock, M.; Marin, G. B., *J. Phys. Chem. B*, (2005) **109**, 2064.
[10] Smith, H. A.; Meriwether, H. T., *J. Am. Chem. Soc.*, (1949) **71**, 413.
[11] Sabatier, P., *Ind. Eng. Chem.*, (1926) **18**, 1005.
[12] Zhu, L.; Zheng, J.; Yu, C.; Zhang, N.; Shu, Q.; Zhou, H.; Li, Y.; Chen, B. H., *RSC Adv.*, (2016) **6**, 13110.
[13] Kang, X.; Liu, H.; Hou, M.; Sun, X.; Han, H.; Jiang, T.; Zhang, Z.; Han, B., *Angew. Chem.*, (2016) **128**, 1092; *Angew. Chem. Int. Ed.*, (2016) **55**, 1080.
[14] Bi, H.; Tan, X.; Dou, R.; Pei, Y.; Qiao, M.; Sun, B.; Zong, B., *Green Chem.*, (2016) **18**, 2216.
[15] Ma, Y.; Huang, Y.; Cheng, Y.; Wang, L.; Li, X., *Appl. Catal., A*, (2014) **484**, 154.
[16] Zahmakıran, M.; Tonbul, Y.; Özkar, S., *J. Am. Chem. Soc.*, (2010) **132**, 6541.
[17] Pellegatta, J.-L.; Blandy, C.; Collière, V.; Choukroun, R.; Chaudret, B.; Cheng, P.; Philippot, K., *J. Mol. Catal. A: Chem.*, (2002) **178**, 55.
[18] Adams, R.; Marshall, J., *J. Am. Chem. Soc.*, (1928) **50**, 1970.
[19] Raney, M., *Ind. Eng. Chem.*, (1940) **32**, 1199.
[20] Osborn, J. A.; Jardine, F. H.; Young, J. F.; Wilkinson, G., *J. Chem. Soc. A*, (1966), 1711.
[21] Young, J. F.; Osborn, J. A.; Jardine, F. H.; Wilkinson, G., *Chem. Commun.*, (1965), 131.
[22] Muetterties, E. L.; Hirsekorn, F. J., *J. Am. Chem. Soc.*, (1974) **96**, 4063.
[23] Russell, M. J.; White, C.; Maitlis, P. M., *J. Chem. Soc., Chem. Commun.*, (1977), 427.
[24] Bennett, M. A.; Huang, T.-N.; Turney, T. W., *J. Chem. Soc., Chem. Commun.*, (1979), 312.
[25] Boxwell, C. J.; Dyson, P. J.; Ellis, D. J.; Welton, T., *J. Am. Chem. Soc.*, (2002) **124**, 9334.
[26] Dyson, P. J.; Ellis, D. J.; Welton, T.; Parker, D. G., *Chem. Commun. (Cambridge)*, (1999), 25.
[27] Dyson, P. J., *Dalton Trans.*, (2003), 2964.
[28] Hagen, C. M.; Vieille-Petit, L.; Laurenczy, G.; Süss-Fink, G.; Finke, R. G., *Organometallics*, (2005) **24**, 1819.
[29] Garcia Fidalgo, E.; Plasseraud, L.; Süss-Fink, G., *J. Mol. Catal. A: Chem.*, (1998) **132**, 5.
[30] Gao, L.; Kojima, K.; Nagashima, H., *Tetrahedron*, (2015) **71**, 6414.
[31] Buil, M. L.; Esteruelas, M. A.; Niembro, S.; Oliván, M.; Orzechowski, L.; Pelayo, C.; Vallribera, A., *Organometallics*, (2010) **29**, 4375.
[32] Moreno-Marrodan, C.; Liguori, F.; Mercadé, E.; Godard, C.; Claver, C.; Barbaro, P., *Catal. Sci. Technol.*, (2015) **5**, 3762.
[33] Das, P.; Sarmah, P. P.; Borah, B. J.; Saikia, L.; Dutta, D. K., *New J. Chem.*, (2016) **40**, 2850.
[34] Kogan, V.; Aizenshtat, Z.; Neumann, R., *New J. Chem.*, (2002) **26**, 272.
[35] Stanislaus, A.; Cooper, B. H., *Catal. Rev.*, (1994) **36**, 75.
[36] Nador, F.; Moglie, Y.; Vitale, C.; Yus, M.; Alonso, F.; Radivoy, G., *Tetrahedron*, (2010) **66**, 4318.
[37] Bresó-Femenia, E.; Chaudret, B.; Castillón, S., *Catal. Sci. Technol.*, (2015) **5**, 2741.
[38] Pan, H.-B.; Wai, C. M., *New J. Chem.*, (2011) **35**, 1649.
[39] Fu, P. P.; Lee, H. M.; Harvey, R. G., *J. Org. Chem.*, (1980) **45**, 2797.
[40] Yalpani, M.; Lunow, T.; Köster, R., *Chem. Ber.*, (1989) **122**, 687.
[41] Segawa, Y.; Stephan, D. W., *Chem. Commun. (Cambridge)*, (2012) **48**, 11963.
[42] Pei, Y.; Zhou, G.; Luan, N.; Zong, B.; Qiao, M.; Tao, F., *Chem. Soc. Rev.*, (2012) **41**, 8140.
[43] Anderson, J. R., *Aust. J. Chem.*, (1957) **10**, 409.
[44] *Metal-Support and Metal-Additive Effects in Catalysis*, Imelik, B.; Naccache, C.; Coudurier, G.; Praliaud, H.; Meriaudeau, P.; Gallezot, P.; Martin, G. A.; Vedrine, J. C., Eds.; Studies in Surface Science and Catalysis Vol. 11; Elsevier: Amsterdam, (1982).

[45] Teichner, S. J.; Hoang-Van, C.; Astier, M., *Stud. Surf. Sci. Catal.*, (1982) **11**, 121.
[46] Galvagno, S.; Donato, A.; Neri, G.; Pietropaolo, D.; Staiti, P., *React. Kinet. Catal. Lett.*, (1988) **37**, 443.
[47] Silveira, E. T.; Umpierre, A. P.; Rossi, L. M.; Machado, G.; Morais, J.; Soares, G. V.; Baumvol, I. J. R.; Teixeira, S. R.; Fichtner, P. F. P.; Dupont, J., *Chem.–Eur. J.*, (2004) **10**, 3734.
[48] Xie, S.; Qiao, M.; Li, H.; Wang, W.; Deng, J.-F., *Appl. Catal., A*, (1999) **176**, 129.
[49] Liu, S.; Liu, Z.; Wang, Z.; Zhao, S.; Wu, Y., *Appl. Catal., A*, (2006) **313**, 49.
[50] Liu, Z.; Xie, S.; Liu, B.; Deng, J.-F., *New J. Chem.*, (1999) **23**, 1057.
[51] Wang, J.; Guo, P.; Yan, S.; Qiao, M.; Li, H.; Fan, K., *J. Mol. Catal. A: Chem.*, (2004) **222**, 229.
[52] Fan, G.-Y.; Li, R.-X.; Li, X.-J.; Chen, H., *Catal. Commun.*, (2008) **9**, 1394.
[53] Zhou, G.; Liu, J.; Tan, X.; Pei, Y.; Qiao, M.; Fan, K.; Zong, B., *Ind. Eng. Chem. Res.*, (2012) **51**, 12 205.
[54] Zhou, G.; Dou, R.; Bi, H.; Xie, S.; Pei, Y.; Fan, K.; Qiao, M.; Sun, B.; Zong, B., *J. Catal.*, (2015) **332**, 119.
[55] Spod, H.; Lucas, M.; Claus, P., *Catalysts*, (2015) **5**, 1756.
[56] Zhou, G.; Tan, X.; Pei, Y.; Fan, K.; Qiao, M.; Sun, B.; Zong, B., *ChemCatChem*, (2013) **5**, 2425.
[57] Wang, W.; Liu, H.; Wu, T.; Zhang, P.; Ding, G.; Liang, S.; Jiang, T.; Han, B., *J. Mol. Catal. A: Chem.*, (2012) **355**, 174.
[58] Zanutelo, C.; Landers, R.; Carvalho, W. A.; Cobo, A. J. G., *Appl. Catal., A*, (2011) **409**, 174.
[59] Liu, J.-L.; Zhu, L.-J.; Pei, Y.; Zhuang, J.-H.; Li, H.; Li, H.-X.; Qiao, M.-H.; Fan, K.-N., *Appl. Catal., A*, (2009) **353**, 282.
[60] Spinacé, E. V.; Vaz, J. M., *Catal. Commun.*, (2003) **4**, 91.
[61] Yan, X.; Zhang, Q.; Zhu, M.; Wang, Z., *J. Mol. Catal. A: Chem.*, (2016) **413**, 85.
[62] Zhang, P.; Wu, T.; Jiang, T.; Wang, W.; Liu, H.; Fan, H.; Zhang, Z.; Han, B., *Green Chem.*, (2013) **15**, 152.
[63] Liu, H.; Liang, S.; Wang, W.; Jiang, T.; Han, B., *J. Mol. Catal. A: Chem.*, (2011) **341**, 35.
[64] Wang, J.; Wang, Y.; Xie, S.; Qiao, M.; Li, H.; Fan, K., *Appl. Catal., A*, (2004) **272**, 29.
[65] Hu, S.-C.; Chen, Y.-W., *Ind. Eng. Chem. Res.*, (1997) **36**, 5153.
[66] Nagahara, H.; Ono, M.; Konishi, M.; Fukuoka, Y., *Appl. Surf. Sci.*, (1997) **121**, 448.
[67] Fang, M.; Machalaba, N.; Sánchez-Delgado, R. A., *Dalton Trans.*, (2011) **40**, 10 621.
[68] Zhao, C.; Wang, H.-z.; Yan, N.; Xiao, C.-x.; Mu, X.-d.; Dyson, P. J.; Kou, Y., *J. Catal.*, (2007) **250**, 33.

1.4 Reduction of Hetarenes

Z.-P. Chen and Y.-G. Zhou

General Introduction

The reduction of hetarenes is an active area of research due to the high interest of synthetic chemists in the construction of heterocycles. Most hetarenes are widely distributed in nature and easily prepared from inexpensive starting materials. Therefore, the direct catalytic reduction of hetarenes would provide an efficient and economic route to the corresponding saturated or partially saturated heterocycles, which are important structural motifs in many biologically active reagents, natural products, synthetic drugs, and materials. In the past decades, this methodology has been extensively developed and now represents a powerful tool in organic synthesis.[1–6] This chapter provides an overview of the most relevant advances in this field, and covers literature from 2000 to mid-2016. Both heterogeneous and homogeneous catalysis approaches involving hydrogenation and transfer hydrogenation with transition-metal catalysts are surveyed, while enantioselective methods are also covered. However, other related works on catalytic reduction of heteroaromatics, including organocatalysis, biocatalysis, and hydrosilylation reactions, are not considered here. An earlier review of asymmetric hydrogenation methods can be found in *Science of Synthesis: Stereoselective Synthesis* (Section 1.6).[7]

1.4.1 Heterogeneous Catalysis

Heterogeneous reduction of hetarenes can be realized without impediments in the presence of palladium, platinum, nickel, ruthenium, or rhodium catalysts. Problems that can be encountered include compatibility with other functional groups that need to be retained, control of regioselectivity in polycyclic aromatics, and control of stereoselectivity. Generally, the complete reduction to saturated heterocyclic rings can be achieved in an efficient manner.

1.4.1.1 Reduction of Quinoline Derivatives

Catalytic reduction of the readily available quinolines provides a convenient and straightforward approach to tetra- or decahydroquinolines, which are very important classes of heterocycles because of their wide range of applications in pharmaceutical research. For this reason, this research area has attracted considerable attention in recent years. In general, hydrogenation of quinolines under moderate conditions proceeds preferentially at the pyridine ring to afford 1,2,3,4-tetrahydroquinolines; for complete reduction to decahydroquinolines, extended reaction times or forcing reaction conditions are required.

Many efficient catalytic systems have been developed for the reduction of quinolines. Among them, palladium on carbon (Pd/C),[8] palladium nanoparticles supported on magnesium oxide (Pd/MgO),[9] and an ordered mesoporous graphitic carbon nitride supported palladium catalyst (Pd@ompg-C_3N_4)[10] are efficient for the regioselective hydrogenation of quinolines to 1,2,3,4-tetrahydroquinolines. A series of functionalized quinolines **1** can be reduced smoothly with excellent conversion under hydrogen atmosphere in methanol by employing rhodium on alumina to furnish 1,2,3,4-tetrahydroquinolines **2** (Scheme 1).[11]

for references see p 191

Scheme 1 Rhodium-on-Alumina Catalyzed Hydrogenation of Quinolines in Methanol[11]

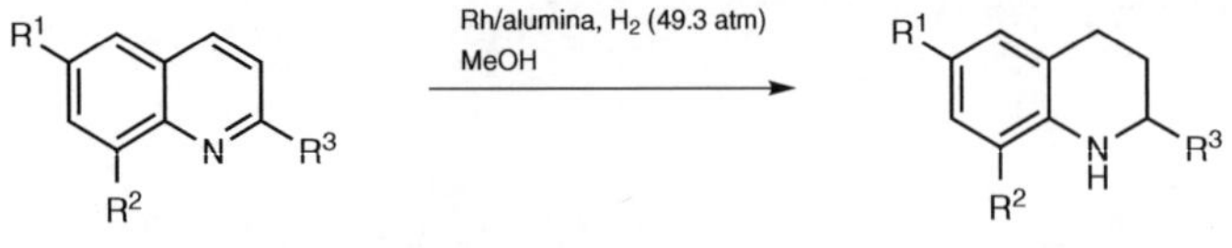

R^1	R^2	R^3	Selectivity[a] (%)	Ref
H	OH	H	100	[11]
H	H	Me	82	[11]
H	H	CO_2H	100	[11]
Br	H	H	100	[11]

[a] Determined by ^{1}H NMR spectroscopy.

Interestingly, when conducting this hydrogenation in 1,1,1,3,3,3-hexafluoropropan-2-ol (HFIP), the decahydroquinoline derivatives **3** are selectively obtained, thus avoiding the use of more severe conditions (Scheme 2).[11]

Scheme 2 Synthesis of Decahydroquinolines by Rhodium-on-Alumina Catalyzed Hydrogenation of Quinolines in 1,1,1,3,3,3-Hexafluoropropan-2-ol[11]

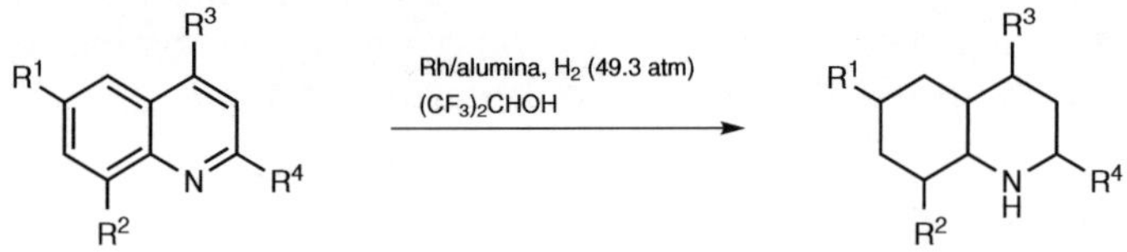

R^1	R^2	R^3	R^4	Selectivity[a] (%)	Ref
H	OH	H	H	100	[11]
H	H	Me	H	60	[11]
H	H	H	CO_2H	100	[11]
Me	H	H	H	100	[11]

[a] Determined by ^{1}H NMR spectroscopy.

Recently, Cao and co-workers reported the chemoselective hydrogenation of quinoline derivatives using supported gold catalysts (Scheme 3).[12] It was demonstrated that gold nanoparticles supported on high-surface-area titanium(IV) oxide (Au/HSA-TiO_2) can catalyze, under mild reaction conditions, the hydrogenation of a variety of quinolines **4** to the corresponding 1,2,3,4-tetrahydroquinolines **5** with high chemoselectivity. Of practical significance is the fact that various valuable functional groups including phenolic hydroxy (entry 4), amino (entry 5), halo (entry 6), ketone (entry 7), and alkene (entry 8) are tolerated under these reaction conditions.

Scheme 3 Hydrogenation of Quinolines Catalyzed by Gold Nanoparticles Supported on High-Surface-Area Titanium(IV) Oxide[12]

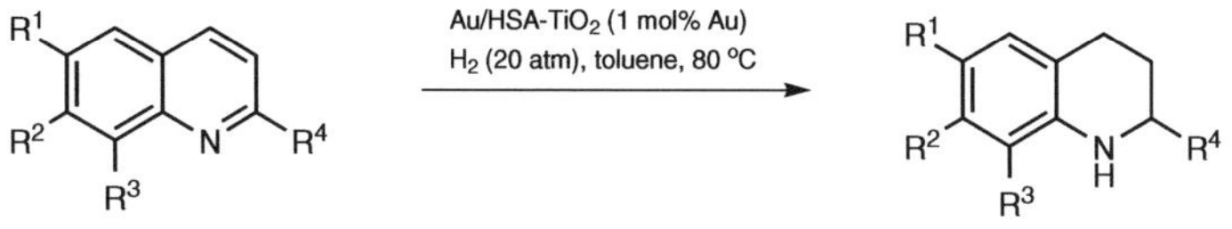

Au/HSA-TiO_2 (1 mol% Au)
H_2 (20 atm), toluene, 80 °C

Entry	R^1	R^2	R^3	R^4	Time (h)	Yield[a] (%)	Ref
1	H	H	H	H	3.5	100	[12]
2	H	H	H	Me	4	100	[12]
3	H	$(CH{=}CH)_2$		H	7	100	[12]
4	H	H	OH	H	10	100	[12]
5	H	H	NH_2	H	24	99	[12]
6	Cl	H	H	H	3	100	[12]
7	Ac	H	H	Me	12	86	[12]
8	H	$CH{=}CH_2$	H	Me	9	80	[12]

[a] Determined by GC.

The heterogeneous reduction of quinolines can also be conducted in an enantioselective manner. Li and co-workers recently synthesized a series of chiral conjugated microporous polymers (CMPs) based on the (*R*)-2,2′-bis(diphenylphosphino)-1,1′-binaphthyl ligand (BINAP-CMPs) with different surface areas.[13] Among them, the iridium/BINAP-CMP-3D-2 catalyst, which can be easily recovered by centrifugation and regular filtration, exhibits high activity and modest enantioselectivity in the hydrogenation of quinolines **6** to provide chiral 1,2,3,4-tetrahydroquinolines **7** (Scheme 4).[14] More importantly, the recovered catalyst can be reused five times without significant deterioration in activity or enantioselectivity.

Scheme 4 Iridium/BINAP-CMP Catalyzed Hydrogenation of Various Quinolines[14]

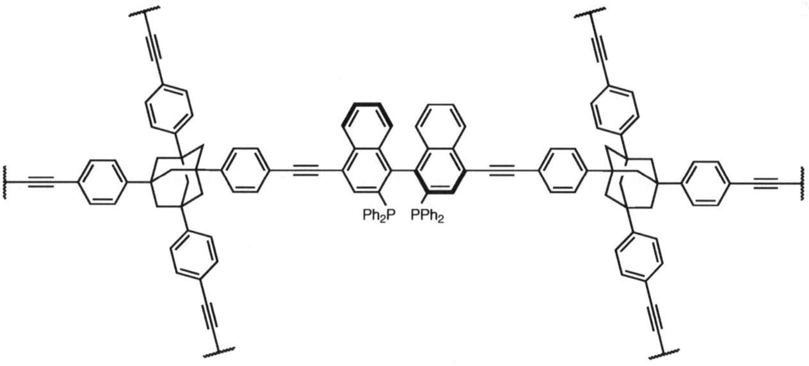

Ph_2P PPh_2

BINAP-CMP-3D-2

for references see p 191

{IrCl(cod)}$_2$ (0.5 mol%)
BINAP-CMP-3D-2 (2 mol%)
I_2 (5 mol%), H_2 (39.5 atm)
CH_2Cl_2, rt

6 → 7

R^1	R^2	Conversion[a] (%)	ee[b] (%)	Ref
Me	H	99	70 (*R*)	[14]
Et	H	99	77 (*R*)	[14]
Pr	H	99	78 (*R*)	[14]
Bu	H	80	77 (*R*)	[14]
iPr	H	83	79 (*R*)	[14]
Ph	H	84	44 (*R*)	[14]
Me	F	99	63 (*S*)	[14]
Me	Me	99	67 (*R*)	[14]
Me	OMe	97	70 (*R*)	[14]

[a] Determined by ^{1}H NMR spectroscopy.
[b] Determined by HPLC analysis using a chiral stationary phase.

1,2,3,4-Tetrahydroquinolines 2; **General Procedure Using Rh/Alumina:**[11]
A quinoline derivative **1** (1 mmol) and Rh/alumina (25 mg, 0.01 mmol Rh) were stirred in MeOH (1.5 mL) under H_2 (49.3 atm) at rt for 1–19 h. The mixture was filtered through Celite and the filtrate was concentrated to afford the crude product.

1,2,3,4-Tetrahydroquinolines 5; **General Procedure Using Au/HSA-TiO$_2$:**[12]
A mixture of a quinoline **4** (0.5 mmol), Au/HSA-TiO$_2$ catalyst (1 mol% Au), and toluene (3 mL) was loaded into a Parr autoclave (25-mL capacity, SS-316). The autoclave was sealed, and charged with H_2 (20 atm) after being completely purged of internal air. The resulting mixture was vigorously stirred at 80 °C for 3–24 h. The product was isolated by chromatography (silica gel, hexanes/EtOAc).

Chiral 1,2,3,4-Tetrahydroquinolines 7; **General Procedure:**[14]
A mixture of {IrCl(cod)}$_2$ (0.85 mg, 1.25 μmol, 0.5 mol%) and BINAP-CMP-3D-2 (5.95 mg, 5.0 μmol, 2 mol%) in CH_2Cl_2 (3 mL) was stirred at rt for 3 h in a glovebox. The mixture was then transferred to a stainless-steel autoclave, which had been charged with I_2 (3.2 mg, 12.5 μmol) and a quinoline **6** (0.25 mmol) beforehand. The hydrogenation was performed at rt under H_2 (39.5 atm) for 2 h. After the H_2 had been carefully released, the mixture was diluted with CH_2Cl_2 (5 mL) and sat. aq Na_2CO_3 (2 mL), and then stirred for 15 min. The layers were separated and the aqueous layer was extracted with CH_2Cl_2 (3 × 5 mL). The combined organic layers were dried (Na_2SO_4), filtered, and concentrated to afford the crude product.

1.4.1.2 Reduction of Isoquinoline Derivatives

Isoquinolines can be conveniently hydrogenated over ruthenium nanoparticles immobilized on poly(4-vinylpyridine),[15] Au/HSA-TiO$_2$,[12] or Adams' catalyst [platinum(IV) oxide][16–18] under a low pressure of hydrogen to afford the corresponding pyridine-ring-saturated heterocycles. Importantly, by using a catalytic amount of platinum(IV) oxide, several 4-substituted isoquinolines **8** containing different kinds of functional groups can be hydrogenated smoothly in acetic acid under an atmosphere of hydrogen to afford 1,2,3,4-tetrahydroisoquinolines **9** (Scheme 5, entries 1–5).[16] By elevating the hydrogen pressure, isoquinoline with a 5-hydroxy group can also be hydrogenated in high yield (entry 6).[17] Isoquinolines containing a 4,4,5,5-tetramethyl-1,3,2-dioxaborolan-2-yl group are also suitable substrates when conducting the reduction in ethanol instead of acetic acid (entry 7).[18]

Scheme 5 Heterogeneous Reduction of Isoquinolines Catalyzed by Adams' Catalyst[16–18]

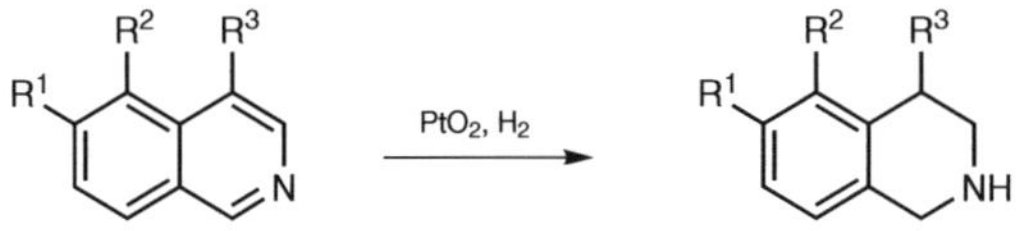

R1 R2 R3 N **8** —PtO_2, H_2→ R1 R2 R3 NH **9**

Entry	R^1	R^2	R^3	Solvent	H_2 (atm)	Yield (%)	Ref
1	H	H	OAc	AcOH	1	71	[16]
2	H	H	NHAc	AcOH	1	73	[16]
3	H	H	NHBz	AcOH	1	87	[16]
4	H	H	$NHCOCF_3$	AcOH	1	83	[16]
5	H	H	CO_2Me	AcOH	1	88	[16]
6	H	OH	H	AcOH	3	93	[17]
7	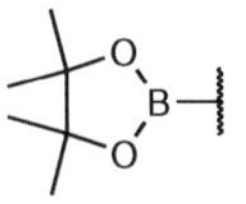 O B O	H	H	EtOH	3	100	[18]

1,2,3,4-Tetrahydroisoquinolin-4-yl Acetate (9, $R^1 = R^2 = H$; $R^3 = OAc$); Typical Procedure:[16]
PtO_2 (25 mg) was added to a soln of isoquinolin-4-yl acetate (**8**, $R^1 = R^2 = H$; $R^3 = OAc$; 250 mg, 1.34 mmol) in AcOH (25 mL) and the resulting suspension was hydrogenated under H_2 (1 atm) at rt. After completion of the reaction (TLC control), the catalyst was removed by filtration through a short pad of Celite. The filtrate was concentrated under reduced pressure to afford the product as an oil; yield: 182 mg (71%).

1.4.1.3 Reduction of Pyridine Derivatives

As an approach for the synthesis of important piperidine products, the reduction of pyridines has attracted considerable interest for a long time. In general, heterogeneous catalytic hydrogenation of pyridine derivatives is usually carried out in acidic media, as protonation activates the pyridine toward hydrogenation and suppresses catalyst poisoning by the resulting products. Adams' catalyst [platinum(IV) oxide] and palladium on carbon are most frequently employed, exhibiting broad substrate scope under low hydrogen pressure, although rhodium on carbon is effective for pyridin-3-ol and nicotinamide.[19] As shown in Table 1, pyridines bearing pyrazole (entry 1)[20] or indole moieties (entry 2)[21] are selectively hydrogenated to the piperidine derivatives using Adams' catalyst either in acidic media or through prior conversion of the pyridines into their hydrochloride salts. Reducible functionalities such as alkynes (entry 3)[22] or nitro groups (entry 4)[23] are reduced during the hydrogenation, whereas a more resistant functionality such as an ester (entry 5)[24] remains intact.

for references see p 191

Table 1 Heterogeneous Hydrogenation of Pyridines Catalyzed by Adams' Catalyst[20–24]

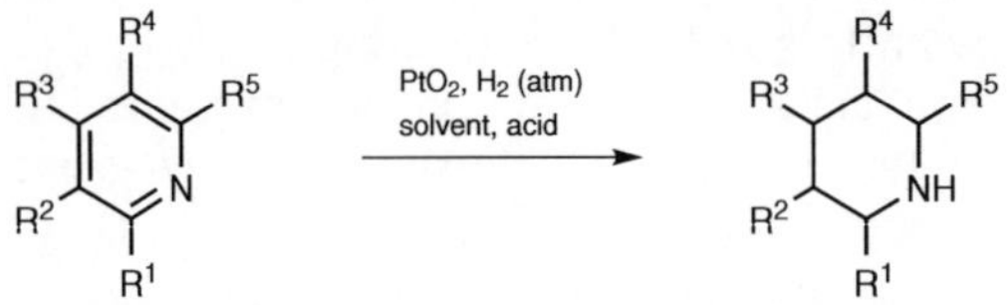

PtO_2, H_2 (atm), solvent, acid

Entry	Substrate	H_2 (atm)	Solvent	Acid	Product	Yield (%)	Ref
1	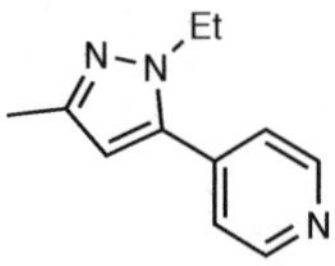	3.4	EtOH	HCl	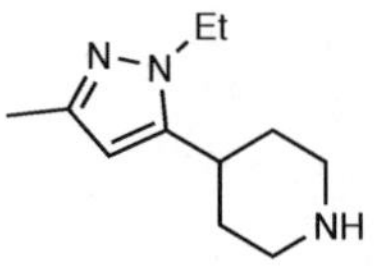	90[a]	[20]
2	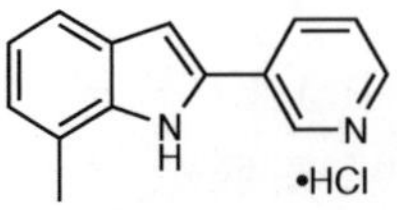 •HCl	3.4	MeOH	–	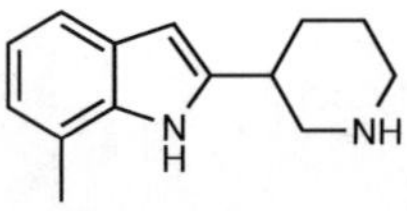	95[a]	[21]
3	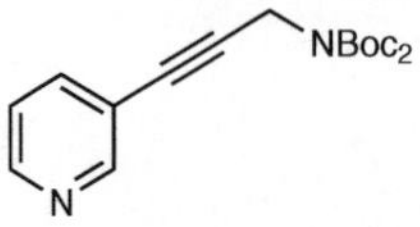	5	EtOH	AcOH	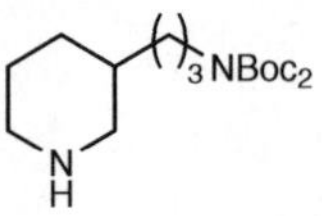	92	[22]
4	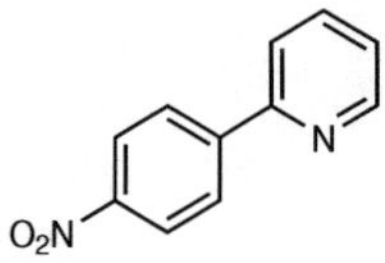	3.4	MeOH	HCl	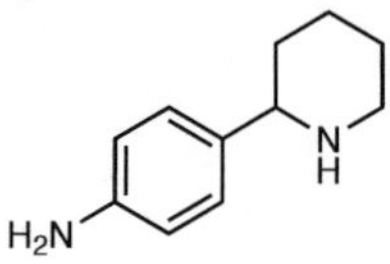	79[a]	[23]
5	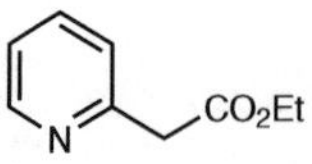	1	EtOH	HCl	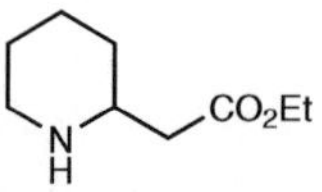	79[a]	[24]

[a] Basic workup.

In 2009, a palladium on carbon catalytic hydrogenation of pyridine carboxamides **10** to the corresponding piperidine carboxamide hydrochlorides **11** was developed by Wang and co-workers (Scheme 6).[25] This method involves the hydrodechlorination of 1,1,2-trichloroethane to release hydrogen chloride, followed by formation of a pyridine carboxamide hydrochloride in situ, which undergoes hydrogenation more easily than pyridine carboxamides.

Scheme 6 Heterogeneous Hydrogenation of Pyridines Catalyzed by Palladium on Charcoal[25]

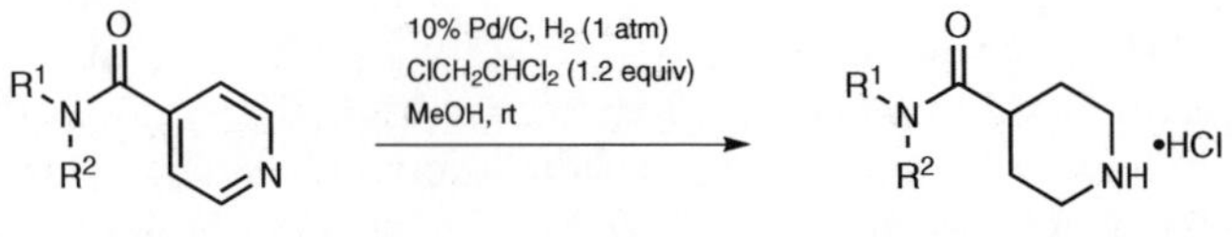

10% Pd/C, H_2 (1 atm), $ClCH_2CHCl_2$ (1.2 equiv), MeOH, rt

10 → **11** •HCl

R^1	R^2	Yield (%)	Ref
H	H	98	[25]
Me	H	99	[25]
Bu	H	99	[25]
$(CH_2)_2OH$	H	98	[25]
Ph	H	98	[25]
Et	Et	96	[25]
$(CH_2)_5$		99	[25]
$(CH_2)_2O(CH_2)_2$		96	[25]

Significantly, Glorius and co-workers have revealed an auxiliary-based method for the asymmetric reduction of substituted pyridines and 5,6,7,8-tetrahydroquinolines using the standard achiral heterogeneous hydrogenation catalysts (Scheme 7, entries 1–7).[26] Thus, the hydrogenation of 2-oxazolidinone-substituted pyridines and 5,6,7,8-tetrahydroquinolines **12** under a high pressure of hydrogen (100 atm) with palladium- and/or rhodium-based catalysts directly affords the corresponding piperidines and decahydroquinolines **13** in good yields and with high stereoselectivity (up to 98% ee) through traceless cleavage of the chiral auxiliary. Additionally, chiral methyl-substituted 5,6,7,8-tetrahydroquinolines can also be hydrogenated and isolated as the corresponding decahydroquinoline hydrochlorides in 99% yield and with a dr of >97:3 (Scheme 7, entry 8).[27]

Scheme 7 Heterogeneous Diastereoselective Reduction of Pyridines[26,27]

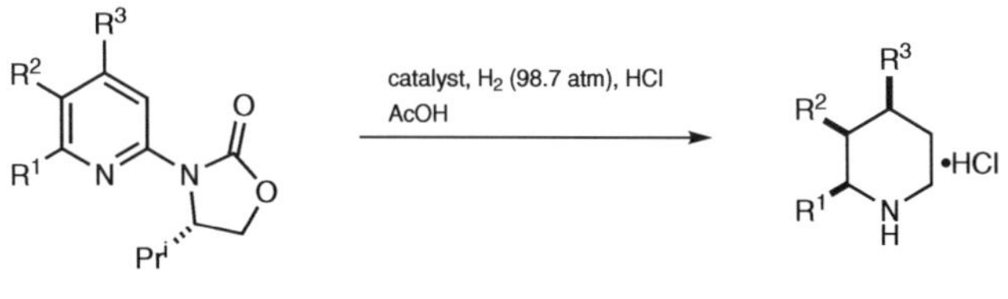

catalyst, H_2 (98.7 atm), HCl
AcOH

12 → **13** •HCl

Entry	R^1	R^2	R^3	Catalyst	ee[a] (%)	Yield (%)	Ref
1	Me	H	H	$Pd(OH)_2/C$	91	93	[26]
2	Pr	H	H	$Pd(OH)_2/C$	95	95	[26]
3	H	Me	H	$Pd(OH)_2/C$	98	90	[26]
4	H	CF_3	H	$Pd(OH)_2/C$	95	93	[26]
5	H	$CONMe_2$	H	$Pd(OH)_2/C$	85	92	[26]
6	$(CH_2)_4$		H	Rh/Pd/C	96	94	[26]
7	$(CH_2)_4$		Me	Rh/C	94	92[b]	[26]
8	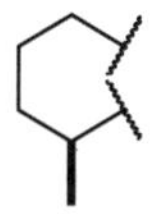		Me	Rh/C	>99[c]	99[c]	[27]

[a] Determined by GC analysis of the *N*-trifluoroacetamide derivative of the crude reaction product.
[b] H_2 pressure was 148 atm.
[c] Determined by chiral GC.

N-Butylpiperidine-4-carboxamide Hydrochloride (11, R^1 = Bu; R^2 = H); **Typical Procedure:**[25]
The suspension of *N*-butylpyridine-4-carboxamide (**10**, R^1 = Bu; R^2 = H; 178 mg, 1.0 mmol), 10% Pd/C (44.5 mg, 25 wt%), and 1,1,2-trichloroethane (160 mg, 1.2 mmol) in MeOH (30 mL) was hydrogenated (using an atmospheric pressure hydrogenator) at rt until absorption of H_2 ceased. The catalyst was removed by filtration and the filtrate was concentrated on a rotary evaporator. The residue was diluted with Et_2O (10 mL), and the product was collected by filtration as white crystals; yield: 220 mg (99%).

(*S*)-2-Propylpiperidine Hydrochloride (13, R^1 = Pr; R^2 = H); **Typical Procedure:**[26]
A mixture of wet 20% $Pd(OH)_2/C$ (w/w; 140 mg), (*S*)-4-isopropyl-3-(6-propylpyridin-2-yl)oxazolidin-2-one (**12**, R^1 = Pr; R^2 = H; 496 mg, 2.0 mmol), and AcOH (15 mL) was stirred in an autoclave under H_2 (98.7 atm) at 40 °C for 22 h. The mixture was filtered through a short pad of Celite, which was subsequently washed with MeOH (15 mL). Concd HCl (333 μL, 4.0 mmol) was added, and the solvent was removed by rotary evaporation until no AcOH was left. The remaining white solid was washed repeatedly with *t*-BuOMe/hexanes to yield a white solid; yield: 310 mg (95%); 95% ee. Concentration of the organic phase afforded (*S*)-4-isopropyloxazolidin-2-one as a white solid; yield: 227 mg (88%).

1.4.2 Homogeneous Catalysis

A few years ago, the concept of homogeneous arene hydrogenation was called into question because the harsh conditions needed always led to decomposition of the transition-metal catalysts, resulting in heterogeneous catalysis.[28] Since then, as a result of rapid developments in the field of catalysis, there has been a great deal of progress in homogeneous hetarene hydrogenation. Although it is not imagined that homogeneous hydrogenation catalysts will replace heterogeneous catalysts in large-scale industrial processes, the use of selective homogeneous hetarene hydrogenation catalysts in the synthesis of fine chemicals can be envisaged, because many structurally diverse hetarenes are readily available and the most convenient route to the corresponding heterocycles could be achieved via direct reduction reactions.

1.4.2.1 Racemic Reductions

1.4.2.1.1 Reduction of Quinoline Derivatives

In 2011, Crabtree, Eisenstein, and co-workers disclosed that iridium catalysts (e.g., **14**), containing N-heterocyclic carbenes as ancillary ligands, are effective for the hydrogenation of quinoline derivatives **15** under mild reaction conditions to furnish 1,2,3,4-tetrahydroquinolines **16** (Scheme 8).[29] The addition of triphenylphosphine greatly increases the reaction rate. In addition, this catalytic system tolerates halo substitution (entries 4 and 6) on the carbocyclic ring, although these substrates require elevated hydrogen pressure (5 atm).

Scheme 8 Iridium–N-Heterocyclic Carbene Catalyzed Hydrogenation of Quinolines[29]

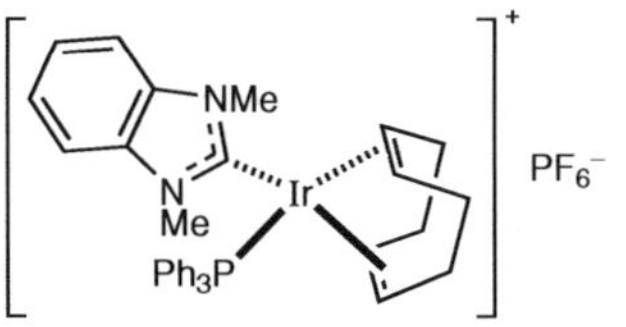

14

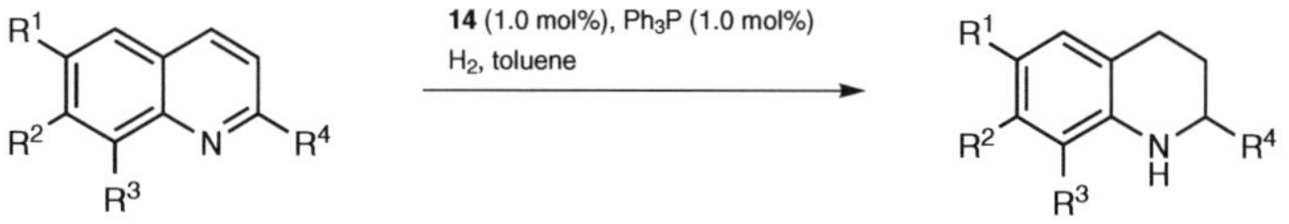

14 (1.0 mol%), Ph_3P (1.0 mol%)
H_2, toluene

15 → 16

Entry	R^1	R^2	R^3	R^4	H_2 (atm)	Conversion[a] (%)	Ref
1	H	H	H	Me	1	>95	[29]
2	H	H	H	Ph	1	>95	[29]
3	H	(CH=CH)$_2$		Me	1	94	[29]
4	H	H	Cl	H	5	>95	[29]
5	H	H	H	H	5	>95	[29]
6	Br	H	H	H	5	>95	[29]

[a] Determined by ^{1}H NMR spectroscopy.

Based on experimental and theoretical results, a stepwise outer-sphere mechanism was proposed (Scheme 9). In hydrogenation cycle I, the cationic iridium dihydrogen complex **19**, which is formed in situ via the reaction of the iridium–N-heterocyclic carbene catalyst with triphenylphosphine under a hydrogen atmosphere, transfers a proton to substrate **17**, resulting in a neutral iridium species **18** and a protonated substrate. In the next step, a hydride of species **18** transfers to the protonated substrate, providing a coordinatively unsaturated cationic species **20**, which then coordinates hydrogen and completes catalytic cycle I. Isomerization of the thus-formed enamine **21** to imine **22** takes place via a classical acid catalysis in the presence of protons liberated by the acidic iridium dihydrogen complex, thus allowing the protonation by complex **19** to form species **18**. Subsequently, another hydride transfer furnishes tetrahydroquinoline product **23** and completes catalytic cycle II.

for references see p 191

Scheme 9 Proposed Mechanism for Iridium–N-Heterocyclic Carbene Catalyzed Hydrogenation of Quinolines[29]

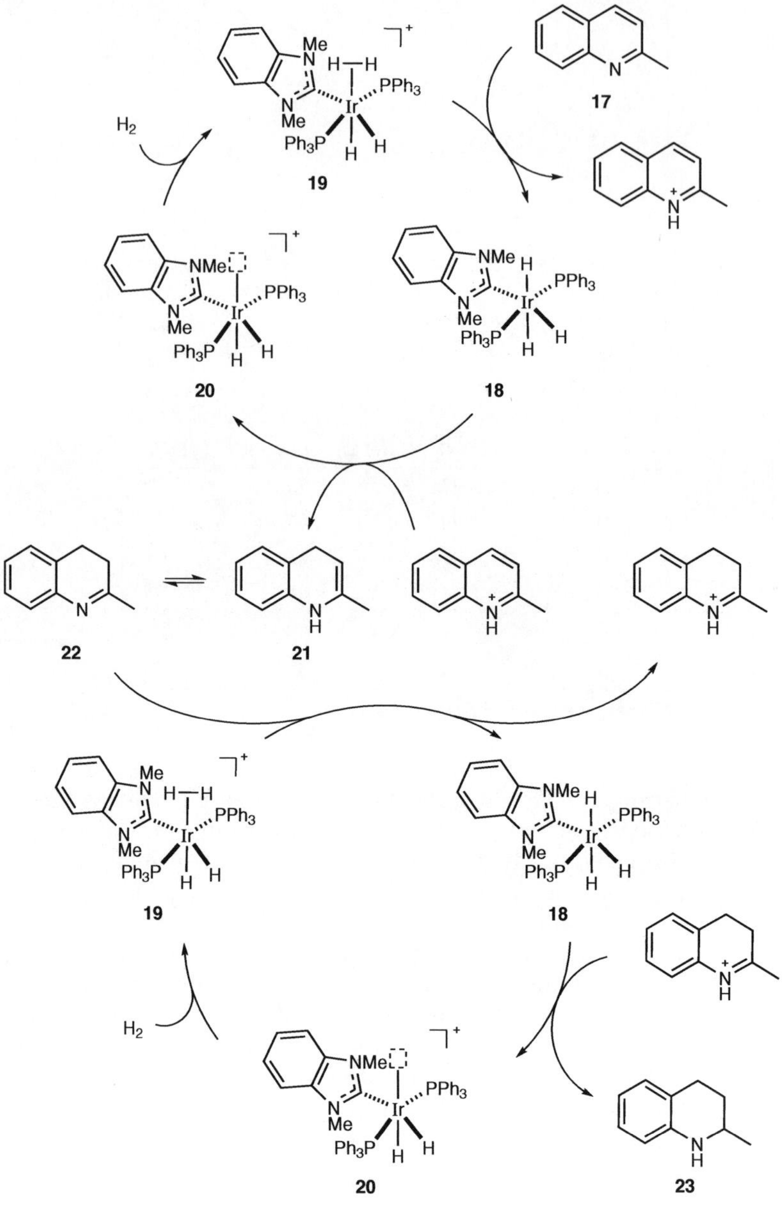

Beside hydrogen, propan-2-ol has also been used as the hydrogen source in iridium-catalyzed reduction of quinolines. Fujita, Yamaguchi, and co-workers have reported that dichloro(η^5-pentamethylcyclopentadienyl)iridium(III) dimer catalyzes the transfer hydrogenation of quinolines **24** with propan-2-ol as hydrogen source (Scheme 10).[30] Notably, the addition of perchloric acid not only greatly accelerates the reaction rate but also suppresses the formation of the undesired *N*-isopropyl-1,2,3,4-tetrahydroquinoline side product.

Scheme 10 Iridium-Catalyzed Transfer Hydrogenation of Quinolines[30]

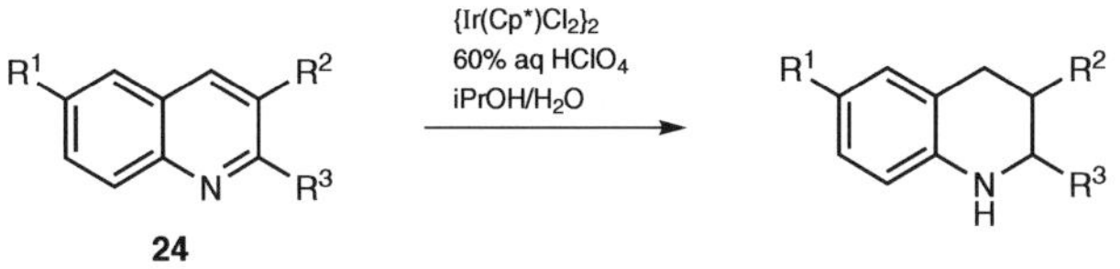

$\{Ir(Cp^*)Cl_2\}_2$, 60% aq $HClO_4$, iPrOH/H_2O; substrate **24**

R^1	R^2	R^3	Ir (mol%)	$HClO_4$ (mol%)	Yield (%)	Ref
H	H	H	1.0	10	93	[30]
H	H	Me	4.0	0	82	[30]
H	Me	H	2.0	10	79	[30]
Me	H	H	2.0	10	78	[30]
NO_2	H	H	2.0	0	94	[30]
Cl	H	H	2.0	10	78	[30]
CO_2H	H	H	4.0	10	64	[30]
OMe	H	H	1.9	11	79	[30]

Xiao and co-workers developed another efficient transfer hydrogenation of 2-substituted quinolines under mild conditions using dichloro(η^5-pentamethylcyclopentadienyl)rhodium(III) dimer as the catalyst and formic acid/triethylamine as the hydrogen source (Scheme 11).[31] By introducing a catalytic amount of potassium iodide as activator, quinolines **25** bearing various substituents including alkyl, aryl, methoxy, and halogens can be reduced successfully to give the corresponding pyridine-ring-reduced products in good to excellent yields.

Scheme 11 Rhodium-Catalyzed Transfer Hydrogenation of Quinolines[31]

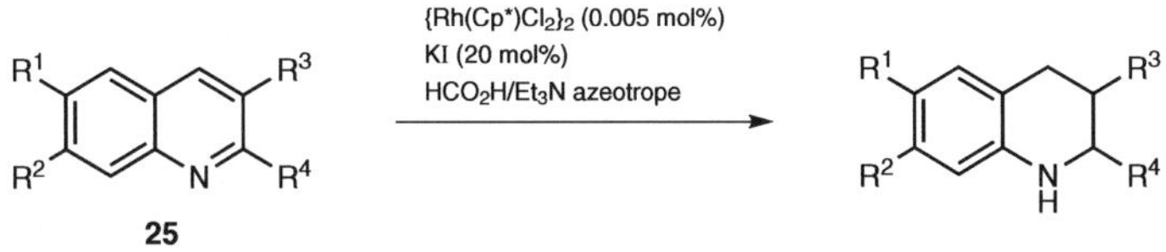

$\{Rh(Cp^*)Cl_2\}_2$ (0.005 mol%), KI (20 mol%), HCO_2H/Et_3N azeotrope; substrate **25**

R^1	R^2	R^3	R^4	Yield (%)	Ref
H	H	H	Me	91	[31]
H	H	H	Et	93	[31]
H	H	H	Bu	96	[31]
H	H	H	*t*-Bu	99	[31]
OMe	H	H	Me	73	[31]
Me	H	H	Me	76	[31]
Cl	H	H	Me	92	[31]
H	F	H	Me	83	[31]
H	H	Me	Me	89	[31]
H	H	$(CH_2)_4$		69	[31]
H	H	H	cyclopropyl	85	[31]

R^1	R^2	R^3	R^4	Yield (%)	Ref
H	H	H	Ph	98[a]	[31]
H	H	H	4-$MeOC_6H_4$	94[a]	[31]
H	H	H	4-FC_6H_4	98[a]	[31]

[a] {Rh(Cp*)Cl_2}$_2$ (0.05 mol%).

Based on control experiments and labelling studies, the authors suggest that this reduction proceeds through the following pathway: First, the rhodium species undergoes ligand displacement by iodide to generate **26** (Scheme 12). Next, reaction with formate leads to the highly active anionic diiodorhodium hydride species **27** and releases carbon dioxide. Then, the hydride of species **27** transfers to the protonated substrate, probably via an initial 1,4-addition pathway. It is noteworthy that addition of an appropriate amount of potassium iodide is crucial for the reactivity, as increasing the iodide concentration promotes the formation of the active species **27** and accelerates the reaction rate; however, excess of iodide favors the formation of the inactive triiodo rhodium species **26**, thus decreasing the concentration of the active catalyst and impeding the reduction.[31]

Scheme 12 Proposed Mechanism for the Rhodium-Catalyzed Transfer Hydrogenation[31]

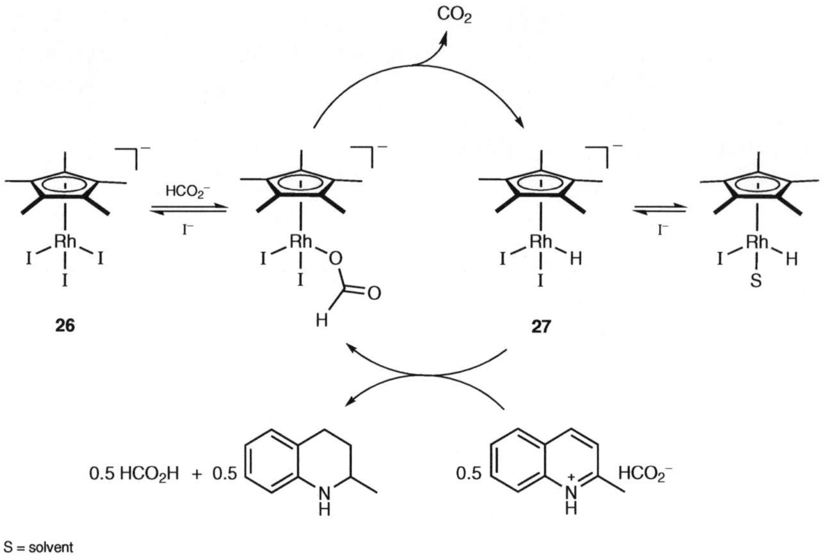

CO_2

HCO_2^- I$^-$ I$^-$

26 27

0.5 HCO_2H + 0.5 0.5 HCO_2^-

S = solvent

1,2,3,4-Tetrahydroquinolines 16; **General Procedure under Atmospheric H_2 Pressure:**[29]
In a flame-dried Schlenk flask under air was added a stirrer bar, iridium–NHC complex **14** (4.4 mg, 5.2 μmol), Ph_3P (1.36 mg, 5.2 μmol), and a quinoline **15** (0.52 mmol). Anhyd degassed toluene (2 mL) was added. The flask was rapidly evacuated and backfilled with H_2 three times, and the flask was kept under ambient H_2 pressure during the reaction. The mixture was then heated to 35 °C and stirred for 18 h. The soln was passed through basic alumina to afford the crude product. The pure product was obtained by chromatography (silica gel, hexanes/EtOAc).

1,2,3,4-Tetrahydroquinolines 16; **General Procedure under Higher H_2 Pressure:**[29]
In a glass reactor sleeve under air was added a stirrer bar, iridium–NHC complex **14** (4.4 mg, 5.2 μmol), Ph_3P (1.36 mg, 5.2 μmol), and a quinoline **15** (0.52 mmol). Anhyd degassed toluene (2 mL) was added. The sleeve was inserted into the reactor vessel, and the reactor was assembled under air. The reactor was purged with H_2 by pressurizing the ap-

paratus to 5 atm and venting three times. The reactor was pressurized to 5 atm with H_2 and heated to 35 °C, and the mixture was stirred for 18 h. The reactor was vented and the soln was passed through basic alumina to afford the crude product. The pure product was obtained by chromatography (silica gel, hexanes/EtOAc).

1.4.2.1.2 Miscellaneous Substrates

Aside from quinolines, other hetarenes such as isoquinolines, quinoxalines, indoles, pyrroles, and (benzo)furans are also suitable substrates for transition-metal-catalyzed homogeneous reduction reactions. In 2012, Xiao and co-workers reported that the dichloro(η^5-pentamethylcyclopentadienyl)rhodium(III) dimer/potassium iodide catalytic system (see Section 1.4.2.1.1, Scheme 11) is also effective in the transfer hydrogenation of isoquinolines **28** and quinoxalines **30**, providing 1,2,3,4-tetrahydroisoquinolines **29** and 1,2,3,4-tetrahydroquinoxalines **31**, respectively (Scheme 13).[31] For some of these challenging transfer hydrogenation substrates, a higher catalyst loading is necessary to obtain high yields.

Scheme 13 Rhodium-Catalyzed Transfer Hydrogenation of Isoquinolines and Quinoxalines[31]

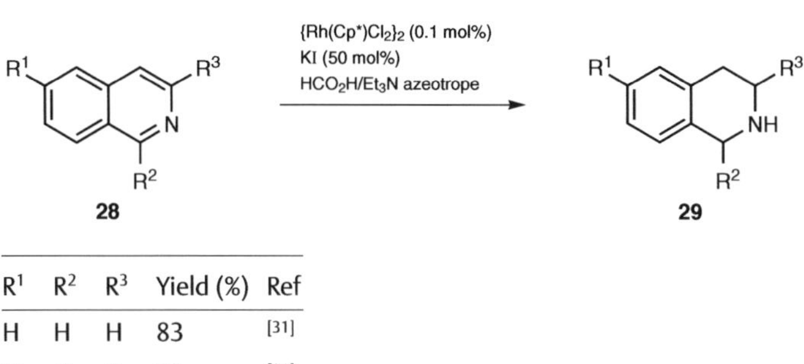

R¹	R²	R³	Yield (%)	Ref
H	H	H	83	[31]
Me	H	H	95	[31]
H	Me	H	92ᵃ	[31]
H	H	Me	91	[31]

ᵃ HCO_2H/Et_3N 3.5:1.0.

{Rh(Cp*)Cl2}2
KI (50 mol%)
HCO2H/Et3N azeotrope

30 → **31**

R¹	R²	R³	R⁴	Catalyst (mol%)	Yield (%)	Ref
Me	H	H	H	0.01	90	[31]
H	H	H	H	0.01	99	[31]
H	H	H	Me	0.01	86	[31]
H	Me	H	H	0.01	92	[31]
H	H	Me	Me	0.1	91	[31]

Dichloro(η^5-pentamethylcyclopentadienyl)iridium(III) dimer is also used as transition-metal precursor to catalyze transfer hydrogenation of quinoxalines **32** with sodium for-

for references see p 191

mate as the hydrogen source (Scheme 14).[32] The activity of this reduction is pH dependent and the utilization of the bidentate amino sulfonamide ligand **33** is vital. With this air-stable iridium catalyst, an array of quinoxalines are smoothly reduced in the acetic acid/sodium acetate buffered solution, affording tetrahydroquinoxalines **34** with good to excellent yields.

Scheme 14 Transfer Hydrogenation of Quinoxalines Catalyzed by an Iridium–Bidentate Ligand Complex[32]

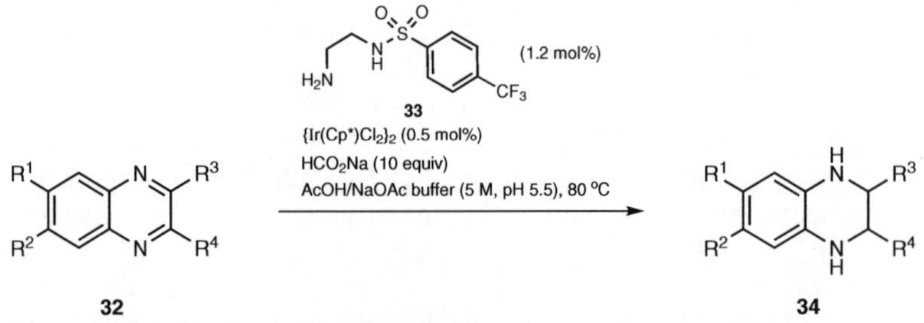

33 (1.2 mol%)
$\{Ir(Cp^*)Cl_2\}_2$ (0.5 mol%)
HCO_2Na (10 equiv)
AcOH/NaOAc buffer (5 M, pH 5.5), 80 °C

32 → **34**

R^1	R^2	R^3	R^4	Yield (%)	Ref
H	H	H	Me	96	[32]
H	H	H	Et	97	[32]
H	H	H	Bu	93	[32]
H	H	H	iBu	96	[32]
H	H	H	Cy	92	[32]
Me	Me	H	Et	96	[32]
H	H	H	Ph	97[a]	[32]
H	H	H	4-Tol	93[a]	[32]
H	H	H	(*E*)-CH=CHPh	95[b]	[32]
H	H	Me	Me	94	[32]

[a] AcOH/NaOAc buffer (5 M, pH 4.3) with EtOAc.
[b] AcOH/NaOAc buffer (5 M, pH 4.5).

Homogeneous hydrogenation with hydrogen gas as the reductant represents an atom economical way to furnish heterocycles from the corresponding hetarenes. Ito and co-workers have reported that the rhodium complex prepared in situ from (acetylacetonato)(1,5-cycloocta-1,5-diene)rhodium(I) and triphenylphosphine exhibits high performance in the hydrogenation of a wide range of five-membered heteroaromatic compounds.[33] A broad range of substrates including *N*-*tert*-butoxycarbonyl-protected indoles **35** to furnish *tert*-butyl indoline-1-carboxylates **36** (Scheme 15), pyrroles (Scheme 16), furans (Scheme 16), and benzo[*b*]furans provides the products in high yields.

Scheme 15 Rhodium-Catalyzed Hydrogenation of Indoles[33]

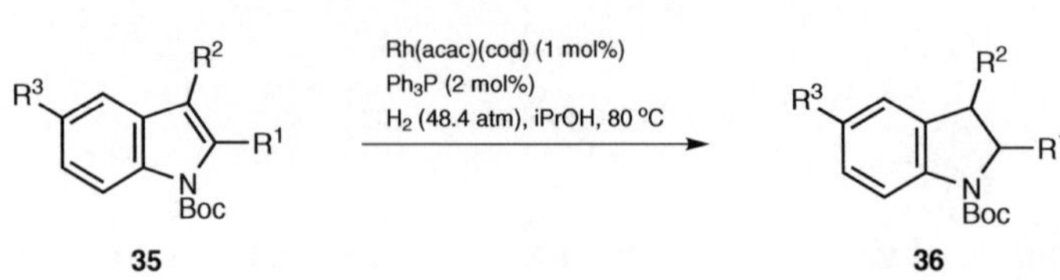

Rh(acac)(cod) (1 mol%)
Ph_3P (2 mol%)
H_2 (48.4 atm), iPrOH, 80 °C

35 → **36**

R^1	R^2	R^3	Yield (%)	Ref
H	H	H	91	[33]
Me	H	H	96	[33]
Ph	H	H	98	[33]
CO_2Me	H	H	85	[33]
H	H	Me	85	[33]
H	H	OMe	99	[33]
H	H	CO_2Me	95	[33]

Scheme 16 Rhodium-Catalyzed Hydrogenation of Pyrroles and Furans[33]

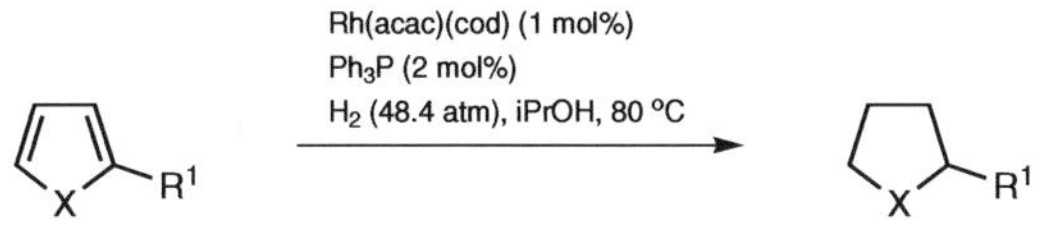

Rh(acac)(cod) (1 mol%)
Ph_3P (2 mol%)
H_2 (48.4 atm), iPrOH, 80 °C

R^1	X	Yield (%)	Ref
H	NBoc	91	[33]
CH_2OH	O	89	[33]

1,2,3,4-Tetrahydroisoquinolines 29; **General Procedure:**[31]

A carousel reaction tube containing a magnetic stirrer bar, $\{Rh(Cp^*)Cl_2\}_2$ (0.31 mg, 0.5 μmol), isoquinoline **28** (0.5 mmol), and KI (42.5 mg, 0.25 mmol) in HCO_2H/Et_3N azeotrope (3 mL) was sealed without degassing and placed in a carousel reactor. The mixture was stirred at 40 °C for 24 h, and then cooled to rt and basified with aq KOH. The resulting mixture was extracted with EtOAc (3 × 5 mL). The extracts were dried (Na_2SO_4), filtered, and concentrated under reduced pressure. The product was purified by flash column chromatography (hexanes/EtOAc).

1,2,3,4-Tetrahydroquinoxalines 34; **General Procedure:**[32]

A carousel reaction tube containing a magnetic stirrer bar, $\{Ir(Cp^*)Cl_2\}_2$ (2 mg, 2.5 μmol), amino sulfonamide ligand **33** (1.6 mg, 6 μmol), quinoxaline substrate **32** (0.5 mmol), and HCO_2Na (340 mg, 5 mmol) in a soln of AcOH/NaOAc was sealed without degassing. The mixture was stirred at 80 °C for 0.25–12 h, and then cooled to rt, basified with aq KOH, and extracted with Et_2O (3 × 5 mL). The extracts were dried (Na_2SO_4), filtered, and concentrated under reduced pressure. The product was purified by flash column chromatography.

***tert*-Butyl Indoline-1-carboxylate** (36, **$R^1 = R^2 = R^3 =$ H); Typical Procedure:**[33]

A mixture of Rh(acac)(cod) (1.6 mg, 5.0 μmol) and Ph_3P (2.6 mg, 10.0 μmol) in iPrOH (1.0 mL) was stirred vigorously at rt for 10 min. The resulting mixture was transferred by cannula to a stainless-steel autoclave filled with N_2, in which *tert*-butyl 1*H*-indole-1-carboxylate (**35**, $R^1 = R^2 = R^3 =$ H; 109 mg, 0.50 mmol) was placed beforehand. H_2 was introduced into the vessel until the pressure gauge indicated 48.4 atm. The mixture was stirred at 80 °C for 2–12 h, and then cooled and concentrated. The residue was purified by flash column chromatography (silica gel, hexanes/EtOAc 20:1); yield: 100 mg (91%).

for references see p 191

1.4.2.2 Enantioselective Reductions

Homogeneous enantioselective reductions require chiral catalysts that commonly consist of a metal-containing precursor and a chiral ligand. Today, many structurally and electronically diverse phosphorus-containing ligands (e.g., **37–55**) as well as phosphorus-free ligands have been designed and applied in enantioselective reduction (Scheme 17).[34]

Scheme 17 Selected Ligands for Enantioselective Reduction of Hetarenes[34]

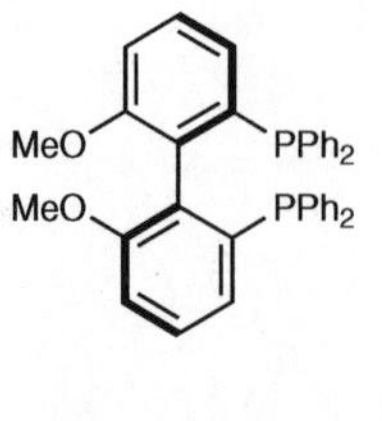

(*R*)-**37** (*R*)-MeO-BIPHEP

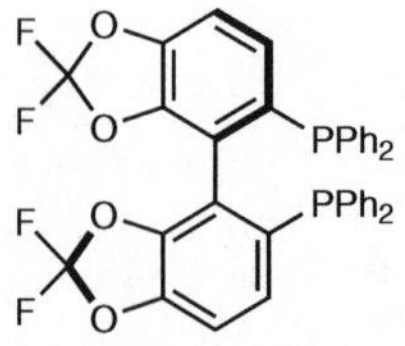

38 (*R*)-DIFLUORPHOS

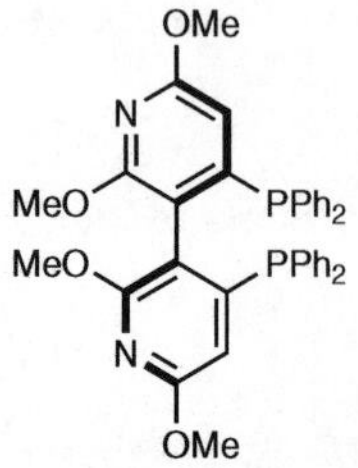

39 (*R*)-P-Phos

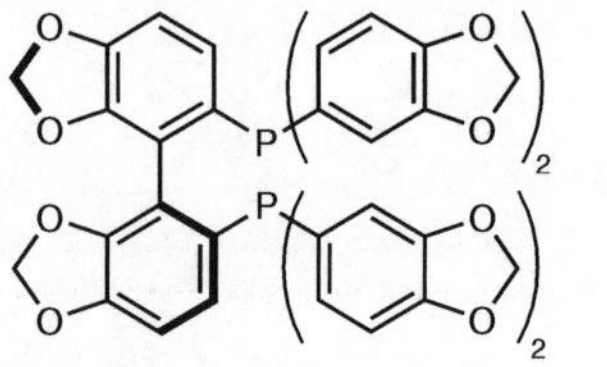

40

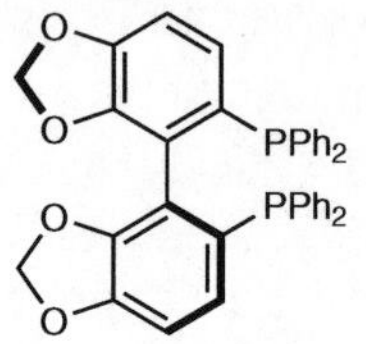

41 (*S*)-SEGPHOS

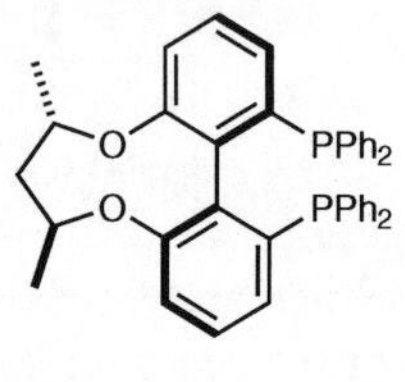

42 (R_{ax},*S*,*S*)-C_3*-TunePhos

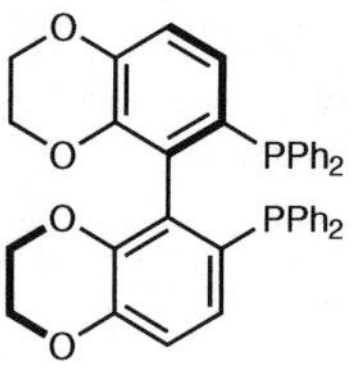

(*R*)-**43** SYNPHOS

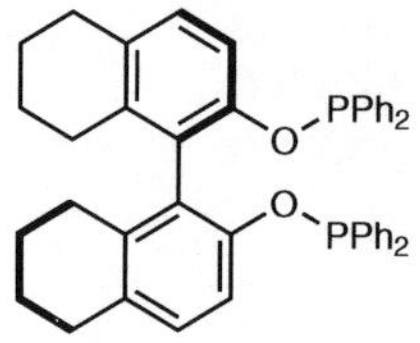

44 (*R*)-H_8-BINAPO

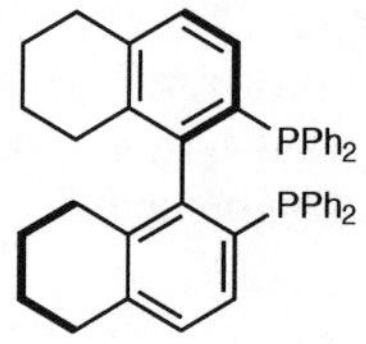

45 (*R*)-H_8-BINAP

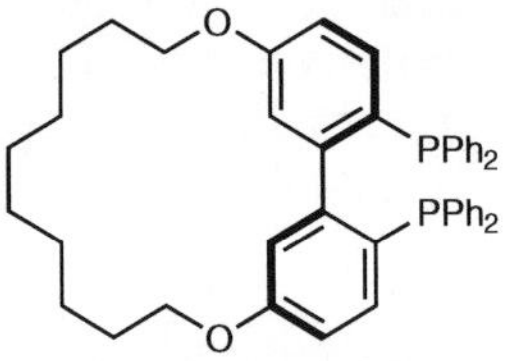

46 (*R*)-C_{10} BridgePHOS

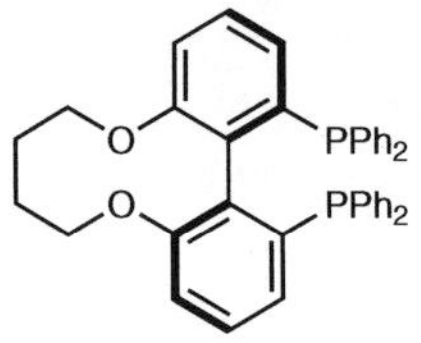

47 (*R*)-C_4-TunePhos

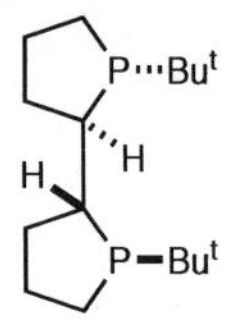

48 (R_C,S_P)-TangPhos

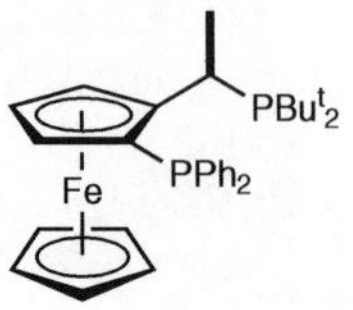

49 Josiphos

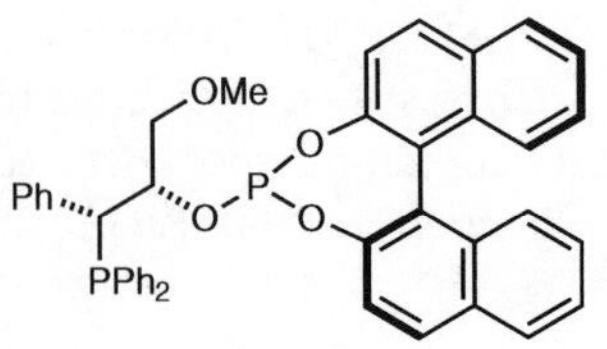

50 P-OP Ligand

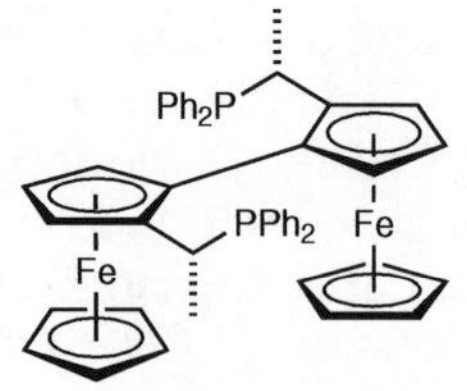

51 (*S*,*S*)-(*R*,*R*)-PhTRAP

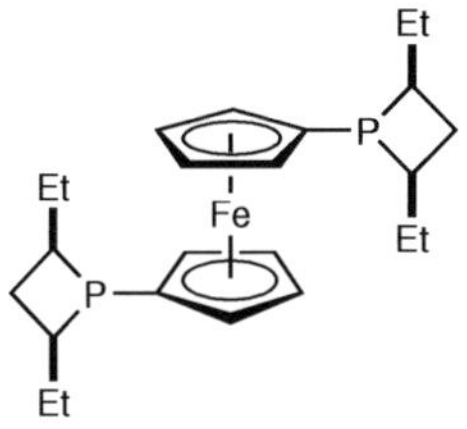

52 (*S*,*S*)-Et-FerroTANE

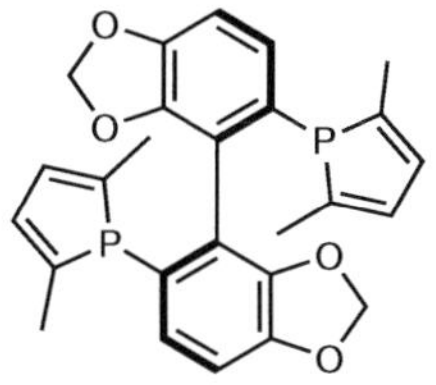

53 (*R*)-MP2-SEGPHOS

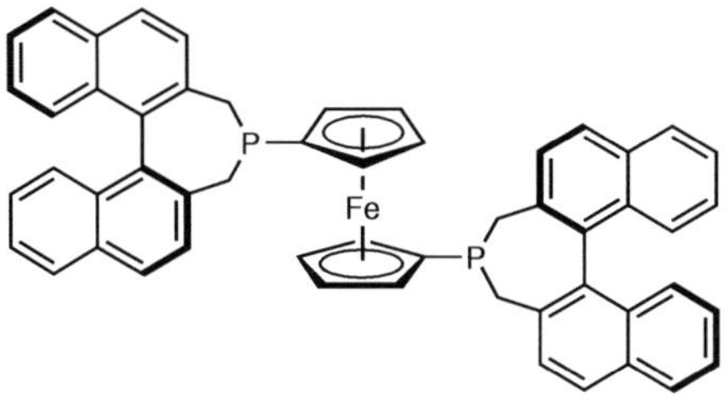

54 (*S*,*S*)-f-Binaphane

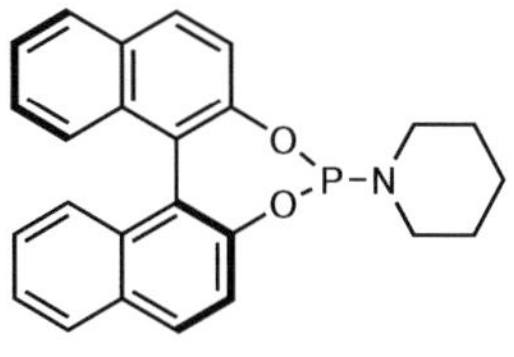

55 (*S*)-PipPhos

Generally, in the reduction of hetarenes the atropisomeric biaryl bisphosphine ligands have been most often used and found to be superior. Other types of phosphorus-containing ligands, including diphosphinite, diphosphonite, phosphoramidite, and *P,N*-ligands, also work well in the reduction of most hetarenes. The readily available and air-stable chiral diamine ligands are highly attractive and also exhibit high efficiency in this transformation, especially for polycyclic aromatic substrates. Various transition-metal complexes, including those based on ruthenium (e.g., **56** and **57**), rhodium (e.g., **58** and **59**), and iridium (e.g., **60–62**), have been developed and successfully applied in these transformations (Scheme 18).[34]

Scheme 18 Selected Transition-Metal Catalysts for the Reduction of Hetarenes[34]

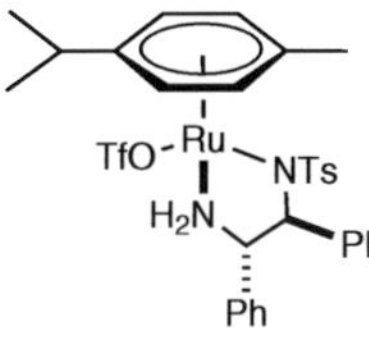

56

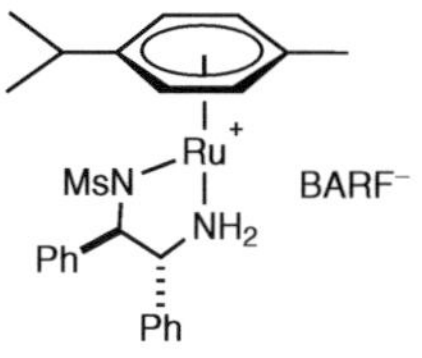

57

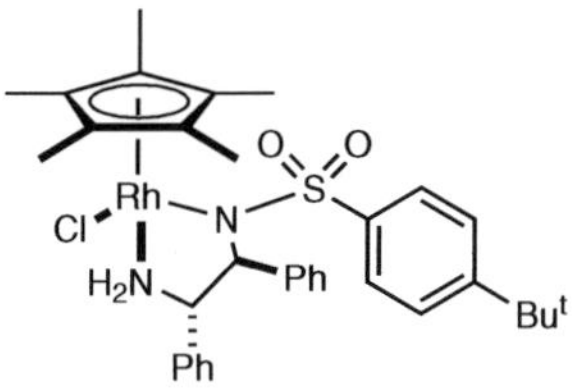

58

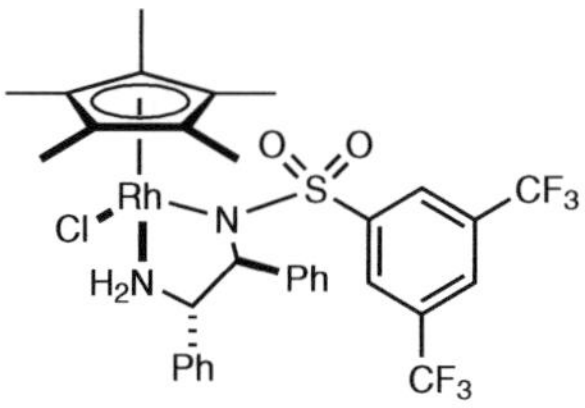

59

for references see p 191

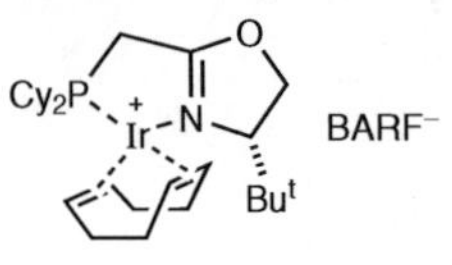

60 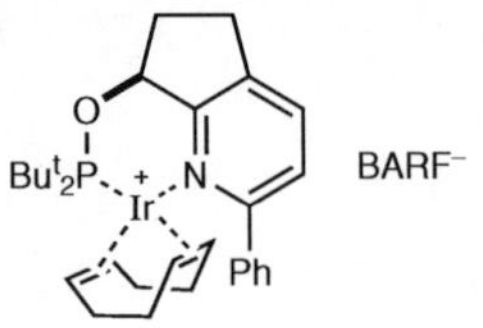 61 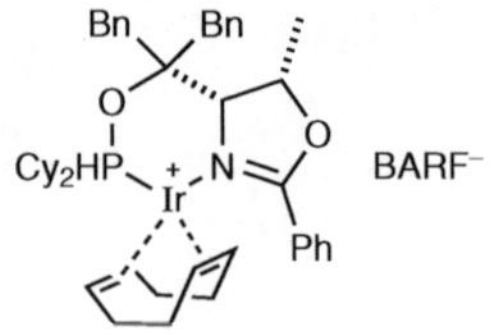 62

BARF = $[3,5-(F_3C)_2C_6H_3]_4B^-$

By using these versatile chiral metallic catalysts, a variety of hetarenes such as quinolines, isoquinolines, quinoxalines, pyridines, indoles, pyrroles, imidazoles, oxazoles, and furans can be reduced with good to excellent enantioselectivities.[35,36] It is necessary to emphasize that additives (such as chloroformates, halogen reagents, Brønsted acids, etc.) often play an important role in most of these catalytic hetarene reductions, where they frequently improve activity and/or stereoselectivity.

1.4.2.2.1 Neutral Iridium Complexes as Catalysts

In one of the first examples of the asymmetric hydrogenation of quinolines, the Zhou group reported the use of a chiral iridium catalyst, prepared in situ by combination of chloro(cycloocta-1,5-diene)iridium(I) dimer with the chiral bisphosphine ligand **37** (MeO-BIPHEP), for the reduction of a variety of 2-alkyl-substituted quinolines **63**, with the aid of a catalytic amount of iodine, to provide chiral 1,2,3,4-tetrahydroquinolines **64** (Scheme 19).[37] Subsequently, this catalytic system was extended to other functionalized quinolines by the same group and others as described in refs[38–40].

Scheme 19 Hydrogenation of Quinolines Catalyzed by Neutral Iridium Complexes[37–40]

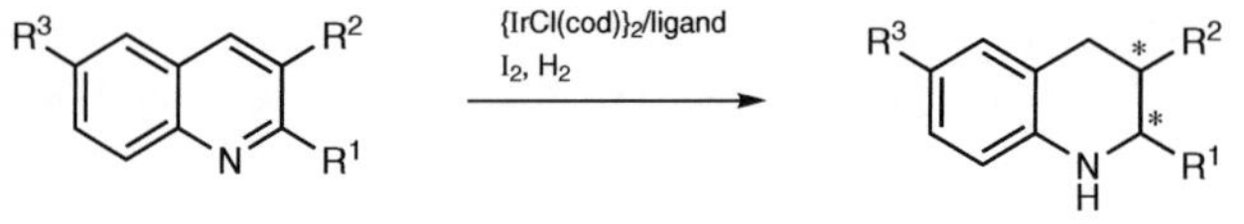

$\{IrCl(cod)\}_2$/ligand, I_2, H_2

63 → 64

R^1	R^2	R^3	Ligand	H_2 (atm)	ee[a] (%) (Config)	Yield (%)	Ref
Me	H	H	(*R*)-**37**	44.2	94 (*R*)	94	[37]
Ph	H	H	(*R*)-**37**	44.2	72 (*S*)	95	[37]
Bn	H	H	(*S*)-**37**	47.6	94 (*R*)	95	[38]
$CH_2C(OH)Me_2$	H	H	(*R*)-**37**	47.6	94 (*S*)	87	[37]
Me	H	Me	(*R*)-**37**	47.6	91 (*R*)	91	[37]
CH_2Ac	H	H	(*S*)-**37**	54.4	90 (*R*)	93	[38]
CH_2CO_2Et	H	H	(*S*)-**37**	54.4	92 (*R*)	93	[38]
CH_2SO_2Ph	H	H	(*S*)-**37**	54.4	90 (*R*)	97	[38]
Et	Me	Me	(*S*)-**37**	2.7	84 (*S*,*S*)	91	[38]
Et	Me	OMe	(*S*)-**37**	2.7	85 (*S*,*S*)	76	[38]
Bu	NPhth	H	**38**	13.6	93 (*R*,*R*)	99	[39]
CH_2NBoc_2	H	H	**39**	49.3	85 (+)	100	[40]

[a] Determined by HPLC analysis with a chiral stationary phase.

Later mechanistic investigations suggested that the phosphine–iridium complex **65** firstly undergoes oxidative addition of molecular iodine, forming an active diiodoiridium(III) species (Scheme 20).[38] Then, σ-bond metathesis with molecular hydrogen proceeds to generate the metal hydride complex **66**. The quinoline substrate coordinates to this active complex to form intermediate **67**, and then 1,4-hydride transfer affords the intermediate **68**. Subsequently, addition of molecular hydrogen to intermediate **68** via heterolytic cleavage generates intermediate **69**, followed by dissociation to furnish an enamine **70** and regenerate complex **66**. Next, the isomerization of the enamine **70** to imine **71** proceeds, catalyzed by the hydrogen iodide Brønsted acid, generated in the initial formation of **66**. Coordination of the imine to complex **66** forms intermediate **72**, and subsequent reduction of the remaining C=N bond provides intermediate **73** with a stereogenic center. Finally, the product is released via hydrogenolysis of intermediate **73**, completing the catalytic cycle.

Scheme 20 Possible Mechanism for Iridium-Catalyzed Hydrogenation of Quinolines[38]

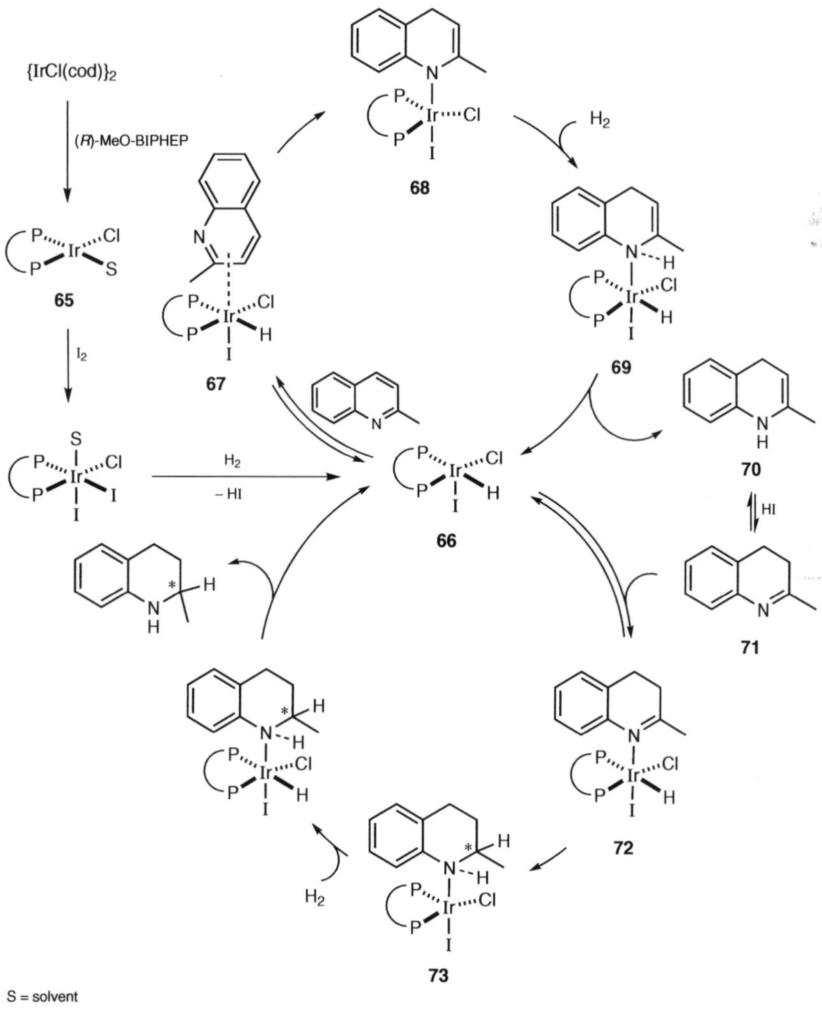

{IrCl(cod)}$_2$

(*R*)-MeO-BIPHEP

65

I_2

H_2

– HI

66

67

68

H_2

69

70

HI

71

72

73

H_2

S = solvent

The chloro(cycloocta-1,5-diene)iridium(I) dimer/bisphosphine ligand system can also be employed to reduce other kinds of hetarenes. As depicted in Table 2, an array of nitrogen-containing arenes, including isoquinolines, quinoxalines, pyridines, indoles, and pyrimidines, are successfully hydrogenated with high enantioselectivity.

When chloroformates are added as the activating reagents, isoquinolines can be hydrogenated to 1,2-dihydroisoquinolines with chloro(cycloocta-1,5-diene)iridium(I) dimer/(*S*)-SEGPHOS (**41**) as catalyst (entry 1).[41] The iridium complex that is generated from (R_{ax},*S*,*S*)-C_3*-TunePhos (**42**) and chloro(cycloocta-1,5-diene)iridium(I) dimer facilitates enantioselective reduction of another type of active isoquinoline derivative, 1,2-disubstituted isoquinolinium bromides (entry 2).[42] 3,4-Disubstituted isoquinolines (entry 3)[43] and 4-fluorinated isoquinolinium salts (entry 4)[44] are also suitable substrates for neutral iridium-catalyzed asymmetric hydrogenation with 3-bromo-1-chloro-5,5-dimethyl-2,4-imidazolidine-2,4-dione (BCDMH) as an additive, the role of which is supposedly to oxidize the low-valent metal to a more active species. The presence of an alkoxycarbonyl group at the C4 position of 3,4-disubstituted isoquinolines is not necessary for the reactivity, yet is crucial for achieving high enantioselectivity. In addition, the use of the hydrochloride salt of a fluorinated isoquinoline as the substrate, instead of the isoquinoline, is important for obtaining high yield and enantiomeric excess.

In 2009, Fan, Chan, and co-workers reported that chiral diphosphinites such as **44**, derived from (*R*)-5,5′,6,6′,7,7′,8,8′-octahydro-1,1′-bi-2-naphthol, in combination with chloro(cycloocta-1,5-diene)iridium(I) dimer induce higher enantioselectivities in the asymmetric hydrogenation of 2-substituted quinoxalines (entry 5).[45] At about the same time, Minnaard, Feringa, de Vries, and co-workers described that the iridium catalyst prepared in situ from chloro(cycloocta-1,5-diene)iridium(I) dimer and monodentate phosphoramidite ligand PipPhos (**55**) also catalyzes this transformation (entry 6).[46]

In 2008, Zhou and co-workers found that their iridium/bisphosphine/molecular iodine catalytic system (see above) is also effective for the asymmetric hydrogenation of pyridines with an electron-withdrawing group (entry 7).[47] Shortly afterwards, they described the asymmetric hydrogenation of pyridinium salts, another type of activated pyridine, affording 2-substituted *N*-benzylpiperidines in good yields and enantiomeric excesses with a similar iridium/bisphosphine catalytic system (entry 8).[48] Subsequently, Chen and Zhang revealed that the use of phosphole-based (*R*)-MP2-SEGPHOS (**53**) can further improve the yields and enantioselectivities (entry 9).[49] Recently, the Zhou group reported that 3-(trifluoromethyl)pyridinium hydrochloride derivatives (entry 10)[50] and 6-substituted 3-hydroxypyridinium salts (entry 11)[51] can be hydrogenated using iridium/bisphosphine catalysts in different solvents.

Vidal-Ferran and co-workers have reported the use of a neutral iridium complex based on P-OP ligand **50** for the hydrogenation of unprotected indoles (entry 12).[52] With a stoichiometric amount of racemic 10-camphorsulfonic acid as additive, various indoles are hydrogenated with moderate yields and high enantioselectivities.

Pyrimidines can be partially hydrogenated with chloro(cycloocta-1,5-diene)iridium(I) dimer/Josiphos (**49**) as the catalyst (entry 13).[53] Generally, by employing ytterbium(III) trifluoromethanesulfonate as an activator, a variety of 2,4-disubstituted pyrimidines are converted into cyclic amidines smoothly with high yields and enantiomeric excesses. However, pyrimidines with an electron-withdrawing group at the C4 position produce amidines with low enantiomeric excesses albeit with high yields.

In 2014, Zhou and co-workers reported an example of iridium-catalyzed asymmetric hydrogenation of *N*-benzylpyrrolo[1,2-*a*]pyrazinium salts using a neutral iridium complex (entry 14).[54] The atropisomeric biaryl bisphosphine ligand (R_{ax},*S*,*S*)-**42** is the most efficient ligand for this hydrogenation, and the *N*-benzyl protecting group remains intact. The presence of cesium carbonate is necessary to prevent racemization of the stereocenter by the in situ generated hydrogen bromide.

Table 2 Hydrogenation of Hetarenes Catalyzed by Neutral Iridium Complexes[41–54]

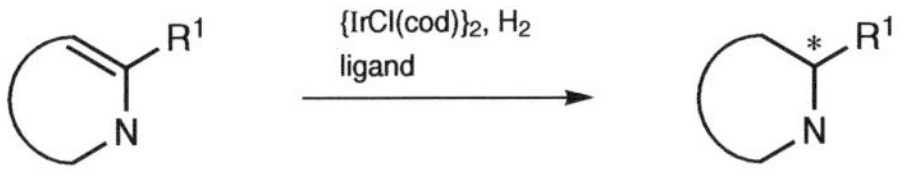

$\{IrCl(cod)\}_2$, H_2
ligand

Entry	Substrate	Ligand	H_2 (atm)	Conditions	Product	ee (%)	Yield (%)	Ref
1	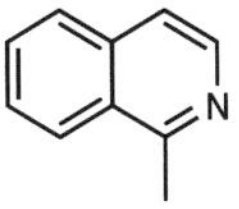	**41**	40.8	THF, $ClCO_2Me$, Li_2CO_3, $LiBF_4$	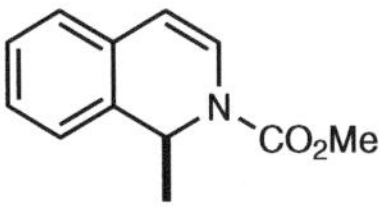	80	85	[41]
2	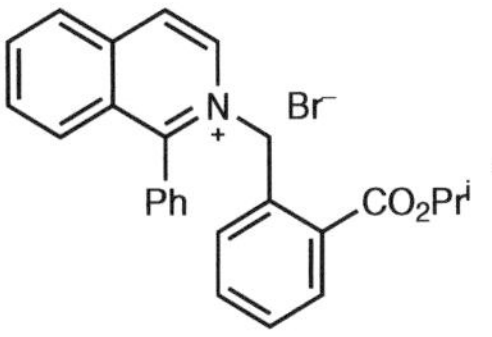	**42**	40.8	THF/CH_2Cl_2	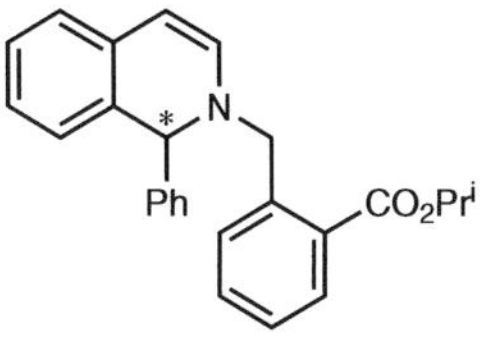	96	99	[42]
3	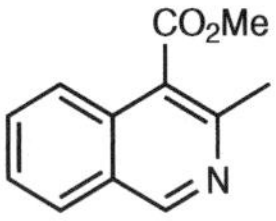	(*R*)-**43**	2.7	BCDMH, toluene	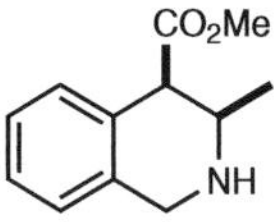	96	99	[43]
4	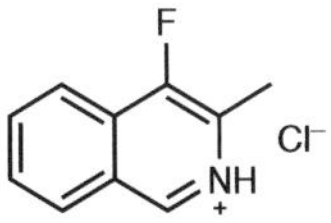	(*R*)-**43**	39.5	BCDMH, dioxane/iPrOH	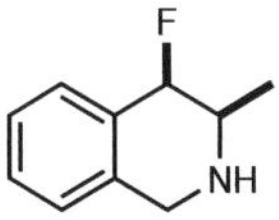	93	92	[44]
5	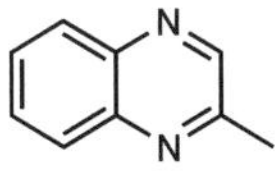	**44**	47.6	I_2, THF	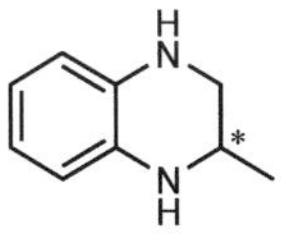	93	>99	[45]
6	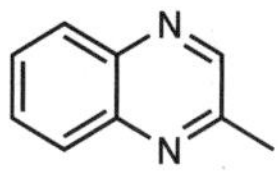	**55**	24.7	THF	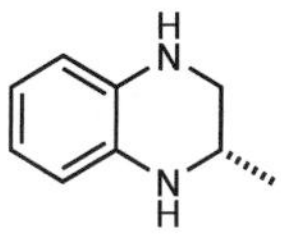	96	85	[46]
7	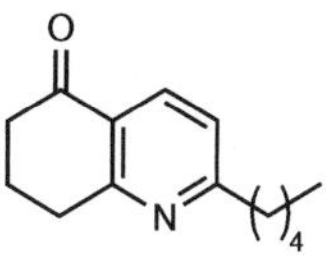	(*S*)-**37**	24.7	I_2, benzene	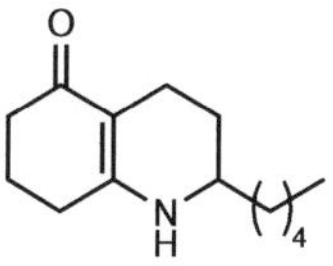	97	98	[47]
8	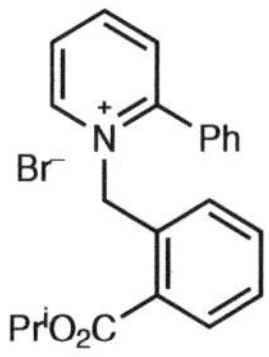	(*R*)-**43**	47.6	toluene/CH_2Cl_2	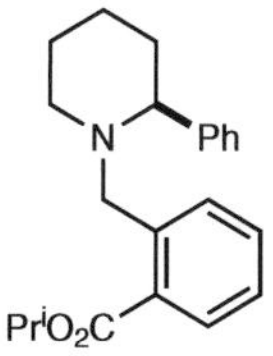	93	99	[48]

for references see p 191

Table 2 (cont.)

Entry	Substrate	Ligand	H_2 (atm)	Conditions	Product	ee (%)	Yield (%)	Ref
9		**53**	34	1,2-dichloroethane/acetone	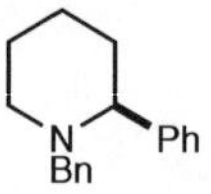	96	97	[49]
10	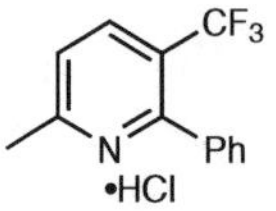	**38**	54.4	TCCA,[a] 1,2-dichloroethane/iPrOH	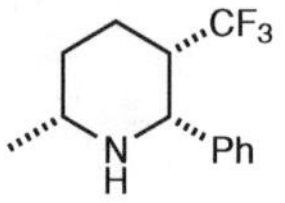	90	95	[50]
11		**54**	40.8	1,2-dichloroethane/acetone	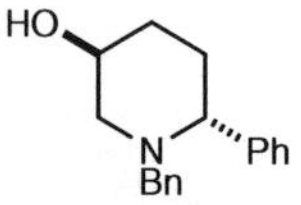	93	98	[51]
12	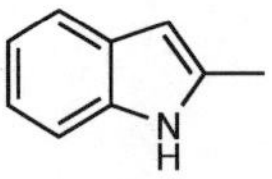	**50**	79.0	CSA, 2-methyltetrahydrofuran/CH_2Cl_2	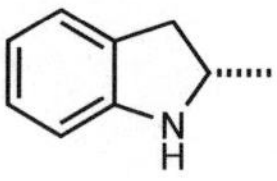	91	78	[52]
13	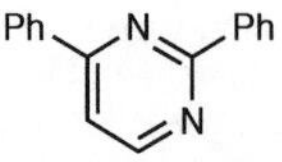	**49**	49.3	I_2, $Yb(OTf)_3$, EtOAc	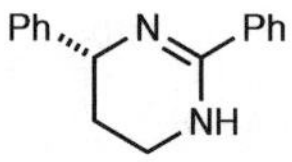	97	92	[53]
14	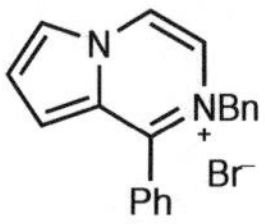	**42**	27.2	THF, Cs_2CO_3	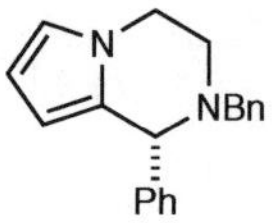	92	97	[54]

[a] TCCA = trichloroisocyanuric acid.

1,2,3,4-Tetrahydroquinolines 64; General Procedure:[37]
A mixture of $\{IrCl(cod)\}_2$ (3.4 mg, 5 μmol) and ligand (*R*)-**37** [(*R*)-MeO-BIPHEP; 6.4 mg, 11 μmol] in toluene (5 mL) was stirred at rt for 30 min in a glovebox. The mixture was transferred by syringe to a stainless-steel autoclave in which I_2 (12.7 mg, 0.05 mmol) and a quinoline **63** (1.0 mmol) had been placed beforehand. The hydrogenation was performed at rt under H_2 (44.2 atm) for 12–15 h. After carefully releasing the H_2, the mixture was diluted with CH_2Cl_2 (20 mL). Sat. aq Na_2CO_3 (5 mL) was added and the mixture was stirred for 15 min. The aqueous layer was extracted with CH_2Cl_2 (3 × 15 mL), and the combined organic layers were dried (Na_2SO_4), filtered, and concentrated. The residue was purified by column chromatography (silica gel, hexanes/EtOAc).

1.4.2.2.2 Cationic Iridium Complexes as Catalysts

In 2006, a series of cationic triply halogen-bridged dinuclear iridium(III) complexes (e.g., **75**) of bisphosphines, including SYNPHOS (**43**), were synthesized and applied in the asymmetric hydrogenation of quinolines, with moderate to good enantioselectivities observed (Table 3, entry 1).[55] The use of a similar catalytic system with triply chloro-bridged iridium(III) complexes **74** containing the DIFLUORPHOS (**38**) ligand extended the enantioselective hydrogenation to quinolinium salts (entries 2 and 3),[56] quinoxalines (entry 4),[57] and pyridinium salts (entry 5).[58]

Table 3 Hydrogenation of Hetarenes Catalyzed by Dinuclear Iridium Complexes[55–58]

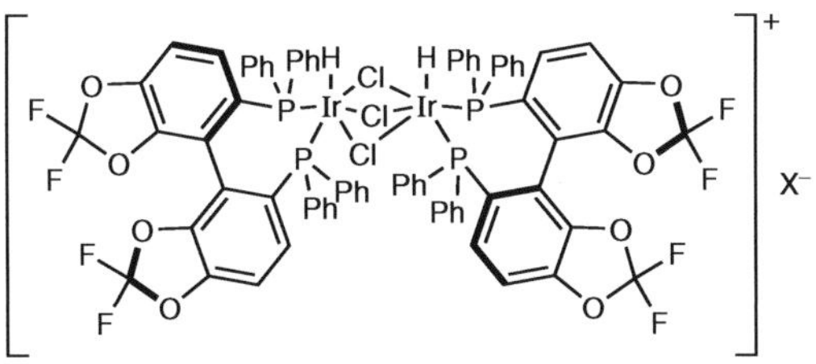

74

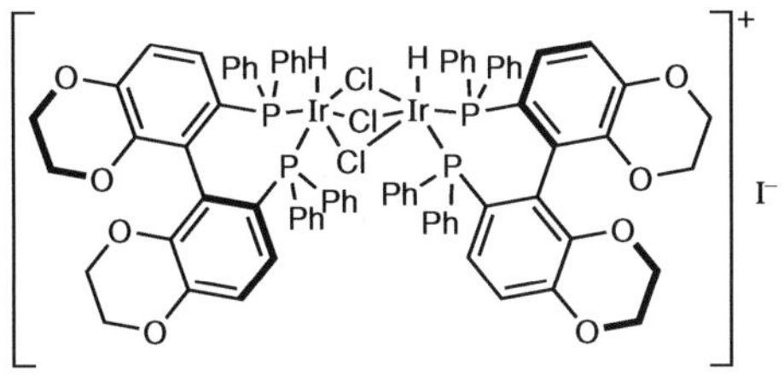

75

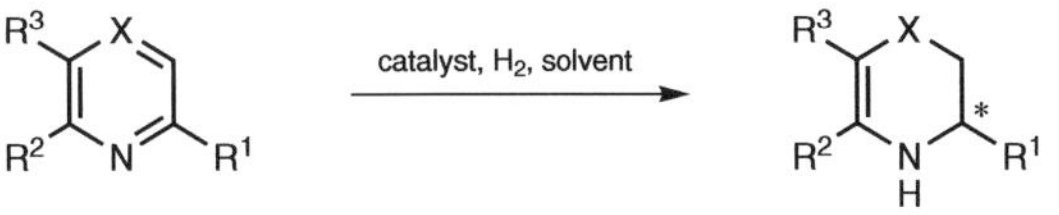

catalyst, H_2, solvent

Entry	Substrate	Catalyst	H_2 (atm)	Solvent	Product	ee (%)	Yield (%)	Ref
1	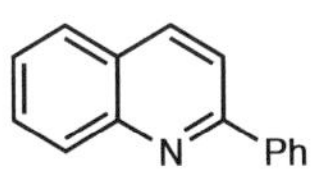	**75**	50	THF	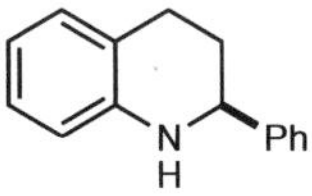	64	42	[55]
2	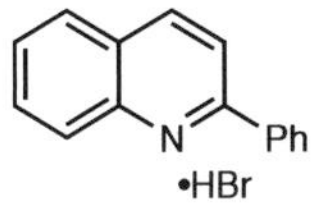 •HBr	**74** (X = Br)	29.6	1,4-dioxane/MeOH	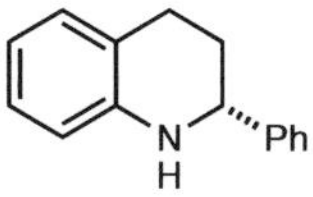	91	>95	[56]
3	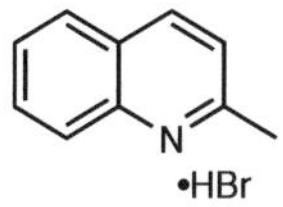 •HBr	**74** (X = Br)	29.6	1,4-dioxane/MeOH	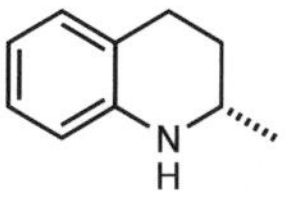	94	81	[56]
4	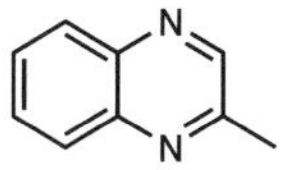	**74** (X = Cl)	29.6	toluene	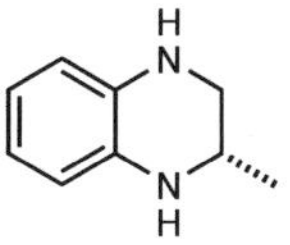	94	99	[57]
5	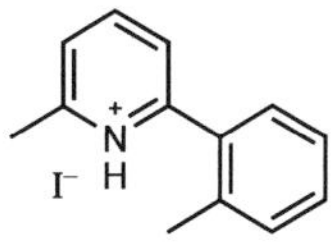	**74** (X = Cl)	49.3	1,4-dioxane	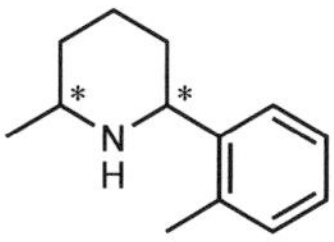	82	>95	[58]

for references see p 191

Significantly, with the catalyst system involving complex **74** (X = Cl), a broad range of isoquinolinium salts **76** are hydrogenated to yield the corresponding 1,2,3,4-tetrahydroisoquinolines **77** with good yields and enantioselectivities (Scheme 21).[59]

Scheme 21 Hydrogenation of Isoquinolinium Salts Catalyzed by a Dinuclear Iridium Complex[59]

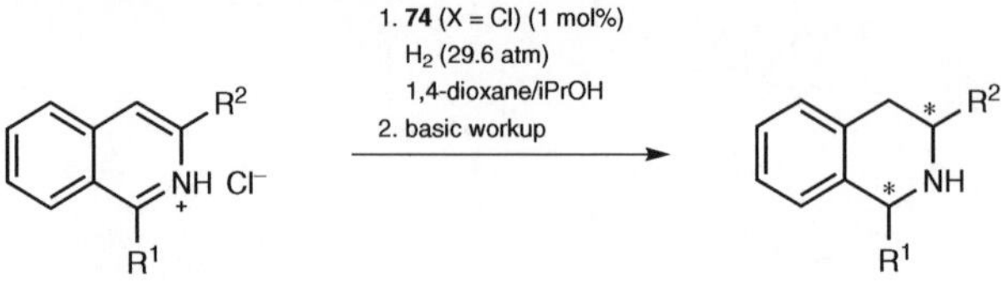

1. **74** (X = Cl) (1 mol%) H_2 (29.6 atm) 1,4-dioxane/iPrOH
2. basic workup

76 → **77**

R^1	R^2	ee[a] (%) (Config)	Yield (%)	Ref
H	Ph	96 (+)	99	[59]
H	2-$MeOC_6H_4$	81 (+)	93	[59]
H	4-$MeOC_6H_4$	95 (+)	99	[59]
H	Cy	91 (+)	98	[59]
Cy	H	79 (−)	99	[59]
iPr	H	91 (*S*)	98	[59]
Ph	H	96 (*S*)	99	[59]
Ph	Ph	98 (*S*,*S*)	98	[59]

[a] Determined by HPLC analysis with a chiral stationary phase.

In 2010, Pfaltz and co-workers demonstrated that the cationic iridium complexes **60–62** derived from phosphine–oxazoline or chiral *P*,*N*-ligands are efficient catalysts for the asymmetric hydrogenation of N-protected indoles **78** (Scheme 22).[60] The protecting group (*N*-Boc-, *N*-acetyl-, or *N*-tosyl) influences both the reactivity and enantioselectivity; thus, the right combination of catalyst and protecting group is vital for achieving high yields and enantiomeric excesses.

Scheme 22 Hydrogenation of Indoles Catalyzed by Cationic Iridium Complexes[60]

Ir catalyst 1–4 mol%) H_2 (100 atm), CH_2Cl_2

78

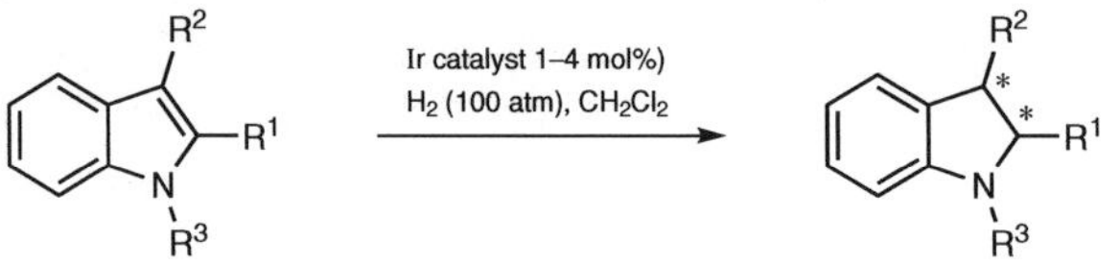

R^1	R^2	R^3	Ir Catalyst	ee[a] (%)	Yield (%)	Ref
Me	H	Ts	**60**	99	94	[60]
Ph	H	Ts	**61**	98	70[b]	[60]
CO_2Et	H	Ac	**61**	99	97	[60]
H	Me	Ts	**61**	98	97	[60]
CO_2Et	H	Boc	**62**	93	88	[60]

[a] Determined by HPLC analysis with a chiral stationary phase.
[b] Chlorobenzene was used as the solvent.

1,2,3,4-Tetrahydroisoquinolines 77; General Procedure:[59]

The Ir complex **74** (X = Cl; 2.4 μmol, 1.0 mol%) and isoquinolinium salt **76** (0.24 mmol) were added to a glass tube in the reactor, and the tube was charged with argon. A mixture of 1,4-dioxane/anhyd iPrOH (10:1; 3 mL) was added to the glass tube in the reactor through the inlet. The glass tube was charged with H_2, and the pressure was increased to 30 atm. The mixture was stirred for 20 h. After the release of H_2 from the tube, the solvent was removed on a rotary evaporator. The residual liquid was poured into sat. aq $NaHCO_3$. The mixture was extracted with EtOAc and the separated organic layer was washed with brine, dried (Na_2SO_4), filtered, and concentrated. The ee was determined by HPLC.

1.4.2.2.3 Ruthenium Complexes as Catalysts

The air-stable chiral diamine–ruthenium catalyst **56** has attracted much attention due to its application in enantioselective hydrogenation of ketones. In 2008, Fan and Chan found that this catalytic system can also be employed at room temperature in methanol or ionic liquids {e.g., 1-butyl-3-methylimidazolium hexafluorophosphate ([bmim]PF_6)} for the efficient catalytic hydrogenation of 2-substituted quinolines (Table 4, entry 1).[61] Following this amazing discovery, the extended application of such diamine ruthenium catalysts in the hydrogenation of 2- or 2,3-substituted quinoxalines (e.g., entry 2),[62] 1,5- or 1,8-naphthyridines (entries 3 and 4),[63,64] and 1,10-phenanthrolines (entries 5 and 6)[65] was also reported.

Table 4 Hydrogenation of Hetarenes Catalyzed by Chiral Diamine Ruthenium Complexes[61–65]

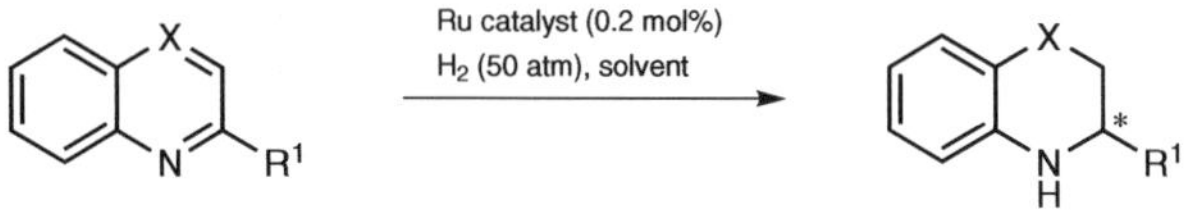

Ru catalyst (0.2 mol%)
H_2 (50 atm), solvent

Entry	Substrate	Ru Catalyst	Solvent	Product	ee[a] (%)	Yield (%)	Ref
1	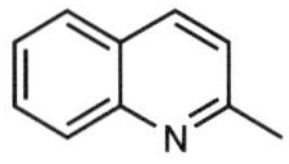	**56**	[bmim]PF_6	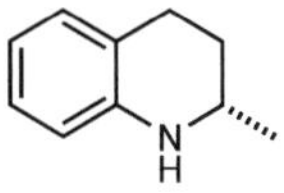	99	96	[61]
2	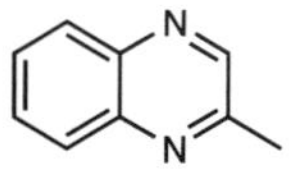	**57**	1,2-dichloroethane	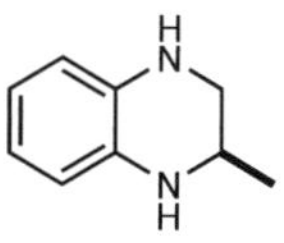	98	95	[62]
3	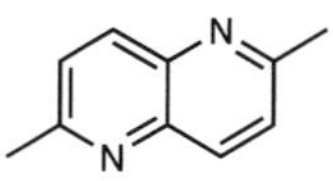	**56**	EtOH	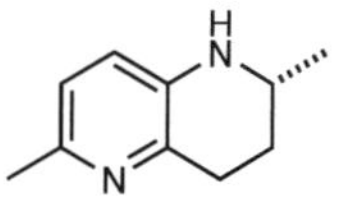	99	95	[63]

for references see p 191

Table 4 (cont.)

Entry	Substrate	Ru Catalyst	Solvent	Product	ee[a] (%)	Yield (%)	Ref
4	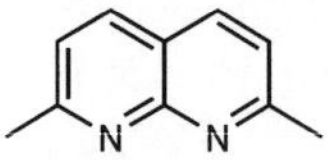	**56**	iPrOH	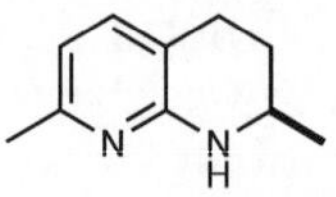	99	99	[64]
5	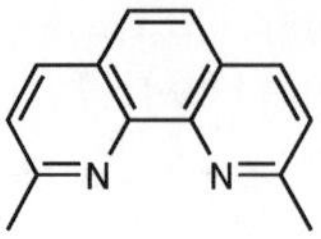	**56**	MeOH	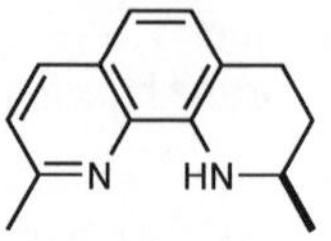	99	100[b]	[65]
6	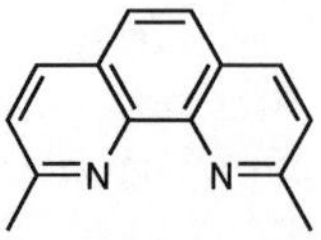	**56**	MeOH	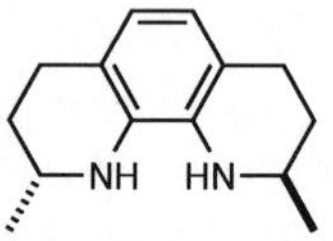	99	90	[65]

[a] Determined by HPLC analysis with a chiral stationary phase.
[b] After releasing H_2, the mixture was stirred for another 12 h at rt under air.

A plausible mechanism for diamine ruthenium catalysis was proposed (Scheme 23):[66] First, molecular hydrogen reversibly coordinates to the ruthenium complex **79** to form dihydrogen complex **80**. Then, complex **80** transfers a proton to the substrate, providing both the active ruthenium hydride species **81** and the protonated substrate. A subsequent 1,4-hydride transfer from complex **80** to the protonated substrate affords the enamine intermediate **21** and regenerates **79**. Then, enamine-to-imine isomerization takes place to generate imine **22**, which serves as a base to deprotonate the dihydrogen ligand of complex **80** resulting in protonated 3,4-dihydroquinoline **82** and ruthenium hydride species **81**. Finally, 1,2-hydride transfer gives the 1,2,3,4-tetrahydroquinoline product and regenerates complex **79** to complete the catalytic cycle.

Scheme 23 Proposed Mechanism for Ruthenium-Catalyzed Hydrogenation of Quinolines[66]

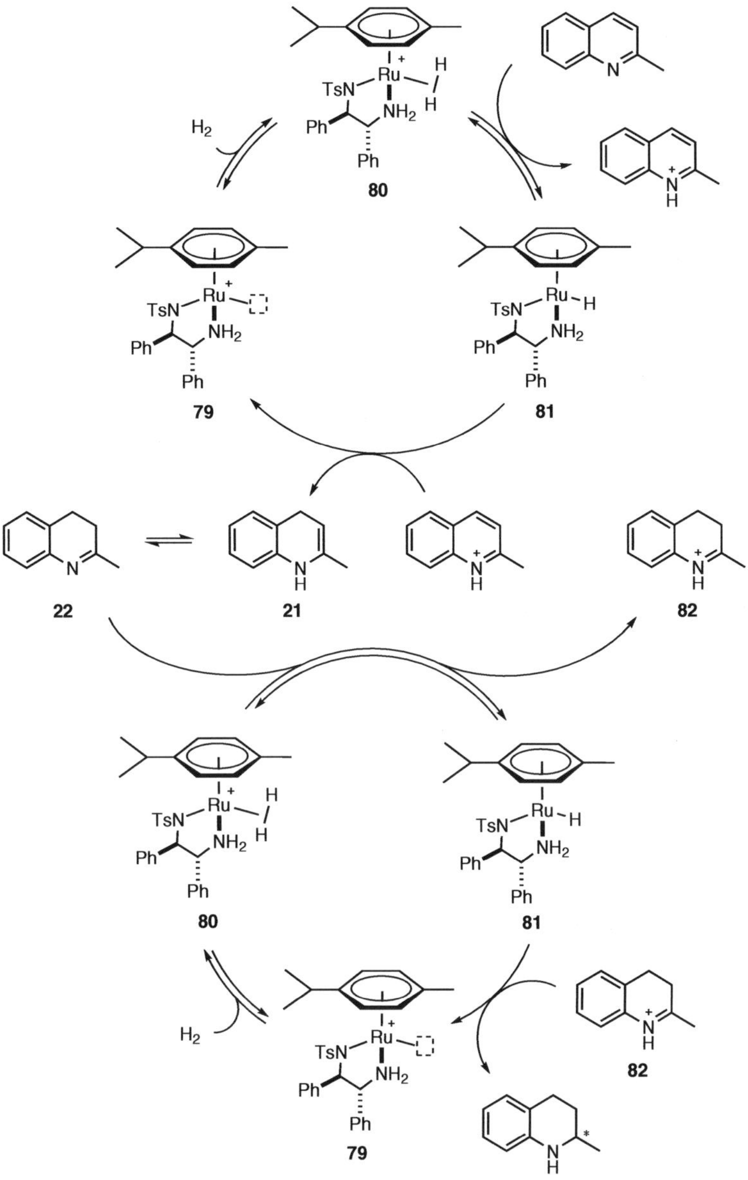

Kuwano and co-workers found that the *trans*-chelating bisphosphine ruthenium complexes perform with high activity and stereoselectivity in the hydrogenation of hetarenes. In 2006, they disclosed that a ruthenium complex bearing PhTRAP ligand (**51**) ([RuCl(*p*-cymene){(*S*,*S*)-(*R*,*R*)-PhTRAP}]Cl) enables enantioselective hydrogenation of a wide range of *N*-(*tert*-butoxycarbonyl)indoles **83** (Scheme 24).[67] Various 2- or 3-substituted indoles are converted into chiral indolines **84** in methanol or propan-2-ol with good enantiomeric excesses. In addition, 2,3-dimethylindoles have also been verified as suitable substrates when bis(2-methylallyl)(cycloocta-1,5-diene)ruthenium(II) [Ru(η^3-methylallyl)$_2$(cod)] is used as the precursor complex, giving *cis*-2,3-dimethylindolines with moderate enantioselectivity.

for references see p 191

Scheme 24 *trans*-Chelating Bisphosphine Ruthenium Catalyzed Hydrogenation of Indoles[67]

[RuCl(*p*-cymeme){(*S,S*)-(*R,R*)-PhTRAP}]Cl (1.0 mol%)
Cs_2CO_3 (10 mol%), H_2 (50 atm)
MeOH or iPrOH

83 → **84**

R^1	R^2	R^3	ee[a] (%)	Yield (%)	Ref
Me	H	H	95	99	[67]
Me	H	OMe	91	97	[67]
Me	H	F	90	96	[67]
CO_2Me	H	H	90	91	[67]
H	Ph	H	87	85	[67]
H	Me	H	94	92	[67]
Me	Me	H	72[b]	59	[67]

[a] Determined by HPLC analysis with a chiral stationary phase.
[b] $Ru(\eta^3\text{-methylallyl})_2(cod)$/**51** was used.

When employing bis(2-methylallyl)(cycloocta-1,5-diene)ruthenium(II) as transition-metal precursor, the *trans*-chelating bisphosphine PhTRAP (**51**) catalyst system also performs well in the enantioselective hydrogenation of *N-tert*-butoxycarbonylpyrroles **85** to furnish the hydrogenated products **86** (Table 5).[68] 2,3,5-Trisubstituted pyrroles bearing a large substituent at the 5-position are hydrogenated with high enantioselectivity to give chiral 4,5-dihydropyrroles in high yields (entries 2 and 3). The regioselectivity depends largely on the structure of the substrates, and the hydrogenation of a triarylpyrrole affords the dihydropyrrole exclusively (entry 4).

Table 5 *trans*-Chelating Bisphosphine Ruthenium Catalyzed Hydrogenation of Pyrroles[68]

85 → 86: $Ru(\eta^3$-2-methylallyl$)_2$(cod) (2.5 mol%), **51** (2.8 mol%), Et_3N (25 mol%), H_2 (50 atm), EtOAc

Entry	Substrate	Product	ee[a] (%)	Yield (%)	Ref
1			79	92[b]	[68]
2			96	85	[68]
3			93	96	[68]
4			99	>99	[68]

[a] Determined by HPLC analysis with a chiral stationary phase.
[b] iPrOH was used as the solvent.

In the presence of a catalytic amount of triethylamine to improve the reactivity, a series of 4-alkyl-2-phenyl-*N*-(*tert*-butoxycarbonyl)imidazoles **87** can also be hydrogenated smoothly to provide the corresponding imidazolines **88** with excellent enantioselectivities and yields (Scheme 25).

Scheme 25 *trans*-Chelating Bisphosphine Ruthenium Catalyzed Hydrogenation of Imidazoles[69]

87 → 88: $Ru(\eta^3$-2-methylallyl$)_2$(cod) (2.5 mol%), **51** (2.8 mol%), Et_3N (25 mol%), H_2 (50 atm), EtOAc

R^1	ee[a] (%)	Yield (%)	Ref
Me	97	97	[69]
CF_3	99	89	[69]

[a] Determined by HPLC analysis with a chiral stationary phase.

for references see p 191

2,4- and 2,5-disubstituted oxazoles **89** are also hydrogenated smoothly by this catalytic system, affording oxazolines **90** with high to excellent enantioselectivity (Scheme 26).[69] For 4-substituted 2-phenyloxazoles, the basic additive *N*,*N*,*N*′,*N*′-tetramethylguanidine (TMG) is needed to accelerate this selective hydrogenation. In contrast, for 5-substituted 2-phenyloxazoles, the additive is not essential and toluene is the best solvent in terms of enantioselectivity. Significantly, hydrogenolysis or acid-facilitated hydrolysis of the obtained chiral imidazolines and oxazolines could provide valuable acyclic chiral 1,2-diamines and β-amino alcohols, respectively, without loss of enantiopurity.

Scheme 26 Hydrogenation of Oxazoles Catalyzed by *trans*-Chelating Bisphosphine Ruthenium Complexes[69]

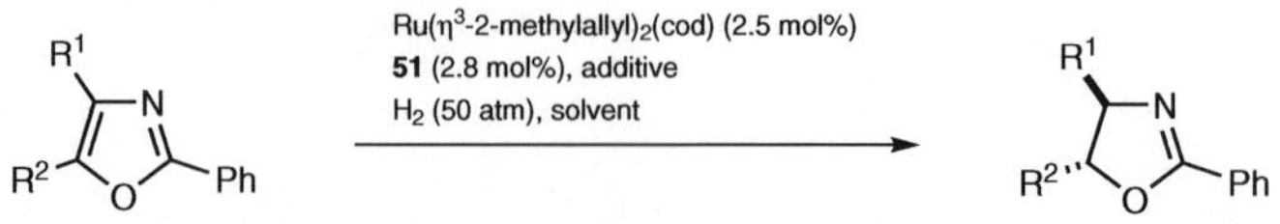

$Ru(\eta^3\text{-2-methylallyl})_2(cod)$ (2.5 mol%)
51 (2.8 mol%), additive
H_2 (50 atm), solvent

89 → **90**

R^1	R^2	Additive	Solvent	ee[a] (%)	Yield (%)	Ref
Ph	H	TMG (25 mol%)	iBuOH	98	>99	[69]
4-FC_6H_4	H	TMG (25 mol%)	iBuOH	97	92	[69]
Me	H	TMG (25 mol%)	iBuOH	91	88	[69]
H	Ph	–	toluene	97	97	[69]
H	4-FC_6H_4	–	toluene	93	>99	[69]
H	Me	–	toluene	86	85	[69]

[a] Determined by HPLC analysis with a chiral stationary phase.

Glorius and co-workers have demonstrated the efficiency of ruthenium–N-heterocyclic carbene catalysts in the reduction of hetarenes. By using bis(2-methylallyl)(cycloocta-1,5-diene)ruthenium(II) as transition-metal precursor and *N*-heterocyclic carbene **92** as ligand, a variety of hetarenes **91** including benzo[*b*]furans (Table 6, entries 1–3),[70,71] furans (entry 4),[72] benzo[*b*]thiophenes (entries 5 and 6),[73] thiophenes (entry 7),[73] and other N-heterocycles (entries 8 and 9)[74] are hydrogenated successfully to the corresponding products **93** with moderate to excellent enantioselectivities.

Table 6 Hydrogenation of Hetarenes Catalyzed by a Chiral Ruthenium–N-Heterocyclic Carbene Complex[70–74]

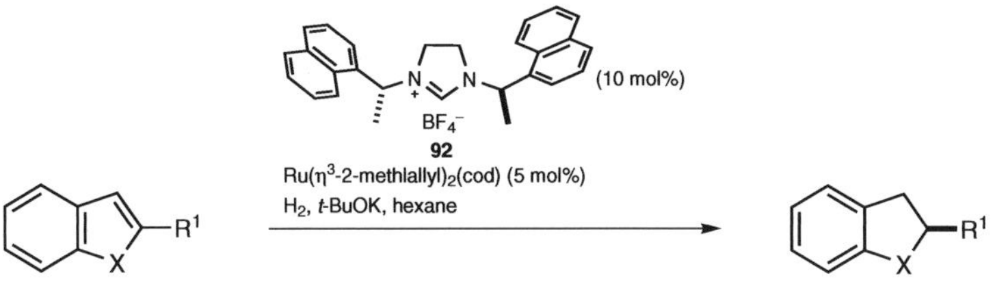

92 (10 mol%)
$Ru(\eta^3$-2-methlallyl$)_2$(cod) (5 mol%)
H_2, t-BuOK, hexane

91 → 93

Entry	Substrate	H_2 (atm)	Product	ee[a] (%)	Yield (%)	Ref
1	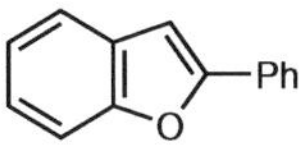	9.9	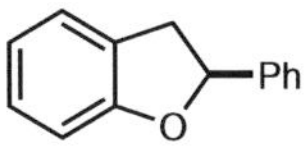	98	99	[70]
2	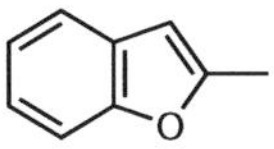	9.9	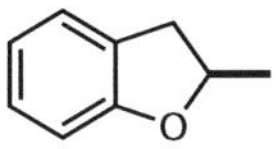	92	99	[70]
3	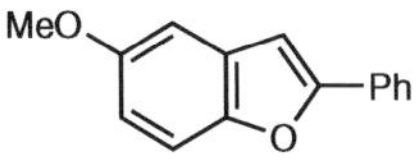	59.2	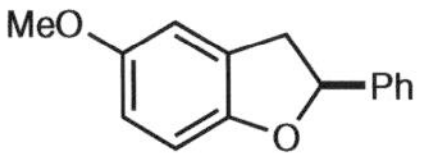	98	>99	[71]
4	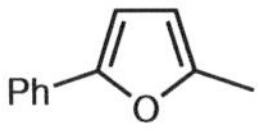	128.3	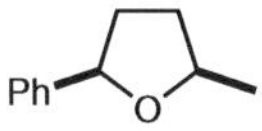	80	99[b]	[72]
5	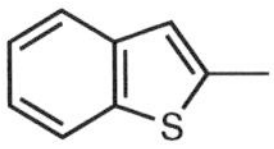	88.8	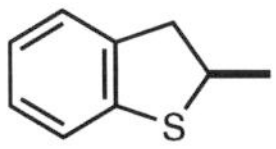	98	98	[73]
6	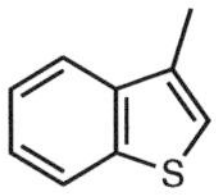	88.8	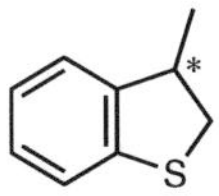	98	79	[73]
7	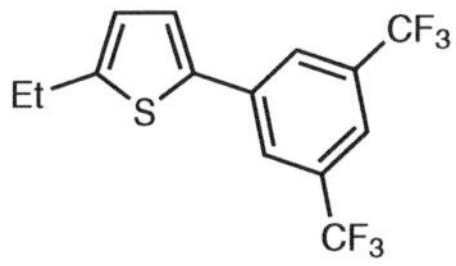	88.8	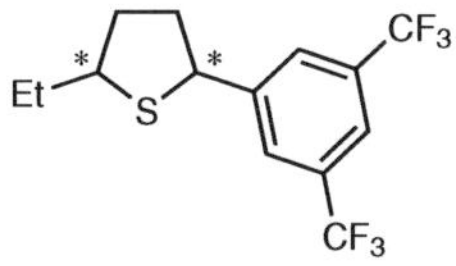	94	>99	[73]
8	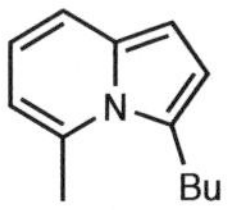	98.7	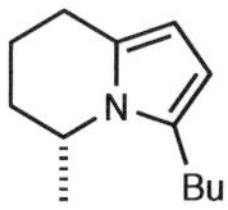	94	99	[74]
9	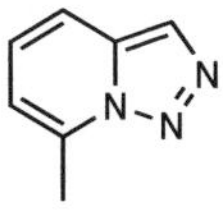	98.7	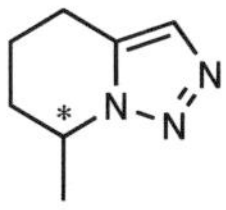	66	99	[74]

[a] Determined by HPLC analysis with a chiral stationary phase.
[b] dr 6.9:1.

(*R*)-2,6-Dimethyl-1,2,3,4-tetrahydro-1,5-naphthyridine (Table 4, Entry 3); Typical Procedure:[63]

A 50-mL glass-lined stainless-steel reactor equipped with a magnetic stirrer bar was charged with 2-methyl-1,5-naphthyridine (0.3 mmol) and Ru catalyst **56** (0.5 mg, 0.6 μmol, 0.2 mol%) in EtOH (1 mL) under N_2 in a glovebox. The autoclave was closed, and the final pressure of the H_2 was adjusted to 50 atm after purging the autoclave with H_2 several times. The mixture was stirred at 20 °C for 10 h and then the H_2 was carefully released. The mixture was filtered through a short pad of silica gel, eluting with CH_2Cl_2, to give the pure product; yield: 95%; 99% ee.

Methyl (*S*)-*N*-(*tert*-Butoxycarbonyl)pyrrolidine-2-carboxylate (Table 5, Entry 1); Typical Procedure:[68]

Under N_2, anhyd iPrOH (1.0 mL) and anhyd Et_3N (5.1 mg, 50 μmol, 25 mol%) were added to a mixture of Ru(η^3-2-methylallyl)$_2$(cod) (1.6 mg, 5.0 μmol, 2.5 mol%) and ligand **51** (4.4 mg, 5.5 μmol, 2.8 mol%). The mixture was stirred at rt for 10 min and then transferred via a cannula into methyl *N*-(*tert*-butoxycarbonyl)-1*H*-pyrrole-2-carboxylate (0.2 mmol) in a test tube. The mixture was treated with 50 atm of H_2, and then the pressure was carefully released to 1 atm. This procedure was repeated twice, and finally the inside of the autoclave was pressurized with H_2 to 50 atm. The mixture was stirred at 80 °C for 24 h. The autoclave was allowed to cool to rt and excess H_2 was released carefully. The resulting mixture was concentrated under reduced pressure. The residue was purified over a short column (silica gel, EtOAc/hexane 1:2) to give a colorless oil; yield: 92%; 79% ee.

(*S*)-2-Phenyl-2,3-dihydrobenzo[*b*]furan (Table 6, Entry 1); Typical Procedure:[70]

To a flame-dried, screw-capped tube equipped with a magnetic stirrer bar was added Ru(η^3-2-methylallyl)$_2$(cod) (4.8 mg, 0.015 mmol, 5 mol%), imidazolium salt **92** (14.1 mg, 0.03 mmol, 10 mol%), and anhyd *t*-BuOK (5.0 mg, 0.045 mmol) in a glovebox. The mixture was suspended in hexane (2 mL), stirred at 70 °C for 12 h under argon, and then transferred under argon to a glass vial containing 2-phenylbenzo[*b*]furan (58 mg, 0.3 mmol) and a magnetic stirrer bar. The glass vial was placed in a 150-mL stainless-steel reactor. The autoclave was carefully pressurized/depressurized with H_2 three times before the pressure was adjusted to 9.9 atm. The hydrogenation was performed at 25 °C for 16 h. After the autoclave was depressurized carefully, the crude mixture was filtered through a plug of silica gel using pentane/EtOAc (9:1), yielding the analytically pure product; yield: 58 mg (99%); 98% ee.

1.4.2.2.4 Rhodium Complexes as Catalysts

A number of cationic rhodium precatalysts combined with chiral ligands have been used for the enantioselective hydrogenation of various heterocycles. In an early study by Studer, bis(norbornadiene)rhodium(I) tetrafluoroborate {[Rh(nbd)$_2$]BF_4} was found to act as an effective precatalyst for the hydrogenation of picolinic acid and its ethyl ester.[75] In 2009, the Xiao group reported that rhodium/*N*-(arylsulfonyl)-1,2-diphenylethylenediamine complexes such as **58** and **59** are effective catalysts for transfer hydrogenation of quinolines **94** in aqueous solution to afford 1,2,3,4-tetrahydroquinolines **95** (Scheme 27).[76] The pH of the reaction medium plays a critical role for high activity, whereas the enantioselectivity is only slightly influenced by the pH. An acetic acid/sodium acetate buffer system (pH 5) is the best choice for a broad range of substrates, giving the desired hydrogenation products with high yields and excellent enantioselectivities.

Scheme 27 Enantioselective Hydrogenation of Quinolines Catalyzed by Chiral Diamine Rhodium Complexes[76]

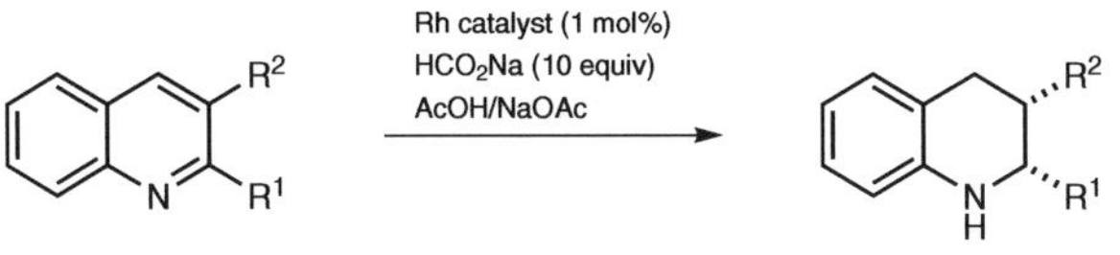

Rh catalyst (1 mol%)
HCO_2Na (10 equiv)
AcOH/NaOAc

94 → **95**

R^1	R^2	Catalyst	ee[a] (%) (Config)	Yield (%)	Ref
Me	H	**58**	97 (*S*)	96	[76]
Et	H	**58**	96 (*S*)	95	[76]
cyclopropyl	H	**58**	98 (*R*)	88	[76]
Me	Me	**58**	92 (*S*,*S*)	89 (dr 4:1)	[76]
Ph	H	**59**	90 (*R*)	96	[76]

[a] Determined by HPLC analysis with a chiral stationary phase.

Kuwano and Ito have reported the combination of bis(norbornadiene)rhodium(I) hexafluoroantimonate {$[Rh(nbd)_2]SbF_6$} and the bisphosphine ligand **51** [(*S*,*S*)-(*R*,*R*)-PhTRAP] for the enantioselective hydrogenation of 2- and 3-substituted indoles (Table 7, entries 1–3).[77,78] Minnaard, Feringa, and de Vries demonstrated a similar efficiency with a bis(cycloocta-1,5-diene)rhodium(I) tetrafluoroborate {$[Rh(cod)_2]BF_4$}/phosphoramidite ligand **55** (PipPhos) system for the hydrogenation of methyl *N*-acetylindole-2-carboxylate (entry 4).[79] Most recently, Yamashita reported the asymmetric hydrogenation of a substituted benzo[*b*]furan using bis(cycloocta-1,5-diene)rhodium(I) trifluoromethanesulfonate {$[Rh(cod)_2]OTf$} and bisphosphino ferrocene ligand **52** [(*S*,*S*)-Et-FerroTANE] (entry 5).[80]

Table 7 Hydrogenation of Hetarenes Catalyzed by Phosphine Rhodium Complexes[77–80]

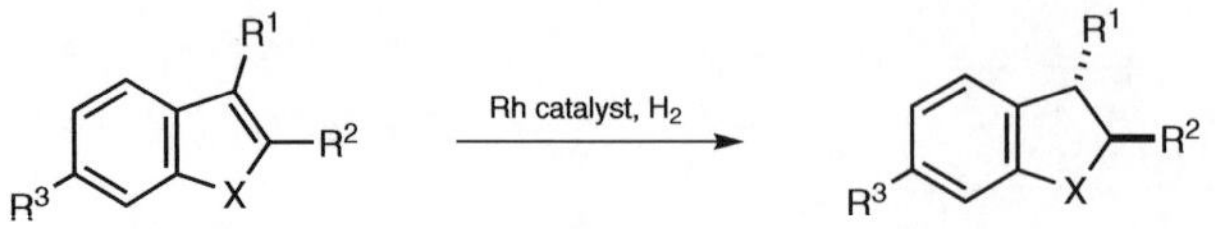

Entry	Substrate	H_2 (atm)	Conditions	Product	ee (%)	Yield (%)	Ref
1	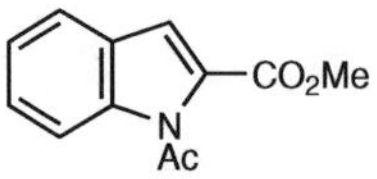	50	$[Rh(nbd)_2]SbF_6$/**51**, Cs_2CO_3, iPrOH	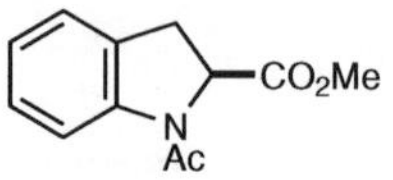	95	95	[77]
2	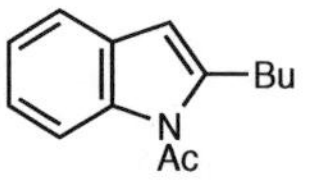	50	$[Rh(nbd)_2]SbF_6$/**51**, Cs_2CO_3, iPrOH	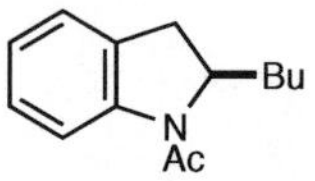	93	100	[78]
3	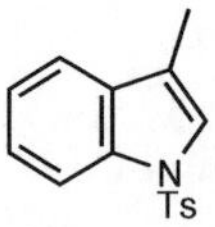	50	$[Rh(nbd)_2]SbF_6$/**51**, Cs_2CO_3, iPrOH	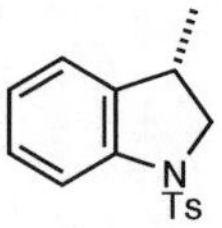	98	100	[78]
4	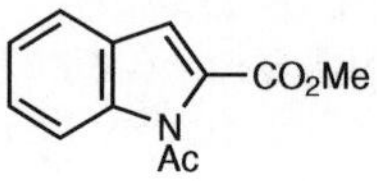	24.7	$[Rh(cod)_2]BF_4$/**55**, Cs_2CO_3, CH_2Cl_2	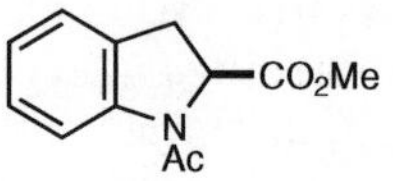	74	100	[79]
5	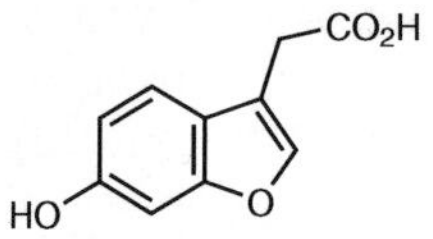	6.9	$[Rh(cod)_2]OTf$/**52**, NaOMe (0.1 equiv), MeOH	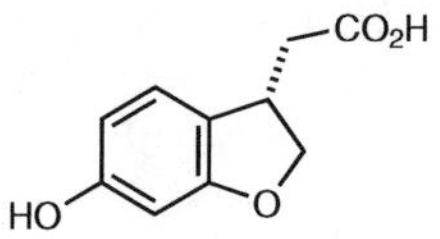	90	95	[80]

1,2,3,4-Tetrahydroquinolines 95; General Procedure:[76]
A carousel reaction tube containing a magnetic stirrer bar, Rh complex **58** or **59** (5 μmol), quinoline **94** (0.5 mmol), and HCO_2Na (0.34 g, 5 mmol) in an aqueous solution of AcOH/NaOAc (2 M, pH 5; 5 mL) was sealed without degassing and placed in a carousel reactor. The mixture was stirred at 40 °C for 6–24 h, and then cooled to rt and basified with aq KOH. The resulting mixture was extracted with Et_2O (3 × 5 mL). The extracts were dried ($MgSO_4$), filtered, and concentrated under reduced pressure. The product was purified by flash column chromatography.

1.4.2.2.5 Palladium Complexes as Catalysts

Palladium is often used in the heterogeneous hydrogenation of unsaturated double bonds such as alkenes, imines, and sometimes arenes. However, the homogeneous palladium-catalyzed asymmetric hydrogenation of hetarenes for a long time remained unexplored. It was not until 2010 that Zhou and Zhang reported the first enantioselective palladium-catalyzed hydrogenation of substituted indoles **96** in the presence of ligand (*R*)-2,2′-bis(diphenylphosphino)-5,5′,6,6′,7,7′,8,8′-octahydro-1,1′-binaphthyl [**45**; (*R*)-H_8-BINAP] and L-10-camphorsulfonic acid (CSA) to furnish indolines **97** (Scheme 28).[81]

Scheme 28 Palladium-Catalyzed Asymmetric Hydrogenation of Indoles[81]

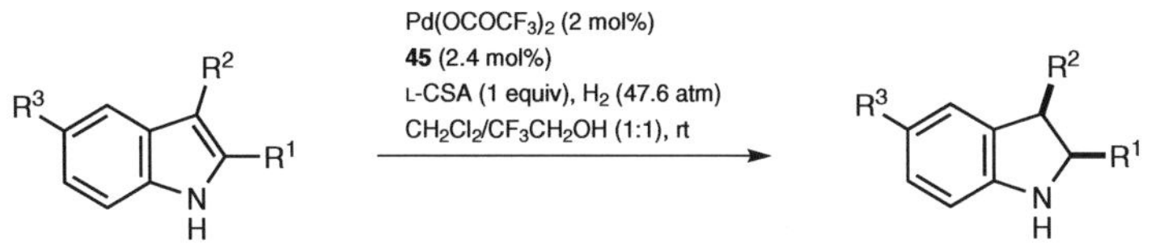

$Pd(OCOCF_3)_2$ (2 mol%)
45 (2.4 mol%)
L-CSA (1 equiv), H_2 (47.6 atm)
CH_2Cl_2/CF_3CH_2OH (1:1), rt

96 → **97**

R^1	R^2	R^3	ee[a] (%)	Yield (%)	Ref
Me	H	H	91	88	[81]
Bu	H	H	93	82	[81]
Cy	H	H	95	90	[81]
$(CH_2)_2Ph$	H	H	93	89	[81]
Bn	H	H	95	99	[81]
Me	H	F	84	81	[81]
Me	H	Me	96	83	[81]
Me	Me	H	92	84[b]	[81]
$(CH_2)_4$		H	91	91	[81]
$(CH_2)_5$		H	90	96	[81]

[a] Determined by HPLC analysis with a chiral stationary phase.
[b] H_2 (1.4 atm), 50 °C.

Subsequently, the Zhou group found that hydroxy (Table 8, entry 1)[82] and tosylamino groups (entry 2)[83] could be employed as the leaving group to form a similar vinylogous iminium intermediate in the palladium-catalyzed asymmetric hydrogenation, yielding *cis*-2,3-disubstituted indolines with high enantioselectivities. The Zhang group developed an axially chiral bisphosphine ligand **46** [(*S*)-C_{10} BridgePHOS] and successfully applied it to the palladium-catalyzed enantioselective hydrogenation of substituted indoles with excellent activity (entry 3).[84] Aside from indoles, other types of hetarenes can also be hydrogenated with palladium catalysts, such as 2,5-disubstituted pyrroles (entry 4),[85] 2-substituted 3-phthalimidoquinolines (entry 5),[86] and fluorinated pyrazol-5-ols (entry 6).[87]

for references see p 191

Table 8 Hydrogenation of Hetarenes Catalyzed by Phosphine Palladium Complexes[82–87]

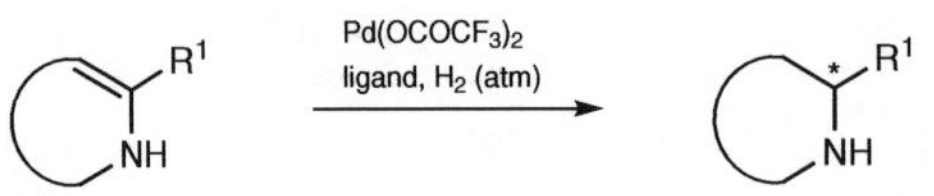

Entry	Substrate	Ligand	H_2 (atm)	Conditions	Product	ee (%)	Yield (%)	Ref
1	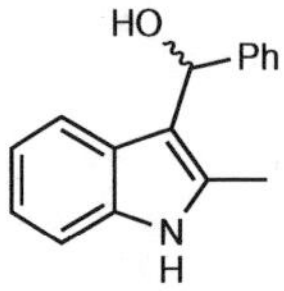	**45**	40.8	$TsOH{\bullet}H_2O$, CH_2Cl_2/CF_3CH_2OH	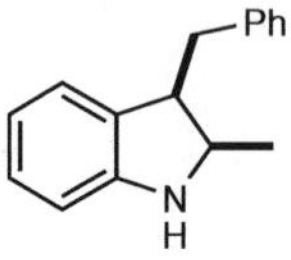	91	96	[82]
2	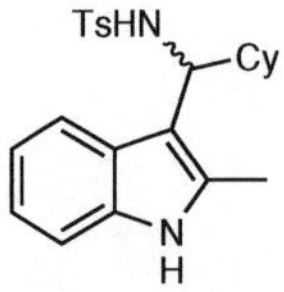	**45**	40.8	$TsOH{\bullet}H_2O$, CH_2Cl_2/CF_3CH_2OH	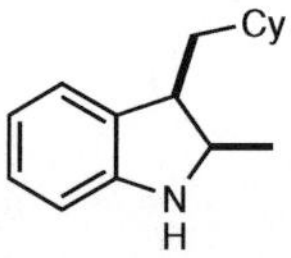	92	97	[83]
3	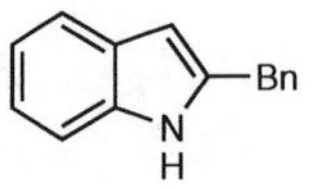	**46**	59.2	D-CSA	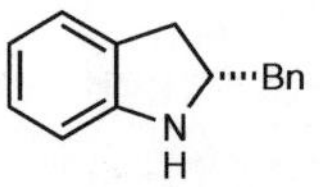	95	>95	[84]
4	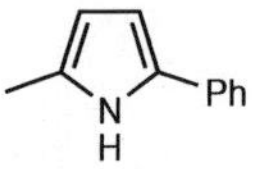	**47**	40.8	$EtSO_3H$, toluene/CF_3CH_2OH	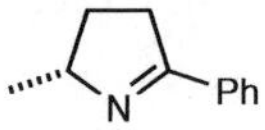	92	80	[85]
5	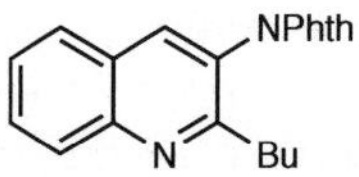	**40**	68.0	L-CSA, CH_2Cl_2/CF_3CH_2OH	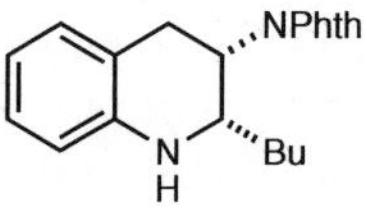	90	91	[86]
6	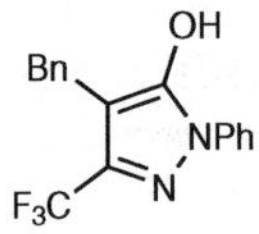	**48**	81.7	L-CSA, CF_3CH_2OH	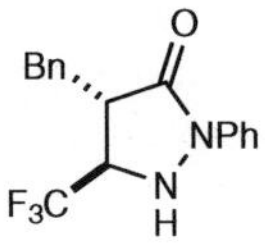	95	94	[87]

Combined experimental and theoretical efforts have revealed a plausible outer-sphere hydrogenation mechanism (Scheme 29).[88] Firstly, the in situ generated bisphosphine–palladium complex **98** is protonated, followed by dissociation of trifluoroacetic acid in the presence of hydrogen to form the palladium dihydrogen complex **99**. Subsequently, the active palladium hydride catalyst **100** is generated by an ionic mechanism with the assistance of trifluoroacetate. Then, the hydride of active catalyst **100** transfers to the protonated indole intermediate, which is generated in situ via protonation by the strong Brønsted acid trifluoroacetic acid; the reduced product is formed along with the coordinatively unsaturated palladium species **102**. The high enantioselectivity is ascribed to the hydrogen-bonding interaction between the NH of the iminium salt and the oxygen atom of the coordinated trifluoroacetate ligand in the eight-membered ring transition state **101** for hydride transfer. In addition, based on deuterium-labeling experiments, it was further elucidated that the stereoselectivity of the reaction arises from reversible protonation of the substrate. This process facilitates a dynamic kinetic resolution by which only the desired enantiomer of the protonated indole is hydrogenated.

Scheme 29 Proposed Mechanism for Palladium-Catalyzed Hydrogenation of Indoles[88]

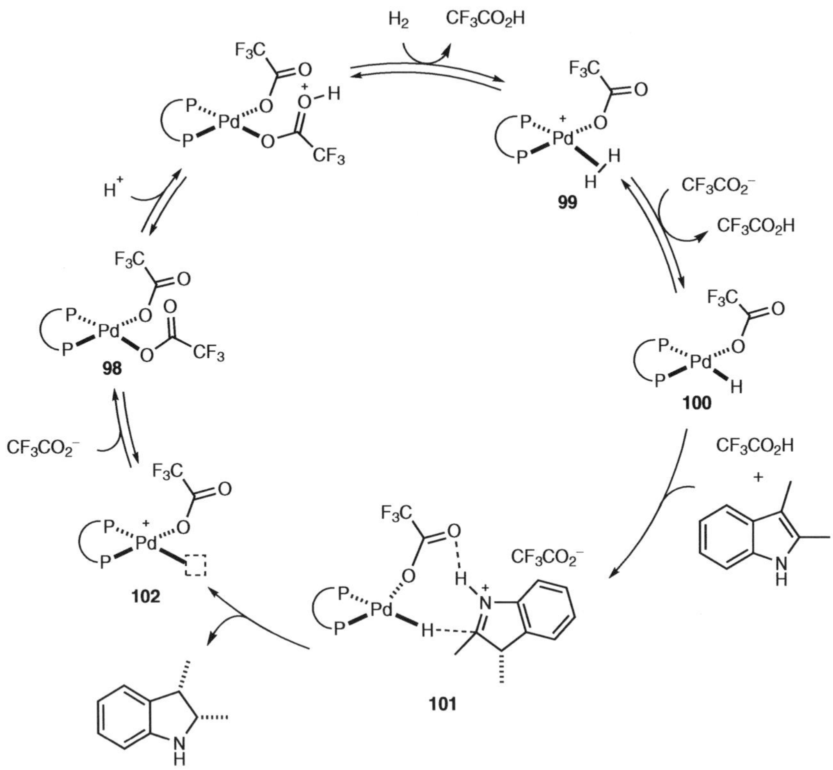

2,3-Dihydro-1*H*-indoles 97; General Procedure:[81]
(*R*)-2,2′-Bis(diphenylphosphino)-5,5′,6,6′,7,7′,8,8′-octahydro-1,1′-binaphthyl (**45**; 3.8 mg, 6 μmol) and $Pd(OCOCF_3)_2$ (1.7 mg, 5 μmol) were placed in a dried Schlenk tube under N_2 and degassed anhyd acetone was added. The mixture was stirred at rt for 1 h. The solvent was removed under reduced pressure to give the catalyst. In a glovebox, L-CSA (0.25 mmol) and substituted indole **96** (0.25 mmol) were stirred in a mixture of CH_2Cl_2 and CF_3CH_2OH (1:1; 1 mL mixed before use) at rt for 5 min. Subsequently, the above catalyst together with a mixture of CH_2Cl_2 and CF_3CH_2OH (1:1; 2 mL) were added. The hydrogenation was performed at rt under H_2 (47.6 atm) in a stainless-steel autoclave for 24 h. After carefully releasing the H_2, the resulting mixture was concentrated under reduced pressure. Sat. aq $NaHCO_3$ (5 mL) was added to the residue. The mixture was stirred for 10 min and extracted with CH_2Cl_2 (3 × 5 mL). The extracts were dried (Na_2SO_4), filtered, and concentrated. The product was purified by chromatography (silica gel, petroleum ether/EtOAc 10:1).

1.4.3 Conclusions and Future Perspectives

This chapter focused on recent advances in transition-metal-catalyzed reduction of hetarenes. Different types of hetarenes, including quinolines, isoquinolines, quinoxalines, pyridines, indoles, pyrroles, furans, thiophenes, imidazoles, and oxazoles, have been successfully subjected to such transformations, providing a facile and economic approach to a variety of heterocyclic compounds. However, despite significant progress, this field is still far from mature. There is a great diversity of hetarenes; however, reduction of a large number of these systems remains unexplored and the enantioselective hydrogenation of such compounds remains full of challenges. In addition, although a number of these methods have already been scaled up for large-scale production, most hetarene hydrogenations are still far removed from the efficiency required for this, especially for stereoselective processes. Given these challenges, the focus on further research in this area is expected to be directed at more efficient homogeneous and heterogeneous catalytic systems and new activation strategies to extend the substrate scope and enhance the catalytic efficiency. Overall, in view of the fast-paced development of catalysis chemistry, we believe that there will be more breakthroughs in the future.

References

[1] Zhou, Y.-G., *Acc. Chem. Res.*, (2007) **40**, 1357.
[2] Kuwano, R., *Heterocycles*, (2008) **76**, 909.
[3] He, Y.-M.; Fan, Q.-H., *Org. Biomol. Chem.*, (2010) **8**, 2497.
[4] Wang, D.-S.; Chen, Q.-A.; Lu, S.-M.; Zhou, Y.-G., *Chem. Rev.*, (2012) **112**, 2557.
[5] Gualandi, A.; Savoia, D., *RSC Adv.*, (2016) **6**, 18 419.
[6] Giustra, Z. X.; Ishibashi, J. S. A.; Liu, S.-Y., *Coord. Chem. Rev.*, (2016) **314**, 134.
[7] Lu, S.-M.; Zhou, Y.-G., In *Science of Synthesis: Stereoselective Synthesis*, de Vries, J. G., Ed.; Thieme: Stuttgart, (2011); Vol. 1, Section 1.6, pp 257–294.
[8] Ohno, M.; Tanaka, Y.; Miyamoto, M.; Takeda, T.; Hoshi, K.; Yamada, N.; Ohtake, A., *Bioorg. Med. Chem.*, (2006) **14**, 2005.
[9] Rahi, R.; Fang, M.; Ahmed, A.; Sánchez-Delgado, R. A., *Dalton Trans.*, (2012) **41**, 14 490.
[10] Gong, Y.; Zhang, P.; Xu, X.; Li, Y.; Li, H.; Wang, Y., *J. Catal.*, (2013) **297**, 272.
[11] Fache, F., *Synlett*, (2004), 2827.
[12] Ren, D.; He, L.; Yu, L.; Ding, R.-S.; Liu, Y.-M.; Cao, Y.; He, H.-Y.; Fan, K.-N., *J. Am. Chem. Soc.*, (2012) **134**, 17 592.
[13] Wang, X.; Lu, S.; Li, J.; Liu, Y.; Li, C., *Catal. Sci. Technol.*, (2015) **5**, 2585.
[14] Wang, X.; Li, J.; Lu, S.; Liu, Y.; Li, C., *Chin. J. Catal.*, (2015) **36**, 1170.
[15] Fang, M.; Machalaba, N.; Sánchez-Delgado, R. A., *Dalton Trans.*, (2011) **40**, 10 621.
[16] Bernabeu, M. C.; Díaz, J. L.; Jiménez, O.; Lavilla, R., *Synth. Commun.*, (2004) **34**, 137.
[17] Fray, M. J.; Allen, P.; Bradley, P. R.; Challenger, C. E.; Closier, M.; Evans, T. J.; Lewis, M. L.; Mathias, J. P.; Nichols, C. L.; Po-Ba, Y. M.; Snow, H.; Stefaniak, M. H.; Vuong, H. V., *Tetrahedron*, (2006) **62**, 6869.
[18] Wang, Q.; Lucien, E.; Hashimoto, A.; Pais, G. C. G.; Nelson, D. M.; Song, Y.; Thanassi, J. A.; Marlor, C. W.; Thoma, C. L.; Cheng, J.; Podos, S. D.; Ou, Y.; Deshpande, M.; Pucci, M. J.; Buechter, D. D.; Bradbury, B. J.; Wiles, J. A., *J. Med. Chem.*, (2007) **50**, 199.
[19] Maegawa, T.; Akashi, A.; Yaguchi, K.; Iwasaki, Y.; Shigetsura, M.; Monguchi, Y.; Sajiki, H., *Chem.–Eur. J.*, (2009) **15**, 6953.
[20] Dirat, O.; Clipson, A.; Elliott, J. M.; Garrett, S.; Jones, A. B.; Reader, M.; Shaw, D., *Tetrahedron Lett.*, (2006) **47**, 1729.
[21] Phillips, J. G.; Jaworska, M.; Lew, W., *Synthesis*, (2010), 714.
[22] Russo, O.; Alami, M.; Brion, J.-D.; Sicsic, S.; Berque-Bestel, I., *Tetrahedron Lett.*, (2004) **45**, 7069.
[23] Wallace, D. J.; Baxter, C. A.; Brands, K. J. M.; Bremeyer, N.; Brewer, S. E.; Desmond, R.; Emerson, K. M.; Foley, J.; Fernandez, P.; Hu, W.; Keen, S. P.; Mullens, P.; Muzzio, D.; Sajonz, P.; Tan, L.; Wilson, R. D.; Zhou, G.; Zhou, G., *Org. Process Res. Dev.*, (2011) **15**, 831.
[24] Fürstner, A.; Kennedy, J. W. J., *Chem.–Eur. J.*, (2006) **12**, 7398.
[25] Cheng, C.; Xu, J.; Zhu, R.; Xing, L.; Wang, X.; Hu, Y., *Tetrahedron*, (2009) **65**, 8538.
[26] Glorius, F.; Spielkamp, N.; Holle, S.; Goddard, R.; Lehmann, C. W., *Angew. Chem.*, (2004) **116**, 2910; *Angew. Chem. Int. Ed.*, (2004) **43**, 2850.
[27] Heitbaum, M.; Fröhlich, R.; Glorius, F., *Adv. Synth. Catal.*, (2010) **352**, 357.
[28] Dyson, P. J., *Dalton Trans.*, (2003), 2964.
[29] Dobereiner, G. E.; Nova, A.; Schley, N. D.; Hazari, N.; Miller, S. J.; Eisenstein, O.; Crabtree, R. H., *J. Am. Chem. Soc.*, (2011) **133**, 7547.
[30] Fujita, K.-i.; Kitatsuji, C.; Furukawa, S.; Yamaguchi, R., *Tetrahedron Lett.*, (2004) **45**, 3215.
[31] Wu, J.; Wang, C.; Tang, W.; Pettman, A.; Xiao, J., *Chem.–Eur. J.*, (2012) **18**, 9525.
[32] Tan, J.; Tang, W.; Sun, Y.; Jiang, Z.; Chen, F.; Xu, L.; Fan, Q.-H.; Xiao, J., *Tetrahedron*, (2011) **67**, 6206.
[33] Kuwano, R.; Sato, K.; Ito, Y., *Chem. Lett.*, (2000), 428.
[34] *Privileged Chiral Ligands and Catalysts*, Zhou, Q.-L., Ed.; Wiley-VCH: Weinheim, Germany, (2011).
[35] Balakrishna, B.; Núñez-Rico, J. L.; Vidal-Ferran, A., *Eur. J. Org. Chem.*, (2015), 5293.
[36] Chen, Z.-P.; Zhou, Y.-G., *Synthesis*, (2016) **48**, 1769.
[37] Wang, W.-B.; Lu, S.-M.; Yang, P.-Y.; Han, X.-W.; Zhou, Y.-G., *J. Am. Chem. Soc.*, (2003) **125**, 10 536.
[38] Wang, D.-W.; Wang, X.-B.; Wang, D.-S.; Lu, S.-M.; Zhou, Y.-G.; Li, Y.-X., *J. Org. Chem.*, (2009) **74**, 2780.
[39] Cai, X.-F.; Guo, R.-N.; Chen, M.-W.; Shi, L.; Zhou, Y.-G., *Chem.–Eur. J.*, (2014) **20**, 7245.
[40] Maj, A. M.; Suisse, I.; Méliet, C.; Hardouin, C.; Agbossou-Niedercorn, F., *Tetrahedron Lett.*, (2012) **53**, 4747.

[41] Lu, S.-M.; Wang, Y.-Q.; Han, X.-W.; Zhou, Y.-G., *Angew. Chem.*, (2006) **118**, 2318; *Angew. Chem. Int. Ed.*, (2006) **45**, 2260.
[42] Ye, Z.-S.; Guo, R.-N.; Cai, X.-F.; Chen, M.-W.; Shi, L.; Zhou, Y.-G., *Angew. Chem.*, (2013) **125**, 3773; *Angew. Chem. Int. Ed.*, (2013) **52**, 3685.
[43] Shi, L.; Ye, Z.-S.; Cao, L.-L.; Guo, R.-N.; Hu, Y.; Zhou, Y.-G., *Angew. Chem.*, (2012) **124**, 8411; *Angew. Chem. Int. Ed.*, (2012) **51**, 8286.
[44] Guo, R.-N.; Cai, X.-F.; Shi, L.; Ye, Z.-S.; Chen, M.-W.; Zhou, Y.-G., *Chem. Commun. (Cambridge)*, (2013) **49**, 8537.
[45] Tang, W.; Xu, L.; Fan, Q.-H.; Wang, J.; Fan, B.; Zhou, Z.; Lam, K.; Chan, A. S. C., *Angew. Chem.*, (2009) **121**, 9299; *Angew. Chem. Int. Ed.*, (2009) **48**, 9135.
[46] Mršić, N.; Jerphagnon, T.; Minnaard, A. J.; Feringa, B. L.; de Vries, J. G., *Adv. Synth. Catal.*, (2009) **351**, 2549.
[47] Wang, X.-B.; Zeng, W.; Zhou, Y.-G., *Tetrahedron Lett.*, (2008) **49**, 4922.
[48] Ye, Z.-S.; Chen, M.-W.; Chen, Q.-A.; Shi, L.; Duan, Y.; Zhou, Y.-G., *Angew. Chem.*, (2012) **124**, 10 328; *Angew. Chem. Int. Ed.*, (2012) **51**, 10 181.
[49] Chang, M.; Huang, Y.; Liu, S.; Chen, Y.; Krska, S. W.; Davies, I. W.; Zhang, X., *Angew. Chem.*, (2014) **126**, 12 975; *Angew. Chem. Int. Ed.*, (2014) **53**, 12 761.
[50] Chen, M.-W.; Ye, Z.-S.; Chen, Z.-P.; Wu, B.; Zhou, Y.-G., *Org. Chem. Front.*, (2015) **2**, 586.
[51] Huang, W.-X.; Yu, C.-B.; Ji, Y.; Liu, L.-J.; Zhou, Y.-G., *ACS Catal.*, (2016) **6**, 2368.
[52] Núñez-Rico, J. L.; Fernández-Pérez, H.; Vidal-Ferran, A., *Green Chem.*, (2014) **16**, 1153.
[53] Kuwano, R.; Hashiguchi, Y.; Ikeda, R.; Ishizuka, K., *Angew. Chem.*, (2015) **127**, 2423; *Angew. Chem. Int. Ed.*, (2015) **54**, 2393.
[54] Huang, W.-X.; Yu, C.-B.; Shi, L.; Zhou, Y.-G., *Org. Lett.*, (2014) **16**, 3324.
[55] Yamagata, T.; Tadaoka, H.; Nagata, M.; Hirao, T.; Kataoka, Y.; Ratovelomanana-Vidal, V.; Genêt, J.-P.; Mashima, K., *Organometallics*, (2006) **25**, 2505.
[56] Tadaoka, H.; Cartigny, D.; Nagano, T.; Gosavi, T.; Ayad, T.; Genêt, J.-P.; Ohshima, T.; Ratovelomanana-Vidal, V.; Mashima, K., *Chem.–Eur. J.*, (2009) **15**, 9990.
[57] Cartigny, D.; Berhal, F.; Nagano, T.; Phansavath, P.; Ayad, T.; Genêt, J.-P.; Ohshima, T.; Mashima, K.; Ratovelomanana-Vidal, V., *J. Org. Chem.*, (2012) **77**, 4544.
[58] Kita, Y.; Iimuro, A.; Hida, S.; Mashima, K., *Chem. Lett.*, (2014) **43**, 284.
[59] Iimuro, A.; Yamaji, K.; Kandula, S.; Nagano, T.; Kita, Y.; Mashima, K., *Angew. Chem.*, (2013) **125**, 2100; *Angew. Chem. Int. Ed.*, (2013) **52**, 2046.
[60] Baeza, A.; Pfaltz, A., *Chem.–Eur. J.*, (2010) **16**, 2036.
[61] Zhou, H.; Li, Z.; Wang, Z.; Wang, T.; Xu, L.; He, Y.; Fan, Q.-H.; Pan, J.; Gu, L.; Chan, A. S. C., *Angew. Chem.*, (2008) **120**, 8592; *Angew. Chem. Int. Ed.*, (2008) **47**, 8464.
[62] Qin, J.; Chen, F.; Ding, Z.; He, Y.-M.; Xu, L.; Fan, Q.-H., *Org. Lett.*, (2011) **13**, 6568.
[63] Zhang, J.; Chen, F.; He, Y.-M.; Fan, Q.-H., *Angew. Chem.*, (2015) **127**, 4705; *Angew. Chem. Int. Ed.*, (2015) **54**, 4622.
[64] Ma, W.; Chen, F.; Liu, Y.; He, Y.-M.; Fan, Q.-H., *Org. Lett.*, (2016) **18**, 2730.
[65] Wang, T.; Chen, F.; Qin, J.; He, Y.-M.; Fan, Q.-H., *Angew. Chem.*, (2013) **125**, 7313; *Angew. Chem. Int. Ed.*, (2013) **52**, 7172.
[66] Wang, T.; Zhuo, L.-G.; Li, Z.; Chen, F.; Ding, Z.; He, Y.; Fan, Q.-H.; Xiang, J.; Yu, Z.-X.; Chan, A. S. C., *J. Am. Chem. Soc.*, (2011) **133**, 9878.
[67] Kuwano, R.; Kashiwabara, M., *Org. Lett.*, (2006) **8**, 2653.
[68] Kuwano, R.; Kashiwabara, M.; Ohsumi, M.; Kusano, H., *J. Am. Chem. Soc.*, (2008) **130**, 808.
[69] Kuwano, R.; Kameyama, N.; Ryuhei, I., *J. Am. Chem. Soc.*, (2011) **133**, 7312.
[70] Ortega, N.; Urban, S.; Beiring, B.; Glorius, F., *Angew. Chem.*, (2012) **124**, 1742; *Angew. Chem. Int. Ed.*, (2012) **51**, 1710.
[71] Ortega, N.; Beiring, B.; Urban, S.; Glorius, F., *Tetrahedron*, (2012) **68**, 5185.
[72] Wysocki, J.; Ortega, N.; Glorius, F., *Angew. Chem.*, (2014) **126**, 8896; *Angew. Chem. Int. Ed.*, (2014) **53**, 8751.
[73] Urban, S.; Beiring, B.; Ortega, N.; Paul, D.; Glorius, F., *J. Am. Chem. Soc.*, (2012) **134**, 15 241.
[74] Ortega, N.; Tang, D.-T. D.; Urban, S.; Zhao, D.; Glorius, F., *Angew. Chem.*, (2013) **125**, 9678; *Angew. Chem. Int. Ed.*, (2013) **52**, 9500.
[75] Studer, M.; Wedemeyer-Exl, C.; Spindler, F.; Blaser, H.-U., *Monatsh. Chem.*, (2000) **131**, 1335.
[76] Wang, C.; Li, C.; Wu, X.; Pettman, A.; Xiao, J., *Angew. Chem.*, (2009) **121**, 6646; *Angew. Chem. Int. Ed.*, (2009) **48**, 6524.
[77] Kuwano, R.; Sato, K.; Kurokawa, T.; Karube, D.; Ito, Y., *J. Am. Chem. Soc.*, (2000) **122**, 7614.

[78] Kuwano, R.; Kaneda, K.; Ito, T.; Sato, K.; Kurokawa, T.; Ito, Y., *Org. Lett.*, (2004) **6**, 2213.
[79] Mršić, N.; Jerphagnon, T.; Minnaard, A. J.; Feringa, B. L.; de Vries, J. G., *Tetrahedron: Asymmetry*, (2010) **21**, 7.
[80] Yamashita, M.; Negoro, N.; Yasuma, T.; Yamano, T., *Bull. Chem. Soc. Jpn.*, (2014) **87**, 539.
[81] Wang, D.-S.; Chen, Q.-A.; Li, W.; Yu, C.-B.; Zhou, Y.-G.; Zhang, X., *J. Am. Chem. Soc.*, (2010) **132**, 8909.
[82] Wang, D.-S.; Tang, J.; Zhou, Y.-G.; Chen, M.-W.; Yu, C.-B.; Duan, Y.; Jiang, G.-F., *Chem. Sci.*, (2011) **2**, 803.
[83] Duan, Y.; Chen, M.-W.; Chen, Q.-A.; Yu, C.-B.; Zhou, Y.-G., *Org. Biomol. Chem.*, (2012) **10**, 1235.
[84] Li, C.; Chen, J.; Fu, G.; Liu, D.; Liu, Y.; Zhang, W., *Tetrahedron*, (2013) **69**, 6839.
[85] Wang, D.-S.; Ye, Z.-S.; Chen, Q.-A.; Zhou, Y.-G., *J. Am. Chem. Soc.*, (2011) **133**, 8866.
[86] Cai, X.-F.; Huang, W.-X.; Chen, Z.-P.; Zhou, Y.-G., *Chem. Commun. (Cambridge)*, (2014) **50**, 9588.
[87] Chen, Z.-P.; Chen, M.-W.; Shi, L.; Yu, C.-B.; Zhou, Y.-G., *Chem. Sci.*, (2015) **6**, 3415.
[88] Duan, Y.; Li, L.; Chen, M.-W.; Yu, C.-B.; Fan, H.-J.; Zhou, Y.-G., *J. Am. Chem. Soc.*, (2014) **136**, 7688.

1.5 Catalytic Reduction of Alkynes and Allenes

W. Bonrath, J. A. Medlock, and M.-A. Müller

General Introduction

Catalytic reductions are among the most important transformations for the chemical industry,[1] and in the field of alkyne and allene reduction the most widely used method is hydrogenation. Since the work of Sabatier approximately 150 years ago using dispersed metals for several catalytic reduction reactions, numerous processes have been developed and implemented in the fine chemical and pharmaceutical industries for the production of a wide variety of chemicals.[2]

The aim of this chapter is to provide an overview of the current best methods for the reduction of alkynes and allenes. The focus will be mainly on methods, processes, and catalysts developed since about 1995. However, a large number of "state-of-the-art" systems were developed many years ago, and are still widely practiced. Therefore these "original" methods will also be discussed in addition to their modern variants.

The least expensive reducing agent is hydrogen gas, and it is clearly favored for large-scale manufacturing. Despite the need for reactors capable of withstanding a pressurized gas and the safety precautions regarding the use of hydrogen, it is widely used in many laboratories throughout the world in a safe and practical manner. An alternative to handling gases are transfer-hydrogenation procedures, where a separate reducing agent is used. This can be converted by the catalyst into the same intermediate as in a hydrogenation reaction, or can operate via an alternative mechanism. Both hydrogenations and transfer hydrogenations will be considered in this section.

Outside of the scope of this review are transformations that proceed via a hydrometalation to an isolated intermediate and then undergo protodemetalation in a separate step. Reactions where this is believed to occur as part of the catalytic cycle but takes place in one process have been included.

The selective semi-hydrogenation of an alkyne or allene to an alkene can result in a number of possible alkene isomers in both the position and geometry (*E* or *Z*) of the resulting C=C bond. The selectivities of the various isomers will be indicated in the text and wherever possible a rational for their formation will be given. Cases where isomerization from one isomer to another also occurs under the reaction conditions will also be discussed. However, the deliberate isomerization of alkenes (both geometric and positional) is beyond the scope of this section.

1.5.1 Reduction of Alkynes

The hydrogenation of alkynes results in the formation of alkenes and/or alkanes. The formation of alkenes, the so-called selective hydrogenation or semi-hydrogenation, is of great industrial importance and needs special types of catalysts, e.g. Lindlar catalysts.[3] The catalyst is especially needed because the dissociation enthalpy of a hydrogen molecule is 434 $kJ \cdot mol^{-1}$.[4] Catalytic hydrogenations can be performed in a homogeneous or heterogeneous manner. Typical catalysts are based on nickel or palladium supported on carbon, silica, titanium(IV) oxide, barium sulfate (Rosenmund catalyst), or calcium carbonate (Lindlar catalyst). Furthermore, platinum, rhodium, ruthenium, cobalt, and iron find applications as hydrogenation catalysts. Copper chromite catalysts (the Adkins cata-

for references see p 225

lyst, $CuCr_2O_4$) are used for the reduction of C=O groups.[5] Zinc chromite ($ZnCr_2O_4$) is a suitable catalyst for the selective catalytic hydrogenation of α,β-unsaturated carbonyl compounds because the C=C bond is not attacked.[6] The hydrogenation of C=C bonds in a homogeneous manner is carried out in the presence of a rhodium catalyst, e.g. the Wilkinson catalyst [$RhCl(PPh_3)_3$].[7] This type of hydrogenation can be performed in an enantioselective manner using chiral ligands.[8]

For the selective hydrogenation of alkynes with heterogeneous catalysts, it is important to inhibit the active sites of the catalyst; non-inhibited catalytic hydrogenation results in alkane formation. The blockage of the active site can be performed by utilizing poisons, such as lead or bismuth, added during the catalyst preparation (e.g., the Lindlar catalyst) and/or the addition of thiols or nitrogen-containing molecules, such as quinoline, to the hydrogenation reaction mixture. The selective hydrogenation of an alkyne in the presence of a Lindlar catalyst stops at the formation of a Z-configured alkene.

1.5.1.1 Total Hydrogenation of Alkynes to Alkanes

In the presence of hydrogen, a non-poisoned heterogeneous palladium catalyst reduces alkynes to alkanes. This is a sequential reaction in which hydrogenation of the alkyne to the alkene is followed by its hydrogenation to the alkane. Here it must be pointed out that the alkane formation during alkyne hydrogenation can be avoided by partial hydrogenation, or blocking of the active sites (see Section 1.5.1.2). In general, the alkyne binds more strongly than the alkene to the catalyst surface; however, if the alkene is present in high concentrations then alkane formation cannot be avoided. A perfectly selective catalyst should first consume all of the alkyne before the hydrogenation of the alkene begins, or, even better, it stops at the alkene stage (Figure 1). However, this is not the case for real life applications. Normally the catalyst starts to consume the alkene before all alkyne is converted and the selectivity drops, especially at the end of the reaction, even for highly selective processes.[9,10]

Figure 1 Different Stages of a Perfect, Selective Alkyne Hydrogenation

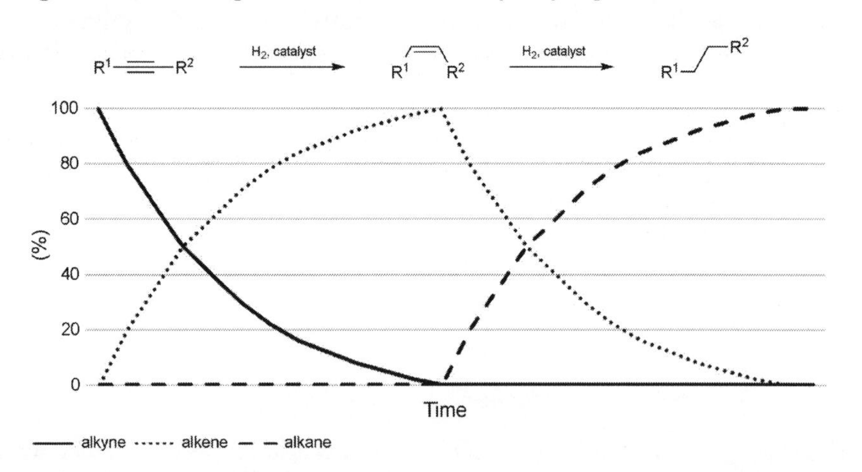

1.5.1.2 Selective Hydrogenation of Alkynes to Alkenes

The reduction of C≡C bonds to C=C bonds is a major topic in alkyne reduction and it is generally carried out by hydrogenation. Critical aspects are avoiding the uptake of a second hydrogen molecule, which results in over-hydrogenation, and the control of the temperature, since alkyne hydrogenation is exothermic (Table 1).[1,11,12]

Table 1 Heats of Hydrogenation of C≡C Bonds[1,11,12]

$R^1{-}C{\equiv}C{-}R^2 \xrightarrow{H_2,\ catalyst} R^1CH{=}CHR^2$

R^1	R^2	Heat of Formation (kJ)	Ref
H	H	177	[1]
Me	H	166	[1]
Me	Me	155	[1]
Bu	H	166	[12]
$(CH_2)_5Me$	H	177	[12]
Ph	H	277	[11]
Ph	Ph	250	[11]

Since the fundamental innovation of Herbert Lindlar, the so-called Lindlar catalyst (Pd-Pb/$CaCO_3$) has been the main choice for the semi-hydrogenation of alkynes.[3,13,14] The addition of modifiers, e.g. nitrogen- or sulfur-containing compounds such as quinoline, is frequently applied with the aim of achieving even greater selectivity.

Selective alkyne hydrogenation is particularly important in the bulk and fine chemical industries, especially in the synthesis of biologically active compounds such as carotenoids, insect pheromones, and vitamins.[15] Electronic and geometric features of the sp-hybridized C≡C bond influence the catalytic activity and selectivity in alkyne hydrogenation.[16] Therefore, the choice of the metal in alkyne hydrogenation is an enormous challenge. Some simplified rules guide the way to finding a suitable catalyst. In general, the order of reactivity is terminal alkynes > internal alkynes.[17]

The catalyst selectivity can be ordered as follows, based on a study for selective pent-2-yne hydrogenation: Pd > Rh > Pt > Ru > Os > Ir.[18] For catalyst activity, the following ranking can be given: Pd > Pt > Rh, Ni > Co > Ir.[19]

The selectivity in alkyne hydrogenation (decrease of over-hydrogenation) is also determined by the concentration of hydrogen, and thus can be influenced by the hydrogen pressure. High selectivity in alkyne hydrogenation can usually be obtained as long as alkyne is still present, because the alkyne is much more strongly bound to the active site of the catalyst than the corresponding alkene.[20] Thus, it is important to stop the reaction once all alkyne has been consumed. The reaction temperature is a key parameter that influences the selectivity in Lindlar-type hydrogenations; in general, decreasing the reaction temperature results in improved selectivity.[21] The role of the support and modifier in selective hydrogenations will be discussed separately.

The first industrial application of semi-hydrogenations in the synthesis of fine chemicals was in the preparation of vitamin A. A key intermediate is hydroxenin (**2**), which was originally prepared from the corresponding alkyne, oxenin (**1**) (Scheme 1).[14,22] The original method was reported by Isler and co-workers and used a palladium on carbon catalyst with an organic modifier/poison to reduce the amount of over-hydrogenation.[23] Although this transformation was successful and allowed the completion of the synthesis of vitamin A, determination of the end point of the reaction (i.e., mono-reduction) was difficult and required careful monitoring of the hydrogen consumption. A huge step forward was made by Lindlar[3,14] by replacing the palladium on carbon catalyst with a less active palladium on calcium carbonate catalyst that had been further modified by lead. Although the subsequent over-hydrogenation was not completely eliminated, its rate was significantly suppressed, resulting in higher selectivity and yield and easier reaction end-point determination. Further improvements were made with the introduction of an organic molecule as an additional modifier/poison,[3] for example quinoline.

for references see p 225

Scheme 1 Hydrogenation of Oxenin to Hydroxenin in the Synthesis of Vitamin A[3,14,22]

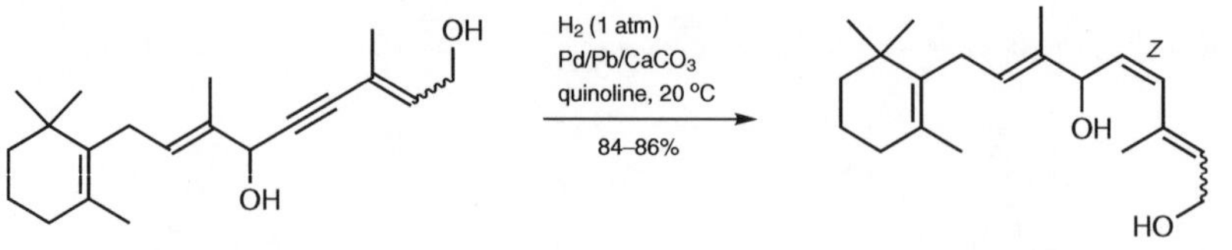

H_2 (1 atm)
Pd/Pb/$CaCO_3$
quinoline, 20 °C
84–86%

1 → 2

Herein, only the semi-hydrogenation of "functionalized" alkynes is discussed. There is also a significant amount of literature on the semi-hydrogenation of acetylene and propyne to give ethene and propene, respectively. This is critically important for the purification of alkene monomers for polymerization reactions;[24] however, these reactions usually take place in the gas phase and the catalysts and conditions used are significantly different to the liquid-phase reactions of larger, functionalized molecules. For further information about the semi-hydrogenation of simple alkynes, please consult a review article.[24–27]

This section has been structured by separating the catalysts and processes that usually result in Z-alkene products (Section 1.5.1.2.1) and those that are usually selective for *E*-alkene products (Section 1.5.1.2.2). Terminal alkynes, as no *E*/*Z*-isomers of the alkene products exist, have been included in the Z-selective section. It can be difficult to accurately determine if a semi-hydrogenation is actually *E* or *Z* selective, because, depending on the reaction conditions, it is possible to get isomerization of the *Z*-isomer to the more thermodynamically stable *E*-isomer. Where possible, hydrogenations that proceed via formation of one isomer followed by isomerization to another isomer have been indicated. From a mechanistic point of view, heterogeneous catalysts generally initially form the *Z*-isomer via *syn*-addition of hydrogen to the triple bond; however, exceptions do exist.

Lead-Modified Palladium on Calcium Carbonate Catalyst (Lindlar Catalyst):[3]

CAUTION: *The Lindlar catalyst is pyrophoric in the presence of solvents. General precautions for handling hydrogenation catalysts should be followed.*

To a mixture of $CaCO_3$ (50 g) and distilled H_2O (400 mL), a soln of $PdCl_2$ (5% Pd) was added and the mixture was stirred at ambient temperature for 5 min, followed by 10 min at 80 °C. The hot suspension was added to a shaken autoclave and shaken under a H_2 atmosphere (pressure not reported) until no more H_2 uptake was detected. The suspension was filtered, and the filter cake was washed thoroughly with distilled H_2O. The filter cake was suspended with distilled H_2O (500 mL), and $Pb(OAc)_2$ (5 g) in distilled H_2O (100 mL) was added. The suspension was stirred for 10 min at 20 °C, followed by 40 min at reflux. The catalyst was collected by filtration, thoroughly washed with distilled H_2O, and dried in vacuo at 40–45 °C.

A checked version of the procedure has also been reported in *Organic Syntheses* using sodium formate instead of hydrogen as the reducing agent.[13]

Hydroxenin (2); General Procedure for the Heterogeneous Z-Selective Semi-hydrogenation of Alkynes:[3]

CAUTION: *The Lindlar catalyst is pyrophoric in the presence of solvents. General precautions for handling hydrogenation catalysts should be followed.*

To a suspension of oxenin (**1**; 50 g) in petroleum ether (bp 80–120 °C; 100 mL) were added the Lindlar catalyst (5 g) and quinoline (2 g). The suspension was shaken under a H_2 atmosphere at 20 °C until almost no more additional H_2 was consumed, after an uptake of approximately 100–105% of the calculated H_2 amount. Oxenin (**1**) becomes soluble during the reaction whereas hydroxenin (**2**) begins to crystallize out. The mixture was warmed up so that the hydroxenin (**2**) dissolved, and the catalyst was filtered off and washed with petroleum ether. Crystallization and drying afforded the product as white plates; yield: 84–86%.

1.5.1.2.1 Semi-hydrogenations of Terminal Alkynes and *Z*-Selective Reduction of Internal Alkynes

1.5.1.2.1.1 Heterogeneous Hydrogenations

Several review articles have been published recently that summarize work on alkyne semi-hydrogenation. A 2012 review covered many catalyst developments for both simple hydrocarbons and also functionalized molecules.[28] An interesting review from 2013 focused on the synthesis of bioactive molecules and sought to examine the suitability and selectivity of a range of semi-reduction methods for industrial use.[29]

As discussed in Section 1.5.1.2, the first highly selective catalyst for alkyne semi-hydrogenation was developed by Lindlar for use in the preparation of vitamin A.[14,22] It used approximately 5 wt% palladium on a calcium carbonate support, doped with lead (usually 1–5 wt%). This was so successful that an *Organic Syntheses* procedure was published and it has become known as a "named catalyst/reaction", the "Lindlar catalyst".[13] Almost all the major precious metal catalyst manufacturers (Evonik, Johnson Matthey, Hindustan Platinum) offer their own versions of the Lindlar catalyst; all of them supply variations of palladium–lead on calcium carbonate. The exception to this rule is BASF, who previously supplied a traditional Lindlar catalyst, but have switched production to their NanoSelect catalyst (see below).

For all newly developed catalysts, their use has been examined with only a small number of substrates and their performance on industrially relevant molecules is generally unknown. The standard substrates are usually taken from the following classes (Scheme 2): hydrocarbons [e.g., hex-1-yne (**3**), hex-3-yne (**4**), phenylacetylene (**5**), diphenylacetylene (**6**)], hex-3-yn-1-ol (**7**), and propargylic alcohols [e.g., 2-methylbut-3-yn-2-ol (**8**)]. The industrially most relevant by volume are dehydroisophytol (**9**) and isophytol intermediates, oxenin (**1**), but-2-yne-1,4-diol (**10**), and aroma compounds [e.g., dehydrolinalool (**11**) in the linalool synthesis].[2,15]

Scheme 2 Typical Substrates Used in Alkyne Semi-hydrogenation Reactions[2,15]

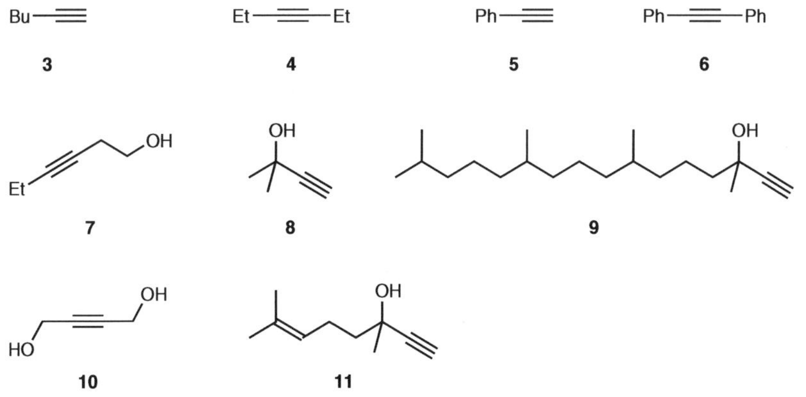

Bu—≡ **3** Et—≡—Et **4** Ph—≡ **5** Ph—≡—Ph **6**

Et, OH **7** OH **8** OH **9**

HO, OH **10** OH **11**

The standard Lindlar catalyst is prepared in a two-step procedure.[3,13] First, a palladium on calcium carbonate catalyst is prepared by reduction of a palladium(II) solution in the presence of calcium carbonate to produce a reduced palladium(0) catalyst. This intermediate

for references see p 225

catalyst is then resuspended in water and treated with an aqueous lead(II) salt, usually lead(II) acetate. After drying, the catalyst is ready to use without further reduction or calcination. (See Section 1.5.1.2 for an experimental procedure.)

The efficient applicability of these catalysts was demonstrated by Lindlar in the hydrogenation of oxenin (**1**) to hydroxenin (**2**), a key intermediate in the synthesis of vitamin A (see Scheme 1, and the experimental procedure in Section 1.5.1.2). Due to its success, there have been a number of investigations into the Lindlar catalyst, to determine why it is so selective and to find possible modifications to improve the selectivity further. In general, there is not a significant difference between these palladium catalysts regarding the relative hydrogenation rates of alkynes and alkenes. However, it is generally accepted that the high selectivities are mainly due to the relative binding coefficients between the substrate and the palladium surface, with alkyne binding in the presence of alkenes being significantly preferred until low alkyne concentrations are reached.[10,30]

The role of the additional metal (lead) poison has been investigated by Maier.[31] Characterization of the Lindlar catalyst has also been reported.[32,33] The use of metals other than lead has been investigated by a number of groups, with the most comprehensive investigations being those performed by Dev and co-workers.[34] The most selective are barium-, cadmium-, manganese-, and tin-doped catalysts; however, none of these have been commercialized. Related to these "doped" catalysts are specifically designed intermetallic catalysts where the two metals are added in known ratios to form stable alloys. A pioneer in this area has been Armbrüster, who has mainly worked on acetylene/propyne.[35] For more functionalized molecules, some success has been achieved with palladium–bismuth intermetallic species, including their use in flow semi-hydrogenation systems.[36,37] More advances in this area can be expected.

In addition to changing the second metal, a number of other variations to catalyst formulation have been investigated. Variations in the palladium deposition method, including the use of poly(1-vinylpyrrolidin-2-one) (PVP) to control the palladium nanoparticle size, have been investigated by an Evonik group.[38] Improvements in the hydrogenation of hex-2-yne were studied, but the catalysts and conditions were not tried with other substrates. The particle size of the calcium carbonate support was found to have a significant effect on the semi-hydrogenation selectivity for an industrially relevant internal alkyne.[39] Controlling the size and shape of the palladium nanoparticles is another area that has been heavily investigated. Some success has been achieved by the group of Kiwi-Minsker for the hydrogenation of 2-methylbut-3-yn-2-ol (**8**).[40] The activity of the semi-hydrogenation depends on the coordination number of the palladium atoms, with edge atoms fourfold less active compared to plane atoms. Over-hydrogenation is believed to only occur on the edge atoms. The selectivity could then be linked to the fraction of edge sites on each type of palladium nanocrystal.

The use of nonconventional heating and dispersion techniques for the synthesis of selective reduction catalysts (including semi-hydrogenation) has been reviewed.[41]

The only other commercial catalysts designed for the semi-hydrogenation of alkynes to alkenes (apart from the Lindlar catalyst) are the palladium NanoSelect catalysts developed by BASF. These catalyst were developed to be an alternative to the Lindlar catalyst that are lead-free but still offer equivalent performance.[42–44] To obtain high selectivity, the size of the palladium nanoparticles is controlled using a surfactant/organic modifier in the catalyst preparation. The surfactant chosen, hexadecyl(2-hydroxyethyl)dimethylammonium dihydrogen phosphate (HHDMA), combines this function with a reducing group so that the palladium(II) species used in the catalyst preparation is both reduced and "templated" at the same time. The palladium nanoparticles formed are in the range of 4–8 nm, with a very narrow size distribution. The supports used are activated carbon and titanium silicate (TiS). Metal loadings are generally 0.5 wt%, which can be considered an additional advantage, due to the high cost of the precious metal.

Excellent results were obtained for the semi-hydrogenation of hex-3-yn-1-ol (**7**) to (Z)-hex-3-en-1-ol [(Z)-**12**], which is an important fragrance compound with a fresh grassy aroma (Scheme 3).[42,44,45] Additionally, other alkyne substrates were tested giving generally >95% alkene selectivity with high Z-selectivity. For certain substrates, the hydrogen uptake decreased to almost zero after consumption of 1 equivalent of hydrogen (i.e., complete consumption of the starting alkyne; Figure 2).

Scheme 3 Hydrogenation of Hex-3-yn-1-ol to (Z)-Hex-3-en-1-ol[42,44]

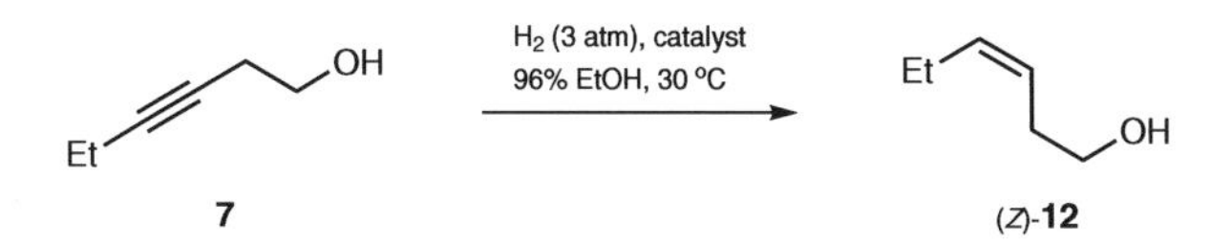

Catalyst	Conversion (%)	Yield (%) of (E/Z)-**12**	Yield (%) of (Z)-**12**	Ref
Pd nanoparticles/TiS	97	99	97	[44]
Lindlar	>99	99	97	[44]

Figure 2 Hydrogen Consumption Curves for the Hydrogenation of Hex-3-yn-1-ol: (a) Pd/C; (b) Pd/C Treated with HHDMA; (c) Pd/TiS NanoSelect Catalyst; (d) Lindlar Catalyst[44]

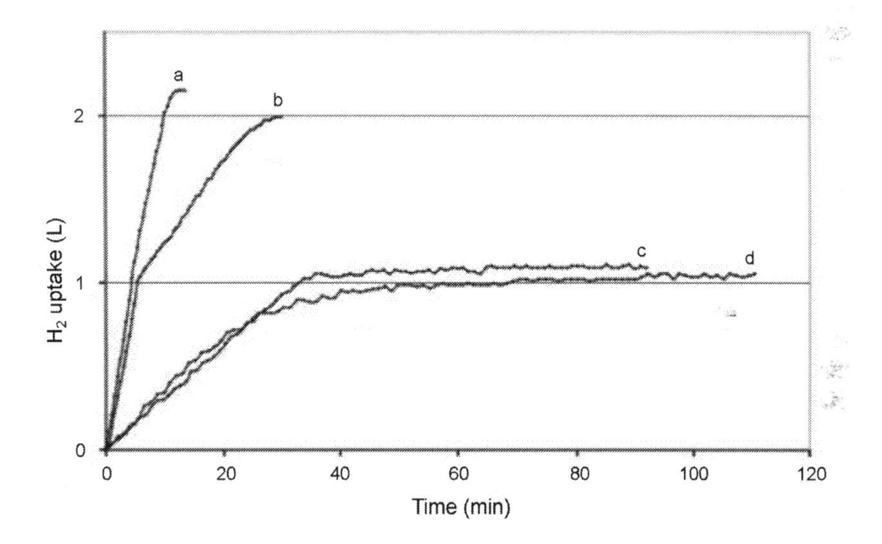

The origin of the selectivity is believed to be due to a combination of the surfactant/organic stabilizer and location of all the palladium nanoparticles on the surface of the support ("egg-shell" distribution).[46] Further investigation of this catalyst system by the Pérez-Ramírez group[47] included the immobilization of the palladium nanoparticles on a monolith for continuous semi-hydrogenations.[48] It is not known if any industrial processes have been implemented using the NanoSelect catalyst to date (2017).

In the field of non-palladium-based catalysts, two systems stand out: (1) the use of cerium oxide catalysts for selective alkyne hydrogenation, and (2) silica-supported copper nanoparticles for alkyne semi-hydrogenation. The cerium oxide catalysts were initially used successfully for acetylene/propyne semi-hydrogenation.[49] However, the use of 20% cerium(IV) oxide supported on titanium(IV) oxide was successful for a range of hydrocarbons with high selectivity (>99%) at 30–100% conversion.[50] More functionalized mol-

for references see p 225

ecules, for instance those with hydroxy groups, give high selectivity, but lower conversion (generally 50%). These cerium(IV) oxide catalysts are significantly less active than palladium catalysts and more forcing conditions (e.g., ~90 atm of H_2, 140 °C) are required.

An examination of narrowly dispersed, silica-supported copper nanoparticles for alkyne semi-hydrogenation[51] utilized a high-throughput testing system to screen a range of ligands (phosphines, carbenes, amines), with ten alkyne substrates studied initially. Tricyclohexylphosphine was the best modifier, with generally high conversion and good selectivity (>90%; ~50 atm of H_2, toluene, 60 °C) (Scheme 4).

Scheme 4 Reduction of Alkynes to Alkenes with a Phosphine-Modified Copper Catalyst[51]

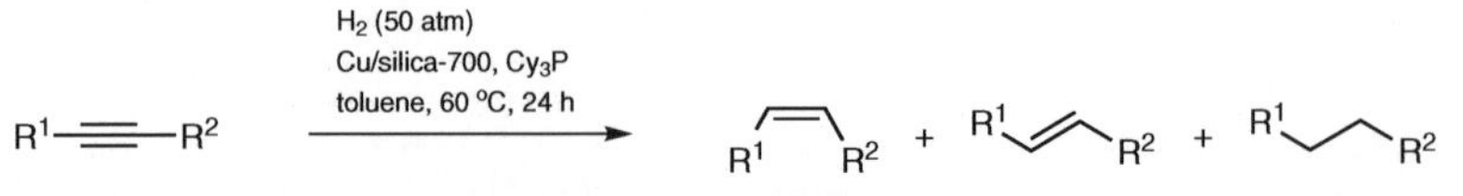

R^1—≡—R^2 → (H2 (50 atm), Cu/silica-700, Cy3P, toluene, 60 °C, 24 h) → (Z)-13 + (E)-13 + 14

R1	R2	Conversion[a] (%)	Selectivity[a] (%)			Ref
			(Z)-13	(E)-13	14	
Bu	Bu	100	99	–	–	[51]
Ph	Bu	100	92	–	–	[51]
Ph	Ph	100	95	1	–	[51]
Ph	1-naphthyl	100[b,c]	94[c]	–	–	[51]
Ph	CO_2Et	100	56	12	32	[51]

[a] Determined by GC.
[b] Reaction temperature was 80 °C.
[c] Determined by NMR.

Almost all of the systems described so far have operated in batch mode. A more recent concept and a highly active area of research is the design of semi-hydrogenation reactors where the catalyst has been coated on the walls.

The most promising approaches are those developed by Kiwi-Minsker using palladium on sintered metal fibers,[52,53] and designed porous structures reported by Rudolf von Rohr.[54,55] Both of these approaches use a metal reactor/support with an inorganic base layer, usually a mixture of zinc oxide and alumina. The active palladium nanoparticles are then prepared and deposited on the support. After activation/calcination, both catalyst systems give long-term high selectivity in the semi-hydrogenation of 2-methylbut-3-yn-2-ol (**8**). Promising results have also been obtained using palladium–intermetallic complexes coated on capillaries.[36,37]

(Z)-Hex-3-en-1-ol [(Z)-12]; Typical Procedure:[42,44]
A 250-mL stainless steel autoclave was charged with Pd/C or Pd nanoparticles/TiS (50 mg) and a soln of hex-3-yn-1-ol (5 mL) in 96% EtOH (100 mL), and the mixture was heated to 30 °C. Without stirring, the autoclave was flushed with H_2 and pressurized with H_2 (3 atm). The reaction was started by stirring (1500 rpm). Using the Lindlar catalyst, the procedure was the same except there was a 15 min pre-hydrogenation step.

1.5.1.2.1.2 Heterogeneous Transfer Hydrogenations

Very few transfer-hydrogenation processes have been reported for the selective semi-hydrogenation of alkynes with heterogeneous catalysts. One of the few reports is the use of commercial palladium on carbon catalysts with ammonium hypophosphite monohydrate ($H_2PO_2NH_4{\cdot}H_2O$) as reductant.[56] A small number of acetylenes bearing at least one phenyl group are successfully reduced to the alkene with 90–99% selectivity; however, a significant amount of isomerization from the *Z*- to the *E*-isomer is observed.

Heck has investigated the use of triethylammonium formate as a hydrogen donor for the reduction of acetylenes (and allylic amines).[57] Eight alkynes were studied, and moderate to good yields of *Z*-alkenes were obtained in five cases, e.g. the formation of (*Z*)-**15** and **18**. A small excess (3–14%) of formate is used. Occasionally, isomerization of the *Z*- to *E*-alkene is observed, e.g. formation of the mixture (*Z*/*E*)-**17**, or the completely reduced compound (e.g., **16**) is obtained as the sole product. Selected results are shown in Scheme 5.

Scheme 5 Transfer Hydrogenation of Alkynes Using Triethylammonium Formate[57]

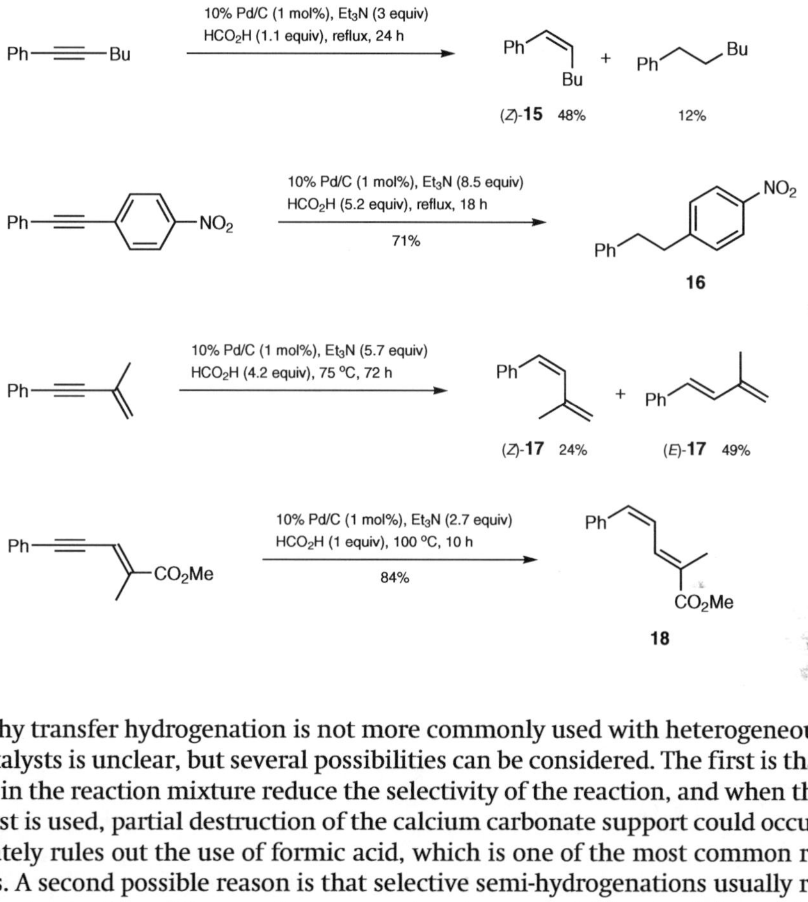

10% Pd/C (1 mol%), Et_3N (3 equiv), HCO_2H (1.1 equiv), reflux, 24 h
(*Z*)-**15** 48%; 12%

10% Pd/C (1 mol%), Et_3N (8.5 equiv), HCO_2H (5.2 equiv), reflux, 18 h; 71%
16

10% Pd/C (1 mol%), Et_3N (5.7 equiv), HCO_2H (4.2 equiv), 75 °C, 72 h
(*Z*)-**17** 24%; (*E*)-**17** 49%

10% Pd/C (1 mol%), Et_3N (2.7 equiv), HCO_2H (1 equiv), 100 °C, 10 h; 84%
18

The reason why transfer hydrogenation is not more commonly used with heterogeneous palladium catalysts is unclear, but several possibilities can be considered. The first is that acids present in the reaction mixture reduce the selectivity of the reaction, and when the Lindlar catalyst is used, partial destruction of the calcium carbonate support could occur. This immediately rules out the use of formic acid, which is one of the most common reducing agents. A second possible reason is that selective semi-hydrogenations usually require careful optimization of the reaction parameters to ensure high selectivity, which could be incompatible with the reducing agent (either by instability, or hindering the transfer of hydrogen to the catalyst surface). In contrast, hydrogen gas can easily be regulated via pressure and generally only one side-reaction has to be considered (over-reduction). A final reason is that the moderately active catalysts needed for selective semi-hydrogenation are not active with most reducing agents. A report on the use of palladium–sepiolite catalysts stated that cyclohexene was not a suitable hydrogen transfer reagent, possibly due to the substrate being more strongly bound to the catalyst than cyclohexene.[58]

In terms of systems that have been reported that do not involve palladium catalysts, unsupported nanoporous gold has been used with both organosilanes/water and formic

for references see p 225

acid for the selective semi-hydrogenation of alkynes.[59,60] Using organosilanes, dimethyl(phenyl)silane is the most effective with either dimethylformamide or pyridine/acetonitrile as solvent. With 2 mol% of AuNPore catalyst, full conversion is achieved in 1–15 hours with high yields (generally >90%) and exclusive formation of the *Z*-alkene. With formic acid and catalytic triethylamine in dimethylformamide, high yields (generally >95%) are obtained, together with *Z*-alkene selectivity of >98%. Although the catalyst is not simple to prepare, it could be reused for at least 5 runs without loss in activity.

Alkenes, e.g. 15, 17, **and** 18; **General Procedure:**[57]
The substrate, 10% Pd/C (1 mol%), and Et_3N (2.7–14.2 equiv) were added to a flask equipped with a reflux condenser and a $CaCl_2$-filled drying tube. The mixture was stirred and 97% HCO_2H (1.0–5.2 equiv) was added slowly at rt. The reaction was stirred at the indicated temperature until GLC analysis showed that all the substrate had been reduced or no further reduction was taking place. Products were obtained by filtering the catalyst from the reaction soln and washing the solid with Et_2O (2 ×). The filtrate and Et_2O washings were combined, washed with distilled H_2O, dried ($MgSO_4$), and concentrated to give the crude product.

1.5.1.2.1.3 Homogeneous Hydrogenations

Besides heterogeneous semi-hydrogenation catalysts (Section 1.5.1.2.1.1), which have dominated the field of alkyne semi-hydrogenation from the very beginning, homogeneous catalysts have also emerged as potential alternatives.[61] Several catalytic systems based on different metals have been developed since the 1980s, and show promising results in the hydrogenation of internal alkynes, resulting in the formation of the *Z*- or *E*-alkene. However, the application of homogeneous systems (especially industrial application) for semi-hydrogenation is relatively rare when compared to the use of the well-established heterogeneous catalysts.

Homogeneous alkyne semi-hydrogenation with the goal of obtaining *Z*-configured alkenes can be performed in several ways. Cationic rhodium-based complexes that have the general formula $[Rh(nbd)L_n]^+X^-$ (nbd = norbornadiene), with L being a phosphine and X^- a weakly coordinating counterion, are among the most prominent catalysts that allow for the *Z*-selective hydrogenation of internal alkynes with high stereoselectivity.[62,63] In the hydrogenation of hex-2-yne, up to 99:1 ratio in favor of (*Z*)-hex-2-ene over (*E*)-hex-2-ene is observed, and hex-1-yne is also converted with high selectivity into hex-1-ene. The presence of an acid is beneficial in certain cases to suppress hydrogenation of the alkene. The selectivity depends directly on the hydrogenation rate of the alkyne and alkene. Best results are obtained by applying $Rh(nbd)(PMe_2Ph)_3^+$ with ClO_4^- or PF_6^- as the counterion, and by conducting the reaction in acetone or 2-methoxyethanol. The reaction is assumed to proceed in an analogous fashion to hydrogenations of C=C bonds in coordinating solvents, via mono- and dihydride complexes as the catalytically active species.[62] In weakly coordinating solvents, these catalysts are normally less effective and show different reaction rates, leading to the postulation of a different mechanism in which the diene (i.e., nbd) remains coordinated to the metal center during the course of the reaction.[64] Furthermore, various strategies have been developed for the recovery of such catalysts, such as the use of fluorinated ligands in a fluorous biphasic system,[65] or by the immobilization of the catalyst on a solid support.[66] (The latter approach was realized for catalysts that generate the alkane from the alkyne.) The application of these catalysts in a numbered-up microchannel reactor has also been demonstrated.[67]

Heller and co-workers reported that the use of a rhodium–TangPhos-derived trinuclear hydride species at 1 atmosphere achieves full conversion with a substrate/catalyst ratio of up to 5000. For example, diphenylacetylene (**6**) undergoes semi-hydrogenation to give (*Z*)-stilbene [(*Z*)-**19**] and (*E*)-stilbene [(*E*)-**19**], but no over-reduction product 1,2-diphenylethane (**20**) (Scheme 6).[68] Good Z selectivity, with little or no over-reduction, is obtained in methanol or tetrahydrofuran. However, the reactions of only a few different substrates using this catalyst were reported; more complex substrates such as 1,4-diphenylbuta-1,4-diyne showed reduction and subsequent dimerization.

Scheme 6 Semi-hydrogenation of Diphenylacetylene Using a Rhodium–TangPhos Catalyst[68]

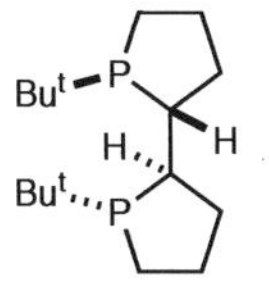

TangPhos

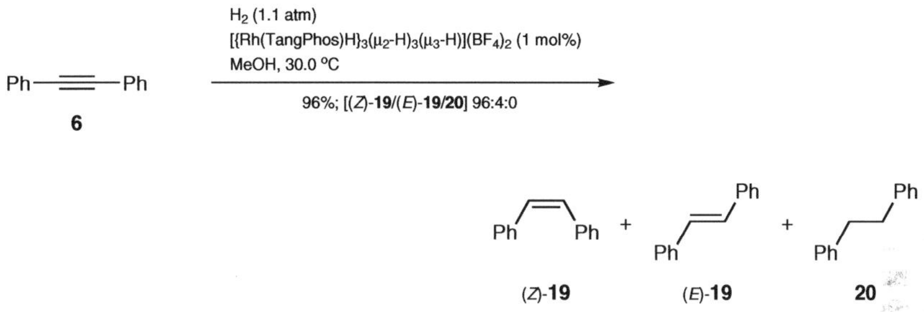

H_2 (1.1 atm)
[{Rh(TangPhos)H}$_3$(μ_2-H)$_3$(μ_3-H)](BF$_4$)$_2$ (1 mol%)
MeOH, 30.0 °C
96%; [(*Z*)-**19**/(*E*)-**19**/**20**] 96:4:0

Ph–≡–Ph **6** → (*Z*)-**19** + (*E*)-**19** + **20**

A different palladium-based catalyst has been utilized for the Z-selective hydrogenation of internal alkynes, as reported by Elsevier and co-workers. Thus, homogeneous palladium(0) complex **21**, bearing a bidentate diimine ligand, shows excellent results for dialkyl-substituted and terminal alkynes under mild reaction conditions (Scheme 7).[69] Enynes are also converted into the corresponding dienes with excellent selectivity. High chemoselectivity is obtained in the presence of a variety of functionalities, such as nitro or alkoxycarbonyl groups. In several cases, the results are superior to the tested Lindlar catalysts and exhibit typical turnover frequencies (TOFs) of 100–200 mol•mol^{-1}•h^{-1}. Aryl-substituted internal alkynes lead to lower selectivities.[70,71]

Scheme 7 Palladium(0)-Catalyzed Hydrogenation of Alkynes To Give Z-Alkenes[69]

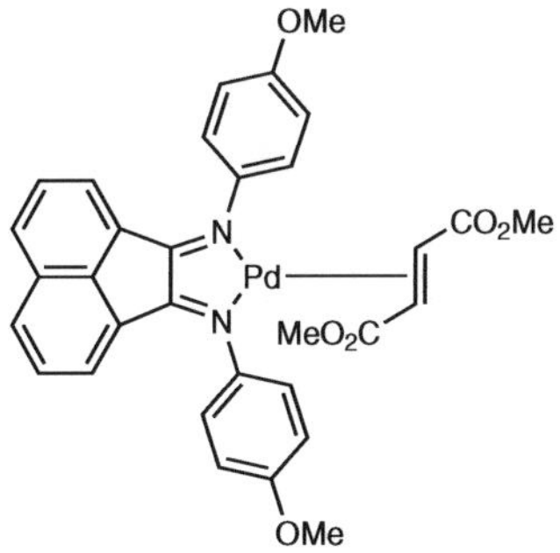

OMe
N
Pd
N
CO_2Me
MeO_2C
OMe
21

for references see p 225

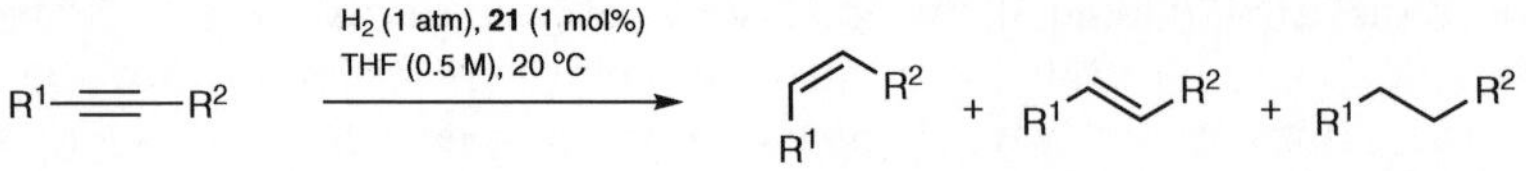

H_2 (1 atm), **21** (1 mol%)
THF (0.5 M), 20 °C

R^1—≡—R^2 → (*Z*)-**22** + (*E*)-**22** + **23**

R^1	R^2	Yield (%)			Ref
		(*Z*)-**22**	(*E*)-**22**	**23**	
Bu	H	100	–	–	[69]
Ph	H	>99[a]	–	–	[69]
Et	Et	>99	–	–	[69]
$(CH_2)_2OH$	Et	>99	–	–	[69]
Ph	Me	92	2	6	[69]
Ph	Ph	87	–	13	[69]
cyclohex-1-enyl	H	100	–	–	[69]

[a] D_2 was used instead of H_2.

A detailed kinetic study with the more reactive palladium/N^1,N^2-bis[3,5-bis(trifluoromethyl)phenyl]acenaphthylene-1,2-diimine/maleic anhydride complex revealed turnover frequencies of up to 16000 for oct-4-yne.[72] Kinetic studies have shown that only mononuclear species are involved in product formation. An alternative palladium/pyridine-2-carbaldimine [Pd(pyca)] complex shows Z selectivity that is as high as the previously described systems; however, these catalysts are less stable, and decomposition was observed in several cases before full conversion of 1-phenylprop-1-yne was reached.[73]

Palladium(0)–N-heterocyclic carbene catalysts **25** have also been utilized for the Z-selective hydrogenation of internal alkynes.[74] These catalysts show superior results for the hydrogenation of 1-phenylprop-1-yne, with selectivities of up to 95% for (Z)-1-phenylprop-1-ene (**26**) (Scheme 8). However, with other substrates such as oct-4-yne less than 1% conversion is observed, and the hydrogenation of, for example, diphenylacetylene results in lower selectivity compared to the previously published systems: (Z)-stilbene (88%), (*E*)-stilbene (5%), and 1,2-diphenylethane (7%).

Scheme 8 *Z*-Selective Hydrogenation of an Internal Alkyne with a Palladium(0)–N-Heterocyclic Carbene Catalyst[74]

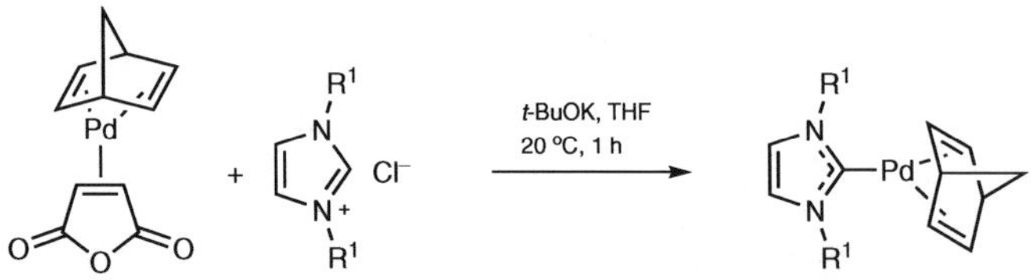

24 + imidazolium chloride (Cl^-) → *t*-BuOK, THF, 20 °C, 1 h → **25**

R^1 = Mes, 2-Tol, 2,6-$iPr_2C_6H_3$, 2,6-$Et_2C_6H_3$

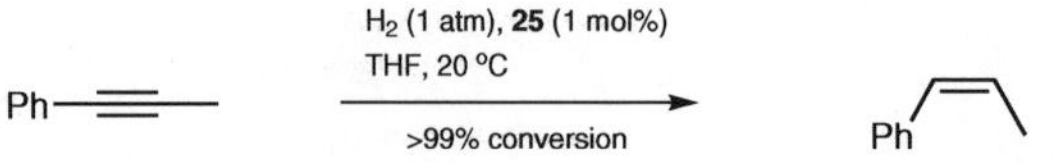

Ph—≡— → H_2 (1 atm), **25** (1 mol%), THF, 20 °C; >99% conversion → **26** up to 95% (*Z*)-alkene

The *Z*-selective hydrogenation of internal alkynes has been accomplished by Matsubara and co-workers using (5,10,15,20-tetraphenylporphyrinato)palladium(II) (**27**) as catalyst.

Unique reaction conditions, with pyridine as solvent and 4-(dimethylamino)pyridine as an additive, allow the hydrogenation of a wide range of functionalized substrates.[75] In general, high selectivity for the *Z*-alkene **28** is obtained, with little over-reduction (Scheme 9). The nature of the catalytically active palladium species is unknown. Therefore, a heterogeneous system, for example involving pyridine-stabilized nanoparticles, cannot be ruled out.[76]

Scheme 9 Palladium–Porphyrin-Catalyzed Hydrogenation of Selected Internal Alkynes[75]

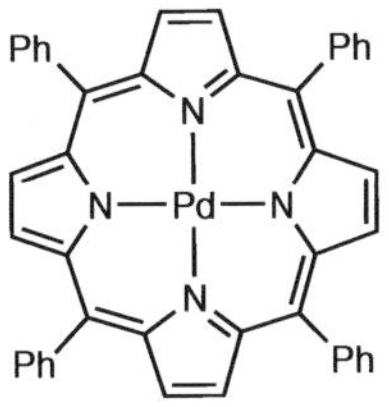

Ph Ph N N—Pd—N N Ph Ph

27

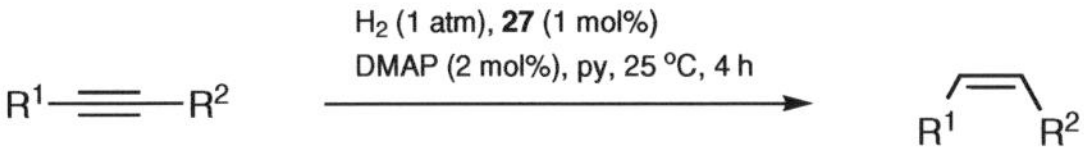

R^1—≡—R^2 → (H_2 (1 atm), **27** (1 mol%), DMAP (2 mol%), py, 25 °C, 4 h) → R^1 R^2 **28**

R^1	R^2	Yield (%)	Ref
Ph	Ph	95[a]	[75]
Ph	CO_2Et	99	[75]
Ph	cyclopropyl	82	[75]
Ph	$(CH_2)_4Me$	85	[75]
$(CH_2)_5Me$	$(CH_2)_5Me$	89[b]	[75]
Bu	$(CH_2)_2OTs$	81	[75]
Bu	$(CH_2)_2OAc$	99	[75]

[a] Together with (*E*)-stilbene (4%) and 1,2-diphenylethane (<1%).
[b] Reaction time was 2 h.

(Arene)tricarbonylchromium(0) complexes are among the most selective catalyst classes for the *Z*-selective hydrogenation of internal alkynes **31**.[77] However, rather forcing conditions with up to 120 °C, 69 atm hydrogen pressure, and 20 mol% catalyst loading are applied (Scheme 10). Excellent selectivity is achieved for nonfunctionalized molecules containing internal triple bonds. These catalysts also show reactivity in the hydrogenation of specific double bonds, such as in Michael acceptors or in the mono-hydrogenation of dienes.[78]

Scheme 10 Semi-hydrogenation of Internal Alkynes Using Chromium-Based Catalysts[77,78]

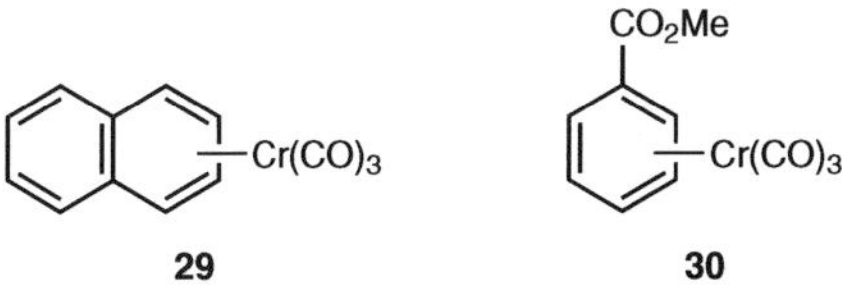

$Cr(CO)_3$ **29**

CO_2Me $Cr(CO)_3$ **30**

for references see p 225

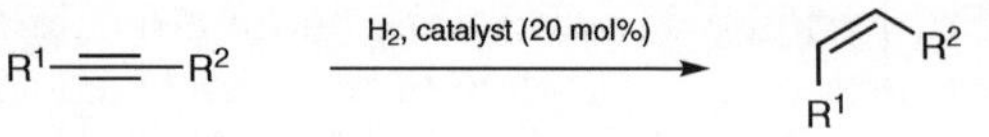

R^1	R^2	Catalyst	H_2 (atm)	Conditions	Yield (%)	Ref
Ph	Me	**29**	20	acetone, 45 °C, 24 h	92	[77,78]
$(CH_2)_5Me$	$(CH_2)_5Me$	**30**	69	THF, 120 °C, 15 h	100	[77,78]
Bu	CH_2OH	**30**	69	THF, 120 °C, 8 h	95	[77,78]

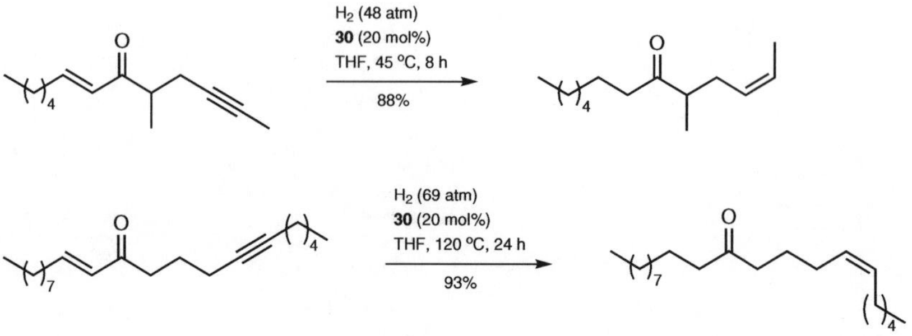

Various copper catalysts have been applied by several research groups for the Z-selective hydrogenation of internal alkynes. These catalyst do not tolerate the presence of acidic protons, and no product is observed for terminal alkynes.[79–81] The use of N-heterocyclic carbene based tethered copper alkoxide complexes, which are generated using butyllithium in the reaction medium prior to alkyne addition, have been reported by Teichert and co-workers. Scheme 11 shows an example for the Z-selective semi-hydrogenation of diphenylacetylene (**6**) using a complex derived from mesitylcopper(I) and 2-hydroxyethyl-tethered N-heterocyclic carbene precursor **32** to give (Z)-stilbene [(Z)-**19**].[79] A variety of functional groups are tolerated, and in general high Z-selectivity is observed using 5–10 mol% catalyst loading in tetrahydrofuran at 40 °C. In most cases, nearly full conversion and yields of up to 93% are achieved. A drawback of this method is the high hydrogen pressure (~100 atm) that is required.

Scheme 11 Semi-hydrogenation of Diphenylacetylene Using a Copper–Carbene Catalyst[79]

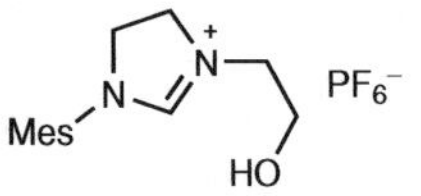

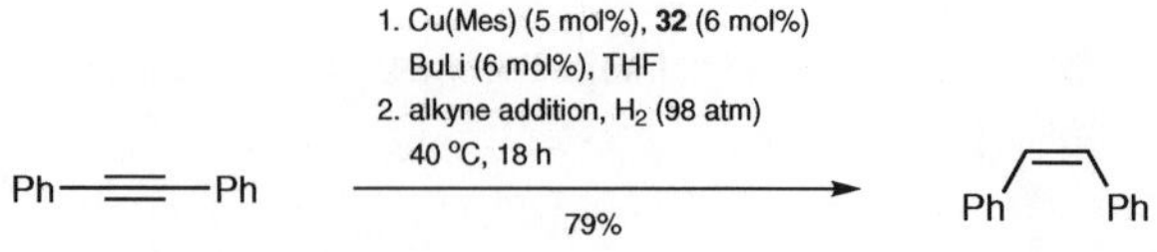

A similar copper-based catalytic system, reported by Nakao and co-workers, utilizes $\{CuCl(PPh_3)\}_4$ (2.0 mol%) in combination with lithium *tert*-butoxide (50 mol%) and propan-2-ol (1.0 equiv), under 5 atm hydrogen pressure in toluene at 100 °C (Scheme 12).[80] Full conversion of internal alkynes to afford *Z*-alkene/*E*-alkene/alkane selectivities of higher than 98:1:1 in up to 95% yield is achieved. Deuteration experiments using iPrOD support the assumption that the reaction occurs via a regioselective copper–hydride addition followed by protonation.

Scheme 12 Semi-hydrogenation of Diphenylacetylene Using a Copper Catalyst[80]

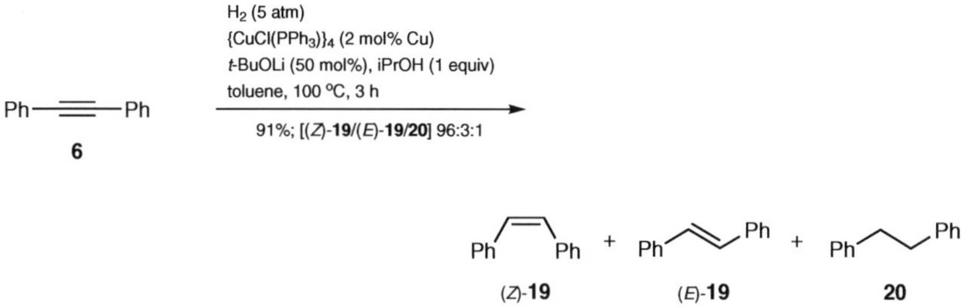

Ph—≡—Ph **6**

H_2 (5 atm), $\{CuCl(PPh_3)\}_4$ (2 mol% Cu), *t*-BuOLi (50 mol%), iPrOH (1 equiv), toluene, 100 °C, 3 h

91%; [(*Z*)-**19**/(*E*)-**19**/**20**] 96:3:1

(*Z*)-**19** + (*E*)-**19** + **20**

Sawamura and co-workers reported that use of a commercially available simple N-heterocyclic carbene ligand [1,3-dimesityl-4,5-dihydroimidazol-2-ylidene (SIMes)] in the copper-catalyzed semi-hydrogenation of internal alkynes gives good to moderate yields of alkenes, with *Z*/*E* ratios of up to >99:1 (Scheme 13).[81] These reactions proceed with 10 mol% catalyst loading and sodium *tert*-butoxide (10 mol%) in an octane/dioxane mixture (4:1) at 100 °C.

Scheme 13 Semi-hydrogenation of Diphenylacetylene Using a Copper–N-Heterocyclic Carbene Complex[81]

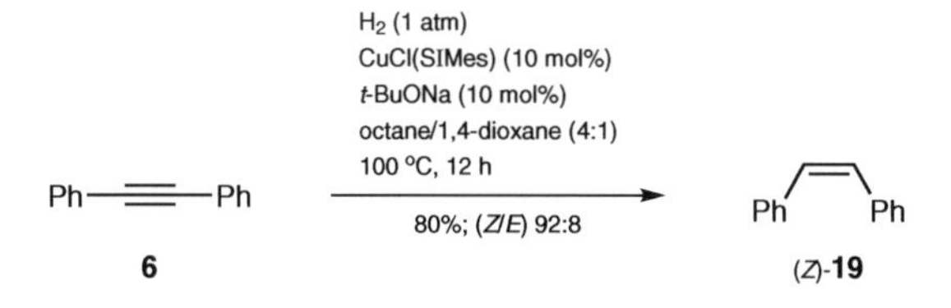

Niobium(III)–imido complexes show interesting performance under homogeneous reaction conditions.[82] The reactions are performed in the presence of small amounts of carbon monoxide, which is assumed to replace the alkene on the metal center and drive catalyst turnover. (*Z*)-1-Phenylprop-1-ene can be synthesized in 2 hours at room temperature. The high catalyst loading (20 mol%) is a major disadvantage; however, up to 85% yield is achieved.

A cationic vanadium–bis(imido) complex has been utilized for the *Z*-selective semi-hydrogenation of internal alkynes with yields of 44–100%, as disclosed by Toste and co-workers.[83] This catalyst has also been applied in the hydrogenation of terminal alkynes; however, significantly lower yields are obtained. The functional-group tolerance of this system is rather limited, with stronger ligands such as acetonitrile resulting in ligand exchange, and aldehydes and ketones leading to decomposition of the catalyst. General reaction conditions were 20 mol% catalyst loading, under 1 atm of hydrogen at 60 °C in (trifluoromethyl)benzene.

The cationic complex $[RuH(PMe_2Ph)_5]PF_6$ has been applied as catalyst in the hydrogenation of terminal (hept-1-yne, phenylacetylene) and internal [but-2-yne, pent-2-yne (**33**)] alkynes (Scheme 14).[84] No over-reduction or isomerization of the *Z*-alkene is observed in the semi-hydrogenation of but-2-yne, even after complete consumption of the alkyne.

for references see p 225

Mild reaction conditions (20 °C, 1 atm H_2) in methanol or acetone are applied. Lower conversions are observed at increased temperatures (>20 °C), which was explained by catalyst decomposition. This drawback can be alleviated by the addition of excess phosphine ligand (5–25 equiv). Under optimized conditions, the hydrogenation of hept-1-yne occurs at hydrogen diffusion controlled rates. The related complex $[RuH(PMe_2Ph)_3]PF_6$ reveals an unusual reactivity order with hex-1-yne < hex-2-yne < hex-3-yne.[85] A similar complex, $[Ru(H)(H_2)(PMe_2Ph)_4]PF_6$, has been reported for the selective hydrogenation of alk-1-ynes to alk-1-enes.[86] However this catalyst is rapidly deactivated during the course of the reaction.

Scheme 14 Semi-hydrogenation of an Alkyne Using a Ruthenium Complex[84]

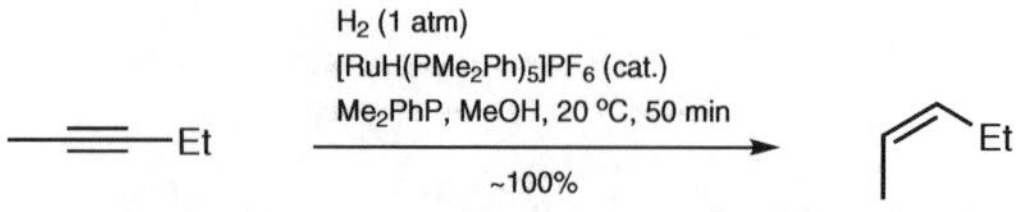

H_2 (1 atm)
$[RuH(PMe_2Ph)_5]PF_6$ (cat.)
Me_2PhP, MeOH, 20 °C, 50 min
~100%

33 → 34

Du and co-workers reported a metal-free protocol for semi-hydrogenation that uses bis(pentafluorophenyl)borane as the catalyst. Either the *Z*- or *E*-alkene can be obtained, depending on the applied reaction conditions. The drawbacks of this method are high catalyst loadings (up to 20 mol%), a high temperature (140 °C), and limited tolerance of functional groups.[87]

Various catalytic systems have been reported for the exclusive hydrogenation of terminal triple bonds. Among these is the application, by Werner and co-workers, of a hetero-bidentate *P,O*-iridium complex ($[Ir(cod)\{P(iPr)_2[(CH_2)_2OMe]\}]BF_4$) for the hydrogenation of phenylacetylene.[88] This study focused mainly on the mechanistic aspects of this reaction. It was assumed that the higher coordination ability of phenylacetylene compared to styrene favors the displacement of the OMe group and, therefore, the alkyne hydrogenation. The coordinating diene of the catalyst [cycloocta-1,5-diene (cod)] is not reduced in the course of the reaction.

Similar results have been obtained for phenylacetylene using Zaragoza–Würzburg catalysts $[OsClH(CO)(PR^1_3)_2$; $PR^1_3 = P(iPr)_3$, $P\text{-}t\text{-}BuMe_2]$.[89] The origin of the high selectivity is believed to be the thermodynamic sink of the vinyl complexes, which suppresses the kinetically favored hydrogenation of styrene; selectivities close to 100% were reached.[90]

Bianchini and co-workers have reported the semi-hydrogenation of terminal alkynes using mild conditions under 1 atm hydrogen in tetrahydrofuran with a *cis*-hydride η^2-dihydrogen complex, $[Fe(H)(H_2)\{P[(CH_2)_2PPh_2]_3\}]BPh_4$.[91] Alkynes containing an alkyl or aryl substituent are converted into the corresponding alkenes chemoselectively. Only in the use of (trimethylsilyl)acetylene did the reaction lead to the formation of a 1,4-disubstituted butadiene via reductive dimerization. Interestingly, the free coordinating site required for alkyne hydrogenation is provided by dissociation of one of the phosphine moieties, not by the remarkably stable dihydrogen ligand. An example of the semi-hydrogenation of phenylacetylene (**5**) to give styrene (**35**) is given in Scheme 15.[92]

Scheme 15 Semi-hydrogenation of Phenylacetylene To Give Styrene Using an Iron Complex[92]

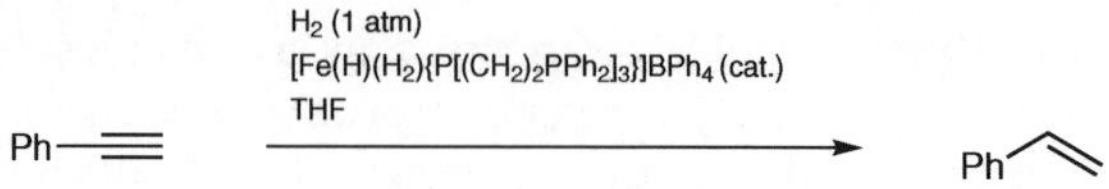

H_2 (1 atm)
$[Fe(H)(H_2)\{P[(CH_2)_2PPh_2]_3\}]BPh_4$ (cat.)
THF

5 → 35

Alkenes 22; **General Procedure:**[69]
Pd catalyst **21** (25 mg, 0.04 mmol) and the alkyne (4.0 mmol) were added to anhyd THF (50 mL) in a Schlenk tube under N_2. The soln was then subjected to a H_2 atmosphere of 1 atm by first flushing with H_2 and then slowly blowing H_2 over the surface while the soln was vigorously stirred at 20 °C. The reaction was monitored by GC and was stopped when all the alkyne had been consumed (conversion 99.5–100%). The composition of the mixture was determined by GC and ^1H NMR analysis.

Alkenes, e.g. 26; **General Procedure:**[74]
In situ generation of the catalyst **25**: THF (30 mL), the imidazolium chloride (0.05 mmol), Pd complex **24** (0.05 mmol), and *t*-BuOK (0.20 mmol) were added to a two-necked Schlenk tube equipped with a septum and a stirrer bar. After being stirred for 1 h, the soln was filtered through Celite to remove traces of palladium black. Then, the appropriate alkyne (5.0 mmol) was added using a syringe, and H_2 (1 atm) was bubbled slowly through the soln. Samples were removed periodically for GC analysis.

Alkenes 28; **General Procedure:**[75]
The reactions were performed in a 20-mL reactor equipped with a Teflon-coated magnetic stirrer bar. Pd complex **27** (3.6 mg, 0.005 mmol), DMAP (1.2 mg, 0.01 mmol), and pyridine (1 mL) were stirred at 25 °C for 0.5 h under H_2 (1 atm). Then, a soln of the alkyne (0.5 mmol) in pyridine (4 mL) was added. The mixture was stirred for 4 h (or the indicated time), and then H_2O was added and the aqueous phase was extracted with EtOAc. The combined organic phases were dried (Na_2SO_4) and concentrated under reduced pressure. The residue was purified by flash column chromatography (silica gel).

(*Z*)-Stilbene [(*Z*)-19**]; Typical Procedure:**[80]
In a glovebox, $\{CuCl(PPh_3)\}_4$ (7.2 mg, 5.0 μmol), toluene (1.0 mL), *t*-BuOLi (40 mg, 0.50 mmol), and iPrOH (60 mg, 1.0 mmol) were added in this order to a vial. The resulting mixture was stirred for 1 min at rt, and then diphenylacetylene (**6**; 178 mg, 1.0 mmol) and toluene (2.0 mL) were added. The vial was placed in an autoclave and the autoclave was taken out of the glovebox. The atmosphere in the autoclave was replaced with H_2 and the mixture was stirred at 100 °C for 3 h under H_2 (5 atm). After cooling to rt, H_2 was released and the mixture was diluted with EtOAc. The conversion was determined by GC. The resulting soln was filtered through a pad of silica gel and concentrated to yield the crude product.

Alkenes, e.g. (*Z*)-Pent-2-ene (34**); General Procedure:**[84]
The Ru catalyst precursor $[RuH(PMe_2Ph)_5]PF_6$, was prepared and purified by established methods.[93] $[RuH(PMe_2Ph)_5]PF_6$ (0.1 mmol), N_2-degassed MeOH (50 mL), and a magnetic stirrer bar were placed in a 300-mL Schlenk flask. The soln was frozen at −196 °C and the reaction flask was evacuated to 0.1 Torr. H_2 was then added to reach 1.0 atm pressure, and the catalyst soln was warmed to the reaction temperature in an oil bath. The alkyne (e.g., **33**; 10 mmol) was added by syringe and the progress of the reaction was monitored at 15-min intervals by GLC.

Alkenes, e.g. Styrene (35**); General Procedure:**[92]
The catalytic reactions were followed by measuring the H_2 consumption as a function of time. The catalyst $[Fe(H)(H_2)\{P(CH_2CH_2PPh_2)_3\}]BPh_4$ was added to a degassed soln of the substrate in 1,2-dichloroethane (or THF; 8 mL) in a 25-mL flask. The system was evacuated and refilled with H_2 (6 ×), and the flask was immersed in a constant-temperature bath. The mixture was vigorously shaken during the reaction. Alternatively, the substrate (2 mmol), THF (12 mL), and a stirrer bar were placed in a reaction vessel fitted with a reflux condenser and with a side arm with a rubber septum under a constant pressure of H_2 (1 atm). The

for references see p 225

vessel was immersed in a constant-temperature oil bath (20 or 63 °C). The catalyst was then added. The soln was sampled after 2 h, and the samples were analyzed by GC and GC/MS.

1.5.1.2.1.4 Homogeneous Transfer Hydrogenations

Only a limited number of transfer hydrogenation catalysts and procedures have been reported that use homogeneous catalysts. The favored hydrogen source is usually formic acid or one of its salts, and palladium catalysts predominate. The most widely investigated palladium system is that reported by Elsevier.[94] Here, palladium–N-heterocyclic carbene (NHC) complexes [e.g., **38** bearing the 1,3-dimesitylimidazol-2-ylidene (IMes) ligand] are used with formic acid–triethylamine as the hydrogen source. In the first report of this protocol, mainly arylalkynes were converted, with little of the overhydrogenated product formed. In the reactions of nonconjugated alkynes, the *Z*-alkene is the major product; however, the selectivity is dependent on both the alkyne structure and also the solvent used (Scheme 16).

Scheme 16 Transfer Hydrogenation of Alkynes Using a Palladium–N-Heterocyclic Carbene Complex[94]

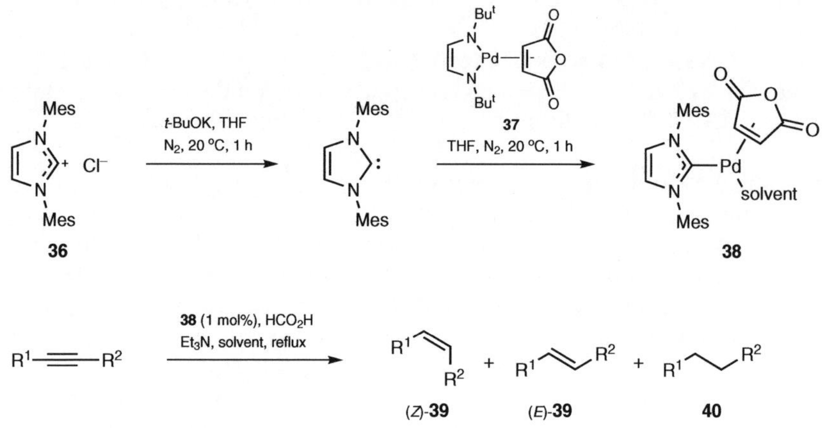

R^1	R^2	Solvent	Time (h)	Conversion (%)	Product Distribution (%)			Ref
					(*Z*)-**39**	(*E*)-**39**	**40**	
Ph	Me	MeCN	12	>99	96	4	1	[94]
Ph	Ph	THF	7	>99	98	–	2	[94]
Pr	Pr	MeCN	<24[a]	>99	93	3	4	[94]
Pr	$(CH_2)_2OH$	MeCN	<24[a]	>99	>99	–	–	[94]
Ph	H	THF	<2	>99	50[b]	–	50	[94]
Ph	H	MeCN	24	>99	>99[b]	–	–	[94]

[a] The exact time for full conversion is not known, but no more starting material was present after 24 h.

[b] *E*/*Z* isomerism does not apply for the alkene product.

Investigations into the mechanism of the reaction show that acetonitrile acts as a hemilabile ligand, whereas maleic anhydride is assumed to be partly coordinated during reaction.[95] Alternative palladium–N-heterocyclic carbene catalysts have been developed that show higher activity, but lower selectivity.[96] A more convenient modification uses the commercially available PdCl(allyl)(IMes) (IMes = 1,3-dimesitylimidazol-2-ylidene) catalyst with triphenylphosphine as an additional ligand.[97]

Another protocol for the palladium-catalyzed Z-selective semi-hydrogenation of internal and terminal alkynes, reported by Han and co-workers, uses formic acid in 1,4-dioxane and tetrakis(triphenylphosphine)palladium(0) as the catalyst (Scheme 17).[98] High Z-selectivity in combination with little over-reduction was obtained for various substrates. Interestingly, when 25% aqueous formic acid is used instead, the E-alkene is formed preferentially (see Section 1.5.1.2.2.2).

Scheme 17 Palladium-Catalyzed Z-Selective Semi-hydrogenation[98]

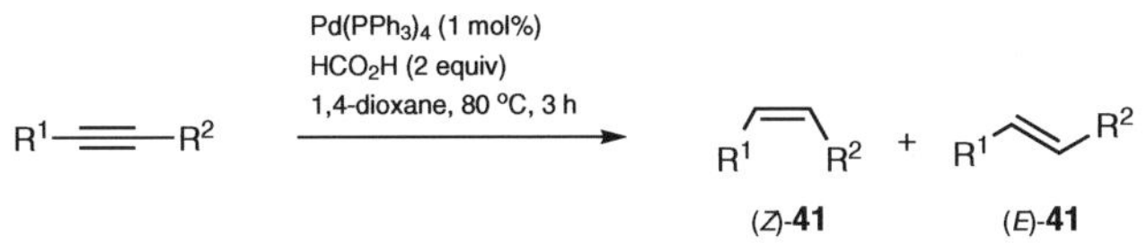

R^1—≡—R^2 → ($Pd(PPh_3)_4$ (1 mol%), HCO_2H (2 equiv), 1,4-dioxane, 80 °C, 3 h) → (Z)-**41** + (E)-**41**

R^1	R^2	Ratio [(Z)-**41**/(E)-**41**]	Yield (%)	Ref
$(CH_2)_{12}Me$	H	–	95[a]	[98]
Ph	H	–	90[a]	[98]
$(CH_2)_5Me$	Ph	95:5	93	[98]
$4-ClC_6H_4$	Ph	99:1	91	[98]
CO_2Me	$(CH_2)_4Me$	98:2	90	[98]
t-Bu	Ph	91:8	55	[98]

[a] Determined by GC.

A simple procedure uses palladium(II) acetate as the catalyst and dimethylformamide and potassium hydroxide as a hydrogen source (Scheme 18).[99] Excellent chemo- and stereoselectivity is reported for both aryl- and alkyl-substituted alkynes. The disadvantages of this method are the relatively high temperature (145 °C) that is required, and that the reported substrates included relatively few types of functional group.

Scheme 18 Palladium-Catalyzed Z-Selective Transfer Hydrogenation with Dimethylformamide/Potassium Hydroxide as the Hydrogen Source[99]

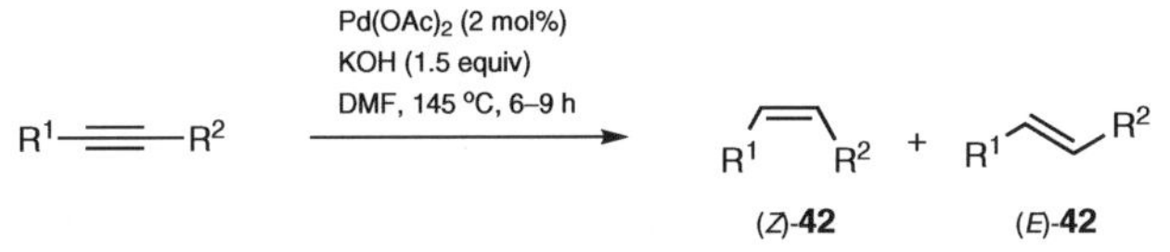

R^1—≡—R^2 → ($Pd(OAc)_2$ (2 mol%), KOH (1.5 equiv), DMF, 145 °C, 6–9 h) → (Z)-**42** + (E)-**42**

R^1	R^2	Ratio [(Z)-**42**/(E)-**42**]	Yield (%)	Ref
Ph	Ph	97:3	96	[99]
Ph	$2-MeOC_6H_4$	95:5	93	[99]
Ph	Me	98:2	88	[99]
Ph	Bu	99:5	92	[99]
Bu	Bu	>99:1	87	[99]

for references see p 225

Z-Alkenes (Z)-39; General Procedure:[94]
The solvent (MeCN or THF; 30 mL), 1,3-dimesitylimidazolium chloride (**36**; 0.055 mmol), and *t*-BuOK (0.22 mmol) were introduced into a two-necked Schlenk tube, equipped with a septum and a stirrer bar. The mixture was stirred for 1 h at 20 °C, and then Pd complex **37** (0.05 mmol) was added. This mixture was stirred for 1 h at 20 °C before adding the alkyne (5 mmol), Et_3N (25 mmol), HCO_2H (25 mmol), and *p*-xylene (5 mmol), and was then heated to the appropriate temperature until the reaction was complete (3–24 h). Samples were taken periodically for GC analyses. At the end of the reaction, the mixture was allowed to cool to rt, filtered through Celite, diluted with pentane (30 mL), washed with H_2O (4 × 50 mL), and dried ($MgSO_4$). The solvent was removed under reduced pressure and the crude mixture was analyzed by ^1H NMR spectroscopy and GC.

1.5.1.2.2 *E*-Selective Semi-hydrogenations of Internal Alkynes

1.5.1.2.2.1 Homogeneous Hydrogenations

There are no general methods known for the *E*-selective catalytic hydrogenation of internal alkynes, and stoichiometric methods, such as the Birch reduction, are usually applied.[100] Alternatively, a two-step protocol using a ruthenium-catalyzed hydrosilylation followed by a desilylation can be used. A disadvantage of this method is the stoichiometric amount of fluoride used in the desilylation step.[101] A similar approach using a hydrostannylation–destannylation approach has been reported;[102] by using stoichiometric amounts of copper(I) thiophene-2-carboxylate (CuTC) and tetrabutylammonium diphenylphosphinate, conventional harsh reagents applied for protodestannylation are avoided and therefore various functional groups are tolerated. A comprehensive review of such reactions was published in 2016 by Fürstner.[103]

Reports of homogeneous systems that achieve the formation of *E*-alkenes from alkyne semi-hydrogenation in one step have been constantly increasing in recent years. However, most of these reactions proceed through *Z*-selective hydrogenation followed by isomerization to generate the thermodynamically favored *E*-isomer.[104] In this case, the *E* selectivity is strongly affected by the substituents on the resulting double bond. Catalysts that achieve this transformation are discussed below.

Examples of catalysts that facilitate the direct generation of the *E*-stereoisomer are relatively rare. A pioneering NMR study by Bargon and co-workers used a cationic (η^5-pentamethylcyclopentadienyl)ruthenium/sorbic acid complex for the semi-hydrogenation of internal alkynes to give *E*-alkenes.[105] By using *para*-hydrogen induced polarization,[106] it was demonstrated that the *E*-selective hydrogenation is intrinsic and is not the result of a hydrogenation/isomerization sequence. A catalyst system that allows the direct *E*-selective hydrogenation of internal alkynes using different (η^5-pentamethylcyclopentadienyl)-ruthenium complexes **43–46** has been reported by Fürstner and co-workers (Scheme 19).[107]

Scheme 19 Ruthenium Complexes Applied for *E*-Selective Hydrogenation of Internal Alkynes[107]

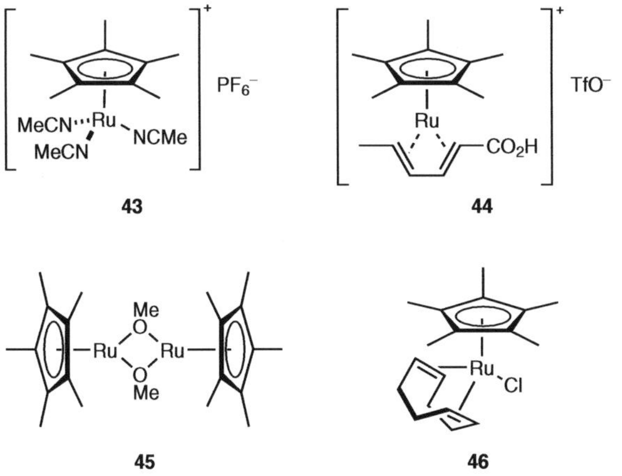

[$Cp^*Ru(MeCN)_3$]$^+$ PF_6^- **43**; [Cp^*Ru(diene)CO_2H]$^+$ TfO$^-$ **44**; [$Cp^*Ru(\mu$-OMe)$]_2$ **45**; Cp^*RuCl(cod) **46**

The hydrogenation of alkynes using chloro(cycloocta-1,5-diene)(η^5-pentamethylcyclopentadienyl)ruthenium(II) (**46**), with moderate reaction times of several hours and under 10 atm of hydrogen pressure in solvents such as dichloromethane or methanol, leads to alkenes **47** with *E*/*Z* ratios of up to 98:2 and yields of up to 96% (Scheme 20).[107] Many functional groups are tolerated under these conditions, e.g. alcohols, carboxylic esters and acids, nitro groups, thioethers, and bromides. The applicability of these catalysts to more complex molecules has been successfully demonstrated in the total synthesis of brefeldin A.[108] Inhibition of the catalyst is observed when diene units or terminal alkynes are present in the substrate. Mechanistic investigations show strong evidence for a *gem*-hydrogen transfer as the origin for the high selectivity of these catalysts.[109]

Scheme 20 *E*-Selective Hydrogenation of Internal Alkynes Using a Commercially Available Ruthenium Catalyst and Silver(I) Trifluoromethanesulfonate[107]

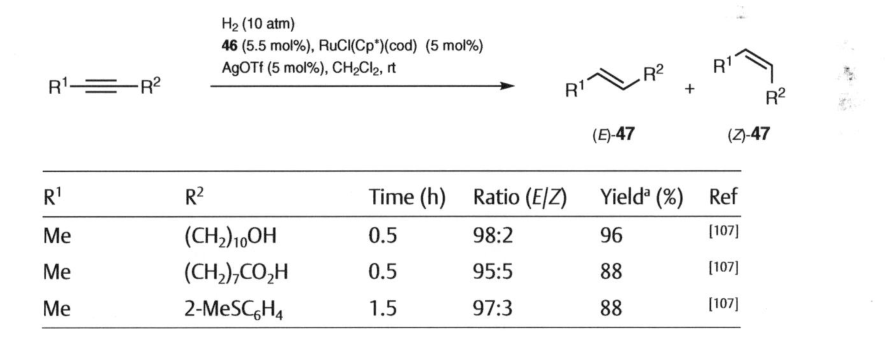

R¹	R²	Time (h)	Ratio (*E*/*Z*)	Yield[a] (%)	Ref
Me	$(CH_2)_{10}OH$	0.5	98:2	96	[107]
Me	$(CH_2)_7CO_2H$	0.5	95:5	88	[107]
Me	$2\text{-}MeSC_6H_4$	1.5	97:3	88	[107]

for references see p 225

R^1	R^2	Time (h)	Ratio (E/Z)	Yield[a] (%)	Ref
Me	2-$MeOC_6H_4$	0.5	95:5	87	[107]
	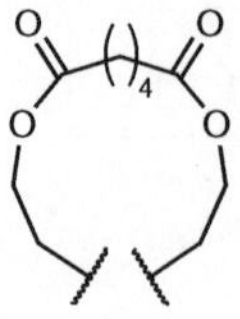	0.5	98:2	89[b]	[107]
Me	cyclohept-1-enyl	–	–	–[c]	[107]

[a] Combined yield of E/Z-alkene and alkane (5–15%) products; determined by GC.
[b] 1.29 mmol scale; isolated yield.
[c] No reaction occurs.

An alternative system, reported by Lindhardt and co-workers, employs a commercially available ruthenium hydride complex [$RuClH(CO)(PPh_3)_3$] to obtain the *E*-alkenes **48** from internal alkynes bearing at least one aryl substituent.[110] In general, high isolated yields of up to 99% are obtained, with tetrahydrofuran or toluene as solvent (Scheme 21). Various functional groups, such as a diene unit, are tolerated and full conversion was obtained after upscaling and optimization of the reaction parameters, resulting in a catalyst loading as low as 0.001 mol%. This protocol has been successfully applied in the semi-hydrogenation or deuteration of terminal alkynes. The reactions were conducted in a flow system or with a two-chamber reactor in which the second chamber is used for the ex situ generation of hydrogen from zinc and hydrochloric acid. For dialkyl-substituted acetylenes, complex product mixtures resulting from double-bond migration are observed. Mechanistic experiments indicate a two-step process that involves *Z*-selective hydrogenation followed by isomerization of the double bond to form the *E*-alkene.

Scheme 21 Hydrogenation of Various Arylalkynes To Give *E*-Arylalkenes[110]

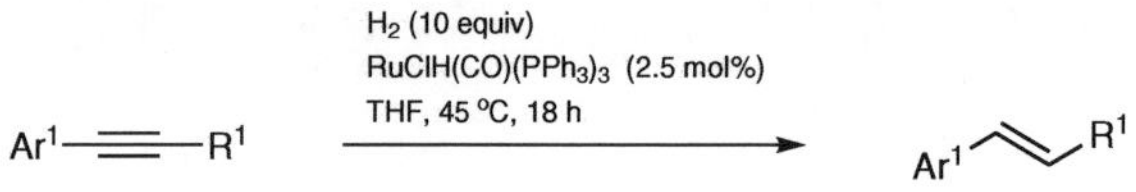

Ar^1—≡—R^1 → (H_2 (10 equiv), $RuClH(CO)(PPh_3)_3$ (2.5 mol%), THF, 45 °C, 18 h) → **48**

Ar^1	R^1	Yield (%)	Ref
Ph	Ph	99[a]	[110]
Ph	4-BrC_6H_4	92	[110]
Ph	4-HOC_6H_4	95	[110]
2-Me-4-$MeOC_6H_3$	2-$MeOC_6H_4$	91[b]	[110]
3-$F_3CC_6H_4$	cyclohex-1-enyl	80	[110]
Ph	Et	89	[110]

[a] Conditions: H_2 (1.2 equiv), $RuClH(CO)(PPh_3)_3$ (0.001 mol%), 90 °C.
[b] Temperature was 90 °C.

A different catalytic system, based on halide-bridged dinuclear iridium complexes **49**, has been reported by Mashima and co-workers (Scheme 22).[111] Using 1 mol% of catalyst **49**, the desired *E*-alkenes (*E*)-**50** are obtained in up to 97% yield, under 1 atm hydrogen pressure in 1,4-dioxane at 80 °C. In general, high *E*/*Z* ratios are observed, with the formation of alkanes **51** as the major byproduct. The substrate scope is limited to alkynes that contain at least one aryl substituent on the triple bond. A mononuclear iridium dihydride species is proposed to catalyze the *Z*-selective alkyne semi-hydrogenation followed by isomerization to form the *E*-alkene.

Scheme 22 Semi-hydrogenation of Internal Alkynes Using a Halide-Bridged Dinuclear Iridium Complex[111]

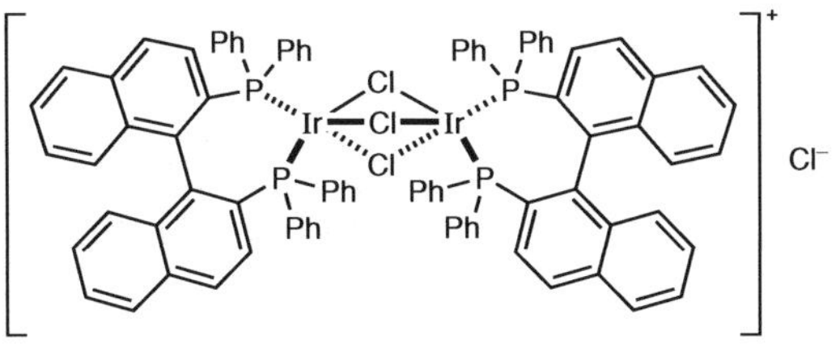

49

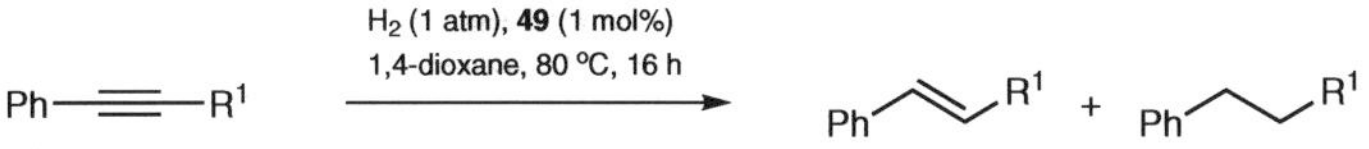

Ph—≡—R[1] → H_2 (1 atm), **49** (1 mol%), 1,4-dioxane, 80 °C, 16 h → (*E*)-**50** + **51**

R^1	Yield[a] (%)			Ref
	(*E*)-**50**	(*Z*)-**50**	**51**	
Ph	97	n.d.	3	[111]
4-$MeOC_6H_4$	94	n.d.	5	[111]
4-BrC_6H_4	92	n.d.	8	[111]
4-NCC_6H_4	95	n.d.	5	[111]
Bu	90	n.d.	10	[111]
CMe_2OH	87	5	6	[111]

[a] Determined by NMR using phenanthrene as an internal standard; n.d. = not determined.

A conceptually interesting system has been reported by Mankad and co-workers that uses a hetero-bimetallic N-heterocyclic carbene catalyst [(IMes)AgRu(Cp)$(CO)_2$; IMes = 1,3-dimesitylimidazol-2-ylidene]. An example is shown for the hydrogenation of diphenylacetylene (**6**) to give (*E*)-stilbene [(*E*)-**19**], (*Z*)-stilbene [(*Z*)-**19**], and 1,2-diphenylethane (**20**) (Scheme 23).[112] Only a few substrates were examined and substantial amounts of the *Z*-isomer are observed in several cases. Various functional groups show no influence on the reactivity of the catalyst, and terminal alkynes are also successfully converted into alkenes. However, the *E*/*Z* selectivity is reduced by the presence of a keto group or a pyridine. Mechanistic investigations indicate that this system also undergoes *Z*-selective hydrogenation followed by isomerization to the *E*-alkene. The hydrogen activation is assumed to be heterolytic.

Scheme 23 Semi-hydrogenation of Diphenylacetylene Using a Ruthenium Catalyst[112]

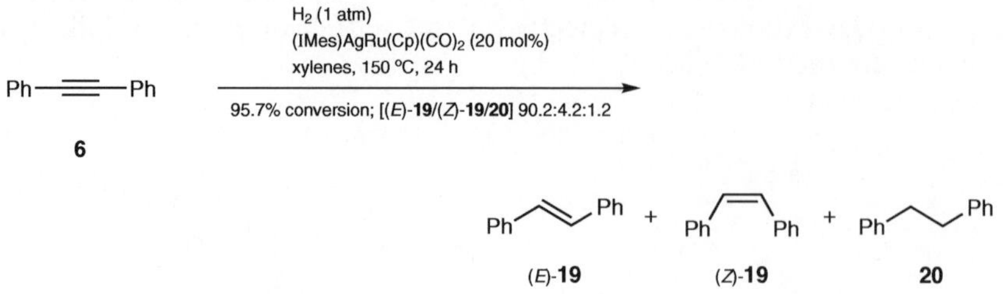

An iron pincer catalyst **52** has been applied by Milstein for the generation of *E*-alkenes from internal alkynes (Scheme 24).[113] This conversion requires long reaction times at 90 °C, but affords stilbenes with conversions of up to 99% and high *E*/*Z* selectivity. Various functional groups, such as carbonyl, nitrile, and chloro substituents, remain intact, and phenylacetylene can be successfully converted into styrene. The high *E* selectivity is due to isomerization of the resultant *Z*-alkene, limiting the substrate scope to arylalkynes or bis(trimethylsilyl)acetylene.

Scheme 24 Semi-hydrogenation of Diphenylacetylene Using an Iron Catalyst[113]

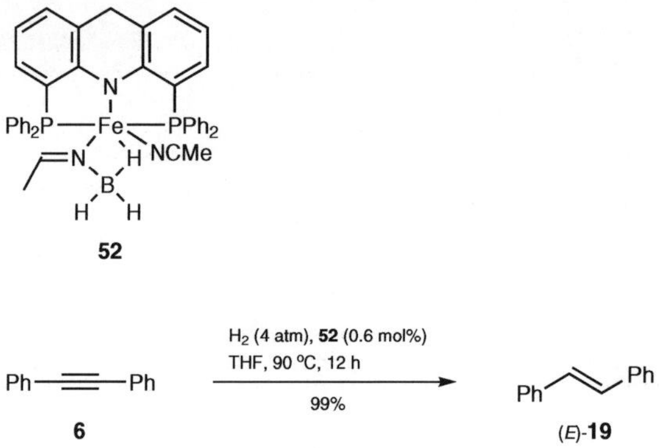

A cobalt dihydrogen complex containing a phenyl-bridged bis(mesitylbenzimidazol-2-ylidene) ligand has been utilized by Fout and Tokmic for the synthesis of *E*-alkenes from internal alkynes.[114] The reactions are performed with 1–2 mol% catalyst at 30 °C under 4 atm hydrogen in tetrahydrofuran, and afford the products in up to 96% yield with high *E*/*Z* ratios for monoaryl-substituted alkynes. One example of a dialkyl-substituted acetylene was given, in which dec-5-yne is converted into dec-5-ene in 69% isolated yield and with an *E*/*Z* ratio of >81:19. Attempts to hydrogenate phenylacetylene result in catalyst decomposition. Mechanistic studies conducted by *para*-hydrogen induced polarization revealed *Z*-selective semi-hydrogenation followed by isomerization.

Almost all of the catalytic systems in this section have been reported since about 2013. These catalysts allow for the one-step generation of *E*-alkenes from an internal triple bond with usually high functional-group tolerance, and they generally occur via a hydrogenation/isomerization sequence. The field is continuing to develop and alternative catalysts are likely to be discovered.

(*E*)-1,8-Dioxacyclotetradec-11-ene-2,7-dione [47, R^1,R^2 = $(CH_2)_2OC(=O)(CH_2)_4C(=O)O(CH_2)_2$]; Typical Procedure:[107]

In a Schlenk tube, a suspension containing RuCl(Cp*)(cod) (**46**; 27.1 mg, 0.071 mmol) and AgOTf (16.5 mg, 0.064 mmol) in CH_2Cl_2 (3.3 mL) was stirred for ca. 1 min under argon. The precipitate was allowed to settle and the supernatant soln was transferred via syringe into a predried autoclave under argon. The autoclave was then pressurized with H_2 (10 atm) and the mixture stirred at rt for 10–15 min. [This pre-hydrogenation of the catalyst prior to the addition of the substrate is not obligatory, but tends to suppress competing isomerization of the alkene product formed in the reaction.] The pressure was carefully released and a soln of 1,8-dioxacyclotetradec-11-yne-2,7-dione (290 mg, 1.29 mmol) in CH_2Cl_2 (3.2 mL) was added under argon. The autoclave was repressurized to 10 atm H_2 and the mixture was stirred for 30 min. After carefully releasing the pressure, the soln was analyzed by GC, which showed complete conversion of the substrate and formation of the alkene [ratio (*E*/*Z*) 98:2] and the corresponding cycloalkane (8%). For workup, the solvent was evaporated and the crude material was passed through a short column of silica gel (EtOAc/hexane 1:4) to give a colorless solid; yield: 260 mg (89%).

1.5.1.2.2.2 Homogeneous Transfer Hydrogenations

Various catalyst systems have been reported that allow the selective synthesis of *E*-alkenes via transfer semi-hydrogenation. The reaction conditions employed in these protocols also enable the isomerization of a *Z*-alkene to its *E*-isomer, which supports the assumption that the reactions of alkynes proceed via a *Z*-selective semi-hydrogenation followed by isomerization. Therefore, the *E* selectivity of these reactions is strongly connected to the groups directly attached to the triple bond.

A method for the transfer hydrogenation of alkynes to *E*-alkenes catalyzed by hydrido(methoxo)iridium(III) complexes was reported by Tani and co-workers (Scheme 25).[115] Good results are obtained in the semi-hydrogenation of diphenylacetylene (**6**) using complex **53** (2 mol%) in a propan-2-ol/toluene mixture, with 100% conversion to give (*E*)-stilbene [(*E*)-**19**; 90% yield] and the product of over-reduction 1,2-diphenylethane (**20**; 10%). The same reaction conducted in methanol/toluene (1:1) affords (*E*)-stilbene in 83% yield and 1,2-diphenylethane in 17% yield at 100% conversion. The reaction of 1-phenylprop-1-yne, i.e. replacement of one phenyl group by a methyl group, results in a significant erosion of selectivity. Hydrogenation of dialkylacetylenes results in mixtures of alkenes and alkanes. Only the semi-hydrogenation of substrates containing alkyl or aryl groups were examined. Competitive studies with a *Z*-alkene showed that isomerization to the *E*-alkene proceeds much faster than the competing hydrogenation to its alkane, which indicates that the overall reaction occurs via *Z*-selective hydrogenation followed by isomerization.

Scheme 25 Semi-hydrogenation of Diphenylacetylene Using an Iridium Catalyst[115]

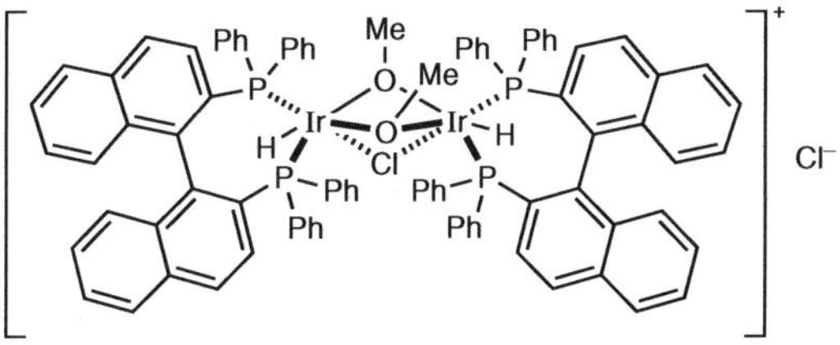

Ph Ph Me Ph Ph O Me P Ir O Ir P H H Cl P Ph Ph P Ph Ph Cl^-

53

53 (2 mol%)
MeOH/toluene (1:1)
80 °C, 18 h
~100% conversion; [(*E*)-19/20] 83:17

Ph—≡—Ph **6** → Ph–CH=CH–Ph (*E*)-**19** + Ph–CH$_2$CH$_2$–Ph **20**

Various functionalized diarylalkynes undergo successful transfer semi-hydrogenation using dodecacarbonyltriruthenium(0) [$Ru_3(CO)_{12}$] in combination with triphenylphosphine and an acid, with dimethylformamide and water as hydrogen donors.[116] The reaction affords the *Z*- or *E*-alkene selectively depending on the applied reaction conditions. The *Z*-alkene is the major product using acetic acid as the additive, whereas with trifluoroacetic acid, the *E*-isomer is obtained almost exclusively: the *E*-alkenes are obtained in up to 93% yield and with *E*/*Z* selectivity >99:1 using trifluoroacetic acid (1.5 equiv) as the additive at 145 °C. The mechanism of this reaction requires further investigation; however, isomerization most likely takes place similarly to the isomerization of (*Z*)-stilbene in the presence of dodecacarbonyltriruthenium(0) and water.

Han and co-workers reported the efficient formation of diaryl-substituted *E*-alkenes by the semi-hydrogenation of diarylalkynes using a nickel–1,3-bis(diphenylphosphino)-propane catalyst, hypophosphorous acid, and acetic acid.[117] Isolated yields of up to 86% and up to >99:1 *E*/*Z* selectivity are obtained. Replacement of one aryl-substituent by an alkyl group results in moderate isolated yields, and large amounts of a hydration side product are formed. Terminal alkynes lead predominantly to the dehydration product under these reaction conditions.

Another protocol for a nickel-catalyzed transfer semi-hydrogenation, disclosed by Moran and Richmond, uses the combination of nickel(II) chloride–ethylene glycol dimethyl ether complex ($NiCl_2$•DME) and bis[2-(diphenylphosphino)ethyl](phenyl)phosphine (triphos) with zinc and formic acid to generate the catalytically active species.[118] High *E* selectivity of up to >95:5 is obtained for diarylalkynes, whereas for dialkylalkynes, or in the absence of triphos, only the *Z*-alkene is observed.

A palladium catalyst has been utilized for the *E*-selective transfer semi-hydrogenation of alkynes.[98] With tetrakis(triphenylphosphine)palladium(0) and aqueous formic acid, high selectivity is obtained for a variety of aryl or carboxylic ester substituted internal alkynes. The *Z*-alkene is obtained if anhydrous formic acid is used (see Section 1.5.1.2.1.4). It is assumed that the first step of the mechanism is *cis*-selective palladium hydride addition. In the presence of water, the *E*-alkene is probably formed by isomerization of the alkenylpalladium intermediate and reductive elimination, rather than formation of the *Z*-alkene and isomerization. Experimentally, (*Z*)-octene is not isomerized under the reaction conditions to the *E*-isomer, which supports this hypothesis. Interestingly, full hydrogenation to the corresponding alkane was observed upon the addition of tricyclohexylphosphine. The use of tricyclohexylphosphine as the ligand prevents β-H elimination in palladium-catalyzed reactions.[119]

1.5.2 Reduction of Allenes

Allenes are compounds that have cumulated C=C bonds. The simplest representative of this family is allene (propadiene, $H_2C=C=CH_2$). Allene synthesis can be performed by copper-catalyzed addition of alkynes to formaldehyde in the presence of a base,[120] dehalogenation of 1,2-dihalocyclopropanes using lithium bases followed by Skattebøl rearrangement,[121] elimination of allylic alcohols containing a tin-substituent at the α-position,[122] or the reduction of propargyl halides or acetates.[123] From an industrial point of view the Saucy–Marbet rearrangement[124] of an alk-1-yn-3-ol–isopropenyl methyl ether adduct to give allene ketones is the most important method for the synthesis of allene derivatives.

Allene ketones and derivatives thereof are produced in several thousand kilotons and are intermediates in the industrial production of fat-soluble vitamins and carotenoids.[15]

Compounds with an allene-type structure are also found in nature, such as carotenoids or pheromones (e.g., in *Acanthoscelides obtectus*). The C39-carotenoid peridinin is found in marine organisms, such as flagellates, and is involved in energy transfer to chlorophyll.[125] Several methods have been described for its synthesis,[126–128] and its physical chemical behavior has been documented.[127]

The heat of hydrogenation of allene was determined to be 297 $kJ \cdot mol^{-1}$ and is comparable to that of buta-1,3-diene (247 $kJ \cdot mol^{-1}$) or penta-1,4-diene (255 $kJ \cdot mol^{-1}$).[129]

In the catalytic hydrogenation reactions of allenes using supported metal catalysts, similar results are obtained to the corresponding reactions of acetylene and buta-1,3-diene. The fundamental work of Bond, Webb, Wells, and Winterbottom shows that the selectivity of the hydrogenation (i.e., formation of the monoalkene versus the alkane at partial conversion) depends on the metal chosen, with the following sequence for selectivity: Pd > Rh > Pt > Ru > Os > Ir.[130] The catalyst activity was found to follow the order: Pt > Pd > Rh > Ir > Ru > Os.[131]

The structure of the substrate and steric interactions have a significant effect on the regio- and stereoselectivity of heterogeneous palladium-catalyzed semi-hydrogenation of allenes. Some substrates show almost exclusive initial reaction at one C=C bond, whereas in other substrates, the reduction of the two different C=C bonds proceeds at similar reaction rates.[132]

The hydrogenation of allene can be performed by applying nickel, palladium, or platinum catalysts. The adsorbed allene is selectively converted into propene. This reaction is first order with respect to hydrogen and zero order with respect to allene. Allene is more strongly adsorbed than propene, and isomerization of allene into propyne could not be detected.[133]

1.5.2.1 Hydrogenation of Functionalized Allenes

In general, there are only a limited number of literature reports on the hydrogenation of functionalized allenes. With the exception of allene ketones (see Section 1.5.2.2), palladium on carbon or platinum oxide catalysts are generally used.

Allenes bearing a neighboring stereogenic center have been successfully hydrogenated using platinum(IV) oxide without loss of stereochemical information (Scheme 26).[134] Platinum(IV) oxide was also applied in the reduction of ionone derivatives containing a trisubstituted allene, e.g. **54** (Scheme 27).[135] Palladium on carbon was utilized as catalyst for the hydrogenation of a related compound **55** bearing an extra methyl group. Here the yield is lower as a result of palladium-catalyzed C=C bond migration and side reactions.

Scheme 26 Platinum Oxide Catalyzed Hydrogenation of an Allene[134]

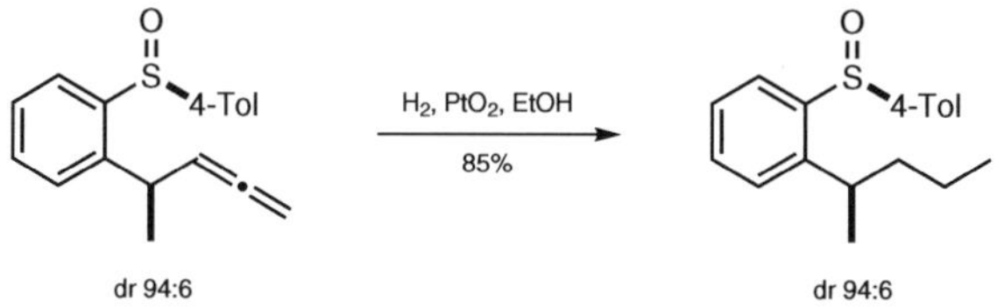

4-Tol
H_2, PtO_2, EtOH
85%
4-Tol
dr 94:6
dr 94:6

Scheme 27 Platinum- and Palladium-Catalyzed Allene Hydrogenation[135]

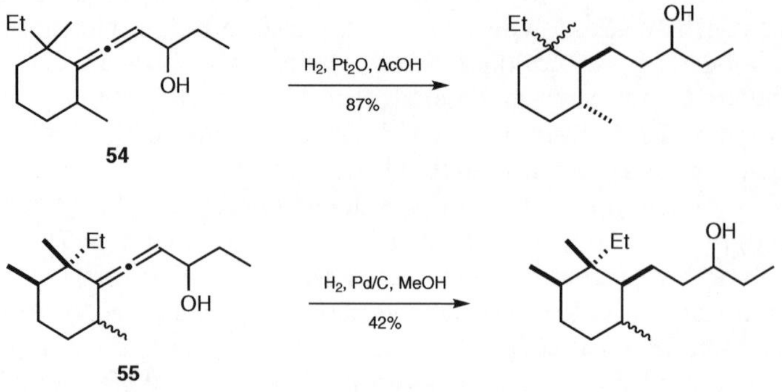

H_2, PtO_2, AcOH
87%

54

H_2, Pd/C, MeOH
42%

55

Palladium on carbon catalysts have also been used in the selective semi-hydrogenation of cyclic trisubstituted allene **56**.[136] The least sterically hindered C=C bond is selectively reduced (Scheme 28). For larger, bicyclic systems with a bridging allene, palladium on carbon leads to a mixture of mono- and direduced compounds **58** and **59** (Scheme 29).[137] Wilkinson's catalyst [$RhCl(PPh_3)_3$] is unreactive with the smaller ring compound **57** (n = 4), but with the larger ring substrate **57** (n = 7) it gives the fully hydrogenated product **59** (n = 7). Wilkinson's catalyst also results in a 10–66% conversion in the hydrogenation of a small number of allene hydrocarbons, with the major product being the Z-alkene.[138]

Scheme 28 Selective Mono-hydrogenation of a Hindered Allene[136]

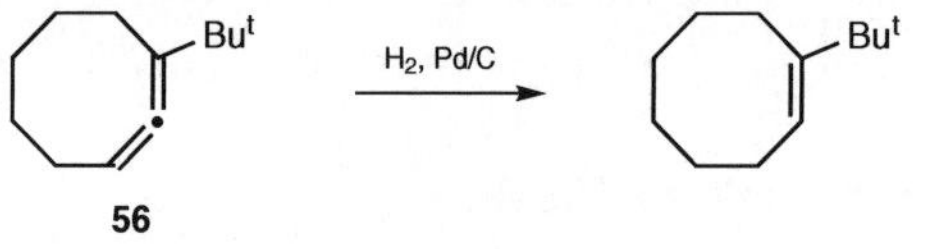

H_2, Pd/C

56

Scheme 29 Hydrogenation of Bicyclic Allenes Using Wilkinson's Catalyst[137]

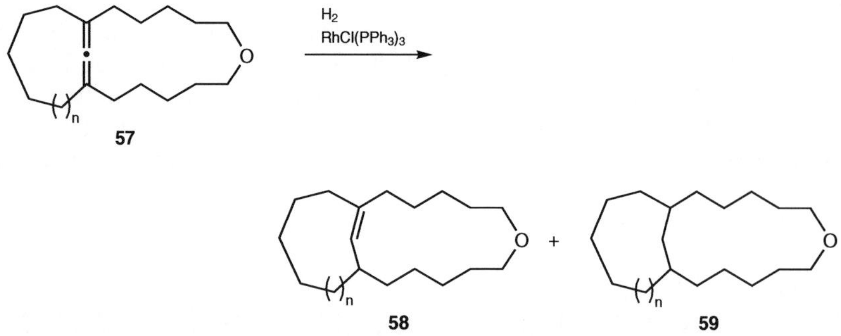

H_2
$RhCl(PPh_3)_3$

57

58 59

n = 4, 7

A comparison of the selectivity and activity of heterogeneous metal catalysts, Wilkinson's catalyst, and stoichiometric reducing agents has been made by Crombie for a small selection of alkynyl and allenyl carboxylic acids and alcohols. Significant differences in selectivity are observed when changing both the substrate and also the catalyst.[139]

1.5.2.2 Hydrogenation of Allene Ketones

Allene ketones, prepared in an efficient manner from alk-1-yn-3-ols by applying the Saucy–Marbet methodology,[124] are of industrial relevance in the manufacture of ionones or isoprenoid building blocks, such as isophytol. The transformation of dehydrolinalool (**11**) into C13 allene ketone **60** followed by base-catalyzed rearrangement results in the formation of a mixture of *E*/*Z*-isomers of pseudoionone **61**. Pseudoionone is a key intermediate in the syntheses of vitamins A and E and several carotenoids. The hydrogenation of pseudoionone (**61**) or C13 allene ketone **60** gives hexahydropseudoionone **62**, a compound of commercial interest in vitamin E synthesis (Scheme 30).[15,140,141] The hydrogenation of β,γ,δ- or α,β,γ,δ-unsaturated carbonyl compounds can be carried out using palladium/carrier catalysts such as palladium on carbon, palladium/alumina, or palladium/silica.[140–142] This type of hydrogenation can be performed either in batch mode or continuously.

Scheme 30 Hydrogenation of Pseudoionone and a C13 Allene Ketone

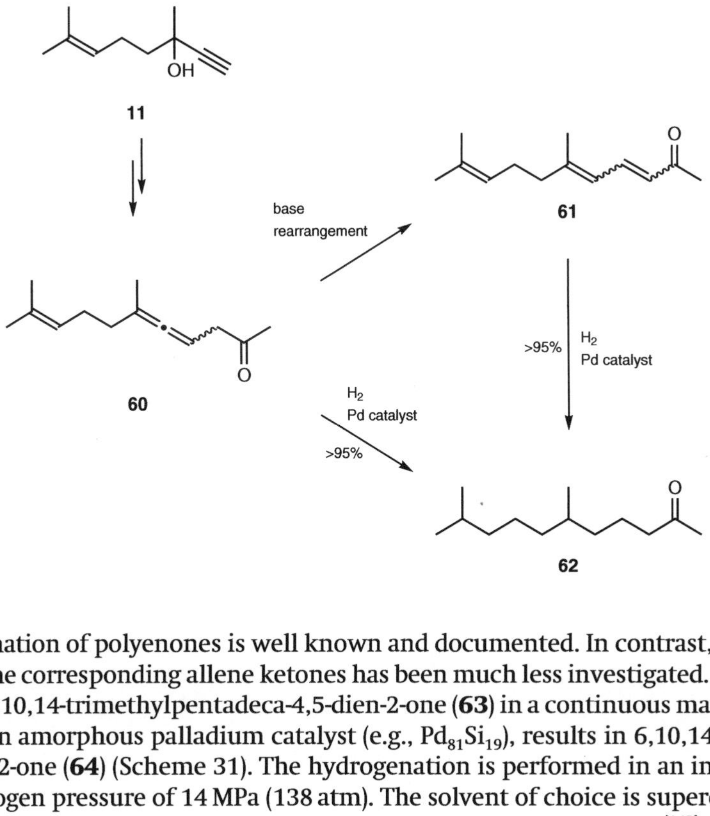

The hydrogenation of polyenones is well known and documented. In contrast, the hydrogenation of the corresponding allene ketones has been much less investigated. The hydrogenation of 6,10,14-trimethylpentadeca-4,5-dien-2-one (**63**) in a continuous manner in the presence of an amorphous palladium catalyst (e.g., $Pd_{81}Si_{19}$), results in 6,10,14-trimethylpentadecane-2-one (**64**) (Scheme 31). The hydrogenation is performed in an inert solvent under a hydrogen pressure of 14 MPa (138 atm). The solvent of choice is supercritical carbon dioxide or near-critical carbon dioxide; yields of ca. 90% are achieved.[143]

Scheme 31 Palladium-Catalyzed Hydrogenation of an Allene Ketone[143]

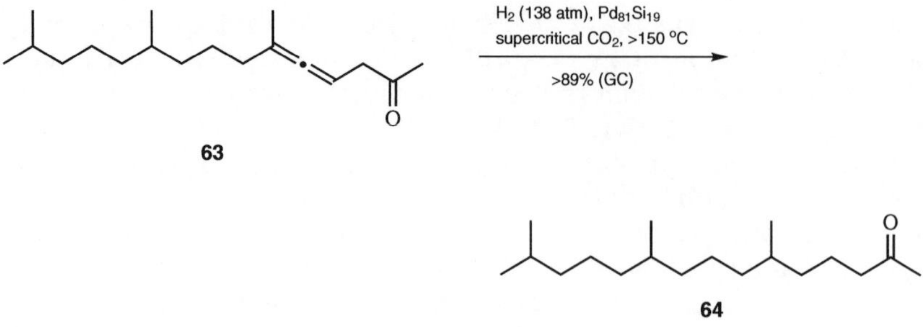

H_2 (138 atm), $Pd_{81}Si_{19}$
supercritical CO_2, >150 °C
>89% (GC)

63

64

The synthesis of isophytol by hydrogenation of the corresponding allene ketone precursor has also been performed in the presence of a nickel alloy catalyst modified with cobalt(II) chloride.[144]

Allene esters, e.g. **65**, are hydrogenated to give the corresponding saturated carbonyl compounds in the presence of palladium/calcium carbonate in excellent yields (Scheme 32),[145] and allenedicarboxylic acids react in methanol in the presence of palladium on carbon to give saturated dicarboxylic acids in 95% yield.[146]

Scheme 32 Palladium-Catalyzed Hydrogenation of an Allene Ester[145]

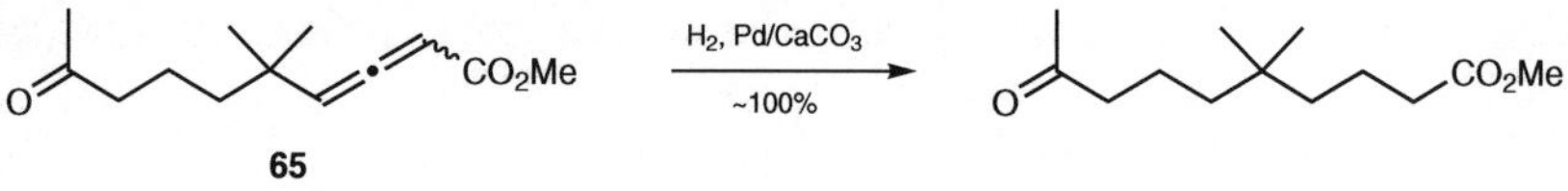

CO_2Me → H_2, $Pd/CaCO_3$, ~100% → CO_2Me

65

1.5.3 Conclusions

The total hydrogenation of alkynes and allenes can be performed in high yield by using a number of metal catalysts, the most preferred being palladium. On the other hand, in semi-hydrogenations it is significantly harder to obtain high selectivities, especially those required for industrial production. The main methods employed still utilize the heterogeneous Lindlar catalyst (and its variations), which was first reported in the 1950s. Since ca. 2010, a number of homogeneous catalytic systems have been reported for alkyne semi-hydrogenation. Promising results have been reported with a range of model substrates and further developments are expected. Whether one broadly applicable homogeneous catalyst system will emerge that can challenge the Lindlar catalyst in terms of versatility remains to be seen.

References

[1] Nishimura, S., *Handbook of Heterogeneous Catalytic Hydrogenation for Organic Synthesis*; Wiley Interscience: New York, (2001).

[2] Bonrath, W.; Medlock, J.; Schütz, J.; Wüstenberg, B.; Netscher, T., In *Hydrogenation*, Karamé, I., Ed.; InTech: Rijeka, Croatia, (2012); Chapter 3; available online at www.intechopen.com/books/hydrogenation/hydrogenation-in-the-vitamins-and-fine-chemicals-industry-an-overview (accessed October 5, 2017); DOI 10.5772/48751.

[3] Lindlar, H., *Helv. Chim. Acta*, (1952) **35**, 446.

[4] Huheey, J. E., *Anorganische Chemie: Prinzipien von Struktur und Reaktivität*; de Gruyter: Berlin, (1988).

[5] Lazier, W. A.; Arnold, H. R., *Org. Synth., Coll. Vol. II*, (1943), 142.

[6] Adkins, H.; Burgoyne, E. E.; Schneider, H. J., *J. Am. Chem. Soc.*, (1950) **72**, 2626.

[7] Johnson, C. R.; Tait, B. D., *J. Org. Chem.*, (1987) **52**, 281.

[8] Knowles, W. S.; Sabacky, M. J., *Chem. Commun.*, (1968), 1445.

[9] Bonrath, W.; Ondruschka, B.; Schmoeger, C.; Stolle, A., WO 2010020671, (2010).

[10] Bruehwiler, A.; Semagina, N.; Grasemann, M.; Renken, A.; Kiwi-Minsker, L.; Saaler, A.; Lehmann, H.; Bonrath, W.; Roessler, F., *Ind. Eng. Chem. Res.*, (2008) **47**, 6862.

[11] Davis, H. E.; Allinger, N. L.; Rogers, D. W., *J. Org. Chem.*, (1985) **50**, 3601.

[12] Rogers, D. W.; Dagdagan, O. A.; Allinger, N. L., *J. Am. Chem. Soc.*, (1979) **101**, 671.

[13] Lindlar, H.; Dubuis, R., *Org. Synth., Coll. Vol. V*, (1973), 880.

[14] Lindlar, H., US 2681938, (1954).

[15] Eggersdorfer, M.; Laudert, D.; Létinois, U.; McClymont, T.; Medlock, J.; Netscher, T.; Bonrath, W., *Angew. Chem.*, (2012) **124**, 13134; *Angew. Chem. Int. Ed.*, (2012) **51**, 12960.

[16] Bond, G. C., *Platinum Met. Rev.*, (1957) **1**, 87.

[17] Dobson, N. A.; Eglinton, G.; Krishnamurti, M.; Raphael, R. A.; Willis, R. G., *Tetrahedron*, (1961) **16**, 16.

[18] Bond, G. C.; Rank, J. S., In *Proceedings of the Third International Congress on Catalysis: Amsterdam, 20–25 July, 1964*, Sachtler, W. M. H.; Schuit, G. C. A.; Zwietering, P., Eds.; North Holland: Amsterdam, (1965); Vol. 2, pp 1225–1236.

[19] Bond, G. C.; Wells, P. B., *Adv. Catal.*, (1964) **15**, 151.

[20] Siegel, S., In *Comprehensive Organic Synthesis*, Trost, B. M.; Fleming, I., Eds.; Pergamon: Oxford, (1991); Vol. 8, pp 417–442.

[21] Henrick, C. A., *Tetrahedron*, (1977) **33**, 1845.

[22] Isler, O., *Pure Appl. Chem.*, (1979) **51**, 447.

[23] Isler, O.; Huber, W.; Ronco, A.; Kofler, M., *Helv. Chim. Acta*, (1947) **30**, 1911.

[24] Borodziński, A.; Bond, G. C., *Catal. Rev.: Sci. Eng.*, (2006) **48**, 91.

[25] McCue, A. J.; Anderson, J. A., *Front. Chem. Sci. Eng.*, (2015) **9**, 142.

[26] Bridier, B.; López, N.; Pérez-Ramírez, J., *Dalton Trans.*, (2010) **39**, 8412.

[27] Borodziński, A.; Bond, G. C., *Catal. Rev.: Sci. Eng.*, (2008) **50**, 379.

[28] Crespo-Quesada, M.; Cárdenas-Lizana, F.; Dessimoz, A.-L.; Kiwi-Minsker, L., *ACS Catal.*, (2012) **2**, 1773.

[29] Oger, C.; Balas, L.; Durand, T.; Galano, J.-M., *Chem. Rev.*, (2013) **113**, 1313.

[30] Vernuccio, S.; Rudolf von Rohr, P.; Medlock, J., *Ind. Eng. Chem. Res.*, (2015) **54**, 11543.

[31] Ulan, J. G.; Kuo, E.; Maier, W. F.; Rai, R. S.; Thomas, G., *J. Org. Chem.*, (1987) **52**, 3126.

[32] Schlägl, R.; Noack, K.; Zbinden, H.; Reller, A., *Helv. Chim. Acta*, (1987) **70**, 627.

[33] Albers, P. W.; Möbus, K.; Frost, C. D.; Parker, S. F., *J. Phys. Chem. C*, (2011) **115**, 24485.

[34] Rajaram, J.; Narula, A. P. S.; Chawla, H. P. S.; Dev, S., *Tetrahedron*, (1983) **39**, 2315.

[35] Armbrüster, M.; Schlögl, R.; Grin, Y., *Sci. Technol. Adv. Mater.*, (2014) **15**, 034803.

[36] Cherkasov, N.; Ibhadon, A. O.; Rebrov, E. V., *Lab Chip*, (2015) **15**, 1952.

[37] Cherkasov, N.; Ibhadon, A. O.; McCue, A. J.; Anderson, J. A.; Johnston, S. K., *Appl. Catal., A*, (2015) **497**, 22.

[38] Klasovsky, F.; Claus, P.; Wolf, D., *Top. Catal.*, (2009) **52**, 412.

[39] Bonrath, W.; Buss, A.; Medlock, J. A.; Mueller, T., WO 2013190076, (2013).

[40] Crespo-Quesada, M.; Yarulin, A.; Jin, M.; Xia, Y.; Kiwi-Minsker, L., *J. Am. Chem. Soc.*, (2011) **133**, 12787.

[41] Wu, Z.; Borretto, E.; Medlock, J.; Bonrath, W.; Cravotto, G., *ChemCatChem*, (2014) **6**, 2762.

[42] Witte, P. T., WO 2009096783, (2009).

[43] Witte, P. T.; de Groen, M.; de Rooij, R. M.; Bakermans, P.; Donkervoort, H. G.; Berben, P. H.; Geus, J. W., *Stud. Surf. Sci. Catal.*, (2010) **175**, 135.

[44] Witte, P. T.; Berben, P. H.; Boland, S.; Boymans, E. H.; Vogt, D.; Geus, J. W.; Donkervoort, J. G., *Top. Catal.*, (2012) **55**, 505; material modified from original article published by Springer (creativecommons.org/licenses/by/3.0/legalcode).

[45] Gildemeister, E.; Hoffmann, F., *Die ätherischen Öle*; Akademie Verlag: Berlin, (1960).

[46] Witte, P. T.; Boland, S.; Kirby, F.; van Maanen, R.; Bleeker, B. F.; de Winter, D. A. M.; Post, J. A.; Geus, J. W.; Berben, P. H., *ChemCatChem*, (2013) **5**, 582.

[47] Albani, D.; Vilé, G.; Mitchell, S.; Witte, P. T.; Almora-Barrios, N.; Verel, R.; López, N.; Pérez-Ramírez, J., *Catal. Sci. Technol.*, (2016) **6**, 1621.

[48] Albani, D.; Vilé, G.; Beltran Toro, M. A.; Kaufmann, R.; Mitchell, S.; Pérez-Ramírez, J., *React. Chem. Eng.*, (2016) **1**, 454.

[49] Vilé, G.; Bridier, B.; Wichert, J.; Pérez-Ramírez, J., *Angew. Chem.*, (2012) **124**, 8748; *Angew. Chem. Int. Ed.*, (2012) **51**, 8620.

[50] Vilé, G.; Wrabetz, S.; Floryan, L.; Schuster, M. E.; Girgsdies, F.; Teschner, D.; Pérez-Ramírez, J., *ChemCatChem*, (2014) **6**, 1928.

[51] Fedorov, A.; Liu, H.-J.; Lo, H.-K.; Copéret, C., *J. Am. Chem. Soc.*, (2016) **138**, 16502.

[52] Semagina, N.; Grasemann, M.; Xanthopoulos, N.; Renken, A.; Kiwi-Minsker, L., *J. Catal.*, (2007) **251**, 213.

[53] Crespo-Quesada, M.; Grasemann, M.; Semagina, N.; Renken, A.; Kiwi-Minsker, L., *Catal. Today*, (2009) **147**, 247.

[54] Elias, Y.; Rudolf von Rohr, P.; Bonrath, W.; Medlock, J.; Buss, A., *Chem. Eng. Process.*, (2015) **95**, 175.

[55] Vernuccio, S.; Goy, R.; Rudolf von Rohr, P.; Medlock, J.; Bonrath, W., *React. Chem. Eng.*, (2016) **1**, 445.

[56] Khai, B. T.; Arcelli, A., *Chem. Ber.*, (1993) **126**, 2265.

[57] Weir, J. R.; Patel, B. A.; Heck, R. F., *J. Org. Chem.*, (1980) **45**, 4926.

[58] Aramendía, M. A.; Borau, V.; Jiménez, C.; Marinas, J. M.; Porras, A.; Urbano, F. J.; Villar, L., *J. Mol. Catal.*, (1994) **94**, 131.

[59] Yan, M.; Jin, T.; Ishikawa, Y.; Minato, T.; Fujita, T.; Chen, L.-Y.; Bao, M.; Asao, N.; Chen, M.-W.; Yamamoto, Y., *J. Am. Chem. Soc.*, (2012) **134**, 17536.

[60] Wagh, Y. S.; Asao, N., *J. Org. Chem.*, (2015) **80**, 847.

[61] Kluwer, A. M.; Elsevier, C. J., In *The Handbook of Homogeneous Hydrogenation*, de Vries, J. G.; Elsevier, C. J., Eds.; Wiley-VCH: Weinheim, Germany, (2007); Vol. 1, pp 375–394.

[62] Schrock, R. R.; Osborn, J. A., *J. Am. Chem. Soc.*, (1976) **98**, 2134.

[63] Schrock, R. R.; Osborn, J. A., *J. Am. Chem. Soc.*, (1976) **98**, 2143.

[64] Esteruelas, M. A.; González, I.; Herrero, J.; Oro, L. A., *J. Organomet. Chem.*, (1998) **551**, 49.

[65] de Wolf, E.; Spek, A. L.; Kuipers, B. W. M.; Philipse, A. P.; Meeldijk, J. D.; Bomans, P. H. H.; Frederik, P. M.; Deelman, B.-J.; van Koten, G., *Tetrahedron*, (2002) **58**, 3911.

[66] Burk, M. J.; Gerlach, A.; Semmeril, D., *J. Org. Chem.*, (2000) **65**, 8933.

[67] Al-Rawashdeh, M.; Zalucky, J.; Müller, C.; Nijhuis, T. A.; Hessel, V.; Schouten, J. C., *Ind. Eng. Chem. Res.*, (2013) **52**, 11516.

[68] Kohrt, C.; Wienhöfer, G.; Pribbenow, C.; Beller, M.; Heller, D., *ChemCatChem*, (2013) **5**, 2818.

[69] van Laren, M. W.; Elsevier, C. J., *Angew. Chem.*, (1999) **111**, 3926; *Angew. Chem. Int. Ed.*, (1999) **38**, 3715.

[70] van Asselt, R.; Elsevier, C. J.; Smeets, W. J. J.; Spek, A. L.; Benedix, R., *Recl. Trav. Chim. Pays-Bas*, (1994) **113**, 88.

[71] van Asselt, R.; Elsevier, C. J.; Smeets, W. J. J.; Spek, A. L., *Inorg. Chem.*, (1994) **33**, 1521.

[72] Kluwer, A. M.; Koblenz, T. S.; Jonischkeit, T.; Woelk, K.; Elsevier, C. J., *J. Am. Chem. Soc.*, (2005) **127**, 15470.

[73] van Laren, M. W.; Duin, M. A.; Klerk, C.; Naglia, M.; Rogolino, D.; Pelagatti, P.; Bacchi, A.; Pelizzi, C.; Elsevier, C. J., *Organometallics*, (2002) **21**, 1546.

[74] Sprengers, J. W.; Wassenaar, J.; Clement, N. D.; Cavell, K. J.; Elsevier, C. J., *Angew. Chem.*, (2005) **117**, 2062; *Angew. Chem. Int. Ed.*, (2005) **44**, 2026.

[75] Nishibayashi, R.; Kurahashi, T.; Matsubara, S., *Synlett*, (2014) **25**, 1287.

[76] Giachi, G.; Oberhauser, W.; Frediani, M.; Passaglia, E.; Capozzoli, L.; Rosi, L., *J. Polym. Sci., Part A: Polym. Chem.*, (2013) **51**, 2518.

[77] Sodeoka, M.; Shibasaki, M., *J. Org. Chem.*, (1985) **50**, 1147.

[78] Sodeoka, M.; Shibasaki, M., *Synthesis*, (1993), 643.
[79] Pape, F.; Thiel, N. O.; Teichert, J. F., *Chem.–Eur. J.*, (2015) **21**, 15 934.
[80] Semba, K.; Kameyama, R.; Nakao, Y., *Synlett*, (2015) **26**, 318.
[81] Wakamatsu, T.; Nagao, K.; Ohmiya, H.; Sawamura, M., *Organometallics*, (2016) **35**, 1354.
[82] Gianetti, T. L.; Tomson, N. C.; Arnold, J.; Bergman, R. G., *J. Am. Chem. Soc.*, (2011) **133**, 14 904.
[83] La Pierre, H. S.; Arnold, J.; Toste, F. D., *Angew. Chem.*, (2011) **123**, 3986; *Angew. Chem. Int. Ed.*, (2011) **50**, 3900.
[84] Albers, M. O.; Singleton, E.; Viney, M. M., *J. Mol. Catal.*, (1985) **30**, 213.
[85] Nkosi, B. S.; Coville, N. J.; Albers, M. O.; Singleton, E., *J. Mol. Catal.*, (1987) **39**, 313.
[86] Lough, A. J.; Morris, R. H.; Ricciuto, L.; Schleis, T., *Inorg. Chim. Acta*, (1998) **270**, 238.
[87] Liu, Y.; Hu, L.; Chen, H.; Du, H., *Chem.–Eur. J.*, (2015) **21**, 3495.
[88] Esteruelas, M. A.; López, A. M.; Oro, L. A.; Pérez, A.; Schulz, M.; Werner, H., *Organometallics*, (1993) **12**, 1823.
[89] Andriollo, A.; Esteruelas, M. A.; Meyer, U.; Oro, L. A.; Sánchez-Delgado, R. A.; Sola, E.; Valero, C.; Werner, H., *J. Am. Chem. Soc.*, (1989) **111**, 7431.
[90] Esteruelas, M. A.; Oro, L. A., *Chem. Rev.*, (1998) **98**, 577.
[91] Bianchini, C.; Meli, A.; Peruzzini, M.; Vizza, F.; Zanobini, F.; Frediani, P., *Organometallics*, (1989) **8**, 2080.
[92] Bianchini, C.; Meli, A.; Peruzzini, M.; Frediani, P.; Bohanna, C.; Esteruelas, M. A.; Oro, L. A., *Organometallics*, (1992) **11**, 138.
[93] Ashworth, T. V.; Singleton, E.; Hough, J. J., *J. Chem. Soc., Dalton Trans.*, (1977), 1809.
[94] Hauwert, P.; Maestri, G.; Sprengers, J. W.; Catellani, M.; Elsevier, C. J., *Angew. Chem.*, (2008) **120**, 3267; *Angew. Chem. Int. Ed.*, (2008) **47**, 3223.
[95] Hauwert, P.; Boerleider, R.; Warsink, S.; Weigand, J. J.; Elsevier, C. J., *J. Am. Chem. Soc.*, (2010) **132**, 16 900.
[96] Hauwert, P.; Dunsford, J. J.; Tromp, D. S.; Weigand, J. J.; Lutz, M.; Cavell, K. J.; Elsevier, C. J., *Organometallics*, (2013) **32**, 131.
[97] Drost, R. M.; Bouwens, T.; van Leest, N. P.; de Bruin, B.; Elsevier, C. J., *ACS Catal.*, (2014) **4**, 1349.
[98] Shen, R.; Chen, T.; Zhao, Y.; Qiu, R.; Zhou, Y.; Yin, S.; Wang, X.; Goto, M.; Han, L.-B., *J. Am. Chem. Soc.*, (2011) **133**, 17 037.
[99] Li, J.; Hua, R.; Liu, T., *J. Org. Chem.*, (2010) **75**, 2966.
[100] Pasto, D. J., In *Comprehensive Organic Synthesis*, Trost, B. M.; Fleming, I., Eds.; Pergamon: Oxford, (1991); Vol. 8, pp 471–488.
[101] Trost, B. M.; Ball, Z. T., *Synthesis*, (2005), 853.
[102] Rummelt, S. M.; Preindl, J.; Sommer, H.; Fürstner, A., *Angew. Chem.*, (2015) **127**, 6339; *Angew. Chem. Int. Ed.*, (2015) **54**, 6241.
[103] Frihed, T. G.; Fürstner, A., *Bull. Chem. Soc. Jpn.*, (2016) **89**, 135.
[104] Michaelides, I. N.; Dixon, D. J., *Angew. Chem.*, (2013) **125**, 836; *Angew. Chem. Int. Ed.*, (2013) **52**, 806.
[105] Schleyer, D.; Niessen, H. G.; Bargon, J., *New J. Chem.*, (2001) **25**, 423.
[106] Bowers, C. R.; Weitekamp, D. P., *Phys. Rev. Lett.*, (1986) **57**, 2645.
[107] Radkowski, K.; Sundararaju, B.; Fürstner, A., *Angew. Chem.*, (2013) **125**, 373; *Angew. Chem. Int. Ed.*, (2013) **52**, 355.
[108] Fuchs, M.; Fürstner, A., *Angew. Chem.*, (2015) **127**, 4050; *Angew. Chem. Int. Ed.*, (2015) **54**, 3978.
[109] Leutzsch, M.; Wolf, L. M.; Gupta, P.; Fuchs, M.; Thiel, W.; Farès, C.; Fürstner, A., *Angew. Chem.*, (2015) **127**, 12 608; *Angew. Chem. Int. Ed.*, (2015) **54**, 12 431.
[110] Neumann, K. T.; Burhardt, M. N.; Skrydstrup, T.; Klimczyk, S.; Lindhardt, A. T.; Bang-Andersen, B., *ACS Catal.*, (2016) **6**, 4710.
[111] Higashida, K.; Mashima, K., *Chem. Lett.*, (2016) **45**, 866.
[112] Karunananda, M. K.; Mankad, N. P., *J. Am. Chem. Soc.*, (2015) **137**, 14 598.
[113] Srimani, D.; Diskin-Posner, Y.; Ben-David, Y.; Milstein, D., *Angew. Chem.*, (2013) **125**, 14 381; *Angew. Chem. Int. Ed.*, (2013) **52**, 14 131.
[114] Tokmic, K.; Fout, A. R., *J. Am. Chem. Soc.*, (2016) **138**, 13 700.
[115] Tani, K.; Iseki, A.; Yamagata, T., *Chem. Commun. (Cambridge)*, (1999), 1821.
[116] Li, J.; Hua, R., *Chem.–Eur. J.*, (2011) **17**, 8462.
[117] Chen, T.; Xiao, J.; Zhou, Y.; Yin, S.; Han, L.-B., *J. Organomet. Chem.*, (2014) **749**, 51.
[118] Richmond, E.; Moran, J., *J. Org. Chem.*, (2015) **80**, 6922.
[119] Netherton, M. R.; Dai, C.; Neuschütz, K.; Fu, G. C., *J. Am. Chem. Soc.*, (2001) **123**, 10 099.

[120] Crabbé, P.; Nassim, B.; Robert-Lopes, M.-T., *Org. Synth., Coll. Vol. VII*, (1990), 276.
[121] Skattebøl, L., *J. Org. Chem.*, (1966) **31**, 2789.
[122] Nativi, C.; Ricci, A.; Taddei, M., *Tetrahedron Lett.*, (1987) **28**, 2751.
[123] Brandsma, L., In *Synthesis of Acetylenes, Allenes and Cumulenes: Methods and Techniques*; Elsevier: Oxford, (2004).
[124] Marbet, R.; Saucy, G., *Helv. Chim. Acta*, (1967) **50**, 2095.
[125] Kajikawa, T.; Aoki, K.; Iwashita, T.; Niedzwiedzki, D. M.; Frank, H. A.; Katsumura, S., *Org. Biomol. Chem.*, (2010) **8**, 2513.
[126] Olpp, T.; Brückner, R., *Angew. Chem.*, (2006) **118**, 4128; *Angew. Chem. Int. Ed.*, (2006) **45**, 4023.
[127] Niedzwiedzki, D. M.; Kajikawa, T.; Aoki, K.; Katsumura, S.; Frank, H. A., *J. Phys. Chem. B*, (2013) **117**, 6874.
[128] Kajikawa, T.; Hasegawa, S.; Iwashita, T.; Kusumoto, T.; Hashimoto, H.; Niedzwiedzki, D. M.; Frank, H. A.; Katsumura, S., *Org. Lett.*, (2009) **11**, 5006.
[129] Kistiakowsky, G. B.; Ruhoff, J. R.; Smith, H. A.; Vaughan, W. E., *J. Am. Chem. Soc.*, (1936) **58**, 146.
[130] Bond, G. C.; Webb, G.; Wells, P. B.; Winterbottom, J. M., *J. Catal.*, (1962) **1**, 74.
[131] Mann, R. S.; To, D. E., *Can. J. Chem.*, (1968) **46**, 161.
[132] Crombie, L.; Jenkins, P. A.; Mitchard, D. A., *J. Chem. Soc., Perkin Trans. 1*, (1975), 1081.
[133] Bond, G. C.; Sheridan, J., *Trans. Faraday Soc.*, (1952) **48**, 658.
[134] García Ruano, J. L.; Marcos, V.; Alemán, J., *Synthesis*, (2009), 3339.
[135] Schulte-Elte, K. H.; Giersch, W.; Winter, B.; Pamingle, H.; Ohloff, G., *Helv. Chim. Acta*, (1985) **68**, 1961.
[136] Price, J. D.; Johnson, R. P., *Tetrahedron Lett.*, (1986) **27**, 4679.
[137] Marshall, J. A.; Rothenberger, S. D., *Tetrahedron Lett.*, (1986) **27**, 4845.
[138] Bhagwat, M. M.; Devaprabhakara, D., *Tetrahedron Lett.*, (1972) **13**, 1391.
[139] Crombie, L.; Jenkins, P. A.; Roblin, J., *J. Chem. Soc., Perkin Trans. 1*, (1975), 1099.
[140] Bonrath, W.; Kircher, T.; Kuenzi, R.; Tschumi, J., WO 2006 029 737, (2006).
[141] Dobler, W.; Bahr, N.; Kindler, A.; Miller, C.; Salden, A., WO 2005 056 508, (2005).
[142] Göbbel, H.-G.; Kaibel, G.; Miller, C.; Dobler, W.; Dirnsteiner, T.; Hahn, T.; Breuer, K.; Aquila, W., WO 2004 007 413, (2004).
[143] Jansen, M.; Rehren, C., EP 841 314, (1998).
[144] Sul'mann, E. M.; Ankudinova, T. V.; Sidorov, A. I., *Khim.-Farm. Zh.*, (1990) **24**, 76.
[145] Alder, A. P.; Wolf, H. R.; Jeger, O., *Helv. Chim. Acta*, (1980) **63**, 1833.
[146] Pover, K. A.; Scheinmann, F., *J. Chem. Soc., Perkin Trans. 1*, (1980), 2338.

1.6 Catalytic Reduction of Phenols, Alcohols, and Diols

S. Tin and J. G. de Vries

General Introduction

The catalytic deoxygenation of organic molecules has attracted a lot of attention in recent years because of interest in the use of biomass-derived fuels and chemicals. The raw materials used may contain up to 50 wt% of oxygen.[1] In this chapter, some practical methods for the selective catalytic hydrodeoxygenation of phenols and alcohols, and deoxydehydration of diols (Scheme 1) by using hydrogen gas or transfer-hydrogenation methods will be described.

Scheme 1 Hydrodeoxygenation of Alcohols and Deoxydehydration of Diols

$$R^1OH \xrightarrow{H_2,\ catalyst} R^1H$$

$$R^1CH(OH)CH(OH)R^2 \xrightarrow[catalyst]{hydrogen\ source} R^1CH{=}CHR^2$$

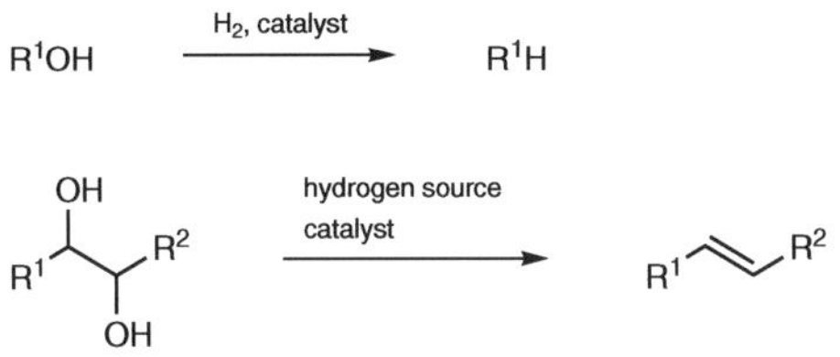

SAFETY: Hydrogen gas is extremely flammable. Benzene, which is formed by the reduction of phenol, is a highly toxic and carcinogenic molecule. Alkenes and alkanes, which are formed in some reactions, are highly flammable. Also, care must be taken regarding the pressure in the reactors if low-molecular-weight hydrocarbons are synthesized.

1.6.1 Hydrodeoxygenation of Phenols To Give Arenes

The drawback of the hydrodeoxygenation of phenols is that many of the existing methods produce either mixtures of aromatic compounds or the corresponding saturated products, with high selectivities for the undesired saturated products.[2–8] For example, the reduction of phenol can result in the formation of benzene, cyclohexanone, cyclohexanol, cyclohexene, cyclohexane, and other products. Examples of the selective reduction of the C—OH bond in phenol (or alkyl-substituted phenols) to a C—H bond with retention of the aromatic moiety by using hydrogen gas, where the selectivity for the desired arene is over 80%, are very rare.[4,6,9,10] The most useful method, to the best of our knowledge, is the one published by Austin, Grabow, and co-workers that employs ruthenium nanoparticles on a titanium(IV) oxide support as catalyst (Scheme 2).[10] Although a method for isolation of the desired product, benzene (**1**), is not provided in the experimental section of the report, it can be easily distilled from remaining phenol and other side products with >95% purity. It was also demonstrated that water is a crucial additive. If no additive is used, only 38% selectivity toward benzene is observed. If a nonpolar additive, such as octane, is used, only 40% selectivity toward the desired product is achieved. In addition, the conversion of phenol is lower in the latter two cases.

for references see p 240

Scheme 2 Selective Hydrogenolysis of Phenol to Benzene Using a Ruthenium Catalyst[10]

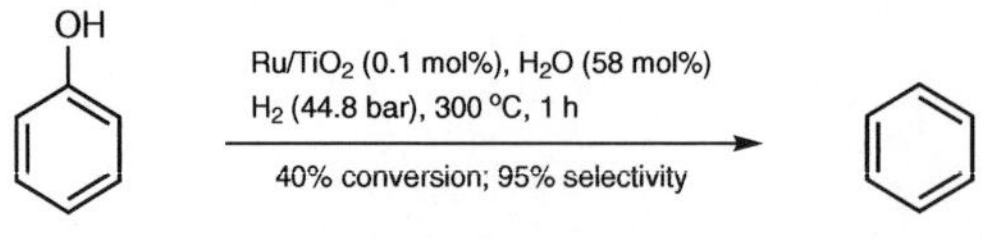

OH

Ru/TiO_2 (0.1 mol%), H_2O (58 mol%)
H_2 (44.8 bar), 300 °C, 1 h

40% conversion; 95% selectivity

1

For the hydrogenolysis of polyhydroxybenzenes, a heterogeneous rhodium/alumina catalyst can be used to cleave only a single C—O bond to give the products, e.g. **2**, in moderate yields (Scheme 3).[11] Although the yields are only moderate, it was shown for one example that a substrate bearing a methoxy group instead of a hydroxy group results in a high yield of the product of mono-deoxygenation at that position (this type of transformation is outside the scope of this chapter).

Scheme 3 Hydrogenolysis of Polyhydroxybenzenes[11]

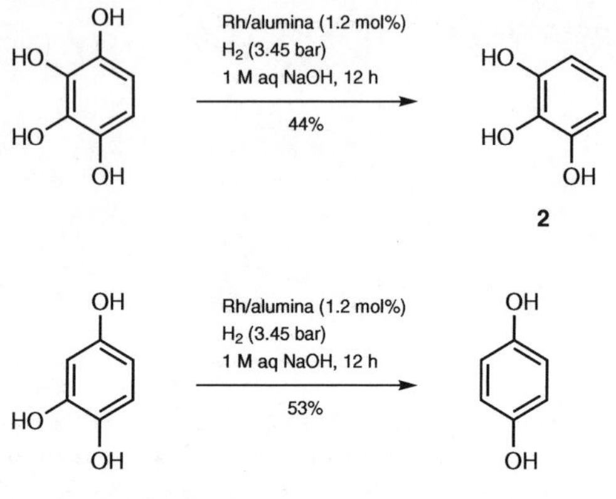

OH HO HO OH

Rh/alumina (1.2 mol%)
H_2 (3.45 bar)
1 M aq NaOH, 12 h

44%

HO HO OH

2

OH HO OH

Rh/alumina (1.2 mol%)
H_2 (3.45 bar)
1 M aq NaOH, 12 h

53%

OH OH

Benzene (1); Typical Procedure:[10]

Ru/TiO_2[6] (~5 wt% Ru; 100 mg, 49.5 μmol) was introduced into a reactor along with liquefied phenol (~10 wt% H_2O; 5 g, 47.8 mmol), and H_2 was bubbled through the resulting soln for 10 min. After this time, the reactor was purged three times with H_2 (5.2 bar). The mixture was heated to 300 °C with stirring, and the pressure was brought to 44.8 bar by pressurization with H_2. During the reaction time, the pressure of H_2 was kept constant. After 1 h, the reactor was cooled to rt, H_2 gas pressure was released, and the sample was frozen.

1,2,3-Trihydroxybenzene (2); Typical Procedure:[11]

5% Rh/alumina (0.125 g, 0.060 mmol of Rh) was added to a soln of 1,2,3,4-tetrahydroxybenzene (0.700 g, 5.00 mmol) in 1.0 M aq NaOH (5 mL), and the resulting suspension was shaken for 12 h under pressure of H_2 (3.45 bar). After this time, the suspension was filtered through Celite, and the pH was adjusted to 6.0 by the addition of 10% HCl. The soln was concentrated under reduced pressure, and the resulting oil was dissolved in 0.5 M H_2SO_4 (20 mL); the soln was heated under reflux under an atmosphere of argon for 12 h. The soln was cooled to rt, the organics were extracted with Et_2O (4 × 50 mL), and the solvent was removed under reduced pressure. Kugelrohr distillation at 90 °C afforded a white crystalline material; yield: 0.277 g (44%).

1.6.2 Hydrodeoxygenation of Aliphatic Alcohols To Give Alkanes

Although the hydrogenolysis of the C—OH bond in aliphatic alcohols to a C—H bond is well explored, with glycerol being the most studied substrate, the current reports in the literature generally do not include isolation methods for the products, and typically only GC or NMR yields are reported.[12,13] Notably, some alternative multistep routes reported for the reduction of aliphatic alcohols to alkanes do include procedures with yields of the isolated products (e.g., via silylation or dehydrogenation/Wolff–Kishner reduction).[14,15] In this section, only those examples of one-pot catalytic hydrogenolysis that at least report GC yields are explored. The selection of examples is based on the conversion of the starting material and the selectivity for the desired deoxygenated product, but it should be kept in mind that in the selected examples, the isolation of the products, for example by distillation, should not be very difficult.

Zhu and co-workers have reported a heterogeneous copper catalyst that can be used to prepare propane-1,2-diol (**3**) from glycerol with high selectivity (Scheme 4).[16] Interestingly, it was shown that the boron content of the catalyst is important; the absence (or overabundance) of boron results in lower selectivities. The optimum catalyst, 3CuB/silica, affords full conversion of glycerol with 98% selectivity for the desired product.

Scheme 4 Highly Selective Deoxygenation of Glycerol to Propane-1,2-diol Using a Copper Catalyst[16]

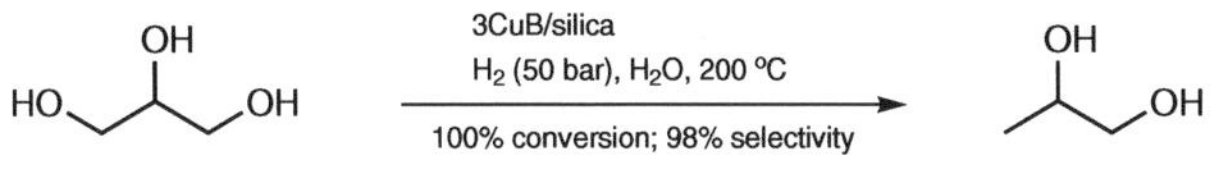

HO–CH₂–CH(OH)–CH₂–OH → (3CuB/silica; H_2 (50 bar), H_2O, 200 °C; 100% conversion; 98% selectivity) → propane-1,2-diol **3**

Another example of the highly selective hydrogenolysis of an alcohol is the formation of propan-1-ol (**4**) from propane-1,3-diol in the presence of 2.0Pt/WZ10(700) as a catalyst [2 wt% platinum, 10 wt% tungsten(VI) oxide on a zirconium(IV) oxide support with a calcination temperature of 700 °C].[17] As shown in Scheme 5, only partial conversion of propane-1,3-diol is achieved at 140 °C, but the selectivity toward the desired product, propan-1-ol, is >99.9%. The reaction was also tested at other temperatures (120 and 130 °C), at which lower conversions are achieved (12 and 22%, respectively), but propan-1-ol remains the only product of the reaction. The same catalyst has been employed to perform the hydrogenolysis of propane-1,2-diol. Whereas the conversion for this reaction is better than that for the hydrogenolysis of propane-1,3-diol under the same conditions, the product is a mixture of propan-1-ol and propan-2-ol.

Scheme 5 Selective Formation of Propan-1-ol by Hydrogenolysis of Propane-1,3-diol[17]

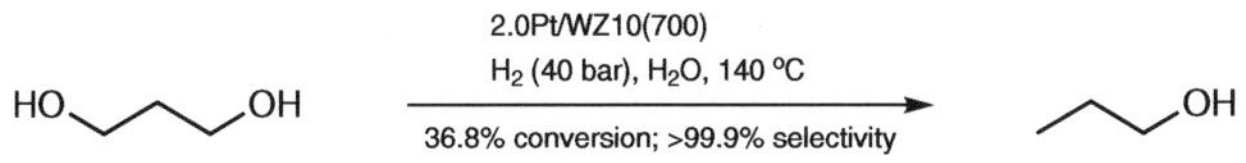

HO–(CH₂)₃–OH → (2.0Pt/WZ10(700); H_2 (40 bar), H_2O, 140 °C; 36.8% conversion; >99.9% selectivity) → propan-1-ol **4**

For the case in which full deoxygenation of glycerol is required, the procedure described by Taher and co-workers affords propane gas (**6**) from glycerol in 100% yield (Scheme 6).[18] In this method, homogeneous ruthenium catalyst **5**, bearing a terpyridine ligand, is employed under acidic conditions at 250 °C. One can safely assume that the catalyst does not

for references see p 240

tolerate these conditions, and that the true catalyst is ruthenium metal or ruthenium nanoparticles. Although the product was not isolated, it can be easily collected as a gas fraction and, owing to the absolute selectivity, does not require any purification.

Scheme 6 Absolute Selectivity for Propane in the Deoxygenation of Glycerol[18]

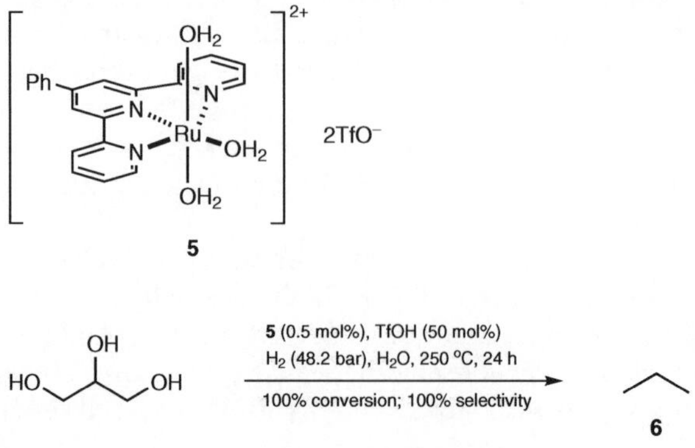

5 (0.5 mol%), TfOH (50 mol%)
H_2 (48.2 bar), H_2O, 250 °C, 24 h
100% conversion; 100% selectivity

Other examples of the hydrogenolysis of alcohols to alkanes have been reported, but these methods usually have either poor selectivities or very low conversions. For example, perfect selectivity is achieved in the one-step conversion of hexane-1,2,6-triol into hexane-1,6-diol, but the conversion is only 8.1% using a Rh–ReO_x catalyst.[19] Because the difference in the boiling points between the starting material and the product is large (hexane-1,2,6-triol: bp 172 °C at 3 Torr[20] or 577 °C at atmospheric pressure; hexane-1,6-diol: bp 253–260 °C at atmospheric pressure[21]), hexane-1,6-diol can be isolated by distillation. Attempts to improve the conversion by increasing the catalyst loading from a catalyst/substrate mass ratio of 1:9 to 2:7 results in improved conversion (59.3%), but the selectivity decreases significantly (61.9% for hexane-1,6-diol). The same reaction has also been described by de Vries, Heeres, and co-workers by using a Rh–Re/silica catalyst. In this case, hexane-1,2,6-triol is converted into hexane-1,6-diol with 17% conversion and absolute selectivity.[22]

Examples of the hydrogenolysis of the vicinal diol groups in glucaric acid, mucic acid, or the diesters of these acids mostly proceed through a two-step process, through which the dialkene is formed and then reduced with hydrogen gas to afford adipic acid (**7**) (multi-pot reactions are outside of the scope of this chapter);[23–27] one-pot processes are rare.[28,29] A good example where the process is performed in one pot (using a rhodium/platinum on silica catalyst) is described in a patented procedure (Scheme 7).[28]

Scheme 7 One-Pot Procedure To Convert Glucaric Acid into Adipic Acid[28]

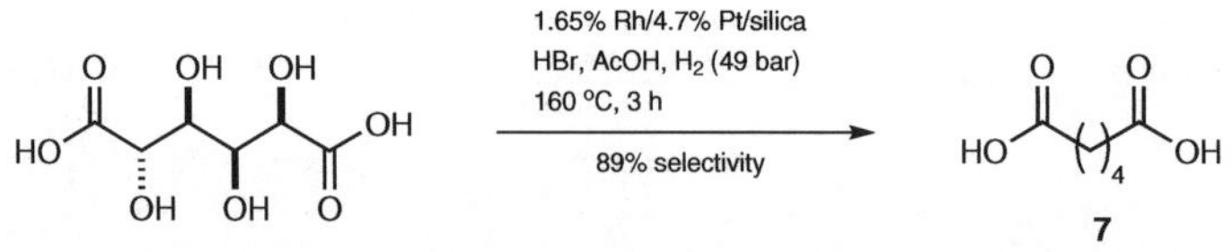

1.65% Rh/4.7% Pt/silica
HBr, AcOH, H_2 (49 bar)
160 °C, 3 h
89% selectivity

Propane-1,2-diol (3); Typical Procedure:[16]

3CuB/SiO_2 (4.0 g) was loaded into a vertical fixed-bed reactor (inner diameter: 12 mm; length: 600 mm) and was reduced in situ by H_2 gas flow (100 $cm^3 \cdot min^{-1}$) at 250 °C for 2 h. After this time, the reactor was cooled to 200 °C and pressurized with H_2 gas (50 bar). A 10 wt% aq soln of glycerol was continuously pumped into the reactor, and the reactor was co-fed with H_2 gas at 150 $cm^3 \cdot min^{-1}$. The products were condensed and collected in a gas–liquid separator at 0 °C. The sample was analyzed by GC. Although a procedure for the isolation of the product was not reported, the obtained product was already 98% pure. It could be purified by distillation from some of the identified side products, such as 1-hydroxypropan-2-one, as the boiling point of propane-1,2-diol is 188 °C[30] and the boiling point of 1-hydroxypropan-2-one is 50 °C at 20 Torr (145 °C at atmospheric pressure).[31]

Propan-1-ol (4); Typical Procedure:[17]
The 2.0Pt/WZ10(700) catalyst (2.0 mL) was loaded into a fixed-bed down-flow reactor (inner diameter: 10.0 mm; length: 340 mm) and was reduced by H_2 gas flow (50 $cm^3 \cdot min^{-1}$) starting at rt with heating to 250 °C at a rate of 0.5 $°C \cdot min^{-1}$. Once the temperature reached 250 °C, activation of the catalyst was continued for 1 h. After this time, the reactor was cooled to 140 °C and pressurized with H_2 gas (40 bar). A 60 wt% aq soln of propane-1,3-diol was continuously fed into the reactor at a rate of 0.5 $mL \cdot h^{-1}$, and the reactor was co-fed with H_2 at a rate of 10 $cm^3 \cdot min^{-1}$ for 24 h. At the outlet of the reactor, the products were condensed by cooling and were collected in a gas–liquid separator. The sample was analyzed by GC. Although a procedure for the isolation of the product was not reported, the resulting mixture contained only propane-1,3-diol and the title compound. The mixture could be separated by distillation, as the difference in the boiling points between the two alcohols is large (propane-1,2-diol: bp 188 °C;[30] propan-1-ol: bp 97.1 °C[32]).

Propane (6); Typical Procedure:[18]
A stock soln containing 0.5 M glycerol, 0.1 M dimethyl sulfone as an internal standard, 2.5 mM **5**, and 0.25 M aq TfOH was prepared. The stock soln (1.0 mL; 0.5 mmol of glycerol, 0.25 mmol of TfOH, 2.5 μmol of Ru) was loaded into a tube, and the tube was loaded into a parallel reactor. The reactor was sealed and evacuated for 2 min using a water pump, and was then pressurized with H_2 (75.8 bar). This evacuation cycle was repeated two more times, and the reactor was brought under H_2 pressure (48.2 bar). The soln was vigorously stirred at 250 °C for 24 h. After this time, the reactor was cooled in an ice bath for 30 min, which was followed by cooling for 5 min in a dry ice/acetone bath. The reactor was vented, warmed to rt, and a change in the pressure was noted. The reactor was opened to air, and a GC sample was taken.

Adipic Acid (7); Typical Procedure:[28]
The 1.65% Rh/4.7% Pt/silica catalyst (8 mg) was loaded into a glass vial, and a soln of 0.2 M glucaric acid/0.2 M HBr in AcOH (250 μL) was added. The vial was covered with a Teflon pin-hole sheet, a silicone pin-hole mat, and a steel gas-diffusion plate. The vial was placed into a pressure vessel; the vessel was pressurized and vented three times with N_2 gas, which was followed by purging three times with H_2. The pressure was brought to 49 bar of H_2, and the vial was heated to 160 °C and shaken for 3 h. After this time, the reactor was cooled to rt, vented, and purged with N_2, and H_2O (750 μL) was added. The vial was shaken to ensure adequate mixing and was placed in a centrifuge to separate the catalyst particles. The mixture was diluted twofold with H_2O and subjected to analysis by HPLC.

1.6.3 Reduction of Diols To Give Alkenes

1.6.3.1 Reduction of Diols by Metal-Free Catalysis

Although the deoxydehydration of vicinal diols often requires a transition-metal catalyst,[33,34] there are examples of methods where the latter is not required. One such practical method involves the use of formic acid as both the source of hydrogen and the catalyst for the reduction of vicinal diol groups.[35] Some examples with good to high yields of products **8** are shown in Table 1. Notably, the side products of this reaction are water

for references see p 240

and carbon dioxide (see Scheme 8 for the proposed mechanism). For cases in which the substrate contains additional hydroxy groups (e.g., glycerol, entries 1 and 2), the formation of allyl formate is also observed, but it can easily be separated from the product (see below for details). The conversion of glycerol into prop-2-en-1-ol can even be performed on a 500-gram scale (5.4 mol of starting material); 265 grams (84% yield) of the desired product was obtained. This proves that the method is scalable. It has also been shown that the configuration of the newly formed double bond depends on the relative stereochemistry of the hydroxy groups in the starting material. In the deoxydehydration of decane-3,4-diol, either the *E*- or *Z*-product is obtained selectively, depending on the stereochemistry of the starting material (entries 9 and 10). This suggests that exclusively *syn*-elimination of the hydroxy groups occurs.

Cyclic diols can also be converted successfully. In all cases, the hydroxy groups are *cis* to each other. All attempts to perform the reactions with cyclic *trans*-diols have been unsuccessful. This observation is consistent with that observed for the deoxygenation of the two different diastereomers of decane-3,4-diol discussed above. Although this property of the transformation enables access to a specific geometry of the alkene product by appropriate choice of the starting material, it also means that the method is limited to *cis*-diols if cyclic alcohols are used as starting materials.

Table 1 Reduction of Vicinal Diols to Alkenes Using Formic Acid[35]

HO OH R1 R4 R2 R3 → HCO_2H (1.44 equiv), 230–240 °C → R1 R4 R3 R2 **8**

Entry	Reactant	Reaction Scale (mmol)	Product	Yield (%)	Ref
1	HO, OH, OH	50	OH	89	[35]
2	HO, OH, OH	150	OH	82	[35]
3	HO, OH, ( )5	20	( )5	91	[35]
4	HO, OH, ( )7	20	( )7	93	[35]
5	OH, OH	20		78	[35]
6	OH, OH	25		96	[35]
7	OH, OH	20		86	[35]

Table 1 (cont.)

Entry	Reactant	Reaction Scale (mmol)	Product	Yield (%)	Ref
8	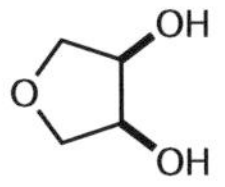	15	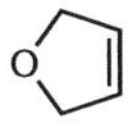	87	[35]
9	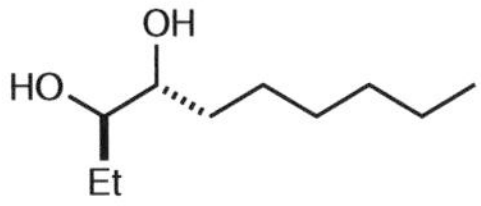	10		83	[35]
10	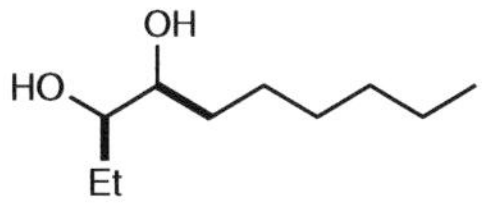	8.1	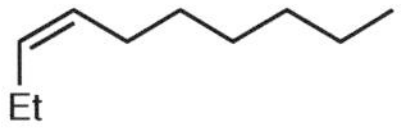	74	[35]

Scheme 8 Proposed Mechanism for the Formic Acid Mediated Deoxygenation of Vicinal Diols[35]

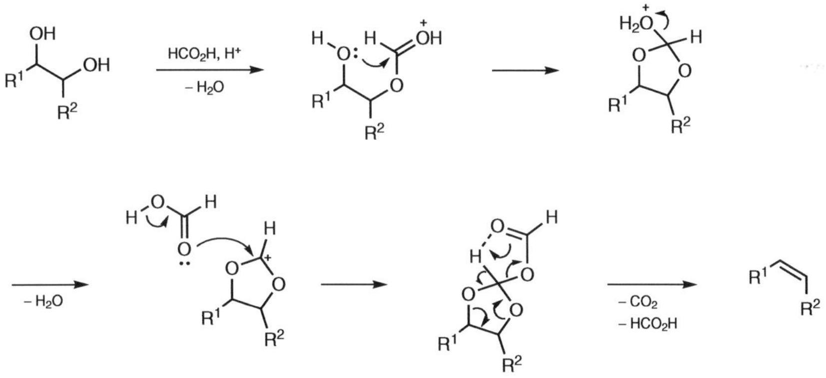

This method has proven to be very useful in multistep processes, such as in the two-step synthesis of penta-1,3-diene from xylitol (Scheme 9).[36] Although the yield of the first step (the formic acid mediated deoxydehydration) was not reported, penta-1,3-diene was isolated in high purity after the second step. Notably, although the selectivity for the desired penta-2,4-dienyl formate is only average, one of the side products is penta-2,4-dien-1-ol, the product of the hydrolysis of this formate.

Scheme 9 Two-Step Synthesis of Penta-1,3-diene via Formic Acid Mediated Deoxydehydration[36]

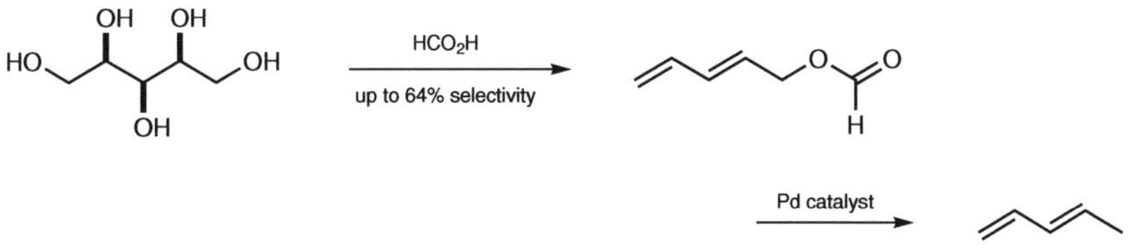

Prop-2-en-1-ol (Table 1, Entry 2); Typical Procedure:[35]

A flask equipped with a fractionating column, a reflux condenser, and a collecting flask was charged with glycerol (13.8 g, 150 mmol) and 99% HCO_2H (3.36 mL, 89 mmol). The mixture was stirred under a N_2 atmosphere for 20 min, and was then heated to 240 °C with collection of the distillate. Once distillation had ceased, the mixture was cooled to rt and more HCO_2H (2.40 mL, 63.5 mmol) was added. The heating, distillation, and cooling process was repeated in the same manner. One more portion of HCO_2H (2.40 mL, 63.5 mmol) was added, and the procedure was repeated again. After this time, the distillates were combined, K_2CO_3 was added, and the organic phase was separated. Distillation delivered the title compound; yield: 7.16 g (82%).

1.6.3.2 Reduction of Diols Using Rhenium Catalysts

The deoxydehydration of diols to give alkenes using a primary or secondary alcohol as the reductant can be performed using catalysts based on various transition metals, such as vanadium and molybdenum;[37–40] however, rhenium-based catalysts are by far the most commonly used.[34,41] In 2010, Bergman and co-workers described a process for the conversion of tetradecane-1,2-diol into tetradec-1-ene (**9**), using decacarbonyldirhenium(0) as the catalyst, in which the product is isolated in high yield (Scheme 10).[42] One of the advantages of this reaction is that it is performed under aerobic conditions. Octan-3-ol plays the role of the reductant, and is oxidized to the corresponding ketone during the process. Sulfuric acid can be used instead of 4-toluenesulfonic acid, but in that case an increase in the catalyst loading to 2.5 mol% is required to achieve the same levels of conversion and yield.

Scheme 10 Rhenium-Catalyzed Deoxydehydration of Tetradecane-1,2-diol To Give Tetradec-1-ene[42]

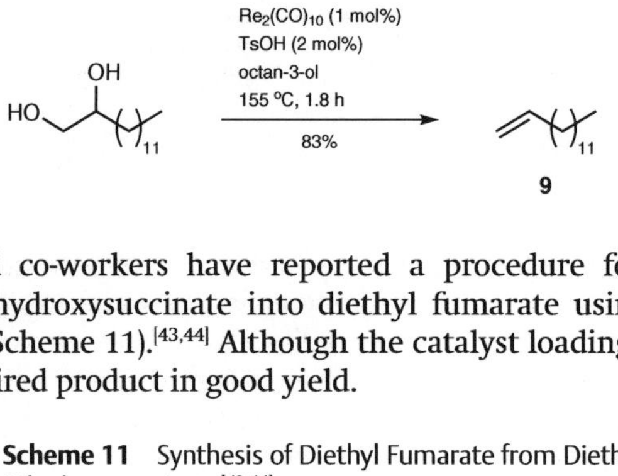

$Re_2(CO)_{10}$ (1 mol%)
TsOH (2 mol%)
octan-3-ol
155 °C, 1.8 h
83%
9

Nicholas and co-workers have reported a procedure for the conversion of diethyl (2*R*,3*R*)-2,3-dihydroxysuccinate into diethyl fumarate using ammonium perrhenate as the catalyst (Scheme 11).[43,44] Although the catalyst loading is very high, this method delivers the desired product in good yield.

Scheme 11 Synthesis of Diethyl Fumarate from Diethyl (2*R*,3*R*)-2,3-Dihydroxysuccinate[43,44]

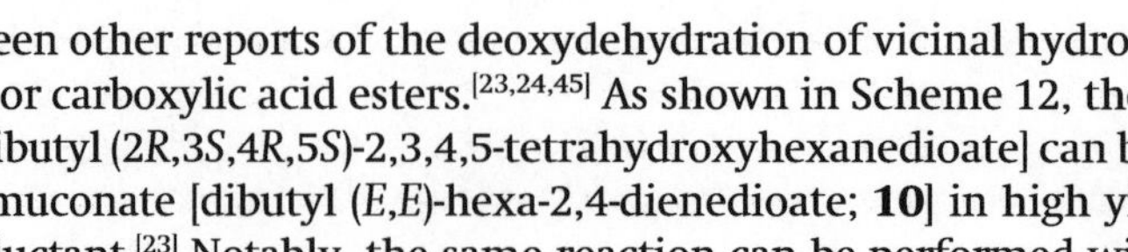

NH_4ReO_4 (10 mol%)
BnOH (1.1 equiv)
benzene, 145 °C, 24 h
85%

There have been other reports of the deoxydehydration of vicinal hydroxy groups in carboxylic acids or carboxylic acid esters.[23,24,45] As shown in Scheme 12, the dibutyl ester of mucic acid [dibutyl (2*R*,3*S*,4*R*,5*S*)-2,3,4,5-tetrahydroxyhexanedioate] can be converted into dibutyl (*E*,*E*)-muconate [dibutyl (*E*,*E*)-hexa-2,4-dienedioate; **10**] in high yield by using butanol as a reductant.[23] Notably, the same reaction can be performed with mucic acid as the starting material. In this case, the product is also dibutyl (*E*,*E*)-muconate, but it is obtained in a lower yield.

Scheme 12 Reduction of the Dibutyl Ester of Mucic Acid To Give a Muconate[23]

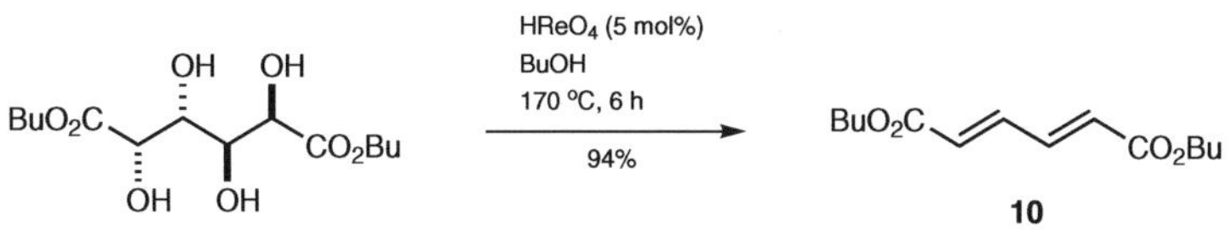

$HReO_4$ (5 mol%)
BuOH
170 °C, 6 h
94%

BuO_2C ... CO_2Bu
10

If muconic acid (**11**) is the target, a method reported by Zhang and co-workers affords the desired non-esterified product from mucic acid in 72% yield (Scheme 13).[24] Attempts to push the conversion further by allowing the reaction to run for a longer time result in partial esterification of the acid groups.

Scheme 13 Rhenium-Catalyzed Synthesis of Muconic Acid from Mucic Acid[24]

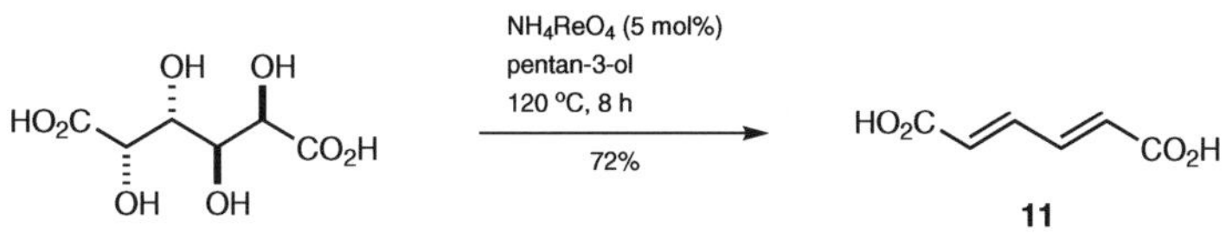

NH_4ReO_4 (5 mol%)
pentan-3-ol
120 °C, 8 h
72%

HO_2C ... CO_2H
11

Another high-yielding example of the deoxydehydration of a polyhydroxy dicarboxylic acid was reported in 2016 by Li and Zhang (Scheme 14).[45] Here, maleic acid (**12**) is prepared by the reduction of tartaric acid.

Scheme 14 High-Yielding Rhenium-Catalyzed Deoxydehydration Process toward Maleic Acid[45]

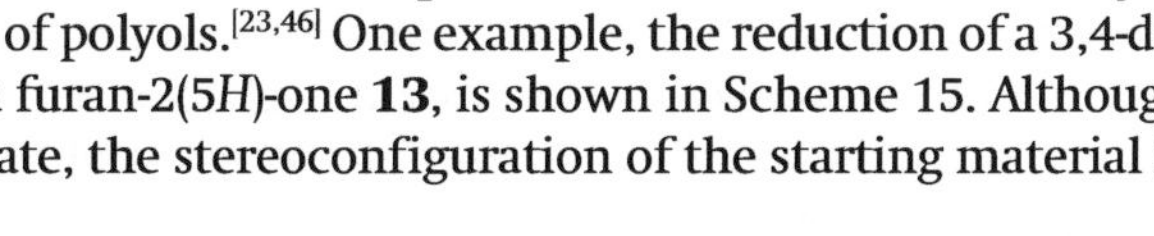

NH_4ReO_4 (5 mol%)
pentan-3-ol
120 °C, 24 h
91%

HO_2C CO_2H
12

A few other practically useful examples of this reaction have been published. For example, Toste and Shiramizu have reported the use of rhenium catalysts to deoxydehydrate a good number of polyols.[23,46] One example, the reduction of a 3,4-dihydroxydihydrofuran-2(3*H*)-one to a furan-2(5*H*)-one **13**, is shown in Scheme 15. Although the yield of the reaction is moderate, the stereoconfiguration of the starting material is retained in the product.[23]

Scheme 15 Rhenium-Catalyzed Preparation of a Furan-2(5*H*)-one from a 3,4-Dihydroxydihydrofuran-2(3*H*)-one[23]

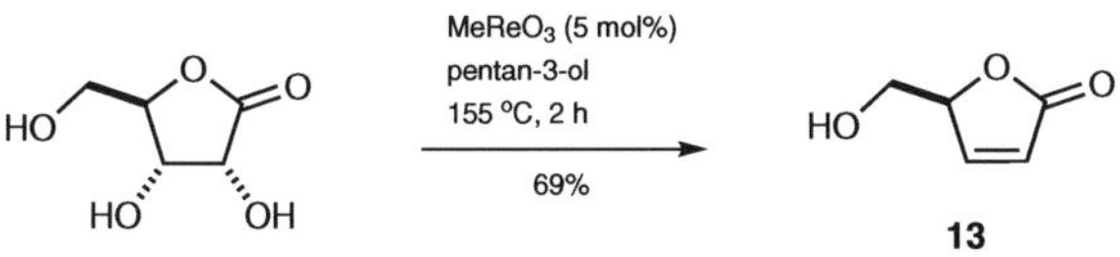

$MeReO_3$ (5 mol%)
pentan-3-ol
155 °C, 2 h
69%

13

Smaller molecules can also be prepared by this route. For example, as shown in Scheme 16, buta-1,3-diene can be formed with high selectivity from *meso*-erythritol.[46] Although the product was not isolated in this study, in principle it can be separated by distillation. The desired molecule can also be prepared from the racemic starting diol, from which the selectivity toward buta-1,3-diene is slightly reduced (81%).

for references see p 240

Scheme 16 Rhenium-Catalyzed Deoxydehydration of *meso*-Erythritol To Give Buta-1,3-diene[46]

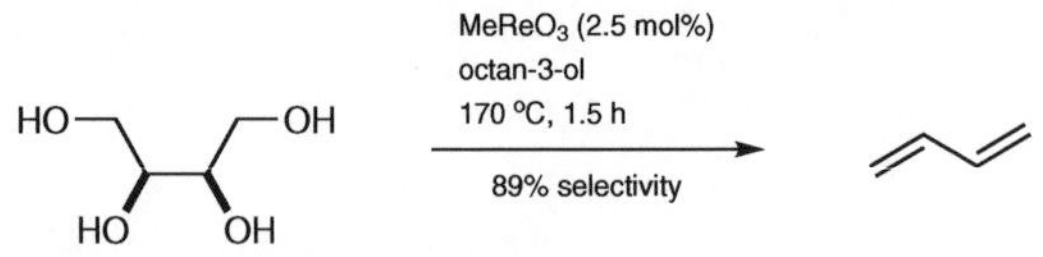

A similar procedure has been employed to prepare (*E*)-hexa-1,3,5-triene from sorbitol.[46] Although the yield of the isolated product is low (mainly as a result of difficulties during separation), the result proves that this type of transformation is achievable. The authors also show that in addition to the deoxydehydration of vicinal diols, it is possible to perform this reaction on nonvicinal 2-ene-1,4-diols and 2,4-diene-1,6-diols.[23] The mechanism for this transformation is thought to proceed through a rhenium-catalyzed 1,3-shift of the hydroxy group of the allylic alcohol, which has previously been reported.[47]

Tetradec-1-ene (9); Typical Procedure:[42]
Tetradecane-1,2-diol (2.5 mmol), $Re_2(CO)_{10}$ (0.025 mmol, 1 mol%), and TsOH (0.05 mmol, 2 mol%) were mixed with octan-3-ol (5 mL), and the resulting mixture was heated under reflux for 1.8 h. After this time, the mixture was cooled to rt and passed through a short plug of silica gel. The silica gel was washed with pentane (5–10 mL), pentane was removed under reduced pressure, and the product was purified by flash column chromatography (silica gel, hexane/EtOAc 40:1); yield: 83%.

Dibutyl (*E*,*E*)-Hexa-2,4-dienedioate (10); Typical Procedure:[23]
A vial was charged with dibutyl (2*R*,3*S*,4*R*,5*S*)-2,3,4,5-tetrahydroxyhexanedioate (97.8 mg, 0.3 mmol), and the vial was purged with N_2. A soln of 76.5% aq $HReO_4$ (4.8 mg, 0.015 mmol, 5 mol%) dissolved in BuOH (6 mL) was added. The vial was immersed into an oil bath preheated at 170 °C and was left in the oil bath for 6 h with magnetic stirring. After this time, the vial was cooled in an ice bath, the volatiles were removed under reduced pressure, and the product was purified by flash column chromatography (hexanes/CH_2Cl_2 4:1 to 2:1) to afford a pale-yellow crystalline solid; yield: 71.2 mg (94%).

Muconic Acid (11); Typical Procedure:[24]
A mixture of mucic acid (420.0 mg, 2.0 mmol), NH_4ReO_4 (26.8 mg, 0.1 mmol), and pentan-3-ol (40 mL) was heated under reflux at 120 °C for 8 h under a N_2 atmosphere. After this time, the mixture was cooled to rt, the soln was filtered, the volatiles were removed under reduced pressure, and the solids were washed with hexane (2 × 20 mL). The hexane layers were combined, and the solvent was removed under reduced pressure; yield: 204 mg (72%).

Maleic Acid (12); Typical Procedure:[45]
A mixture of L-tartaric acid (150.0 mg, 1.0 mmol), NH_4ReO_4 (13.4 mg, 0.05 mmol), and pentan-3-ol (20 mL) was heated under reflux at 120 °C for 24 h under a N_2 atmosphere. After this time, the mixture was cooled to rt and filtered through Celite; the volatiles were removed under reduced pressure, and the solids were washed with hexane (10 mL) and then separated by centrifugation; yield: 106 mg (91%).

(*S*)-5-(Hydroxymethyl)furan-2(5*H*)-one (13); Typical Procedure:[23]
A vial was charged with D-(+)-ribono-1,4-lactone [(3*R*,4*S*,5*R*)-3,4-dihydroxy-5-(hydroxymethyl)dihydrofuran-2(3*H*)-one; 44.5 mg, 0.3 mmol] and $MeReO_3$ (4.0 mg, 0.016 mmol), and the vial was purged with N_2. Pentan-3-ol (3 mL) was added, and the vial was immersed into an oil bath preheated at 155 °C for 2 h with magnetic stirring. After this time, the vial

was cooled in an ice bath, and the resulting black mixture was filtered through Celite. The Celite pad was washed with CH_2Cl_2. The organic phases were combined and the solvent was removed under reduced pressure to afford a colorless oil; yield: 23.5 mg (69%).

1.6.4 Conclusions

This chapter describes methods that are currently the best for the (transfer) hydrogenation of phenols, alcohols, and diols. Regarding the conversion of the C—OH bonds in phenols into C—H bonds (Section 1.6.1), reports on selective synthetically useful procedures are rare. This is in part because the C—O bond energy in phenol is very high,[48] and therefore, it is difficult to cleave such bonds selectively by hydrogenation. Usually, a mixture of products is obtained. Alternative routes for this reaction can be considered, and these would first involve activation of the O—H bond.[49,50]

In the conversion of aliphatic alcohols into alkanes with hydrogen gas (Section 1.6.2), the current catalyst systems that have been reported provide no information on methods for the isolation of the final products. Although the products should be easily separable by distillation, as discussed, other catalytic routes (e.g., via silylation[14] or dehydrogenation/Wolff–Kishner reduction[15]) might also be considered.

For the deoxydehydration of vicinal diols, some practically applicable examples are given in Section 1.6.3. The two major ways to achieve this transformation involve the use of harsh conditions and formic acid as both the catalyst and the source of hydrogen,[35,36] or the use of a rhenium catalyst and an alcohol as the reductant.[23,24,42–46]

for references see p 240

References

[1] Furimsky, E., *Appl. Catal., A*, (2000) **199**, 147.
[2] Dickinson, J. G.; Savage, P. E., *ACS Catal.*, (2014) **4**, 2605.
[3] de Souza, P. M.; Rabelo-Neto, R. C.; Borges, L. E. P.; Jacobs, G.; Davis, B. H.; Sooknoi, T.; Resasco, D. E.; Noronha, F. B., *ACS Catal.*, (2015) **5**, 1318.
[4] Wang, W.; Yang, Y.; Luo, H.; Hu, T.; Liu, W., *Catal. Commun.*, (2011) **12**, 436.
[5] Horáček, J.; Šťávová, G.; Kelbichová, V.; Kubička, D., *Catal. Today*, (2013) **204**, 38.
[6] Newman, C.; Zhou, X.; Goundie, B.; Ghampson, T. I.; Pollock, R. A.; Ross, Z.; Wheeler, M. C.; Meulenberg, R. W.; Austin, R. N.; Frederick, B. G., *Appl. Catal., A*, (2014) **477**, 64.
[7] Huynh, T. M.; Armbruster, U.; Pohl, M.-M.; Schneider, M.; Radnik, J.; Hoang, D.-L.; Phan, B. M. Q.; Nguyen, D. A.; Martin, A., *ChemCatChem*, (2014) **6**, 1940.
[8] Shafaghat, H.; Rezaei, P. S.; Daud, W. M. A. W., *RSC Adv.*, (2015) **5**, 33 990.
[9] Hong, Y.; Zhang, H.; Sun, J.; Ayman, K. M.; Hensley, A. J. R.; Gu, M.; Engelhard, M. H.; McEwen, J.-S.; Wang, Y., *ACS Catal.*, (2014) **4**, 3335.
[10] Nelson, R. C.; Baek, B.; Ruiz, P.; Goundie, B.; Brooks, A.; Wheeler, M. C.; Frederick, B. G.; Grabow, L. C.; Austin, R. N., *ACS Catal.*, (2015) **5**, 6509.
[11] Hansen, C. A.; Frost, J. W., *J. Am. Chem. Soc.*, (2002) **124**, 5926.
[12] Ruppert, A. M.; Weinberg, K.; Palkovits, R., *Angew. Chem. Int. Ed.*, (2012) **51**, 2564.
[13] Wang, Y.; Zhou, J.; Guo, X., *RSC Adv.*, (2015) **5**, 74 611.
[14] Yasuda, M.; Onishi, Y.; Ueba, M.; Miyai, T.; Baba, A., *J. Org. Chem.*, (2001) **66**, 7741.
[15] Dai, X.-J.; Li, C.-J., *J. Am. Chem. Soc.*, (2016) **138**, 5433.
[16] Zhu, S.; Gao, X.; Zhu, Y.; Zhu, Y.; Zheng, H.; Li, Y., *J. Catal.*, (2013) **303**, 70.
[17] Qin, L.-Z.; Song, M.-J.; Chen, C.-L., *Green Chem.*, (2010) **12**, 1466.
[18] Taher, D.; Thibault, M. E.; Di Mondo, D.; Jennings, M.; Schlaf, M., *Chem.–Eur. J.*, (2009) **15**, 10 132.
[19] Chia, M.; Pagán-Torres, Y. J.; Hibbitts, D.; Tan, Q.; Pham, H. N.; Datye, A. K.; Neurock, M.; Davis, R. J.; Dumesic, J. A., *J. Am. Chem. Soc.*, (2011) **133**, 12 675.
[20] Zelinski, R.; Eichel, H. J., *J. Org. Chem.*, (1958) **23**, 462.
[21] Gong, J.; Lou, X.-J.; Li, W.-D.; Jing, X.-K.; Chen, H.-B.; Zeng, J.-B.; Wang, X.-L.; Wang, Y.-Z., *J. Polym. Sci., Part A: Polym. Chem.*, (2010) **48**, 2828.
[22] Buntara, T.; Noel, S.; Phua, P. H.; Melián-Cabrera, I.; de Vries, J. G.; Heeres, H. J., *Angew. Chem. Int. Ed.*, (2011) **50**, 7083.
[23] Shiramizu, M.; Toste, F. D., *Angew. Chem. Int. Ed.*, (2013) **52**, 12 905.
[24] Zhang, H.; Li, X.; Su, X.; Ang, E. L.; Zhang, Y.; Zhao, H., *ChemCatChem*, (2016) **8**, 1500.
[25] Zhang, Y.; Li, X., WO 2016 032 403, (2016).
[26] Li, X.; Wu, D.; Lu, T.; Yi, G.; Su, H.; Zhang, Y., *Angew. Chem. Int. Ed.*, (2014) **53**, 4200.
[27] Shin, N.; Kwon, S.; Moon, S.; Hong, C. H.; Kim, Y. G., *Tetrahedron*, (2017) **73**, 4758.
[28] Boussie, T. R.; Dias, E. L.; Fresco, Z. M.; Murphy, V. J.; Shoemaker, J.; Archer, R.; Jiang, H., WO 2010 144 862, (2010).
[29] Zhang, Y.; Li, X., WO 2015 084 265, (2015).
[30] Sevonkaev, I. V.; Kumar, A.; Pal, A.; Goia, D. V., *RSC Adv.*, (2014) **4**, 3653.
[31] Matsumoto, T.; Ohishi, M.; Inoue, S., *J. Org. Chem.*, (1985) **50**, 603.
[32] Wang, Q.-Y.; Zeng, H.; Song, H.; Liu, Q.-S.; Yao, S., *J. Chem. Eng. Data*, (2010) **55**, 5271.
[33] Sousa, S. C. A.; Fernandes, A. C., *Coord. Chem. Rev.*, (2015) **284**, 67.
[34] Raju, S.; Moret, M.-E.; Klein Gebbink, R. J. M., *ACS Catal.*, (2015) **5**, 281.
[35] Arceo, E.; Marsden, P.; Bergman, R. G.; Ellman, J. A., *Chem. Commun. (Cambridge)*, (2009), 3357.
[36] Sun, R.; Zheng, M.; Li, X.; Pang, J.; Wang, A.; Wang, X.; Zhang, T., *Green Chem.*, (2017) **19**, 638.
[37] Chapman, G.; Nicholas, K. M., *Chem. Commun. (Cambridge)*, (2013) **49**, 8199.
[38] Gopaladasu, T. V.; Nicholas, K. M., *ACS Catal.*, (2016) **6**, 1901.
[39] Hills, L.; Moyano, R.; Montilla, F.; Pastor, A.; Galindo, A.; Álvarez, E.; Marchetti, F.; Pettinari, C., *Eur. J. Inorg. Chem.*, (2013), 3352.
[40] Sandbrink, L.; Beckerle, K.; Meiners, I.; Leffmann, R.; Rahimi, K.; Okuda, J.; Palkovits, R., *ChemSusChem*, (2017) **10**, 1375.
[41] Dethlefsen, J. R.; Fristrup, P., *ChemSusChem*, (2015) **8**, 767.
[42] Arceo, E.; Ellman, J. A.; Bergman, R. G., *J. Am. Chem. Soc.*, (2010) **132**, 11 408.
[43] Boucher-Jacobs, C.; Nicholas, K. M., *ChemSusChem*, (2013) **6**, 597.
[44] McClain, J. M.; Nicholas, K. M., *ACS Catal.*, (2014) **4**, 2109.
[45] Li, X.; Zhang, Y., *ChemSusChem*, (2016) **9**, 2774.

[46] Shiramizu, M.; Toste, F. D., *Angew. Chem. Int. Ed.*, (2012) **51**, 8082.
[47] Herrmann, A. T.; Saito, T.; Stivala, C. E.; Tom, J.; Zakarian, A., *J. Am. Chem. Soc.*, (2010) **132**, 5962.
[48] Mortensen, P. M.; Grunwaldt, J.-D.; Jensen, P. A.; Jensen, A. D., *ACS Catal.*, (2013) **3**, 1774.
[49] Chen, Q.-Y.; He, Y.-B., *Synthesis*, (1988), 896.
[50] Wang, X.-Y.; Leng, J.; Wang, S.-M.; Asiri, A. M.; Marwani, H. M.; Qin, H.-L., *Tetrahedron Lett.*, (2017) **58**, 2340.

1.7 Hydrogenolysis of Ethers

Y. Nakagawa, M. Tamura, and K. Tomishige

General Introduction

Ethers are generally less reactive than other functionalized compounds such as alcohols. Therefore, cleavage of one of the C—O bonds in an ether group is not common in organic synthesis, except in the case of very reactive ethers such as benzyl ethers, *tert*-butyl ethers, and epoxides. The presence of two C—O bonds in each ether group causes another problem for the utilization of ether transformations in organic syntheses: the conversion of unsymmetrical ethers can produce two different sets of products. However, the recent focus on biomass conversion has renewed interest in ether conversions, because biomass-derived materials often contain ether groups. An example of this is lignin,[1] which is a complex polyether of propylbenzene units. Furan derivatives are also important biomass-derived compounds.[2,3] Biomass resources generally contain large amounts of oxygen atoms and many functional groups. Removal of ether oxygen is one of the main reactions in biomass conversion, and ether hydrogenolysis (Scheme 1) achieves this end.[4,5]

Scheme 1 Hydrogenolysis of Ethers[4,5]

$$R^1{-}O{-}R^2 + H_2 \longrightarrow R^1H + R^2OH$$

Research in the field of ether hydrogenolysis has been rapidly growing since around 2010. In this chapter, selected examples of ether hydrogenolysis that show yields high enough to be used in organic syntheses are discussed. The chapter is divided into three subsections: hydrogenolysis of aryl ethers (Section 1.7.1), hydrogenolysis of saturated ethers (Section 1.7.2), and hydrogenolysis of furans (Section 1.7.3). Note that the hydrogenolysis of benzyl ethers (removal of benzyl ether protecting groups) is discussed separately in Section 2.4.2.3, and was also covered in *Science of Synthesis*, Vol. 37 (Section 37.8.2.2). The hydrogenolysis of epoxides is discussed in *Science of Synthesis*, Vol. 36 (Section 36.1.2.2.3).

1.7.1 Hydrogenolysis of Aryl Ethers

There are more methods for the hydrogenolysis of diaryl ethers or alkyl aryl ethers than for dialkyl ethers. The difference in the electronic states between C_{aryl}—O and C_{alkyl}—O bonds enables the selective C—O dissociation of alkyl aryl ethers. Most reported systems preferentially dissociate the C_{aryl}—O bond in alkyl aryl ethers, producing the corresponding arene and alcohol. Consequently, hydrogenolysis of diaryl ethers is generally easier than that of dialkyl ethers. However, if the reducing activity is too strong this may lead to the hydrogenation of aromatic rings, decreasing the yield of the desired arene (and phenol, in the case of diaryl ethers).

1.7.1.1 Homogeneous Nickel Catalysts for the Hydrogenolysis of Diaryl Ethers and Alkyl Aryl Ethers

Sergeev and Hartwig pioneered the field of hydrogenolysis of aryl ethers in their paper published in 2011.[6] They used N-heterocyclic carbenes as ligands for a homogeneous nickel catalyst, and found that high yields in the hydrogenolysis of diaryl ethers are obtained with bis(η^4-cycloocta-1,5-diene)nickel(0) as the active metal source, 1,3-bis(2,6-diisopropylphenyl)-4,5-dihydroimidazolium chloride (**1**•HCl; SIPr•HCl) as the ligand precursor, 0.1 MPa molecular hydrogen, and base (*t*-BuONa). The ligand precursor is also commercially available in the tetrafluoroborate form **1**•HBF_4. This catalyst can convert various diaryl ethers **2** (Scheme 2) and alkyl aryl ethers **5** (Scheme 3) into the corresponding arenes **3** and **6** and alcohols **4** and **7**. In the case of unsymmetrical diaryl ethers, the C—O bond on the more electron-deficient ring is preferably dissociated. Hydrogenolysis using **1**•HBF_4 as the ligand precursor leads to identical yields of the products. Reaction of a C_{aryl}—O—C_{aryl} group proceeds even when the aryl moieties contain methoxy substituents. Hydride reagents, such as diisobutylaluminum hydride and triethylsilane, can be employed as the hydrogen source instead of molecular hydrogen.

Scheme 2 Hydrogenolysis of Diaryl Ethers Catalyzed by a Nickel Complex with an N-Heterocyclic Carbene Ligand[6]

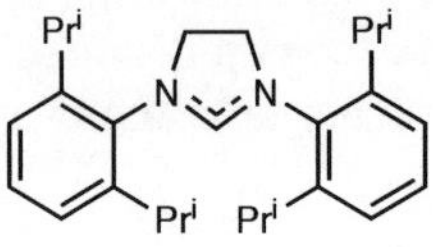

Pri Pri N N Pri Pri

1

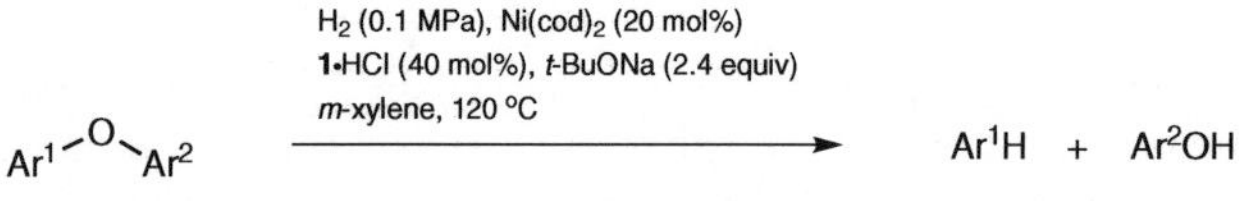

Ar^1–O–Ar^2 (**2**) → H_2 (0.1 MPa), $Ni(cod)_2$ (20 mol%), **1**•HCl (40 mol%), *t*-BuONa (2.4 equiv), *m*-xylene, 120 °C → Ar^1H (**3**) + Ar^2OH (**4**)

Ar^1	Ar^2	Time (h)	Conversion (%)	Yield[a] (%)		Ref
				3	**4**	
Ph	Ph	16	100	99	99	[6]
3-Tol	3-Tol	16	100	96	99	[6]
3-$MeOC_6H_4$	3-$MeOC_6H_4$	16	94	65[b]	83	[6]
4-Tol	4-Tol	48	100	97	99	[6]
4-*t*-BuC_6H_4	4-*t*-BuC_6H_4	48	74	72	73	[6]
4-$F_3CC_6H_4$	Ph	16	100[c]	64[d]	99	[6]
4-$F_3CC_6H_4$	4-$MeOC_6H_4$	16	100[c]	68[e]	92	[6]
Ph	4-$MeOC_6H_4$	16	100	88[f]	80	[6]
2-Tol	2-Tol	32	85	85	85	[6]

[a] Determined by GC.
[b] Benzene (23% yield) and phenol (3% yield) were also obtained.
[c] Conditions: $Ni(cod)_2$ (10 mol%), **1**•HCl (20 mol%), 100 °C.
[d] Toluene was also obtained in 23% yield.
[e] Toluene was also obtained in 19% yield.
[f] Anisole (4% yield) and phenol (17% yield) were also obtained.

Scheme 3 Hydrogenolysis of Alkyl Aryl Ethers Catalyzed by a Nickel Complex with an N-Heterocyclic Carbene Ligand[6]

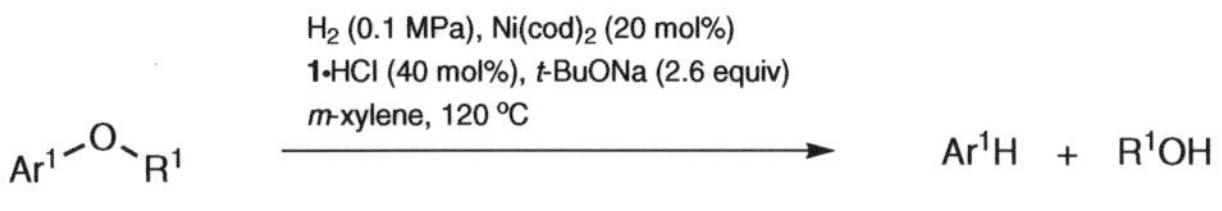

H_2 (0.1 MPa), Ni(cod)$_2$ (20 mol%)
1•HCl (40 mol%), *t*-BuONa (2.6 equiv)
m-xylene, 120 °C

Ar^1–O–R^1 (**5**) → Ar^1H (**6**) + R^1OH (**7**)

Ar^1	R^1	Time (h)	Conversion (%)	Yield[a] (%)		Ref
				6	7	
2-naphthyl	$(CH_2)_5Me$	16	100	95	98	[6]
2-naphthyl	Me	16	89	89	n.d.	[6]
1-naphthyl	Me	16	72	72	n.d.	[6]
4-PhC_6H_4	$(CH_2)_5Me$	32	85	85	85	[6]
4-PhC_6H_4	Me	32	60	59	n.d.	[6]

[a] Yield determined by GC; n.d. = not determined.

The mechanism of this nickel-based reaction system is proposed to involve oxidative addition of the C_{aryl}–O bond to nickel(0) to form an arylnickel(II) alkoxide/phenoxide species such as **8** (Scheme 4).[7]

Scheme 4 Proposed Mechanism for Nickel-Catalyzed Hydrogenolysis of Aryl Ethers[7]

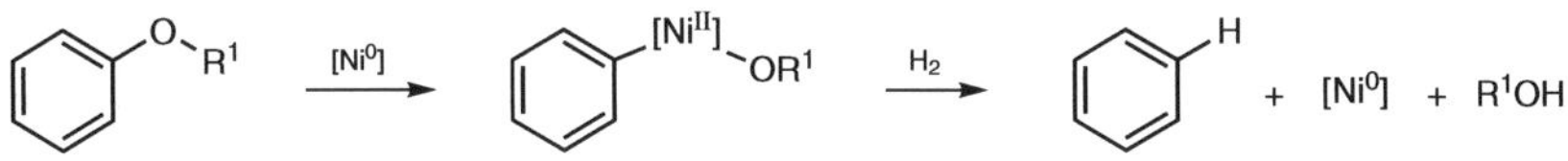

PhOR^1 —[Ni^0]→ Ph–[Ni^{II}]–OR^1 (**8**) —H_2→ PhH + [Ni^0] + R^1OH

For alkyl aryl ethers, the nickel complex **8** does not react with hydrogen, but rather undergoes hydride abstraction from the nickel alkoxide and subsequent reductive elimination to give an arene and an aldehyde (Scheme 5), as was deduced by deuterium-labelling studies. Therefore, removal of an alkoxy group from alkoxyarenes is possible without the use of an additional reducing agent, if recovery of the removed alkyl group is not necessary.[8]

Scheme 5 Mechanism of Nickel-Catalyzed Cleavage of a C_{aryl}–O–C_{alkyl} Unit without a Reducing Agent[8]

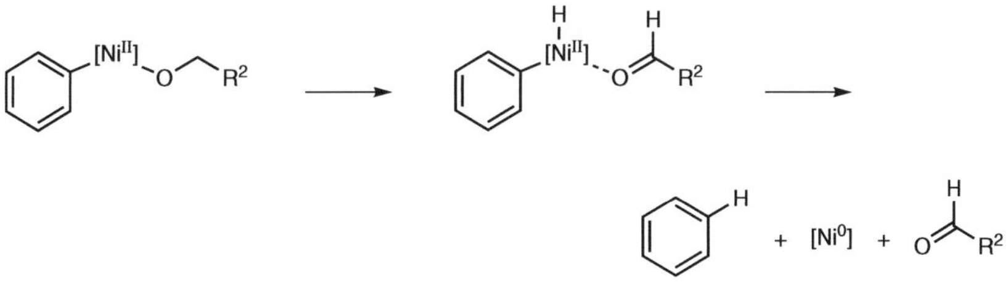

Ph–[Ni^{II}]–O–CH$_2$–R^2 → Ph–[Ni^{II}](H)···O=CH–R^2 → PhH + [Ni^0] + O=CH–R^2

1,3-Di(2-adamantyl)imidazolium hydrochloride (**9**•HCl) is the precursor to an effective NHC ligand in the catalysis of the cleavage of aryl alkyl ethers **10** (Scheme 6).[8] The aryl group can contain substituents such as a C=C bond, an amide, or pyridine. The alkyl group needs to be a straight-chain alkyl, such as methyl, ethyl, or octyl. The ligand is synthesized from adamantan-2-amine, glyoxal, *N*,*N*,*N'*,*N'*-tetramethylmethanediamine, and acetyl chloride via an ethane-1,2-diimine.

Scheme 6 Cleavage of C_{aryl}–O Bonds Catalyzed by a Nickel Complex without a Reducing Agent[8]

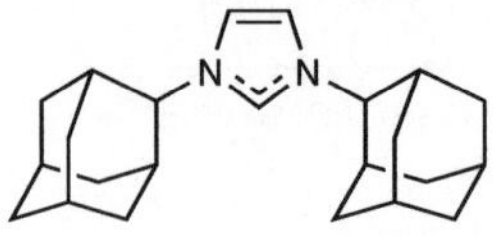

9

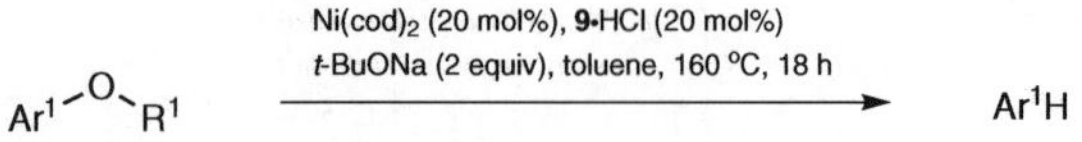

Ar^1–O–R^1 (10) → Ar^1H

Ni(cod)$_2$ (20 mol%), **9**•HCl (20 mol%)
t-BuONa (2 equiv), toluene, 160 °C, 18 h

Ar^1	R^1	Yield[a] (%)	Ref
2-naphthyl	Me	84	[8]
2-naphthyl	Et	91	[8]
2-naphthyl	$(CH_2)_7Me$	90	[8]
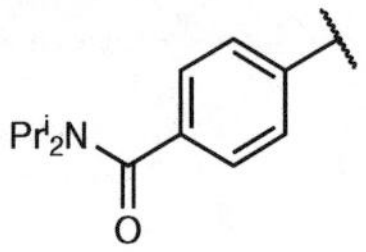 4-(Pr^i_2NCO)C$_6$H$_4$	Et	67	[8]
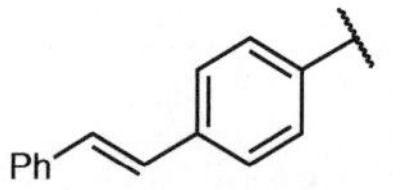 4-(PhCH=CH)C$_6$H$_4$	Et	78	[8]
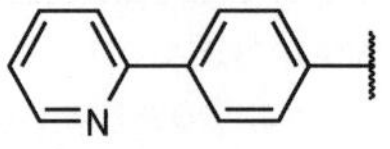 4-(2-pyridyl)C$_6$H$_4$	Et	75	[8]

[a] Determined by GC.

Benzene (3, Ar[1] = Ph) and Phenol (4, Ar[2] = Ph); Typical Procedure:[6]

Ni(cod)$_2$ (8.3 mg, 0.03 mmol), 1,3-bis(2,6-diisopropylphenyl)-4,5-dihydroimidazolium chloride (**1**•HCl; 27.8 mg, 0.065 mmol), and *t*-BuONa (37.2 mg, 0.387 mmol) as well as a magnetic stirrer bar were added to a 15-mL Schlenk tube in a glovebox. Then, a soln of diphenyl ether (**2**, $Ar^1 = Ar^2 = Ph$; 26.4 mg, 0.155 mmol) in anhyd *m*-xylene (0.8 mL) was added, and the mixture was stirred for 3 min, degassed, and pressurized with H_2 (0.1 MPa) at rt. The color of the mixture changed slightly from dark brown to dark red brown. The tube was heated at 120 °C for 16 h. The mixture was cooled to rt and diluted with Et_2O (1 mL). 1.5 M aq HCl (1 mL) was added and the organic layer was separated. The aqueous layer was extracted with Et_2O (1 mL) and the combined organic layers were passed through a short pad of Celite. The filtrate was subjected to GC analysis.

1.7.1.2 Solid Catalysts for the Hydrogenolysis of Aryl Ethers

Diaryl ethers are relatively reactive unless strongly electron-donating groups are present. Therefore, the hydrogenolysis of diaryl ethers can also proceed without an added ancillary ligand (Scheme 7; cf. Scheme 2).[9] It was found that in the absence of an added ancillary ligand, heterogeneous nickel particles form in situ from the soluble precursor bis(η^4-cycloocta-1,5-diene)nickel(0), in the presence of sodium *tert*-butoxide as base. The nickel particles catalyze the cleavage of C—O bonds in diaryl ethers, in some cases at low catalyst loadings.

Scheme 7 Hydrogenolysis in the Presence and Absence of an N-Heterocyclic Carbene Ligand[6,9]

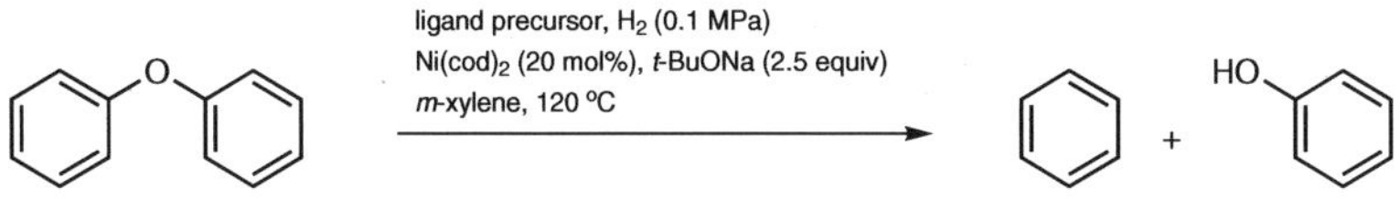

ligand precursor, H_2 (0.1 MPa)
Ni(cod)$_2$ (20 mol%), *t*-BuONa (2.5 equiv)
m-xylene, 120 °C

HO

Ligand Precursor	Time (h)	Yield (%) of Benzene	Yield (%) of Phenol	Ref
1•HCl	16	99	99	[6]
none	48	99	99	[9]

The deposition of nickel on an activated carbon support increases the activity further.[10] The resulting nickel-on-carbon (Ni/C) catalyst can even cleave diaryl ethers with many electron-donating substituents, such as a model substrate of lignin [5-ethyl-1,2-dimethoxy-3-(2-methoxy-4-propylphenoxy)benzene (**11**)], giving 4-ethyl-1,2-dimethoxybenzene (**12**), 2-methoxy-4-propylphenol (**13**), 5-ethyl-2,3-dimethoxyphenol (**14**), and 1-methoxy-3-propylbenzene (**15**) in 94% GC yield (Scheme 8). The hydrogenolysis of **11** does not proceed with soluble bis(η^4-cycloocta-1,5-diene)nickel(0) [Ni(cod)$_2$] or Ni(CH$_2$SiMe$_3$)$_2$(TMEDA). The nickel-on-carbon catalyst is prepared by impregnating activated carbon with bis(η^4-cycloocta-1,5-diene)nickel(0) in tetrahydrofuran. A nickel boride on carbon (Ni-B/C) catalyst, prepared from a less expensive nickel precursor and sodium borohydride, is also active; however, larger amounts of the catalyst are required to obtain a similar yield.

Scheme 8 Hydrogenolysis of a Lignin Model Dimer over Nickel on Carbon[10]

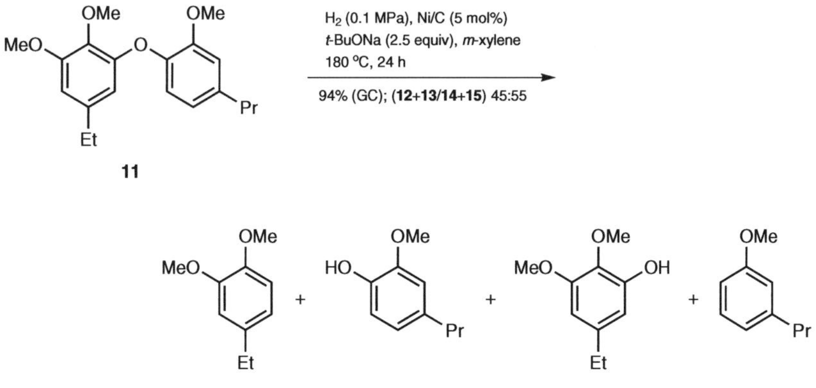

OMe OMe MeO O Pr Et **11**

H_2 (0.1 MPa), Ni/C (5 mol%)
t-BuONa (2.5 equiv), *m*-xylene
180 °C, 24 h

94% (GC); (**12**+**13**/**14**+**15**) 45:55

MeO OMe Et **12** + HO OMe Pr **13** + MeO OMe OH Et **14** + OMe Pr **15**

Iron[11] and cobalt[12] species are also active in dissociations of C$_{aryl}$—O bonds in diaryl ethers and alkyl aryl ethers with base additive providing arenes **16** and phenols **17** {and

[1,1′-biphenyl]-2-ol (**18**)} (Scheme 9). Inexpensive and air-stable acetylacetonate complexes of cobalt(II) and iron(III) can be used as catalysts. However, lithium aluminum hydride is used as a reducing agent for both the iron- and cobalt-based systems. The use of molecular hydrogen as a reducing agent is not effective. In this process, the metal complexes are converted into solid metal species in the reaction media, in a similar fashion to the nickel-catalyzed system.

Scheme 9 Reductive Cleavage of C_{aryl}–O Bonds Using Tris(acetylacetonato)iron(III) or Bis(acetylacetonato)cobalt(II) Catalysts and Lithium Aluminum Hydride[11,12]

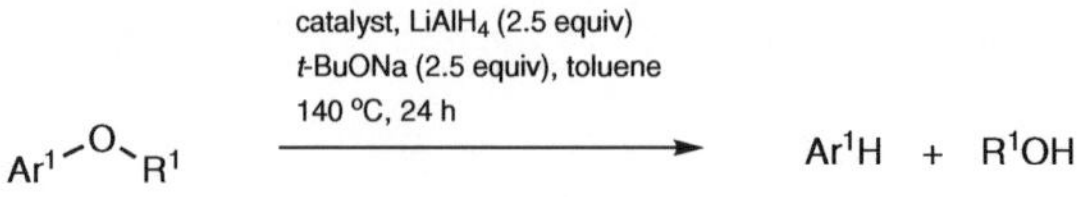

Ar^1–O–R^1 → (catalyst, $LiAlH_4$ (2.5 equiv), t-BuONa (2.5 equiv), toluene, 140 °C, 24 h) → Ar^1H (**16**) + R^1OH (**17**)

Ar[1]	R[1]	Catalyst (mol%)	Yield[a] (%)		Ref
			16	**17**	
Ph	Ph	$Fe(acac)_3$ (20)	97	100 (81)	[11]
Ph	Ph	$Co(acac)_2$ (15)	92	93	[12]
4-Tol	4-Tol	$Fe(acac)_3$ (20)	98	99 (88)	[11]
4-Tol	4-Tol	$Co(acac)_2$ (15)	100	98 (88)	[12]
1-naphthyl	Me	$Fe(acac)_3$ (20)	95	–	[11]
1-naphthyl	Me	$Co(acac)_2$ (15)	49	–	[12]
Ph	Me	$Fe(acac)_3$ (20)	trace	–	[11]
Ph	Me	$Co(acac)_2$ (15)	trace	–	[12]

[a] Determined by GC; isolated yield in parentheses.

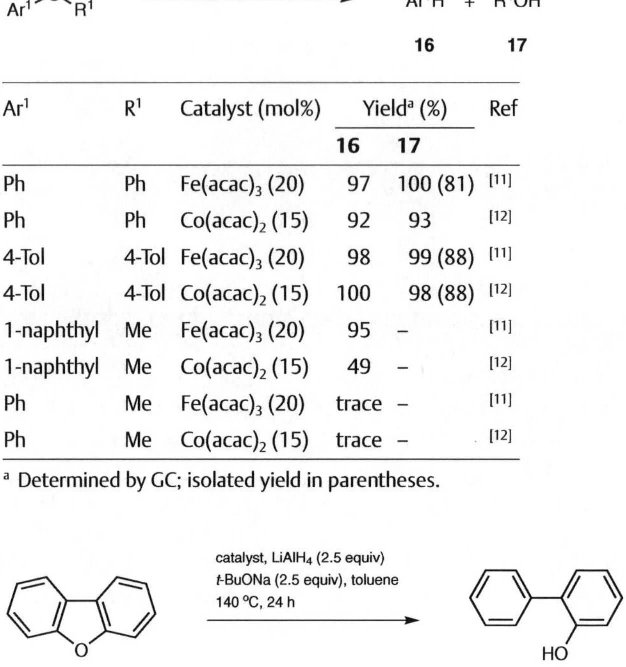

Dibenzofuran → (catalyst, $LiAlH_4$ (2.5 equiv), t-BuONa (2.5 equiv), toluene, 140 °C, 24 h) → **18**

Catalyst (mol%)	Yield[a] (%)	Ref
$Fe(acac)_3$ (20)	(80)	[11]
$Co(acac)_2$ (15)	93 (81)	[12]

[a] Determined by GC; isolated yield in parentheses.

4-Ethyl-1,2-dimethoxybenzene (12), 2-Methoxy-4-propylphenol (13), 5-Ethyl-2,3-dimethoxyphenol (14), and 1-Methoxy-3-propylbenzene (15); Typical Procedure:[10]

Preparation of the Ni/C catalyst: Activated carbon (500 mg) was added in one portion to a soln of $Ni(cod)_2$ (70 mg, 0.254 mmol) in THF (8 mL) in a glovebox. The suspension was stirred at rt for 1 h, and the solid was collected by vacuum filtration in the glovebox. The black solid was washed with additional THF (5 × 2 mL) and dried under reduced pressure. The solid was then exposed to more severe drying conditions (~1 Pa at 120 °C for 12 h). The catalyst should not be exposed to air, and should be stored in a freezer inside a glovebox.

Catalytic hydrogenolysis: Ni/C (6 μmol Ni, 5 mol%), *t*-BuONa (29 mg, 0.3 mmol), and a magnetic stirrer bar were added to a 15-mL Schlenk tube in a glovebox. Then, a soln of 5-ethyl-1,2-dimethoxy-3-(2-methoxy-4-propylphenoxy)benzene (**11**; 40 mg, 0.12 mmol) in *m*-xylene (0.8 mL) was added, and the mixture was stirred for 1 min, and then degassed and pressurized with H_2 (0.1 MPa) at rt. The tube was heated at 180 °C for 24 h. The resulting dark black mixture was cooled to rt, diluted with Et_2O (1 mL), and quenched with 1.5 M HCl (1 mL). The organic layer was separated and the aqueous layer was extracted with Et_2O (1 mL). The combined organic layers were passed through a short pad of Celite and anhyd Na_2SO_4. The soln was analyzed by GC.

Arenes 16 and Phenols 17; General Procedure Using Fe(acac)$_3$:[11]

CAUTION: *Solid lithium aluminum hydride reacts vigorously with a variety of substances, and can ignite on rubbing or vigorous grinding.*

The aryl ether (0.2 mmol), $LiAlH_4$ (19 mg, 0.5 mmol), Fe(acac)$_3$ (14 mg, 0.04 mmol), *t*-BuONa (48 mg, 0.5 mmol), and toluene (1.5 mL) were added to a dried tube (ca. 40 mL volume) equipped with a magnetic stirrer bar under N_2. The tube was sealed and heated to 140 °C. After 24 h, the tube was cooled to rt, and the mixture was acidified to pH 5~6 with 2 M aq HCl. The products were purified by column chromatography.

1.7.1.3 Hydrogenolysis of the β-O-4 Linkage in Lignin Model Compounds

Lignin is composed of oxygenated propylbenzene units linked to each other with ether and C—C bonds. The most abundant linkages are β-O-4 (ether linkage of the β-position of one propylbenzene unit and the phenol hydroxy group at the 4-position of another propylbenzene unit), α-O-4, and 4-O-5 (Scheme 10).[5]

Scheme 10 Linkages between Propylbenzene Units in Lignin

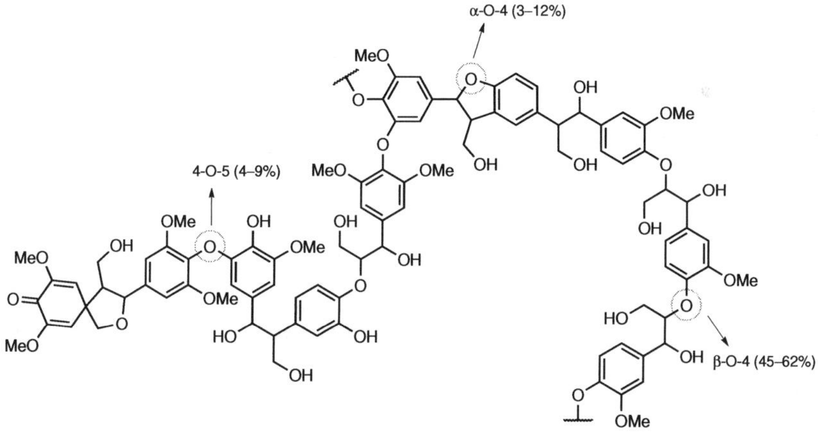

The 4-O-5 linkage is a diaryl ether and is mimicked by model compounds **11**, **19**, and **20** (Scheme 11). The α-O-4 linkage is a benzyl ether and therefore can be hydrogenolyzed relatively easily; it is mimicked by model compound **21**. The β-O-4 linkage can be mimicked by model compounds **22–24**.

for references see p 265

Scheme 11 Lignin Model Compounds

4-O-5 model compounds: 19, 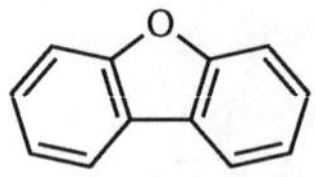 20

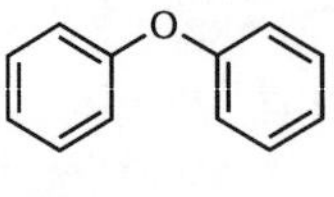

α-O-4 model compound: 21

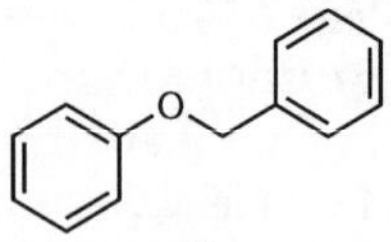

β-O-4 model compounds: 22, 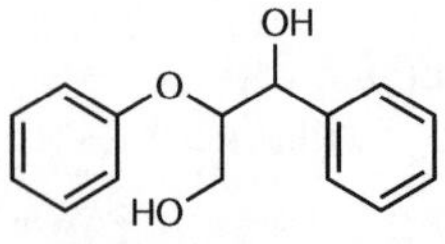 23, 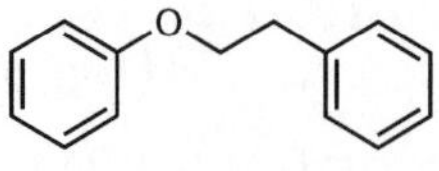 24

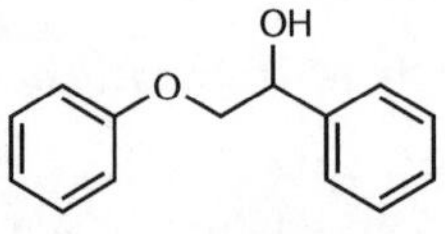

Several methods for breaking the latter bond have been developed. Typical β-O-4 linkages in lignin have a hydroxy group at the α-position of the propylbenzene unit with the ether linkage at the β-position; this type of structure is mimicked by 2-phenoxy-1-phenylethanol (**22**). The reactivity of this compound is relatively high; even a simple base, without any reducing agent, can decompose **22** into phenol and acetophenone.[13] However, strongly basic conditions convert the ketone product (acetophenone) into a complex mixture by aldol condensation. In the presence of a reducing agent, the aldol condensation can be suppressed by rapid reduction of acetophenone, and a good yield of hydrogenolysis products is obtained (Scheme 12).[11,14] Overreduction of 1-phenylethanol (to give ethylbenzene) can take place under more harsh conditions (Scheme 12).[14]

Scheme 12 Reductive Cleavage of the β-O-4 Linkage in 2-Phenoxy-1-phenylethanol in the Presence of Base[11,14]

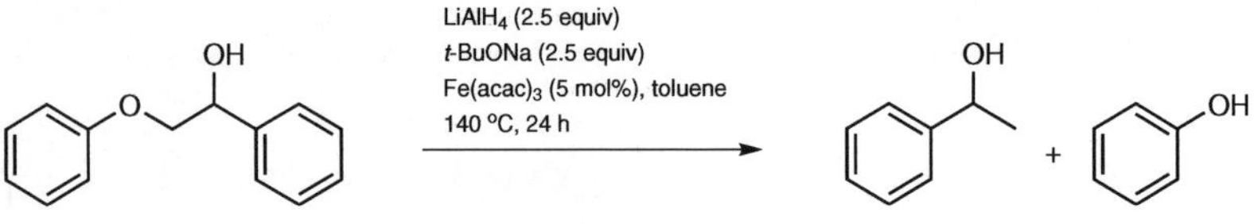

22 → ($LiAlH_4$ (2.5 equiv), t-BuONa (2.5 equiv), $Fe(acac)_3$ (5 mol%), toluene, 140 °C, 24 h) → 1-phenylethanol 41% (GC) + phenol 83%

22 → ($LiAlH_4$ (5 equiv), t-BuOK (5 equiv), toluene, 180 °C, 24 h) → ethylbenzene 74% (GC) + phenol 94%

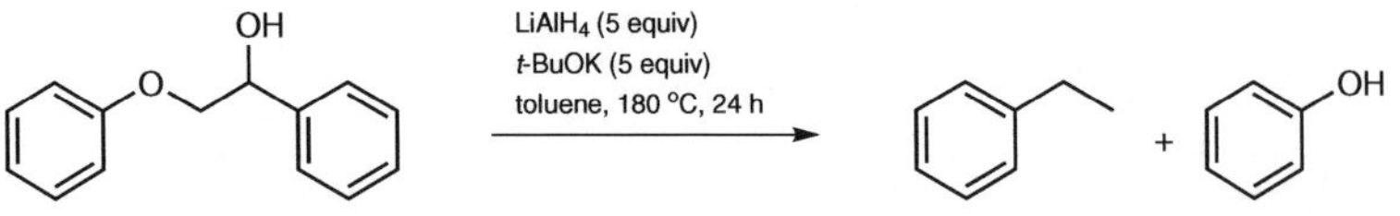

Hydrogenolysis of the β-O-4 linkage in another lignin model dimer has been reported using a Zn/Pd/C catalyst (Scheme 13).[15] However, cleavage of the β-O-4 linkage can also be achieved with an appropriate metal catalyst without a reducing agent via a borrowing hydrogen type reaction. Following pioneering work by Bergman and Ellman using homogeneous ruthenium catalysts,[16] various homogeneous and heterogeneous catalysts were investigated for nonreductive cleavage of the β-O-4 linkage. Even the commercially available palladium-on-carbon catalyst is effective in the decomposition of 2-phenoxy-1-phenylethanol (**22**) to give acetophenone and phenol.[17]

Scheme 13 Hydrogenolysis of the β-O-4 Linkage in a Lignin Model Dimer with a Zn/Pd/C Catalyst[15]

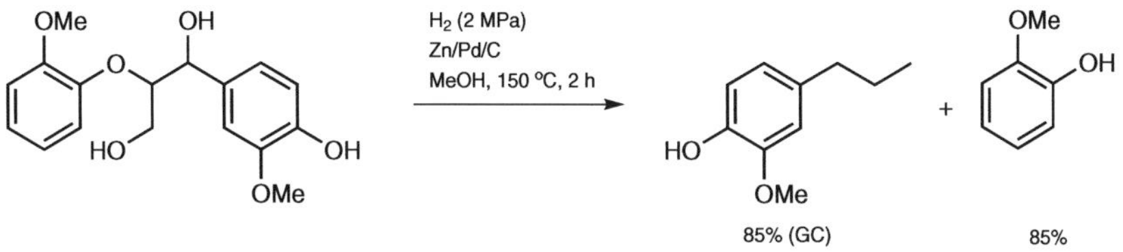

H_2 (2 MPa)
Zn/Pd/C
MeOH, 150 °C, 2 h

85% (GC) 85%

The reactivity of the β-O-4 linkages is much lower when an α-hydroxy group is not present.[11] Similarly to the case with an α-hydroxy group, the C_{alkyl}—O bond is preferably dissociated in the reduction of phenyl 2-phenylethyl ether (**24**), affording ethylbenzene (**25**) and phenol (**26**) (Scheme 14).[11,18–21] This is in contrast to the reactions of simple alkyl aryl ethers, which are usually converted into arenes and alkyl alcohols via C_{aryl}—O dissociation. Hydrogenation of the aromatic ring is a side reaction when high hydrogen pressure and high temperature are employed (Scheme 14, entries 5 and 6).

Scheme 14 Hydrogenolysis of the β-O-4 Linkage Model Compound Phenyl 2-Phenylethyl Ether[11,18–21]

catalyst

24 **25** **26**

Entry	Catalyst	Conditions	Yield[a] (%) **25**	Yield[a] (%) **26**	Ref
1	$Fe(acac)_3$	$LiAlH_4$ (2.5 equiv), *t*-BuOK (2.5 equiv), toluene, 140 °C, 24 h	trace	7	[11]
2	Ni@SiC	*t*-BuOK (1 equiv), TBAB (0.3 equiv), H_2 (0.6 MPa), H_2O, 120 °C, 6 h	–	96	[18]
3	Raney Ni	H_2 (1 MPa), *p*-xylene, 120 °C, 16 h	62	57	[19]
4	Ni/metal–organic framework	H_2 (1 MPa), *p*-xylene, 120 °C, 16 h	79	80	[19]
5	Pd-Fe/mesoporous carbon	H_2 (1 MPa), hexadecane, 250 °C, 3 h	21[b]	16	[20]
6	FeMoP[c]	H_2 (4.2 MPa), decane, 400 °C, 0.5 h	50[d]	6	[21]

[a] Determined by GC.
[b] Benzene (20% yield) and saturated compounds (11% yield) were also obtained.
[c] Bimetallic Fe-Mo phosphide catalyst.
[d] Benzene (32% yield) and cyclohexane (12% yield) were also obtained.

1.7.1.4 Hydrogenolysis of Methoxy Groups in Lignin Model Monomers

In addition to the linkages between the propylbenzene units (see Section 1.7.1.3), lignin contains a large number of methoxy groups bound to the aromatic rings. Because methoxybenzenes are in lower demand than arenes and phenols, the hydrogenolysis of methoxybenzenes has been investigated in recent years. 2-Methoxyphenol and derivatives are frequently used as model substrates. However, most studies use very harsh conditions (e.g., >300 °C) and aim at the production of small hydrocarbons for biofuel.[22,23] Although demethoxylation (hydrogenolysis of the C_{aryl}—OMe bond) of substrates without other oxygen atoms can be achieved by using the systems introduced above (Sections 1.7.1.1 and 1.7.1.2), methods for the regioselective hydrogenolysis of methoxybenzenes are rather limited.

One example of the regioselective hydrogenolysis at a methoxy position of a lignin model monomer is the ruthenium-catalyzed demethoxylation directed by a phenolic hydroxy group (Scheme 15).[24,25] Thus, methoxyphenols **27**, such as 2-methoxyphenol and 2,6-dimethoxyphenol are converted into cyclohexanol (**28**) and methanol using a commercially available ruthenium-on-carbon (Ru/C) catalyst in combination with magnesium oxide. Demethoxylation of 2-methoxycyclohexan-1-ol is very slow with this system, indicating that the demethoxylation proceeds before hydrogenation of the aromatic ring to a cyclohexane. If no phenolic hydroxy group is present in the substrate (i.e., methoxybenzenes), the yield of cyclohexane (<20%) is much lower than the yield of cyclohexanol from methoxyphenols.

Scheme 15 Hydrogenolysis of Methoxyphenols Using Ruthenium on Carbon and Magnesium Oxide[24]

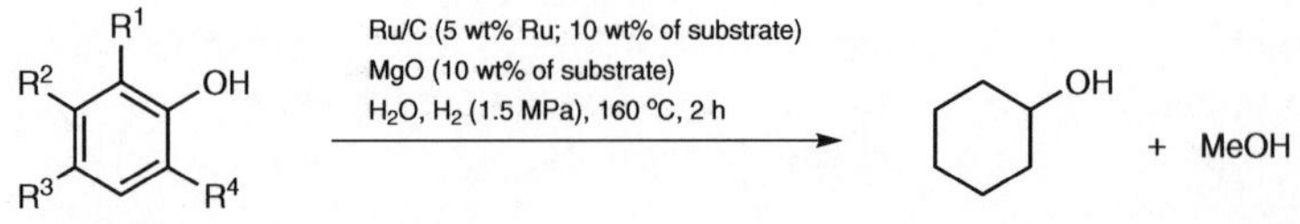

27 → 28 + MeOH

Ru/C (5 wt% Ru; 10 wt% of substrate)
MgO (10 wt% of substrate)
H_2O, H_2 (1.5 MPa), 160 °C, 2 h

R^1	R^2	R^3	R^4	Yield[a] (%) 28	Yield[a] (%) MeOH	Ref
OMe	H	H	H	79	85	[24]
H	OMe	H	H	85	85	[24]
H	H	OMe	H	55	52	[24]
OMe	H	H	OMe	69	84[b]	[24]

[a] Determined by GC.
[b] Based on the total number of methoxy groups in the substrate.

Reports of the selective demethylation of methoxybenzenes to give phenols (i.e., hydrogenolysis of a $H_3C—OC_{aryl}$ bond) are limited in comparison with demethoxylation. Kusumoto and Nozaki have reported that (hydroxycyclopentadienyl)iridium complexes (e.g., **29**) catalyze demethylation (Scheme 16).[26] Although the selectivity is high at low conversions, this catalyst also shows activity for the hydrogenolysis of phenols to arenes; therefore, it is difficult to obtain a high yield of the phenol (e.g., **30**).

Scheme 16 H_3C–O Hydrogenolysis Catalyzed by an Iridium Complex[26]

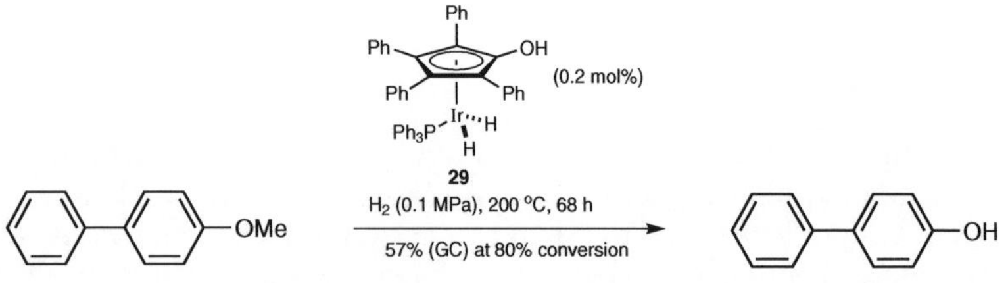

29 (0.2 mol%)
H_2 (0.1 MPa), 200 °C, 68 h
57% (GC) at 80% conversion

30

1.7.2 Hydrogenolysis of Saturated Ethers

1.7.2.1 Ethers Bearing Directing Hydroxy Groups

The hydrogenolysis of saturated ethers is more difficult than that of aryl ethers. Selective hydrogenolysis of one of the two C—O bonds is also rather difficult. However, if a hydroxymethyl group is attached to one carbon of the ether group [R^1O—$C(R^2)$—CH_2OH structure], the O—C bond can be selectively dissociated. This reaction was first reported in 2009,[27] and typical substrates are tetrahydrofurfuryl alcohol (**31**, n = 1),[27] (tetrahydro-2*H*-pyran-2-yl)methanol (**31**, n = 2),[28] and 2-alkoxyethanols **33**.[29] The products from these substrates are pentane-1,5-diol (**32**, n = 1), hexane-1,6-diol (**32**, n = 2), and a 1:1 mixture of ethanol and alcohols **34**, respectively.

Active catalysts for this hydrogenolysis have the general formula of M^1-M^2O_x/support (M^1 = Rh, Ir; M^2 = V, Mo, W, Re; support = silica gel or carbon).[27–43] Monometallic M^1/support or M^2O_x/support catalysts have much lower activity and selectivity. Rh-ReO_x/silica gel,[27,30,31,33,35,36] Rh-ReO_x/carbon,[28,34,37] Rh-MoO_x/silica gel,[30,31,34,38] and Ir-ReO_x/silica gel[32,43] catalysts have been investigated intensively. These catalysts also exhibit some activity in the C—O hydrogenolysis of simple alcohols. In order to obtain high yields of the target alcohols, it is necessary to make an appropriate selection of the catalyst and reaction conditions (particularly the temperature and time). Generally speaking, temperatures higher than 120 °C decrease the yield because of over-hydrogenolysis of the produced alcohol. Water is by far the best solvent. Among the reported catalysts, Rh-ReO_x/carbon gives the highest yield of pentane-1,5-diol (**32**, n = 1; 94% GC yield) from tetrahydrofurfuryl alcohol (**31**, n = 1).[28] The Rh-ReO_x/C catalyst is easily prepared from a commercially available rhodium-on-carbon (Rh/C) catalyst and ammonium perrhenate. Because the catalyst is poisoned by iron or nickel species, it should not come into contact with metal apparatus during preparation, storage, or use. Sulfur compounds or base also deactivate the catalyst.

The reported data for this type of reaction is summarized in Scheme 17. It should be noted that the low conversion values in some cases do not mean that the catalytic reaction had ceased: Deactivation of these types of catalysts during the reaction is usually not severe, and prolonging the reaction time generally leads to an increase in conversion; however, the selectivity is seen to decrease due to over-hydrogenolysis of the produced alcohol.

for references see p 265

Scheme 17 Hydrogenolysis of Ethers Directed by a Hydroxymethyl Group with M^1-M^2O_x-Type Catalysts[27–30,32,34,41]

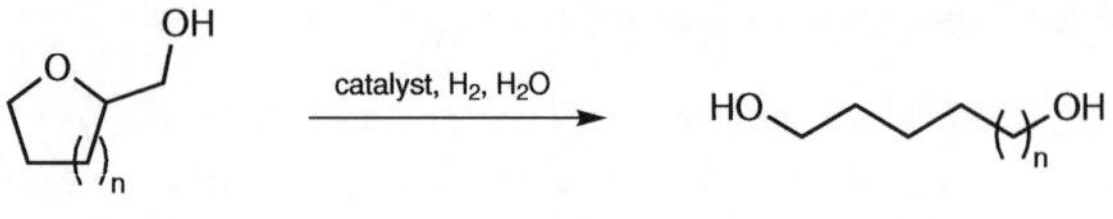

catalyst, H_2, H_2O

31 → **32**

n	Catalyst	H_2 Pressure (MPa)	Conditions	Conversion (%)	Yield[a] (%)	Ref
1	Rh-ReO$_x$/C	8	100 °C, 24 h	99	94	[28]
1	Rh-ReO$_x$/silica gel	8	120 °C, 24 h	96	80	[27]
1	Rh-MoO$_x$/silica gel	8	100 °C, 24 h	94	85	[30]
1	Rh-WO$_x$/silica gel	8	120 °C, 4 h	30	26	[30]
1	Rh-MoO$_x$/C	3.4	100 °C, 4 h	52	47	[34]
1	Ir-ReO$_x$/silica gel	8	120 °C, 8 h	92	82	[32]
1	Ir-MoO$_x$/silica gel	8	120 °C, 2 h	31	28	[32]
1	Ir-WO$_x$/silica gel	8	120 °C, 2 h	5.1	4.8	[32]
1	Ir-VO$_x$/silica gel	6	80 °C, flow	57	50	[41]
2	Rh-ReO$_x$/C	8	100 °C, 24 h	97	84	[28]

[a] Determined by GC.

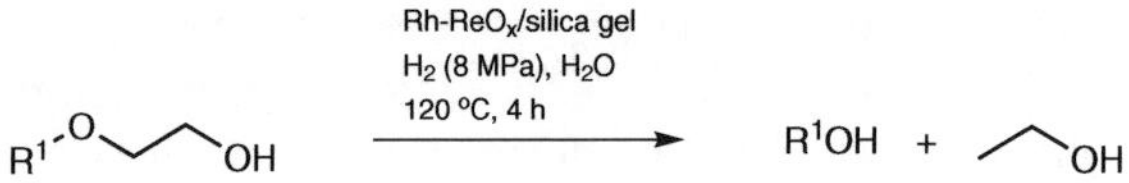

Rh-ReO$_x$/silica gel
H_2 (8 MPa), H_2O
120 °C, 4 h

33 → R^1OH (**34**) + EtOH

R^1	Conversion (%)	Yield[a] (%)		Ref
		34	EtOH	
Et	32	31[b]		[29]
iPr	42	40	39	[29]
Bu	41	39	40	[29]

[a] Determined by GC.
[b] Yield based on the formation of two molecules of EtOH from one substrate molecule.

The mechanism of the hydrogenolysis of ethers bearing a hydroxymethyl group over catalysts of the type M^1-M^2O_x has been investigated in detail by employing catalyst characterization, kinetic studies, and labeling studies with molecular deuterium and deuterium oxide.[29,31,32] The proposed mechanism is composed of four steps: (i) adsorption of the substrate with a hydroxymethyl group at the M^2 site; (ii) dissociation of molecular hydrogen at the M^1 site to give a hydride-like species and a proton; (iii) S_N2-like attack of the hydride-like species at the M^1-M^2 interface site; and (iv) desorption of the product (Scheme 18). The formation of a hydride-like active species is also supported by the fact that this type of catalyst shows very high activity and selectivity in the hydrogenation of α,β-unsaturated aldehydes to unsaturated alcohols,[44,45] which typically proceeds with hydride reagents such as sodium borohydride.

Scheme 18 Proposed Mechanism for Hydrogenolysis Using M^1-M^2O_x-Type Catalysts[29,31,32]

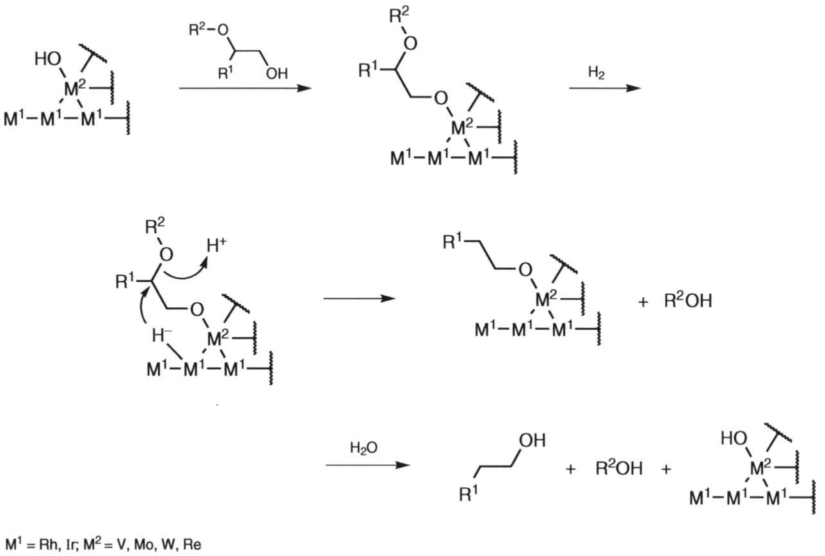

M^1 = Rh, Ir; M^2 = V, Mo, W, Re

The adsorption of secondary alcohols onto a metal oxide surface is more difficult than that of primary alcohols because of steric effects. Consequently, the directing effect of the hydroxy group on the hydrogenolysis of the neighboring C—O bond is weaker in secondary alcohols than in primary alcohols. The hydrogenolysis of tetrahydrofuran-3-ol (**37**) and tetrahydro-2*H*-pyran-3-ol (**39**) over Ir-ReO_x/silica gel gives butane-1,3-diol (**38**) and pentane-1,4-diol (**40**), respectively, with ca. 80% selectivity. However, the selectivity in the transformation of tetrahydrofurfuryl alcohol (**35**) into pentane-1,5-diol (**36**) is over 90% in the same conversion range using the same catalyst (Scheme 19).[32] Rhodium-based catalysts are less selective in this hydrogenolysis directed by a secondary hydroxy group, because of their relatively high activity in the hydrogenolysis of C—O bonds without directing groups (see Section 1.7.2.2).

Scheme 19 Hydrogenolysis of Ethers Directed by a Secondary Hydroxy Group Using an Ir-ReO_x/Silica Gel Catalyst[32]

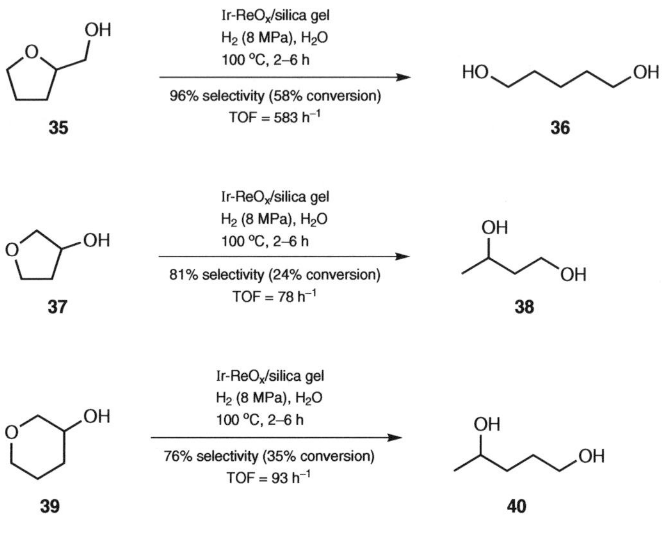

35 → **36**: Ir-ReO_x/silica gel, H_2 (8 MPa), H_2O, 100 °C, 2–6 h; 96% selectivity (58% conversion), TOF = 583 h^{-1}

37 → **38**: Ir-ReO_x/silica gel, H_2 (8 MPa), H_2O, 100 °C, 2–6 h; 81% selectivity (24% conversion), TOF = 78 h^{-1}

39 → **40**: Ir-ReO_x/silica gel, H_2 (8 MPa), H_2O, 100 °C, 2–6 h; 76% selectivity (35% conversion), TOF = 93 h^{-1}

for references see p 265

Pentane-1,5-diol (32, **n = 1**)**; Typical Procedure:**[28]

Preparation of the Rh-ReO$_x$/C catalyst: NH_4ReO_4 (33 mg, 0.12 mmol) was dissolved in H_2O (1.3 mL). A part of this soln (ca. 0.5 mL) was added to a commercial Rh/C catalyst (1 g; 5 wt% Rh, 0.48 mmol) in a 50-mL beaker, and the reagents were well mixed using a glass or plastic spatula. The beaker was placed on a hot plate (ca. 70 °C), and the solvent was evaporated with constant mixing. During the evaporation, the rest of the NH_4ReO_4 soln was added slowly to keep the catalyst wet. After completion of addition of the NH_4ReO_4 soln, the catalyst was dried over the hot plate with mixing, and then further dried in an oven at 110 °C overnight under air.

Catalytic hydrogenolysis: The reaction was performed in a stainless-steel autoclave with an inserted glass vessel. A 5% aq soln of tetrahydrofurfuryl alcohol (**31**, n = 1; 20 mL), the catalyst Rh-ReO$_x$/C (100 mg), and a stirrer bar were added to the autoclave. The reactor was sealed and the air content was purged by flushing three times with H_2 (1 MPa). The autoclave was pressurized with H_2 (1 MPa) and then heated to 100 °C. After 1 h (for reduction of the catalyst), the pressure of H_2 was increased to 8 MPa. After the appropriate reaction time (24 h), the reactor was cooled in a water bath. The autoclave contents were filtered and transferred into vials. The products were analyzed by GC. To calculate the selectivity values accurately, the products in the gas phase were also analyzed.

1.7.2.2 Ethers without Directing Hydroxy Groups

For substrates without a hydroxy group at the adjacent carbon of the ether site, catalysts of the type M^1-M^{2}O$_x$/support show much lower activity. The unmodified rhodium on silica gel catalyst has a higher activity in the hydrogenolysis of simple ethers, such as 2-methyltetrahydrofuran (**41**), than Rh-M^{2}O$_x$/silica gel catalysts.[27,29] The selectivity is simply governed by steric effects: a $C_{primary}$—O bond is preferentially dissociated over a $C_{secondary}$—O bond (Scheme 20).

Scheme 20 Hydrogenolysis of 2-Methyltetrahydrofuran Using Rhodium on Silica Gel Catalyst[29]

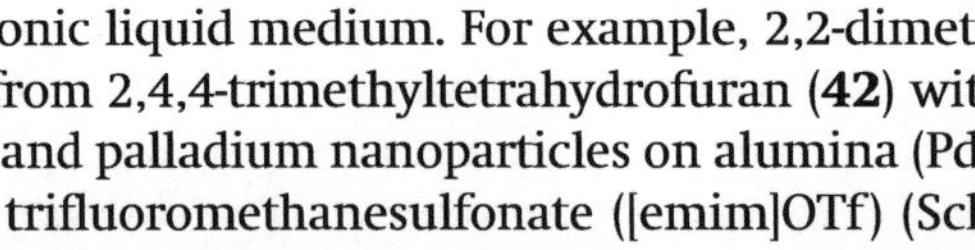

Rh/silica gel, H_2 (8 MPa)
H_2O, 120 °C
initial selectivity 90%
OH
41

Marks and co-workers have reported the selective hydrogenolysis of saturated ethers using a combination of a lanthanide trifluoromethanesulfonate, palladium nanoparticles, and an ionic liquid medium. For example, 2,2-dimethylpentan-1-ol (**43**) is obtained in 77% yield from 2,4,4-trimethyltetrahydrofuran (**42**) with ytterbium(III) trifluoromethanesulfonate and palladium nanoparticles on alumina (Pd@alumina) in 1-ethyl-3-methylimidazolium trifluoromethanesulfonate ([emim]OTf) (Scheme 21).[46] In contrast to the rhodium-catalyzed hydrogenolysis, the $C_{secondary}$—O bond is selectively dissociated. The mechanism of this system is proposed to involve the Lewis acid (lanthanide) catalyzed elimination of alkoxide to form an alkene and alcohol, followed by hydrogenation of the alkene by the palladium catalyst.

Scheme 21 Selective Hydrogenolysis of a $C_{secondary}$—O Bond Using Ytterbium(III) Trifluoromethanesulfonate and a Palladium Catalyst in an Ionic Liquid[46]

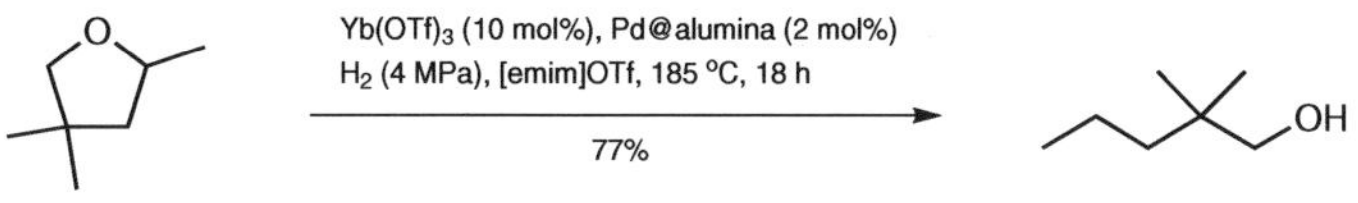

Yb(OTf)$_3$ (10 mol%), Pd@alumina (2 mol%)
H_2 (4 MPa), [emim]OTf, 185 °C, 18 h
77%

42 → 43

2,2-Dimethylpentan-1-ol (43); Typical Procedure:[46]
Yb(OTf)$_3$ (40 mg, 0.065 mmol), Pd@alumina (2 wt% Pd; 0.013 mmol Pd) [prepared by atomic layer deposition of the bis(1,1,1,5,5,5-hexafluoroacetylacetonato)palladium(II) precursor], 1-ethyl-3-methylimidazolium trifluoromethanesulfonate ([emim]OTf; 7.5 g, 26 mmol), and 2,4,4-trimethyltetrahydrofuran (**42**; 74 mg, 0.65 mmol) were added to a 100-mL Teflon sleeve in a glovebox. The Teflon sleeve was then put into an autoclave in the glovebox and sealed. The autoclave was purged with H_2 (0.7 MPa), and then the reactor was heated to 185 °C. The heated reactor was pressurized with H_2 (4 MPa), and the mixture was stirred for 18 h. Upon reaction completion, the reactor was cooled to rt and depressurized. The mixture was extracted with small portions of Et_2O and the combined extracts were passed through a plug of silica gel to remove residual ionic liquid, affording the product as a pale-yellow liquid; yield: 77%.

1.7.2.3 Total Hydrodeoxygenation of Cyclic Ethers to Alkanes

The hydrogenolysis of both C—O bonds in cyclic ethers (e.g., **44**,[47] **45** and **46**,[48] and **47**[49]) produces alkanes (hydrodeoxygenation). This reaction has been intensively investigated in the field of biofuel production.[50] The system involves the use of a combination of acid and noble metal catalysts (Scheme 22). Most other functional groups, such as alcohols and ketones, are completely reduced during the ether hydrogenolysis (total hydrodeoxygenation).[48] Because severe reaction conditions are applied to obtain oxygen-free hydrocarbons, C—C dissociation and isomerization of the carbon chain are sometimes also observed.[49] Dissociation of C—C bonds is always an undesirable reaction, and should be suppressed. Although a certain degree of isomerization of the carbon chain is desirable in biofuel production, because it increases the octane number (for gasoline) or the freezing point (for jet or diesel fuels), isomerization is also unfavorable from the point of view of organic synthesis. The use of acids that are too strong is not suitable for total hydrodeoxygenation if C—C dissociation or isomerization needs to be avoided.

Scheme 22 Total Hydrodeoxygenation of Saturated Ethers[47–49]

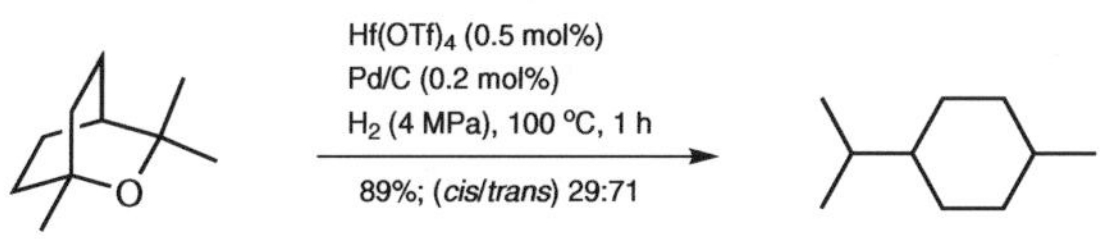

Hf(OTf)$_4$ (0.5 mol%)
Pd/C (0.2 mol%)
H_2 (4 MPa), 100 °C, 1 h
89%; (*cis/trans*) 29:71

44

Pt/silica gel•alumina
H_2 (6 MPa), THF
260 °C, flow system

45 43:57 46

11 + 10

73% 16%

Pt/ZrP, H_2
350 °C, flow system

Bu

4 4

+ other C_{15} hydrocarbons

47 47% 28%

Tetrahydrofuran derivatives, as the precursors of biofuels, are produced by hydrogenation of the corresponding furan derivatives. Noble metal catalysts have activity in hydrogenation, and therefore recent research in biofuel production tends to directly convert furan derivatives into hydrocarbons without isolation of the tetrahydrofuran derivatives. These systems are described in Section 1.7.3.1. The catalysts introduced in Section 1.7.3.1 (Scheme 24) can probably also be applied to the total hydrodeoxygenation of saturated ethers.

1.7.3 Hydrogenolysis of Furans

Furan derivatives are one of the major classes of biomass-derived compounds. Dehydration of hemicellulose and hexoses produces furan-2-carbaldehyde (furfural) and 5-(hydroxymethyl)furan-2-carbaldehyde, respectively.[2,3] In particular, furan-2-carbaldehyde is currently produced on a large scale, and various compounds containing furan or tetrahydrofuran rings are synthesized from this precursor. Interest in the reactions of furan derivatives has grown rapidly in recent years. In this section, recent achievements in the hydrogenolysis of the C—O bond in the furan ring are summarized.

1.7.3.1 Hydrogenolysis via Tetrahydrofuran Derivatives

Many hydrogenolysis catalysts are also active in hydrogenation. Using such catalysts, furan derivatives can be directly converted into saturated hydrogenolysis products. For example, Pd-Ir-ReO_x/silica gel[51] and Rh-Ir-ReO_x/silica gel[52] are active in both the hydrogenation of furan-2-carbaldehyde (**48**) to (tetrahydrofuran-2-yl)methanol, and the hydrogenolysis of (tetrahydrofuran-2-yl)methanol to give pentane-1,5-diol (**36**). Furan-2-carbaldehyde can be converted into pentane-1,5-diol (**36**) using these catalysts under two-step temperature conditions (Scheme 23).[52] Hydrogenation and hydrogenolysis proceed in the low-temperature and high-temperature steps, respectively.

Scheme 23 Conversion of Furan-2-carbaldehyde into Pentane-1,5-diol Using a Rh-Ir-ReO_x/Silica Gel Catalyst[52]

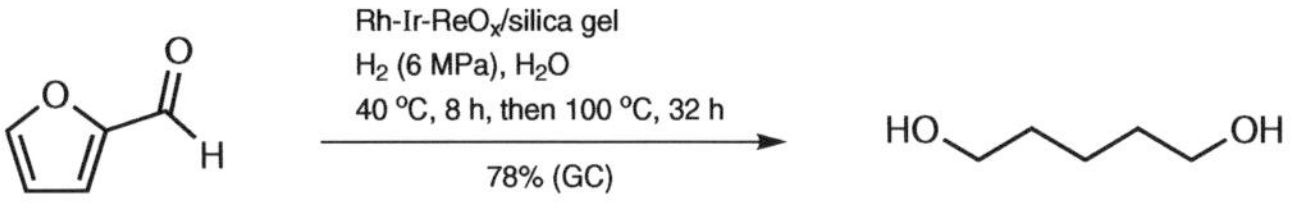

Hydrogenation/hydrogenolysis of furan derivatives (e.g., **49**,[53–56] **50**,[57] **51**,[58] and **52**[59]) is often applied in biofuel production. Condensation of furan-2-carbaldehyde (**48**) or 5-(hydroxymethyl)furan-2-carbaldehyde with another molecule such as acetone produces compounds (e.g., **49**) with a carbon number of 8–15; total hydrodeoxygenation then leads to jet or diesel range hydrocarbons (Scheme 24).[53–59] Combination of a noble metal catalyst and an acid catalyst with moderate acid strength, such as niobium phosphate, is generally effective in the total hydrodeoxygenation of furan derivatives. As described in Section 1.7.2.3, the use of reaction conditions that are too severe can lead to C—C dissociation and isomerization. The catalyst combination of a noble metal and niobium phosphate has also been applied in the total hydrodeoxygenation of other biomass-derived oxygenates, such as sorbitol[60–62] and triglycerides.[63]

Scheme 24 Total Hydrodeoxygenation of Furan Derivatives Using Noble Metal and Acid Catalysts[53–59]

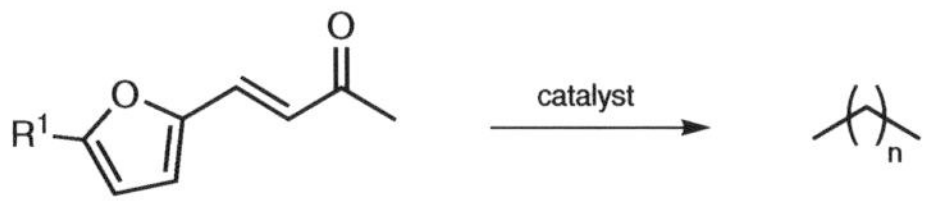

R^1	n	Catalyst	Conditions	Yield[a] (%)	Ref
CH_2OH	7	$Hf(OTf)_4$ (2 mol%), Pd/C (5 mol%)	H_2 (6 MPa), octane, 60 °C, 8 h, then 180 °C, 20 h	90	[53]
H	6	Pd/Nb_2O_5	H_2 (2 MPa), cyclohexane, 170 °C, 24 h	96	[54]
H	6	$Pd/NbOPO_4$	H_2 (2 MPa), cyclohexane, 170 °C, 24 h	94	[54]
H	6	Pd/Al-MCM-41	H_2 (4 MPa), CO_2 (14 MPa), 80 °C, 20 h	99	[55]
H	6	Pd/Nb_2O_5/silica gel	H_2 (2.5 MPa), cyclohexane, 170 °C, 24 h	96	[56]

[a] Determined by GC.

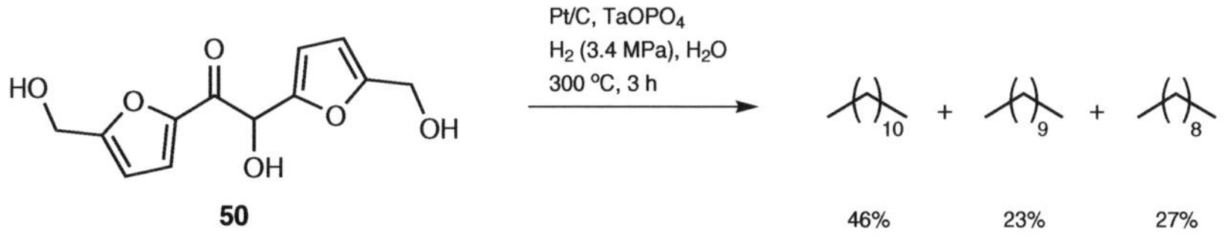

for references see p 265

Pr
51
Ni/H-β-zeolite
H_2 (6 MPa), neat
230 °C, flow
42%
Pr
4 4

Et
O
H
52
Pt/$NbOPO_4$
H_2 (3.4 MPa)
octane
250 °C, 4 h
4 33%
6 20%
7 13%
Et Et 12%

1.7.3.2 Ring Opening of Furan Derivatives not Involving a Tetrahydrofuran Intermediate

The furan ring is a reactive group in substitution, ring-opening, and rearrangement reactions, as well as hydrogenations.[64] Rearrangement reactions and subsequent hydrogenation or hydrogenolysis can give products different to those obtained by hydrogenolysis of tetrahydrofuran derivatives (see Section 1.7.3.1). Ring-opening reaction and subsequent hydrogenation also has a different selectivity pattern to ring opening (hydrogenolysis) that occurs after hydrogenation to a tetrahydrofuran derivative. Furfuryl alcohol is a key reactant in rearrangement or ring-opening reactions because of the high reactivity of the benzyl-like hydroxymethyl group. Furfuryl alcohol is easily produced from furan-2-carbaldehyde under reductive conditions, and the latter has often been used as a precursor for the rearrangement/reduction of furfuryl alcohol.

In Table 1, results for the reduction of furan derivatives that do not result in tetrahydrofuran formation are summarized.[33,65–77] The formation of a cyclopentane ring (entries 1–8), first reported by Hronec and Fulajtarová in 2012,[65] is distinctive. This reaction occurs in near-neutral water as solvent. Acidic conditions lead to the formation of a colored polymerized solid. A possible mechanism for the cyclopentane ring formation from furan-2-carbaldehyde derivatives is shown in Scheme 25.[68,69,72] The key steps are dehydration/hydration of the furfuryl alcohol and subsequent intramolecular aldol condensation. Conversion of furfuryl alcohol into 4-hydroxycyclopent-2-en-1-one in hot water has indeed been reported.[78] Reduction of (5-hydroxymethyl)furan-2-carbaldehyde to 1-hydroxyhexane-2,5-dione is a reaction that is related to the cyclopentane ring formation: hydrogenation of the C=C bond after dehydration/hydration gives a dione product (entries 9–11). Production of polyketones using a similar reaction to the one for 1-hydroxyhexane-2,5-dione formation has also been reported (entries 12 and 13).[76,79]

On the other hand, reduction of furan-2-carbaldehyde or furfuryl alcohol using a platinum/hydrotalcite catalyst can give pentane-1,2-diol (entries 14 and 15).[77] Hydrotalcite is a basic material, and the use of a magnesium oxide support, which is a typical basic support, gives similar results (68% yield of pentane-1,2-diol from furan-2-carbaldehyde). Other products include pentane-1,5-diol, as well as the hydrogenation product tetrahydrofurfuryl alcohol. In contrast to M^1-M^2O_x-type catalyst systems, tetrahydrofurfuryl alcohol does not react at all in this platinum/hydrotalcite catalyst system, indicating that tetrahydrofurfuryl alcohol is not an intermediate in the formation of the pentanediols. Other catalysts that have been reported for the production of pentane-1,2- and pentane-1,5-diols from furfuryl alcohol are metal catalysts on basic supports such as Pt/Li-Co_2AlO_4,[80] Cu/$Mg_3AlO_{4.5}$,[81] and Ru/MnO_x;[82] however, the yield is lower in these cases. The mechanism of pentane-1,2- and pentane-1,5-diol formation has not yet been elucidated.

Table 1 Reduction of Furan Derivatives without Tetrahydrofuran Ring Formation[33,65–77,79]

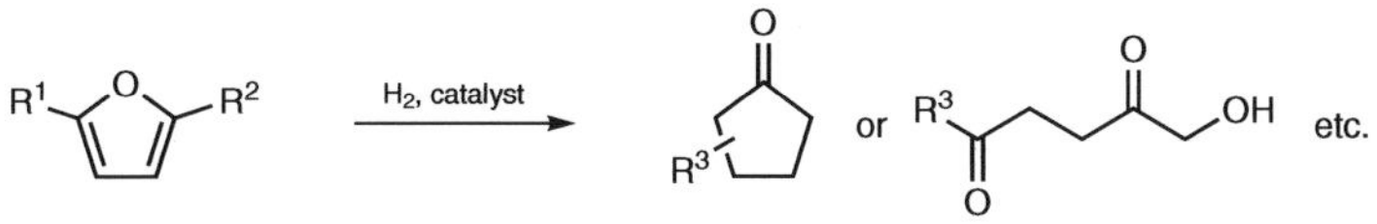

Entry	Substrate	Catalyst	Conditions	Product	Yield[a] (%)	Ref
1	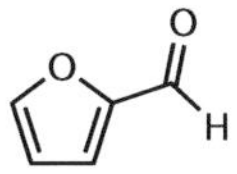	Pt/C	H_2 (8 MPa), H_2O, 160 °C, 0.5 h	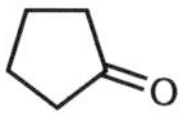	77	[65]
2	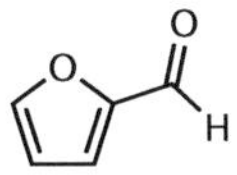	Ni/CNT[b]	H_2 (5 MPa), H_2O, 140 °C, 10 h	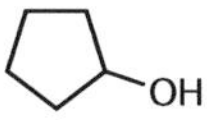	84	[66]
3	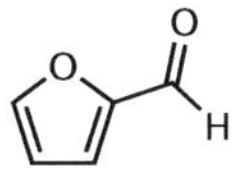	Cu/Zn/Al mixed oxide	H_2 (4 MPa), H_2O, 150 °C, 10 h	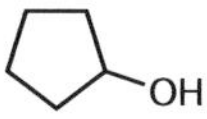	84	[67]
4	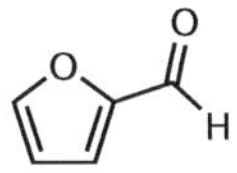	Pd-Cu/C	H_2 (3 MPa), H_2O, 160 °C, 1 h	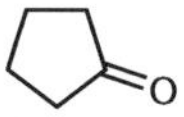	92	[68]
5	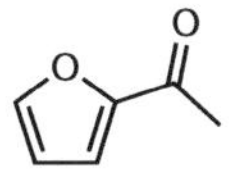	Ru/MIL-101[c]	H_2 (4 MPa), H_2O, 160 °C, 2.5 h	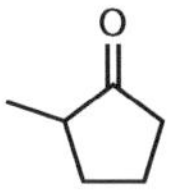	82	[69]
6	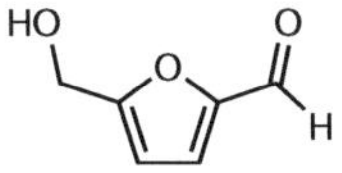	Au/Nb_2O_5	H_2 (8 MPa), H_2O, 140 °C, 12 h	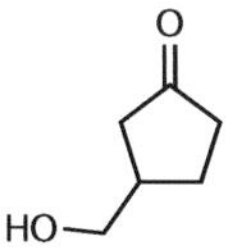	86	[70]
7	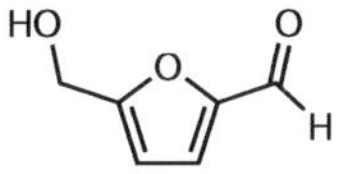	Ni/alumina	H_2 (2 MPa), H_2O, 140 °C, 6 h	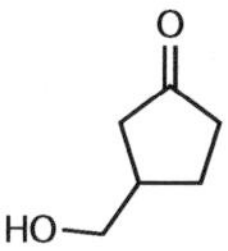	81	[71]
8	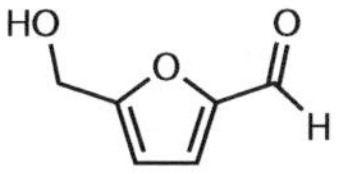	Ta_2O_5, Pt/silica gel	H_2 (3 MPa), H_2O, 140 °C, 12 h	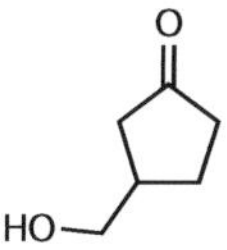	81	[72]
9	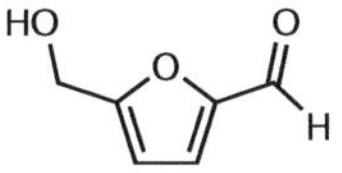	Rh-Re/silica gel	H_2 (1 MPa), H_2O, 120 °C, 1 h; H_2 (8 MPa), 17 h	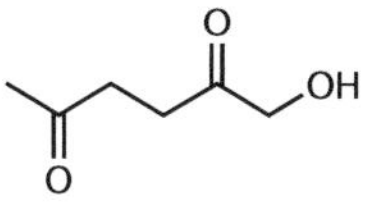	81	[33]
10	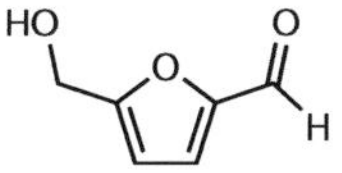	Pd/C, Amberlyst 15	H_2 (5 MPa), THF, 80 °C, 15 h	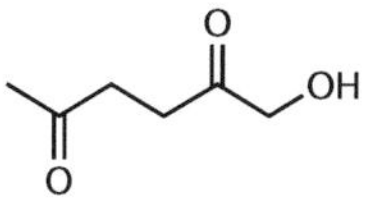	77	[73]

for references see p 265

Table 1 (cont.)

Entry	Substrate	Catalyst	Conditions	Product	Yield[a] (%)	Ref
11	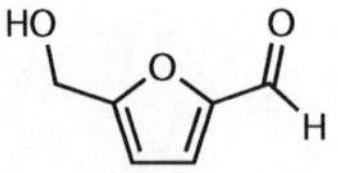	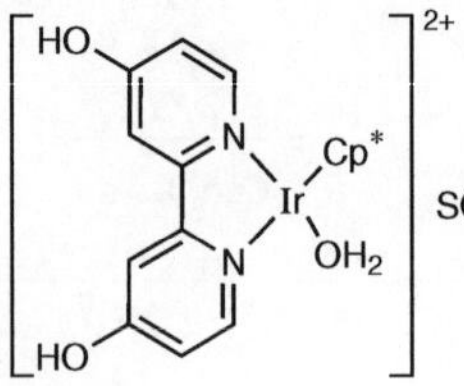 SO_4^{2-}	formate buffer (pH 2.5), 130 °C, 2 h	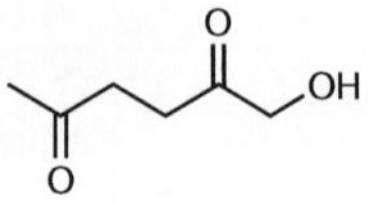	92	[74]
12	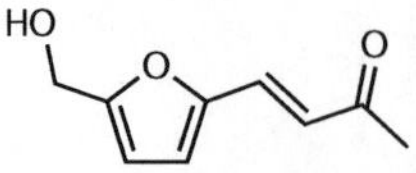	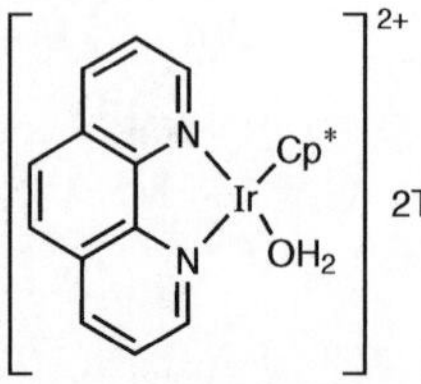 2TfO⁻	H_2 (1 MPa), H_2O, 120 °C, 2 h	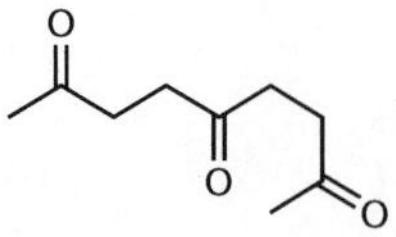	72[d]	[79]
13	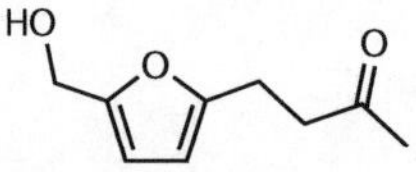	Pd/C	50% aq AcOH, trace H_2, air, 100 °C, 3 h	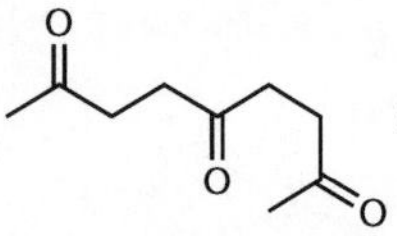	96[d]	[76]
14	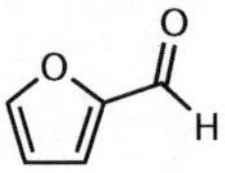	Pt/hydrotalcite	H_2 (3 MPa), iPrOH, 150 °C, 4 h	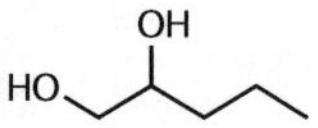	73	[77]
15	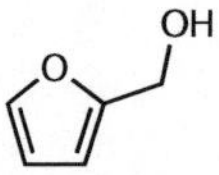	Pt/hydrotalcite	H_2 (3 MPa), iPrOH, 150 °C, 4 h	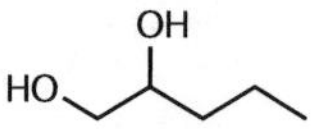	80	[77]

[a] Determined by GC, unless otherwise noted.
[b] CNT = carbon nanotubes.
[c] MIL-101 = materials of Institut Lavoisier No. 101.
[d] Isolated yield.

Scheme 25 Mechanism of Reduction of 2-Furyl Carbonyl Compounds to Cyclopentanones or Diketones[68,69,72]

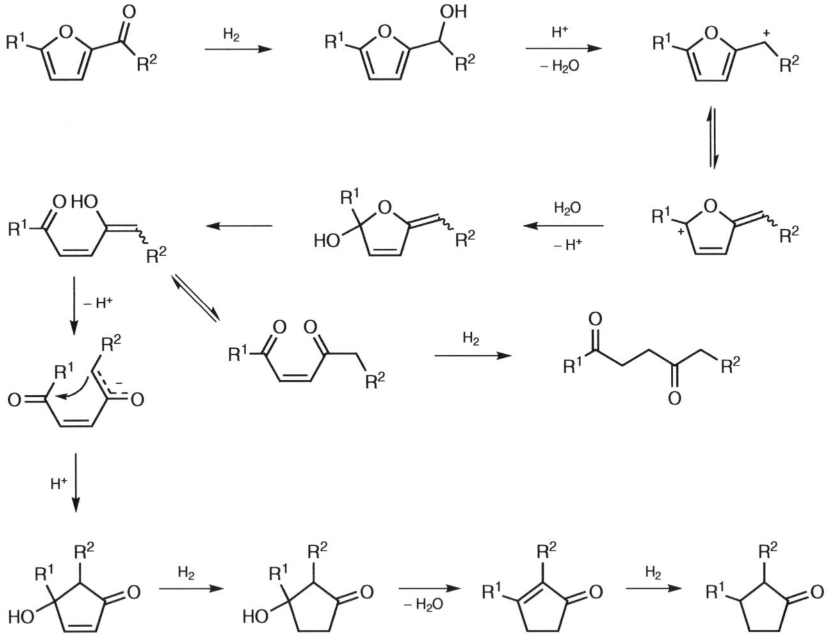

Although good yields of a single product are obtained in the transformations described in Table 1, the product compositions vary widely in response to small changes in the reaction conditions such as hydrogen pressure, temperature, and pH value. C5 ring formation and diketone formation are competing reactions. In addition, hydrogenation of the furan ring and polymerization of furfuryl alcohols also compete with target C5 ring or diketone formation. Careful optimization of reaction conditions is necessary to obtain good yields.

Cyclopentanone (Table 1, Entry 1):[65]
Furan-2-carbaldehyde (1 g, 10.4 mmol), H_2O (20 mL), and commercially available Pt/C catalyst (5 wt% Pt, 0.05 g) were added to a 100-mL, stainless-steel autoclave equipped with a stirrer and temperature controller. The reactor was sealed and flushed several times with low pressure H_2 and then pressurized with H_2 (8 MPa) at rt. The reactor was then heated to 160 °C. After 30 min, the reactor was quickly cooled, and the contents of the reactor were transferred to a vial. The catalyst was removed by centrifugation. The aqueous phase was analyzed by GC; yield: 77%.

Nonane-2,5,8-trione (Table 1, Entry 12):[79]
The catalyst was prepared by the reaction of $\{IrCl_2(Cp^*)\}_2$ with the ligand 1,10-phenanthroline in MeOH and subsequent chloride removal with AgOTf in H_2O.[75]

The reduction was conducted in a 15-mL autoclave. (*E*)-4-[5-(Hydroxymethyl)furan-2-yl]but-3-en-2-one (83 mg, 0.5 mmol), H_2O (2 mL), and the catalyst (0.625 μmol; 0.13 mol%) were added to the reactor. The reactor was flushed three times with H_2 (1 MPa) and then heated to 120 °C. After 2 h, the reactor was cooled to rt, and the mixture was extracted with CH_2Cl_2 (3 × 2 mL). The extract was dried (Na_2SO_4), filtered, and concentrated; yield: 61.3 mg (72%).

for references see p 265

1.7.4 Conclusions

A number of catalyst systems for the hydrogenolysis of ethers have been reported since 2010. Selective hydrogenolysis of the C_{aryl}—O bond in alkyl aryl ethers is now possible. Catalysts of the type M^1-M^2O_x/support promote highly selective hydrogenolysis of saturated ethers that contain a hydroxymethyl group. Methods for the formation of cyclopentane rings and diketones have been discovered that use furan-2-carbaldehyde, furfuryl alcohol, and their substituted derivatives as substrates. The production of alkanes that retain the carbon chain of the substrate obtained by aldol condensation of furan-2-carbaldehyde or 5-(hydroxymethyl)furan-2-carbaldehyde with ketones has also been intensively investigated.

To date, most of the recently discovered systems are oriented toward the production of biofuels or biochemicals, and not toward the synthesis of fine chemicals. The substrate scope of these reactions has not been well-explored, and the optimization of many parameters of the reaction conditions is often required for each substrate. Further development of these catalyst systems will encourage their application for the preparation of a variety of fine chemicals.

References

[1] Zakzeski, J.; Bruijnincx, P. C. A.; Jongerius, A. L.; Weckhuysen, B. M., *Chem. Rev.*, (2010) **110**, 3552.
[2] Lange, J.-P.; van der Heide, E.; van Buijtenen, J.; Price, R., *ChemSusChem*, (2012) **5**, 150.
[3] van Putten, R.-J.; van der Waal, J. C.; de Jong, E.; Rasrendra, C. B.; Heeres, H. J.; de Vries, J. G., *Chem. Rev.*, (2013) **113**, 1499.
[4] Ruppert, A. M.; Weinberg, K.; Palkovits, R., *Angew. Chem.*, (2012) **124**, 2614; *Angew. Chem. Int. Ed.*, (2012) **51**, 2564.
[5] Zaheer, M.; Kempe, R., *ACS Catal.*, (2015) **5**, 1675.
[6] Sergeev, A. G.; Hartwig, J. F., *Science (Washington, D. C.)*, (2011) **332**, 439.
[7] Kelley, P.; Lin, S.; Edouard, G.; Day, M. W.; Agapie, T., *J. Am. Chem. Soc.*, (2012) **134**, 5480.
[8] Tobisu, M.; Morioka, T.; Ohtsuki, A.; Chatani, N., *Chem. Sci.*, (2015) **6**, 3410.
[9] Sergeev, A. G.; Webb, J. D.; Hartwig, J. F., *J. Am. Chem. Soc.*, (2012) **134**, 20 226.
[10] Gao, F.; Webb, J. D.; Hartwig, J. F., *Angew. Chem.*, (2016) **128**, 1496; *Angew. Chem. Int. Ed.*, (2016) **55**, 1474.
[11] Ren, Y.; Yan, M.; Wang, J.; Zhang, Z. C.; Yao, K., *Angew. Chem.*, (2013) **125**, 12 906; *Angew. Chem. Int. Ed.*, (2013) **52**, 12 674.
[12] Ren, Y.; Tian, M.; Tian, X.; Wang, Q.; Shang, H.; Wang, J.; Zhang, Z. C., *Catal. Commun.*, (2014) **52**, 36.
[13] Huo, W.; Li, W.; Zhang, M.; Fan, W.; Chang, H.; Jameel, H., *Catal. Lett.*, (2014) **144**, 1159.
[14] Xu, H.; Yu, B.; Zhang, H.; Zho, Y.; Yang, Z.; Xu, J.; Han, B.; Liu, Z., *Chem. Commun. (Cambridge)*, (2015) **51**, 12 212.
[15] Parsell, T. H.; Owen, B. C.; Klein, I.; Jarrell, T. M.; Marcum, C. L.; Haupert, L. J.; Amundson, L. M.; Kenttämaa, H. I.; Ribeiro, F.; Miller, J. T.; Abu-Omar, M. M., *Chem. Sci.*, (2013) **4**, 806.
[16] Nichols, J. M.; Bishop, L. M.; Bergman, R. G.; Ellman, J. A., *J. Am. Chem. Soc.*, (2010) **132**, 12 554.
[17] Zhou, X.; Mitra, J.; Rauchfuss, T. B., *ChemSusChem*, (2014) **7**, 1623.
[18] Zaheer, M.; Hermannsdörfer, J.; Kretschmer, W. P.; Motz, G.; Kempe, R., *ChemCatChem*, (2014) **6**, 91.
[19] Stavila, V.; Parthasarathi, R.; Davis, R. W.; El Gabaly, F.; Sale, K. L.; Simmons, B. A.; Singh, S.; Allendorf, M. D., *ACS Catal.*, (2016) **6**, 55.
[20] Kim, J. K.; Lee, J. K.; Kang, K. H.; Lee, J. W.; Song, I. K., *J. Mol. Catal. A: Chem.*, (2015) **410**, 184.
[21] Rensel, D. J.; Rouvimov, S.; Gin, M. E.; Hicks, J. C., *J. Catal.*, (2013) **305**, 256.
[22] Wang, H.; Male, J.; Wang, V., *ACS Catal.*, (2013) **3**, 1047.
[23] Saidi, M.; Samimi, F.; Karimipourfard, D.; Nimmanwudipong, T.; Gates, B. C.; Rahimpour, M. R., *Energy Environ. Sci.*, (2014) **7**, 103.
[24] Nakagawa, Y.; Ishikawa, M.; Tamura, M.; Tomishige, K., *Green Chem.*, (2014) **16**, 2197.
[25] Ishikawa, M.; Tamura, M.; Nakagawa, Y.; Tomishige, K., *Appl. Catal., B*, (2016) **182**, 193.
[26] Kusumoto, S.; Nozaki, K., *Nat. Commun.*, (2015) **6**, 6296.
[27] Koso, S.; Furikado, I.; Miyazawa, T.; Kunimori, K.; Tomishige, K., *Chem. Commun. (Cambridge)*, (2009), 2035.
[28] Chen, K.; Koso, S.; Kubota, T.; Nakagawa, Y.; Tomishige, K., *ChemCatChem*, (2010) **2**, 547.
[29] Koso, S.; Nakagawa, Y.; Tomishige, K., *J. Catal.*, (2011) **280**, 221.
[30] Koso, S.; Ueda, N.; Shinmi, Y.; Okumura, K.; Kuzuka, T.; Tomishige, K., *J. Catal.*, (2009) **267**, 89.
[31] Koso, S.; Watanabe, H.; Okumura, K.; Nakagawa, Y.; Tomishige, K., *Appl. Catal., B*, (2012) **111–112**, 27.
[32] Chen, K.; Mori, K.; Watanabe, H.; Nakagawa, Y.; Tomishige, K., *J. Catal.*, (2012) **294**, 171.
[33] Buntara, T.; Noel, S.; Phua, P. H.; Melián-Cabrera, I.; de Vries, J. G.; Heeres, H. J., *Angew. Chem.*, (2011) **123**, 7221; *Angew. Chem. Int. Ed.*, (2011) **50**, 7083.
[34] Chia, M.; Pagán-Torres, Y. J.; Hibbitts, D.; Tan, Q.; Pham, H. N.; Datye, A. K.; Neurock, M.; Davis, R. J.; Dumesic, J. A., *J. Am. Chem. Soc.*, (2011) **133**, 12 675.
[35] Buntara, T.; Noel, S.; Phua, P. H.; Melián-Cabrera, I.; de Vries, J. G.; Heeres, H. J., *Top. Catal.*, (2012) **55**, 612.
[36] Buntara, T.; Melián-Cabrera, I.; Tan, Q.; Fierro, J. L. G.; Neurock, M.; de Vries, J. G.; Heeres, H. J., *Catal. Today*, (2013) **210**, 106.
[37] Chia, M.; O'Neill, B. J.; Alamillo, R.; Dietrich, P. J.; Ribeiro, F. H.; Miller, J. T.; Dumesic, J. A., *J. Catal.*, (2013) **308**, 226.
[38] Guan, J.; Peng, G.; Cao, Q.; Mu, X., *J. Phys. Chem. C*, (2014) **118**, 25 555.

[39] Wang, Z.; Pholjaroen, B.; Li, M.; Dong, W.; Li, N.; Wang, A.; Wang, X.; Cong, Y.; Zhang, T., *J. Energy Chem.*, (2014) **23**, 427.
[40] Hakim, S. H.; Sener, C.; Alba-Rubio, A. C.; Gostanian, T. M.; O'Neill, B. J.; Ribeiro, F. H.; Miller, J. T.; Dumesic, J. A., *J. Catal.*, (2015) **328**, 75.
[41] Pholjaroen, B.; Li, N.; Huang, Y.; Li, L.; Wang, A.; Zhang, T., *Catal. Today*, (2015) **245**, 93.
[42] Alba-Rubio, A. C.; Sener, C.; Hakim, S. H.; Gostanian, T. M.; Dumesic, J. A., *ChemCatChem*, (2015) **7**, 3881.
[43] Xiao, B.; Zheng, M.; Li, X.; Pang, J.; Sun, R.; Wang, H.; Pang, X.; Wang, A.; Wang, X.; Zhang, T., *Green Chem.*, (2016) **18**, 2175.
[44] Tamura, M.; Tokonami, K.; Nakagawa, Y.; Tomishige, K., *Chem. Commun. (Cambridge)*, (2013) **49**, 7034.
[45] Tamura, M.; Tokonami, K.; Nakagawa, Y.; Tomishige, K., *ACS Catal.*, (2016) **6**, 3600.
[46] Atesin, A. C.; Ray, N. A.; Stair, P. C.; Marks, T. J., *J. Am. Chem. Soc.*, (2012) **134**, 14682.
[47] Li, Z.; Assary, R. S.; Atesin, A. C.; Curtiss, L. A.; Marks, T. J., *J. Am. Chem. Soc.*, (2014) **136**, 104.
[48] Xing, R.; Subrahmanyam, A. V.; Olcay, H.; Qi, W.; van Walsum, G. P.; Pendse, H.; Huber, G. W., *Green Chem.*, (2010) **12**, 1933.
[49] Li, G.; Li, N.; Wang, Z.; Li, C.; Wang, A.; Wang, X.; Cong, Y.; Zhang, T., *ChemSusChem*, (2012) **5**, 1958.
[50] Nakagawa, Y.; Liu, S.; Tamura, M.; Tomishige, K., *ChemSusChem*, (2015) **8**, 1114.
[51] Liu, S.; Amada, Y.; Tamura, M.; Nakagawa, Y.; Tomishige, K., *Green Chem.*, (2014) **16**, 617.
[52] Liu, S.; Amada, Y.; Tamura, M.; Nakagawa, Y.; Tomishige, K., *Catal. Sci. Technol.*, (2014) **4**, 2535.
[53] Song, H.-J.; Deng, J.; Cui, M.-S.; Li, X.-L.; Liu, X.-X.; Zhu, R.; Wu, W.-P.; Fu, Y., *ChemSusChem*, (2015) **8**, 4250.
[54] Xia, Q.-N.; Cuan, Q.; Liu, X.-H.; Gong, X.-Q.; Lu, G.-Z.; Wang, Y.-Q., *Angew. Chem.*, (2014) **126**, 9913; *Angew. Chem. Int. Ed.*, (2014) **53**, 9755.
[55] Chatterjee, M.; Matsushima, K.; Ikushima, Y.; Sato, M.; Yokoyama, T.; Kawanami, H.; Suzuki, T., *Green Chem.*, (2010) **12**, 779.
[56] Shao, Y.; Xia, Q.; Liu, X.; Lu, G.; Wang, Y., *ChemSusChem*, (2015) **8**, 1761.
[57] Liu, D.; Chen, E. Y.-X., *ChemSusChem*, (2013) **6**, 2236.
[58] Li, G.; Li, N.; Yang, J.; Li, L.; Wang, A.; Wang, X.; Cong, Y.; Zhang, T., *Green Chem.*, (2014) **16**, 594.
[59] Sreekumar, S.; Balakrishnan, M.; Goulas, K.; Gunbas, G.; Gokhale, A. A.; Louie, L.; Grippo, A.; Scown, C. D.; Bell, A. T.; Toste, F. D., *ChemSusChem*, (2015) **8**, 2609.
[60] West, R. M.; Tucker, M. H.; Braden, D. J.; Dumesic, J. A., *Catal. Commun.*, (2009) **10**, 1743.
[61] Li, N.; Tompsett, G. A.; Huber, G. W., *ChemSusChem*, (2010) **3**, 1154.
[62] Xi, J.; Xia, Q.; Shao, Y.; Ding, D.; Yang, P.; Liu, X.; Lu, G.; Wang, Y., *Appl. Catal., B*, (2016) **181**, 699.
[63] Xia, Q.; Zhuang, X.; Li, M. M.-J.; Peng, Y.-K.; Liu, G.; Wu, T.-S.; Soo, Y.-L.; Gong, X.-Q.; Wang, Y.; Tsang, S. C. E., *Chem. Commun. (Cambridge)*, (2016) **52**, 5160.
[64] Nakagawa, Y.; Tamura, M.; Tomishige, K., *ACS Catal.*, (2013) **3**, 2655.
[65] Hronec, M.; Fulajtarová, K., *Catal. Commun.*, (2012) **24**, 100.
[66] Zhou, M.; Zhu, H.; Niu, L.; Xiao, G.; Xiao, R., *Catal. Lett.*, (2014) **144**, 235.
[67] Wang, Y.; Zhou, M.; Wang, T.; Xiao, G., *Catal. Lett.*, (2015) **145**, 1557.
[68] Hronec, M.; Fulajtarová, K.; Vávra, I.; Soták, T.; Dobročka, E.; Mičušík, M., *Appl. Catal., B*, (2016) **181**, 210.
[69] Fang, R.; Liu, H.; Luque, R.; Li, Y., *Green Chem.*, (2015) **17**, 4183.
[70] Ohyama, J.; Kanao, R.; Esaki, A.; Satsuma, A., *Chem. Commun. (Cambridge)*, (2014) **50**, 5633.
[71] Perret, N.; Grigoropoulos, A.; Zanella, M.; Manning, T. D.; Claridge, J. B.; Rosseinsky, M. J., *ChemSusChem*, (2016) **9**, 521.
[72] Ohyama, J.; Kanao, R.; Ohira, Y.; Satsuma, A., *Green Chem.*, (2016) **18**, 676.
[73] Liu, F.; Audemar, M.; Vigier, K. D. O.; Clacens, J.; Campo, F. D.; Jérôme, F., *Green Chem.*, (2014) **16**, 4110.
[74] Wu, W.-P.; Xu, Y.-J.; Zhu, R.; Cui, M.-S.; Li, X.-L.; Deng, J.; Fu, Y., *ChemSusChem*, (2016) **9**, 1209.
[75] Xu, Z.; Yan, P.; Li, H.; Liu, K.; Liu, X.; Jia, S.; Zhang, Z. C., *ACS Catal.*, (2016) **6**, 3784.
[76] Sutton, A. D.; Waldie, F. D.; Wu, R.; Schlaf, M.; Silks, L. A. P., III; Gordon, J. C., *Nat. Chem.*, (2013) **5**, 428.
[77] Mizugaki, T.; Yamakawa, T.; Nagatsu, Y.; Maeno, Z.; Mitsudome, T.; Jitsukawa, K.; Kaneda, K., *ACS Sustainable Chem. Eng.*, (2014) **2**, 2243.
[78] Li, G.; Li, N.; Zheng, M.; Li, S.; Wang, A.; Cong, Y.; Wang, X.; Zhang, T., *Green Chem.*, (2016) **18**, 3607.
[79] Xu, Z.; Yan, P.; Xu, W.; Liu, X.; Xia, Z.; Chung, B.; Jia, S.; Zhang, Z. C., *ACS Catal.*, (2015) **5**, 788.

[80] Xu, W.; Wang, H.; Liu, X.; Ren, J.; Wang, Y.; Lu, G., *Chem. Commun. (Cambridge)*, (2011) **47**, 3924.
[81] Liu, H.; Huang, Z.; Zhao, F.; Cui, F.; Liu, X.; Xia, X.; Chen, J., *Catal. Sci. Technol.*, (2016) **6**, 668.
[82] Zhang, B.; Zhu, Y.; Ding, G.; Zheng, H.; Li, Y., *Green Chem.*, (2012) **14**, 3402.

1.8 Catalytic Reduction of Carbonates

Y. Li and K. Ding

General Introduction

This chapter focuses on the reduction of carbonates and is categorized according to the major products of the reductive transformations: formates, methanol, methane, and other products. Given that in the literature carbonates are referred to as either metal (bi)carbonates or organic carbonate esters, catalytic reduction involving both types of substrates, including (bi)carbonate salts, cyclic carbonates, acyclic carbonates, and polycarbonates, will be covered herein.

Carbonates are basic chemicals that are widely used in both industry and academia, and the annual production of various organic carbonates, including dimethyl carbonate, ethylene carbonate, propylene carbonate, and diphenyl carbonate, exceeds one million metric tons. Generally, these compounds are thermodynamically very stable, and their reduction is challenging, even though the reduction of metal bicarbonates has been known for over 100 years. In fact, organic carbonates are still viewed as "inert" compounds and are often used as green reaction media, fuel additives, and electrolytes.

As carbonates can be prepared relatively easily from carbon dioxide through low-barrier nonreductive transformations (for the reduction of carbon dioxide, see Section 1.9), the reduction of carbonates represents a viable route for the valorization of carbon dioxide to valuable chemicals and fuels, such as formic acid, formaldehyde, methanol, and their derivatives. Formate metal salts, accessible by hydrogenation of the corresponding bicarbonates, are considered potential chemical carriers in hydrogen storage and transportation.[1] In this regard, combining the hydrogenation of metal bicarbonates and the dehydrogenation of metal formates will probably lead to a practical hydrogen battery system.

The development of efficient catalytic systems for carbonate reduction is also attractive in the area of synthetic chemistry. The aim of this chapter is to provide an overview of the important methodological developments in the field of carbonate reduction by catalytic hydrogenation and transfer hydrogenation, whereas the photo- and electroreduction as well as the enzymatic reduction of carbonates is discussed only briefly.

SAFETY: Hydrogen is flammable. It poses an immediate fire and explosive hazard at very low concentrations. High concentrations cause suffocation. High-pressure reactors should be used in properly ventilated hoods.

1.8.1 Catalytic Reduction of Carbonates to Formates

For the reduction of carbonates to formates, almost all reports so far have been focused on the catalytic hydrogenation of metal bicarbonates. Early reports are based on heterogeneous catalysts, and many homogeneous catalytic systems based on transition-metal complexes have been developed in the last couple of decades. For related reductions of carbon dioxide to formates, see Section 1.9.1.

1.8.1.1 Homogeneous Reductions

The reduction of metal carbonates to formates using homogeneous catalytic systems has been extensively studied in recent decades, and many transition-metal complexes have been shown to be effective as catalysts for this reaction, mostly involving rhodium, ruthenium, iridium, or palladium as the active centers. However, reports on the hydrogenation of bicarbonate without the addition of carbon dioxide gas are rare, and the efficiencies of such reactions are generally much lower than those for hydrogenation performed in the presence of carbon dioxide.[2]

In 2010, Beller and Laurenczy disclosed an important finding that the iron complex formed in situ from iron(II) tetrafluoroborate hexahydrate and tetradentate phosphine ligand **1** could be used for the hydrogenation of bicarbonates (Scheme 1).[3] In the hydrogenation of sodium bicarbonate to sodium formate (**3**) in methanol under hydrogen pressure (6.0 MPa), turnover numbers of up to 610 are obtained with a turnover frequency of 30.5 h^{-1}. In a more recent report, new tetradentate ligand **2**, with a triphenylphosphine scaffold, was used with iron(II) tetrafluoroborate hexahydrate to reduce sodium bicarbonate to formate with a turnover number of up to 7546.[4]

Scheme 1 Hydrogenation of Sodium Bicarbonate to Sodium Formate Using Iron/Tetradentate Phosphine Ligands[3,4]

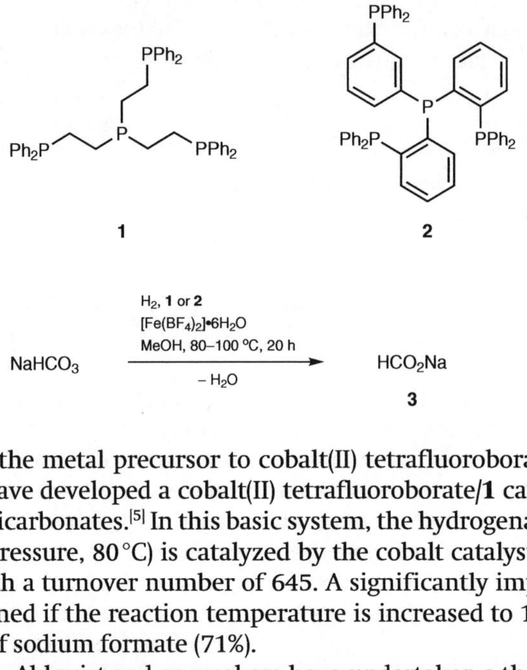

$NaHCO_3$ → (H_2, **1** or **2**, $[Fe(BF_4)_2]\cdot 6H_2O$, MeOH, 80–100 °C, 20 h; $-H_2O$) → HCO_2Na (**3**)

By changing the metal precursor to cobalt(II) tetrafluoroborate hexahydrate, Beller and co-workers have developed a cobalt(II) tetrafluoroborate/**1** catalyst system for the hydrogenation of bicarbonates.[5] In this basic system, the hydrogenation of sodium bicarbonate (6.0 MPa H_2 pressure, 80 °C) is catalyzed by the cobalt catalyst to give sodium formate in 94% yield with a turnover number of 645. A significantly improved turnover number of 3877 is obtained if the reaction temperature is increased to 120 °C, albeit at the expense of the yield of sodium formate (71%).

Recently, Ahlquist and co-workers have undertaken a theoretical study of the mechanism for the hydrogenation of bicarbonate catalyzed by tetradentate phosphine/iron complexes where they discussed the critical role of solvents.[6] In hydrogen-bond-donor solvents (protic solvents), the bicarbonate substrate is well solvated, and its reaction with the active $[FeH(\eta^2\text{-}H_2)(\mathbf{2})]^+$ complex results in an intermediate with a high free energy of formation. Accordingly, it is suggested that the rate of hydrogenation of bicarbonate can be increased simply by using a protic solvent with a lower ability to solvate the bicarbonate anion. Specifically, the use of solvents with a low dielectric constant will make the interaction between the substrate and the iron–hydride species more efficient, with a lower free-energy barrier for the entire reaction. In the case of a non-hydrogen-bonding solvent (aprotic solvent), the polarity of the solvent has a reverse effect. The solubility of metal bicarbonates in a medium with a low dielectric constant (e.g., tetrahydrofuran) is poor, and the solubility of the iron–formate intermediate is much better, which leads to a more stable intermediate. This will create an overall higher free-energy barrier. Therefore, it might be advantageous to use an aprotic solvent with a high dielectric constant.

In 2015, Gonsalvi and co-workers described the use of a linear tetradentate phosphine ligand, 1,1,4,7,10,10-hexaphenyl-1,4,7,10-tetraphosphadecane (**4**), in the iron-catalyzed hydrogenation of sodium bicarbonate (Scheme 2).[7] Interestingly, the *meso*-**4**/iron(II) tetrafluoroborate combination has rather poor reactivity, with a turnover number of 62 and a 6% yield of sodium formate (**3**) after 24 hours (80 °C) under hydrogen pressure (6.0 MPa). Upon using *rac*-**4**/iron(II) tetrafluoroborate as the catalyst, sodium formate is formed in 58% yield with a turnover number of 575 under otherwise identical conditions. Under the conditions mentioned above, well-defined iron(II) complex **5** is more active: a turnover number of 762 is obtained, and sodium formate is obtained in yields of up to 76%. If the catalyst loading is lowered to 0.01 mol%, the turnover number increases to 1229, although sodium formate is obtained in a lower yield of 12%.

Scheme 2 Iron-Catalyzed Hydrogenation of Sodium Bicarbonate to Sodium Formate Using a Linear Tetradentate Phosphine[7]

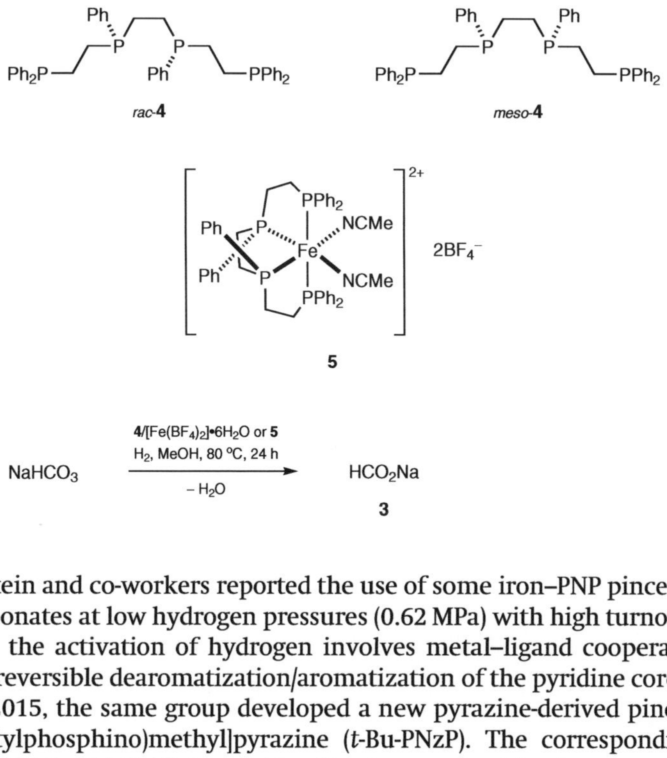

rac-**4** *meso*-**4**

5

$NaHCO_3$ $\xrightarrow[-\ H_2O]{\mathbf{4}/[Fe(BF_4)_2]\cdot 6H_2O \text{ or } \mathbf{5};\ H_2,\ MeOH,\ 80\,^{\circ}C,\ 24\,h}$ HCO_2Na (**3**)

In 2011, Milstein and co-workers reported the use of some iron–PNP pincer complexes to reduce bicarbonates at low hydrogen pressures (0.62 MPa) with high turnover numbers.[8] Significantly, the activation of hydrogen involves metal–ligand cooperation, which is based on the reversible dearomatization/aromatization of the pyridine core on the ligand skeleton. In 2015, the same group developed a new pyrazine-derived pincer ligand, 2,6-bis[(di-*tert*-butylphosphino)methyl]pyrazine (*t*-Bu-PNzP). The corresponding iron complex, FeClH(CO)(*t*-Bu-PNzP) (**6**), is capable of hydrogenating bicarbonates at low pressures under mild conditions.[9] The optimal results (turnover numbers up to 149) for the hydrogenation of sodium bicarbonate are obtained at 45 °C with 0.65 MPa of hydrogen and a 0.1 mol% catalyst loading in a mixture of tetrahydrofuran/water (1:10) (Scheme 3).

for references see p 287

Scheme 3 Hydrogenation of Sodium Bicarbonate Using an Iron–PNP Complex[9]

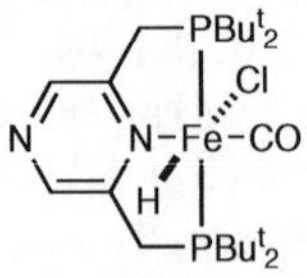

$NaHCO_3$ → HCO_2Na (3); conditions: H_2 (0.65 MPa), **6** (0.1 mol%), H_2O/THF (10:1), 45 °C, 16 h; – H_2O

Knölker's iron complex **8** is a well-known catalyst for the transfer hydrogenation of carbonyl groups.[10] This complex and some derivatives have been tested in the catalytic reduction of sodium bicarbonate by Zhou and co-workers.[11] Of all the complexes studied, complex **8** was the most active, and the hydrogenation occurs at low hydrogen pressures of 0.1 to 0.5 MPa; the highest reactivity is obtained with a hydrogen gas pressure of 4.0 MPa. A turnover number up to 447 is achieved in a water/ethanol solvent mixture. It is proposed that, similar to the Hieber base reaction, in this basic system, the carbonyl group of iron precursor **7** is attacked by a hydroxide ion to form the active hydride complex, and this is accompanied by the release of carbon dioxide (Scheme 4). The Renaud group has also described a phosphine-free, air- and moisture-tolerant iron catalyst system that promotes the reduction of bicarbonate into formate.[12] Complex **9**, developed by Renaud and bearing a cyclopentadienone framework that is more electron-rich than the framework of Knölker's iron complex, also shows high activities. The best activity is obtained in a 1:1 mixture of dimethyl sulfoxide/water, and a turnover number of 1246 is achieved at 100 °C under a hydrogen pressure of 5.0 MPa.

Scheme 4 Hydrogenation of Sodium Bicarbonate to Sodium Formate Using Iron Complexes[11,12]

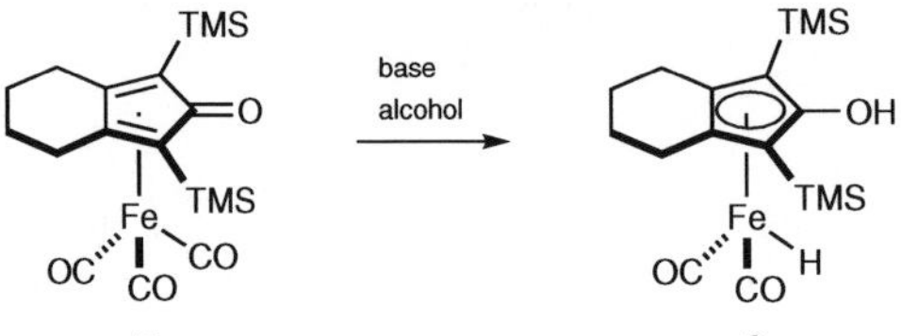

7 → **8** (base, alcohol)

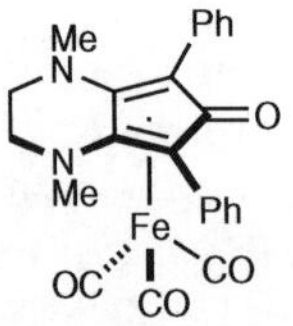

$NaHCO_3$ → HCO_2Na (3); conditions: H_2 (3.0–5.0 MPa), **7** or **9**, 100–120 °C, 20–24 h; – H_2O

Recently, Kühn and co-workers have reported the use of some rhodium- and iridium-based N-heterocyclic carbene complexes as catalysts for the efficient hydrogenation of bicarbonates to formates.[13] The synthesis of these water-soluble bis(N-heterocyclic carbene) complexes of rhodium and iridium is shown in Scheme 5. The complexes **11** (M = Rh, Ir) are obtained in good yields in a one-pot procedure from the reaction of bis(imidazolium salt) **10** and silver(I) oxide with the corresponding metal precursors.

Scheme 5 Synthesis of Metal–Bis(carbene) Complexes[13]

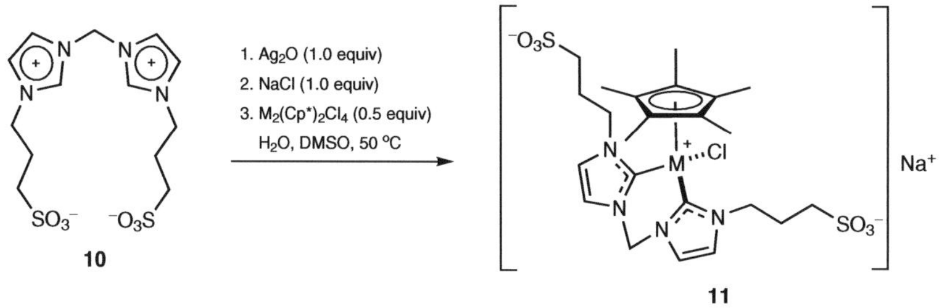

1. Ag_2O (1.0 equiv)
2. NaCl (1.0 equiv)
3. $M_2(Cp^*)_2Cl_4$ (0.5 equiv) H_2O, DMSO, 50 °C

10 → **11**

M	Yield (%)	Ref
Rh	93	[13]
Ir	90	[13]

Using bis(N-heterocyclic carbene) complexes **11** as the catalyst, the smooth reduction of metal bicarbonates (potassium bicarbonate, water as the solvent in the presence of 5.0 MPa hydrogen) can be realized, and no inert gas conditions are required in preparing the hydrogenation reactions. The yields, with turnover numbers up to 3600, can be conveniently analyzed by integrating the area of the signal for the formyl proton in the ^{1}H NMR spectrum, with dimethyl sulfoxide serving as an external standard (Scheme 6). DFT calculations suggest that two catalyst molecules are needed for the reactivity, wherein one species carries the bicarbonate group and the second molecule provides an external hydride as the reductant. Under similar conditions, the rhodium complex is more active than the iridium complex in the hydrogenation of bicarbonates.

Scheme 6 Metal–N-Heterocyclic Carbene Complexes in the Hydrogenation of Metal Bicarbonates[13]

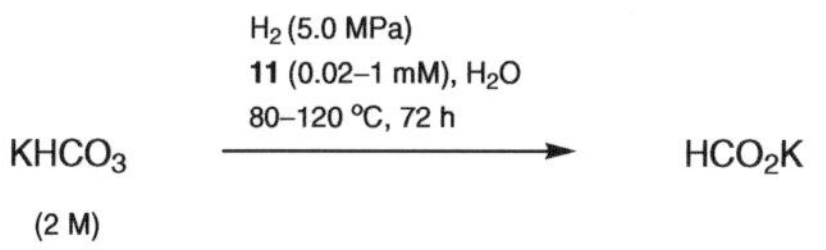

$KHCO_3$ (2 M) → HCO_2K

H_2 (5.0 MPa)
11 (0.02–1 mM), H_2O
80–120 °C, 72 h

As an alternative to hydrogenation, transfer hydrogenation has also been utilized in bicarbonate reduction, with the advantage that the reducing reagent is easier to handle. In 2014, Beller and co-workers developed a transfer-hydrogenation system to reduce sodium and potassium bicarbonates to the corresponding formate salts **13**.[14] Using ruthenium–pincer complex **12** as a catalyst, the reaction proceeds smoothly in methanol, which also serves as the source of hydrogen (Scheme 7). This protocol is recognized as a green and sustainable process, in which both the metal bicarbonate and methanol are converted into the formate in the presence of potassium hydroxide. Of the several catalysts screened, ruthenium complex **12** was the most effective, and potassium bicarbonate is

for references see p 287

more reactive than sodium bicarbonate. Under relatively mild conditions (150 °C) in methanol/water (25:5, v/v), high efficiency is achieved for the production of potassium formate (92% yield) from the reduction of the bicarbonate and the dehydrogenation of methanol with a turnover number up to 18422 in 20 hours. Notably, higher concentrations of potassium bicarbonate do not favor the efficiency of the reaction, probably as a result of higher pH values of the reaction system.

Scheme 7 Ruthenium-Catalyzed Transfer Hydrogenation of Metal Bicarbonates[14]

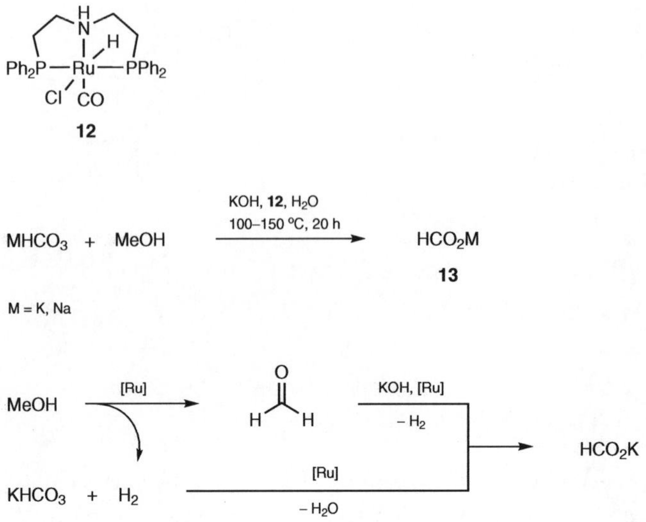

Sodium Formate (3); General Procedure:[7]

A magnetically stirred 40-mL stainless-steel autoclave was charged with $NaHCO_3$ (typically 840 mg, 10 mmol) and the Fe catalyst **5** (0.01–0.001 mmol as solid or stock soln in propylene carbonate) under an inert atmosphere. The autoclave was then sealed and thoroughly purged through several vacuum–argon cycles. MeOH (20.0 mL) was then added to the autoclave by suction, followed by the introduction of H_2 to the desired pressure. The autoclave was then placed in an oil bath preheated to the desired temperature, and the mixture was stirred for the set reaction time. The autoclave was then cooled and depressurized, which was followed by transfer of the mixture to a flask and removal of the solvent at rt. The yield was determined by analyzing aliquots (ca. 30 mg) of the solid mixture dissolved in D_2O (0.5 mL) by ^1H NMR spectroscopy (relaxation delay of 20 s), with the use of dry THF (20 μL) as an internal standard.

Potassium Formate (13, M = K); General Procedure:[14]

$KHCO_3$ (20 mmol), KOH (60 mmol), and Ru complex **12** (5 mmol) were placed in a 100-mL autoclave, and MeOH (25 mL) and H_2O (5 mL) were then added under an argon atmosphere. The mixture was stirred at 150 °C for 20 h. Then, the autoclave was cooled in an ice–water bath and the pressure was carefully released. The mixture was fully concentrated under reduced pressure, and the white product was analyzed by ^1H NMR spectroscopy (relaxation delay of 20 s) in D_2O, using THF as an internal standard.

1.8.1.2 Heterogeneous Reductions

In 1914, two German chemists reported the catalytic hydrogenation of bicarbonates to prepare formates by using heterogeneous catalysts. In one of the earliest works involving the use of a palladium-based heterogeneous catalyst, the catalytic hydrogenation of potassium bicarbonate was reported by Bredig and Carter from the Technische Hochschule Karlsruhe.[15] The palladium-catalyzed reaction was performed in water, and potassium formate was obtained in 74.7% yield after 23 hours at 70 °C under 6.0 MPa hydrogen pressure. However, not much further progress was reported until the 1980s. In 1983, Wrighton and co-workers reported the catalytic hydrogenation of sodium bicarbonate at room temperature and 0.1 MPa hydrogen pressure in the presence of supported palladium catalysts, such as palladium on carbon, palladium on γ-alumina, and palladium on a heterogeneous tungsten-containing redox-active polymer.[16,17] This work demonstrated that, under appropriate conditions, palladium-based catalysts can exhibit good activity for both the hydrogenation of bicarbonates and the dehydrogenation of formates. Accordingly, under a hydrogen atmosphere (0.1–0.17 MPa) at 298 K, the equilibrium of the hydrogen/aqueous sodium bicarbonate/aqueous sodium formate system reaches a formate/bicarbonate ratio ranging from 1:1 to 1.5:1.

In 1988, Wiener and co-workers studied the kinetics of the palladium-on-carbon catalyzed hydrogenation of metal bicarbonates to formates in aqueous solution. For this catalytic system, the hydrogenation rate increases upon increasing the hydrogen pressure. Furthermore, it was observed that as the concentration of bicarbonate is increased, the rate eventually reaches a maximum.[18]

1.8.2 Catalytic Reduction of Dialkyl Carbonates to Methanol

Even though the reduction of carbonates to methanol represents a method for the indirect reduction of carbon dioxide to a liquid fuel, breakthroughs in this area were only achieved after 2010. The hydrogenation of carbon dioxide to methanol is discussed in Section 1.9.2.

1.8.2.1 Homogeneous Reductions

In 2011, Milstein and co-workers reported the first general procedure for the hydrogenation of organic carbonates for the synthesis of methanol.[19] By using ruthenium–PNN pincer complexes **14–17** as the catalysts, turnover numbers of up to 4400 could be achieved in the reduction of dimethyl carbonate to methanol at 145 °C with hydrogen (5.0 MPa) (Scheme 8). In this system, 3 moles of methanol are produced through the reduction of 1 mole of dimethyl carbonate with consumption of 3 moles of hydrogen.

Scheme 8 Ruthenium-Catalyzed Hydrogenation of Dimethyl Carbonate[19]

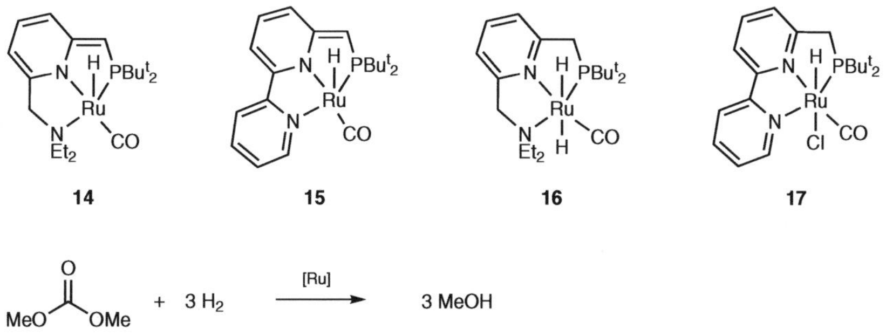

for references see p 287

The proposed mechanism involves metal–ligand cooperation during the critical steps and is supported by stoichiometric reactions of dimethyl carbonate and methyl formate with dihydride complex **16** (Scheme 9). Insertion of the C=O bond of dimethyl carbonate into the Ru—H bond in **16** leads to the formation of intermediate **18**, in which the methoxy group is protonated by proton transfer from the ligand to form formate-coordinated species **19**. By a similar hydrogenation procedure, acetal intermediate **20**, formaldehyde intermediate **21**, and methoxide intermediate **22** are formed. The release of methanol from **22** is realized by activation of one hydrogen atom with generation of the key sixteen-electron species **14**.

Scheme 9 Proposed Mechanism for Milstein's Catalytic Hydrogenation of Dimethyl Carbonate[19]

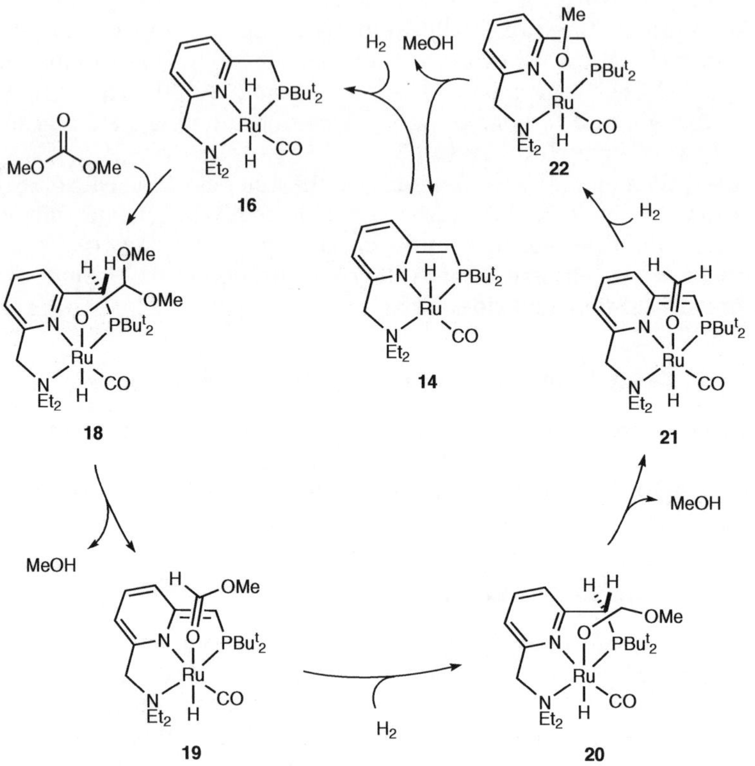

Several computational studies have sought to provide detailed mechanistic understanding of Milstein's method for the hydrogenation of organic carbonates to alcohols that employs pincer-ligated ruthenium complexes as catalysts. In 2012, Yang[20] and Wang[21] independently proposed a direct pathway with three sequential stages for the hydrogenation of dimethyl carbonate to methyl formate, formaldehyde, and finally methanol. DFT calculations have revealed the essential role of the non-innocent pincer ligand in the hydrogen splitting, C—O bond breaking, and O—H bond formation steps. Later, Hasanayn and coworkers[22] found that transfer of a hydride ion from ruthenium to the carbonyl carbon atom has a surprisingly low barrier, which leads to an activated ion pair species ($[Ru]^+$/$[C{-}O]^-$). Hence, the authors propose a metathesis-type transformation involving a procedure in which a hydride ion of the metal hydride species and an alkoxide of the carbonate molecule are exchanged through an outer-sphere interaction mode.

In contrast with the hydrogenation of dimethyl carbonate to form methanol, in the reduction of cyclic carbonates, the corresponding diol is cogenerated together with methanol without the involvement of water (Scheme 10).[23] This reaction might potentially constitute an interesting route for the production of vicinal diols, especially considering the high energy cost to remove water in both the traditional oxirane hydration process and the ethylene carbonate (1,3-dioxolan-2-one) mediated Shell OMEGA process for the production of ethane-1,2-diol.

Scheme 10 Comparison of the Reduction of Cyclic Carbonates with the Shell OMEGA Process for Ethane-1,2-diol Production[23]

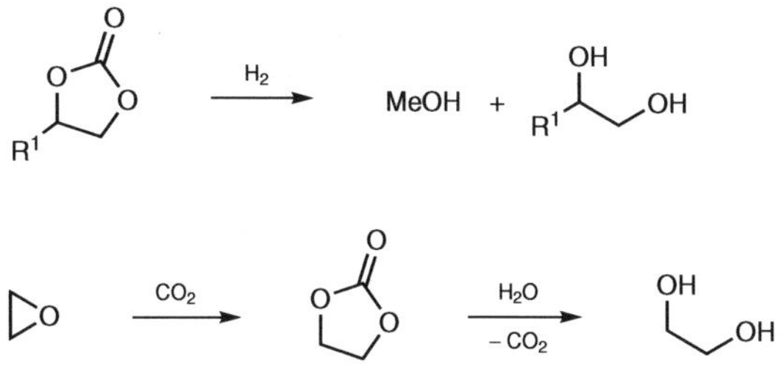

Ding and co-workers have disclosed the highly efficient hydrogenation of cyclic carbonates by using ruthenium complex **12** to produce diols **23** and **24** and methanol (Scheme 11).[23] In the presence of 0.1 mol% of complex **12**, both methanol and ethane-1,2-diol are obtained in >99% yield within 1 hour at 140 °C under 5.0 MPa hydrogen pressure. Upon lowering the catalyst loading to 0.001 mol%, the reaction proceeds smoothly with a turnover number of 87000 and a turnover frequency of up to 1200 h^{-1}. Under similar conditions, poly(propylene carbonate) ($M_w/M_n = 1.77$, >99% carbonate linkages) is reduced quantitatively to the corresponding diol and methanol in the presence of 0.1 mol% catalyst **12**. It is proposed that the Ru—H and N–H moieties of the complex are cooperatively involved in the outer-sphere hydrogenation of the carbonate.

Scheme 11 Ruthenium-Catalyzed Hydrogenation of Cyclic Carbonates[23]

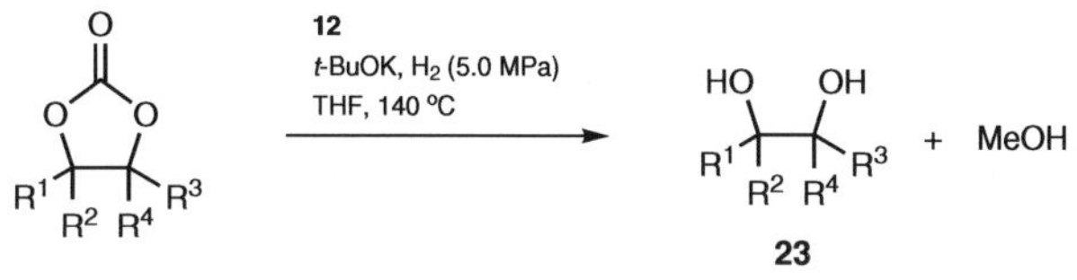

R^1	R^2	R^3	R^4	Catalyst (mol%)	Time (h)	Yield (%)		Ref
						23	MeOH	
H	H	H	H	0.1	0.5	>99	>99	[23]
H	H	H	H	0.01	48	>99	>99	[23]
H	Me	H	Me	0.05	10	99	98	[23]
Me	Me	H	H	0.05	12	97	>99	[23]
Me	Me	Me	Me	0.1	20	96	95	[23]
H	Bu	H	H	0.05	4	99	>99	[23]
H	Ph	H	H	0.05	4	99	>99	[23]

for references see p 287

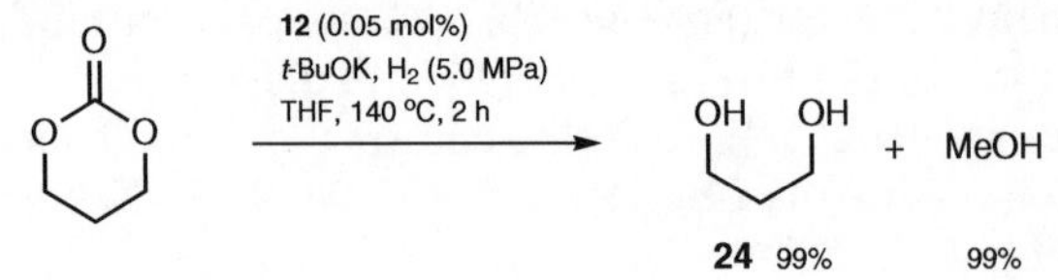

12 (0.05 mol%)
t-BuOK, H_2 (5.0 MPa)
THF, 140 °C, 2 h

OH OH + MeOH

24 99% 99%

Later, Gao and co-workers described the use of lutidine-bridged N-heterocyclic carbene complexes of ruthenium as catalysts for the hydrogenation of ethylene carbonate.[24] A series of air- and moisture-stable lutidine-bridged (and one 1,3-phenylene-bridged) N-heterocyclic carbene ligands are prepared as their hydrobromide salts in yields of 83–93% (Scheme 12). By combining *N*-butyl-substituted CNC pincer ligand **25** with 0.1 mol% of the metal complex carbonyl(chloro)hydridotris(triphenylphosphine)ruthenium(II) [$RuClH(CO)(PPh_3)_3$], full conversion of ethylene carbonate is achieved under 5 MPa hydrogen at 130 °C, to give ethane-1,2-diol in 92% yield and methanol in 42% yield. Here, the low yield of methanol is thought to be caused by the formation of byproducts such as 2-hydroxyethyl formate and carbon dioxide from the decomposition of ethylene carbonate, as well as carbon monoxide from both the decarbonylation of ethylene carbonate and the hydrogenation of carbon dioxide.

Scheme 12 Bis(N-Heterocyclic Carbene) Ligands Used in Gao's Work[24]

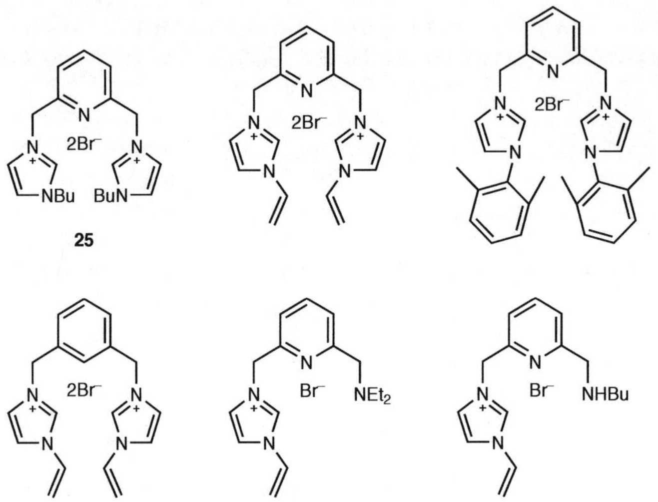

N 2Br⁻ NBu BuN **25**
N 2Br⁻
N 2Br⁻
2Br⁻
Br⁻ NEt_2
Br⁻ NHBu

The transfer hydrogenation of organic carbonates has also been investigated recently. In 2014, Kim and co-workers reported the use of ruthenium complex **12** in the transfer hydrogenation of methyl formate and cyclic carbonates by using propan-2-ol as the reductant with yields of the corresponding diols **26** of up to 99% (Scheme 13).[25] The reduction of propylene carbonate proceeds smoothly at 140 °C in propan-2-ol in the presence of 0.05 mol% catalyst to give quantitative yields of methanol and propane-1,2-diol (**26**, R^1 = Me) in 5 hours. Under similar conditions, the reduction of linear carbonates (e.g., diethyl carbonate) is much less efficient with this system, and only 6% yield of methanol is obtained. It is thought that the ring opening of the cyclic carbonates, with release of strain, is helpful for the hydrogenation and the formation of the 2-hydroxyethyl formate intermediates.

Scheme 13 Ruthenium-Catalyzed Transfer Hydrogenation of Cyclic Carbonates[25]

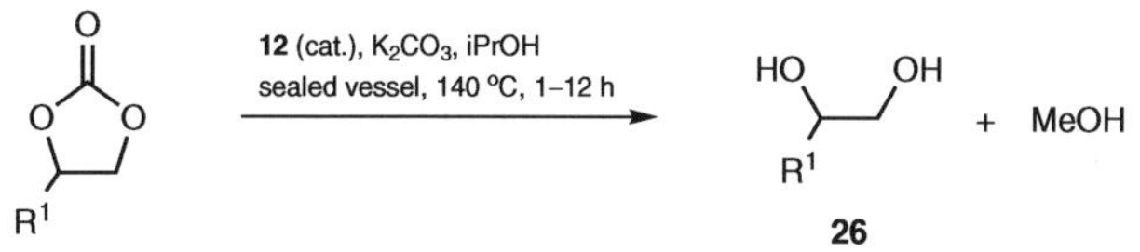

12 (cat.), K_2CO_3, iPrOH
sealed vessel, 140 °C, 1–12 h

HO OH R1 **26** + MeOH

R^1	**12** (mol%)	K_2CO_3 (mol%)	Yield (%)		Ref
			26	MeOH	
H	0.1	0.1	91	91	[25]
Me	0.05	0.05	>99	>99	[25]
Et	0.1	0.1	94	94	[25]
Bu	0.1	0.1	97	>99	[25]
Ph	0.1	0.1	95	93	[25]

Ethane-1,2-diol (23, $R^1 = R^2 = R^3 = R^4 = H$); Typical Procedure:[23]
In a glovebox, a 125-mL Parr autoclave was charged with Ru complex **12** (17.4 mg, 28.6 μmol), *t*-BuOK (3.2 mg, 28.6 μmol,), THF (20 mL), and ethylene carbonate (2.52 g, 28.6 mmol). *p*-Xylene (50 μL) was added to the mixture as an internal standard. The reaction vessel was sealed and then purged three times with H_2. The pressure of H_2 in the autoclave was finally adjusted to 5.0 MPa. The vessel was heated at 140 °C for 0.5 h with stirring and was then cooled in an ice–water bath. The residual H_2 gas was released carefully, and the mixture was analyzed by GC to determine the yield.

Propane-1,2-diol (26, R^1 = Me); General Procedure:[25]
A pressure-tolerating reaction vessel was charged with complex **12** (1.7 mg, 2.8 μmol), K_2CO_3 (0.39 mg, 2.8 μmol), iPrOH (20 mL), and propylene carbonate (2.8 mmol) under inert conditions. The vessel was heated to 140 °C for 2 h. The generated H_2 was released carefully in a hood. The yield was determined by GC analysis, using *p*-xylene as an internal standard.

1.8.2.2 Heterogeneous Reductions

Heterogeneous catalysts are favorable in industrial applications from the perspective of recycling and reusing the catalyst. For the hydrogenation of carbon dioxide on the surface of a catalyst (see Section 1.9.1.2), it was proposed in pioneering studies that (bi)carbonates could be intermediates.[26,27] In the reverse water–gas shift reaction, carbonates can be formed by the insertion of carbon dioxide into hydroxide sites, and the ensuing reduction of the carbonate moieties will result in the formation of formate species; these formates undergo decomposition to carbon monoxide and water.

Copper-based heterogeneous catalysts, such as copper chromite, have been investigated in the hydrogenation of organic carbonates. Li and co-workers have reported a study involving the use of a heterogeneous Cu/CrO_x-based catalyst for the hydrogenation of cyclic carbonates (e.g., ethylene carbonate) to diols and methanol.[28] Moderate selectivity (60%) for methanol is achieved, along with the generation of methane, carbon monoxide, and carbon dioxide as byproducts. Although the hydrogenolysis of C—O bonds on the surfaces of transition metals is well known and overreduction to ethanol or ethane may occur, ethane-1,2-diol is obtained in up to 93% yield. The catalyst can be reused over four runs with some loss in catalytic activity (about 20% lower conversion).

Considering that chromium-based materials are toxic and are harmful to the environment, a better option for the development of catalyst supports is the use of less-toxic metals supported on silica zeolites. It has been well known for a long time that silica-supported copper/hexagonal mesoporous silica catalysts are effective for the catalytic hydrogenation of esters.[29] Recently, Xia, Chen, Ding, and co-workers reported the use of a copper–silica nanocomposite for the hydrogenation of cyclic carbonates to methanol and the corresponding diols.[30] The catalyst is prepared by a precipitation–gel method and possesses a uniform copper dispersion on silica. Under relatively mild conditions, the catalyst demonstrates remarkable activity, selectivity, and stability in both batch and fixed-bed continuous-flow reactors, thus providing a highly practical approach for the transformation of carbon dioxide (via conversion into the cyclic carbonate substrates) into methanol with the coproduction of diols (up to >97% yield).

In this work, the heterogeneous hydrogenation of carbonates to alcohols using different metals and supports was systematically studied. Several different catalysts, including Ru/C (5 wt%), NiO/silica (10 wt%), Co_3O_4/silica (10 wt%), CuO/silica (10 wt%), and 10% Cu–silica modified with 5 wt% WO_3 or 5 wt% K_2O, were prepared by either an impregnation–drying–calcination or a precipitation–gel method. Catalyst screening revealed that the combination of copper and silica is optimal. The selectivity for methanol improves gradually upon increasing the copper content, and maximum selectivity for methanol and ethane-1,2-diol (both 97%) at >99% ethylene carbonate conversion is obtained in the presence of 70 wt% copper under relatively mild conditions (6 MPa H_2, 160 °C, 10 h) in tetrahydrofuran. Remarkably, neat ethylene carbonate can be efficiently hydrogenated with >99% conversion, with selectivities for methanol and ethane-1,2-diol of 98 and 97%, respectively. The stability of the catalyst has been tested over six hydrogenation runs; the yields of methanol remain essentially constant at 95–97% with about 100% conversion.

High reactivity is achieved for copper particles 8–10 nm in size, and this suggests that the high efficiency is a result of the edge sites of copper(0)–copper(1+) species and copper–silica interfacial sites. Silica, with weak acid–base properties, has the highest reactivity, whereas other metal oxide supports with a high concentration of acidic and basic sites do not promote the reaction efficiently. In a control experiment, the use of pure copper metal offers a rather low conversion (26%) and low selectivity for methanol (8%) in the hydrogenation of ethylene carbonate. Similar to other hydrogenation systems based on copper–silica, it is proposed that the copper(0) sites dissociatively adsorb hydrogen and function as metal hydrides, and that the copper(1+) sites function as Lewis acid sites to activate the carbonyl groups of the carbonates. Notably, the copper(1+) species interacting with the surface of silica are strong Lewis acids. On the basis of the characterization data, the authors conclude that the copper particle size, the surface acidity–basicity balance, and the copper valence state of the catalyst all play important roles in determining the activity and selectivity. In this catalytic system, a synergetic effect between the copper(0) and copper(1+) sites is considered the most critical factor for achieving high yields of methanol and diols. Taking the hydrogenation of dimethyl carbonate as an example, the reaction pathway is shown in Scheme 14.

Scheme 14 Proposed Mechanism for the Copper-Catalyzed Hydrogenation of Dimethyl Carbonate[30]

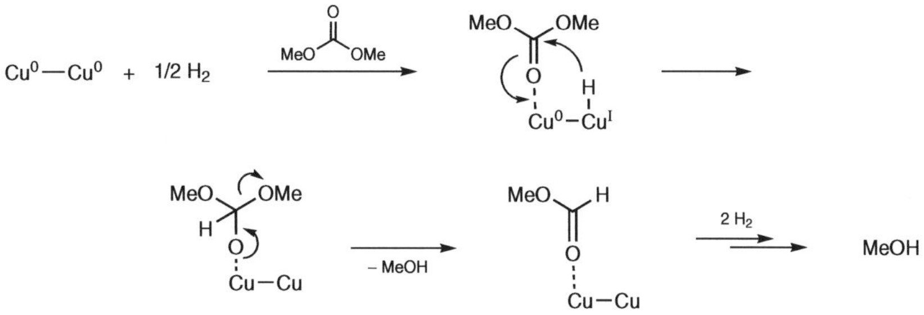

Dai and co-workers have reported the copper–silica-based catalytic hydrogenation of ethylene carbonate to methanol and ethane-1,2-diol on a continuous fixed bed. The copper/hexagonal mesoporous silica (Cu/HMS) catalyst, prepared by the ammonia evaporation method, enables a conversion of 100% and a selectivity for methanol of 74% in the gas-phase fixed-bed heterogeneous hydrogenation. The reactions are conducted at 180 °C under 3 MPa of hydrogen pressure on a fixed-bed tubular reactor. The 50Cu/HMS catalyst, which contains 50 wt% copper after calcination, shows the highest copper(1+) ratio {copper(1+)/[copper(1+) + copper(0)]; determined by X-ray diffraction measurements} and gives high hydrogenation reactivity. The yield of methanol correlates well with the surface area of copper(0) metal in the catalysts, which indicates that the copper(0) surface area plays an important role in the catalytic performance.[31]

Later, the same group developed a gas–solid-phase fixed-bed copper/silica catalyst for the catalytic hydrogenation of dimethyl carbonate to methanol.[32] Vapor-phase hydrogenation under solvent-free conditions makes this process promising for practical use.

Recently, the use of copper/cerium(IV) oxide sub-nanoparticles as effective catalysts for the synthesis of methanol from the hydrogenation of organic carbonates has been disclosed.[33] At 160 °C and 2.5 MPa hydrogen pressure, the hydrogenation of dimethyl carbonate delivers 94% methanol (98% yield based on the carbonyl moiety and total produced methanol). The reduction of copper/cerium(IV) oxide produces copper metal particles with a size of <1.0 nm. Under similar conditions, copper/cerium(IV) oxide is the most effective catalyst among the various metal–support combinations tested, including metals such as copper, platinum, silver, gold, iridium, nickel, rhodium, cobalt, palladium, and ruthenium and supports such as carbon, zirconium(IV) oxide, magnesium oxide, titanium(IV) oxide, γ-alumina, silica–alumina, and silica. Interestingly, it was found that pure cerium(IV) oxide shows comparatively high selectivity (81%) for methanol, albeit with very low activity. In the initial stage of the hydrogenation reaction, the formation of methyl formate is observed. The X-ray absorption near-edge structure spectrum of the active catalyst exhibits no peaks for copper(2+) or copper(1+) sites. The Cu—Cu bond with a coordination number of 5.5 suggests that copper is in the metallic state. Notably, copper/cerium(IV) oxide shows no activity for the hydrogenation of carbon dioxide, and consequently, the catalyst surface interacts more effectively with carbonates such as dimethyl carbonate and diethyl carbonate. However, it should be kept in mind that in the hydrogenation of carbon dioxide, a large copper loading is generally used and larger copper metal particles with sizes ≥3 nm are often formed.

Generally, the type of support used for catalyst immobilization is crucial for both catalytic reactivity and selectivity. In this context, Li and co-workers studied the influence of the support on the performance of copper catalysts for the hydrogenation of ethylene carbonate to ethane-1,2-diol and methanol.[34] Thermodynamic considerations of the hydrogenation of ethylene carbonate indicate that the reaction is exothermic, with a standard molar enthalpy of the reaction of $-71.59\ kJ \cdot mol^{-1}$ and a standard molar Gibbs energy of

for references see p 287

the reaction of −25.62 kJ·mol^{-1}. The ammonia evaporation method was used to prepare mesoporous materials as copper supports, e.g. KIT-6, MCM-41, and SBA-15 (three types of ordered mesoporous molecular sieves), and the observed activity for the hydrogenation of ethylene carbonate follows the sequence Cu/SBA-15 > Cu/KIT-6 > Cu/MCM-41. The highest activity is obtained with Cu/SBA-15, which gives turnover numbers to ethane-1,2-diol and methanol of 22.0 and 11.4, respectively. The authors propose that this correlates with the larger surface area of metallic copper, with a higher dispersion on SBA-15. Under the optimized conditions, 100% conversion of ethylene carbonate with yields of 94.7% for ethane-1,2-diol and 62.3% for methanol are achieved within 4 hours.

Chorkendorff and co-workers have monitored the hydrogenation of carbonate (produced in the presence of 0.03 MPa carbon dioxide) on copper overlayers deposited on a platinum(111) single crystal by X-ray photoelectron spectroscopy and temperature-programmed desorption. Upon subjecting carbonate to hydrogen gas pressure (0.02 MPa), a significant loss is observed at room temperature, which suggests that the hydrogenation of carbonate may constitute a relevant pathway for the generation of methanol.[35]

1.8.3 Catalytic Reduction of Carbonates to Methane

One of the earliest examples of the hydrogenation of solid metal carbonates to methane was reported in 1992.[36] The reduction of nickel(II) carbonate and cobalt(II) carbonate proceeds in good yields in the presence of high pressures of hydrogen at 200–250 °C without additional catalysts. The reduction of alkali and alkaline-earth metal carbonates takes place at higher temperatures (300–400 °C) in the presence of commercially available metal powder catalysts such as cobalt, nickel, and palladium. It is presumed that in situ formed metallic nickel and cobalt work as the hydrogenation catalysts.

For the reduction of carbon dioxide to methane, see Section 1.9.3.

1.8.4 Other Reductive Transformations

Other than the reduction of carbonates to formates, methanol, or methane in the presence of hydrosilanes, borane, or hydrogen as a reductant, reductive transformations such as the methylation of amines have recently been discovered. In addition, the reduction of carbonates by enzymatic methods will be briefly discussed.

1.8.4.1 Catalytic Reductive Methylation

Methylation of N or C nucleophiles is important in organic synthesis, especially for the synthesis of bioactive N-methylated compounds (for reductive methylation using carbon dioxide, see Section 1.9.4). Recently, the reductive methylation of amines with hydrogen and carbonates has emerged as a new route for the indirect utilization of carbon dioxide (Scheme 15).

Scheme 15 Reductive N-Methylation Using Carbonates

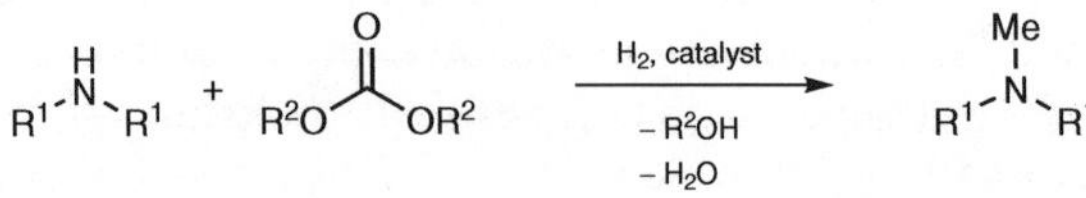

$$R^1NHR^1 + R^2OC(O)OR^2 \xrightarrow[-\,R^2OH,\ -\,H_2O]{H_2,\ \text{catalyst}} R^1N(Me)R^1$$

In 2016, Beller and co-workers described the general and selective ruthenium/bis[2-(diphenylphosphino)ethyl]phenylphosphine (triphos)/trifluoromethanesulfonimide catalyzed reductive N-methylation of both primary and secondary amines by employing dimethyl carbonate as a carbon source and molecular hydrogen as the reducing agent.[37]

Isotopic-labeling experiments performed to elucidate the mechanism revealed that the formed N-methyl group of the final product originates from the carbonyl group and not from the methoxy moiety of dimethyl carbonate. As shown in Scheme 16, it is proposed that the main reaction pathway proceeds through the initial hydrogenation of dimethyl carbonate to methyl formate, which reacts with the aniline to give a formanilide. Then, the dimethylated product can be formed from the reaction of the formanilide.

Scheme 16 Proposed Reaction Pathways for Aniline Dimethylation Using Dimethyl Carbonate and Molecular Hydrogen[37]

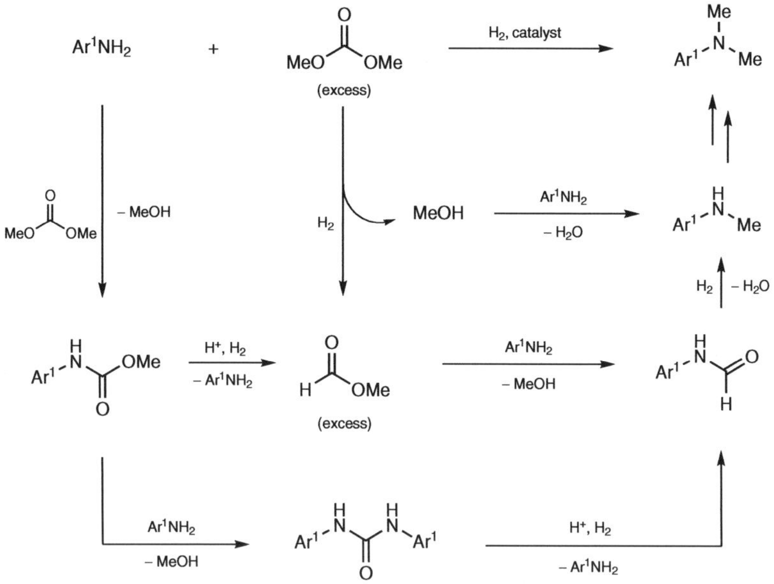

1.8.4.2 Photo-/Electro- and Enzymatic Reduction of Carbonates

In addition to the chemical reduction by hydrogen or silanes, carbonates can also be reduced under photoreduction, electroreduction, or enzymatic reduction conditions. In 1983, Chandrasekaran and Thomas reported the photochemical reduction of carbonate to formaldehyde on titanium(IV) oxide powder.[38]

In 1989, Willner and Mandler studied the photosensitized reduction of bicarbonate to formate in an aqueous medium composed of deazariboflavin as a photosensitizer, N,N′-dimethyl-4,4′-bipyridinium as a primary electron acceptor, and sodium oxalate as a sacrificial electron donor. A palladium colloid stabilized by β-cyclodextrin (Pd-β-CD) was used as a catalyst.[39] Importantly, the procedure to prepare the Pd-β-CD colloid has a dramatic effect on the catalytic activity; whereas preparing the palladium colloid at 60 °C results in an active catalyst for the reduction of bicarbonate to formate, the Pd-β-CD colloid prepared at 90 °C is inactive toward bicarbonate reduction.

In 2015, Heagy and co-workers reported the distinct photocatalytic behavior of the two crystal forms of zinc sulfide (wurtzite and sphalerite) on productivity in the photoreduction of bicarbonate in propan-2-ol or glycerol as the scavenger with respect to the parameters of size [micrometer (mP) and nanoscale (np)], crystal lattice, surface area, and band gap.[40] The largest sized catalyst, mP-Sphal, was shown to have the poorest performance, whereas the wurtzite crystal form had the highest apparent quantum efficiency of 0.9% upon using propan-2-ol as the hole scavenger. Among the three examined hy-

for references see p 287

droxylic positive hole scavengers (i.e., ethane-1,2-diol, propan-2-ol, and glycerol), glycerol is the best sacrificial hole scavenger under the described photochemical conditions, and an apparent quantum efficiency of 3.2% is achieved for the nP-Wurtz particles.

Organic photocatalysts have also been studied in the reduction of carbonates. Ameta and co-workers have developed a method for the photocatalytic reduction of aqueous sodium and potassium carbonates in the presence of toluidine blue.[41] The formation of formic acid and formaldehyde is detected. The efficiency of the reduction depends on several parameters, including pH, amount of the photocatalyst, concentration of the carbonate, and light intensity.

Wrighton and co-workers have reported the electrochemical reduction of aqueous bicarbonate to formate with a high current efficiency near the thermodynamic potential at chemically derivatized electrodes (Scheme 17).[42] The electrode surface contains metallic palladium dispersed on a supported polymer-coated palladium cathode. Such chemically derivatized electrodes are prepared by using monomer **27** for the construction of the surface-confined polymer, which is redox active and provides a pathway to transfer the reducing equivalents to the high-surface-area palladium catalyst sites.

Scheme 17 Electroreduction of Metal Bicarbonates Using Polymer-Coated Palladium Electrodes[42]

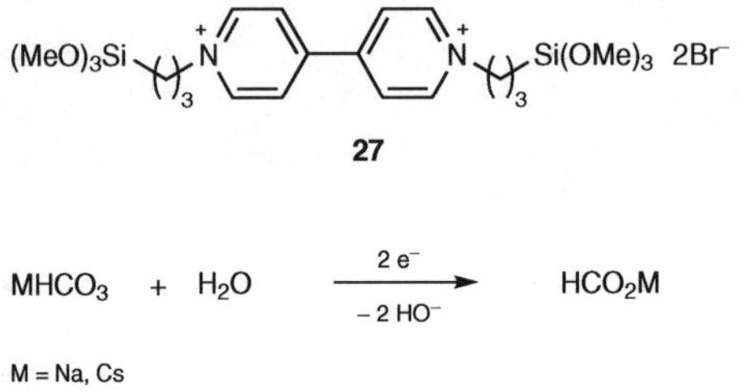

$(MeO)_3Si$ ... N^+ ... N^+ ... $Si(OMe)_3$ $2Br^-$

27

$$MHCO_3 + H_2O \xrightarrow[-\,2\,HO^-]{2\,e^-} HCO_2M$$

M = Na, Cs

In this electroreduction system, the reduction of aqueous carbon dioxide takes place near the thermodynamic potential (80 mV of the standard potential) with a reasonable rate and with high current efficiency (turnover number on palladium atom $\approx$ 1000). Notably, under the same conditions, a palladium-impregnated electrode is an effective catalyst for the reduction of bicarbonate with hydrogen, for which the analogous platinum electrode shows negligible activity.

In a report from 1985, for which a palladium cathode was used, the influence of formaldehyde and formate on the electrochemical reduction of bicarbonate ions in aqueous solution is discussed.[43] If formaldehyde is added to a solution of aqueous sodium bicarbonate, it selectively affects the reduction of bicarbonate, and the partial current corresponding to the reduction of bicarbonate is severely retarded. In contrast, the addition of formate has almost no effect on the shape of the polarization plot and the amount of the current. Moreover, different heterogeneous palladium materials affect the equilibration of hydrogen/bicarbonate and formate/water, which can be monitored by exchange of hydrogen and carbon isotopes in the water/hydrogen and bicarbonate/formate redox couples.[17]

In 1986, Augustynski and co-workers reported that the rate of the cathodic reduction of bicarbonate anions at palladium could be enhanced by the presence of cesium cations.[44] The net current densities for the electroreduction of bicarbonate anions in cesium bicarbonate solution are up to 9 times larger than those observed in sodium bicarbonate solution. This result is interpreted on the basis of the involvement of cesium(1+) cations in the reaction at the cathode through the formation of ion pairs with the bicarbonate or formate anions.

Rophael and co-workers have studied the photocatalyzed reduction of aqueous carbonate using chemically treated semiconductors.[45] In general, there are two ways to extend the photoresponse of a semiconductor to light absorption: doping with a suitable foreign element or cation, or sensitizing by coating with a photoactive dye. In Rophael's work, powders of titanium(IV) oxide, strontium titanate, zinc oxide, zinc sulfide, molybdenum(IV) sulfide, and cadmium(II) sulfide with diameters smaller than 48 μm (by passing through a 300-mesh sieve) are used in their bare forms or after treatment with phthalocyanine dyes [iron(2+)–phthalocyanine or cobalt(2+)–phthalocyanine] with 1–3% coating. The reductions are performed in 1 M aqueous sodium carbonate solution containing the catalyst as a powder suspension. Under a nitrogen atmosphere, the suspension is irradiated with 254-nm light. For the coated catalyst, >80% of the 254-nm radiation is absorbed by the phthalocyanine dyes to effect sensitization of the semiconductor. In many cases, both methanol and formaldehyde are produced, but methanol is the major product. For dye-coated titanium(IV) oxide, by using cobalt(2+)–phthalocyanine, which has a redox potential different to that of iron(2+)–phthalocyanine, the yield of methanol increases by about 10%. Moreover, the bare molybdenum(IV) sulfide photocatalyst gives the highest yield of methanol, owing to the characteristic behavior of semiconducting layer-type disulfides. It is proposed that the photo-produced electrons reduce carbonate initially to formate, then to formaldehyde, and finally to methanol.

Generally, the photoreduction of (bi)carbonates can give methanol, formate, or formaldehyde as the final product. On the other hand, it has been shown that aqueous sodium carbonates can also be reduced photocatalytically to carbon by using platinized titania.[46,47] A low yield of formaldehyde is obtained in the presence of ultraviolet to visible light. In fact, titania both with and without platinum doping results in the formation of carbon and formaldehyde.

Related to the reduction of carbonates, the radical transformation of dialkyl carbonates, which are used as solvents in batteries, is also briefly discussed here. Given that organic carbonates are used as electrolytes in lithium-ion batteries, the electrochemical reduction of carbonate molecules is possible and might play an important role in the long-term performance of batteries, as evidenced by numerous experimental and computational studies. For example, in a paper published in 2013, radical intermediates generated by the one-electron reduction of carbonates are observed by spectroscopic methods. Importantly, such radical intermediates are proposed to be involved in the formation of the solid–electrolyte interface in lithium-ion batteries.[48] In this work, radiolysis and laser photoionization of carbonate electrolytes are used to observe and identify these reaction intermediates by using electron paramagnetic resonance (EPR) spectroscopy. For example, through the one-electron reduction of propylene carbonate in solvent or solid phase, the radical anion of propylene carbonate is formed, as shown by EPR spectroscopy and quantum-chemistry calculations. It is suggested that branching and the formation of a polymer network is favored, and these lead to the formation of different solid–electrolyte interfaces.

In natural photosynthesis processes, carbon dioxide usually associates with its aqueous bicarbonate and carbonate ions. It is clear that the evolution of biosystems should lead to enzymes suitable for charge or electron transfer to and from bicarbonate or carbonate ions. For a long time, it was unclear whether the enzymatic reduction of carbon dioxide was a direct process or a bicarbonate-mediated one. The first report with solid evidence of a direct process is based on the use of a flavin enzyme to reduce carbon dioxide directly with NADH (reduced form of nicotinamide adenine dinucleotide), which was published in 1976.[49]

Immobilized cells have also been studied for the catalytic hydrogenation of sodium bicarbonate to formate.[50] *Alcaligenes eutrophus* cells containing formate dehydrogenase have been immobilized in κ-carrageenan. Specifically, immobilized cells and hydrogen gas are introduced to a solution of sodium bicarbonate in phosphate buffer (pH 7.2), and

for references see p 287

this is followed by shaking at room temperature. After 22 hours, 11 mM of formates is produced from 0.3 M of bicarbonate. Interestingly, it was found that by using palladium on activated carbon as the catalyst, an enhanced productivity of formate (13 mM) is obtained under similar conditions.

1.8.5 Conclusions

Within the last 30 years, many elegant catalytic systems have been developed for the reduction of metal bicarbonates to metal formates with excellent reactivity and selectivity. Some of the catalysts accelerate the equilibrium between bicarbonates and formates, and this has created the possibility to set up a chemical battery based on hydrogen storage and release (the charge–discharge cycle). The challenge here, in order to have a high potential for application, is to have a catalyst that is efficient for both the hydrogenation of bicarbonates and the dehydrogenation of formates. Learning from nature should be helpful for catalyst design.

The catalytic hydrogenation of cyclic carbonates has been investigated as a viable route to both diols and methanol. Clearly, this process brings a new approach that contrasts the direct hydrogenation of carbon dioxide to methanol for the chemical fixation of carbon dioxide. With a sustainable supply of hydrogen, the reductive degradation of polycarbonates might also prove to be a feasible process.

At the same time, the reductive transformation of carbonates for the methylation of amines has been shown to be interesting for the utilization of carbon dioxide. Nonetheless, higher reactivity and selectivity are still needed. The photo/electroreduction of (bi)carbonates or organic carbonates might also allow for the possibility of a practical production of formates or methanol. In addition, coupling other reactions with such direct reductions may lead to the discovery of new useful methodologies for organic synthesis.

In summary, it is clear that carbonate reduction is an interesting topic for organic synthesis, photo/electrochemistry, and energy research, and the development of more advanced catalysts and catalytic systems is desired in order to achieve final applications.

References

[1] Zaidman, B.; Wiener, H.; Sasson, Y., *Int. J. Hydrogen Energy*, (1986) **11**, 341.
[2] Wang, W.-H.; Himeda, Y.; Muckerman, J. T.; Manbeck, G. F.; Fujita, E., *Chem. Rev.*, (2015) **115**, 12936.
[3] Federsel, C.; Boddien, A.; Jackstell, R.; Jennerjahn, R.; Dyson, P. J.; Scopelliti, R.; Laurenczy, G.; Beller, M., *Angew. Chem. Int. Ed.*, (2010) **49**, 9777.
[4] Ziebart, C.; Federsel, C.; Anbarasan, P.; Jackstell, R.; Baumann, W.; Spannenberg, A.; Beller, M., *J. Am. Chem. Soc.*, (2012) **134**, 20701.
[5] Federsel, C.; Ziebart, C.; Jackstell, R.; Baumann, W.; Beller, M., *Chem.–Eur. J.*, (2012) **18**, 72.
[6] Marcos, R.; Xue, L.; Sánchez-de-Armas, R.; Ahlquist, M. S. G., *ACS Catal.*, (2016) **6**, 2923.
[7] Bertini, F.; Mellone, I.; Ienco, A.; Peruzzini, M.; Gonsalvi, L., *ACS Catal.*, (2015) **5**, 1254.
[8] Langer, R.; Diskin-Posner, Y.; Leitus, G.; Shimon, L. J. W.; Ben-David, Y.; Milstein, D., *Angew. Chem. Int. Ed.*, (2011) **50**, 9948.
[9] Rivada-Wheelaghan, O.; Dauth, A.; Leitus, G.; Diskin-Posner, Y.; Milstein, D., *Inorg. Chem.*, (2015) **54**, 4526.
[10] Knölker, H.-J.; Baum, E.; Goesmann, H.; Klauss, R., *Angew. Chem. Int. Ed.*, (1999) **38**, 2064.
[11] Zhu, F.; Zhu-Ge, L.; Yang, G.; Zhou, S., *ChemSusChem*, (2015) **8**, 609.
[12] Thai, T.-T.; Mérel, D. S.; Poater, A.; Gaillard, S.; Renaud, J.-L., *Chem.–Eur. J.*, (2015) **21**, 7066.
[13] Jantke, D.; Pardatscher, L.; Drees, M.; Cokoja, M.; Herrmann, W. A.; Kühn, F. E., *ChemSusChem*, (2016) **9**, 2849.
[14] Liu, Q.; Wu, L.; Gülak, S.; Rockstroh, N.; Jackstell, R.; Beller, M., *Angew. Chem. Int. Ed.*, (2014) **53**, 7085.
[15] Bredig, G.; Carter, S. R., *Ber. Dtsch. Chem. Ges.*, (1914) **47**, 541.
[16] Stalder, C. J.; Chao, S.; Summers, D. P.; Wrighton, M. S., *J. Am. Chem. Soc.*, (1983) **105**, 6318.
[17] Chao, S.; Stalder, C. J.; Summers, D. P.; Wrighton, M. S., *J. Am. Chem. Soc.*, (1984) **106**, 2723.
[18] Wiener, H.; Blum, J.; Feilchenfeld, H.; Sasson, Y.; Zalmanov, N., *J. Catal.*, (1988) **110**, 184.
[19] Balaraman, E.; Gunanathan, C.; Zhang, J.; Shimon, L. J. W.; Milstein, D., *Nat. Chem.*, (2011) **3**, 609.
[20] Yang, X., *ACS Catal.*, (2012) **2**, 964.
[21] Li, H.; Wen, M.; Wang, Z.-X., *Inorg. Chem.*, (2012) **51**, 5716.
[22] Hasanayn, F.; Baroudi, A.; Bengali, A. A.; Goldman, A. S., *Organometallics*, (2013) **32**, 6969.
[23] Han, Z.; Rong, L.; Wu, J.; Zhang, L.; Wang, Z.; Ding, K., *Angew. Chem. Int. Ed.*, (2012) **51**, 13041.
[24] Wu, X.; Ji, L.; Ji, Y.; Elageed, E. H. M.; Gao, G., *Catal. Commun.*, (2016) **85**, 57.
[25] Kim, S. H.; Hong, S. H., *ACS Catal.*, (2014) **4**, 3630.
[26] Deluzarche, A.; Hindermann, J. P.; Kieffer, R., *J. Chem. Res., Synop.*, (1981), 72.
[27] Deluzarche, A.; Hindermann, J. P.; Kieffer, R., *J. Chem. Res., Miniprint.*, (1981), 934.
[28] Lian, C.; Ren, F.; Liu, Y.; Zhao, G.; Ji, Y.; Rong, H.; Jia, W.; Ma, L.; Lu, H.; Wang, D.; Li, Y., *Chem. Commun. (Cambridge)*, (2015) **51**, 1252.
[29] Yin, A.; Wen, C.; Dai, W.-L.; Fan, K., *J. Mater. Chem.*, (2011) **21**, 8997.
[30] Liu, H.; Huang, Z.; Han, Z.; Ding, K.; Liu, H.; Xia, C.; Chen, J., *Green Chem.*, (2015) **17**, 4281.
[31] Chen, X.; Cui, Y.; Wen, C.; Wang, B.; Dai, W.-L., *Chem. Commun. (Cambridge)*, (2015) **51**, 13776.
[32] Cui, Y.; Chen, X.; Dai, W.-L., *RSC Adv.*, (2016) **6**, 69530.
[33] Tamura, M.; Kitanaka, T.; Nakagawa, Y.; Tomishige, K., *ACS Catal.*, (2016) **6**, 376.
[34] Li, F.; Wang, L.; Han, X.; He, P.; Cao, Y.; Li, H., *RSC Adv.*, (2016) **6**, 45894.
[35] Schumacher, N.; Andersson, K. J.; Nerlov, J.; Chorkendorff, I., *Surf. Sci.*, (2008) **602**, 2783.
[36] Tsuneto, A.; Kudo, A.; Saito, N.; Sakata, T., *Chem. Lett.*, (1992), 831.
[37] Cabrero-Antonino, J. R.; Adam, R.; Junge, K.; Beller, M., *Catal. Sci. Technol.*, (2016) **6**, 7956.
[38] Chandrasekaran, K.; Thomas, J. K., *Chem. Phys. Lett.*, (1983) **99**, 7.
[39] Willner, I.; Mandler, D., *J. Am. Chem. Soc.*, (1989) **111**, 1330.
[40] Leonard, D. P.; Pan, H.; Heagy, M. D., *ACS Appl. Mater. Interfaces*, (2015) **7**, 24543.
[41] Jain, S.; Vardia, J.; Sharma, A.; Ameta, S. C., *Int. J. Energy Res.*, (2001) **25**, 107.
[42] Stalder, C. J.; Chao, S.; Wrighton, M. S., *J. Am. Chem. Soc.*, (1984) **106**, 3673.
[43] Spichiger-Ulmann, M.; Augustynski, J., *J. Chem. Soc., Faraday Trans. 1*, (1985) **81**, 713.
[44] Spichiger-Ulmann, M.; Augustynski, J., *Helv. Chim. Acta*, (1986) **69**, 632.
[45] Khalil, L. B.; Youssef, N. S.; Rophael, M. W.; Moawad, M. M., *J. Chem. Technol. Biotechnol.*, (1992) **55**, 391.
[46] Rophael, M. W.; Malati, M. A., *J. Chem. Soc., Chem. Commun.*, (1987), 1418.
[47] Raphael, M. W.; Malati, M. A., *J. Photochem. Photobiol., A*, (1989) **46**, 367.

[48] Shkrob, I. A.; Zhu, Y.; Marin, T. W.; Abraham, D., *J. Phys. Chem. C*, (2013) **117**, 19255.
[49] Rusching, U.; Müller, U.; Willnow, P.; Höpner, T., *Eur. J. Biochem.*, (1976) **70**, 325.
[50] Klibanov, A. M.; Alberti, B. N.; Zale, S. E., *Biotechnol. Bioeng.*, (1982) **24**, 25.

1.9 Hydrogenation of Carbon Dioxide

F. Nahra and C. S. J. Cazin

General Introduction

Carbon dioxide is an economical, safe, and renewable C_1 source. In recent years, it has become an attractive C_1 building block for the synthesis of organic chemicals, materials, and carbohydrates.[1–7] The utilization of carbon dioxide as a feedstock to produce chemicals and potential fuel derivatives in the near future will most certainly contribute to countering the damaging effects of global climate change.[8,9] In addition, it will provide challenging opportunities to explore new concepts for catalytic and industrial developments. However, carbon dioxide is not used extensively as a source of carbon in current laboratory and industrial practices, and its use is limited to a few industrial processes, including the synthesis of urea and its derivatives, salicylic acid, and carbonates.[10] Its main use is in the synthesis of inorganic carbonates and pigments.[11] This is considered a main consequence of the thermodynamic stability of carbon dioxide; therefore, high-energy substances or electroreductive processes are usually required for any transformation involving this material.[12–14] Hydrogen, which is a high-energy material, is typically used for the reduction of carbon dioxide.

The main products of the hydrogenation or reduction of carbon dioxide fall into two categories: fuels and chemicals. The products resulting from the hydrogenation of carbon dioxide, including methanol and hydrocarbons, which are considered to be excellent fuels for internal combustion engines, can be easily stored and transported.[15,16] Furthermore, methanol and formic acid are raw materials that can be used as important building blocks in many chemical industries.[16,17] Recent investigations in the context of catalysis, surface science, biology, nanotechnology, and environmental science have sparked renewed interest in the hydrogenation of carbon dioxide, mainly as a result of its fundamental and practical significance. Both homogeneous and heterogeneous catalysts have been used to hydrogenate carbon dioxide.[1,2,7,15,18–21] Homogeneous catalysts show satisfactory activity and selectivity, but the recovery and regeneration of the catalysts are still problematic, which thus renders the procedures less attractive for industrial applications. In contrast, high potential for the stability and convenient separation, handling, and reuse of heterogeneous catalysts make them desirable for large-scale productions.[2,10,12,13,17,19,22,23] With the aim of combining the high reactivity of homogeneous catalysts with the recyclability and stability of heterogeneous catalysts, several research groups have made significant contributions to the field of supported catalysis by means of immobilizing homogeneous catalysts on solid supports. In addition, there has been ongoing research into novel heterogeneous catalysts and the use of greener solvents such as ionic liquids and supercritical carbon dioxide.[19,24–26]

The catalytic hydrogenation of carbon dioxide is mainly driven by the development and implementation of various homogeneous and heterogeneous metal systems. The development of new catalytic conversions of carbon dioxide requires knowledge of the properties of these metal systems. In this context, the following sections will highlight the most important advancements in this field (Scheme 1). Related discussion on the reduction of carbonates can be found in Section 1.8.

for references see p 311

Scheme 1 Compounds Generated by the Hydrogenation of Carbon Dioxide and Highlighted in this Chapter

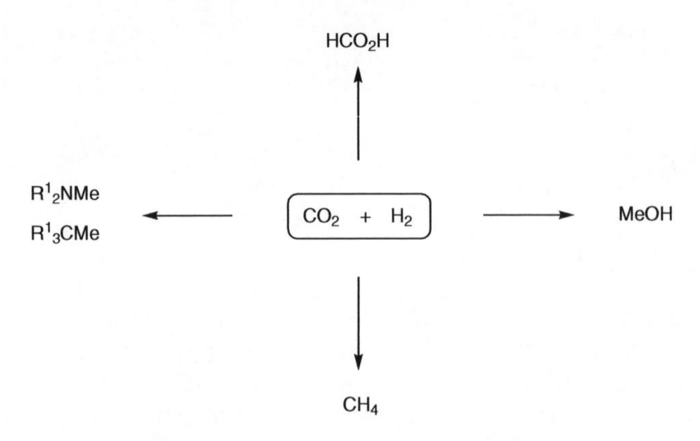

1.9.1 Hydrogenation of Carbon Dioxide to Formic Acid or Formate Salts

One very important chemical among the diverse products that are derived from the hydrogenation of carbon dioxide is formic acid, which is a valuable basic chemical used in various industries.[27,28] In addition, it plays a major role in synthetic chemistry as an acid, a reductant, and a synthetic precursor. Furthermore, its potential use as a hydrogen-storage material in energy industries has attracted much interest in recent years, mainly as a result of its ease of storage and transportation (Scheme 2).[21,29] More recently, formic acid has emerged as a promising fuel source in direct liquid fuel-cell systems.[30,31]

Scheme 2 Reversible Synthesis of Formic Acid from Hydrogen and Carbon Dioxide and Its Potential Use as a Hydrogen-Storage Material[21,29]

$$CO_2(g) + H_2(g) \rightleftharpoons HCO_2H$$

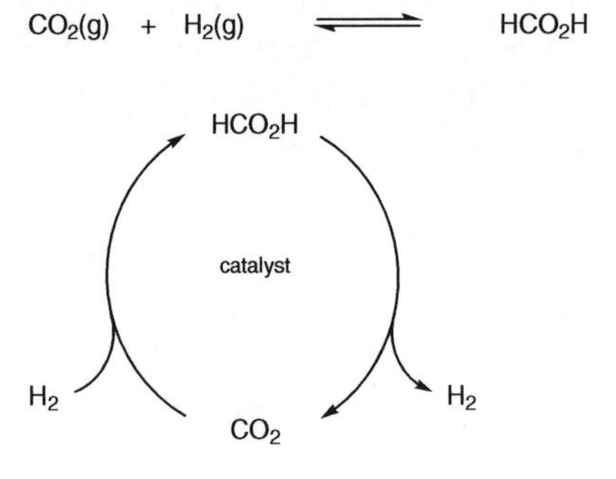

Owing to the countless applications of formic acid as well as the utilization of carbon dioxide, the direct synthesis of formic acid/formates from carbon dioxide and hydrogen catalyzed by transition-metal systems has attracted much attention from the scientific and technological communities and has been widely investigated in the last 20 years.

1.9.1.1 Homogeneous Systems

Following the pioneering work of Inoue and co-workers,[32] who highlighted the high activity of Wilkinson's catalyst [$RhCl(PPh_3)_3$] and its ruthenium analogue [$RuCl_2(PPh_3)_3$] relative to that of other catalysts of the time (e.g., palladium, nickel, and iridium), a variety of rhodium and ruthenium catalysts have been developed and applied in the hydrogenation of carbon dioxide to formic acid/formate. Recent work has extended the success of these early findings to include various phosphine,[33–37] nitrogen-based,[38–45] PNP,[46–50] and N-heterocyclic carbene[51,52] ligands **1–13** (Scheme 3).

Scheme 3 A Selection of Ligands Involved in the Hydrogenation of Carbon Dioxide to Formic Acid/Formate Salts[33–52]

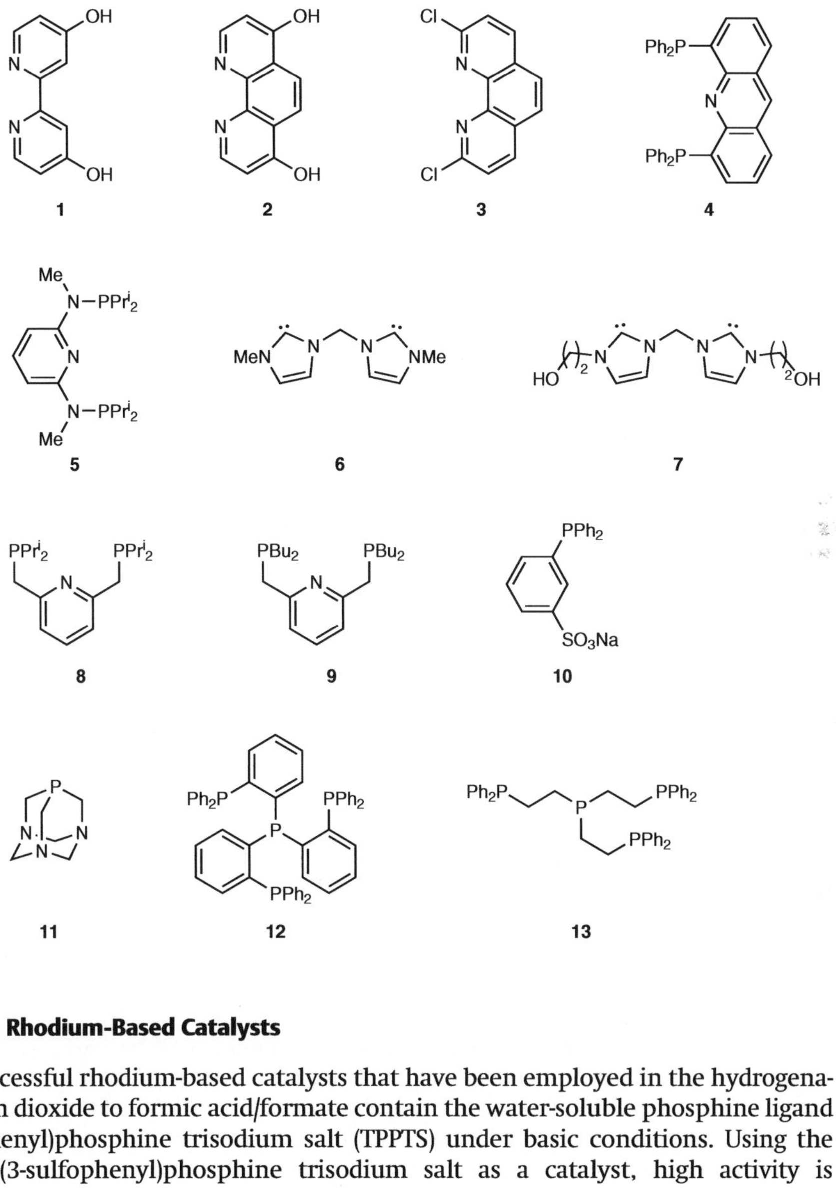

1.9.1.1.1 Rhodium-Based Catalysts

The most successful rhodium-based catalysts that have been employed in the hydrogenation of carbon dioxide to formic acid/formate contain the water-soluble phosphine ligand tris(3-sulfophenyl)phosphine trisodium salt (TPPTS) under basic conditions. Using the rhodium–tris(3-sulfophenyl)phosphine trisodium salt as a catalyst, high activity is achieved, as expressed by a turnover number (TON) of 3439 and a turnover frequency (TOF) of 7260 h^{-1} (Scheme 4).[33,53] Production of formic acid (**14**, X = H) at room temperature and ambient pressure is also observed, although at a much lower rate (a turnover frequency of 3 per day).

More recent work has shown that rhodium-based catalysts bearing bipyridine- and phenanthroline-derived ligands [4,4′-dihydroxy-2,2′-bipyridine (**1**) and 4,7-dihydroxy-1,10-phenanthroline (**2**)] can also achieve high activity in the aforementioned reaction; a rhodium-based catalyst with ligand **2** exhibits high activity and reaches a turnover number of 2400 in 1 M potassium hydroxide solution under 40 bar of hydrogen/carbon dioxide (1:1) at 80 °C after 32 hours.[45]

Scheme 4 Selected Examples of the Hydrogenation of Carbon Dioxide to Formic Acid/Formate Using Rhodium-Based Catalysts[33,45,53]

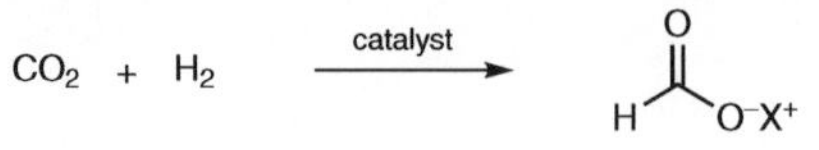

$$CO_2 + H_2 \xrightarrow{\text{catalyst}} HCO_2^-X^+$$

14

Catalyst[a]	Conditions	X	TON	TOF (h^{-1})	Ref
$RhCl(TPPTS)_3$	Me_2NH, H_2O, H_2/CO_2 (1:1, 40 bar), rt, 12 h	Me_2NH_2	3439	7260[b]	[33,53]
[Rh(Cp*)Cl(**1**)]Cl	KOH, H_2O, H_2/CO_2 (1:1, 40 bar), 80 °C, 12 h	K	1800	790	[45]
[Rh(Cp*)Cl(**2**)]Cl	KOH, H_2O, H_2/CO_2 (1:1, 40 bar), 80 °C, 32 h	K	2400	270	[45]

[a] For the structures of ligands, see Scheme 3, Section 1.9.1.1.

[b] Initial turnover frequency, determined under the same conditions at 81 °C after 0.5 h.

Potassium Formate (14, X = K); Typical Procedure:[45]

A degassed soln of [Rh(Cp*)Cl(**2**)]Cl (5 μmol) in 1 M aq KOH (50 mL) in a 100-mL stainless-steel reactor equipped with a sampling device was saturated with CO_2. The reactor was heated to 80 °C and was then pressurized with CO_2/H_2 (1:1, 40 bar). The concentration of formate was monitored by HPLC on an anion-exclusion column [Tosoh TSKgel SCX(H^+)] with 2 mM aq phosphate as an eluent and a UV detector (λ 210 nm). After 32 h, the final concentration of the title compound was 0.24 M with a turnover number of 2400 and a turnover frequency of 270 h^{-1}.

1.9.1.1.2 Ruthenium-Based Catalysts

In the hydrogenation of carbon dioxide to formic acid/formate, ruthenium-based catalysts are usually much more efficient than their rhodium counterparts and, therefore, have been widely studied (Scheme 5). Using a sodium bicarbonate solution (0.3 M) and a ruthenium complex bearing water-soluble ligand **10**, a turnover frequency (TOF) of 9600 h^{-1} is achieved, which at the time of the study was the highest rate in pure aqueous solution.[34] In non-water-based systems, the turnover frequency of this reaction increases to 95 000 h^{-1} using triethylamine and 2,3,4,5,6-pentafluorophenol as additives, in supercritical carbon dioxide at 50 °C and a carbon dioxide/hydrogen (70:120) pressure of 190 bar.[54] This increase in the turnover frequency by one order of magnitude means that this is considered to be the highest turnover frequency for a ruthenium-based homogeneous system. Despite these results, water-based systems have still found more widespread use. In this context, considerably high turnover numbers (TON up to 15 400) and turnover frequencies (up to 4400 h^{-1}) are achieved upon using water-soluble bipyridine- and phenanthroline-derived ligands **1** and **2**.[45]

A ruthenium complex with bis(N-heterocyclic carbene) **6** as the ligand achieves a high turnover number of 23 000 at 40 bar hydrogen/carbon dioxide (1:1) and 200 °C in 1 M potassium hydroxide solution in 75 hours.[51] Notably, the same catalyst can also be used to achieve the transfer hydrogenation of carbon dioxide with propan-2-ol to generate potassium formate; a turnover number of 874 is obtained, which is the highest turnover number so far reported for this type of reaction.[51]

Scheme 5 Selected Examples of the Hydrogenation of Carbon Dioxide to Formic Acid/Formate Using Ruthenium-Based Catalysts[34,35,45,46,51,54,55]

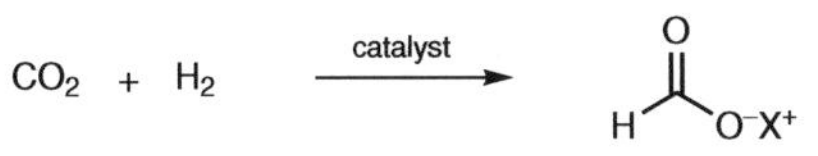

$$CO_2 + H_2 \xrightarrow{\text{catalyst}} HC(O)O^-X^+ \quad \textbf{14}$$

Catalyst[a]	Conditions	X	TON	TOF (h^{-1})	Ref
$\{RuCl_2(\mathbf{10})_2\}_2$	$NaHCO_3$, H_2O, H_2/CO_2 (60:35, 95 bar), 80 °C	Na	–	9600	[34]
$[Ru(\mathbf{3})_2(OH_2)_2](OTf)_2$	Et_3N, EtOH, H_2/CO_2 (1:1, 60 bar), 150 °C, 8 h	Et_3NH	5000	625	[55]
$RuCl(OAc)(PMe_3)_4$	Et_3N/C_6F_5OH, $scCO_2$,[b] H_2/CO_2 (70:120, 190 bar), 50 °C, 0.33 h	Et_3NH	–	95 000	[54]
$[Ru(\eta^6\text{-}C_6Me_6)Cl(\mathbf{1})]Cl$	KOH, H_2O, H_2/CO_2 (1:1, 60 bar), 120 °C, 8 h	K	13 620	4400	[45]
$[Ru(\eta^6\text{-}C_6Me_6)Cl(\mathbf{2})]Cl$	KOH, H_2O, H_2/CO_2 (1:1, 60 bar), 120 °C, 24 h	K	15 400	3600	[45]
$[Ru(\eta^6\text{-}C_6Me_6)Cl(\mathbf{6})]PF_6$	KOH, H_2O, H_2/CO_2 (1:1, 40 bar), 200 °C, 75 h	K	23 000	306	[51]
$[Ru(\eta^6\text{-}C_6Me_6)Cl(\mathbf{7})]Cl$	KOH, H_2O, H_2/CO_2 (1:1, 40 bar), 200 °C, 20 h	K	9500	475	[51]
$RuCl_2(\mathbf{11})_4$	DMSO/H_2O (9:1), H_2/CO_2 (1:1, 100 bar), 50 °C, 120 h	H	475	4	[35]
$RuCl(OBz)(PPh_3)(\mathbf{4})$	DMSO/H_2O (95:5), H_2/CO_2 (2:1, 120 bar), 60 °C, 16 h	H	4200	260	[46]
$RuCl(OBz)(PPh_3)(\mathbf{4})$	AcOH/NaOAc, DMSO/H_2O (95:5), H_2/CO_2 (2:1, 120 bar), 60 °C, 16 h	H	16 310	1019	[46]

[a] For the structures of ligands, see Scheme 3, Section 1.9.1.1.
[b] $scCO_2$ = supercritical CO_2.

A catalytic system that allows for the formation of high concentrations of free formic acid (**14**, X = H) in dimethyl sulfoxide in the absence of a base has recently been developed.[35] This has sparked renewed interest in ruthenium-based systems, and a more efficient version achieves a high turnover number of 16 310 and a turnover frequency of 1019 h^{-1} at 120 bar hydrogen/carbon dioxide (2:1) and 60 °C in 16 hours (Scheme 5).[46] Dimethyl sulfoxide, a well-known hydrogen-bond acceptor, was identified as a favorable solvent for such hydrogenations of carbon dioxide on the basis that hydrogen bonding plays an important role in the stabilization of formic acid. A density functional theory model of a 1:1 complex of dimethyl sulfoxide/formic acid was identified and analyzed in the same study, and it provides a theoretical rationale for the thermodynamic stabilization of formic acid in dimethyl sulfoxide and a dimethyl sulfoxide/water mixture.[46] However, no adduct has been isolated experimentally. Furthermore, the authors show that the increase in acidity of the media owing to the production of formic acid is responsible for protonation of the catalyst and its consequent deactivation. Therefore, to counteract this process, an acetate buffer can be used and this leads to a significant increase in both the turnover number and turnover frequency of the hydrogenation reaction.[46]

Potassium Formate (14, X = K); Typical Procedure:[51]
The reaction was performed in a Hastelloy Autoclave Mini-Reactor system equipped with a 50-mL cylinder. $[RuCl(\eta^6\text{-}C_6Me_6)(\mathbf{6})]PF_6$ was dissolved in a degassed soln of 1 M aq KOH (10 mL). The reactor was pressurized with CO_2/H_2 (1:1, 40 bar) and heated at 200 °C for 75 h. After reducing the pressure to 1 bar and cooling to rt, the solvent was removed by

for references see p 311

evaporation, and the residue was dissolved in D_2O. The yield of the title compound was determined by ^{1}H NMR spectroscopy in D_2O, using isonicotinic acid as an internal standard; yield: 0.46 mmol; turnover number: 23 000; turnover frequency: 306 h^{-1}.

Formic Acid (14, X = H); Typical Procedure:[46]
The ruthenium complex $RuCl(OBz)(PPh_3)$(**4**) (0.5 mL of a stock soln of 2.0 mg in 4.0 mL DMSO; 0.25 mg, 0.23 μmol) was mixed with DMSO (2.2 mL) and 4 M aq AcOH/NaOAc buffer (0.3 mL) in a Schlenk tube. The bright blue soln was transferred into a 20-mL steel autoclave. The autoclave was subsequently pressurized with CO_2 (40 bar) and H_2 (80 bar) at rt, and then placed in a preheated heating cone. The mixture was stirred at 60 °C for 16 h. Afterwards, the autoclave was cooled to rt and carefully vented. The mixture was analyzed directly by ^{1}H NMR spectroscopy in DMSO-d_6, using mesitylene as an internal standard; yield: 1.27 M; turnover number: 16 310; turnover frequency: 1019 h^{-1}.

1.9.1.1.3 Iridium-Based Catalysts

For years, the use of iridium-based complexes in carbon dioxide reductions was associated with low conversions and rates.[56,57] Himeda and co-workers finally succeeded in achieving high catalytic activity by using well-designed half-sandwich complexes. The water-soluble bipyridine- and phenanthroline-derived ligands **1** and **2**, which have successfully been applied to rhodium and ruthenium complexes, have proven to be even more efficient if combined with iridium (Scheme 6).[40,43–45] Using iridium complexes of ligands **1** and **2**, maximum catalytic activity [turnover frequency (TOF) up to 42 000 h^{-1} and turnover number (TON) up to 222 000] for the formation of potassium formate is obtained at 60 bar hydrogen/carbon dioxide (1:1) and 120 °C in 1 M aqueous potassium hydroxide (Scheme 7).[45] In addition, these impressive catalysts allow the reaction to proceed at atmospheric pressure.

Scheme 6 Ligands Involved in the Iridium-Catalyzed Hydrogenation of Carbon Dioxide to Formic Acid/Formate Salts[45,48,52,58]

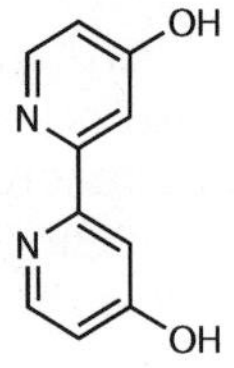

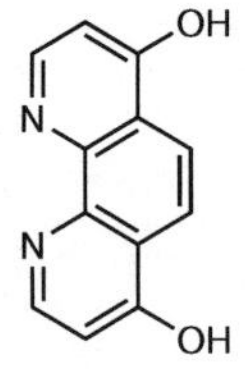

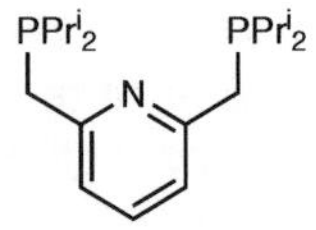

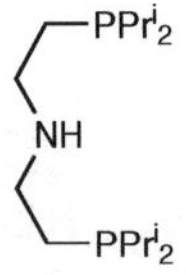

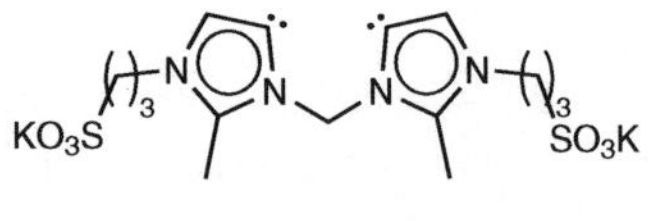

Scheme 7 Selected Examples of the Hydrogenation of Carbon Dioxide to Formate Using Iridium-Based Catalysts[45,48,52,58]

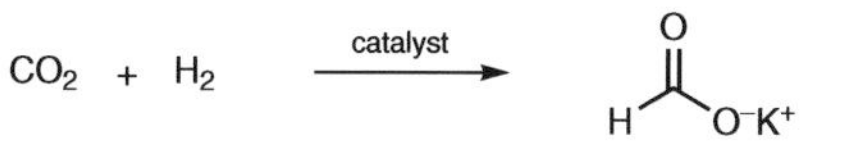

$$CO_2 + H_2 \xrightarrow{\text{catalyst}} HCO_2^-K^+$$

17

Catalyst	Conditions	TON	TOF (h^{-1})	Ref
[Ir(Cp*)Cl(**1**)]Cl	KOH, H_2O, H_2/CO_2 (1:1, 60 bar), 120 °C, 57 h	190000	42000[a]	[45]
[Ir(Cp*)Cl(**2**)]Cl	KOH, H_2O, H_2/CO_2 (1:1, 60 bar), 120 °C, 48 h	222000	33000[a]	[45]
IrH_3(**8**)	KOH, H_2O/THF, H_2/CO_2 (1:1, 60 bar), 120 °C, 48 h	3500000	73000	[48]
$Ir(O_2CH)H_2$(**15**)	KOH, H_2O, H_2/CO_2 (1:1, 55 bar), 185 °C, 24 h	348000	14500	[58]
IrI_2(OAc)(**16**)	KOH, H_2O, H_2/CO_2 (1:1, 60 bar), 200 °C, 75 h	190000	2533	[52]

[a] Initial turnover frequency calculated by nonlinear least-squares fit of the experimental data from the initial part of the reaction.

Using an iridium(III) complex bearing pincer PNP ligand **8**, an unprecedented turnover number of 3.5 million is observed along with a high turnover frequency of 73000 h^{-1} at 60 bar hydrogen/carbon dioxide (1:1) and 120 °C in tetrahydrofuran and 1 M aqueous potassium hydroxide.[48] A simpler PNP-based iridium catalyst with ligand **15** can still efficiently catalyze the reaction, albeit with a lower turnover number (up to 348000) and a lower turnover frequency (up to 14500 h^{-1}) relative to that observed for the previous system.[58] A series of N-heterocyclic-carbene-based iridium complexes can also catalyze the hydrogenation of carbon dioxide to formate, and for the complex bearing water-soluble N-heterocyclic carbene ligand **16**, a decent turnover number of 190000 is obtained.[52] Notably, these N-heterocyclic-carbene-based iridium complexes can also be used for the transfer hydrogenation of carbon dioxide to formate upon using propan-2-ol as the hydrogen donor.[52]

Potassium Formate (17); Typical Procedure:[48]
The iridium complex IrH_3(**8**) (20 μmol) was dissolved in THF (4.00 mL) and diluted to 100 μmol•L^{-1}. A 50-mL stainless steel autoclave was charged with the catalyst soln (100 μL) and degassed 1 M aq KOH (5.00 mL), and was pressurized with CO_2/H_2 (1:1, 60 bar). The mixture was stirred at 120 °C for 48 h. Sodium 3-(trimethylsilyl)propane-1-sulfonate (11.3 mg, 51.8 μmol) was added to the mixture as an internal standard, and an aliquot of the mixture was dissolved in D_2O to estimate the yield using ^1H NMR spectroscopy; yield: 70% (based on the added internal standard); turnover number: 3500000; turnover frequency: 73000 h^{-1}. The reaction was repeated at least two times to confirm reproducibility.

1.9.1.1.4 Other Metal-Based Catalysts

Catalysts based on other metals have been less frequently investigated owing to their low efficiency, and therefore the development of nonprecious-metal-based homogeneous catalytic systems for the conversion of carbon dioxide into formates is still limited (Scheme 8). Most advances in this area have been achieved with iron-based complexes.[36,37,47,50] Nickel- and copper-based complexes have also been used as catalysts in this transformation;[59,60] however, these complexes are much less efficient than their precious-metal counterparts. Upon using [1,2-bis(dicyclohexylphosphino)ethane]dichloronickel(II) {[$NiCl_2$(dcpe)], dcpe = 1,2-bis(dicyclohexylphosphino)ethane}, a maximum turnover number (TON) of 4400 is obtained along with a very low turnover frequency (TOF) of 20 h^{-1}.[59]

for references see p 311

The promising activity of iron complexes is showcased by the encouraging turnover number of 9840 and an average turnover frequency of 469 h^{-1} that is obtained using a complex derived from ligand **5**, at 80 bar hydrogen/carbon dioxide (1:1) and 80 °C in ethanol, using 1,8-diazabicyclo[5.4.0]undec-7-ene as the base (Scheme 8).[47] Under these conditions, formate is obtained in excellent yield (98%).

Scheme 8 Selected Examples of the Hydrogenation of Carbon Dioxide to Formic Acid/Formate Using Non-precious-Metal-Based Catalysts[36,37,47,50,59,60]

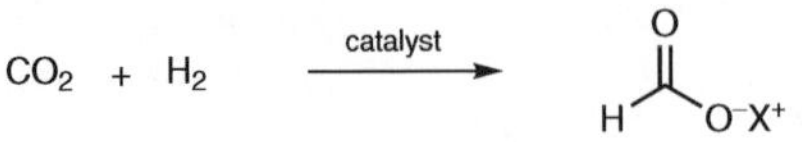

$$CO_2 + H_2 \xrightarrow{\text{catalyst}} HCO_2^- X^+$$

14

Catalyst[a]	Conditions	X	TON	TOF (h^{-1})	Ref
$FeH_2(CO)$(**9**)	NaOH, H_2O/THF, H_2/CO_2 (6.7:3.3, 10 bar), 80 °C, 5 h	Na	788	156	[50]
$NiCl_2$(dcpe)	DBU, DMSO, H_2/CO_2 (1:4, 200 bar), 50 °C, 216 h	DBUH	4400	20	[59]
$Cu(OAc)_2$•H_2O	DBU, dioxane, H_2/CO_2 (1:1, 80 bar), 100 °C, 21 h	DBUH	165	8	[60]
[FeF(**12**)]BF_4	Et_3N, MeOH/H_2O (88:12), H_2/CO_2 (1:1, 60 bar), 100 °C, 20 h	Et_3NH	1897	95	[36]
$FeCl_2$/**13**	Et_3N, MeOH, H_2/CO_2 (2:1, 90 bar), 100 °C, 21 h	Et_3NH	256	12	[37]
FeBrH(CO)(**5**)	DBU, EtOH, H_2/CO_2 (1:1, 80 bar), 80 °C, 21 h	DBUH	9840	469	[47]

[a] For the structures of ligands, see Scheme 3, Section 1.9.1.1.

Formate–1,8-Diazabicyclo[5.4.0]undec-7-ene Complex (14, X = DBUH); Typical Procedure:[47]

DBU (10 mmol) was added to a soln of FeBrH(CO)(**5**) (1 μmol) in distilled and dried EtOH (25 mL). The soln was then pressurized with CO_2/H_2 (1:1, 80 bar). The mixture was stirred at 80 °C for 21 h. The yield was determined by ^{1}H NMR spectroscopy by using DMF as an internal standard; yield: 98% (based on the added internal standard); turnover number: 9840; average turnover frequency: 469 h^{-1}.

1.9.1.2 Heterogeneous Systems

Despite the fact that homogeneous catalysts can exhibit excellent efficiencies for the hydrogenation of carbon dioxide to formic acid/formate, there might be reluctance in industry to use them for large-scale production owing to the perceived difficulty of separating the catalyst from the final mixture. Because of such limitations, diverse heterogeneous catalysts have been reported.[19] These heterogeneous catalysts are mainly made from precious metals, and ruthenium- and iridium-based catalysts are the most catalytically active. Various supports for the selected metal have been developed for this transformation [e.g., titanium(IV) oxide, alumina, silica];[19] amine-functionalized silica (**18** and **19**)[61,62] as well as a bipyridine-based covalent triazine framework (**21**)[63] and Tröger's base derived microporous organic polymers (e.g., **20**)[64] have proven to be the most successful, as they afford high turnover numbers (TON) and turnover frequencies (TOF) (Scheme 9). The most-active catalyst reported (**21**) affords an initial turnover frequency of 5300 h^{-1} at 80 bar hydrogen/carbon dioxide (1:1) and 120 °C after 0.25 hours (Scheme 10).[63] The reaction reaches a remarkably high turnover number of 5000 after 2 hours. This catalyst shows good resistance to metal leaching and can be recycled several times.

Scheme 9 Heterogeneous Catalysts for the Hydrogenation of Carbon Dioxide to Formic Acid/Formate[61–64]

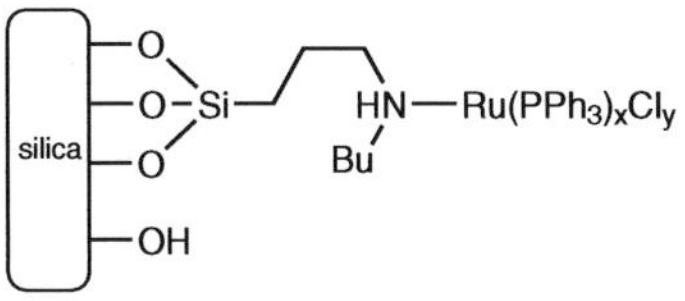

silica O O Si O OH HN Ru(PPh3)xCly Bu

18

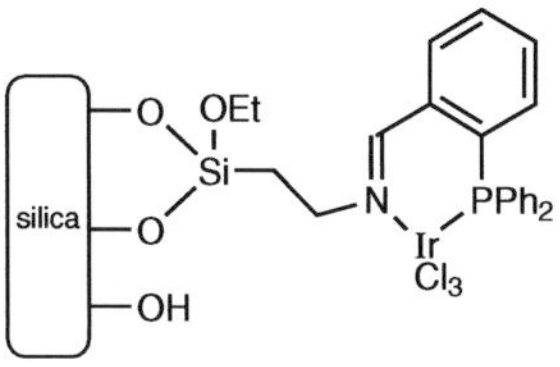

silica O OEt Si O OH N PPh2 Ir Cl3

19

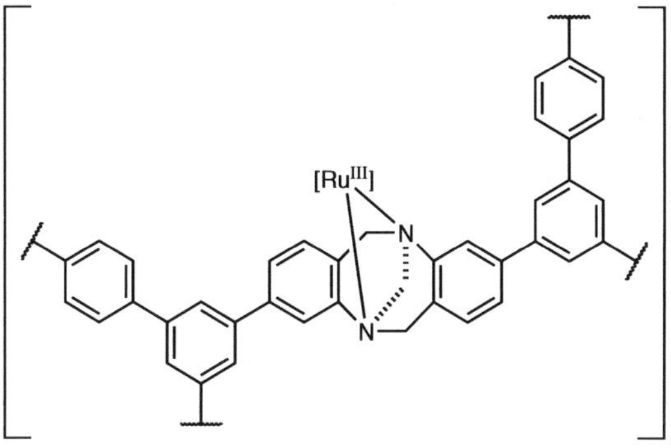

[RuIII] N N n

20

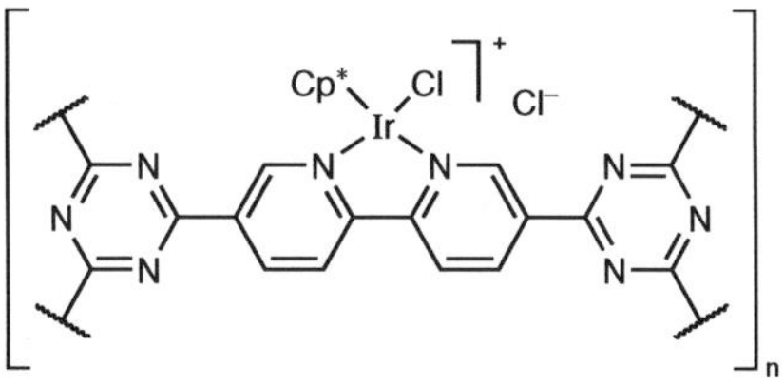

Cp* Cl Ir + Cl− N N N N N N N N n

21

for references see p 311

Scheme 10 Selected Examples of the Hydrogenation of Carbon Dioxide to Formic Acid/Formate Using Heterogeneous Catalysts[61–68]

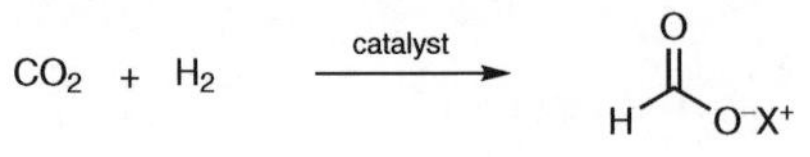

$$CO_2 + H_2 \xrightarrow{\text{catalyst}} HCO_2^-X^+$$

14

Catalyst[a]	Conditions	X	TON	TOF (h^{-1})	Ref
AUROlite	Et_3N, H_2/CO_2 (1:1, 180 bar), 40 °C, 52 h	Et_3NH	855	16.4	[65]
AuNPs/alumina	Et_3N, EtOH, H_2/CO_2 (1:1, 40 bar), 70 °C, 20 h	Et_3NH	215	11	[66]
AuNPs/TiO_2	Et_3N, EtOH, H_2/CO_2 (1:1, 40 bar), 70 °C, 20 h	Et_3NH	111	5.5	[67]
Ru/γ-alumina (nanorods)	Et_3N, EtOH, H_2/CO_2 (50:85, 135 bar), 80 °C, 1 h	Et_3NH	731	731	[68]
18	Ph_3P/Et_3N, EtOH, H_2/CO_2 (1:3, 160 bar), 80 °C, 1 h	Et_3NH	1384	1384	[61]
19	Et_3N, H_2O, H_2/CO_2 (1:1, 40 bar), 80 °C, 2 h	Et_3NH	1300	620	[62]
19	Et_3N, H_2O, H_2/CO_2 (1:1, 40 bar), 120 °C, 2 h	Et_3NH	2300	1200	[62]
19	Et_3N, H_2O, H_2/CO_2 (1:1, 40 bar), 60 °C, 20 h	Et_3NH	2700	140	[62]
20	Ph_3P, Et_3N, H_2/CO_2 (1:1, 120 bar), 60 °C, 24 h	Et_3NH	2254	94	[64]
21	Et_3N, H_2O, H_2/CO_2 (1:1, 80 bar), 120 °C, 2 h	Et_3NH	5000	2500	[63]
21	Et_3N, H_2O, H_2/CO_2 (1:1, 80 bar), 120 °C, 0.25 h	Et_3NH	1300	5300	[63]

[a] AUROlite = 1 wt% Au on TiO_2 extrudates (Mintek); NPs = nanoparticles.

Triethylammonium Formate (14, X = Et_3NH); Typical Procedure:[63]
Hydrogenation was performed in a homemade 100-mL stainless-steel reactor with a glass vessel insert. In a typical run, complex **21** (0.01 g) was added to a CO_2-saturated aq soln of Et_3N, and the vessel was closed and flushed with CO_2. The reactor was pressurized first with CO_2 and then with H_2 (1:1) to the targeted pressure (80 bar) at rt, and was then heated at 120 °C. The mixture was cooled to rt after the appropriate time, and the pressure was slowly released. The concentration of the title compound was analyzed by HPLC using 0.005 M H_2SO_4 as the eluent. An initial turnover frequency of 5300 h^{-1} was observed after 0.25 h and a maximum turnover number of 5000 was attained after 2 h. In the recycling experiment, the catalyst was recovered by filtration, washed with H_2O, and dried under vacuum. The collected material was reused following the above procedure.

Pure and Anhydrous Formic Acid (14, X = H) from Triethylammonium Formate Complex by the Amine-Exchange Method; Typical Procedure:[65]
Trihexylamine (133 g, 474 mmol) was added to $HCO_2H{\bullet}NEt_3$ adduct (acid/amine molar ratio of 1.715:1; 100 g, 555 mmol). Two liquid phases arose and were analyzed by ^1H NMR spectroscopy: the upper phase was composed of Et_3N/trihexylamine in a 1:10 molar ratio with only a trace amount of HCO_2H, and the heavier one was composed of HCO_2H/Et_3N/trihexylamine in a 22.5:12:1 molar ratio. The biphasic system was fractioned under reduced pressure (90 Torr) with a 10-cm Vigreux column. Once the oil bath temperature had reached 120 °C, Et_3N began to distil within the 35–40 °C range. Pure Et_3N (50.5 g, 499 mmol, 90%) was recovered in a Schlenk flask cooled at 0 °C, whereas the residual liquid became clear (first fraction). The vessel containing the amine was removed, and the temperature was increased again (90 Torr). At 160 °C (oil bath temperature), a second fraction began to distil within the 125–135 °C range, and an acid liquid (48.8 g) was collected. The ^1H NMR spectrum showed HCO_2H/Et_3N/trihexylamine in a 148:9:1 molar ratio (85 wt% HCO_2H purity contaminated by 11.5 wt% Et_3N and 3.5 wt% trihexylamine). Once the high-boiling fraction started to distil, the residue again became opaque, affording at the end, quantitatively, clear trihexylamine (third fraction). Redistillation of the second

fraction at 100 °C under atmospheric pressure gave the title compound; yield: 36.350 g (790 mmol; 83% yield with respect to HCO_2H in the HCO_2H/Et_3N adduct). The residue was composed of HCO_2H/Et_3N/trihexylamine in a 19:8.7:1 molar ratio (11.7 g).

1.9.2 Hydrogenation of Carbon Dioxide to Methanol

Methanol is a common solvent, an alternative fuel, and a raw material for the chemical industry. As an alternative feedstock, carbon dioxide has the promising potential to replace carbon monoxide in methanol production; this alternative pathway has received much attention, as it can be considered a more efficient way to utilize carbon dioxide.[12] Other products can be formed during the hydrogenation of carbon dioxide, such as carbon monoxide, hydrocarbons, and higher alcohols.[69] Therefore, in the synthesis of methanol, a highly selective catalyst is required to avoid the formation of these undesired byproducts. Typically, catalysts used in the hydrogenation of carbon dioxide are the same as those used in the synthesis of methanol by the hydrogenation of carbon monoxide. In this context, a number of investigations have addressed the effects of active components, supports, promoters, preparation methods, and surface morphology on reactivity.

1.9.2.1 Homogeneous Systems

The use of homogeneous catalysis by means of transition-metal complexes for the synthesis of methanol through hydrogenation of carbon dioxide is still in its early stages. The highest activities for this transformation have mainly been exhibited by ruthenium catalysts (Scheme 11).[70,71] However, in recent years, a nickel hydride catalyst has emerged as an efficient alternative with the use of a borane as a reducing reagent.[72] Several other metals have also been investigated.[20] Little to no methanol is obtained upon using tungsten-, molybdenum-, or cobalt-based catalysts, and rhodium-, iridium-, and iron-based catalysts have also been reported to show inferior activity.

Scheme 11 Selected Homogeneous Catalysts for the Hydrogenation of Carbon Dioxide to Methanol[70–72]

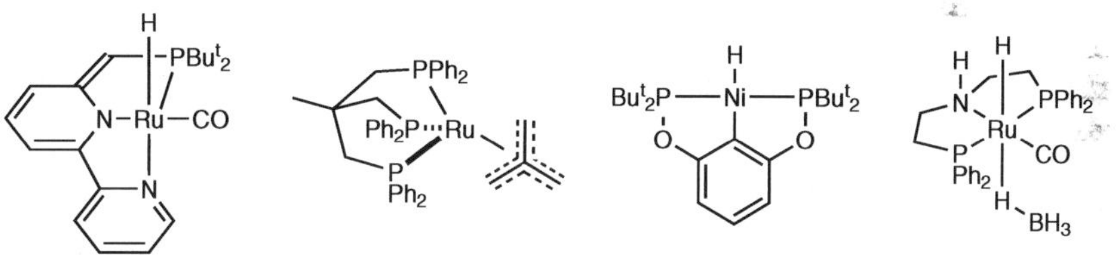

A cascade reaction, involving the hydrogenation of carbon dioxide to formic acid [catalyzed by chlorotetrakis(trimethylphosphine)ruthenium(II) acetate], followed by esterification of formic acid to a formate ester [catalyzed by scandium(III) trifluoromethanesulfonate] and subsequent hydrogenation of the ester (catalyzed by **22**) successfully delivers desired methanol (**26**).[70] The combination of all three processes in one pot delivers methanol with a very low turnover number (TON) of 2.5 at 40 bar hydrogen/carbon dioxide (3:1) and 135 °C in deuterated methanol (Scheme 12); however, if the last step (hydrogenation of the ester) is performed sequentially, albeit separately, the turnover number increases to 21.[70]

A strategy developed later, involving the use of a single-component catalyst, i.e. **23**, improves the above reaction.[71] In this case, the hydrogenation of carbon dioxide liberates methanol with a turnover number up to 221 at 80 bar hydrogen/carbon dioxide (3:1) and 140 °C in the presence of bis(trifluoromethylsulfonyl)amine. Efficient nickel hydride catalysis with a borane as a reductant, under extremely mild conditions of 1 bar carbon dioxide at room temperature, has also been accomplished.[72] Thus, PCP–nickel(II) hydride complex **24** reacts rapidly with carbon dioxide under these conditions to afford methanol with a relatively high turnover number of 495. Notably, the byproduct [2,2′-oxybis(benzo-[*d*][1,3,2]dioxaborole)] precipitates out of the benzene-d_6 solution and thus can be easily removed.

Scheme 12 Selected Examples of the Hydrogenation of Carbon Dioxide to Methanol Using Homogeneous Catalysts[70–74]

$$CO_2 + H_2 \xrightarrow{\text{catalyst}} \underset{\mathbf{26}}{\text{MeOH}}$$

Catalyst	Conditions	Co-products	TON	Ref
$Ru_3(CO)_{12}$	KI, NMP, H_2/CO_2 (3:1, 80 bar), 240 °C, 3 h	CO, CH_4, C_2H_6	94.5	[73]
$RuCl(OAc)(PMe_3)_4$/ $Sc(OTf)_3$/**22**	CD_3OH, H_2/CO_2 (3:1, 40 bar), 135 °C, 16 h	HCO_2CD_3	2.5	[70]
23	Tf_2NH, THF/EtOH, H_2/CO_2 (3:1, 80 bar), 140 °C, 24 h	HCO_2Me	221	[71]
24	catecholborane (then H_2O), benzene-d_6, pCO_2 (1 bar), rt, 1 h	2,2′-oxybis(benzo[*d*]-[1,3,2]dioxaborole)	495	[72]
25	Me_2NH, K_3PO_4, THF, H_2/CO_2 (20:1, 52.5 bar), 95 °C, 18 h, then 155 °C, 36 h	DMF, $HCO_2H•Me_2NH$	550	[74]

In an improved cascade strategy, a tandem dimethylamine/ruthenium complex **25** system affords methanol with a relatively high turnover number of 550 at 52.5 bar hydrogen/carbon dioxide (20:1) in tetrahydrofuran under basic conditions (Scheme 12).[74] Although high quantities of dimethylformamide are also produced, the carbon efficiency of this procedure is quite high (up to 96% conversion of carbon dioxide into dimethylformamide and methanol).

Methanol (26); Typical Procedure:[74]

Under a N_2 atmosphere in a drybox, Ru complex **25** (1.0 mg, 1.7 μmol, 0.03 mol%) was dissolved in THF (2 mL) and a 3.8 M soln of Me_2NH (7.6 mmol, 4470 equiv relative to Ru). The resulting soln was added to a prechilled (in a drybox freezer at −33 °C) metal well of a pressure vessel containing K_3PO_4 (0.250 mmol, 5 mol%) and an octagonal magnetic stirrer bar (5/16 × 1/2 inch). The vessel was sealed and removed from the drybox. The vessel was pressurized with CO_2 (2941 equiv relative to Ru) and then immediately with H_2 (50 bar) at rt. The mixture was then heated using a temperature ramp (18 h at 95 °C and 36 h at 155 °C) with a stir rate of 800 rpm. After 54 h, the mixture was cooled to rt. The pressure vessel was placed in a bath (EtOAc/liq N_2) at −84 °C for 15 min, and was then carefully vented using a metering valve. THF (0.5 mL) was added through the venting valve of the pressure vessel to wash any residual liquids/solids into the vessel. The vessel was then opened, a 0.593 M soln of 1,3,5-trimethoxybenzene in DMSO-d_6 (300 μL, 0.178 mmol) was added as a ^{1}H NMR spectroscopy standard, and the contents of the vessel were diluted with DMSO-d_6. An aliquot (50 μL) of the resulting soln was added into an NMR tube, diluted further with DMSO-d_6, and acidified with 12 M HCl to pH 2. The sample was then analyzed by ^{1}H NMR spectroscopy with solvent suppression. The turnover number was determined to be 550.

1.9.2.2 Heterogeneous Systems

Although many types of metal-based heterogeneous catalysts have been examined for the synthesis of methanol (Scheme 13), copper-based complexes remain the most active catalyst components, together with various metal-based modifiers (e.g., zinc, zirconium, cerium, aluminum, silicon, gallium, and chromium).[75–77] An appropriate support effects the formation and stabilization of the active phase of the catalyst and is also capable of tuning the interactions between the major component and the promoter as well as the properties of the corresponding catalyst.[78]

A highly efficient system using copper/zinc(II) oxide/alumina delivers an impressive space-time yield of 7729 $g_{MeOH}\cdot kg_{cat}^{-1}\cdot h^{-1}$ at a carbon dioxide conversion of 65.8% with selectivity to methanol (**26**) of 77.3%.[79] Higher conversion and selectivity can be achieved using this system, albeit at the expense of a decrease in the space-time yield of methanol. Other metals, such as gold,[80] or other supports, such as zirconium(IV) oxide,[81] can also be used to achieve high conversion and selectivity; however, the yields generated by these systems are still far from optimal.

Scheme 13 Selected Examples of the Hydrogenation of Carbon Dioxide to Methanol Using Heterogeneous Catalysts[79–89]

$$CO_2 + H_2 \xrightarrow{\text{catalyst}} \underset{\mathbf{26}}{\text{MeOH}}$$

Catalyst	Conditions	Conversion[a] (%) of CO_2	Selectivity[b] (%) for MeOH	STY[c]	Ref
Cu/ZnO/alumina	H_2/CO_2 (327:33, 360 bar), 260 °C	65.8	77.3	7729	[79]
Cu/Zn/Ga	H_2/CO_2 (23:7, 30 bar), 270 °C	15.9	29.7	136	[82]
Cu/Ba/γ-alumina	H_2/CO_2 (79:21, 100 bar), 280 °C	25.2	9.3	71	[83]
Ga/Cu/ZnO/ZrO_2	H_2/CO_2 (52:18, 70 bar), 250 °C	22	72	704	[84]
Cu/ZnO/ZrO_2	H_2/CO_2 (37:13, 50 bar), 240 °C	9.7	62	1200	[81]
Pd/Ga/CNT[d]	H_2/CO_2 (37:13, 50 bar), 250 °C	16.5	52.5	512	[85]
Pd/ZnO/CNT[d]	H_2/CO_2 (37:13, 50 bar), 270 °C	19.6	35.5	343	[86]
Au/ZnO/ZrO_2	H_2/CO_2 (3:1, 80 bar), 220 °C	2	100	19	[80]
Cu@ZnO (core–shell)	H_2/CO_2 (23:7, 30 bar), 250 °C	2.3	100	147	[87]
$LaCr_{0.5}Cu_{0.5}O_3$	H_2/CO_2 (3:1, 20 bar), 250 °C	10.4	90.8	278	[88]
Cu/Ga/ZnO	H_2/CO_2 (3:1, 20 bar), 270 °C	6	88	378	[89]

[a] Conversion (%) = (mol CO_2 in – mol CO_2 out)/mol CO_2 in.
[b] Selectivity (%) = mol MeOH produced/mol CO_2 converted.
[c] STY = space-time yield in $g_{MeOH}\cdot kg_{cat}^{-1}\cdot h^{-1}$.
[d] CNT = carbon nanotubes.

Methanol (26); Typical Procedure:[79]

CO_2 hydrogenation to MeOH was performed using a fixed-bed microreaction system. A tubular reactor (inner diameter: 1.74 mm; outer diameter: 3.17 mm) was used. Before charging the Cu/ZnO/alumina catalyst to the reactor, it was pelletized, crushed, and sieved to a size of 100–300 μm. The catalyst (42.5 mg) was then loaded into the reactor (bed length of ca. 10 cm) and reduced in a stream of argon (10%) and H_2 (90%) at 330 °C for 2 h under atmospheric pressure. After prereduction, the catalyst bed was cooled to rt. The reaction mixture was then introduced into the reactor at a gas hourly space velocity of 182 000 h^{-1} and a total pressure of 360 bar H_2/CO_2 (10:1). The reaction was performed at 360 bar and 260 °C. Data points were collected during the steady-state operation of the reaction at specified temperature, pressure, and flow conditions. No clear catalyst deactivation was observed during data collection throughout the catalyst testing. The feed mixture was composed of CO_2 (20%), H_2 (72.5%), and argon (7.5%) as an internal standard for GC analysis. The effluent stream was analyzed by an online GC instrument (Bruker 450), equipped with a Porapak Q+ molecular sieves column and a thermal conductivity detector for analysis of gaseous products, and with a CP wax 52 CB capillary column and a flame ionization detector for analysis of MeOH and other oxygenates. The product transfer line from the reactor to the GC was heated to 150 °C to avoid condensation of the products. CO_2 conversion was determined directly from the CO_2 concentration measured by the thermal conductivity detector. The detection limits for MeOH and CO_2 were 10 and 200 ppm, respectively, on the basis of the signal-to-noise ratio of the chromatograms. The conversion (65.8%) and selectivity (77.3%) values were calculated by averaging over several injections after product concentrations were stabilized. The tendency and accuracy of the catalytic performance were ensured by a minimum of two runs performed on different days. The standard deviations for CO_2 conversion and product selectivities were <2.2%. Methanol was obtained with a space-time yield of 7729 $g_{MeOH} \cdot kg_{cat}^{-1} \cdot h^{-1}$.

1.9.3 Methanation of Carbon Dioxide

Catalytic hydrogenation of carbon dioxide to methane, also called the Sabatier reaction, is an important catalytic reaction. The methanation of carbon dioxide has a range of applications, including the production of syngas and the formation of compressed natural gas. The methanation of carbon dioxide is thermodynamically favorable ($\Delta G_{298K} = -130.8\ kJ \cdot mol^{-1}$); however, reduction of the fully oxidized carbon to methane is an eight-electron process with significant kinetic limitations, and it thus requires a suitable catalyst to achieve acceptable rates and selectivities.[90] Extensive studies have been conducted on metal-based catalytic systems for the hydrogenation of carbon dioxide to methane, mainly with the use of heterogeneous protocols.[91–102]

1.9.3.1 Heterogeneous Systems

Even though several metals, such as nickel, palladium, cobalt, and rhodium, have demonstrated some interesting reactivity as the main active components, ruthenium is still considered the principal component for developing these heterogeneous systems, because it has higher activity and is less expensive than other precious metals.[91] The method for the preparation of the catalyst significantly influences the yield and selectivity of this reaction. The methanation of carbon dioxide with the use of a commercial 3% ruthenium/alumina catalyst is an excellent example of ruthenium activity; this catalyst gives a methane yield of 93% with no coproduction of carbon monoxide at 350 °C, with a gas hourly space velocity of 55 000 h^{-1} and a hydrogen/carbon dioxide ratio of 5:1 (Scheme 14).[101] Recent examples of nickel-based catalysts on cerium(IV) oxide, titanium(IV) oxide, and alumina show promising results.[91] Nickel-based catalysts have been widely investigated for industrial purposes owing to their low cost and abundance. Using 15 wt% nickel/titanium(IV) oxide as the catalyst, a carbon dioxide conversion of 96% and a selectivity to methane of 99% are achieved with a turnover frequency of 1.22×10^{-3} s^{-1} (determined at 200 °C) at a relatively low temperature of 260 °C.[93]

Scheme 14 Selected Examples of the Hydrogenation of Carbon Dioxide to Methane Using Heterogeneous Catalysts[92–102]

$$CO_2 + H_2 \xrightarrow{\text{catalyst}} CH_4$$

27

Catalyst	Catalyst Preparation Method	Conditions[a]	Conversion[b] (%) of CO_2	Selectivity[c] (%) for CH_4	Ref
4.29% Ni/RHA-alumina[d]	ion exchange	H_2/CO_2 (4:1), 500 °C, flow rate = 25 mL•min^{-1}	34	56	[92]
4.09% Ni/silica	ion exchange	H_2/CO_2 (4:1), 500 °C, flow rate = 25 mL•min^{-1}	25	45	[92]
15 wt% Ni/TiO_2	precipitation–deposition	H_2/CO_2 (4:1), 260 °C, GHSV = 2400 h^{-1}	96	99	[93]
15% Ni/RHA-alumina[d]	incipient wetness impregnation	H_2/CO_2 (4:1), 500 °C, flow rate = 25 mL•min^{-1}	63	90	[94]
NiFeAl/$(NH_4)_2CO_3$	coprecipitation	H_2/CO_2 (4:1), 220 °C, GHSV = 9600 mL•g^{-1}•h^{-1}	58.5	99.5	[95]
NiFeAl/NH_4OH	coprecipitation	H_2/CO_2 (4:1), 220 °C, GHSV = 9600 mL•g^{-1}•h^{-1}	54.5	99.4	[95]
25% Ni/alumina	coprecipitation	H_2/CO_2 (9:1), 235 °C, GHSV = 22 250 mL•g^{-1}•h^{-1}	99	99.7	[96]
10% Ni/La_2O_3	impregnation	H_2/CO_2 (4:1), 380 °C, GHSV = 11 000 h^{-1}	100[e]	100	[97]
35Ni5Fe0.6RuAX[f]	sol–gel	H_2/CO_2 (4:1), 220 °C, GHSV = 9600 mL•g^{-1}•h^{-1}	68.2	98.9	[98]
Pd-Mg/silica	reverse microemulsion	H_2/CO_2 (4:1), 450 °C, GHSV = 3273 h^{-1}	59.2	95.3	[99]
Pd-Ni/silica	reverse microemulsion	H_2/CO_2 (4:1), 450 °C, GHSV = 3273 h^{-1}	50.5	89	[99]
Co/meso-silica	excess impregnation	H_2/CO_2 (4.6:1), 280 °C, GHSV = 22 000 mL•g^{-1}•h^{-1}	40	94.1	[100]
Co/KIT-6	excess impregnation	H_2/CO_2 (4.6:1), 300 °C, GHSV = 22 000 mL•g^{-1}•h^{-1}	51	98.9	[100]
3% Ru/alumina	commercially available	H_2/CO_2 (5:1), 350 °C, GHSV = 55 000 h^{-1}	93	100	[101]
$Ce_{0.95}Ru_{0.05}O_2$	combustion	H_2/CO_2 (4:1), 450 °C, GHSV = 10 227 h^{-1}	55	99	[102]

[a] GHSV = gas hourly space velocity.
[b] Conversion (%) = (mol CO_2 in – mol CO_2 out)/mol CO_2 in.
[c] Selectivity (%) = mol CH_4 produced/mol CO_2 converted.
[d] RHA = rice husk ash.
[e] Space-time yield (STY) = 1180 g•kg$_{cat}^{-1}$•h^{-1}
[f] AX = alumina xerogel.

for references see p 311

Methane (27); Typical Procedure:[93]

The catalytic performance of the nickel-based catalyst 15 wt% Ni/TiO_2 was evaluated at atmospheric pressure in a fixed-bed quartz reactor with an interior diameter of 8 mm. The reactor was heated in a tube furnace equipped with a temperature controller. All gases were monitored by calibrated mass-flow controllers. Prior to the catalytic performance test, the catalyst (1.0 g) was pretreated in situ in a gaseous mixture of H_2/N_2 (2:3, v/v) for 4 h with a total gas flow of 100 mL$\cdot$min^{-1} at 450 °C with a heating rate of 5 °C min^{-1}, and it was then cooled to 150 °C under a N_2 atmosphere. Subsequently, a mixture of H_2, CO_2, and argon (an internal standard) in a molar ratio (H_2/CO_2/Ar) of 12:3:5 was introduced into the reactor, and the total flow rate was set to 40 mL min^{-1} (with a gas hourly space velocity of 2400 h^{-1}). The composition of the outlet gases was analyzed online using a GC-2014C gas chromatograph with a TDX-01 column and a thermal conductivity detector. The CO_2 conversion (96%) and selectivity (99%) for CH_4 were determined on the basis of the concentrations of CO_2 and CH_4. The turnover frequency ($1.22 \times 10^{-3} \cdot s^{-1}$) was determined as follows: turnover frequency = mol CH_4 produced per second/mol surface nickel sites.

1.9.4 Reductive Methylation Using Carbon Dioxide

Methylations constitute the most important form of alkylation reactions, and they are widely used in chemical transformations. However, more sustainable methylation reagents are only scarcely applied in the synthesis of functionalized products.[103–105] Hence, the development of new and improved catalytic methods, especially those involving the use of the green gas carbon dioxide, as the methylation agent is highly desired. Recent work has been directed toward the methylation of amines and carbon-based nucleophiles, as illustrated in the following sections. Some of the catalysts and ligands involved in such transformations are depicted in Scheme 15. The related reductive methylation using carbonates is discussed in Section 1.8.4.1.

Scheme 15 Selected Catalysts and Ligands Involved in the Reductive Methylation of Carbon Dioxide Using Homogeneous Systems[106–111]

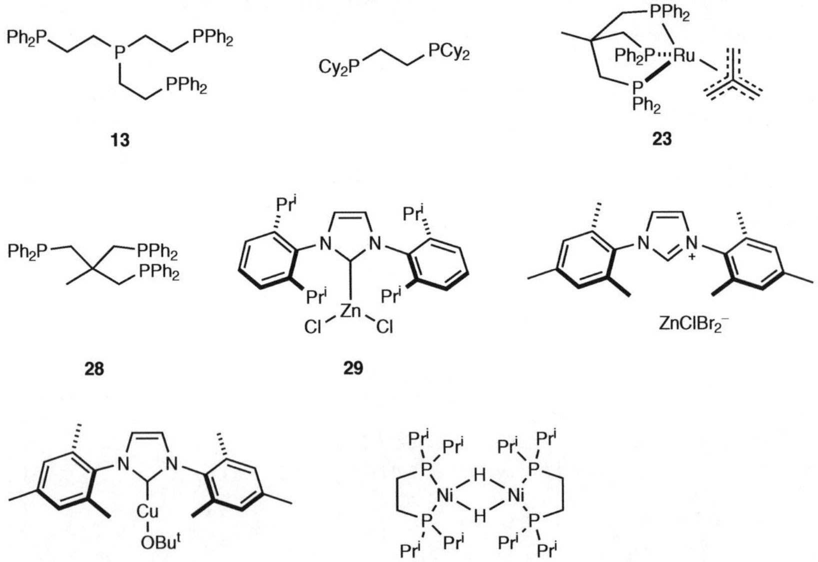

1.9.4.1 Methylation of Amines

1.9.4.1.1 Homogeneous Systems

Several homogeneous systems have emerged in recent years for the reductive methylation of amines. At first, hydrosilanes were used as the reducing agent. Hydrosilanes are well known in the reduction of carbon dioxide.[112] In this context, carbon dioxide is readily captured by primary and secondary amines and is then reduced to give monomethylated amines **30** and **32** or dimethylated amine **31**, depending on the method used (Scheme 16).[106–109,111,113] The most prominent examples of such reactions make use of zinc catalyst **29**[106] or the dichlorotetrakis(dimethylsulfoxide)ruthenium(II)/di(1-adamantyl)butylphosphine system[113] to deliver these products in good yields with high selectivity. Later on, a more sophisticated and atom-economical procedure came to light that makes use of hydrogen as the reductant, which renders this field even more interesting (see Section 1.9.4.1.1.1).

Scheme 16 Methylation of Amines Starting from Hydrosilanes and Carbon Dioxide Using Homogeneous Systems[106–109,111,113]

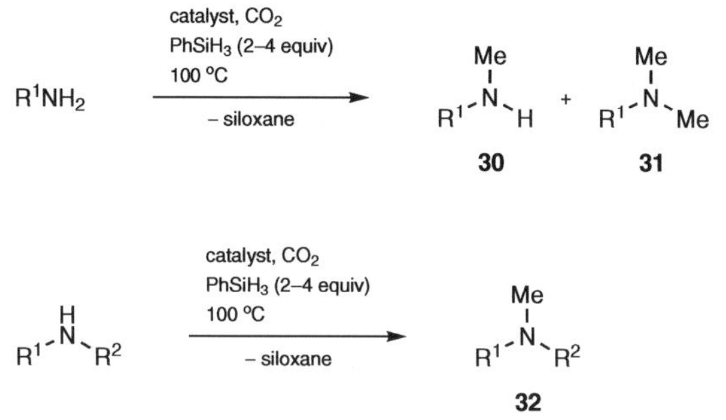

R^1NH_2 → (catalyst, CO_2; $PhSiH_3$ (2–4 equiv); 100 °C; – siloxane) → $R^1N(Me)H$ **30** + $R^1N(Me)Me$ **31**

R^1NHR^2 → (catalyst, CO_2; $PhSiH_3$ (2–4 equiv); 100 °C; – siloxane) → $R^1N(Me)R^2$ **32**

1.9.4.1.1.1 Methylation of Amines Using Hydrogen as Reductant

The use of hydrogen as a reductant in the reductive methylation of amines is a natural transition from hydrosilane-based systems toward a more atom-economical strategy. The synthesis of disubstituted anilines **35** and trisubstituted anilines **33** and **34** is readily achieved by using acid cocatalysts and ruthenium-based systems such as tris(acetylacetonato)ruthenium(III)/triphosphine **28** (triphos)[114] and ruthenium/triphos complex **23**,[110] respectively (Scheme 17).

Scheme 17 Synthesis of Di- and Trisubstituted Amines by the Ruthenium-Catalyzed Methylation of Anilines Using Hydrogen and Carbon Dioxide[110,114]

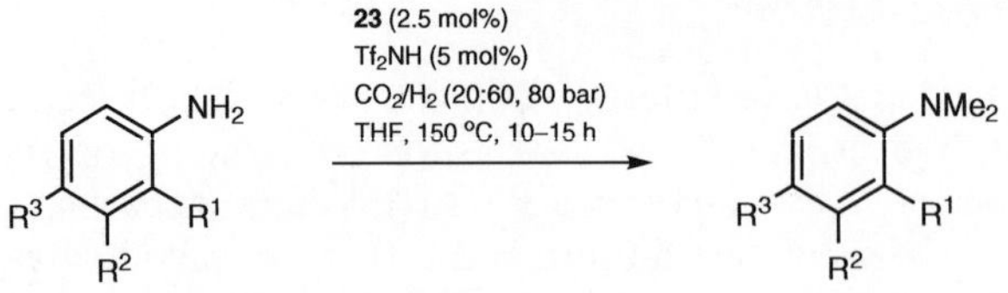

23 (2.5 mol%), Tf_2NH (5 mol%), CO_2/H_2 (20:60, 80 bar), THF, 150 °C, 10–15 h → 33

R^1	R^2	R^3	Yield (%)	Ref
H	H	H	94	[110]
H	H	Cl	93	[110]
F	H	H	93	[110]
H	CF_3	Me	94	[110]

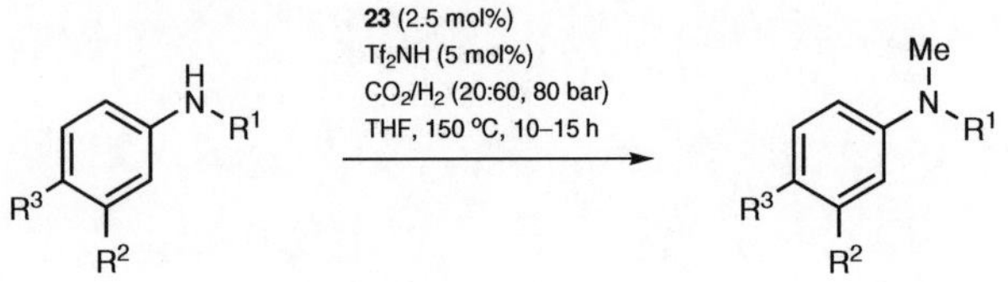

23 (2.5 mol%), Tf_2NH (5 mol%), CO_2/H_2 (20:60, 80 bar), THF, 150 °C, 10–15 h → 34

R^1	R^2	R^3	Yield (%)	Ref
Me	F	H	90	[110]
Me	H	Cl	90	[110]
Me	H	OMe	35	[110]
Cy	H	H	64	[110]
Ph	H	H	27	[110]

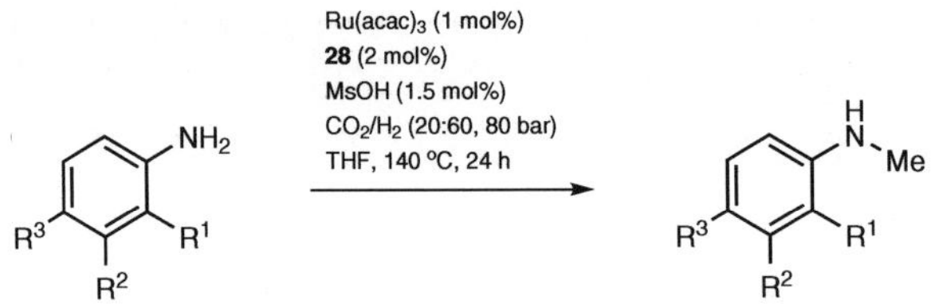

$Ru(acac)_3$ (1 mol%), 28 (2 mol%), MsOH (1.5 mol%), CO_2/H_2 (20:60, 80 bar), THF, 140 °C, 24 h → 35

R^1	R^2	R^3	Yield (%)	Ref
H	H	H	88	[114]
OBn	H	H	70	[114]
H	OCH_2O		71	[114]
H	H	CO_2Me	78	[114]
Cl	H	H	85	[114]

N,N-Dimethylanilines 33; **General Procedure:**[110]
Under an argon atmosphere, Ru catalyst **23** (0.019 g, 0.025 mmol, 2.5 mol%) and Tf_2NH (0.014 g, 0.05 mmol, 5 mol%) were weighed into a Schlenk tube. After dissolving in THF (1.0 mL), the mixture was transferred by cannula to an autoclave, which was followed by the addition of a soln of the aniline substrate (1.0 mmol) in THF (1.0 mL). The autoclave was then pressurized with CO_2 (20 bar), and then H_2 was added up to a total pressure of 80 bar. The mixture was stirred and heated to 150 °C in an oil bath. After 10 h, the autoclave was cooled in an ice bath and was then carefully vented. The reaction soln was analyzed by ^1H NMR spectroscopy with mesitylene as an internal standard, and the results were confirmed by GC, using dodecane as an internal standard.

N-Methylanilines 35; **General Procedure:**[114]
Inside a 300-mL autoclave, a 4-mL glass vial containing a stirrer bar was charged with triphos (**28**; 6.2 mg, 10 μmol, 2 mol%), $Ru(acac)_3$ (2.0 mg, 5 μmol, 1 mol%), and MsOH (0.51 mL, 7.5 μmol, 1.5 mol%). Anhyd THF (2.0 mL) and the aniline substrate (0.5 mmol) were added to the vial, which was sealed by a septum penetrated with a syringe needle. The autoclave was sealed, purged (30 bar CO_2, 2 ×), and pressurized with CO_2 (20 bar) and H_2 (60 bar). Then, the autoclave was placed in an aluminum block on a stirring machine and heated to 140 °C for 24 h. After that, the mixture was cooled in cold H_2O, and the gas was carefully released. The mixture was analyzed by GC/MS and GC, using hexadecane as an internal standard, or was purified through silica gel columns.

1.9.4.1.2 Heterogeneous Systems

Heterogeneous systems have also been used in the reductive methylation of amines using carbon dioxide, mainly with the use of hydrogen as a reductant. In this manner, primary anilines can be monomethylated using a 2 wt% palladium-based catalyst to give products **36** in good to moderate yields (Scheme 18).[115] The monomethylation of a wide range of secondary amines can also be achieved in very good yields using a platinum-based catalyst; with the use of this system, pyrrolidine is readily methylated to give 1-methylpyrrolidine (**37**) in 99% yield (Scheme 18).[116]

Scheme 18 Monomethylation of Primary and Secondary Amines Using Hydrogen and Carbon Dioxide[115,116]

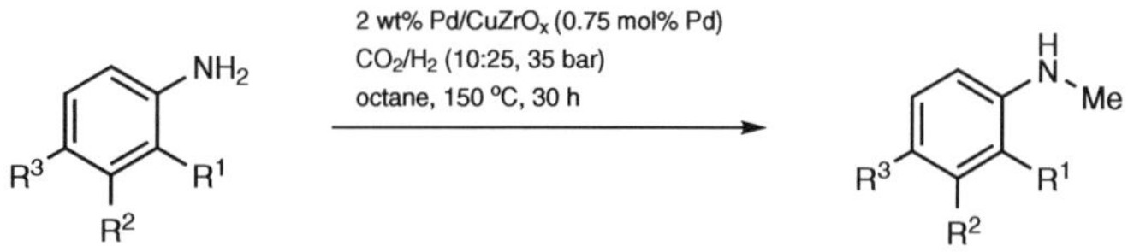

2 wt% $Pd/CuZrO_x$ (0.75 mol% Pd)
CO_2/H_2 (10:25, 35 bar)
octane, 150 °C, 30 h

36

R^1	R^2	R^3	Yield (%)	Ref
H	H	H	75	[115]
H	H	Me	73	[115]
Me	H	H	75	[115]
H	Me	H	67	[115]
H	H	OMe	73	[115]
H	H	Cl	49	[115]
Ph	H	H	79	[115]

for references see p 311

Pt-MoO_x/TiO_2 (2 mol% Pt)
CO_2/H_2 (5:50, 55 bar)
neat, 200 °C, 24 h
99%

N–Me

37

The dimethylation of primary amines can also be achieved using a gold-based catalyst to generate dimethylated products **38** in excellent yields with a relatively high turnover frequency of 45 h^{-1} (Scheme 19).[117] In the presence of aldehydes and under similar conditions, primary amines are converted into unsymmetrical tertiary amines **39** in a one-pot procedure in very good yields.

Scheme 19 Dimethylation and Methylation/Alkylation of Primary Amines Using Hydrogen and Carbon Dioxide[117]

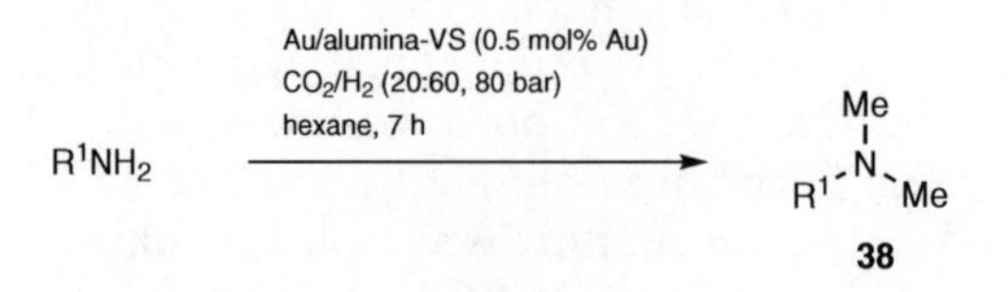

R^1	Temp (°C)	Yield (%)	Ref
Ph	140	92	[117]
2-Tol	140	96	[117]
3-Tol	140	94	[117]
4-$MeOC_6H_4$	140	99	[117]
Cy	170	93	[117]
$(CH_2)_5Me$	170	95	[117]
cyclopentyl	170	92	[117]

R^1NH_2 + R^2CHO

Au/alumina-VS (0.5 mol% Au)
CO_2/H_2 (20:60, 80 bar)
hexane, 170 °C, 7 h

39

R^1	R^2	Yield (%)	Ref
Bn	$(CH_2)_4Me$	92	[117]
Bn	4-Tol	94	[117]
Bn	4-FC_6H_4	98	[117]
$(CH_2)_5Me$	4-$MeOC_6H_4$	97	[117]
$(CH_2)_5Me$	Bu	91	[117]
Ph	Ph	85	[117]

Nitrobenzenes and aromatic nitriles can also be used in this transformation (Scheme 20).[118] The copper-based reduction of these groups with hydrogen followed by in situ capture of carbon dioxide and subsequent reduction of the formed carbonyl moiety directly delivers desired trisubstituted amines **40** and **41** in moderate to good yields.

Scheme 20 Dimethylation of Nitrobenzenes and Aromatic Nitriles Using Hydrogen and Carbon Dioxide[118]

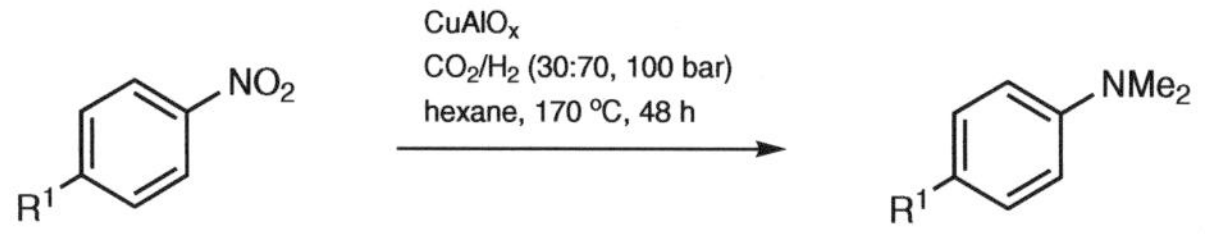

CuAlO$_x$
CO_2/H_2 (30:70, 100 bar)
hexane, 170 °C, 48 h

40

R^1	Yield (%)	Ref
H	86	[118]
Me	79	[118]
OMe	86	[118]

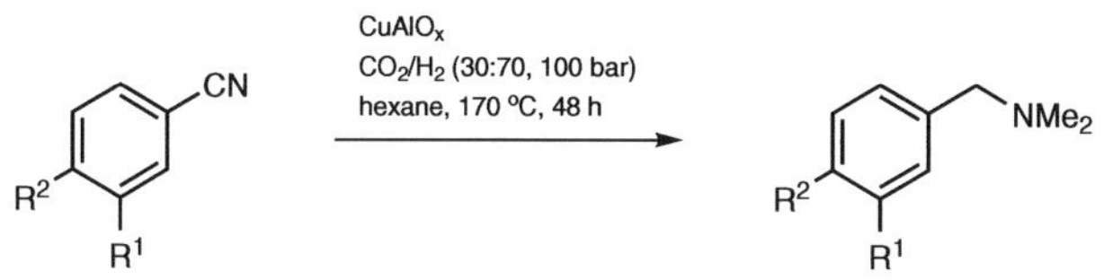

CuAlO$_x$
CO_2/H_2 (30:70, 100 bar)
hexane, 170 °C, 48 h

41

R^1	R^2	Yield (%)	Ref
H	H	55	[118]
H	OMe	45	[118]
OCH_2O		51	[118]

N-Methylanilines 36**; General Procedure:**[115]
The aniline substrate (1.0 mmol), 2 wt% Pd/CuZrO$_x$ (40 mg, 0.75 mol% Pd), and octane (2 mL) were added into an 80-mL autoclave. CO_2 (10 bar) and H_2 (25 bar) were then introduced. The reaction was performed at 150 °C for 30 h under magnetic stirring. Subsequently, the autoclave was cooled to rt. Biphenyl (internal standard) and EtOH (10 mL) were then added for quantitative analysis by GC-FID (Agilent 6890A).

1-Methylpyrrolidine (37)**; Typical Procedure:**[116]
Pt-MoO$_x$/TiO$_2$ (5 wt% Pt, 7 wt% Mo) was first prereduced in 100% H_2 (20 $cm^3 \cdot min^{-1}$) for 0.5 h at 300 °C. After prereduction, the catalyst in a closed glass tube sealed with a septum inlet was cooled to rt under a H_2 atmosphere. A mixture of pyrrolidine (1 mmol, 1 equiv) and dodecane (0.5 mmol) was injected into the prereduced catalyst inside the glass tube through the septum inlet. Then, the septum was removed under air, and a magnetic stirrer was put in the tube, which was followed by insertion of the tube inside a stainless autoclave with a dead space of 14 cm^3. Soon after sealing, the reactor was flushed with CO_2 from a high-pressure gas cylinder and was charged with CO_2 (5 bar) and then H_2 (50 bar) at rt. Then, the reactor was heated at 200 °C under stirring (400 rpm). The yield was determined by GC, using dodecane as an internal standard; yield: 99%.

Trisubstituted Amines 38 **or** 39**; General Procedure:**[117]
A mixture containing the amine (1.0 mmol, 1 equiv), aldehyde (to form products **39**; 1.0 mmol, 1 equiv), hexane (10 mL), and the Au/alumina catalyst (0.5 mol% Au) was charged into a 50-mL Hastelloy-C high-pressure Parr reactor. The atmosphere inside the reactor was exchanged with H_2, and then CO_2 (20 bar) and H_2 (60 bar) were introduced.

for references see p 311

The reaction was performed at 170 °C for 7 h under magnetic stirring. Subsequently, the autoclave was cooled to rt, and octane (1 mmol, internal standard) was added before quantitative analysis by GC-FID (Agilent 7820 A) to determine the yield of **39**. In the absence of an aldehyde, the reaction follows the same procedure (heating at 140 °C or 170 °C for 7 h) to deliver dimethylated amine **38**.

N,N-Dimethylanilines 40; **General Procedure:**[118]
The nitrobenzene substrate (1 mmol), the $CuAlO_x$ catalyst (50 mg), and hexane (2 mL) were added into an 80-mL autoclave. The atmosphere in the autoclave was then exchanged with CO_2, and CO_2 (30 bar) and H_2 (70 bar) were introduced. The reaction took place at 170 °C for 48 h under magnetic stirring. Subsequently, the autoclave was cooled to rt. Biphenyl (internal standard) and EtOH (10 mL) were added for quantitative analysis by GC-FID (Agilent 6890A).

1.9.4.2 Methylation of C—H Bonds

The reductive methylation of C—H bonds with carbon dioxide can be accomplished for several indole, pyrrole, and arene substrates by using ruthenium-based homogeneous systems (Scheme 21).[119] In this manner, indole **42** and 1,3,5-trimethoxybenzene (**44**) are converted into monomethylated products **43** and **45**, respectively, in very good yields; a combination of the tris(acetylacetonato)ruthenium(III)/triphos (**28**) catalyst system and an acid cocatalyst [methanesulfonic acid or aluminum tris(trifluoromethanesulfonate)] is used in these cases.

Scheme 21 Monomethylation of an Indole and Arene Using Hydrogen and Carbon Dioxide with a Ruthenium-Based Homogeneous Catalyst System[119]

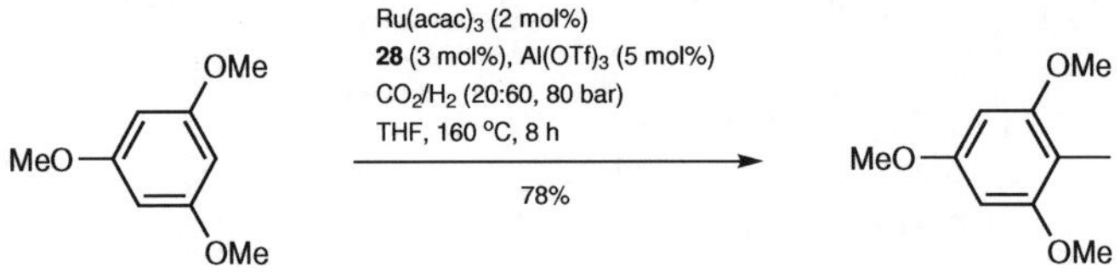

$Ru(acac)_3$ (4 mol%)
28 (6 mol%), MsOH (6 mol%)
CO_2/H_2 (20:60, 80 bar)
THF, 140 °C, 24 h
81%

42 → **43**

$Ru(acac)_3$ (2 mol%)
28 (3 mol%), $Al(OTf)_3$ (5 mol%)
CO_2/H_2 (20:60, 80 bar)
THF, 160 °C, 8 h
78%

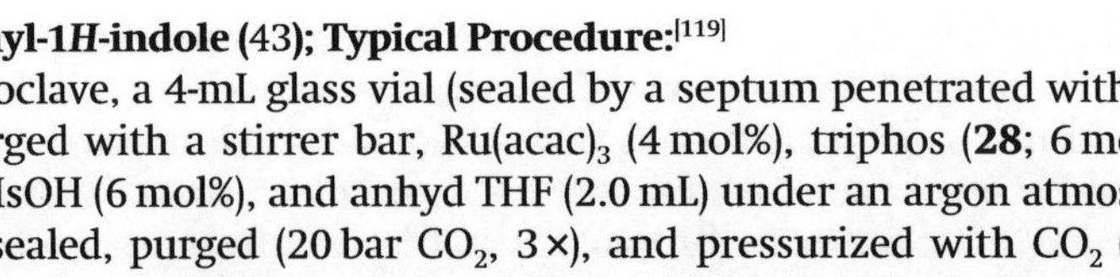

44 → **45**

2,3,5-Trimethyl-1*H*-indole (43); **Typical Procedure:**[119]
Inside an autoclave, a 4-mL glass vial (sealed by a septum penetrated with a syringe needle) was charged with a stirrer bar, $Ru(acac)_3$ (4 mol%), triphos (**28**; 6 mol%), indole **42** (0.5 mmol), MsOH (6 mol%), and anhyd THF (2.0 mL) under an argon atmosphere. The autoclave was sealed, purged (20 bar CO_2, 3 ×), and pressurized with CO_2 (20 bar) and H_2 (60 bar). Then, the autoclave was placed in an aluminum block on a stirring machine and was heated at 140 °C for 24 h. After that, the autoclave was cooled to rt in cold H_2O, and the gas was carefully released. The mixture was purified by column chromatography (silica gel); yield: 81%.

References

[1] Jessop, P. G.; Joó, F.; Tai, C.-C., *Coord. Chem. Rev.*, (2004) **248**, 2425.
[2] Omae, I., *Catal. Today*, (2006) **115**, 33.
[3] Sakakura, T.; Choi, J. C.; Yasuda, H., *Chem. Rev.*, (2007) **107**, 2365.
[4] Aresta, M.; Dibenedetto, A., *Dalton Trans.*, (2007) **28**, 2975.
[5] Sakakura, T.; Kohno, K., *Chem. Commun. (Cambridge)*, (2009), 1312.
[6] Aresta, M., In *Activation of Small Molecules: Organometallic and Bioinorganic Perspectives*, Tolman, W. B., Ed.; Wiley-VCH: Weinheim, Germany, (2006); p 1.
[7] Wang, W.; Wang, S.; Ma, X.; Gong, J., *Chem. Soc. Rev.*, (2011) **40**, 3703.
[8] Dell'Amico, D. B.; Calderazzo, F.; Labella, L.; Marchetti, F.; Pampaloni, G., *Chem. Rev.*, (2003) **103**, 3857.
[9] Riduan, S. N.; Zhang, Y., *Dalton Trans.*, (2010) **39**, 3347.
[10] Centi, G.; Perathoner, S., *Catal. Today*, (2009) **148**, 191.
[11] Ricci, M., In *Carbon Dioxide Recovery and Utilization*, Aresta, M., Ed.; Kluwer: Dordrecht, The Netherlands, (2003); p 395.
[12] Ma, J.; Sun, N.; Zhang, X.; Zhao, N.; Xiao, F.; Wei, W.; Sun, Y., *Catal. Today*, (2009) **148**, 221.
[13] Baiker, A., *Appl. Organomet. Chem.*, (2000) **14**, 751.
[14] Chueh, W. C.; Falter, C.; Abbott, M.; Scipio, D.; Furler, P.; Haile, S. M.; Steinfeld, A., *Science (Washington, D. C.)*, (2010) **330**, 1797.
[15] Younas, M.; Kong, L. L.; Bashir, M. J. K.; Nadeem, H.; Shehzad, A.; Sethupathi, S., *Energy Fuels*, (2016) **30**, 8815.
[16] Olah, G. A.; Goeppert, A.; Surya Prakash, G. K., *Beyond Oil and Gas: The Methanol Economy*, 2nd ed.; Wiley-VCH: Weinheim, Germany, (2009).
[17] Porosoff, M. D.; Yan, B.; Chen, J. G., *Energy Environ. Sci.*, (2016) **9**, 62; and references cited therein.
[18] Centi, G.; Perathoner, S., *Stud. Surf. Sci. Catal.*, (2004) **153**, 1.
[19] Gunasekar, G. H.; Park, K.; Jung, K.-D.; Yoon, S., *Inorg. Chem. Front.*, (2016) **3**, 882; and references cited therein.
[20] Li, Y.-N.; Ma, R.; He, L.-N.; Diao, Z.-F., *Catal. Sci. Technol.*, (2014) **4**, 1498; and references cited therein.
[21] Wang, W.-H.; Himeda, Y., In *Hydrogenation*, Karamé, I., Ed.; InTech: Rijeka, Croatia, (2012); Chapter 10, p 249; 3; available online at www.intechopen.com/books/hydrogenation/recent-advances-in-transition-metal-catalysed-homogeneous-hydrogenation-of-carbon-dioxide-in-aqueous; DOI 10.5772/48 658.
[22] Mikkelsen, M.; Jorgensen, M.; Krebs, F. C., *Energy Environ. Sci.*, (2010) **3**, 43.
[23] Dai, W.-L.; Luo, S.-L.; Yin, S.-F.; Au, C.-T., *Appl. Catal., A*, (2009) **366**, 2.
[24] Zhang, S.; Chen, Y.; Li, F.; Lu, X.; Dai, W.; Mori, R., *Catal. Today*, (2006) **115**, 61.
[25] Sun, J. M.; Fujita, S.; Arai, M., *J. Organomet. Chem.*, (2005) **690**, 3490.
[26] Jessop, P. G., *J. Supercrit. Fluids*, (2006) **38**, 211.
[27] Hietala, J.; Vuori, A.; Johnsson, P.; Pollari, I.; Reutemann, W.; Kieczka, H., *Ullmann's Encyclopedia of Industrial Chemistry*; Wiley-VCH: Weinheim, Germany, (2016); DOI 10.1002/14 356 007.a12_013.pub3.
[28] Grasemann, M.; Laurenczy, G., *Energy Environ. Sci.*, (2012) **5**, 8171.
[29] Jiang, H.-L.; Singh, S. K.; Yan, J.-M.; Zhang, X.-B.; Xu, Q., *ChemSusChem*, (2010) **3**, 541.
[30] For an example of the application of formic acid fuel cells, see www.teamfast.nl/.
[31] Uhm, S.; Lee, H. J.; Lee, J., *Phys. Chem. Chem. Phys.*, (2009) **11**, 9326.
[32] Inoue, Y.; Izumida, H.; Sasaki, Y.; Hashimoto, H., *Chem. Lett.*, (1976), 863.
[33] Leitner, W.; Dinjus, E.; Gassner, F., In *Aqueous-Phase Organometallic Catalysis: Concepts and Applications*, Cornils, B.; Herrmann, W. A., Eds.; Wiley-VCH: Weinheim, Germany, (1998); p 486.
[34] Elek, J.; Nádasdi, L.; Papp, G.; Laurenczy, G.; Joó, F., *Appl. Catal., A*, (2003) **255**, 59.
[35] Moret, S.; Dyson, P. J.; Laurenczy, G., *Nat. Commun.*, (2014) **5**, 4017.
[36] Ziebart, C.; Federsel, C.; Anbarasan, P.; Jackstell, R.; Baumann, W.; Spannenberg, A.; Beller, M., *J. Am. Chem. Soc.*, (2012) **134**, 20 701.
[37] Drake, J. L.; Manna, C. M.; Byers, J. A., *Organometallics*, (2013) **32**, 6891.
[38] Hayashi, H.; Ogo, S.; Abura, T.; Fukuzumi, S., *J. Am. Chem. Soc.*, (2003) **125**, 14 266.
[39] Hayashi, H.; Ogo, S.; Fukuzumi, S., *Chem. Commun. (Cambridge)*, (2004), 2714.
[40] Himeda, Y., *Eur. J. Inorg. Chem.*, (2007), 3927.
[41] Himeda, Y.; Miyazawa, S.; Hirose, T., *ChemSusChem*, (2011) **4**, 487.

[42] Himeda, Y.; Onozawa-Komatsuzaki, N.; Sugihara, H.; Arakawa, H.; Kasuga, K., *Organometallics*, (2004) **23**, 1480.
[43] Himeda, Y.; Onozawa-Komatsuzaki, N.; Sugihara, H.; Kasuga, K., *J. Am. Chem. Soc.*, (2005) **127**, 13118.
[44] Himeda, Y.; Onozawa-Komatsuzaki, N.; Sugihara, H.; Kasuga, K., *J. Photochem. Photobiol., A*, (2006) **182**, 306.
[45] Himeda, Y.; Onozawa-Komatsuzaki, N.; Sugihara, H.; Kasuga, K., *Organometallics*, (2007) **26**, 702.
[46] Rohmann, K.; Kothe, J.; Haenel, M. W.; Englert, U.; Hölscher, M.; Leitner, W., *Angew. Chem. Int. Ed.*, (2016) **55**, 8966.
[47] Bertini, F.; Gorgas, N.; Stöger, B.; Peruzzini, M.; Veiros, L. F.; Kirchner, K.; Gonsalvi, L., *ACS Catal.*, (2016) **6**, 2889.
[48] Tanaka, R.; Yamashita, M.; Nozaki, K., *J. Am. Chem. Soc.*, (2009) **131**, 14168.
[49] Tanaka, R.; Yamashita, M.; Chung, L. W.; Morokuma, K.; Nozaki, K., *Organometallics*, (2011) **30**, 6742.
[50] Langer, R.; Diskin-Posner, Y.; Leitus, G.; Shimon, L. J. W.; Ben-David, Y.; Milstein, D., *Angew. Chem. Int. Ed.*, (2011) **50**, 9948.
[51] Sanz, S.; Azua, A.; Peris, E., *Dalton Trans.*, (2010) **39**, 6339.
[52] Azua, A.; Sanz, S.; Peris, E., *Chem.–Eur. J.*, (2011) **17**, 3963.
[53] Gassner, F.; Leitner, W., *J. Chem. Soc., Chem. Commun.*, (1993), 1465.
[54] Munshi, P.; Main, A. D.; Linehan, J. C.; Tai, C.-C.; Jessop, P. G., *J. Am. Chem. Soc.*, (2002) **124**, 7963.
[55] Lau, C. P.; Chen, Y. Z., *J. Mol. Catal. A: Chem.*, (1995) **101**, 33.
[56] Joó, F.; Laurenczy, G.; Nádasdi, L.; Elek, J., *Chem. Commun. (Cambridge)*, (1999), 971.
[57] Erlandsson, M.; Landaeta, V. R.; Gonsalvi, L.; Peruzzini, M.; Phillips, A. D.; Dyson, P. J.; Laurenczy, G., *Eur. J. Inorg. Chem.*, (2008), 620.
[58] Schmeier, T. J.; Dobereiner, G. E.; Crabtree, R. H.; Hazari, N., *J. Am. Chem. Soc.*, (2011) **133**, 9274.
[59] Tai, C.-C.; Chang, T.; Roller, B.; Jessop, P. G., *Inorg. Chem.*, (2003) **42**, 7340.
[60] Watari, R.; Kayaki, Y.; Hirano, S.-i.; Matsumoto, N.; Ikariya, T., *Adv. Synth. Catal.*, (2015) **357**, 1369.
[61] Zhang, Y.; Fei, J.; Yu, Y.; Zheng, X., *Catal. Commun.*, (2004) **5**, 643.
[62] Xu, Z.; McNamara, N. D.; Neumann, G. T.; Schneider, W. F.; Hicks, J. C., *ChemCatChem*, (2013) **5**, 1769.
[63] Park, K.; Gunasekar, G. H.; Prakash, N.; Jung, K.-D.; Yoon, S., *ChemSusChem*, (2015) **8**, 3410.
[64] Yang, Z.; Zhang, H.; Yu, B.; Zhao, Y.; Ji, G.; Liu, Z., *Chem. Commun. (Cambridge)*, (2015) **51**, 1271.
[65] Preti, D.; Resta, C.; Squarcialupi, S.; Fachinetti, G., *Angew. Chem. Int. Ed.*, (2011) **50**, 12551.
[66] Preti, D.; Squarcialupi, S.; Fachinetti, G., *ChemCatChem*, (2012) **4**, 469.
[67] Filonenko, G. A.; Vrijburg, W. L.; Hensen, E. J. M.; Pidko, E. A., *J. Catal.*, (2016) **343**, 97.
[68] Liu, N.; Du, R. J.; Li, W., *Adv. Mater. Res. (Durnten-Zurich, Switz.)*, (2013) **821–822**, 1330; DOI 10.4028/www.scientific.net/AMR.821–822.1330.
[69] Inui, T.; Takeguchi, T., *Catal. Today*, (1991) **10**, 95.
[70] Huff, C. A.; Sanford, M. S., *J. Am. Chem. Soc.*, (2011) **133**, 18122.
[71] Wesselbaum, S.; vom Stein, T.; Klankermayer, J.; Leitner, W., *Angew. Chem. Int. Ed.*, (2012) **51**, 7499.
[72] Chakraborty, S.; Zhang, J.; Krause, J. A.; Guan, H., *J. Am. Chem. Soc.*, (2010) **132**, 8872.
[73] Tominaga, K.-I.; Sasaki, Y.; Kawai, M.; Watanabe, T.; Saito, M., *J. Chem. Soc., Chem. Commun.*, (1993), 629.
[74] Rezayee, N. M.; Huff, C. A.; Sanford, M. S., *J. Am. Chem. Soc.*, (2015) **137**, 1028.
[75] Liaw, B. J.; Chen, Y. Z., *Appl. Catal., A*, (2001) **206**, 245.
[76] Arena, F.; Barbera, K.; Italiano, G.; Bonura, G.; Spadaro, L.; Frusteri, F., *J. Catal.*, (2007) **249**, 185.
[77] Saito, M.; Murata, K., *Catal. Surv. Asia*, (2004) **8**, 285.
[78] Liu, X.-M.; Lu, G. Q.; Yan, Z.-F.; Beltramini, J., *Ind. Eng. Chem. Res.*, (2003) **42**, 6518.
[79] Bansode, A.; Urakawa, A., *J. Catal.*, (2014) **309**, 66.
[80] Słoczyński, J.; Grabowski, R.; Kozłowska, A.; Olszewski, P.; Stoch, J.; Skrzypek, J.; Lachowska, M., *Appl. Catal., A*, (2004) **278**, 11.
[81] Arena, F.; Mezzatesta, G.; Zafarana, G.; Trunfio, G.; Frusteri, F.; Spadaro, L., *J. Catal.*, (2013) **300**, 141.
[82] Cai, W. J.; de la Piscina, P. R.; Toyir, J.; Homs, N., *Catal. Today*, (2015) **242**, 193.
[83] Bansode, A.; Tidona, B.; Rudolf von Rohr, P.; Urakawa, A., *Catal. Sci. Technol.*, (2013) **3**, 767.
[84] Ladera, R.; Pérez-Alonso, F. J.; González-Carballo, J. M.; Ojeda, M.; Rojas, S.; Fierro, J. L. G., *Appl. Catal., B*, (2013) **142**, 241.

[85] Kong, H.; Li, H.-Y.; Lin, G.-D.; Zhang, H.-B., *Catal. Lett.*, (2011) **141**, 886.
[86] Liang, X.-L.; Xie, J.-R.; Liu, Z.-M., *Catal. Lett.*, (2015) **145**, 1138.
[87] Le Valant, A.; Comminges, C.; Tisseraud, C.; Canaff, C.; Pinard, L.; Pouilloux, Y., *J. Catal.*, (2015) **324**, 41.
[88] Jia, L.; Gao, J.; Fang, W.; Li, Q., *Catal. Commun.*, (2009) **10**, 2000.
[89] Toyir, J.; de la Piscina, P. R.; Fierro, J. L. G.; Homs, N., *Appl. Catal., B*, (2001) **34**, 255.
[90] Wang, W.; Gong, J., *Front. Chem. Sci. Eng.*, (2011) **5**, 2; and references cited therein.
[91] Aziz, M. A. A.; Jalil, A. A.; Triwahyono, S.; Ahmada, A., *Green Chem.*, (2015) **17**, 2647; and references cited therein.
[92] Chang, F.-W.; Tsay, M.-T.; Liang, S.-P., *Appl. Catal., A*, (2001) **209**, 217.
[93] Liu, J.; Li, C.; Wang, F.; He, S.; Chen, H.; Zhao, Y.; Wei, M.; Evans, D. G.; Duan, X., *Catal. Sci. Technol.*, (2013) **3**, 2627.
[94] Chang, F.-W.; Kuo, M.-S.; Tsay, M.-T.; Hsieh, M.-C., *Appl. Catal., A*, (2003) **247**, 309.
[95] Hwang, S.; Hong, U. G.; Lee, J.; Seo, J. G.; Baik, J. H.; Koh, D. J.; Lim, H.; Song, I. K., *J. Ind. Eng. Chem. (Amsterdam, Neth.)*, (2013) **19**, 2016.
[96] Aksoylu, A.; Onsan, Z., *Appl. Catal., A*, (1997) **164**, 1.
[97] Song, H.; Yang, J.; Zhao, J.; Chou, L., *Chin. J. Catal.*, (2010) **31**, 21.
[98] Hwang, S.; Lee, J.; Hong, U. G.; Baik, J. H.; Koh, D. J.; Lim, H.; Song, I. K., *J. Ind. Eng. Chem. (Amsterdam, Neth.)*, (2013) **19**, 698.
[99] Park, J. N.; McFarland, E. W., *J. Catal.*, (2009) **266**, 92.
[100] Zhou, G.; Wu, T.; Xie, H.; Zheng, X., *Int. J. Hydrogen Energy*, (2013) **38**, 10012.
[101] Garbarino, G.; Bellotti, D.; Riani, P.; Magistri, L.; Busca, G., *Int. J. Hydrogen Energy*, (2015) **40**, 9171.
[102] Sharma, S.; Hu, Z.; Zhang, P.; McFarland, E. W.; Metiu, H., *J. Catal.*, (2011) **278**, 297.
[103] Chan, L. K. M.; Poole, D. L.; Shen, D.; Healy, M. P.; Donohoe, T. J., *Angew. Chem. Int. Ed.*, (2014) **53**, 761.
[104] Sorribes, I.; Junge, K.; Beller, M., *Chem.–Eur. J.*, (2014) **20**, 7878.
[105] Savourey, S.; Lefèvre, G.; Berthet, J.-C.; Cantat, T., *Chem. Commun. (Cambridge)*, (2014) **50**, 14033.
[106] Jacquet, O.; Frogneux, X.; Das Neves Gomes, C.; Cantat, T., *Chem. Sci.*, (2013) **4**, 2127.
[107] Santoro, O.; Nahra, F.; Cordes, D. B.; Slawin, A. M. Z.; Nolan, S. P.; Cazin, C. S. J., *J. Mol. Catal. A: Chem.*, (2016) **423**, 85.
[108] Santoro, O.; Lazreg, F.; Minenkov, Y.; Cavallo, L.; Cazin, C. S. J., *Dalton Trans.*, (2015) **44**, 18138.
[109] González-Sebastián, L.; Flores-Alamo, M.; García, J. J., *Organometallics*, (2015) **34**, 763.
[110] Beydoun, K.; vom Stein, T.; Klankermayer, J.; Leitner, W., *Angew. Chem. Int. Ed.*, (2013) **52**, 9554.
[111] Frogneux, X.; Jacquet, O.; Cantat, T., *Catal. Sci. Technol.*, (2014) **4**, 1529.
[112] Tlili, A.; Blondiaux, E.; Frogneux, X.; Cantat, T., *Green Chem.*, (2015) **17**, 157; and references cited therein.
[113] Li, Y.; Fang, X.; Junge, K.; Beller, M., *Angew. Chem. Int. Ed.*, (2013) **52**, 9568.
[114] Li, Y.; Sorribes, I.; Yan, T.; Junge, K.; Beller, M., *Angew. Chem. Int. Ed.*, (2013) **52**, 12156.
[115] Cui, X.; Zhang, Y.; Deng, Y.; Shi, F., *Chem. Commun. (Cambridge)*, (2014) **50**, 13521.
[116] Kon, K.; Siddiki, S. M. A. H.; Onodera, W.; Shimizu, K., *Chem.–Eur. J.*, (2014) **20**, 6264.
[117] Du, X. L.; Tang, G.; Bao, H. L.; Jiang, Z.; Zhong, X. H.; Su, D. S.; Wang, J. Q., *ChemSusChem*, (2015) **8**, 3489.
[118] Cui, X.; Dai, X.; Zhang, Y.; Deng, Y.; Shi, F., *Chem. Sci.*, (2014) **5**, 649.
[119] Li, Y.; Yan, T.; Junge, K.; Beller, M., *Angew. Chem. Int. Ed.*, (2014) **53**, 10476.

1.10 Reduction of Peroxo Compounds, Ozonides, and Molozonides

P. Poechlauer and A. Zimmermann

General Introduction

This chapter focuses on heterogeneous catalytic hydrogenation of peroxo compounds, ozonides (1,2,4-trioxolanes), and molozonides (1,2,3-trioxolanes). The reduction of these compounds is of particular interest for various reasons: first of all, these motifs can be introduced by many different reagents and reactions. The synthesis of such compounds in general has been previously covered in *Science of Synthesis*, Vol. 38 (Peroxides), and *Houben–Weyl*, Vol. E 13 and Vol. VIII, pp 1–74. Thus, structurally diverse compounds containing a peroxo (O—O) group are accessible via well-established methods. The combination of these methods with catalytic hydrogenation leads to concise, selective reaction sequences delivering structural motifs that are frequently not easily and effectively accessible by other routes. Consequently, such sequences, e.g. a sequence consisting of ozonolysis plus hydrogenation, have been used in large-scale manufacturing of fine chemicals. Secondly, the growing knowledge in the field of catalytic hydrogenation allows conscious fine-tuning of reaction parameters and reagents to effect highly selective and high-yielding transformations. This chapter comprises typical examples of heterogeneous catalytic hydrogenations of peroxo compounds. It first focuses on aspects of hydrogenation catalysts and new mechanistic insights, and then examples of selective hydrogenations are described.

SAFETY: Organic peroxides, ozonides, and molozonides are highly reactive, combustible, and thermally unstable substances which may undergo self-accelerating decomposition. They also possess oxidizing characteristics and will react, often violently, with organic matter and chemical reducing agents. Initiation is by heating or by catalytic contaminations with transition-metal compounds (e.g., vanadium, chromium, manganese, iron, and cobalt), amines, strong acids, or alkalis. Peroxides spilled onto combustible materials (e.g., wood, paper, or cotton) may result in a fire. Some organic peroxides can separate out of solution if they become too cool, or by concentration of peroxide-containing solutions resulting in a concentrated peroxide which may be explosive and sensitive to shock.

Reactions and subsequent operations involving peroxy compounds should be run behind a safety shield. New or unfamiliar reactions, particularly those run at elevated temperatures, should be run first on a small scale. Reaction products should never be recovered from the final reaction mixture by distillation until all residual active oxygen compounds (including unreacted peroxy compounds) have been destroyed. Decomposition of active oxygen compounds may be accomplished by the procedure described in ref[1].

for references see p 337

1.10.1 Reduction of Peroxo Compounds

1.10.1.1 Mechanistic Aspects and New Types of Catalysts

A detailed mechanistic understanding is the basis for the development of a robust and commercially attractive process. Here, recent kinetic studies of the palladium-catalyzed hydroperoxide hydrogenation of pinane hydroperoxide[2,3] and cumene hydroperoxide[4] have contributed to the understanding of mechanistic aspects.

For the pinane hydroperoxide system, a careful elucidation of kinetic data allowed a closer insight into elemental catalytic steps of the catalytic hydrogenation comprising the hydroperoxide/hydrogen adsorption on the palladium surface, the hydrogen activation, and the surface reaction itself. The authors describe a retardation of the hydrogenation rate with increasing concentration of the starting material. Thereby, an initial increase in the reaction rate was observed with increasing pinane hydroperoxide concentrations up to 0.6 M. A further increase of the pinane hydroperoxide concentration resulted in a zero-order-type reaction rate. Interestingly, this concentration limit does not depend on the hydrogenation temperature, thus indicating a temperature-independent adsorption ratio for hydrogen and pinane hydroperoxide. Moreover, the zero-order-type reaction rate at concentrations >0.6 M could indicate a monomolecular decomposition mechanism that becomes dominant at elevated peroxide concentrations.

In a further aspect, it was demonstrated that the hydrogenation rate for pinane hydroperoxide increases uniformly with increasing hydrogen pressure up to 2 bar. A further increase in the hydrogen pressure resulted in an approximately three-fold steeper reaction rate, again uniformly increasing with the hydrogen pressure up to 12 bar. Based on these observations, different modes of hydrogen activation were proposed depending on the applied hydrogen pressure (Scheme 1).[2]

Scheme 1 Proposed Mechanisms for Hydrogen Activation in the Hydrogenation of Pinane Hydroperoxide[2]

hydrogen activation at high temperatures and high H_2 pressure

oxidative cleavage

hydrogen activation at low temperatures and low H_2 pressure

heterolytic cleavage

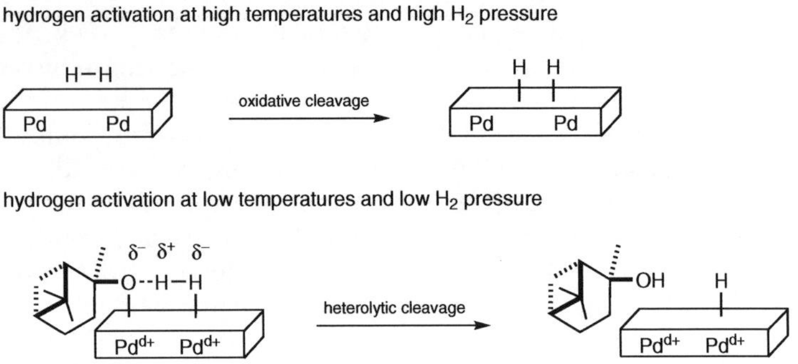

The first mechanism shown in Scheme 2 was assumed to take place at high hydrogen pressures and high temperatures via a classical homolytic hydrogen cleavage at the surface of the palladium catalyst, followed by the reduction of the hydroperoxide. As an alternative mechanism, a heterolytic hydrogen cleavage in the presence of the pinane hydroperoxide oxidized palladium surface was proposed at low hydrogen pressures and low temperatures.[2]

Scheme 2 Proposed Mechanisms for the Surface Reaction at High and Low Hydrogen Pressure[2]

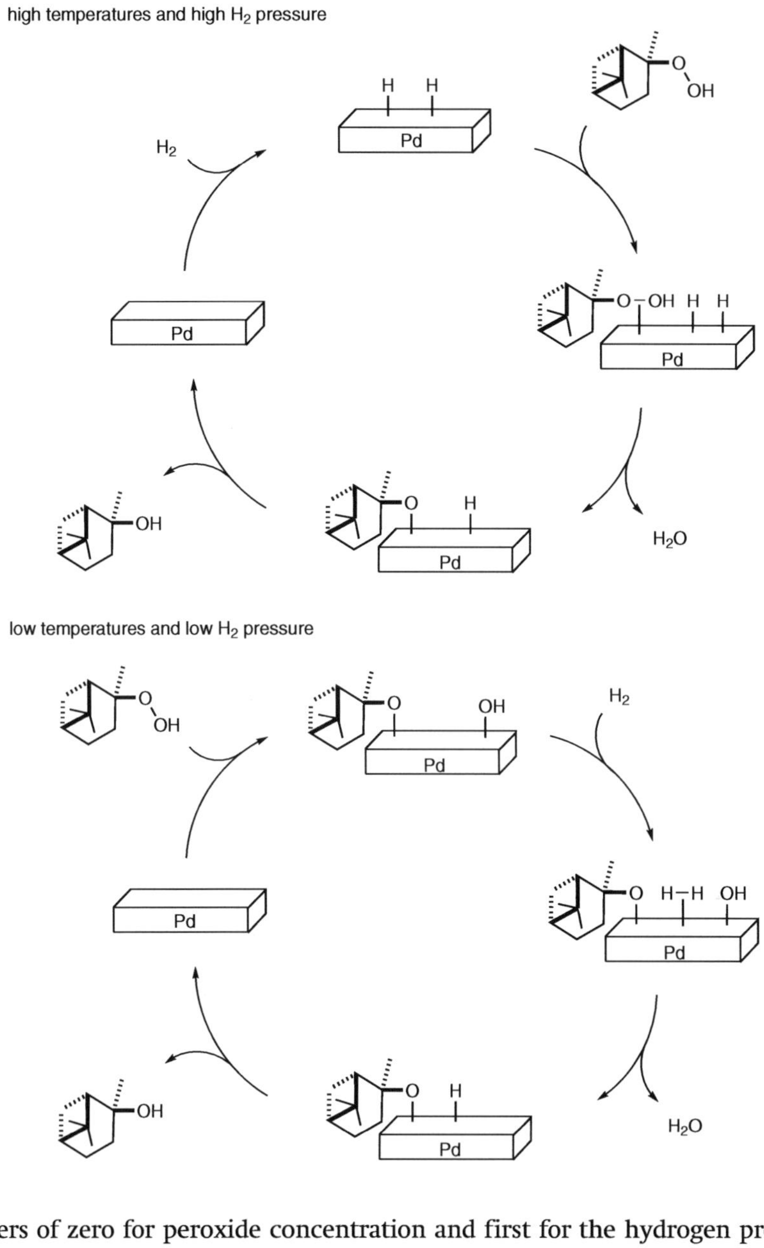

high temperatures and high H_2 pressure

H H Pd H_2 O OH Pd O–OH H H Pd OH O H Pd H_2O

low temperatures and low H_2 pressure

O OH O OH Pd H_2 Pd O H–H OH Pd OH O H Pd H_2O

Reaction orders of zero for peroxide concentration and first for the hydrogen pressure were also found for the catalytic hydrogenation of cumene hydroperoxide over palladium on alumina in a trickle-bed reactor.[4] Hydrogenation readily occurs at 30–60 °C by applying a hydrogen pressure of 0.5–1.5 MPa.

The oxidation of the catalyst surface in the presence of the strongly oxidizing hydroperoxide starting material is also described for the palladium-catalyzed hydrogenation of cumene hydroperoxide in the trickle-bed reactor. The oxidation of the palladium catalyst was demonstrated to be the root cause for the catalyst deactivation at the end of the hydrogenation reaction. Oxidation of the catalyst was supported by thermogravimetric analysis (TGA). However, it was demonstrated that 90% of the hydrogenation activity of the catalyst could be restored by treating the oxidized catalyst with hydrogen for 2 hours.

for references see p 337

For the pinane hydroperoxide system, the effect of the catalyst support and catalyst preparation on the reactivity and selectivity of the palladium-catalyzed hydroperoxide hydrogenation was investigated.[4,5] The reductive preparation of the palladium-on-charcoal system was thereby performed either with formaldehyde or hydrogen-reduced potassium tetrachloropalladate(II) (K_2PdCl_4) precursors. For both species, the trend of improved activity and chemoselectivity at higher temperatures and hydrogen pressure described in the studies mentioned above was confirmed. Furthermore, kinetic studies with both catalysts again showed zero-order reaction rates with respect to the pinane hydroperoxide concentration and first-order reaction rates for the hydrogen pressure. Based on these kinetic data, no significant change in the mechanism is expected for the carbon-supported palladium catalyst.

1.10.1.2 Reduction of Peroxo Compounds in the Presence of Other Reducible Functional Groups

The O—O bond in peroxides is easily reduced under mild conditions.[6] It is hydrogenated at ambient to moderately increased pressure, by moderately active catalysts. This allows selective hydrogenation in the presence of other reducible functional groups.

1.10.1.2.1 Reduction of Peroxo Compounds in the Presence of Alkenes, Esters, Lactones, and Lactams

Due to the easy access to dialkyl peroxy compounds, their hydrogenation in the presence of an alkenic double bond and an ester group has been studied. The unsaturated peroxy acetal ester **1** yields the saturated peroxy ester **2** (hydrogenation of the double bond only), the saturated hydroxy ester **3** (hydrogenation of the double bond and peroxy group), and the saturated oxo ester **4** (hydrogenation of the double bond and rearrangement of the peroxide). The selectivity is determined by the nature of the catalyst (Scheme 3).[7]

Scheme 3 Products of Hydrogenation of a Multifunctional Substrate Depending on the Catalyst[7]

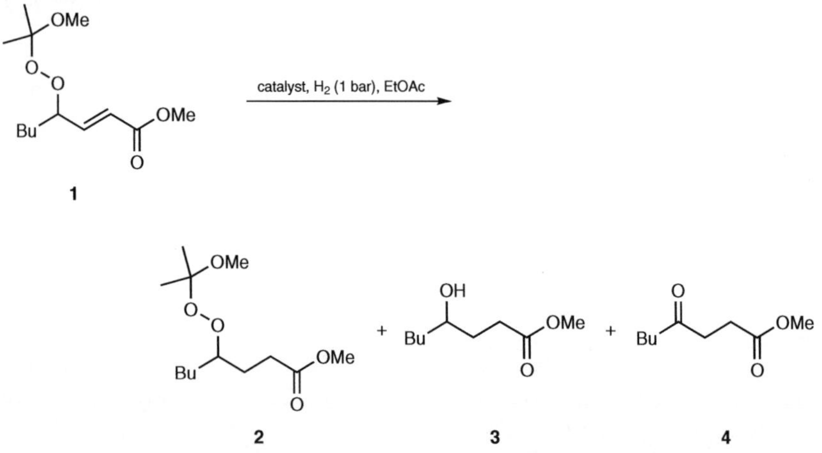

OMe
O
O
Bu
OMe
O
1

catalyst, H_2 (1 bar), EtOAc

OMe
O
O
Bu
OMe
O
2

+

OH
Bu
OMe
O
3

+

O
Bu
OMe
O
4

Catalyst (mol%)	Time (min)	Yield[a] (%)			Ref
		2	3	4	
5% Rh/C (5)	30	49	18	–	[7]
10% Pd/C (5)	15	39	22	28	[7]
10% Pt/C (5)	15	43	17	–	[7]

[a] Isolated yield after purification by flash chromatography; the reaction temperature was not given.

Further examples of the catalytic hydrogenation of peroxides in the presence of ester groups are described in Section 1.10.2. Thus, esters with an α-alkoxy hydroperoxy substituent are hydrogenated to yield hemiacetals without affecting the ester group.

A synthetically attractive sequence consisting of bisperoxidation with subsequent palladium-catalyzed C—H activation, C—C bond formation, and hydrogenation delivers 3-hydroxy-3-(hydroxymethyl)indolin-2-ones (e.g., **7**) from acrylanilides (e.g., **5**) via diperoxide intermediates (e.g., **6**) (Scheme 4).[8]

Scheme 4 Formation of an Indolin-2-one Derivative from an Acrylanilide[8]

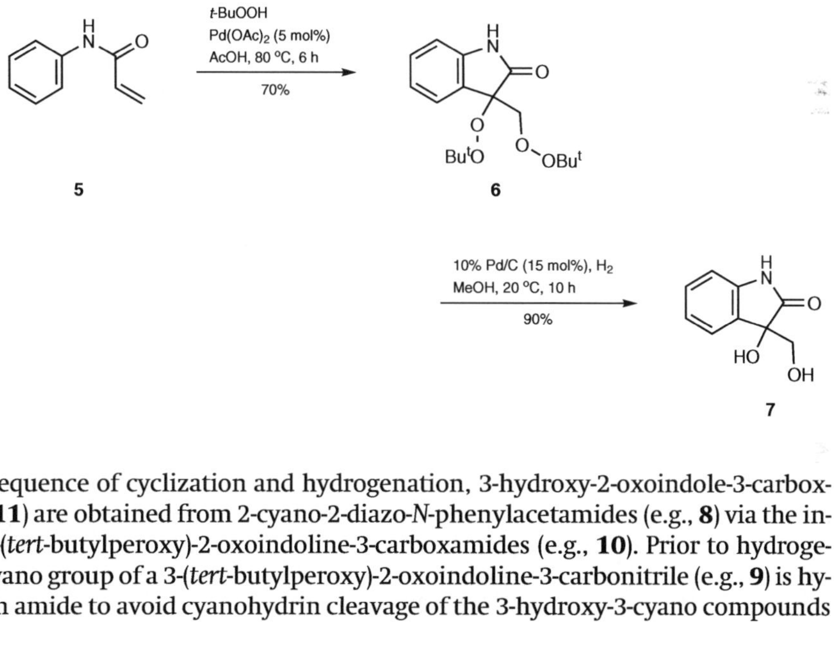

In a similar sequence of cyclization and hydrogenation, 3-hydroxy-2-oxoindole-3-carboxamides (e.g., **11**) are obtained from 2-cyano-2-diazo-*N*-phenylacetamides (e.g., **8**) via the intermediate 3-(*tert*-butylperoxy)-2-oxoindoline-3-carboxamides (e.g., **10**). Prior to hydrogenation, the cyano group of a 3-(*tert*-butylperoxy)-2-oxoindoline-3-carbonitrile (e.g., **9**) is hydrolyzed to an amide to avoid cyanohydrin cleavage of the 3-hydroxy-3-cyano compounds (Scheme 5).[9]

Scheme 5 Formation of a 1-Alkyl-3-hydroxy-2-oxoindoline-3-carboxamide from a 2-Cyano-2-diazo-*N*-phenylacetamide[9]

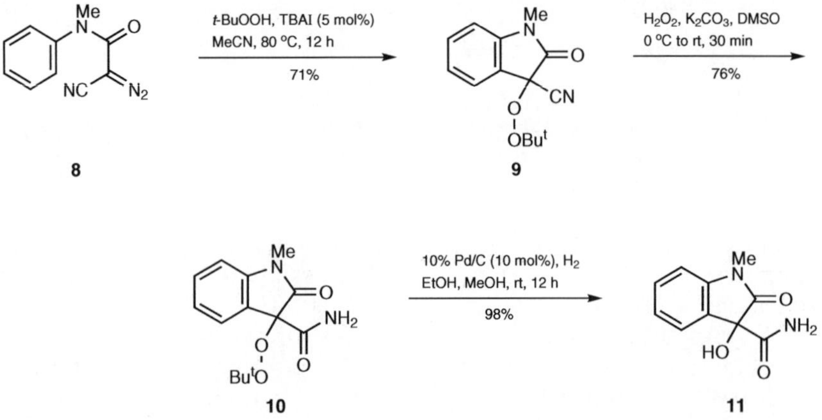

Methyl 4-[(2-Methoxypropan-2-yl)peroxy]octanoate (2) and Methyl 4-Hydroxyoctanoate (3):[7]

CAUTION: *Please read the safety information in the General Introduction (Section 1.10).*

To a soln of methyl (*E*)-4-[(2-methoxypropan-2-yl)peroxy]oct-2-enoate (**1**; 104 mg, 0.40 mmol) in EtOAc (2.0 mL) was added the 5% Rh/C catalyst (38 mg, 5 mol%). The mixture was stirred under H_2 until no further UV activity could be detected by TLC (30 min), whereupon the catalyst was removed by filtration through Celite with EtOAc. The filtrate was concentrated in the presence of 3,5-di-*tert*-butyl-4-hydroxytoluene (BHT) and the crude product was directly subjected to flash chromatography (EtOAc/hexane 1:9) to furnish methyl 4-[(2-methoxypropan-2-yl)peroxy]octanoate (**2**); yield: 51 mg (49%); and methyl 4-hydroxyoctanoate (**3**); yield: 12 mg (18%).

3-(*tert*-Butylperoxy)-3-[(*tert*-butylperoxy)methyl]indolin-2-one (6); Typical Procedure:[8]

CAUTION: *Please read the safety information in the General Introduction (Section 1.10).*

A flask equipped with a reflux condenser was charged with *N*-phenylacrylamide (**5**; 147 mg, 1.0 mmol) and $Pd(OAc)_2$ (11.2 mg, 5 mol%). AcOH (12 mL) was added, followed by a 5–6 M soln of *t*-BuOOH in decane (1.6 mL) or a 4 M soln of *t*-BuOOH in benzene (2.0 mL) (**CAUTION:** *carcinogen*). The mixture was placed in an 80 °C oil bath, stirred for 6 h, cooled to rt, and concentrated. The residue was purified by column chromatography; yield: 70%.

3-Hydroxy-3-(hydroxymethyl)indolin-2-one (7); Typical Procedure:[8]

CAUTION: *Please read the safety information in the General Introduction (Section 1.10).*

3-(*tert*-Butylperoxy)-3-[(*tert*-butylperoxy)methyl]indolin-2-one (**6**; 323 mg, 1 mmol) was dissolved in MeOH (50 mL) at rt and 10% Pd/C (15 mol%) was added. The mixture was stirred under H_2 for 10 h and then passed through Celite. The Celite was washed with Et_2O and the combined organic phases were concentrated. The resulting crude product was purified by column chromatography affording a white solid; yield: 90%.

3-(*tert*-Butylperoxy)-1-methyl-2-oxoindoline-3-carbonitrile (9); Typical Procedure:[9]

CAUTION: *Please read the safety information in the General Introduction (Section 1.10).*

In a Schlenk tube, 2-cyano-2-diazo-*N*-methyl-*N*-phenylacetamide (**8**; 50 mg, 0.25 mmol, 1.0 equiv) was dissolved in MeCN (0.25 M) and a 5.5 M soln of *t*-BuOOH in decane (0.23 mL, 1.25 mmol, 5.0 equiv) was added in one portion under argon. TBAI (4.6 mg, 12.5 μmol, 5 mol%) was added and the tube was sealed and the mixture was stirred at 80 °C for 12 h. The solvent was removed under reduced pressure and the resulting residue was purified by flash chromatography (petroleum ether/EtOAc); yield: 71%.

3-(*tert*-Butylperoxy)-1-methyl-2-oxoindoline-3-carboxamide (10); Typical Procedure:[9]

CAUTION: *Please read the safety information in the General Introduction (Section 1.10).*

To a soln of 3-(*tert*-butylperoxy)-1-methyl-2-oxoindoline-3-carbonitrile (**9**; 65 mg, 0.25 mmol, 1.0 equiv) and K_2CO_3 (0.13 mmol, 0.5 equiv) in DMSO (0.25 mL) was added dropwise 9.8 M aq H_2O_2 (50 μL, 0.50 mmol, 2.0 equiv) at 0 °C within 5 min. The mixture was allowed to warm up to rt and stirred at the same temperature for 30 min. H_2O (3 mL) was added and the mixture was extracted with CH_2Cl_2 (3 × 3 mL). The combined organic layers were dried ($MgSO_4$), filtered, and concentrated under reduced pressure. The resulting residue was purified by flash chromatography (petroleum ether/EtOAc 1:1), affording a colorless solid; yield: 76%.

3-Hydroxy-1-methyl-2-oxoindoline-3-carboxamide (11); Typical Procedure:[9]

CAUTION: *Please read the safety information in the General Introduction (Section 1.10).*

A soln of 3-(*tert*-butylperoxy)-1-methyl-2-oxoindoline-3-carboxamide (**10**; 70 mg, 0.25 mmol, 1.0 equiv) and 10% Pd/C (26.5 mg, 10 mol%) in EtOH/MeOH (3:1; 8.0 mL) was stirred under H_2 (1 bar) at rt for 12 h. The mixture was filtered through Celite. The filtrate was concentrated under reduced pressure and the resulting residue was purified by flash chromatography (petroleum ether/EtOAc 1:1), affording a colorless solid; yield: 51 mg (98%).

1.10.1.2.2 Reduction of Peroxo Compounds in the Presence of Oxo Groups

1.10.1.2.2.1 Hydrogenation of α-Peroxy Carbonyl Compounds

Hydrosilylation of alkynes yields silyl-substituted alkenes (e.g., **12**). Ozonolysis of these compounds does not cleave the double bond, but instead furnishes α-silylperoxy carbonyl compounds (e.g., **13**) in high yield. Catalytic hydrogenation of these compounds over palladium on charcoal at room temperature delivers α-hydroxy carbonyl compounds (acyloins, e.g. **14**) (Scheme 6).[10] The method is most convenient for symmetrically substituted alkynes, delivering symmetrical acyloins, and for terminal alkynes, delivering α-hydroxy aldehydes.

Scheme 6 Ozonolysis and Hydrogenation of Silyl-Substituted Alkenes[10]

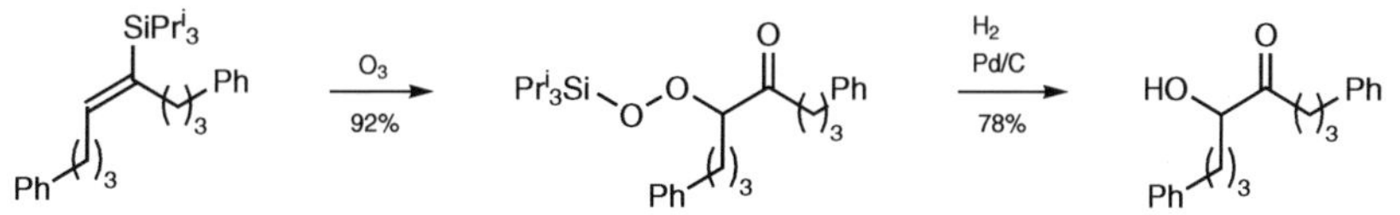

SiPr$^i_3$ Ph Ph 3 3 **12** —(O_3, 92%)→ Pr$^i_3$SiOO–C(=O)Ph Ph 3 3 **13** —(H_2, Pd/C, 78%)→ HO Ph Ph 3 3 **14**

for references see p 337

5-Hydroxy-1,8-diphenyloctan-4-one (14); Typical Procedure:[10]

CAUTION: *Please read the safety information in the General Introduction (Section 1.10).*

CAUTION: *Ozone irritates mucous membranes and the lungs and is highly explosive in the liquid or solid states.*

A stream of ozone (1.2 v/v% in O_2, 150 mL/min) was bubbled through a soln of (*E*)-(1,8-diphenyloct-4-en-4-yl)triisopropylsilane (**12**; 133 mg, 0.32 mmol) in EtOAc at −78 °C. After 30 min, the soln turned pale blue, indicating complete oxidation. Dissolved ozone was removed by bubbling argon through the soln for 5 min. The temperature was allowed to rise to ambient. The solvent was removed and the residue was purified by chromatography (silica gel, hexane/Et_2O 40:1) to afford 1,8-diphenyl-5-[(triisopropylsilyl)peroxy]octan-4-one (**13**) as a colorless oil; yield: 136 mg (92%).

To a soln of 1,8-diphenyl-5-[(triisopropylsilyl)peroxy]octan-4-one (**13**; 415 mg, 0.88 mmol) in EtOAc was added 10% Pd/C (213 mg). The resulting mixture was stirred for 6 h under H_2 at atmospheric pressure and ambient temperature, and then filtered through a pad of Celite. The filtrate was concentrated under reduced pressure and the residue was purified by chromatography (silica gel, hexane/Et_2O 10:1), affording a colorless oil; yield: 205 mg (78%).

1.10.1.2.2.2 Hydrogenation of β-Peroxy Carbonyl Compounds

Peroxidation of α,β-unsaturated ketones can be directed to deliver epoxides or β-peroxy ketones. Suitable chiral catalysts render this reaction chemo- and enantioselective. The resulting homochiral β-peroxy ketones **15** can be catalytically hydrogenated at ambient temperature and pressure to deliver β-hydroxy ketones **16** with unchanged enantiomeric excess in good yield (Scheme 7).[11]

Scheme 7 Synthesis of β-Hydroxy Ketones by Reduction of Homochiral β-Peroxy Ketones[11]

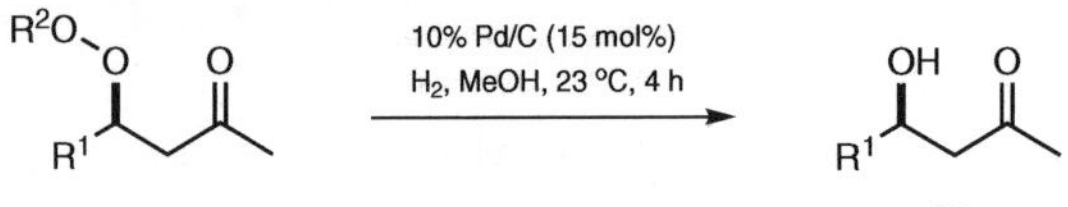

R^2O O O R^1 **15** → 10% Pd/C (15 mol%), H_2, MeOH, 23 °C, 4 h → OH O R^1 **16**

R^1	R^2	Yield[a] (%)	Ref
Pr	*t*-Bu	85	[11]
$(CH_2)_4Me$	CMe_2Ph	84	[11]

[a] Isolated yield after purification by flash chromatography.

(*R*)-4-Hydroxyheptan-2-one (16, R^1 = Pr); Typical Procedure:[11]

CAUTION: *Please read the safety information in the General Introduction (Section 1.10).*

To a soln of (*R*)-4-(*tert*-butylperoxy)heptan-2-one (**15**, R^1 = Pr; R^2 = *t*-Bu; 20 mg, 0.1 mmol) in MeOH (5 mL) at 23 °C was added 10% Pd/C (15 mol%). The mixture was kept under H_2 for 4 h, and then passed through Celite. The Celite was washed with Et_2O. The filtrate was concentrated under reduced pressure, and the residue was purified by flash chromatography (silica gel, hexanes/EtOAc 3:1) affording a colorless oil; yield: 85%; 90% ee (determined by GC on a chiral stationary phase).

1.10.1.2.2.3 Hydrogenation of γ-Peroxy Carbonyl Compounds

γ-Peroxy ketones are elegantly accessible by acid-catalyzed oxidative radical addition of ketones to alkenes. Thus, the reaction of acetone, styrene, and *tert*-butyl hydroperoxide in the presence of catalytic amounts of 4-toluenesulfonic acid delivers 5-(*tert*-butylperoxy)-5-phenylpentan-2-one (**17**) in 74% yield. If this γ-peroxy ketone is catalytically hydrogenated using palladium on carbon in acetonitrile, the reaction proceeds in a completely chemoselective way: neither the oxo group nor the benzylic alcohol formed by hydrogenation of the benzylic *tert*-butylperoxy group is attacked and the corresponding homoaldol product **18** is obtained in 93% yield (Scheme 8).[12]

Scheme 8 Reduction of a γ-Peroxy Ketone To Produce a γ-Hydroxy Ketone[12]

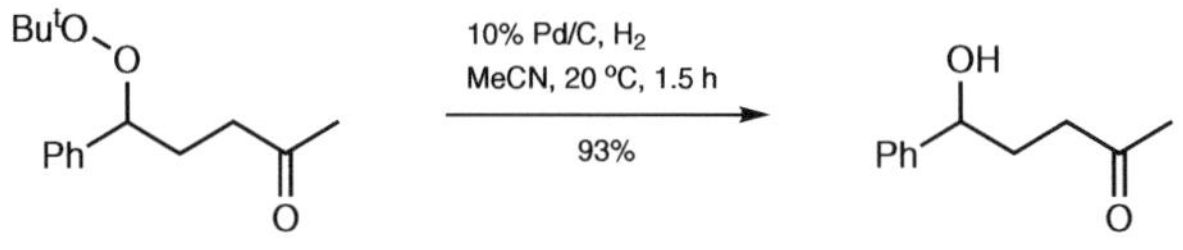

10% Pd/C, H_2
MeCN, 20 °C, 1.5 h
93%

17 → **18**

If the hydrogenation of 5-(*tert*-butylperoxy)-5-phenylpentan-2-one (**17**) is performed in methanol, complete reduction of the benzylic position is observed, giving product **20**. The reaction proceeds via acetal **19**, which can be isolated if the reaction is not allowed to go to completion (Scheme 9).[12]

Scheme 9 Reduction of a γ-Peroxy Ketone with Complete Reduction at the Benzylic Position[12]

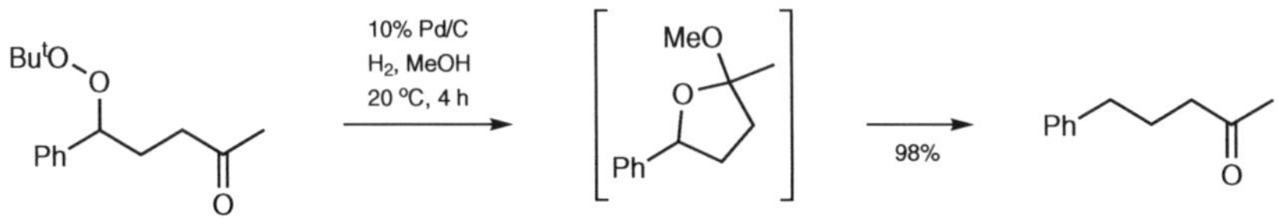

10% Pd/C
H_2, MeOH
20 °C, 4 h

17 → [**19**] → (98%) **20**

5-Hydroxy-5-phenylpentan-2-one (18):[12]

CAUTION: *Please read the safety information in the General Introduction (Section 1.10).*

Into a two-necked flask, 10% Pd/C (20 mg) was introduced and the atmosphere was replaced with argon. A soln of 5-(*tert*-butylperoxy)-5-phenylpentan-2-one (**17**; 125 mg, 0.5 mmol, 1 equiv) in MeCN (5 mL) was added and the atmosphere was then replaced with H_2. The mixture was stirred vigorously at rt for 1.5 h, and then filtered through Celite (CH_2Cl_2). The filtrate was concentrated and the residue was purified by column chromatography (hexanes/EtOAc 4:1), affording the product as a clear oil; yield: 83 mg (93%).

5-Phenylpentan-2-one (20):[12]

CAUTION: *Please read the safety information in the General Introduction (Section 1.10).*

Into a two-necked flask, 10% Pd/C (20 mg) was introduced and the atmosphere was replaced with argon. A soln of 5-(*tert*-butylperoxy)-5-phenylpentan-2-one (**17**; 125 mg, 0.5 mmol, 1 equiv) in MeOH (5 mL) was added and the atmosphere was then replaced with H_2. The mixture was stirred vigorously at rt for 4 h, and then filtered through Celite (MeOH). The filtrate was concentrated and the residue was purified by column chromatography (hexanes/EtOAc 95:5), affording the product as a clear oil; yield: 80 mg (98%).

for references see p 337

1.10.1.2.3 Reduction of Peroxo Compounds in the Presence of Nitro Groups

In an analogous manner to its reactions with α,β-unsaturated ketones, *tert*-butyl hydroperoxide epoxidizes α,β-unsaturated nitroalkenes or adds to the double bond to form β-*tert*-butylperoxy-substituted nitro compounds (e.g., **21**). If the alkene substrate has an additional homochiral allylic ether or silyl ether group, both reactions proceed diastereoselectively. Furthermore, the β-*tert*-butylperoxy-substituted nitro compound **21** can be hydrogenated chemoselectively using palladium on charcoal at ambient temperature and pressure to produce the nitro alcohol **22** in 66% yield. Neither the nitro group nor the benzylic silyl ether in **21** is reduced (Scheme 10).[13] Catalytic hydrogenation of both the epoxide and the β-*tert*-butylperoxy-substituted product proceeds smoothly under mild conditions.

Scheme 10 Chemoselective Reduction of a β-*tert*-Butylperoxy-Substituted Nitro Compound[13]

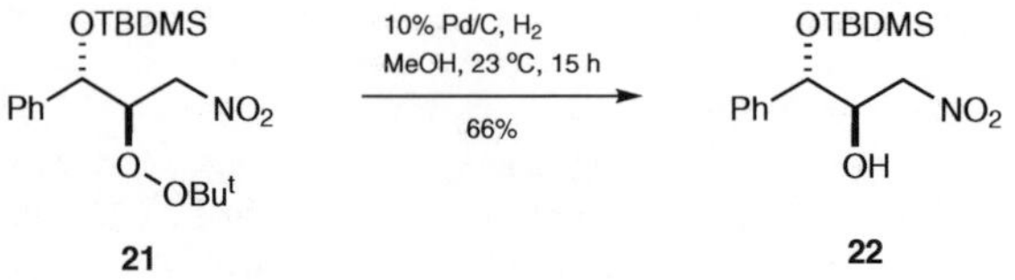

OTBDMS; Ph; NO_2; O; OBu^t; **21** → 10% Pd/C, H_2; MeOH, 23 °C, 15 h; 66% → OTBDMS; Ph; NO_2; OH; **22**

(1*R*,2*S*)-1-(*tert*-Butyldimethylsiloxy)-3-nitro-1-phenylpropan-2-ol (22):[13]

CAUTION: *Please read the safety information in the General Introduction (Section 1.10).*

A soln of (1*R*,2*S*)-1-(*tert*-butyldimethylsiloxy)-2-(*tert*-butylperoxy)-3-nitro-1-phenylpropane (**21**; 383 mg, 1 mmol) in anhyd MeOH (2 mL) was added to a prehydrogenated suspension of 10% Pd/C (40 mg) in anhyd MeOH (20 mL). The resulting mixture was stirred under H_2 at 23 °C for 15 h, and then filtered through Celite. The filtrate was concentrated under reduced pressure and the resulting light-yellow oil was purified by chromatography (silica gel, hexanes/EtOAc 7:3 to 6:4); yield: 66%.

1.10.1.3 Reduction of Peroxides as Part of Multistep Syntheses

A number of multistep syntheses make use of the sequence of introduction of a peroxide moiety and then its reduction. Thus, one or two hydroxy groups can be attached regio- and diastereoselectively. This also creates options for the realization of dihydroxylations by formal addition of oxygen and hydrogen in catalytic processes. This concept is highly attractive in light of the principles of green chemistry.[14]

The following general strategies are most common: (1) hydrogenation of a hydroperoxide formed by an ene reaction of singlet oxygen with a C=C bond to effect a net hydration of this double bond; (2) hydrogenation of a dialkyl peroxide formed by addition of an alkyl hydroperoxide to a C=C bond to again accomplish a net hydration of this double bond; (3) hydrogenation of 1,2-dioxin derivatives formed by [2+4] cycloaddition of singlet oxygen to 1,3-dienes to achieve 1,4-dihydroxylation; and (4) hydrogenation of endoperoxides formed by [2+4] cycloaddition of singlet oxygen to cyclic 1,3-dienes to cause *cis*-1,4-dihydroxylation.

1.10.1.3.1 Hydrogenation of Hydroperoxides Formed by an Ene Reaction of Singlet Oxygen with a C=C Bond

The sequence of an ene reaction of singlet oxygen to form a hydroperoxide and subsequent hydrogenation of this hydroperoxide results in a net "hydration" and has been used in various syntheses of tetracyclines.

For example, Aureomycin (chlortetracycline; **25**, $R^1 = Cl$) is synthesized by photooxygenation of 7-chloroanhydrotetracycline (**23**, $R^1 = Cl$) and hydrogenation of the hydroperoxide **24** ($R^1 = Cl$),[15] while Terramycin (oxytetracycline; **25**, $R^1 = H$) is obtained by a sequence of dye-sensitized photooxygenation of anhydrotetracycline (**23**, $R^1 = H$) and hydrogenation of the resulting hydroperoxide **24** ($R^1 = H$) (Scheme 11).[16]

Scheme 11 Sequence of Ene Reaction To Form a Hydroperoxide and Catalytic Hydrogenation in the Synthesis of Tetracyclines[15,16]

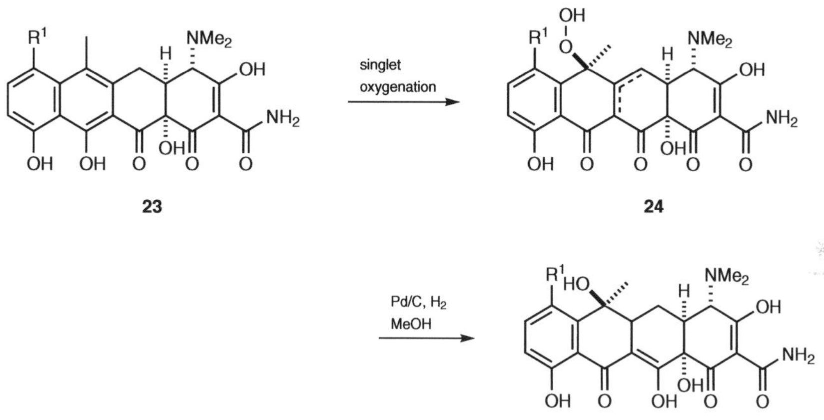

R1 NMe2 H OH NH2 OH OH O OH O O **23** → singlet oxygenation → OH O R1 NMe2 H OH NH2 OH O O OH O O **24** → Pd/C, H2 MeOH → R1 HO NMe2 H OH NH2 OH O OH OH O O **25**

R^1	Conditions for Step 1	Yield (%)		Ref
		24	**25**	
Cl	O_2, benzene, *hν*, 5 d	70[a]	quant	[15]
H	5,10,15,20-tetraphenylporphyrin, O_2, $CHCl_3$, *hν*, 10 min	97	49	[16]

[a] After recycling.

Terramycin (25, $R^1 = H$):[16]

CAUTION: *Please read the safety information in the General Introduction (Section 1.10).*

To a photolysis well was added a soln of anhydrotetracycline (**23**, $R^1 = H$; 100 mg, 0.235 mmol) and 5,10,15,20-tetraphenylporphyrin (5 mg) as the dye sensitizer in $CHCl_3$ (30 mL). The mixture was irradiated with a 650-W halogen tungsten lamp in a closed system at cooling water temperature for 10 min while O_2 was bubbled through. TLC (C18 reversed phase, $BuOH/AcOH/H_2O$ 1:1:8) showed that the starting material was oxidized completely. The cloudy mixture was filtered to remove a very small amount of insoluble material (pale-yellow solid). The filtrate gave a positive peroxide test and was concentrated to about 1 mL. MeOH (10~15 mL) was added to precipitate the dye, which was then removed by filtration. MeOH was removed under reduced pressure to give the crude hydroperoxide **24** ($R^1 = H$) as a pale-yellow solid; yield: 104 mg (97%).

6-Deoxy-6-hydroperoxy-5a(11a)-dehydrotetracycline (**24**, $R^1 = H$; 40 mg, 0.087 mmol) and 10% Pd/C (40 mg) in MeOH (15 mL) were subjected to hydrogenation (48 psi) for 2.5 h

for references see p 337

at rt. The reaction was monitored by taking NMR spectra of reaction aliquots. TLC (C18 reversed phase, MeCN/AcOH/H_2O 8:1:11) showed that there was a spot identical in R_f to that of Terramycin (**25**, R^1 = H) which was UV (365 nm) active. Purification by chromatography (polyamide column) gave the pure product as a pale-yellow solid; yield: 19 mg (49%).

1.10.1.3.2 Hydrogenation of Dialkyl Peroxides Formed by Addition of an Alkyl Hydroperoxide to a C=C Bond To Effect a Net Hydration of this Double Bond

A strategy similar to the one described above (Section 1.10.1.3.1) has been employed in synthetic studies toward a unified total synthesis of various pterocarpans, members of a large family of plant-derived isoflavonoids. Again, a sequence of peroxidation and catalytic reduction effects net hydration of phenol **26**, with incorporation of oxygen at the *para* position relative to the phenolic hydroxy group. In the first step, *tert*-butyl hydroperoxide is employed as terminal oxidant in combination with (diacetoxyiodo)benzene to introduce a *tert*-butylperoxy group and dearomatize **26** to form *tert*-butyl peroxide **27** in 48% yield as a 3:1 mixture of diastereomers. In the second step, the *tert*-butylperoxy group is reduced to form the desired hydroxy derivative **28** (Scheme 12).[17] According to the authors, reductions of similar resorcinol-derived dienones are known to be problematic, leading to reductive rearomatization so as to return the pre-dearomatized material.[18] The authors applied what could be called in situ generated hydrogen or transfer hydrogenation by using ammonium formate in combination with palladium on carbon and microwave irradiation.

Scheme 12 Sequence of Peroxidation and Catalytic Hydrogenation in the Synthesis of Pterocarpans[17]

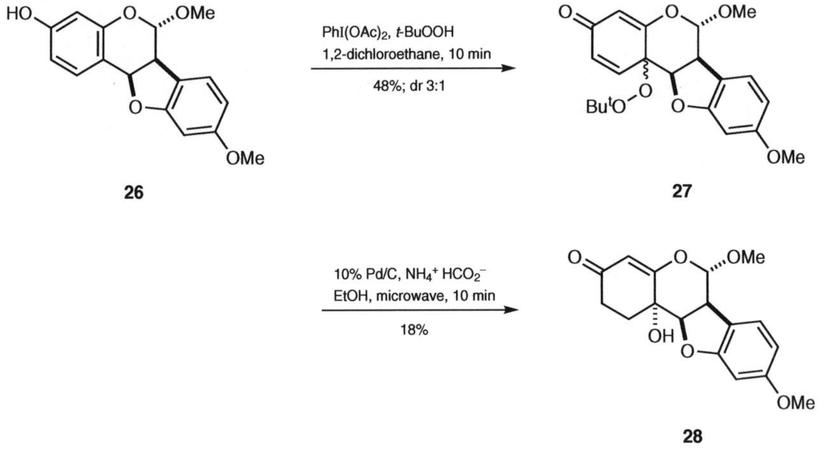

HO, O, OMe, O, OMe — **26**

$PhI(OAc)_2$, t-BuOOH
1,2-dichloroethane, 10 min
48%; dr 3:1

O, O, OMe, Bu^tO, O, O, OMe — **27**

10% Pd/C, $NH_4^+ HCO_2^-$
EtOH, microwave, 10 min
18%

O, O, OMe, OH, O, OMe — **28**

(6*R,6a*R**,11a*S**,11b*R**)-11b-Hydroxy-6,9-dimethoxy-1,2,6,6a,11a,11b-hexahydro-3*H*-benzofuro[3,2-*c*]chromen-3-one (28):**[17]

CAUTION: *Please read the safety information in the General Introduction (Section 1.10).*

To a soln of (6*R**,6a*R**,11a*S**)-6,9-dimethoxy-6a,11a-dihydro-6*H*-benzofuro[3,2-*c*]chromen-3-ol (**26**; 5 mg, 16 μmol) in 1,2-dichloroethane (0.2 mL) was added a 5.5 M soln of *t*-BuOOH in decane (29 μL). The mixture was stirred for 5 min at rt, followed by addition of $PhI(OAc)_2$ (10.7 mg, 0.033 mmol). The resulting mixture was stirred for 10 min, and then concentrated under reduced pressure. The residue was purified by TLC (silica gel, EtOAc/hexane 1:4) to afford (6*R**,6a*R**,11a*S**)-11b-(*tert*-butylperoxy)-6,9-dimethoxy-6,6a,11a,11b-tetrahydro-3*H*-benzofuro[3,2-*c*]chromen-3-one (**27**) as a mixture of diastereomers; yield: 3.4 mg (48%); dr 3:1.

In a microwave tube (0.5–2.0 mL) equipped with a stirrer bar were placed *tert*-butyl peroxide **27** (5 mg, 12.9 μmol), $NH_4^+HCO_2^-$ (5 mg), 10% Pd/C (1 mg), and EtOH (1 mL). The tube was purged with N_2, sealed with an aluminum cap at 1 atm pressure, and then placed in the Biotage Initiator microwave synthesizer (prestir time 30 s, 120 °C, time 10 min). Another 3 batches of the same reaction were run and all 4 batches were combined. The resulting mixture was filtered through Celite. The filtrate was concentrated and the residue was purified by TLC (EtOAc/hexane 3:7); yield: 3.0 mg (18%).

1.10.1.3.3 Hydrogenation of 1,2-Dioxin Derivatives Formed by [2+4] Cycloaddition of Singlet Oxygen to 1,3-Dienes To Effect 1,4-Dihydroxylation

Syntheses employing 1,2-dioxin derivatives are attractive for a number of reasons: Firstly, they allow derivatization of the cyclic double bond with good control of diastereoselectivity. Secondly, the peroxide group in 1,2-dioxins and their saturated derivatives is sufficiently stable to mask functionalities such as the 1,4-diol or the γ-hydroxy carbonyl function. Thus, peroxide generation and hydrogenation need not follow each other immediately in a multistep synthesis.[19] This has been impressively demonstrated in the synthesis of the marine natural product 6-epiplakortolide E.[20]

1.10.1.3.3.1 Dihydroxylation and Hydrogenation of 4-Substituted 1,2-Dioxins

Branched erythro sugars **31** are concisely synthesized from 4-substituted 1,2-dioxins **29** via *cis*-dihydroxylation and hydrogenation.[21] Chiral dihydroxylation yielding 4-substituted 1,2-dioxane-4,5-diols **30** is effected by using chiral Sharpless catalysts, e.g. dihydroquinidine phthalazine-1,4-diyl diether [$(DHQD)_2$PHAL]. Benzylic alcohol groups are not affected by the hydrogenation. Thus, the sequence is rather generally applicable, giving access to a variety of rare or unnatural sugar derivatives. The hydrogenation of dioxins **30** over palladium on carbon returns high yields of tetraols **31** (Scheme 13).[21]

for references see p 337

Scheme 13 Dihydroxylation and Catalytic Hydrogenation To Form Rare or Unnatural Sugar Derivatives[21]

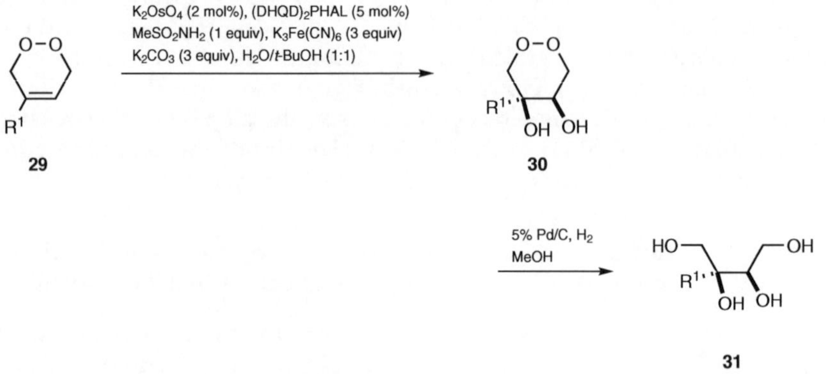

K_2OsO_4 (2 mol%), $(DHQD)_2PHAL$ (5 mol%)
$MeSO_2NH_2$ (1 equiv), $K_3Fe(CN)_6$ (3 equiv)
K_2CO_3 (3 equiv), H_2O/t-BuOH (1:1)

29 → **30**

5% Pd/C, H_2
MeOH

→ **31**

R^1	Yield (%)		ee (%) of **30**	Ref
	30	**31**		
Ph	70	71	93	[21]
Cy	90	80	85	[21]
1-adamantyl	91	84	80	[21]
Me	63	73	10	[21]
Bn	65	74	40	[21]
$(CH_2)_5Me$	83	78	93	[21]

4-Substituted 1,2-Dioxane-4,5-diols 30; **General Procedure:**[21]

CAUTION: *Please read the safety information in the General Introduction (Section 1.10).*

To a stirred mixture of t-BuOH (5 mL) and H_2O (5 mL) was added K_2OsO_4 (2 mol%), chiral amine ligand $(DHQD)_2PHAL$ (5 mol%), $K_3Fe(CN)_6$ (3 mmol), K_2CO_3 (3 mmol), and $MeSO_2NH_2$ (1 mmol), followed by a 1,2-dioxin **29** (1 mmol). The orange mixture was stirred rapidly at ambient temperature until the reaction was complete (ca. 1–2 d, TLC monitoring). The reaction was quenched with solid Na_2SO_3 (3 mmol). The mixture was then stirred for 30 min and extracted with CH_2Cl_2 or EtOAc (4 × 50 mL). The extracts were dried (Na_2SO_4), filtered, and concentrated under reduced pressure. The crude material was purified by flash chromatography.

Erythritols 31; **General Procedure:**[21]

CAUTION: *Please read the safety information in the General Introduction (Section 1.10).*

To a stirred soln of a 4-substituted 1,2-dioxane-4,5-diol **30** (1 mmol) in MeOH (5 mL) was added 5% Pd/C (10% w/w), and the mixture was stirred overnight under H_2. The suspension was then filtered through a small pad of Kenite, washing with MeOH. The filtrate was concentrated under reduced pressure and the crude product was purified by flash chromatography or recrystallized (EtOAc).

1.10.1.3.4 Hydrogenation of Endoperoxides Formed by [2+4] Cycloaddition of Singlet Oxygen to Cyclic 1,3-Dienes To Effect *cis*-1,4-Dihydroxylation

1.10.1.3.4.1 Cyclohexanetetraols by Hydrogenation of 2,3-Dioxabicyclo[2.2.2]octane-5,6-diols

A sequence of dihydroxylation of endoperoxides (e.g., **32**) to furnish diols (e.g., **33**) and catalytic hydrogenation has been used to synthesize carbocyclic tetraols (e.g., **34**) with defined relative stereochemistry in excellent yields (Scheme 14).[22] The authors mention the importance of providing sufficient hydrogen to avoid peroxide rearrangement at the catalyst surface, which leads to the dicarbonyl compounds **35**.

Scheme 14 Dihydroxylation and Catalytic Hydrogenation To Form *cis*-1,4-Disubstituted Carbocyclic Tetraols[22]

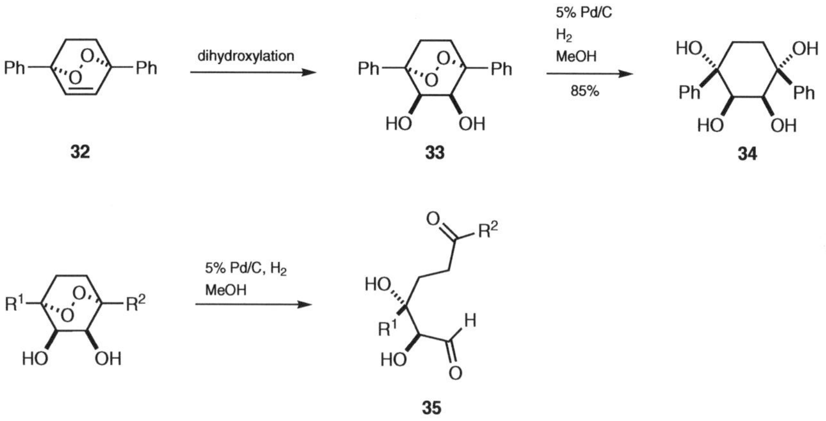

(1*R,2*S**,3*R**,4*S**)-1,4-Diphenylcyclohexane-1,2,3,4-tetraol (34); Typical Procedure:**[22]

CAUTION: *Please read the safety information in the General Introduction (Section 1.10).*

To a stirred soln of (1*R**,4*S**,5*R**,6*S**)-1,4-diphenyl-2,3-dioxabicyclo[2.2.2]octane-5,6-diol (**33**; 298 mg, 1 mmol) in MeOH (5 mL) was added 5% Pd/C (10% w/w), and the mixture was stirred under H_2 until hydrogenation was complete according to TLC. The suspension was then filtered through Kenite, washing with MeOH. The filtrate was concentrated under reduced pressure and the crude residue was purified by crystallization (CH_2Cl_2), providing the product as a white solid; yield: 85%; mp 162–164 °C (CH_2Cl_2).

1.10.1.3.4.2 4-Hydroxycyclohexenones by Hydrogenation of 1-Alkoxy-2,3-dioxabicyclo[2.2.2]oct-5-enes

A sequence of peroxide hydrogenation and selective alkene hydrogenation has been used for derivatization of opium alkaloids to synthesize pharmaceutically useful semisynthetic opiates such as naltrexone derivatives. For example, oripavine, an opioid naturally occurring in poppy, can be N-alkylated and subjected to a sequence of singlet oxygenation, to deliver endoperoxide **36**, and catalytic hydrogenation to furnish enone **37**. The chosen solvent is a 1:1:1 mixture of propan-2-ol, water, and formic acid. To avoid overreduction of the enone group in **37** to the alcohol, the endoperoxide is hydrogenated over 0.05 g/g substrate of 10% palladium on charcoal after addition of 0.04 g/g substrate of thiourea to

for references see p 337

poison the catalyst. Enone **37** is then isolated and hydrogenated over 0.1 g/g substrate of 10% palladium on charcoal in methanol as solvent to yield 74% of (*R*)-methylnaltrexone (**38**) based on endoperoxide **36** (Scheme 15).[23]

Scheme 15 Peroxide and Alkene Hydrogenation in the Synthesis of Opiates[23]

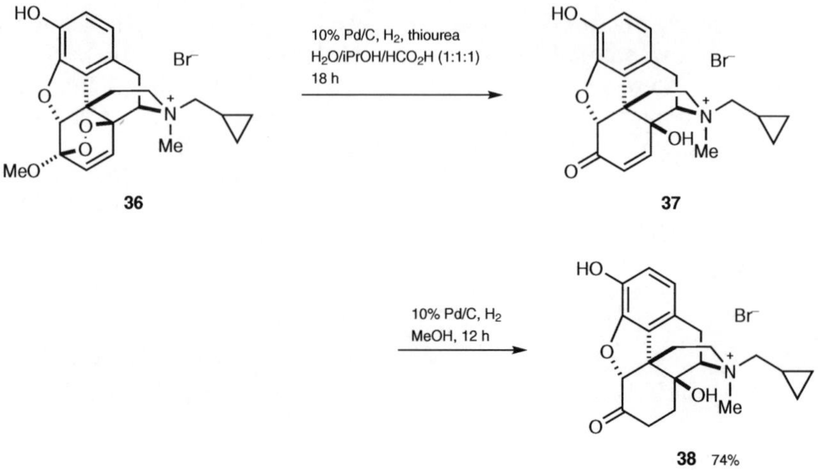

10% Pd/C, H_2, thiourea
H_2O/iPrOH/HCO_2H (1:1:1)
18 h

10% Pd/C, H_2
MeOH, 12 h

38 74%

The strategy of forming an endoperoxide by singlet oxygenation and subsequent hydrogenation can also be employed in the *cis*-1,4-dihydroxylation of 4a,9b-dihydrodibenzo[*b*,*d*]furans.[24]

Enone 37:[23]

CAUTION: *Please read the safety information in the General Introduction (Section 1.10).*

To a soln of the endoperoxide **36** (1.02 g, 2.2 mmol) in a mixture of H_2O/iPrOH/HCO_2H (1:1:1; 5 mL) were added thiourea (0.04 g, 0.52 mmol) and 10% Pd/C (0.05 g). The mixture was flushed with H_2 (3 ×) and then stirred under 1 atm of H_2 for 18 h. The suspension was filtered through a short plug of Celite, washing with MeOH. The filtrate was concentrated under reduced pressure and the residue was purified by flash column chromatography (silica gel, CH_2Cl_2/MeOH 5:1).

(*R*)-Methylnaltrexone (38):[23]
To a soln of the enone **37** (0.20 g, 0.46 mmol) in MeOH (3 mL) was added 10% Pd/C (0.02 g; 10 wt%). The mixture was hydrogenated in a Parr shaker at 40 psi for 12 h. The suspension was filtered through a short plug of Celite and the plug was washed with MeOH. The filtrate was concentrated, furnishing the essentially pure product (checked by HPLC and ^{1}H NMR spectroscopy); yield: 0.19 g (74% over 2 steps from **36**).

1.10.1.3.4.3 2,7-Dioxabicyclo[2.2.1]heptanes by Hydrogenation of 2,3,5-Trioxabicyclo[2.2.2]oct-7-enes

This sequence constitutes a concise strategy to construct the 2,7-dioxabicyclo[2.2.1]heptane moiety (an epoxy-bridged tetrahydropyran frequently occurring in natural products) from readily available 1,2,4-trioxanes. In this case, the hydrogenation over platinum (starting from PtO_2) at a hydrogen pressure of 1 bar contracts the bicyclic peroxy acetal **39** to a bicyclic acetal and saturates the cyclic double bond to form **40** (Scheme 16).[25] In a later study, the same authors used different hydrogenation catalysts. Some of them, such as Lindlar's catalyst, do not hydrogenate the ring double bond, but, as well as peroxide cleavage, catalyze rearrangement to form the furan derivative **41**.[26]

Scheme 16 Synthesis of a 2,7-Dioxabicyclo[2.2.1]heptane from a 1,2,4-Trioxane[25,26]

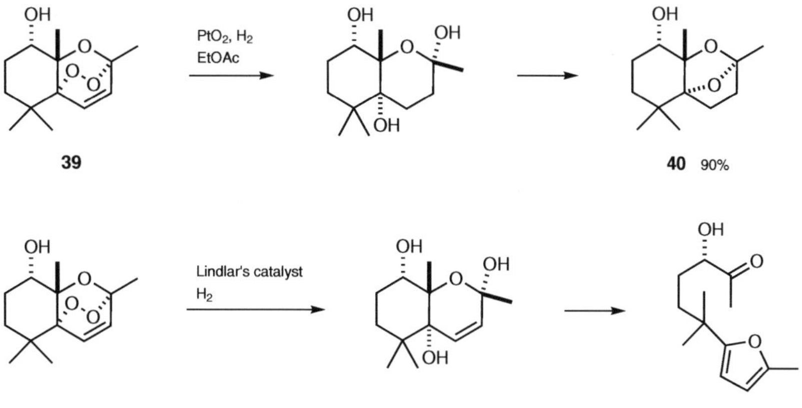

Similarly, peroxy-bridged indolizinones (e.g., **42**) are hydrogenated to yield hydroxy derivatives (e.g., **43**), as shown in Scheme 17; the intermediate aminal supposedly forms an imine and is hydrogenated to yield the saturated lactam structure.[27]

Scheme 17 Synthesis of a Hydroxyindolizinone from a Peroxy-Bridged Indolizinone[27]

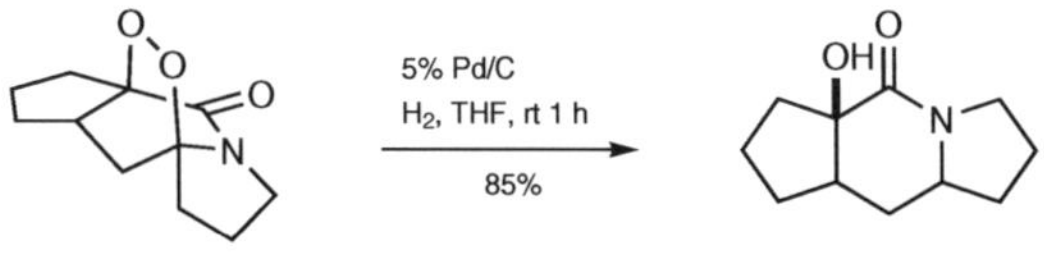

5a-Hydroxydecahydro-5*H*-cyclopenta[*f*]indolizin-5-one (43); Typical Procedure:[27]

CAUTION: *Please read the safety information in the General Introduction (Section 1.10).*

Into a 50-mL round-bottomed flask were placed 5% Pd/C (5 mg) and hexahydro-1*H*,5*H*,6*H*-5a,9a-epidioxycyclopenta[*f*]indolizin-5-one (**42**; 29 mg, 0.14 mmol) in THF (10 mL). The mixture was first flushed with argon and then it was hydrogenated for 1 h at rt and 1 atm of H_2. The catalyst was removed by filtration by pouring the mixture onto a silica gel plug, which was washed and eluted with CH_2Cl_2 (100 mL). The residue was purified by flash column chromatography (silica gel, EtOAc/hexanes 15:1) to give a white solid; yield: 23 mg (85%); mp 145–147 °C.

1.10.2 Reduction of Molozonides, Ozonides, and Hydroxy and Alkoxy Hydroperoxides

Alkenes react with ozone to form molozonides ("primary ozonides", 1,2,3-trioxolanes) as primary reaction products and ozonides ("secondary ozonides", 1,2,4-trioxolanes), their rearrangement products. If the reaction is performed in the presence of water or methanol, 1,2,4-trioxolanes are not obtained, but hydroxy or alkoxy hydroperoxides are formed instead, as water or methanol react with the intermediate carbonyl oxide. These compounds are unstable and are, with few exceptions, not synthetic targets themselves. However, they are versatile intermediates to produce carboxylic acids (by further oxidation), carbonyl compounds (by rearrangement or reduction), or alcohols (by reduction). This chemistry, including reductions and a few hydrogenations, has been described in *Houben–Weyl*, Vol. IV/1 c, p 484 and *Science of Synthesis*, Vol. 25 [Aldehydes (Section 25.1.1)] and Vol. 36 [Alcohols (Sections 36.1.2.1.9.2 and 36.1.4.4)]. This chapter focuses on catalytic hydrogenation of molozonides, ozonides, and hydroxy and alkoxy hydroperoxides as products of the ozonolysis of alkenes, as depicted in Scheme 18.

Scheme 18 Ozonolysis in Methanol and Hydrogenation of Symmetrical 1,2-Disubstituted Alkenes

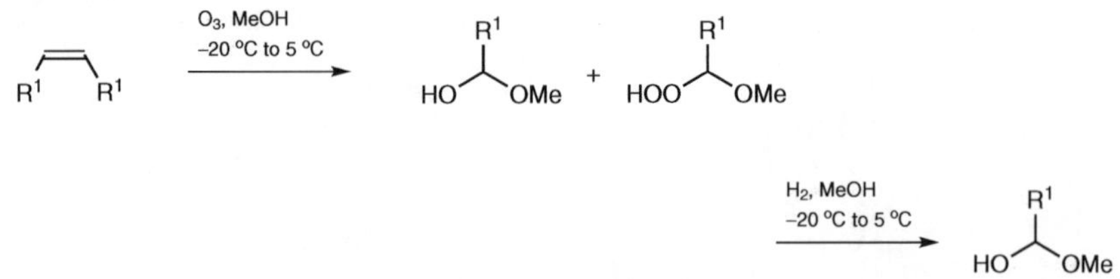

O_3, MeOH; −20 °C to 5 °C

R^1 R^1 → HO(R^1)OMe + HOO(R^1)OMe

H_2, MeOH; −20 °C to 5 °C → HO(R^1)OMe

Whereas reductive workup of the ozonolysis solution on a laboratory scale using sulfide- or phosphine-based reagents is quite common, surprisingly few reports on catalytic hydrogenation of ozonolysis products have entered the literature after initial successful applications as described in *Houben–Weyl* in 1954 (Vol. VII/1, p 339). In contrast, large-scale applications of the reaction sequence "ozonolysis plus hydrogenation" have been developed to produce aldehydes and alcohols from the respective alkenes on an industrial scale. The reason for this discrepancy appears to be the easy handling on a laboratory scale of the above reducing agents that are unacceptable as stoichiometric reagents on a large scale. The catalytic hydrogenation of ozonolysis products, while state-of-the-art on large scale, requires attention on laboratory scale: noble metal catalysts on carriers are used and, to avoid their deactivation by the peroxide solution, this solution is slowly added to the reaction system containing the catalyst and hydrogen. Furthermore, conditions must be carefully chosen to avoid peroxide decomposition or rearrangement at the noble metal catalyst surface prior to hydrogenation. To reach high atom efficiency, either symmetrical 1,2-disubstituted or monosubstituted alkenes (vinyl derivatives) are chosen. An example of this reaction sequence is the synthesis of glyoxylic acid methyl ester hemiacetal (**45**) from dimethyl maleate (**44**) by ozonolysis and hydrogenation in methanol (Scheme 19).[28]

Scheme 19 Ozonolysis of Dimethyl Maleate in Methanol and Subsequent Hydrogenation[28]

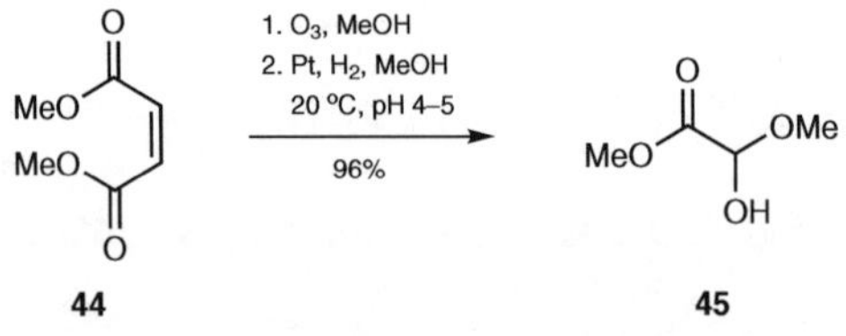

44 → 1. O_3, MeOH; 2. Pt, H_2, MeOH; 20 °C, pH 4–5; 96% → **45**

Much work has been dedicated to the ozonolysis of naphthalene, with the goal of producing *ortho*-phthalaldehyde by a sequence of ozonolysis and hydrogenation. The ozonolysis of one of the two aromatic rings in participating solvents delivers cyclic peroxides, as first recognized by Bailey et al.,[29] but they are not easily reduced by common reducing agents such as iodide.[30] These peroxides crystallize easily and thus pose a significant explosion hazard once they are formed and precipitate. Laboratory procedures to reduce them with dimethyl sulfide[31] or phosphines[32] have been developed. A method for the catalytic hydrogenation of the cyclic peroxides resulting from ozonolysis has been developed and used on an industrial scale for many substrates (Table 1).[33]

Table 1 Reduction of Ozonolysis Products of Alkenes in Methanol[33]

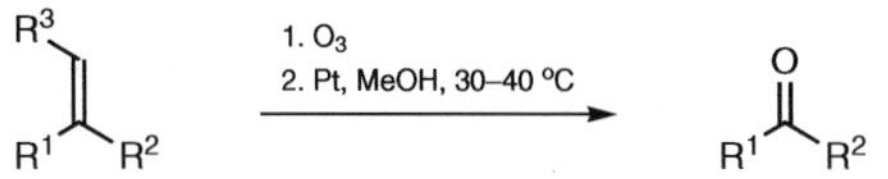

R^3, R^1, R^2 — 1. O_3; 2. Pt, MeOH, 30–40 °C → O, R^1, R^2

Substrate	Ratio (g Catalyst/g Substrate)	Catalyst (mol%)	Product	Yield (%)	Ref
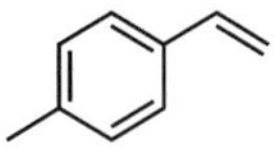	1:177	0.34	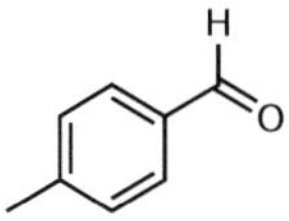 H, O	94	[33]
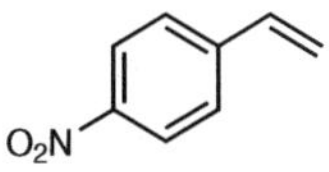 O_2N	1:224	0.34	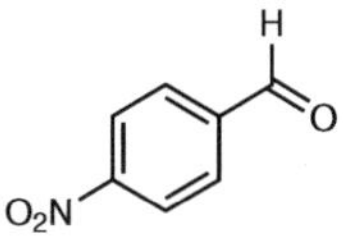 O_2N, H, O	95	[33]
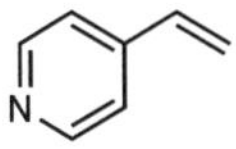 N	1:158	0.34	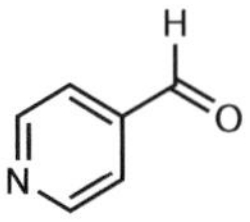 N, H, O	91	[33]
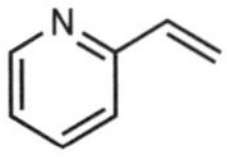 N	1:158	0.34	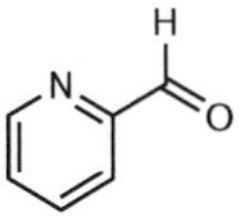 N, H, O	89	[33]
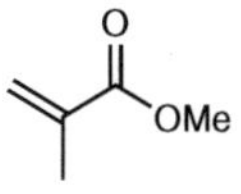 O, OMe	1:150	0.34	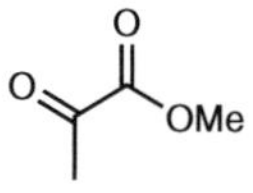 O, O, OMe	91[a]	[33]
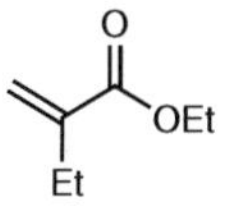 O, OEt, Et	1:192	0.34	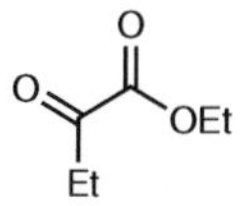 O, O, OEt, Et	88[b]	[33]
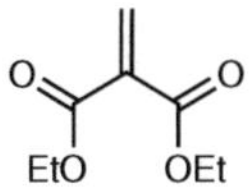 O, O, EtO, OEt	5:258	0.18	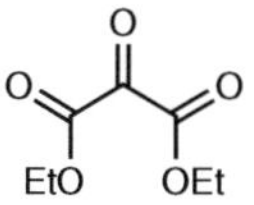 O, O, O, EtO, OEt	82[c]	[33]
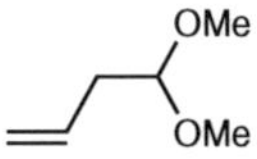 OMe, OMe	1:348	0.18	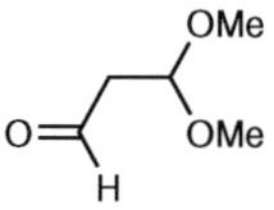 OMe, O, OMe, H	89[d]	[33]
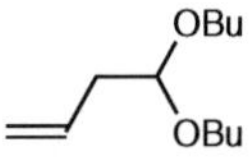 OBu, OBu	1:600	0.18	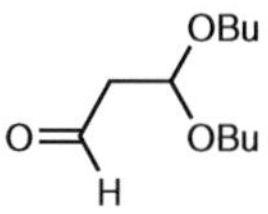 OBu, O, OBu, H	94	[33]

for references see p 337

Table 1 (cont.)

Substrate	Ratio (g Catalyst/g Substrate)	Catalyst (mol%)	Product	Yield (%)	Ref
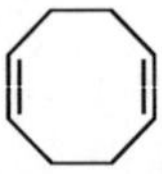	1:81	0.68	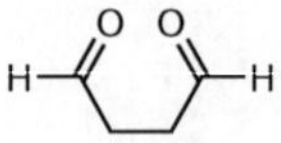	96[e]	[33]
	1:121	0.22	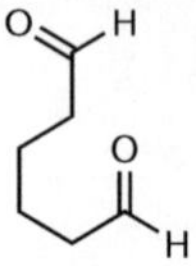	89	[33]
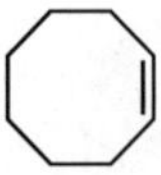	1:165	0.34	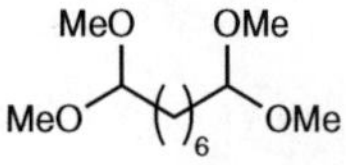	92	[33]
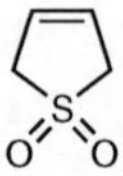	1:177	0.34	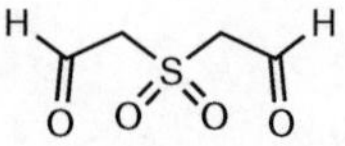	95[f]	[33]
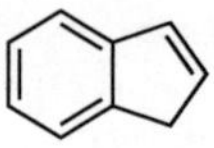	1:87	0.68	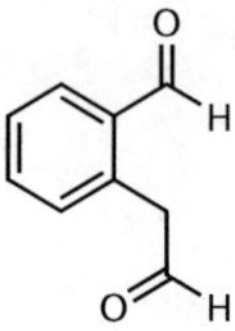	94[f]	[33]
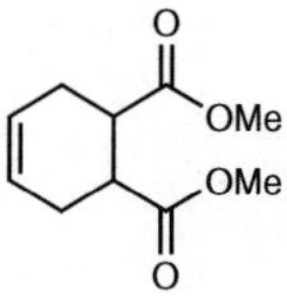	1:297	0.34	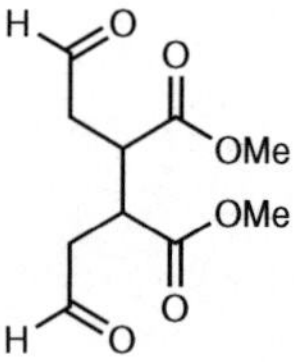	94[g]	[33]
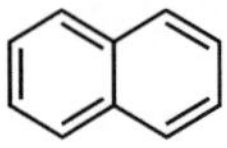	1:96	0.68	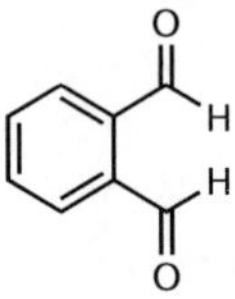	87[h]	[33]
	1:105	0.34	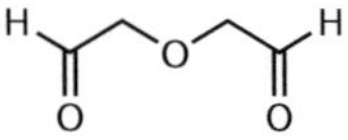	95	[33]

[a] The catalyst was recycled 5 times.
[b] EtOH was used as solvent.
[c] 10% Pd/C was used as catalyst.
[d] The catalyst was recycled twice.
[e] The catalyst was recycled 10 times.
[f] Determined by oxime titration.
[g] Isolated as the corresponding dicarboxylic acid.
[h] Substrate partially dissolved in MeOH.

Ozonolysis of oleic acid (**46**, $R^1 = H$) has been used to produce azelaic acid (**48**, nonanedioic acid) and pelargonic acid (**47**, nonanoic acid) for decades (Scheme 20). Ozonolysis of methyl oleate (**46**, $R^1 = Me$) followed by catalytic hydrogenation over 10% palladium on carbon in participating and nonparticipating solvents is used to synthesize pelargonic aldehyde (**49**, nonanal) and azelaic ester aldehyde **50** as well as 9-hydroxynonanoic acid (**52**) (Scheme 20).[36]

Scheme 20 Synthetic Approaches toward Pelargonic and Azelaic Acid and Aldehydes via Ozonolysis of Oleic Acid Derivatives and Subsequent Oxidative or Reductive Workup of the Ozonides[34–36]

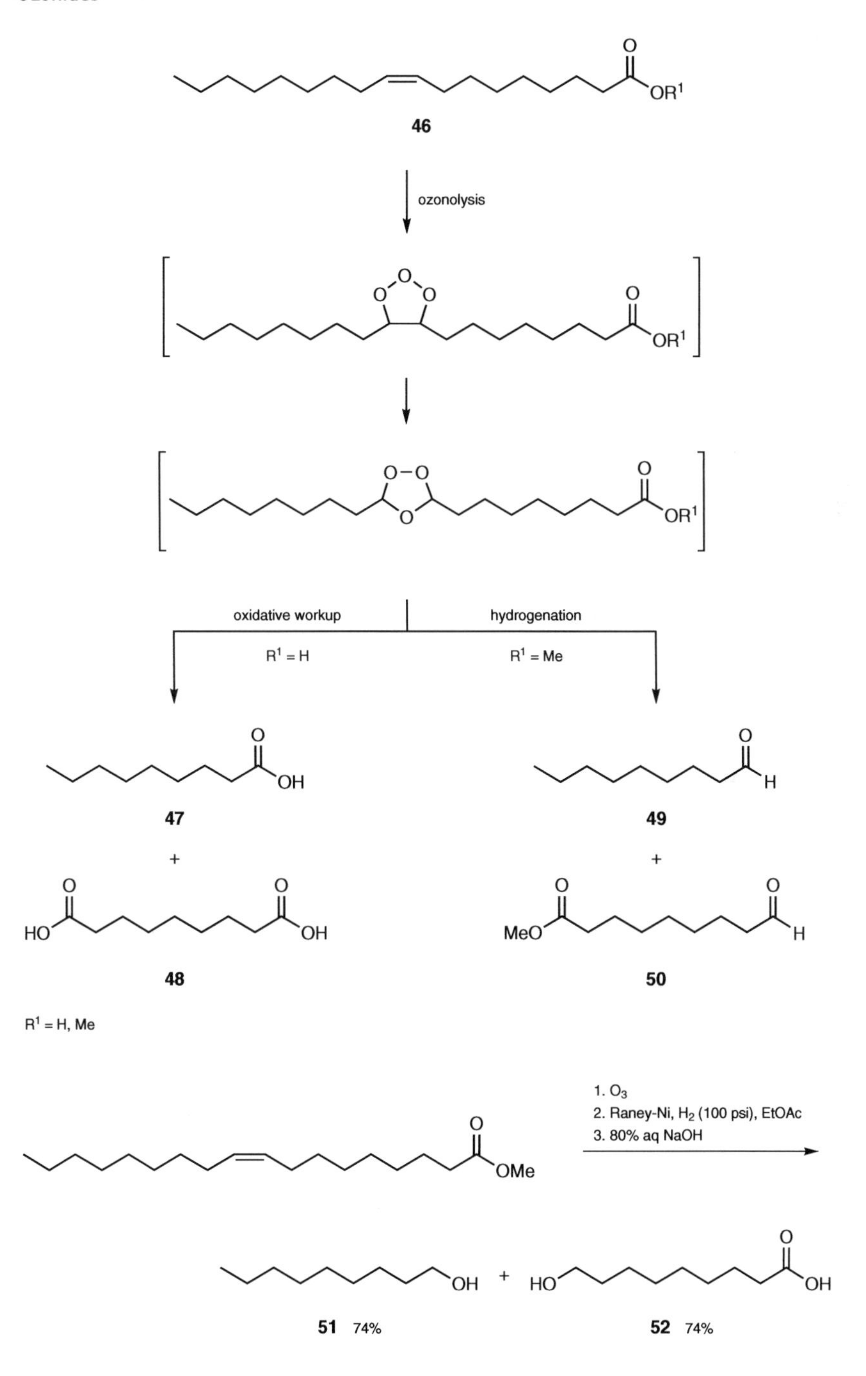

for references see p 337

With the increased interest in producing base chemicals from renewable sources, such as vegetable oils containing unsaturated fatty acids, researchers are turning to modified reductive workup procedures of the respective ozonolysis solutions to produce aldehydes,[35] hydroxy acids, and lactones.[34]

Methyl 2-Hydroxy-2-methoxyacetate (45):[28]

CAUTION: *Please read the safety information in the General Introduction (Section 1.10).*

CAUTION: *Ozone irritates mucous membranes and the lungs and is highly explosive in the liquid or solid states.*

A soln of dimethyl maleate (**44**; 21.6 g, 150 mmol) in MeOH (100 mL) was placed in the reactor and the ozonization was carried out with an O_2/O_3 mixture (60 L/h of O_2 containing 27.8 mg/L of O_3) for 260 min, with cooling to 0–3 °C. O_3 was taken up quantitatively, and a stoichiometric amount of dimethyl maleate (**44**) was consumed. Ozonolysis was continued until the residual content of dimethyl maleate (**44**) was less than 1% of the amount originally present. The obtained soln was placed, with stirring and while passing in H_2, into a hydrogenation reactor, into which a suspension of platinum (0.1 g) in MeOH had been introduced, via a metering vessel in doses such that the peroxide content in the hydrogenation reactor at the start and in the course of the hydrogenation did not exceed 0.02 M. The mixture was kept at 20 °C by external cooling, and a pH value of 4 to 5 was established by addition of a soln of NaOH in MeOH. When the addition of the ozonolysis soln had ended, the mixture was peroxide-free within 5 min. The catalyst was then removed by filtration and used for further hydrogenation reactions. The yield was determined by oxime titration and polarography; yield: 96%.

Nonanol (51) **and 9-Hydroxynonanoic Acid** (52):[36]

CAUTION: *Please read the safety information in the General Introduction (Section 1.10).*

CAUTION: *Ozone irritates mucous membranes and the lungs and is highly explosive in the liquid or solid states.*

Methyl oleate (**46**, R^1 = Me; 20.0 g, 67.4 mmol) and anhyd EtOH (100 mL) were added to a three-necked flask. The flask was fitted with a magnetic stirrer, inlet for O_3, and outlet for gas. O_3 was produced by an ozone generator (Azcozon model RMV16–16 from Azco Industries Ltd., Canada) using cylinder oxygen as a feeding gas. The reaction was performed at –4 °C (ice–salt bath) at a flow rate of 5 L/min of O_2 with an agitation rate of 100 rpm for 27 min. The concentration of O_3 was 62.0 g/m^3. After 27 min, the ozone generator was stopped and the vessel was purged with N_2 for 3 min to remove any residue of O_3. EtOAc (800 mL) was added and the ozonide product was used for hydrogenation.

A slurry of Raney Ni catalyst (5.0 g) in H_2O was added to the ozonolysis product in a hydrogenation vessel (2 L, Parr Instrument Co.) fitted with a magnetic drive. The vessel was charged with H_2 at 100 psi and heated to 70 °C. After 3 h, H_2 flow and heating were stopped and the vessel was allowed to cool to rt and finally purged with N_2 to remove any residue of H_2. The resulting mixture was filtered through a Büchner funnel. The filtrate was then transferred to a flask, and the solvent was removed by rotary evaporation, yielding an oily residue; yield: 20.2 g.

The oily residue (20.2 g) was saponified using 8.0% aq NaOH (100 mL) for 3 h. Afterward, the soln was cooled to rt. The resulting mixture was then washed with Et_2O (3 × 50 mL). The combined Et_2O layers were concentrated to give nonan-1-ol (**51**); yield: 7.2 g (74%). The aqueous layer was cooled to 0 °C, acidified with concd HCl (10.5 mL), and extracted with Et_2O (3 × 50 mL). The organic layers were combined and washed with brine until neutral. The soln was finally dried (Na_2SO_4), filtered, and concentrated with a rotary evaporator. The crude 9-hydroxynonanoic acid (**52**) was obtained as a white solid (10.4 g) and further purified by recrystallization (Et_2O) to give pure 9-hydroxynonanoic acid (**52**); yield: 8.7 g (74%).

References

[1] Korach, M.; Nielsen, D. R.; Rideout, W. H., *Org. Synth., Coll. Vol. V*, (1973), 414.
[2] Semikolenov, V. A.; Ilyna, I. I.; Simakova, I. L., *Appl. Catal., A*, (2001) **211**, 91.
[3] Ilyna, I. I.; Simakova, I. L.; Semikolenov, V. A., *Kinet. Catal.*, (2002) **43**, 652.
[4] Ma, Y.; Zhu, Q.-C., *Chem. Eng. Technol.*, (2012) **35**, 1849.
[5] Zhu, Q.-c.; Shen, B.-x.; Ling, H.; Gu, R., *J. Hazard. Mater.*, (2010) **175**, 646.
[6] Hudlický, M., *Reductions in Organic Chemistry*, Wiley: New York, (1984); p 83.
[7] Dussault, P. H.; Kreifels, S.; Lee, I. Q., *Synth. Commun.*, (1995) **25**, 2613.
[8] An, G.; Zhou, W.; Zhang, G.; Sun, H.; Han, J.; Pan, Y., *Org. Lett.*, (2010) **12**, 4482.
[9] Kischkewitz, M.; Daniliuc, C.-G.; Studer, A., *Org. Lett.*, (2016) **18**, 1206.
[10] Igawa, K.; Kawasaki, Y.; Tomooka, K., *Chem. Lett.*, (2011) **40**, 233.
[11] Lu, X.; Liu, Y.; Sun, B.; Cindric, B.; Deng, L., *J. Am. Chem. Soc.*, (2008) **130**, 8134.
[12] Schweitzer-Chaput, B.; Demaerel, J.; Engler, H.; Klussmann, M., *Angew. Chem. Int. Ed.*, (2014) **53**, 8737.
[13] Jain, A.; Rodríguez, S.; López, I.; González, F. V., *Tetrahedron*, (2009) **65**, 8362.
[14] Krabbe, S. W.; Do, D. T.; Johnson, J. S., *Org. Lett.*, (2012) **14**, 5932.
[15] Scott, A. I.; Bedford, C. T., *J. Am. Chem. Soc.*, (1962) **84**, 2271.
[16] Wasserman, H. H.; Lu, T.-J.; Scott, A. I., *J. Am. Chem. Soc.*, (1986) **108**, 4237.
[17] Feng, Z.-G.; Bai, W.-J.; Pettus, T. R. R., *Angew. Chem. Int. Ed.*, (2015) **54**, 1864.
[18] Wenderski, T. A.; Hoarau, C.; Mejorado, L.; Pettus, T. R. R., *Tetrahedron*, (2010) **66**, 5873.
[19] Zvarec, O.; Avery, T. D.; Taylor, D. K.; Tiekink, E. R. T., *Tetrahedron*, (2010) **66**, 1007.
[20] Jung, M.; Ham, J.; Song, J., *Org. Lett.*, (2002) **4**, 2763.
[21] Robinson, T. V.; Pedersen, D. S.; Taylor, D. K.; Tiekink, E. R. T., *J. Org. Chem.*, (2009) **74**, 5093.
[22] Valente, P.; Avery, T. D.; Taylor, D. K.; Tiekink, E. R. T., *J. Org. Chem.*, (2009) **74**, 274.
[23] Machara, A.; Werner, L.; Leisch, H.; Carroll, R. J.; Adams, D. R.; Haque, D. M.; Cox, D. P.; Hudlicky, T., *Synlett*, (2015) **26**, 2101.
[24] Wang, Z.-Y.; Wong, W.-T.; Yang, D., *Org. Lett.*, (2013) **15**, 4980.
[25] Riveira, M. J.; La-Venia, A.; Mischne, M. P., *J. Org. Chem.*, (2008) **73**, 8678.
[26] Riveira, M. J.; La-Venia, A.; Mischne, M. P., *Tetrahedron Lett.*, (2010) **51**, 804.
[27] Rostami, A.; Wang, Y.; Arif, A. M.; McDonald, R.; West, F. G., *Org. Lett.*, (2007) **9**, 703.
[28] Sajtos, A., US 5 015 760, (1991); *Chem. Abstr.*, (1984) **100**, 156 238.
[29] Bailey, P. S.; Garcia-Sharp, F. J., *J. Org. Chem.*, (1957) **22**, 1008.
[30] Bailey, P. S.; Bath, S. S.; Dobinson, F.; Garcia-Sharp, F. J.; Johnson, C. D., *J. Org. Chem.*, (1964) **29**, 697.
[31] Pappas, J. J.; Keaveney, W. P.; Gancher, E.; Berger, M., *Tetrahedron Lett.*, (1966), 4273.
[32] Pappas, J. J.; Keaveney, W. P.; Berger, M.; Rush, R. V., *J. Org. Chem.*, (1968) **33**, 787.
[33] Sajtos, A.; Wechsberg, M.; Roithner, E.; Pollhammer, S.; Mahringer, A., EP 147 593, (1985); *Chem. Abstr.*, (1986) **104**, 68 447.
[34] Liu, G.; Kong, X.; Wan, H.; Narine, S., *Biomacromolecules*, (2008) **9**, 949.
[35] Omonov, T. S.; Kharraz, E.; Foley, P.; Curtis, J. M., *RSC Adv.*, (2014) **4**, 53 617.
[36] Pryde, E. H.; Anders, D. E.; Teeter, H. M.; Cowan, J. C., *J. Org. Chem.*, (1960) **25**, 618.

1.11 Reduction of Sulfur Compounds Using Metal Catalysts

K. Kaneda and T. Mitsudome

General Introduction

The reductions of sulfur compounds such as sulfoxides, sulfones, and disulfides represent fundamental and important transformations in organic synthesis[1–6] and biochemistry.[7–10] For example, in the asymmetric synthesis of carbinols, chiral sulfoxides are introduced into carbonyl compounds as chiral auxiliaries and, following the asymmetric transformation of the carbonyl group into an alcohol, the sulfoxide moiety is removed from the parent molecule by deoxygenation followed by desulfuration.[11]

Stoichiometric reduction of sulfur compounds has been carried out using excess amounts of reducing reagents such as metal hydrides,[12–16] 1,3-dithiane in the presence of electrophilic halogens as catalyst,[17] hydrogen halides,[18–21] thiols,[17,22,23] and phosphines.[24–31] However, these reactions have serious drawbacks such as the use of highly toxic reagents, the production of large amounts of waste, and low yields of products. Thus, much effort has been devoted to the replacement of these stoichiometric reactions with catalytic ones, and many catalytic systems employing molybdenum, rhenium, copper, or gold combined with phosphorus compounds,[32–35] hydrosilanes,[36–44] or borane[45–47] have been proposed. These catalytic systems, however, still suffer from low atom efficiency.

Based on the current demand for atom economical and environmentally benign chemical reactions, much attention has been directed toward the development of promising catalyst systems that employ molecular hydrogen or an alcohol as a clean reductant. In these systems, the use of harmful reductants can be avoided and water or carbonyl compounds are produced as the byproducts. Several reviews and many reports on the reduction of sulfur compounds using the above-mentioned silanes, boranes, and phosphorus reagents as reductants have already appeared.[11,48] Therefore, the focus in this chapter is on the recently advanced and highly efficient heterogeneous and homogeneous catalysts (not including biocatalysts or organocatalysts) for the deoxygenation of sulfoxides and sulfones using alcohols (Section 1.11.1) or molecular hydrogen (Section 1.11.2) under liquid-phase conditions, with a focus on developments in the past ten years. Furthermore, a rare example of the catalytic hydrogenation/hydrogenolysis of disulfides is described (Section 1.11.3).

1.11.1 Deoxygenation of Sulfoxides to Thioethers Using Alcohols

The Sanz group reported for the first time that dioxomolybdenum(VI) complexes catalyze the deoxygenation of sulfoxides to thioethers using pinacol as a green reducing agent (Scheme 1).[49] For example, di-4-tolyl sulfoxide (**1**, $R^1 = R^2 =$ 4-Tol) is deoxygenated to di-4-tolyl sulfide (**2**, $R^1 = R^2 =$ 4-Tol) in 85% yield using the dimethylformamide adduct of dichlorodioxomolybdenum(VI) [$MoCl_2O_2(DMF)_2$] (2 mol%) in the presence of 2 equivalents of pinacol as an oxygen acceptor at 90 °C in acetonitrile for 4 hours. The dioxomolybdenum catalyst can efficiently promote the deoxygenation under solvent-free conditions, and the process is scalable to the multigram synthesis of 9.4 g of di-4-tolyl sulfide (90% isolated yield) from 11.5 g (50 mmol) of di-4-tolyl sulfoxide. A remarkable advantage of this catalytic system is its high chemoselectivity for the reduction of sulfoxides bearing other poten-

for references see p 353

tially reducible functional groups. Various sulfoxides are chemoselectively reduced to the corresponding thioethers **2** without affecting functional groups such as alkenes, alkynes, carbonyls, nitriles, halides, and hydroxy groups.

Scheme 1 Molybdenum-Catalyzed Deoxygenation of Sulfoxides Using Pinacol as a Reductant[49]

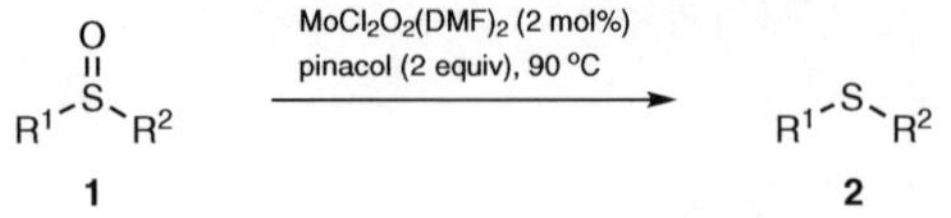

$MoCl_2O_2(DMF)_2$ (2 mol%)
pinacol (2 equiv), 90 °C

1 → **2**

R^1	R^2	Yield (%)	Ref
4-Tol	4-Tol	91	[49]
Ph	Ph	89	[49]
Ph	Me	92	[49]
Ph	Bn	91	[49]
Ph	*s*-Bu	90	[49]
Ph	*t*-Bu	89	[49]
Bu	Bu	79	[49]
$4-ClC_6H_4$	$4-ClC_6H_4$	99[a]	[49]
$4-BrC_6H_4$	Me	91	[49]
Ph	$(CH_2)_3CH{=}CH_2$	82	[49]
Ph	$(CH_2)_3C{\equiv}CH$	91	[49]
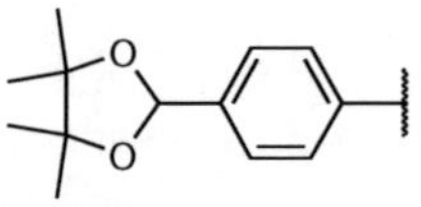	Me	89	[49]
$4-OHCC_6H_4$	Me	88[b]	[49]
Ph	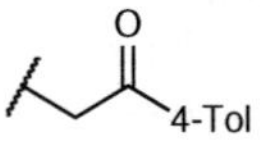 4-Tol	78	[49]
$2-HO_2CC_6H_4$	Me	90[a]	[49]
$2-HO_2CC_6H_4$	Me	85[c]	[49]
Ph	CH_2CO_2Me	79	[49]
Ph	CH_2CN	83	[49]
$4-O_2NC_6H_4$	Me	80[c,d]	[49]
$4-O_2NC_6H_4$	Me	86[c,e]	[49]

[a] Reaction performed using 4 equivalents of pinacol.
[b] HCl and THF were added to the crude mixture, which was then stirred under reflux.
[c] Reaction conducted in MeCN.
[d] Partial reduction (ca. 7%) to 4-aminophenyl methyl sulfide was observed.
[e] Reaction conducted with 1 equivalent of pinacol.

The proposed catalytic cycle is shown in Scheme 2. Thus, $MoCl_2O_2(DMF)_2$ reacts with pinacol, leading to water and the molybdenum(VI) pinacolato complex **5**. Subsequently, molybdenum(VI) is reduced by the pinacolate ligand to the oxomolybdenum(IV) species **6**, bearing a weakly coordinated acetone that is easily displaced by the sulfoxide to give the unstable oxomolybdenum(IV) adduct **3**. Finally, the sulfide is released from the catalyst, completing the catalytic cycle. It was impossible to isolate complex **6**. Instead, a dinuclear oxomolybdenum(V) complex, $Mo_2O_3Cl_4(DMF)_4$ (**4**), was isolated. This complex is the result of a comproportionation reaction between the molybdenum(IV) and molybdenum(VI) complex.

Scheme 2 Mechanism of Molybdenum-Catalyzed Deoxygenation of Sulfoxides Using Pinacol[49]

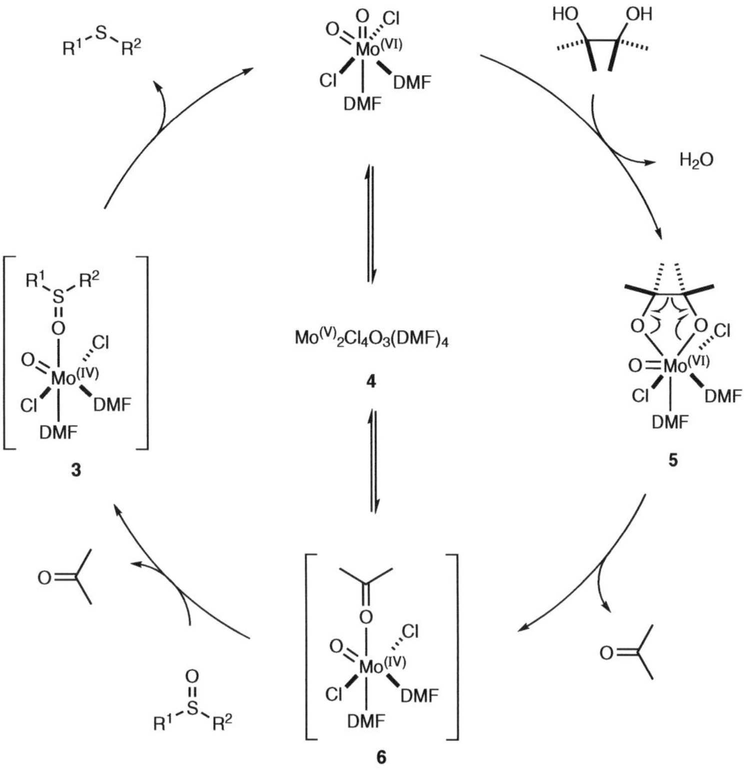

The Sanz group has also reported that the biomass-derived chemical feedstock glycerol can be used in the molybdenum-catalyzed deoxygenation of sulfoxides.[50] Large amounts of glycerol are formed as a byproduct during the production of biodiesel by the transesterification of vegetable oils and animal fats. The rapid increase in demand for biodiesel has caused an oversupply of glycerol. Therefore, the development of catalytic processes for the utilization of abundant glycerol has attracted much attention from both academia and industry as a promising solution to this problem. Although a higher temperature (170 °C) is required compared to when using the molybdenum/pinacol catalyst system, various sulfoxides can be reduced to thioethers in high yield. Furthermore, the deoxygenation can be performed under air, without any solvent.

From an environmental and practical synthetic point of view, the development of highly efficient heterogeneous catalysts for the deoxygenation of sulfoxides is highly desired because of the advantage of easy workup procedures, such as separation of the catalysts from the reaction mixtures and reuse of the catalysts. In this context, the Kaneda group has developed a highly efficient heterogeneous catalyst system for the deoxygenation of sulfoxides that uses alcohols as reducing reagents (Scheme 3). Hydroxyapatite (HAP) supported ruthenium nanoparticles (RuNPs/HAP) act as a reusable catalyst for the deoxygenation of sulfoxides.[51] This is the first report of the catalytic deoxygenation of

for references see p 353

sulfoxides using heterogeneous catalysts with alcohols as reagents. Thus, the ruthenium nanoparticles show high catalytic activity in the deoxygenation of diphenyl sulfoxide (**7**) as a model substrate using propan-2-ol as a reducing reagent in toluene at 110 °C under argon, affording the corresponding diphenyl sulfide (**8**) in >99% yield within 1 hour. In sharp contrast to ruthenium, other metal nanoparticles supported on hydroxyapatite, such as palladium, platinum, rhodium, and gold, do not exhibit any catalytic activity, demonstrating the uniqueness of the ruthenium nanoparticles in the deoxygenation. Further optimization with the hydroxyapatite-supported ruthenium nanoparticles reveals that toluene is the best solvent and polar solvents are not suitable, resulting in low yields of diphenyl sulfide. The use of other alcohols such as ethanol and benzyl alcohol instead of propan-2-ol is also applicable. Hydroxyapatite-supported ruthenium nanoparticles promote the deoxygenation even at lower temperature (40 °C), although a longer reaction time is required.

Scheme 3 Deoxygenation of Diphenyl Sulfoxide Using Various Hydroxyapatite-Supported Metal Nanoparticle Catalysts[51]

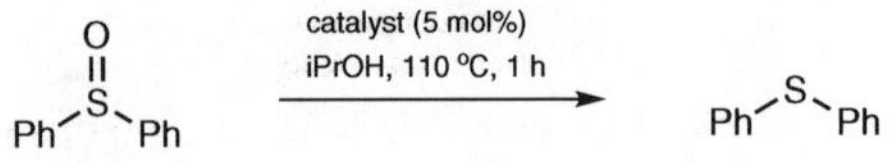

Ph–S(=O)–Ph (**7**) → catalyst (5 mol%), iPrOH, 110 °C, 1 h → Ph–S–Ph (**8**)

Catalyst	Solvent	Conversion[a] (%)	Yield[a] (%)	Ref
RuNPs/HAP	toluene	>99	>99	[51]
PdNPs/HAP	toluene	<1	<1	[51]
PtNPs/HAP	toluene	<1	<1	[51]
RhNPs/HAP	toluene	<1	<1	[51]
AuNPs/HAP	toluene	<1	<1	[51]
RuNPs/HAP	(trifluoromethyl)benzene	84	84	[51]
RuNPs/HAP	heptane	83	83	[51]
RuNPs/HAP	1,4-dioxane	21	21	[51]
RuNPs/HAP	EtOAc	12	12	[51]
RuNPs/HAP	toluene	>99[b]	>99	[51]
RuNPs/HAP	toluene	>99[c]	>99	[51]
RuNPs/HAP	toluene	>99[d]	>99	[51]

[a] Determined by GC using an internal standard.
[b] EtOH was used instead of iPrOH.
[c] BnOH (0.5 mmol) was used instead of iPrOH.
[d] Conditions: catalyst (0.1 g), 40 °C, 12 h.

The hydroxyapatite-supported ruthenium nanoparticles catalyze the selective deoxygenation of a diverse range of sulfoxides **9** to give the corresponding sulfides **10** (Scheme 4). Various aromatic and aliphatic sulfoxides are efficiently converted into the corresponding sulfides in excellent yields. Sulfoxides with reducible functional groups are chemoselectively deoxygenated, whereas halogen and carbonyl groups remain intact during the deoxygenation. One of the advantages of this heterogeneous system is its reusability; after the reaction, the catalyst is easily recoverable by simple filtration and can be reused without loss of activity, but a longer reaction time is required. Transmission electron microscopy (TEM) analysis of the used hydroxyapatite-supported ruthenium nanoparticle catalyst reveals that the size of the nanoparticles does not significantly change compared to the fresh catalyst. These observations are consistent with the high durability of the catalyst. Furthermore, this catalyst system is applicable under scale-up conditions. For example, in a 10-mmol-scale reaction of diphenyl sulfoxide, diphenyl sulfide is obtained in 90% isolated yield by simple removal of the catalyst by filtration and evaporation of the mesitylene solvent.[51]

Scheme 4 Deoxygenation of Sulfoxides Using Propan-2-ol Catalyzed by Hydroxyapatite-Supported Ruthenium Nanoparticles[51]

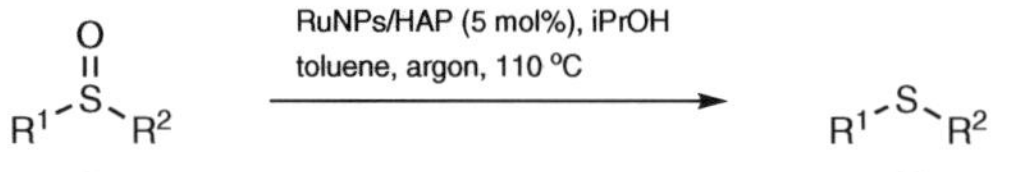

RuNPs/HAP (5 mol%), iPrOH
toluene, argon, 110 °C

$R^1S(O)R^2$ **9** → R^1SR^2 **10**

R^1	R^2	Time (h)	Conversion[a]	Yield[a] (%)	Ref
Ph	Ph	1	>99	>99	[51]
Ph	Ph	3	>99	>99[b]	[51]
Ph	Ph	3	>99	>99[c]	[51]
Ph	Ph	3	>99	>99[d]	[51]
Ph	Ph	20	>99	90[e]	[51]
4-ClC_6H_4	4-ClC_6H_4	1	>99	>99	[51]
4-BrC_6H_4	Me	2	>99	>99	[51]
4-AcC_6H_4	Me	2	>99	>99	[51]
Pr	Pr	1	>99	>99	[51]
Me	$(CH_2)_7Me$	1	>99	>99	[51]

[a] Determined by GC using an internal standard.
[b] 1st reuse.
[c] 2nd reuse.
[d] 3rd reuse.
[e] Isolated yield; reaction performed on 10-mmol scale in mesitylene at 180 °C.

Sulfides 2; General Procedure:[49]
A mixture of a sulfoxide **1** (2 mmol), pinacol (472 mg, 4 mmol, 2 equiv), and the catalyst $MoCl_2O_2(DMF)_2$ (14 mg, 0.04 mmol) was stirred at 90 °C until the sulfoxide was consumed (as determined by GC/MS or TLC). The mixture was cooled to rt, treated with 0.5 M aq NaOH (25 mL), and extracted with Et_2O (3 × 20 mL). The combined organic layers were washed with brine, dried (Na_2SO_4), filtered, and concentrated under reduced pressure. The corresponding sulfide was obtained in pure form without further purification.

Diphenyl Sulfide (10, $R^1 = R^2 = Ph$); Typical Procedure Using RuNPs/HAP:[51]
Preparation of the hydroxyapatite-supported ruthenium nanoparticles: Hydroxyapatite (1.0 g) was stirred in 5 mM aq $RuCl_3$ (100 mL) at rt for 12 h. The resulting dark-brown solid was collected by filtration, washed with deionized H_2O, and dried at 110 °C. The obtained solid was treated with H_2 (1 atm) at 110 °C for 1 h, affording RuNPs/HAP. The Ru loading on RuNPs/HAP was estimated to be 4.76 wt% by elemental analysis. The Ru K-edge X-ray absorption near-edge structure (XANES) spectrum of RuNPs/HAP showed the formation of metallic Ru^0 species. Transmission electron microscopy (TEM) confirmed that highly dispersed Ru nanoparticles with a mean diameter of 3.0 nm were formed on the surface of hydroxyapatite.

Catalytic deoxygenation: The RuNPs/HAP catalyst (50 mg, 25 μmol Ru, 5 mol%) and iPrOH (0.5 mL) were added to a 0.1 M soln of diphenyl sulfoxide (0.5 mmol) in toluene. The mixture was stirred at 110 °C for 1 h. The yield of the product was determined by GC using naphthalene as internal standard; yield: >99%.

1.11.2 Deoxygenation of Sulfoxides and Sulfones to Thioethers Using Molecular Hydrogen

The deoxygenation of sulfoxides to sulfides using molecular hydrogen is a promising method that offers high atom efficiency because molecular hydrogen is an ideal reductant due to the production of water as the sole byproduct. However, in early research on the hydrogenation of sulfoxides, the use of rhodium(III) chloride for the hydrogenation of dimethyl sulfoxide resulted in low yields of dimethyl sulfide together with low catalytic activity and a limited range of substrates.[52]

The Royo group found that high-valent oxomolybdenum(VI) and oxorhenium(V) derivatives, such as dichlorodioxomolybdenum(VI) ($MoCl_2O_2$), bis(*N*,*N*-diethyldithiocarbamate)dioxomolybdenum(VI) [$MoO_2\{SC(=S)NEt_2\}_2$], iododioxobis(triphenylphosphine)rhenium(V) [$ReIO_2(PPh_3)_2$], and trichlorooxobis(triphenylphosphine)rhenium(V) [$ReCl_3O(PPh_3)_2$], have catalytic activity for the deoxygenation of sulfoxides **11** to sulfides **12** using molecular hydrogen as a reducing agent (Scheme 5).[53] These deoxygenations were performed using methyl phenyl sulfoxide or dibutyl sulfoxide as substrates at 120 °C under 50 atm of molecular hydrogen, affording high yields of the corresponding sulfides. DFT calculations reveal that the dioxomolybdenum complexes are readily deoxygenated to an active monooxo molybdenum species by molecular hydrogen through H—H addition to the Mo=O bond, followed by hydride migration to yield a water complex.

Scheme 5 Deoxygenation of Sulfoxides with Molecular Hydrogen Using Homogeneous Molybdenum and Rhenium Catalysts[53]

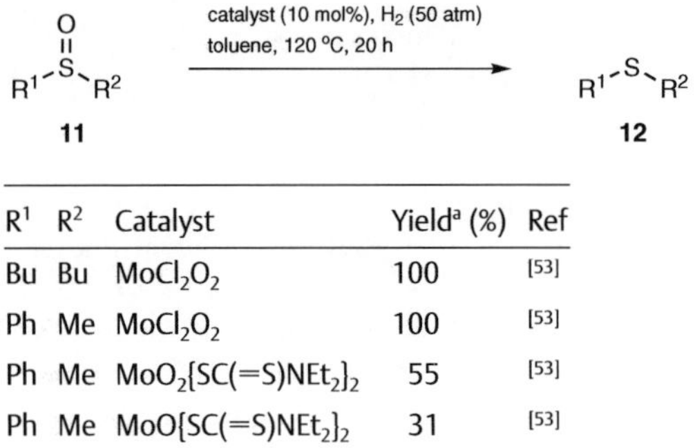

R^1S(=O)R^2 **11** → catalyst (10 mol%), H_2 (50 atm), toluene, 120 °C, 20 h → R^1SR^2 **12**

R^1	R^2	Catalyst	Yield[a] (%)	Ref
Bu	Bu	$MoCl_2O_2$	100	[53]
Ph	Me	$MoCl_2O_2$	100	[53]
Ph	Me	$MoO_2\{SC(=S)NEt_2\}_2$	55	[53]
Ph	Me	$MoO\{SC(=S)NEt_2\}_2$	31	[53]
Bu	Bu	$ReIO_2(PPh_3)_2$	100	[53]
Ph	Me	$ReIO_2(PPh_3)_2$	87	[53]
Bu	Bu	$ReCl_3O(PPh_3)_2$	100	[53]
Ph	Me	$ReCl_3O(PPh_3)_2$	83	[53]

[a] Determined by GC.

The severe conditions (i.e., high pressure of H_2 and high temperature) often preclude the use of these catalysts for the selective deoxygenation of functionalized sulfoxides containing other reducible or thermally labile functional groups, thus restricting the applicability of these processes. Therefore, the development of an efficient catalytic system for the deoxygenation of diverse sulfoxides to sulfides under mild conditions would represent a significant advance. Additionally, heterogeneous catalysts are advantageous over homogeneous catalysts from a viewpoint of separation and reuse of the catalyst. Therefore, an ideal method for the green sustainable deoxygenation of sulfoxides would involve the utilization of a reusable heterogeneous catalyst with atmospheric pressure of molecular hydrogen as a reductant under mild reaction conditions.

The Kaneda group reported the first deoxygenation of sulfoxides using a heterogeneous titanium(IV) oxide supported ruthenium nanoparticle catalyst under atmospheric pressure of molecular hydrogen.[54] Screening of various combinations of metal nanoparticles with supports as catalysts for the hydrogenation of diphenyl sulfoxide (**7**) at 100 °C under atmospheric pressure of molecular hydrogen revealed that ruthenium nanoparticles (RuNPs) show significantly higher catalytic activity than other metal nanoparticles, and titanium(IV) oxide was found to be the best support, providing a quantitative yield of diphenyl sulfide (**8**) (Scheme 6). The catalytic activity of the ruthenium nanoparticles increases with decreasing particle size (entry 1 versus entries 12 and 13).

Scheme 6 Deoxygenation of Diphenyl Sulfoxide Using Various Supported Metal Nanoparticle Catalysts[54]

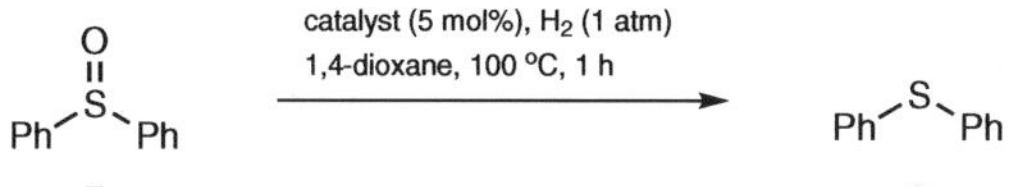

Ph–S(=O)–Ph (**7**) → catalyst (5 mol%), H_2 (1 atm), 1,4-dioxane, 100 °C, 1 h → Ph–S–Ph (**8**)

Entry	Catalyst[a]	Particle Size[b] (nm)	Conversion[c] (%)	Yield[c] (%)	Ref
1	RuNPs/TiO_2	1.6	>99	>99	[54]
2	RhNPs/TiO_2	n.d.	18	16	[54]
3	PtNPs/TiO_2	n.d.	12	11	[54]
4	PdNPs/TiO_2	n.d.	12	10	[54]
5	CuNPs/TiO_2	n.d.	2	2	[54]
6	NiNPs/TiO_2	n.d.	2	2	[54]
7	AgNPs/TiO_2	n.d.	<1	<1	[54]
8	AuNPs/TiO_2	n.d.	<1	<1	[54]
9	RuNPs/SiO_2	2.1	73	72	[54]
10	RuNPs/HAP	3.0	51	50	[54]
11	RuNPs/Al_2O_3	4.2	26	23	[54]
12	RuNPs/TiO_2	2.0	86	85	[54]
13	RuNPs/TiO_2	2.7	69	69	[54]

[a] HAP = hydroxyapatite.
[b] n.d. = not determined.
[c] Determined by GC using an internal standard.

Ruthenium nanoparticles on titanium(IV) oxide (RuNPs/TiO_2) show high catalytic activity for the selective hydrogenation of various sulfoxides to sulfides under atmospheric pressure of molecular hydrogen (Scheme 7). A wide range of aromatic and aliphatic sulfoxides **13** are efficiently converted into the corresponding sulfides **14** in excellent yield. Under these mild reaction conditions, the catalyst enables chemoselective hydrogenation of sulfoxides, with other reducible functional groups, such as halogens, acetals, carbonyls, cyano groups, esters, and amides, remaining intact. Ruthenium nanoparticles on titanium(IV) oxide also promote the deoxygenation of methionine sulfoxide [**13**, R^1 = $(CH_2)_2CH(NH_2)CO_2H$; R^2 = Me], which occurs naturally in metabolic systems in animal livers or plant cells, in aqueous media, giving methionine as the sole product. This catalyst

for references see p 353

system is also applicable to a 50-mmol-scale process. The product is readily isolated in 96% yield by simple filtration of the catalyst followed by evaporation of the filtrate. Turnover numbers (TON) as high as 500 are achieved.

Scheme 7 Deoxygenation of Various Sulfoxides Using Ruthenium Nanoparticles on Titanium(IV) Oxide with Molecular Hydrogen[54]

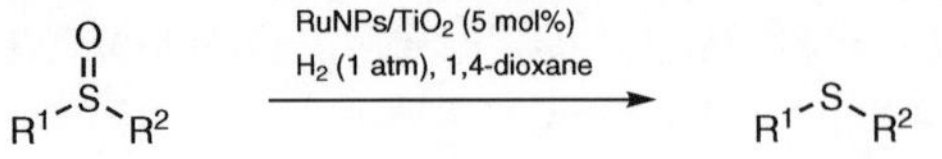

$R^1S(O)R^2$ (**13**) → R^1SR^2 (**14**); RuNPs/TiO_2 (5 mol%), H_2 (1 atm), 1,4-dioxane

R^1	R^2	Temperature (°C)	Time (h)	Conversion[a] (%)	Yield[a] (%)	Ref
Ph	Ph	100	1	>99	>99 (93)	[54]
Ph	Ph	160	36	>99	>99[b] (96)	[54]
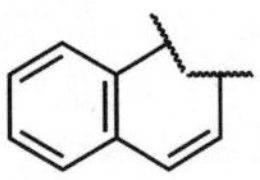		100	1	>99	>99 (95)	[54]
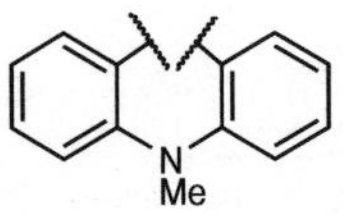		100	1	>99	>99 (94)	[54]
Ph	Bn	80	12	>99	98[c] (90)	[54]
Bn	Bn	60	12	>99	98[c] (94)	[54]
Pr	Pr	100	1	>99	>99 (84)	[54]
iPr	iPr	100	1	>99	>99 (81)	[54]
Me	$(CH_2)_7Me$	100	1	>99	>99 (90)	[54]
4-$H_2NC_6H_4$	Me	100	2	>99	>99 (92)	[54]
4-ClC_6H_4	4-ClC_6H_4	100	1	>99	>99 (95)	[54]
4-BrC_6H_4	Me	80	3	>99	>99 (93)	[54]
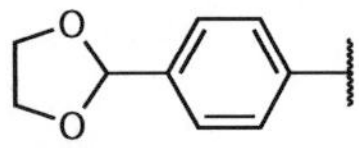	Me	100	5	>99	>99 (95)	[54]
4-AcC_6H_4	Me	100	1	>99	>99 (97)	[54]
4-NCC_6H_4	Me	100	8	>99	>99 (96)	[54]
4-$MeO_2CC_6H_4$	Me	100	2	>99	>99 (97)	[54]
3-$AcHNC_6H_4$	Me	100	3	>99	>99 (94)	[54]
$(CH_2)_2CO(CH_2)_2$		100	8	>99	>99 (91)	[54]
$(CH_2)_2CH(NH_2)CO_2H$	Me	100	12	>99[d]	(90)[c,e]	[54]

[a] Determined by GC using an internal standard; isolated yields in parentheses.
[b] 50-mmol scale using 200 mg of the catalyst and mesitylene (100 mL) as solvent.
[c] 0.20 g of catalyst was used.
[d] Determined by NMR spectroscopy.
[e] H_2O was used as solvent.

The catalyst system also works well in a column flow reactor (Scheme 8). Diphenyl sulfoxide (**7**) in 1,4-dioxane was passed through a column reactor packed with ruthenium nanoparticles on titanium(IV) oxide, along with molecular hydrogen at atmospheric pressure, giving diphenyl sulfide (**8**) in high yield. Inductively coupled plasma (ICP) analysis of the filtrate showed no leaching of the ruthenium species into the reaction mixture, demonstrating the high durability of the catalyst.

Scheme 8 Continuous Flow Reactor System Using Ruthenium Nanoparticles on Titanium(IV) Oxide for the Deoxygenation of Diphenyl Sulfoxide[54]

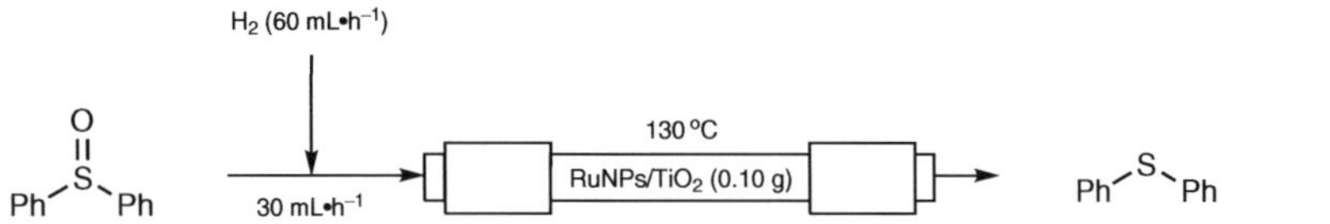

H_2 (60 mL•h^{-1})

130 °C

RuNPs/TiO_2 (0.10 g)

30 mL•h^{-1}

7 10 mmol (2.02 g) in 1,4-dioxane (200 mL)

8 1.73 g (93% isolated yield)

The versatility of the ruthenium/titanium(IV) oxide system has been demonstrated in the asymmetric synthesis of (*S*)-1-phenylethanol (**17**), and in the one-carbon homologation reaction of 4-methoxybenzaldehyde to give (4-methoxyphenyl)acetaldehyde (**20**) (Scheme 9). The deoxygenation of both the β-hydroxy sulfoxide **15** and the styryl sulfoxide **18** occurs efficiently under mild reaction conditions, affording the desired products **16** and **19**, respectively, in excellent yields. The reductive desulfurization of β-hydroxy sulfide **16** was carried out by using Raney nickel,[55] affording (*S*)-1-phenylethanol (**17**) in 93% yield with 81% enantiomeric excess.

Scheme 9 Asymmetric Synthesis of a Carbinol and a One-Carbon Homologation Involving Hydrogenation of a Sulfoxide Catalyzed by Ruthenium Nanoparticles on Titanium(IV) Oxide[54]

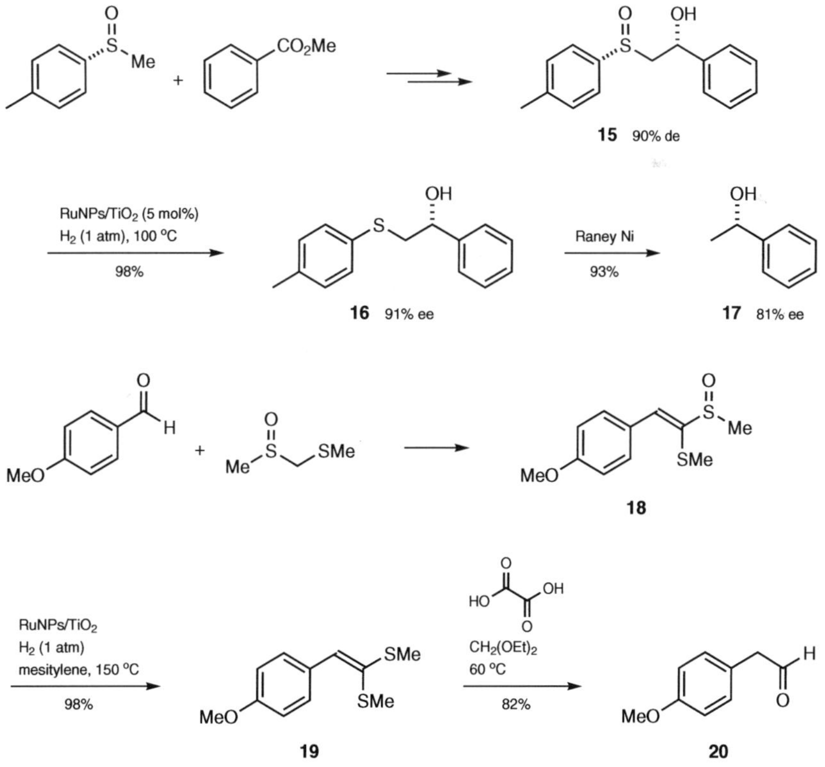

15 90% de

RuNPs/TiO_2 (5 mol%), H_2 (1 atm), 100 °C, 98%

16 91% ee

Raney Ni, 93%

17 81% ee

18

RuNPs/TiO_2, H_2 (1 atm), mesitylene, 150 °C, 98%

19

$CH_2(OEt)_2$, 60 °C, 82%

20

for references see p 353

X-ray absorption fine structure (XAFS) analysis of ruthenium/titanium(IV) oxide revealed the origin of the superior catalysis. The k^3-weighted extended X-ray absorption fine structure (EXAFS) and X-ray absorption near-edge structure (XANES) analyses show that the partially reduced ruthenium(0) species on the surface of the RuO_x nanoparticles directly deoxygenates sulfoxides to sulfides, after which the corresponding oxidized ruthenium species generated in situ are readily reduced by molecular hydrogen back to the original ruthenium nanoparticles.

Additionally, the treatment of diphenyl sulfoxide (**7**) with ruthenium/titanium(IV) oxide under argon leads to the formation of diphenyl sulfone via disproportionation, thus strongly supporting the in situ generation of a Ru=O intermediate during the reaction of the ruthenium/titanium(IV) oxide catalyst with sulfoxides. The oxygen transfer from diphenyl sulfoxide (**7**) to the reduced ruthenium species affords the Ru=O intermediate and then the Ru=O species transfers oxygen to the sulfoxide, affording the corresponding sulfone. The efficient oxygen-transfer behavior of the reduced ruthenium species on RuO_x nanoparticles via the Ru=O intermediate enables catalytic hydrogenation of various sulfoxides under mild conditions (Scheme 10).

Scheme 10 Proposed Reaction Pathway for the Deoxygenation of Sulfoxides Using Molecular Hydrogen Catalyzed by Ruthenium Nanoparticles on Titanium(IV) Oxide

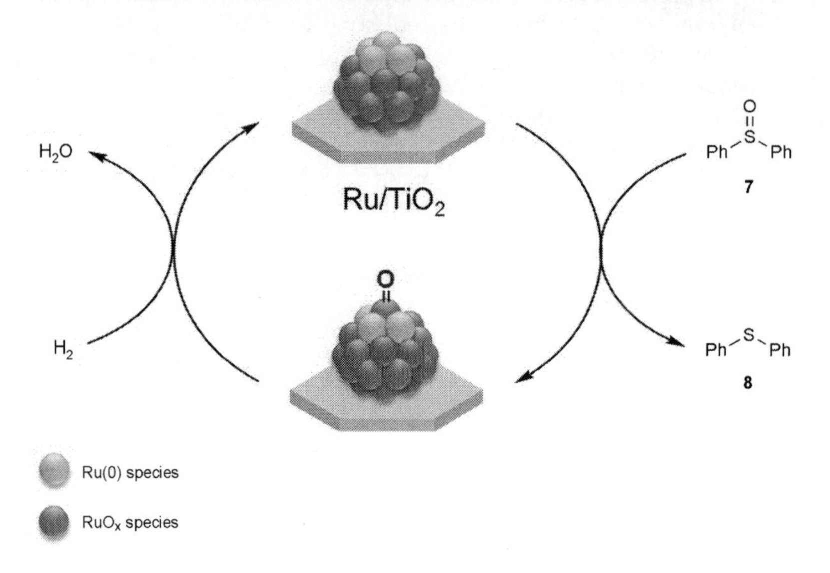

Recently, the Shimizu group reported titanium(IV) oxide supported platinum and MoO_x catalysts for the deoxygenation of sulfoxides using molecular hydrogen.[56] The Pt-MoO_x/TiO_2 catalyst is comprised of Mo^{4+} species and metallic platinum nanoparticles with a mean diameter of 4.7 nm (± 1.1 nm). Pt-MoO_x/TiO_2 promotes the deoxygenation of diphenyl sulfoxide at 7 atm of molecular hydrogen under solvent-free conditions, giving diphenyl sulfide in 97% yield. Among various platinum catalysts, including bimetal-loaded titanium(IV) oxide compounds (Pt-MO_x/TiO_2; M = Mo, V, Nb, W, Re), and various metal-loaded M-MoO_x/TiO_2 catalysts (M = Rh, Pd, Re, Ru, Ni, Cu), Pt-MoO_x/TiO_2 showed the best activity. The yield in the deoxygenation of diphenyl sulfoxide using Pt-MoO_x/TiO_2 is much higher than using MoO_3/TiO_2 or Pt/TiO_2, revealing a synergistic effect between platinum and MoO_x.

The Pt-MoO_x/TiO_2 catalyzed deoxygenation of various sulfoxides **21** into sulfides **22** proceeds at 7 atm of molecular hydrogen under solvent-free conditions (Scheme 11). The method shows chemoselective hydrogenation of sulfoxides without the conversion of other reducible functional groups such as chloro, bromo, and carbonyl groups. The

$Pt\text{-}MoO_x/TiO_2$ catalyst exhibits a high turnover number (TON) of 4400 with respect to the total number of platinum atoms in the catalyst. This value is higher than that of previously reported catalysts.

Scheme 11 Deoxygenation of Sulfoxides To Give Sulfides Using a Titanium(IV) Oxide Supported Platinum/MoO_x Catalyst[56]

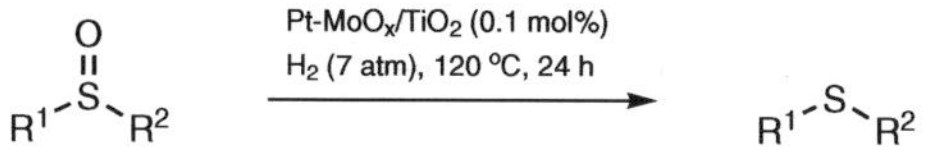

$Pt\text{-}MoO_x/TiO_2$ (0.1 mol%)
H_2 (7 atm), 120 °C, 24 h

$R^1S(O)R^2$ **21** → R^1SR^2 **22**

R^1	R^2	Yield (%)	Ref
Ph	Ph	97[a]	[56]
Ph	Ph	88[b]	[56]
4-Tol	4-Tol	92	[56]
$4\text{-}ClC_6H_4$	$4\text{-}ClC_6H_4$	88	[56]
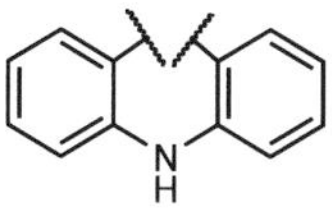		87	[56]
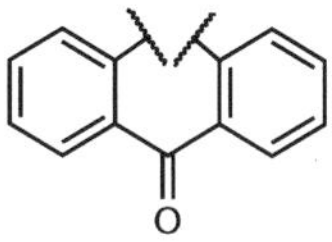		85	[56]
Ph	Me	94	[56]
$4\text{-}BrC_6H_4$	Me	91	[56]
Bn	Bn	96[c]	[56]
Bu	Bu	98	[56]
$(CH_2)_{11}Me$	$(CH_2)_{11}Me$	91	[56]

[a] 50 °C.
[b] Conditions: sulfoxide (5 mmol), H_2 (1 atm), $Pt\text{-}MoO_x/TiO_2$ (0.02 mol%); TON = 4400.
[c] 36 h.

In general, sulfur atoms in sulfides are strongly adsorbed onto the surface of active metals, resulting in the deactivation of the catalyst. However, the addition of 4-(methylsulfanyl)aniline does not significantly affect the initial formation rate of diphenyl sulfide in the hydrogenation of diphenyl sulfoxide by $Pt\text{-}MoO_x/TiO_2$, showing a sulfur tolerance of the $Pt\text{-}MoO_x/TiO_2$ catalyst. This sulfur tolerance may be the reason for the high TON of this catalytic system.

A unique characteristic feature of the $Pt\text{-}MoO_x/TiO_2$ catalyst is its activity for the deoxygenation of sulfones. Thus, benzyl phenyl sulfone (**23**) is deoxygenated under 7 atm of molecular hydrogen in the presence of 0.1 mol% of $Pt\text{-}MoO_x/TiO_2$ to give benzyl phenyl sulfide (**24**) in 85% yield after 40 hours (Scheme 12). This result represents the first example of catalytic hydrogenation of a sulfone to a sulfide by molecular hydrogen.

Scheme 12 Deoxygenation of a Sulfone To Give a Sulfide by Molecular Hydrogen Using a Titanium(IV) Oxide Supported Platinum/MoO_x Catalyst[56]

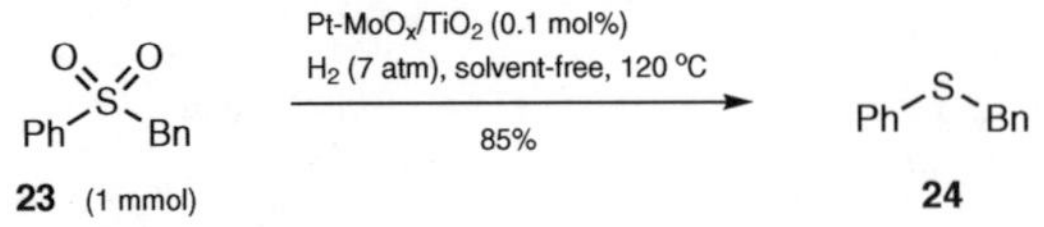

Ph–$S(O)_2$–Bn **23** (1 mmol) → [Pt-MoO_x/TiO_2 (0.1 mol%), H_2 (7 atm), solvent-free, 120 °C; 85%] → Ph–S–Bn **24**

Sulfides 12; General Procedure:[53]
The deoxygenation reaction was carried out in a stainless-steel autoclave, which was charged with $MoCl_2O_2$, $MoO_2\{SC(=S)NEt_2\}_2$, $ReIO_2(PPh_3)_2$, or $ReCl_3O(PPh_3)_2$ (10 mol%) and a magnetic stirrer bar, and then flushed with N_2. A soln of a sulfoxide **11** (1.0 mmol) in toluene (5 mL) was placed into the autoclave, which was pressurized with H_2 (50 atm). The mixture was then stirred at 120 °C. After 20 h, the autoclave was cooled to rt, degassed, and opened. The mixture was filtered and analyzed by GC.

Sulfides 14; General Procedure Using RuNPs/TiO_2:[54]
Preparation of ruthenium nanoparticles on titanium(IV) oxide: An aliquot of 17 wt% aq $TiCl_4$ (20 g) and urea (15 g) were added to deionized H_2O (200 mL) and the mixture was stirred at 100 °C for 12 h. The resulting white gel was collected by filtration and washed with deionized H_2O until a neutral filtrate was obtained, and then dried overnight at 110 °C. The obtained crystals were milled and calcined at 300 °C for 3 h to yield TiO_2 as a light-yellow powder (5.8 g). The TiO_2 (1.0 g) was stirred with 5 mM aq $RuCl_3$ (100 mL) and urea (0.50 g) at 95 °C for 6 h. The resulting dark-brown solid was collected by filtration, washed with deionized H_2O (1 L), and dried at 110 °C. Further treatment of the resulting solid under 1 atm of H_2 at 150 °C for 1 h gave RuNPs/TiO_2 with a mean diameter of 1.6 nm (Ru content: 0.48 mmol • g^{-1}).

Catalytic deoxygenation: The RuNPs/TiO_2 (50 mg, 0.025 mmol Ru, 5 mol%) was placed in a reaction vessel connected to a gas bag filled with H_2 (1 atm). A 0.10 M soln of a sulfoxide **13** in 1,4-dioxane (5 mL, 0.50 mmol) was added and the mixture was stirred at 100 °C. The yield of the product was determined by GC using naphthalene as internal standard.

Diphenyl Sulfide (8); Typical Procedure Using a Continuous-Flow Reactor:[54]
A stainless-steel column (inner diameter: 5.0 mm; length: 200 mm) was filled with RuNPs/TiO_2 (0.50 g) and silica gel (CARiACT Q-3: 75–150 μm, 1.0 g). A soln of diphenyl sulfoxide (**7**; 2.02 g, 10 mmol) in 1,4-dioxane (200 mL) was introduced to the column (30 mL • h^{-1}), which was maintained at 130 °C, along with a flow of H_2 (60 mL • h^{-1}). The resulting soln was concentrated under reduced pressure to afford the product as a colorless oil; yield: 1.73 g (93%).

(*R*)-1-Phenyl-2-(4-tolylsulfanyl)ethan-1-ol (16); Typical Procedure:[54]
RuNPs/TiO_2 (100 mg, 0.05 mmol Ru, 5 mol%) was placed in a reaction vessel connected to a gas bag filled with H_2 (1 atm). A 0.10 M soln of β-hydroxy sulfoxide **15** in 1,4-dioxane (10 mL, 1.0 mol) was added and the mixture was stirred at 100 °C for 3 h. The RuNPs/TiO_2 catalyst was removed by filtration and the filtrate was concentrated; yield: 98%; 91% ee.

(4-Methoxyphenyl)acetaldehyde (20); Typical Procedure:[54]
Sulfoxide **18** (242 mg, 1 mmol) was stirred in mesitylene (10 mL) under H_2 (1 atm) at 150 °C with RuNPs/TiO_2 (10 mol%) for 3 h. The RuNPs/TiO_2 catalyst was removed by filtration and the filtrate was concentrated to afford 1-(4-methoxyphenyl)-2,2-bis(methylsulfanyl)ethene (**19**); yield: 98%. The obtained dithioacetal product was added to a mixture of oxalic acid and diethoxymethane and the mixture was stirred at 60 °C. When the reaction was

complete, the mixture was poured into H_2O and the layers were separated. The organic phase was dried (Na_2SO_4), filtered, and concentrated. The residue was purified by column chromatography (silica gel); yield: 82%.

Sulfides 22**; General Procedure Using Pt-MoO_x/TiO_2:**[56]
Preparation of the Pt-MoO_x/TiO_2 catalyst: TiO_2 (5 g, 62.5 mmol) and $(NH_4)_6Mo_7O_{24}$•$4H_2O$ (1.09 g, 0.88 mmol) were added to H_2O (50 mL) at 50 °C, followed by evaporation at 50 °C. The obtained solid was dried at 90 °C for 12 h, and then calcined in air at 500 °C for 3 h to obtain MoO_3-loaded TiO_2 (MoO_3/TiO_2). MoO_3/TiO_2 (5.0 g) was added to a 13.5 mM soln of $Pt(NH_3)_2(NO_3)_2$ in aq HNO_3 (100 mL), followed by evaporation at 50 °C, and drying at 90 °C for 12 h.

Catalytic deoxygenation: The Pt-MoO_x/TiO_2 was pre-reduced in a Pyrex tube under a flow of H_2 (20 mL•min^{-1}) at 300 °C for 0.5 h and then cooled to rt. Dodecane (0.05 g) was injected to the pre-reduced catalyst inside the glass tube through the septum inlet, and then the septum was removed under air and a sulfoxide **21** (1.0 mmol) and a stirrer bar were charged to the tube. The tube was then inserted into a stainless-steel autoclave with a dead space of 28 mL. After being sealed, the reactor was flushed with H_2 and then charged with H_2 to 7 atm at rt. Then, the reactor was heated at 50 or 120 °C under stirring (180 rpm) for 24 h.

1.11.3 Reduction of Disulfides to Thiols Using Molecular Hydrogen

The synthesis of thiols (mercaptans) is important because they are often used as precursors for pharmaceuticals, agrochemicals, cosmetics, and petrochemical products. Currently, aliphatic thiols are efficiently produced via the direct thiolation of alcohols with hydrogen sulfide over alkali metal catalysts.[57] However, this method is not efficient for the production of aromatic thiols, with low yields obtained. One possible alternative route to the production of aromatic thiols is through reductive cleavage of the corresponding disulfides.[58–67]

The Hardacre group achieved, for the first time, the hydrogenation/hydrogenolysis of a range of disulfides over a supported palladium catalyst using molecular hydrogen under relatively benign conditions (Scheme 13).[68] Thus, the disulfides **25** are converted into the corresponding thiols **26** with high selectivity in tetrahydrofuran at 75 °C under 5 or 50 atm of molecular hydrogen, in the presence of 10 wt% palladium on charcoal (detailed yields and selectivities are not reported in the literature). The aliphatic disulfides, dimethyl disulfide, dibutyl disulfide, and dibenzyl disulfide, exhibit surprisingly high rates of hydrogenolysis. For the bis(nitrophenyl) disulfide derivatives, reduction of the nitro group is also observed. Furthermore, the catalyst activity can be reestablished using pretreatment with molecular hydrogen, but there is still a slight loss in activity upon recycling. Other effective methods for the hydrogenation of disulfides use cobalt sulfide catalysts on solid supports[69,70] or hydridotetrakis(triphenylphosphine)rhodium [$RhH(PPh_3)_4$].[71]

for references see p 353

Scheme 13 Hydrogenation/Hydrogenolysis of Disulfides Using Palladium on Charcoal[68]

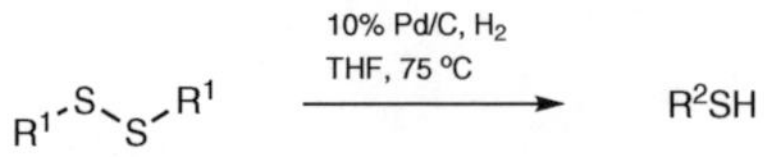

R^1–S–S–R^1 (**25**) → 10% Pd/C, H_2, THF, 75 °C → R^2SH (**26**)

R^1	R^2	Ref
Me	Me	[68]
Bu	Bu	[68]
Bn	Bn	[68]
Ph	Ph	[68]
4-Tol	4-Tol	[68]
4-$MeOC_6H_4$	4-$MeOC_6H_4$	[68]
4-$H_2NC_6H_4$	4-$H_2NC_6H_4$	[68]
4-$O_2NC_6H_4$	4-$H_2NC_6H_4$	[68]
3-$O_2NC_6H_4$	3-$H_2NC_6H_4$	[68]
2-$O_2NC_6H_4$	2-$H_2NC_6H_4$	[68]

Thiols 26; **General Procedure:**[68]
A disulfide **25** (5 mmol) in THF (200 mL) was placed in a high-pressure 300-mL Parr autoclave reactor with Pd/C (200 mg) and heated to 75 °C under N_2 before the desired pressure of H_2 was introduced. Samples were withdrawn at regular intervals and analyzed in an off-line Perkin-Elmer Clarus 500 gas chromatograph fitted with a ZB-5 capillary column.

1.11.4 Conclusions and Future Perspectives

In this review, recently developed efficient metal catalysts for the reduction of sulfur compounds using alcohols and molecular hydrogen as green reducing reagents were explored. Some homogeneous molybdenum and rhenium complexes display activity for the deoxygenation of sulfoxides to afford the corresponding thioethers. In particular, dioxomolybdenum complexes show high chemoselectivity for sulfoxides bearing potentially reducible functional groups. Several advanced heterogeneous catalysts, including ruthenium and platinum active species, have high activity and selectivity for the deoxygenation of a wide range of sulfoxides using alcohols or an atmospheric pressure of molecular hydrogen. These catalysts have also been proven to be reusable, without any loss of activity or selectivity.

Although metal catalysts with high activity have been disclosed and great advances have been made in the reduction of sulfur compounds, several issues must still be addressed. With regard to the reducing reagents, molecular hydrogen is an ideal reagent due to its high atom-efficiency and the production of water as the only byproduct. However, the existing homogeneous catalysts require a high pressure of hydrogen. Therefore, the utilization of an atmospheric pressure of hydrogen is the next challenging issue, which, when solved, will lead to greener catalytic processes.

Great advances have been made recently in the development of heterogeneous catalysts that show high activity for the deoxygenation of sulfoxides under 1 atmosphere of molecular hydrogen. However, precious metals are required; thus, the replacement of precious metals such as ruthenium and platinum with base metals is necessary for the development of cost-effective processes.

In addition, some catalysts have shown activity for the deoxygenation of sulfones and hydrogenation/hydrogenolysis of disulfides. However, in this area the activities and substrate scopes are not broad enough, and alternative catalysts need to be developed. We believe that future research in the above direction will provide new avenues for the green reductions of sulfur compounds using metal catalysts.

References

[1] Kunieda, N.; Nokami, J.; Kinoshita, M., *Chem. Lett.*, (1974), 369.
[2] Solladié, G., *Synthesis*, (1981), 185.
[3] Kosugi, H.; Konta, H.; Uda, H., *J. Chem. Soc., Chem. Commun.*, (1985), 211.
[4] Solladié, G.; Demailly, G.; Greck, C., *Tetrahedron Lett.*, (1985) **26**, 435.
[5] Kosugi, H.; Watanabe, Y.; Uda, H., *Chem. Lett.*, (1989), 1865.
[6] Carreño, M. C., *Chem. Rev.*, (1995) **95**, 1717.
[7] Debaun, J. R.; Menn, J. J., *Science (Washington, D. C.)*, (1976) **191**, 187.
[8] Ejiri, S.-I.; Weissbach, H.; Brot, N., *J. Bacteriol.*, (1979) **139**, 161.
[9] Yoneyama, K.; Matsumura, F., *Pestic. Biochem. Physiol.*, (1981) **15**, 213.
[10] Brot, N.; Weissbach, H., *Trends Biochem. Sci.*, (1982) **7**, 137.
[11] Madesclaire, M., *Tetrahedron*, (1988) **44**, 6537.
[12] Chasar, D. W., *J. Org. Chem.*, (1971) **36**, 613.
[13] Drabowicz, J.; Mikołajczyk, M., *Synthesis*, (1976), 527.
[14] Kano, S.; Tanaka, Y.; Sugino, E.; Hibino, S., *Synthesis*, (1980), 695.
[15] Harrison, D. J.; Tam, N. C.; Vogels, C. M.; Langler, R. F.; Baker, R. T.; Decken, A.; Westcott, S. A., *Tetrahedron Lett.*, (2004) **45**, 8493.
[16] Zhang, J.; Gao, X.; Zhang, C.; Luan, J.; Zhao, D., *Synth. Commun.*, (2010) **40**, 1794.
[17] Iranpoor, N.; Firouzabadi, H.; Shaterian, H. R., *J. Org. Chem.*, (2002) **67**, 2826.
[18] Gazdar, M.; Smiles, S., *J. Chem. Soc.*, (1910) **97**, 2248.
[19] Aida, T.; Furukawa, N.; Oae, S., *Tetrahedron Lett.*, (1973), 3853.
[20] Landini, D.; Maia, A. M.; Rolla, F., *J. Chem. Soc., Perkin Trans. 2*, (1976), 1288.
[21] Doi, J. T.; Musker, W. K., *J. Am. Chem. Soc.*, (1981) **103**, 1159.
[22] Wallace, T. J.; Mahon, J. J., *J. Am. Chem. Soc.*, (1964) **86**, 4099.
[23] Karimi, B.; Zareyee, D., *Synthesis*, (2003), 1875.
[24] Castrillón, J. P. A.; Szmant, H. H., *J. Org. Chem.*, (1965) **30**, 1338.
[25] Still, I. W. J.; Hasan, S. K.; Turnbull, K., *Can. J. Chem.*, (1978) **56**, 1423.
[26] Kikuchi, S.; Konishi, H.; Hashimoto, Y., *Tetrahedron*, (2005) **61**, 3587.
[27] Iranpoor, N.; Firouzabadi, H.; Jamalian, A., *Synlett*, (2005), 1447.
[28] Hua, G.; Woollins, J. D., *Tetrahedron Lett.*, (2007) **48**, 3677.
[29] Bahrami, K.; Khodaei, M. M.; Khedri, M., *Chem. Lett.*, (2007) **36**, 1324.
[30] Pandey, L. K.; Pathak, U.; Rao, A. N., *Synth. Commun.*, (2007) **37**, 4105.
[31] Jang, Y.; Kim, K. T.; Jeon, H. B., *J. Org. Chem.*, (2013) **78**, 6328.
[32] Zhu, Z.; Espenson, J. H., *J. Mol. Catal. A: Chem.*, (1995) **103**, 87.
[33] Arterburn, J. B.; Perry, M. C., *Tetrahedron Lett.*, (1996) **37**, 7941.
[34] Sanz, R.; Escribano, J.; Aguado, R.; Pedrosa, M. R.; Arnáiz, F. J., *Synthesis*, (2004), 1629.
[35] Bagherzadeh, M.; Haghdoost, M. M.; Amini, M.; Derakhshandeh, P. G., *Catal. Commun.*, (2012) **23**, 14.
[36] Fernandes, A. C.; Romão, C. C., *Tetrahedron*, (2006) **62**, 9650.
[37] Sousa, S. C. A.; Fernandes, A. C., *Tetrahedron Lett.*, (2009) **50**, 6872.
[38] Cabrita, I.; Sousa, S. C. A.; Fernandes, A. C., *Tetrahedron Lett.*, (2010) **51**, 6132.
[39] Enthaler, S., *Catal. Sci. Technol.*, (2011) **1**, 104.
[40] Enthaler, S., *ChemCatChem*, (2011) **3**, 666.
[41] Krackl, S.; Company, A.; Enthaler, S.; Driess, M., *ChemCatChem*, (2011) **3**, 1186.
[42] Enthaler, S.; Weidauer, M., *Catal. Lett.*, (2011) **141**, 833.
[43] Mikami, Y.; Noujima, A.; Mitsudome, T.; Mizugaki, T.; Jitsukawa, K.; Kaneda, K., *Chem.–Eur. J.*, (2011) **17**, 1768.
[44] Thiel, K.; Zehbe, R.; Roeser, J.; Strauch, P.; Enthaler, S.; Thomas, A., *Polym. Chem.*, (2013) **4**, 1848.
[45] Fernandes, A. C.; Romão, C. C., *Tetrahedron Lett.*, (2007) **48**, 9176.

[46] Enthaler, S.; Krackl, S.; Irran, E.; Inoue, S., *Catal. Lett.*, (2012) **142**, 1003.
[47] Enthaler, S., *Catal. Lett.*, (2012) **142**, 1306.
[48] Sousa, S. C. A.; Fernandes, A. C., *Coord. Chem. Rev.*, (2015) **284**, 67.
[49] García, N.; García-García, P.; Fernández-Rodríguez, M. A.; Rubio, R.; Pedrosa, M. R.; Arnáiz, F. J.; Sanz, R., *Adv. Synth. Catal.*, (2012) **354**, 321.
[50] García, N.; García-García, P.; Fernández-Rodríguez, M. A.; García, D.; Pedrosa, M. R.; Arnáiz, F. J.; Sanz, R., *Green Chem.*, (2013) **15**, 999.
[51] Takahashi, Y.; Mitsudome, T.; Mizugaki, T.; Jitsukawa, K.; Kaneda, K., *Chem. Lett.*, (2014) **43**, 420.
[52] James, B. R.; Ng, F. T. T.; Rempel, G. L., *Can. J. Chem.*, (1969) **47**, 4521.
[53] Reis, P. M.; Costa, P. J.; Romão, C. C.; Fernandes, J. A.; Calhorda, M. J.; Royo, B., *Dalton Trans.*, (2008), 1727.
[54] Mitsudome, T.; Takahashi, Y.; Mizugaki, T.; Jitsukawa, K.; Kaneda, K., *Angew. Chem. Int. Ed.*, (2014) **53**, 8348.
[55] Node, M.; Nishide, K.; Shigeta, Y.; Obata, K.; Shiraki, H.; Kunishige, H., *Tetrahedron*, (1997) **53**, 12 883.
[56] Touchy, A. S.; Siddiki, S. M. A. H.; Onodera, W.; Kon, K.; Shimizu, K., *Green Chem.*, (2016) **18**, 2554.
[57] Hoyos, L. J.; Primet, M.; Praliaud, H., *J. Chem. Soc., Faraday Trans.*, (1992) **88**, 113.
[58] Brown, H. C.; Nazer, B.; Cha, J. S., *Synthesis*, (1984), 498.
[59] Overman, L. E.; Matzinger, D.; O'Connor, E. M.; Overman, J. D., *J. Am. Chem. Soc.*, (1974) **96**, 6081.
[60] Maiti, S. N.; Spevak, P.; Singh, M. P.; Micetich, R. G.; Narender Reddy, A. V., *Synth. Commun.*, (1988) **18**, 575.
[61] Calais, C.; Lacroix, M.; Geantet, C.; Breysse, M., *J. Catal.*, (1993) **144**, 160.
[62] Calais, C.; Lacroix, M.; Geantet, C.; Breysse, M., *Appl. Catal., A*, (1994) **115**, 303.
[63] Krishnamurthy, S.; Aimino, D., *J. Org. Chem.*, (1989) **54**, 4458.
[64] Ookawa, A.; Yokohama, S.; Soai, K., *Synth. Commun.*, (1986) **16**, 819.
[65] Happer, D. A. R.; Mitchell, J. W.; Wright, G. J., *Aust. J. Chem.*, (1973) **26**, 121.
[66] Overman, L. E.; Petty, S. T., *J. Org. Chem.*, (1975) **40**, 2779.
[67] Maiti, S. N.; Singh, M. P.; Spevak, P.; Micetich, R. G.; Narender Reddy, A. V., *J. Chem. Res., Synop.*, (1988) **1**, 256.
[68] Novakova, E. K.; McLaughlin, L.; Burch, R.; Crawford, P.; Griffin, K.; Hardacre, C.; Hu, P.; Rooney, D. W., *J. Catal.*, (2007) **249**, 93.
[69] Mashkina, A. V.; Krivorucho, O. P.; Khairulina, L. N., *Kinet. Catal.*, (2008) **49**, 103.
[70] Mashkina, A. V.; Khairulina, L. N., *Kinet. Catal.*, (2007) **48**, 125.
[71] Arisawa, M.; Sugata, C.; Yamaguchi, M., *Tetrahedron Lett.*, (2005) **46**, 6097.

1.12 Catalytic Hydrodehalogenation Reactions

B. Ghosh and R. E. Maleczka, Jr.

General Introduction

Hydrodehalogenation (or reductive dehalogenation) of organic halides is an important transformation in organic synthesis and industrial detoxification processes.[1] Numerous methods to effect this transformation have been developed, including the use of microorganisms,[2] stoichiometric amounts of hydride reducing agents,[3–7] metal–halogen exchanges,[8,9] metal mediation (or catalysis),[1,10,11] photocatalysis,[12] and electron-transfer[13] and non-metal-mediated reactions.[14,15] Reviews of metal-mediated hydrodehalogenation reactions were published in 2002[10] and 2008.[11] This review of catalytic hydrodehalogenation reactions is restricted to selected synthetic protocols that involve catalytic hydrogenation by molecular hydrogen or transfer hydrogenations from other reagents (e.g., alcohols, formate, etc.) reported in the last 15 years. Furthermore, page limitations prohibit a full accounting by way of citation of all such methods reported during this time period. The new methods that are discussed fall into two categories of hydrodehalogenation processes: metal-catalyzed (heterogeneous and homogeneous) (Section 1.12.1) and photo-induced (Section 1.12.2). Notably, all methods chosen for this chapter report isolated and spectroscopically characterized reaction products.

1.12.1 Metal-Catalyzed Hydrodehalogenation

1.12.1.1 Heterogeneous Catalysis

Organo bromides and iodides generally undergo hydrodehalogenations more readily than fluorides or chlorides. As such, many methods for the chemoselective reductions of these halides have been reported over the years. Nonetheless, new examples of note continue to emerge.

Mandal and co-workers have reported the synthesis of N-protected α-methylalkylamines **3** under mild conditions (Scheme 1).[16] In this method, the carboxylic acid groups of protected amino acids **1** are converted into the hydroxymethyl derivatives, followed by formation of the iodomethyl compounds **2**. Hydrodeiodination of these iodomethyl groups is achieved using palladium on charcoal (10 wt%), molecular hydrogen, and *N*,*N*-diisopropylethylamine (1.5 equiv) in tetrahydrofuran/methanol (1:2) at room temperature to produce hydrodeiodinated products in good to excellent yields. This chemoselective procedure also allows for hydrodeiodination of an iodomethyl group in the presence of hydrogenolysis-labile benzyloxycarbonyl and 9-fluorenylmethoxycarbonyl protecting groups. Wullschleger, Gertsch, and Altmann have employed this hydrodeiodination during the synthesis and biological study of analogues of the polyketide-derived macrolide peruloside A.[17]

Scheme 1 Hydrodeiodination of Alkyl Iodides Under Base-Mediated Hydrogenation Conditions[16]

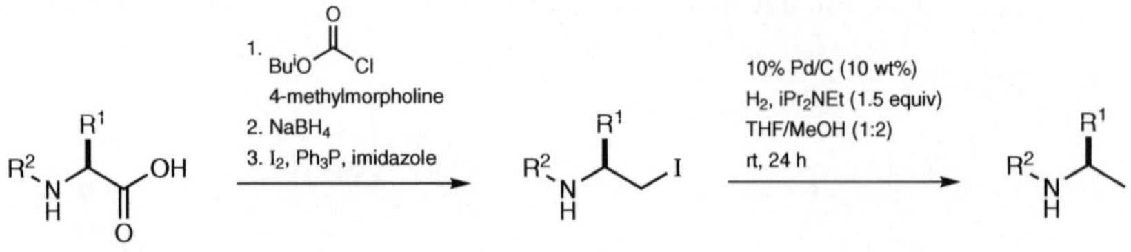

1 → 2: 1. $Bu^tOC(O)Cl$, 4-methylmorpholine; 2. $NaBH_4$; 3. I_2, Ph_3P, imidazole

2 → 3: 10% Pd/C (10 wt%), H_2, iPr_2NEt (1.5 equiv), THF/MeOH (1:2), rt, 24 h

R^1	R^2	Yield (%) of **3** from **2**	Ref
iPr	Boc	91	[16]
iPr	Cbz	64	[16]
Bn	Boc	91	[16]
Bn	Cbz	81	[16]
Bn	Fmoc	89	[16]
CH_2OBn	Fmoc	92	[16]
$(CH_2)_4NHBoc$	Cbz	83	[16]
$(CH_2)_2CO_2Bn$	Fmoc	90	[16]
$4-BnOC_6H_4CH_2$	Cbz	78	[16]
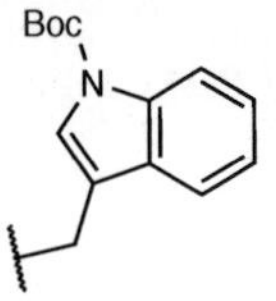 (N-Boc-indol-3-yl)methyl	Fmoc	77	[16]
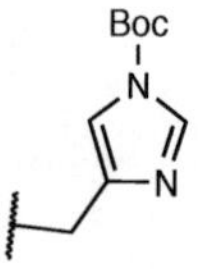 (N-Boc-imidazol-4-yl)methyl	Fmoc	85	[16]
(*R*)-CH(Me)OBn	Boc	88	[16]
(*S*)-CH(Me)OBn	Boc	84	[16]

In another example of sp^3 hydrodehalogenation, Chan and co-workers investigated the reduction of 4-halotetrahydropyrans (Scheme 2).[18] Depending on the halogen, their method allows for the chemoselective one-pot dehalogenation (X = I), debenzylation (X = Cl), or tandem dehalogenation/debenzylation (X = Br) of substituted pyrans. Best results are achieved when palladium on charcoal (40 mol%), methanol/ethyl acetate (9:1), and sodium hydrogen carbonate are used under hydrogenation conditions. The applicability of this methodology to multistep synthesis has been demonstrated by synthesizing a key intermediate of (+)-SCH 351448, a natural product that functions as a low-density lipoprotein receptor activator.

Scheme 2 Chemoselective Hydrodehalogenation/Debenzylation of Tetrahydropyrans[18]

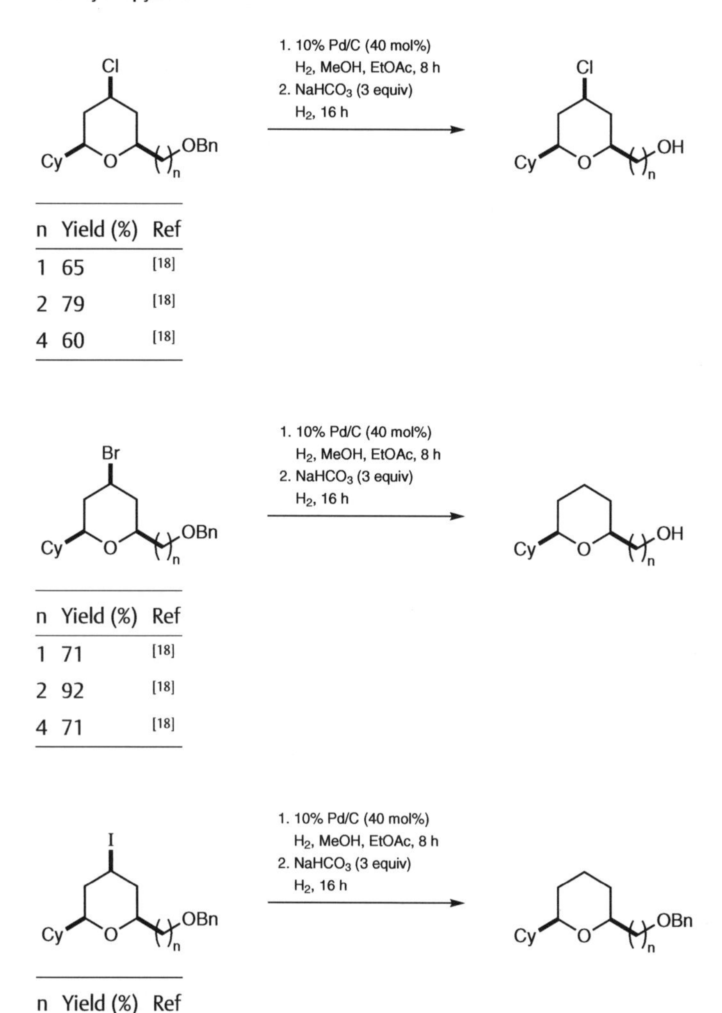

n	Yield (%)	Ref
1	65	[18]
2	79	[18]
4	60	[18]

n	Yield (%)	Ref
1	71	[18]
2	92	[18]
4	71	[18]

n	Yield (%)	Ref
1	47	[18]
2	74	[18]
4	66	[18]

Advances in hydrodehalogenation of sp^2 halides, including aryl chlorides and fluorides, have also emerged in recent years. Monguchi and co-workers demonstrated that the presence of triethylamine accelerates palladium on charcoal catalyzed hydrodehalogenations of aryl chlorides **4** (Scheme 3).[19] With 10 wt% palladium on charcoal and a hydrogen atmosphere maintained by a balloon, the addition of triethylamine (1.2 equiv) increases the rate of hydrodechlorination by >70-fold compared to the same reaction performed without triethylamine. It is suggested that triethylamine serves as both an HCl scavenger and a single-electron donor in the dechlorination reaction. Application of this chemistry to nine substrates affords the reduced arene products **5** in moderate to excellent yields. Substrates bearing alkenes or alkynes are not among the aryl chlorides investigated, so it is unclear whether such groups remain unsaturated. That said, Khan, Iyer, and co-workers

for references see p 373

chose to use this chemistry in the last step of their syntheses of three 8-aminoquinoline–pyrazolopyrimidine hybrids, which were further evaluated as potent antimalarial compounds.[20]

Scheme 3 Palladium on Charcoal Catalyzed Triethylamine-Mediated Hydrodechlorination of Aryl Chlorides[19]

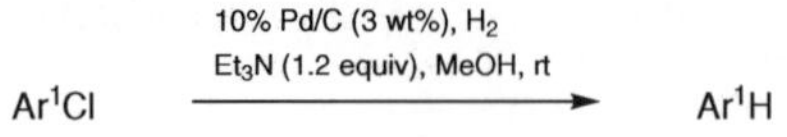

Ar^1Cl (**4**) → [10% Pd/C (3 wt%), H_2, Et_3N (1.2 equiv), MeOH, rt] → Ar^1H (**5**)

Ar^1	Time (h)	Yield (%)	Ref
$2\text{-}HO_2CC_6H_4$	6	99	[19]
$4\text{-}HO_2CC_6H_4$	3	100	[19]
$4\text{-}BzC_6H_4$	1	65	[19]
$4\text{-}HOC_6H_4$	3	93	[19]
$2\text{-}Me\text{-}5\text{-}O_2NC_6H_3$	2	90[a]	[19]
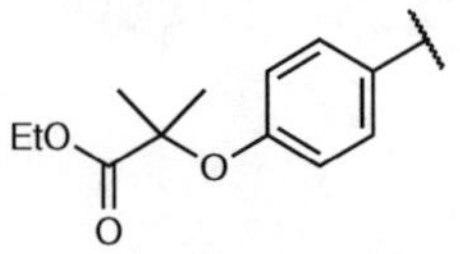 EtO, O	27	71	[19]
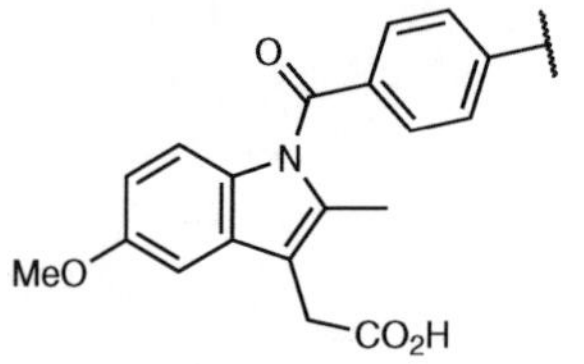 O, N, MeO, CO_2H	25	44	[19]

[a] Reduction of the nitro group also occurs; the product is *p*-toluidine.

Zhang et al. investigated the synthesis of 2-chloropyrimidines **7** from 2,4-dichloropyrimidines **6** via a regioselective palladium on charcoal catalyzed hydrodechlorination. The reaction proceeds under atmospheric hydrogen pressure and at room temperature, with sodium hydrogen carbonate serving as a base and ethanol as solvent (or co-solvent) (Scheme 4).[21] The method appears to be limited to pyrimidines with an electron-rich fused aromatic or heteroaromatic ring, as reactions of electron-poor or neutral substrates often give complex mixtures of products. Nonetheless, Wipf and co-workers relied on this regioselective dechlorination in their preparation of a thiazepinothiophenopyrimidinone analogue, which was later shown to inhibit the PMA-induced PKD1 autophosphorylation at S^{916}.[22] Similar reductions have also been used to remove *para*-situated chlorides or bromides that had served as blocking groups during a selective *ortho* functionalization of aromatic compounds with *ortho*/*para*-directing groups.[23]

Scheme 4 Palladium on Charcoal Catalyzed Regioselective Dechlorination of 2,4-Dichloropyrimidines[21]

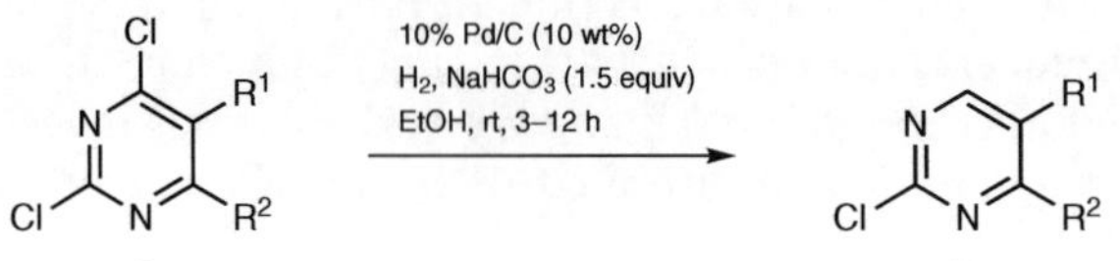

Cl, N, R^1, Cl, N, R^2 (**6**) → [10% Pd/C (10 wt%), H_2, $NaHCO_3$ (1.5 equiv), EtOH, rt, 3–12 h] → N, R^1, Cl, N, R^2 (**7**)

R^1 R^2	Time (h)	Yield (%)	Ref
SCH=CH	12	86[a]	[21]
NHCH=CH	3	53[b]	[21]
CH=C(OMe)C(OMe)=CH	3	78	[21]

[a] 20 wt% of Pd/C was used.
[b] 1.0 equiv of $NaHCO_3$ was used.

In another investigation, Arcadi and co-workers reported palladium on charcoal (5 mol%) catalyzed reductions of aryl chlorides using sodium formate (5 equiv) as a transfer hydrogenation agent (Scheme 5).[24] The temperature plays a key role in controlling the extent of the reduction: at ambient temperature, the hydrodechlorinated arene **8** is generated, but at 100 °C the arene is fully hydrogenated to give cyclohexanes **9**. This method also enables hydrodefluorinations. For example, when 4-fluorophenol is subjected to the reaction conditions at 100 °C, cyclohexanol is obtained in 75% yield.

Scheme 5 Palladium on Charcoal Catalyzed Transfer Reduction of Aryl Chlorides with Sodium Formate[24]

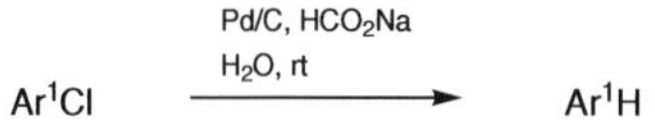

Ar^1Cl —[Pd/C, HCO_2Na, H_2O, rt]→ Ar^1H **8**

Ar^1	Time (h)	Yield (%)	Ref
$4-BzC_6H_4$	24	63	[24]
$3-HOC_6H_4$	3	96	[24]
$4-HOC_6H_4$	3	100	[24]
$2-HOC_6H_4$	5	92	[24]
$4-MeOC_6H_4$	13	95	[24]
$4-AcC_6H_4$	24	100	[24]
$2,4-(HO)_2C_6H_3$	2.5	95	[24]
$2-H_2NC_6H_4$	12	89	[24]

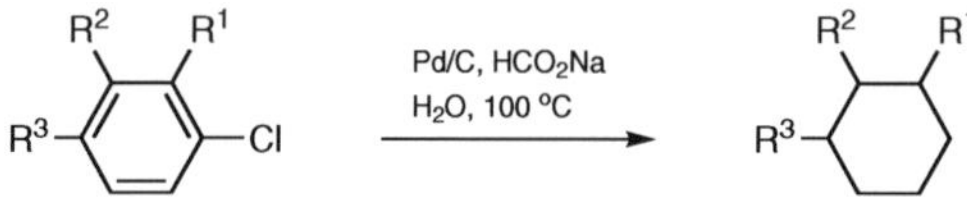

Pd/C, HCO_2Na, H_2O, 100 °C

9

R^1	R^2	R^3	Time (h)	Yield (%)	Ref
H	OH	H	22	97	[24]
H	H	OH	3	85	[24]
OH	H	H	16	100	[24]
H	H	OMe	24	75	[24]

A number of other methods for the hydrodefluorination of fluoroarenes have been recently described. Sawama and co-workers demonstrated a platinum on charcoal catalyzed hydrodefluorination protocol where, interestingly, the ease of dehalogenation was observed to be F > Cl > Br > I (Scheme 6).[25] It was also observed that the addition of sodium

for references see p 373

carbonate as a base accelerates the reaction and that the conditions tolerate the presence of carbonyl, ester, and carboxylic acid functional groups on the fluoroarenes to provide arenes **10**.

Scheme 6 Platinum on Charcoal Catalyzed Hydrodefluorination of Fluoroarenes[25]

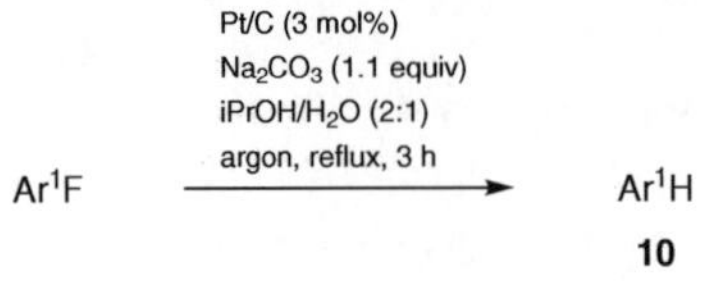

Ar^1F → Ar^1H (**10**)

Pt/C (3 mol%), Na_2CO_3 (1.1 equiv), iPrOH/H_2O (2:1), argon, reflux, 3 h

Ar^1	Time (h)	Yield (%)	Ref
2-PhC_6H_4	6	94	[25]
4-(4-FC_6H_4)C_6H_4	15	88[a]	[25]
4-BnC_6H_4	9	98	[25]
4-HOC_6H_4	9	80	[25]
4-$MeOC_6H_4$	12	83	[25]
4-$H_2NC_6H_4$	6	85	[25]
4-BzC_6H_4	15	89	[25]
4-AcC_6H_4	15	87	[25]
4-$EtO_2CC_6H_4$	6	89	[25]
4-$HO_2CC_6H_4$	6	89	[25]
2-$HO_2CC_6H_4$	6	91	[25]
2-$HO_2CC_6F_4$	16	88[b]	[25]

[a] Hydrodefluorination at both fluorine-substituted positions occurs; the product is 1,1′-biphenyl.

[b] Hydrodefluorination at all fluorine-substituted positions occurs; the product is benzoic acid.

Uozumi and co-workers have described the hydrodechlorination of aryl chlorides **11** with an amphiphilic polymer (polystyrene-PEG) supported nanopalladium catalyst (ARP-Pd) that facilitates ammonium formate assisted transfer hydrogenations (Scheme 7).[26,27] When subjected to these reaction conditions, a number of aryl chlorides with electron-donating, electron-withdrawing, or neutral substituents undergo reduction to give the corresponding arenes **12** with excellent isolated yields. Highlights of the reactions presented include examples that illustrate ketone tolerance and the complete hydrodechlorination of pentachloroaniline.

Scheme 7 Hydrodechlorination Using an Amphiphilic Polymer Supported Nanopalladium Catalyst[26]

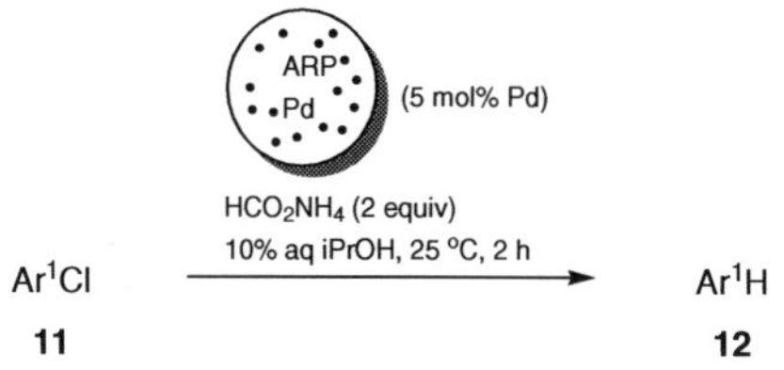

ARP-Pd (5 mol% Pd), HCO_2NH_4 (2 equiv), 10% aq iPrOH, 25 °C, 2 h

Ar^1Cl **11** → Ar^1H **12**

Ar^1	Yield (%)	Ref
Ph	99[a]	[26]
4-Tol	99[a]	[26]
4-$HOCH_2C_6H_4$	99	[26]
4-$MeOC_6H_4$	89	[26]
4-HOC_6H_4	93	[26]
4-$H_2NC_6H_4$	91	[26]
4-AcC_6H_4	97	[26]
4-$H_2NC(O)C_6H_4$	99	[26]
4-$HO_2CC_6H_4$	>99	[26]
4-$MeO_2CC_6H_4$	94	[26]
3-pyridyl	>99[a]	[26]

[a] GC yield.

Arenes 5; **General Procedure:**[19]

After two vacuum/H_2 cycles to remove air from a round-bottomed flask, a suspension of an aryl chloride **4** (100 mg), 10% Pd/C (3.0 mg), and Et_3N (1.2 equiv) in MeOH (10 mL) was vigorously stirred using a stirrer bar under H_2 (balloon) at ambient temperature (ca. 20 °C). After the starting aryl chloride had disappeared, the mixture was filtered through a 0.2-mL Millipore membrane filter and concentrated under reduced pressure. The residue was partitioned between Et_2O (10 mL) and H_2O (10 mL) and the organic layer was washed with brine (10 mL), dried ($MgSO_4$), filtered, and concentrated. The residue was purified by column chromatography (silica gel or reverse phase), if necessary.

Arenes 10; **General Procedure:**[25]

A mixture of a fluoroarene (0.25 mmol), 10% Pt/C (3 mol%), Na_2CO_3 (1.1 equiv), iPrOH (2 mL), and H_2O (1 mL) was heated at reflux in a 100 °C heating bath under an argon balloon in a Chemistation (EYELA) organic synthesizer. After completion of the reaction, the mixture was passed through a membrane filter to remove Pt/C. The filtrate was partitioned between with Et_2O (10 mL) and H_2O (10 mL). The aqueous layer was further extracted with Et_2O (10 × 2 mL). The combined organic layers were dried ($MgSO_4$), filtered, and concentrated under reduced pressure to afford the pure dry product.

Methyl Benzoate (12, **Ar^1 = 4-$MeO_2CC_6H_4$); Typical Procedure:**[26]

To a mixture of ARP-Pd (60 mg, 24 μmol Pd) and methyl 4-chlorobenzoate (**11**, Ar^1 = 4-$MeO_2CC_6H_4$; 85 mg, 0.5 mmol) in 10% aq iPrOH (1.0 mL) was added a soln of HCO_2NH_4 (63 mg, 1.0 mmol) in H_2O (0.2 mL) at 25 °C. The mixture was stirred at 25 °C for 120 min and then filtered. The catalyst beads were extracted with EtOAc (3 × 5 mL), and the com-

for references see p 373

bined extract was washed with brine, dried (Na_2SO_4), filtered, and concentrated (100% conversion; >99% GC yield). The residue was purified by chromatography (silica gel); yield: 94%.

1.12.1.2 Homogeneous Catalysis

Hydrodehalogenations under homogeneous conditions continue to be developed as useful synthetic tools. Logan et al. demonstrated that the use of palladium(II) acetate (2 mol%) with biphenyl-2-yldi-*tert*-butylphosphine (JohnPhos; 4 mol%) and sodium formate (2 equiv) in methanol can accomplish a homogeneous hydrodechlorination of aryl chlorides **13** to furnish arenes **14** (Scheme 8).[28] Here, sodium formate, and not methanol, is the hydrogen donor in the transfer hydrogenation. Five examples of the successful application of this method, with substrates bearing electron-donating, electron-withdrawing, or neutral functional groups, were reported; however, isolated yields were only reported for two of these examples. Despite this limited detail, Grøtli and co-workers employed a microwave-assisted modification of the method in their synthesis and study of chromone-based MEK1/2 modulators.[29]

Scheme 8 Homogeneous Dechlorination of Aryl Chlorides Using Palladium(II) Acetate and Sodium Formate[28]

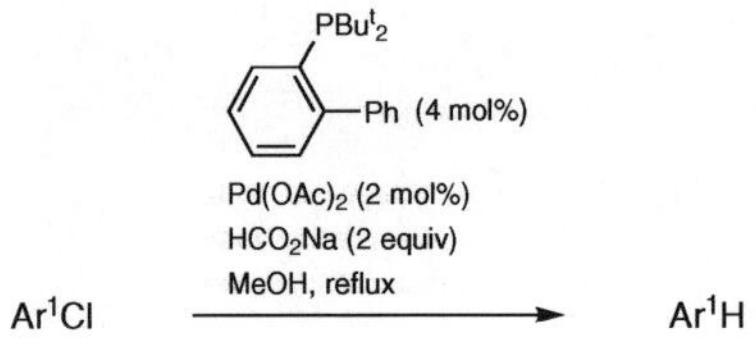

PBu^t_2 / —Ph (4 mol%)
$Pd(OAc)_2$ (2 mol%)
HCO_2Na (2 equiv)
MeOH, reflux

Ar^1Cl **13** → Ar^1H **14**

Ar^1	Yield (%)	Ref
4-$AcHNC_6H_4$	89	[28]
1-naphthyl	90	[28]

Another type of homogeneous catalytic transfer hydrogenation protocol was developed by Bhattacharjya and co-workers, namely a mild palladium-catalyzed (2 mol% $PdCl_2$) dehalogenation protocol for functionalized aryl halides **15** in aqueous media (Scheme 9). Here, tetramethyldisiloxane is used as the transfer hydrogenation source; it is proposed that molecular hydrogen is generated by metathesis between two palladium(II) species formed by the oxidative addition of palladium(0) to the silane.[30] Aryl bromides and iodides **15** (X = Br, I) react at room temperature, whereas aryl chlorides **15** (X = Cl) require elevated temperatures for the reactions to go to completion. Esters, carbamates, amides, nitriles, and the endocyclic double bond of a cholesterol derivative are all tolerated under these reaction conditions, whereas nitro groups, aldehydes, ketones, and exocyclic double bonds do not survive. Notably though, switching to [1,1′-bis(di-*tert*-butylphosphino)-ferrocene]dichloropalladium(II) [$PdCl_2$(dtbpf)] as catalyst in the presence of triethylamine (3 equiv) prevents the reduction of aldehydes and ketones and exocyclic double bonds. Heterocyclic halides also undergo facile dehalogenation reactions. Finally, related palladium-catalyzed transfer hydrodehalogenations of aryl bromides, iodides, and chlorides using hypercoordinated polymethylhydrosiloxane have also been reported.[31–33]

Scheme 9 Hydrodehalogenation of Aryl Halides in Water by In Situ Generated Molecular Hydrogen[30]

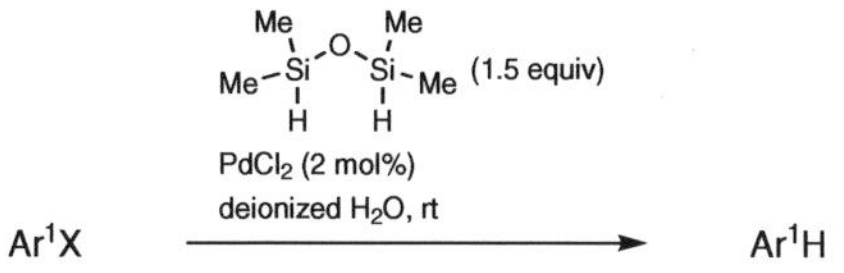

Me–SiH(Me)–O–SiH(Me)–Me (1.5 equiv)
$PdCl_2$ (2 mol%)
deionized H_2O, rt

Ar^1X **15** → Ar^1H

Ar^1	X	Yield (%)	Ref
$4\text{-}MeOC_6H_4$	Cl	75[a]	[30]
$4\text{-}NCC_6H_4$	I	89	[30]
quinolin-3-yl	Br	87	[30]
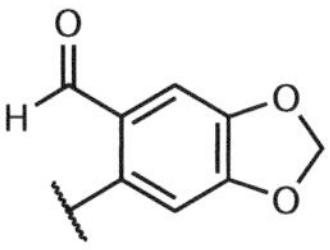 6-formyl-1,3-benzodioxol-5-yl (structure)	Br	88[b]	[30]
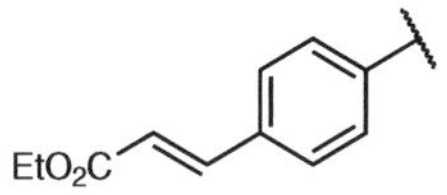 4-[(E)-2-(EtO_2C)vinyl]phenyl (structure)	Cl	97[b]	[30]
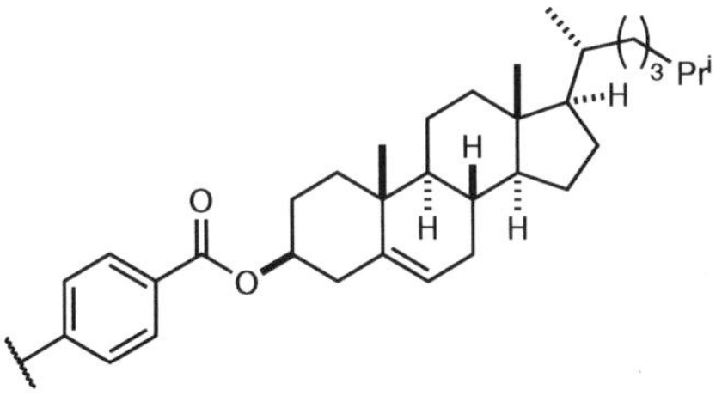 4-(cholesteryloxycarbonyl)phenyl (structure)	Br	86	[30]

[a] At 50 °C.
[b] $PdCl_2(dtbpf)$ (2 mol%) was used as catalyst in the presence of Et_3N (3 equiv).

As demonstrated by Li and co-workers, cobalt catalysis can allow for the selective hydrodefluorination of polyfluoroarenes. With tetrakis(trimethylphosphine)cobalt(0) [$Co(PMe_3)_4$] as the catalyst and sodium formate as the reductant, the selective hydrodefluorination of aryl fluorides **16** is achievable in moderate to excellent yields (Scheme 10).[34] When the reactions are performed in acetonitrile or dimethyl sulfoxide, reductions of polyfluorinated aryl and hetaryl fluorides tend to give monodefluorinated products **17**. Defluorination occurs selectively at the *para*-position to the nitrogen atom of fluorinated pyridines, or at the position *para* to the trifluoromethyl substituent in perfluorotoluene. It is suggested that these reactions proceed via oxidative addition of cobalt(0) into the C—F bond. Substitution of the Co—F bond by hydrogen occurs upon formate decarboxylation leading to a Co—H intermediate. Subsequent reductive elimination produces the final monodefluorinated compound.

Scheme 10 Tetrakis(trimethylphosphine)cobalt(0)-Catalyzed Selective Defluorination of Polyfluoroarenes[34]

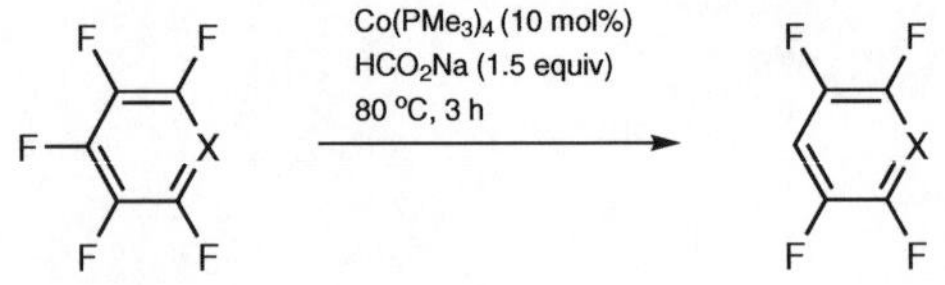

$Co(PMe_3)_4$ (10 mol%)
HCO_2Na (1.5 equiv)
80 °C, 3 h

16 → **17**

X	Solvent	Yield (%)	Ref
CCF_3	MeCN	70	[34]
N	MeCN	61	[34]
CF	MeCN	9[a]	[34]
CF	DMSO	11[a]	[34]
CH	MeCN	7[a]	[34]
CH	DMSO	51[a]	[34]
CC_6F_5	MeCN	76	[34]

[a] Determined by ^{19}F NMR spectroscopy.

Alcohols can also serve as hydrogen donors in a variety of hydrodehalogenation reactions. Such protocols have been used to great effect in isotopic labeling experiments. An example of such a process can be seen in the work of Janni and co-workers. Their homogeneous deuterodehalogenation of aryl halides occurs in a methanol-d_4/toluene (1:4) solution at 80 °C containing palladium(II) acetate (1 mol%), di(1-adamantyl)(butyl)phosphine [$(Ad)_2BuP$; 2 mol%], and potassium phosphate (1.5 equiv) (Scheme 11).[35] Both brominated aromatic and heteroaromatic compounds **18** (X = Br) undergo the bromine–deuterium exchange, affording the deuterated products **19** in moderate to excellent yields and with high deuterium incorporation. The CD_3 group of methanol-d_4 is the deuterium donor in the reaction.

Scheme 11 Catalytic Deuterodehalogenation of Aryl Bromides and Chlorides[35]

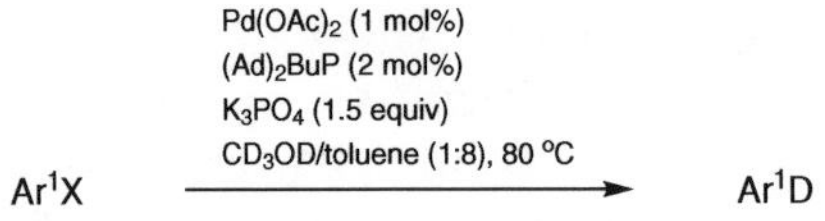

$Pd(OAc)_2$ (1 mol%)
$(Ad)_2BuP$ (2 mol%)
K_3PO_4 (1.5 equiv)
CD_3OD/toluene (1:8), 80 °C

Ar^1X **18** → Ar^1D **19**

Ar^1	X	Time (h)	Yield (%)	Ref
anthracen-9-yl	Br	16	99	[35]
4-NCC_6H_4	Br	10.5	58	[35]
4-BzC_6H_4	Cl	10.5	96	[35]

Ar^1	X	Time (h)	Yield (%)	Ref
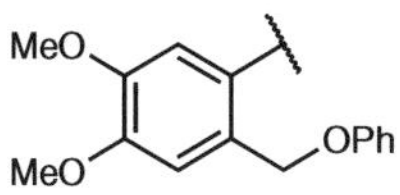 [structure: 4,5-dimethoxy-2-(phenoxymethyl)phenyl]	Br	8	90	[35]
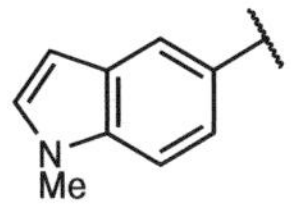 [structure: 1-methylindol-5-yl]	Br	7	68	[35]
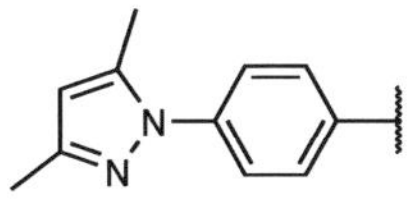 [structure: 4-(3,5-dimethylpyrazol-1-yl)phenyl]	Br	10.5	91	[35]

Chen and co-workers documented a palladium(II) acetate catalyzed transfer hydrogenation of aryl bromides and iodides **20**. The reaction is performed in the presence of triphenylphosphine and potassium carbonate as a base, while butanol (or propan-2-ol for the reduction of halophenols) serves as the solvent and hydrogen donor (Scheme 12).[36] The chemoselective reduction of a bromide in the presence of chloride is possible. The reduction also proceeds in excellent yields when applied to α-bromo ketones **21**. α-Chloro ketones can also be reduced this way, but with diminished yields. The Alami group designed a synthesis of biologically interesting 5-aryldihydrobenzoxepins that employs a slightly modified version of this chemistry.[37]

Scheme 12 Palladium(II) Acetate Catalyzed Hydrodehalogenation of Aryl Halides and α-Bromo Ketones[36]

Ar^1X **20** → Ar^1H [$Pd(OAc)_2$ (1 mol%), Ph_3P (4 mol%), K_2CO_3 (2 equiv), BuOH, 100 °C, 1 h]

Ar^1	X	Yield (%)	Ref
2-Me-4-MeOC_6H_3	I	93	[36]
quinolin-3-yl	Br	99	[36]
4-AcC_6H_4	Br	99	[36]
3-Tol	Br	99	[36]
4-ClC_6H_4	Br	99	[36]

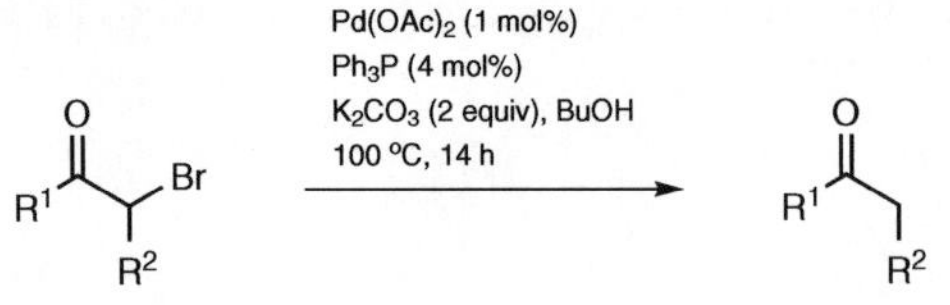

Pd(OAc)$_2$ (1 mol%)
Ph$_3$P (4 mol%)
K$_2$CO$_3$ (2 equiv), BuOH
100 °C, 14 h

21

R^1	R^2	Yield (%)	Ref
Ph	Ph	90	[36]
4-MeOC$_6$H$_4$	H	92	[36]
2-naphthyl	H	94	[36]
t-Bu	H	95	[36]

Hydrodechlorination as a means of remediating polychlorinated biphenyls has long been an area of interest. Among the recent advances in this area is the work of the Akzinnay group. They have demonstrated the hydrodechlorination of polychlorinated biphenyls **23** in the presence of very low loadings (0.04–1.00 mol% Pd) of the commercially available NHC–palladium complex dichloro(di-μ-chloro)bis[1,3-bis(2,6-diisopropylphenyl)imidazol-2-ylidene]dipalladium(II) (**22**), with propan-2-ol serving as the hydrogen donor and sodium hydroxide as a base (Scheme 13).[38] In this report, four examples of polychlorinated biphenyls with varied numbers of chlorine substituents were subjected to the reaction conditions. All of the substrates examined undergo exhaustive dechlorination in excellent isolated yields and high catalyst turnovers.

Scheme 13 Catalytic Hydrodechlorination of Polychlorinated Biphenyls[38]

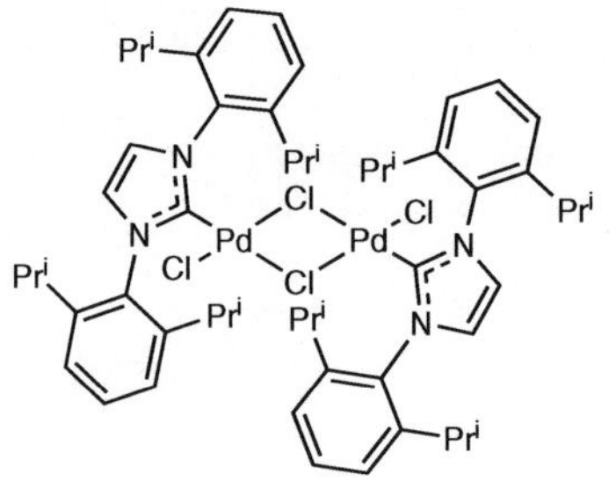

22

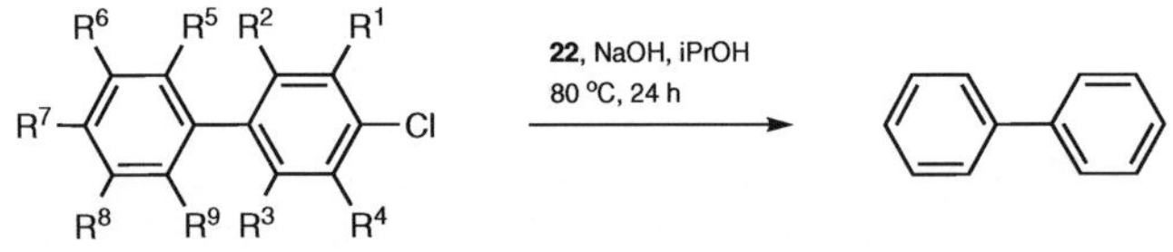

22, NaOH, iPrOH
80 °C, 24 h

23

R^1	R^2	R^3	R^4	R^5	R^6	R^7	R^8	R^9	Pd (mol%)	NaOH^a (Equiv)	Yield (%)	Ref
H	H	H	H	H	H	H	H	H	0.04	1.1	92	[38]
Cl	H	H	H	Cl	H	H	H	H	0.04	3.3	91	[38]
Cl	H	H	H	H	Cl	Cl	H	H	0.04	4.4	95	[38]
Cl	Cl	Cl	Cl	Cl	Cl	Cl	Cl	Cl	1	11	84	[38]

^{a} 10% excess with respect to the number of chlorine atoms.

Transfer hydrodechlorinations of aryl (phenyl and naphthyl) chlorides in the presence of butan-2-ol as a hydrogen donor have been achieved using rhodium catalysis. Specifically, Fujita and co-workers used (η^5-pentamethylcyclopentadienyl)rhodium complexes [e.g., $Rh(Cp^*)(OAc)_2{\bullet}H_2O$] to efficiently reduce a number of functionalized chloroarenes **24** to provide the corresponding arenes **25** (Scheme 14).[39] This method tolerates ester, amide, alkoxide, hydroxy, amine, and carboxylic acid functional groups, which are not reduced. The catalytic cycle is suggested to proceed via an (η^5-pentamethylcyclopentadienyl)rhodium butan-2-olate intermediate that undergoes a β-hydrogen elimination from the butoxy ligand to form an (η^5-pentamethylcyclopentadienyl)rhodium hydride [Rh(Cp*)H], which then reduces the aryl chloride. The resultant chloro(η^5-pentamethylcyclopentadienyl)rhodium [RhCl(Cp*)] intermediate reacts with butan-2-ol and the base to form an (η^5-pentamethylcyclopentadienyl)rhodium butan-2-olate, thereby completing the catalytic cycle.

Scheme 14 Rhodium-Catalyzed Hydrodechlorination of Aryl Chlorides[39]

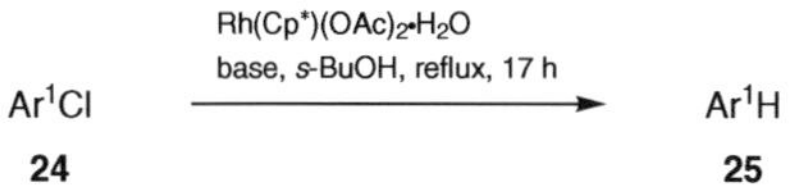

Ar^1Cl (**24**) → [$Rh(Cp^*)(OAc)_2{\bullet}H_2O$, base, *s*-BuOH, reflux, 17 h] → Ar^1H (**25**)

Ar^1	Catalyst (mol%)	Base (Equiv)	Yield (%)	Ref
4-PhC_6H_4	1	KOH (2)	100	[39]
4-$MeO_2CC_6H_4$	5	Cy_2NMe (1)	95	[39]
4-AcC_6H_4	5	Cs_2CO_3 (1)	15[a]	[39]
4-$H_2NC(O)C_6H_4$	5	Cy_2NMe (1)	95	[39]
4-$HO_2CC_6H_4$	5	Cy_2NMe (2)	96	[39]
1-naphthyl	1	KOH (2)	98	[39]
2-naphthyl	1	KOH (2)	96	[39]
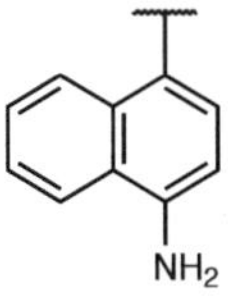 4-amino-1-naphthyl (NH_2)	2	KOH (2)	79	[39]
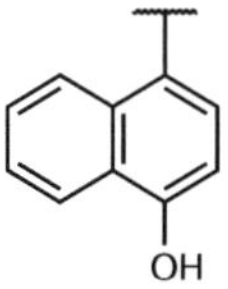 4-hydroxy-1-naphthyl (OH)	1	KOH (3.1)	88	[39]

[a] 1-Phenylethanol (85%) was the main product.

Hydrodehalogenation by concurrent tandem catalysis (CTC) is also possible, as reported by Cannon and co-workers (Scheme 15).[40] In the first concurrent tandem catalytic step of the reaction cycle, an aryl chloride or bromide (e.g., **26**) undergoes metal-catalyzed halogen exchange with an iodide source to generate a reactive (but commercially less available) aryl iodide. The in situ generated aryl iodide (e.g., **27**) then undergoes a facile hydrodeiodination reaction to form the dehalogenated product **28**. The Cannon method calls for copper(I) iodide (20 mol%) and sodium iodide (2 equiv) as the iodide source. An added diamine (1.5 equiv) presumably acts as a ligand for copper, but is also proposed to be the likely source of the hydrogen in the hydrodehalogenation. Under microwave irradiation, the reaction runs at 200 °C to effect the facile hydrodebromination of aryl bromides in good isolated yields. The authors claim the dechlorination of aryl chlorides under the

for references see p 373

same conditions, but no isolated yields are reported. Furthermore, the presence of heteroatom-containing functional groups has a deleterious effect on the reaction, producing unwanted side products.

Scheme 15 Hydrodebromination of Aryl Bromides under Microwave-Assisted Copper-Catalyzed Concurrent Tandem Catalysis[40]

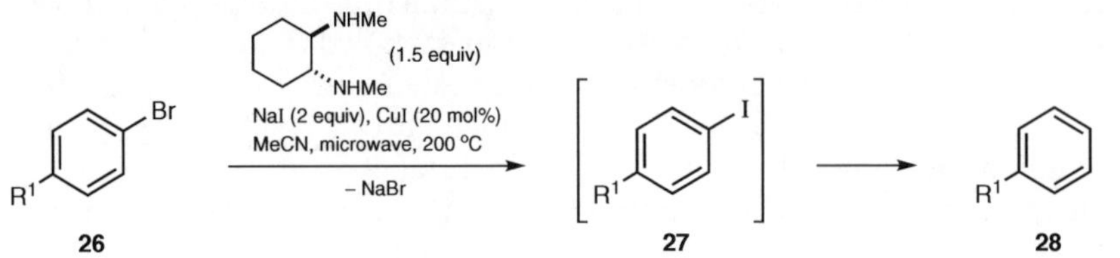

NHMe (1.5 equiv) NHMe
NaI (2 equiv), CuI (20 mol%)
MeCN, microwave, 200 °C
– NaBr

26 **27** **28**

R^1	Time (h)	Yield (%)	Ref
Ph	2	77	[40]
4-BrC_6H_4	2	70[a]	[40]
Bz	1	77	[40]

[a] Hydrodebromination of both bromo groups occurs; the product is 1,1′-biphenyl.

Arenes 14; General Procedure:[28]

An oven-dried, 25-mL, two-necked, round-bottomed flask was equipped with a magnetic stirrer bar, a rubber septum, and a ground-glass stopper. $Pd(OAc)_2$ (2 mol%) and biphenyl-2-yldi-*tert*-butylphosphine (JohnPhos, 4 mol%) were added, followed by HCO_2Na (4 mmol). The mixture was put under argon. MeOH (1 mL) was added by syringe through the septum, and the mixture was stirred for 5 min prior to addition of an aryl chloride **13** (2 mmol). The aryl chloride was added by syringe if it was a liquid. If the aryl chloride was a solid, it was added quickly through the neck of the flask. A second portion of MeOH (1 mL) was added. The mixture was heated at reflux, and the reaction progress was monitored by GC, with or without an internal standard. Upon completion of the reaction, the contents of the reaction flask were transferred using EtOAc (2 × 5 mL). The EtOAc washes were vacuum-filtered through diatomaceous earth (3 mL) in a 15-mL fritted funnel, and the diatomaceous earth was washed with EtOAc (5 mL). The filtrate was washed with H_2O (10 mL), and the aqueous layer was back-extracted with EtOAc (2 × 5 mL). The combined extracts were washed with sat. aq NaCl (10 mL) and dried (Na_2SO_4). EtOAc was removed under reduced pressure to yield the crude product, which was purified by chromatography [silica gel (10 mL)]. The desired fractions were combined, and the solvent was removed under reduced pressure to yield the purified product.

Arenes 17; General Procedure:[34]

A fluoroarene **16** (1 mmol) was dissolved in anhyd (dried by known procedures or distillation under N_2) DMSO or MeCN (3 mL) together with HCO_2Na (1.5 equiv) by stirring for 1 min. After the addition of $Co(PMe_3)_4$ (10 mol%), the suspension was stirred for 3 h at 80 °C under oxygen-free N_2. The soln turned from brown to orange. The yield was determined from the crude soln by ^{19}F NMR spectroscopy [(trifluoromethyl)benzene as external standard]. Where isolation of the final product was possible, the mixture was quenched with 1 M aq HCl and extracted with Et_2O. The extract was dried ($MgSO_4$) and filtered. The pure product was obtained by distillation or column chromatography.

Deuteroarenes 19; General Procedure:[35]

An oven-dried 8-mL vial was charged with $Pd(OAc)_2$ (1 mol%), $(Ad)_2BuP$ (2 mol%), K_3PO_4 (0.75 mmol), and a (het)aryl halide **18** (0.5 mmol) in CD_3OD (0.25 mL) and toluene (2.0 mL). The contents were stirred at 80 °C for 7–19 h and the reaction progress was monitored by TLC or GC/MS analysis. After the starting material had been completely consumed, the mixture was allowed to cool to rt and then concentrated. The residue was purified by chromatography (silica gel).

1.12.2 Photoinduced Hydrodehalogenation

The use of photoredox catalysis in organic synthesis has grown considerably over the past several years. Hydrodehalogenations are no exceptions to this trend. Narayanam and coworkers recently reported a photocatalytic procedure for the reductive hydrodehalogenation of alkyl bromides and chlorides **29**, **31**, and **33** to furnish the corresponding dehalogenated products **30**, **32**, and **34**, respectively (Scheme 16).[41] The reactions are carried out using 2.5 mol% of the metal complex tris(2,2′-bipyridyl)ruthenium(II) chloride hexahydrate [$RuCl_2(bipy)_3$•$6H_2O$] and *N*,*N*-diisopropylethylamine (2–10 equiv). Formic acid (10 equiv) or the Hantzsch ester (1.1 equiv) complete the reagents used. The reaction is believed to proceed via a radical mechanism with *N*,*N*-diisopropylethylamine being a hydrogen donor. This method allows for the reduction of carbon—halogen bonds in the presence of other reducible functional groups such as alkenes and alkynes and is tolerant of hydroxy, silyl ether, and carbamate groups. Excellent chemoselectivity for the reduction of halogens α to an electron-withdrawing group over the reduction of vinyl iodides or aryl bromides and iodides is observed.

Scheme 16 Photocatalyzed Reductive Dehalogenations[41]

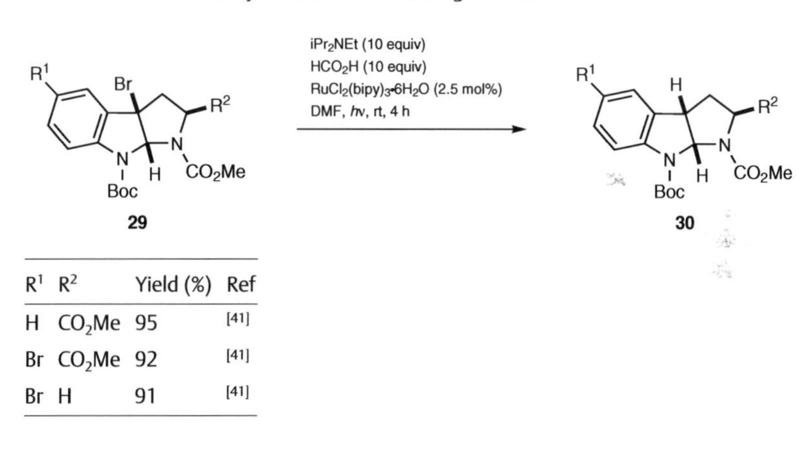

R^1	R^2	Yield (%)	Ref
H	CO_2Me	95	[41]
Br	CO_2Me	92	[41]
Br	H	91	[41]

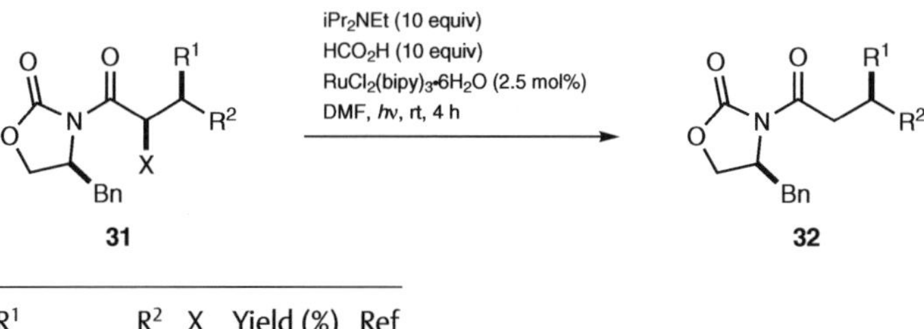

R^1	R^2	X	Yield (%)	Ref
OH	Ph	Br	99	[41]
OTBDMS	iPr	Br	79	[41]
OH	Ph	Cl	80	[41]

for references see p 373

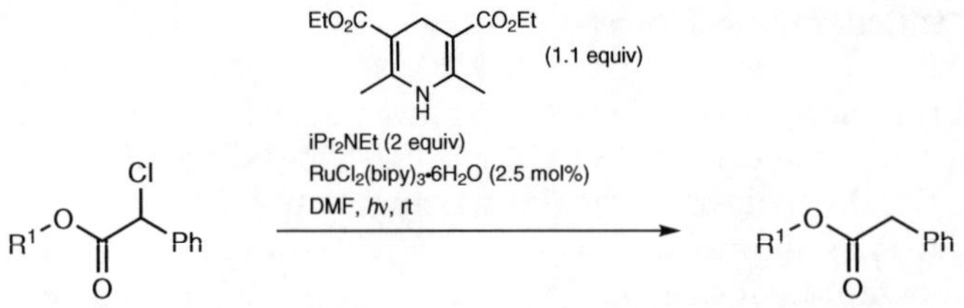

R^1	Yield (%)	Ref
$4\text{-BrC}_6\text{H}_4\text{CH}_2$	78	[41]
$2\text{-IC}_6\text{H}_4\text{CH}_2$	88	[41]
$(E)\text{-(CH}_2)_2\text{CH=CHI}$	81	[41]
$(\text{CH}_2)_2\text{C}\equiv\text{CH}$	89	[41]

Photoredox catalysis can also be used in hydrodefluorinations. Senaweera et al. have reported photoredox catalysis affording partially fluorinated aromatic compounds (Table 1).[42] Here, tris[2-phenylpyridinato-C_2,N]iridium(III) is used as catalyst and *N*,*N*-diisopropylethylamine is the electron and hydrogen-atom donor. Blue light is used to provide the photons. The reaction is again assumed to proceed via a single-electron donation to the aromatic substrate by the amine, followed by hydrogen atom abstraction from the latter by the aromatic compound. A number of fluorinated aromatic compounds have been efficiently subjected to these reaction conditions (Table 1). Certain functional-group modifications facilitate the hydrodefluorination. For example, the reactions of thiol- or amine-substituted substrates proceed better after alkoxycarbonylation and acylation, respectively. Steric effects also play a role in reaction kinetics and evidence points to the reaction having pseudo-zero-order rate dependency on substrate. For most examples provided, regioselective hydrodefluorination is observed.

Table 1 Photocatalytic Hydrodefluorination of Fluoroarenes[42]

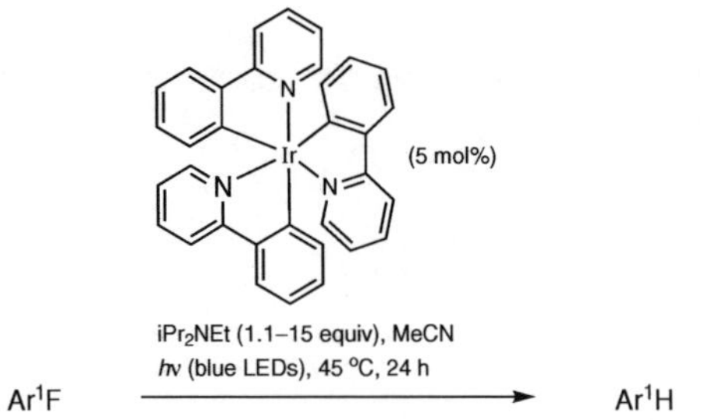

Substrate	iPr_2NEt (Equiv)	Product	Yield (%)	Ref
	1.1	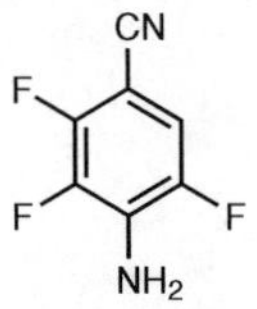	82	[42]

Table 1 (cont.)

Substrate	iPr_2NEt (Equiv)	Product	Yield (%)	Ref
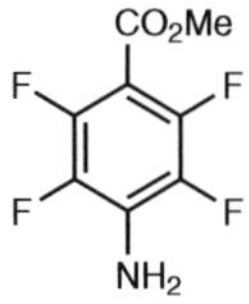	3.3	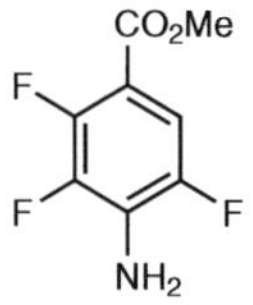	99	[42]
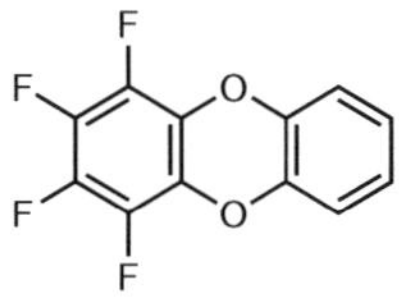	3.3	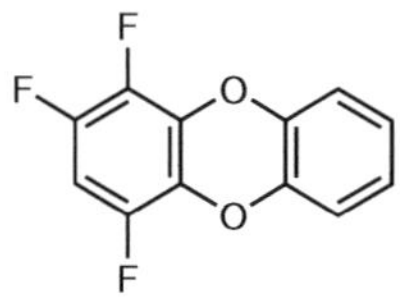	45	[42]
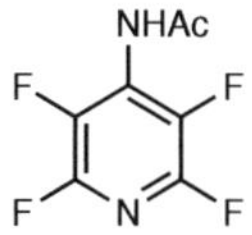	3.3	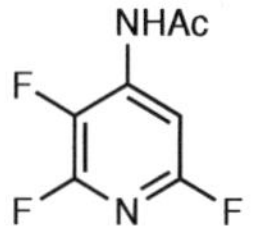	97	[42]
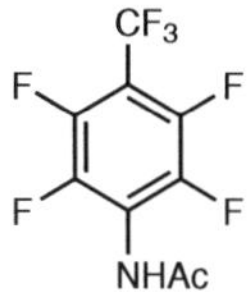	6	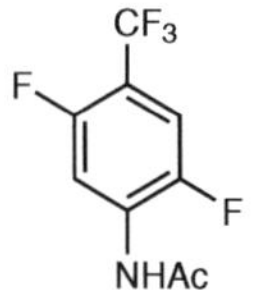	74[a]	[42]
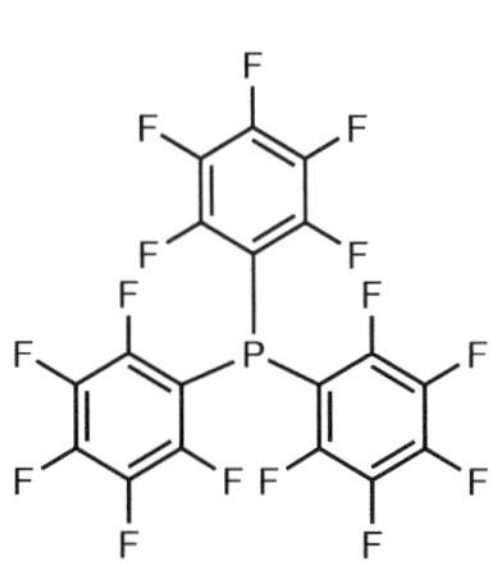	15	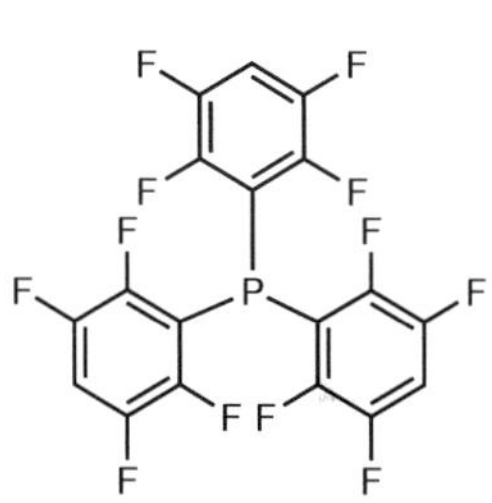	78	[42]

[a] 42 h reaction time.

In an example of photoredox heterogeneous catalysis, McTiernan and co-workers have shown that iodoarenes can be efficiently dehalogenated in the presence of *N*,*N*-diisopropylethylamine with platinum nanoparticles (0.2%) on titanium(IV) oxide (PtNPs@TiO_2) as catalyst under UVA/visible light irradiation (Scheme 17).[43] Here, the amine is thought to serve as a single-electron donor. This method is tolerant of esters, anilines, allylic alcohols, and alkenes. Notably, high chemoselectivity is observed for reductive deiodination in the presence of chlorine or bromine. An added advantage of this system is that the heterogeneous catalyst can be easily separated from the reaction mixture by centrifugation.

for references see p 373

Scheme 17 Heterogeneous Light-Mediated Reductive Deiodination of Arenes and an Alkene[43]

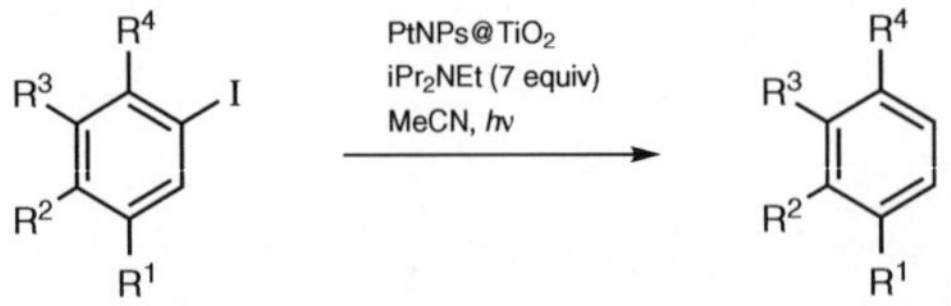

PtNPs@TiO_2
iPr_2NEt (7 equiv)
MeCN, hν

R^1	R^2	R^3	R^4	Time (h)	Yield (%)	Ref
H	CO_2Et	H	H	15	74	[43]
H	CO_2Me	H	H	15	70	[43]
CO_2Me	H	Br	H	15	87	[43]
H	H	Cl	NH_2	19	71	[43]

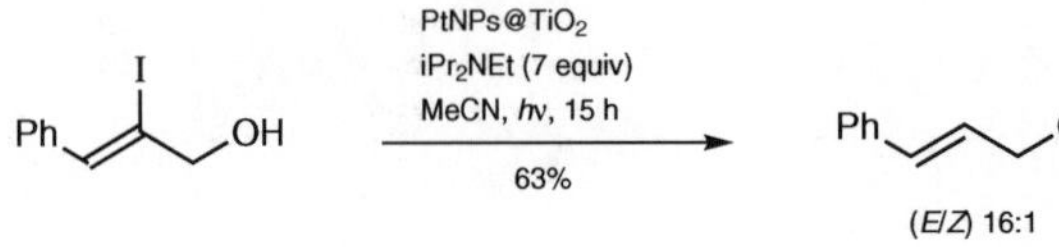

PtNPs@TiO_2
iPr_2NEt (7 equiv)
MeCN, hν, 15 h
63%

(E/Z) 16:1

Dehalogenated Products 30, 32, and 34; General Procedure:[41]
A flame-dried 10-mL round-bottomed flask equipped with a rubber septum and a magnetic stirrer bar was charged with tris(2,2′-bipyridyl)ruthenium(II) chloride hexahydrate (2.5 μmol, 0.025 equiv), the halide (0.10 mmol), iPr_2NEt (1.0 mmol, 10 equiv for formic acid mediated reactions; 0.20 mmol, 2 equiv for Hantzsch ester mediated reactions) and formic acid (1.0 mmol, 10 equiv) or Hantzsch ester (0.11 mmol, 1.1 equiv). The flask was purged with a stream of N_2 and anhyd DMF (1.0 mL) was added. The resultant mixture was degassed for 20 min by sparging with N_2 and then placed at a distance of ~8–10 cm in front of a 15-W fluorescent lamp. After the reaction was complete (as judged by TLC analysis), the mixture was poured into a separatory funnel containing Et_2O (25 mL) and H_2O (25 mL). The layers were separated and the aqueous layer was extracted with Et_2O (2 × 50 mL). The combined organic layers were dried (Na_2SO_4), filtered, and concentrated. The residue was purified by chromatography (silica gel).

1.12.3 Conclusions

As illustrated in this chapter, new catalytic methods for transforming organic halides into their hydrodehalogenated products continue to emerge. Many of these new heterogeneous and homogeneous catalytic systems are coupled with photoredox and nanoparticle chemistry. Lastly, the synthetic utility of several of the hydrodehalogenations described has been validated by chemists, uninvolved with the development of the chemistry, applying those methods to a variety of their own research needs.

The authors thank Dr. Steven P. Tanis for valuable discussions.

References

[1] Lunin, V.; Lokteva, E., *Russ. Chem. Bull.*, (1996) **45**, 1519.
[2] Mohn, W. W.; Tiedje, J. M., *Microbiol. Rev.*, (1992) **56**, 482.
[3] Jefford, C.; Kirkpatrick, D.; Delay, F., *J. Am. Chem. Soc.*, (1972) **94**, 8905.
[4] Guillaumet, G.; Mordenti, L.; Caubere, P., *J. Organomet. Chem.*, (1975) **92**, 43.
[5] Boukherroub, R.; Chatgilialoglu, C.; Manuel, G., *Organometallics*, (1996) **15**, 1508.
[6] Hutchins, R. O.; Hoke, D.; Keogh, J.; Koharski, D., *Tetrahedron Lett.*, (1969), 3495.
[7] Kuivila, H. G.; Menapace, L. W., *J. Org. Chem.*, (1963) **28**, 2165.
[8] Knochel, P.; Dohle, W.; Gommermann, N.; Kneisel, F. F.; Kopp, F.; Korn, T.; Sapountzis, I.; Vu, V. A., *Angew. Chem. Int. Ed.*, (2003) **42**, 4302.
[9] Bailey, W. F.; Patricia, J. J., *J. Organomet. Chem.*, (1988) **352**, 1.
[10] Alonso, F.; Beletskaya, I. P.; Yus, M., *Chem. Rev.*, (2002) **102**, 4009.
[11] Sisak, A.; Simon, O. B., In *The Handbook of Homogeneous Hydrogenation*, de Vries, J. G.; Elsevier, C. J., Eds.; Wiley-VCH: Weinheim, Germany, (2008); p 513.
[12] Yin, H.; Wada, Y.; Kitamura, T.; Yanagida, S., *Environ. Sci. Technol.*, (2001) **35**, 227.
[13] Bunnett, J. F., *Acc. Chem. Res.*, (1992) **25**, 2.
[14] Kwok, W. M.; Zhao, C.; Li, Y.-L.; Guan, X.; Wang, D.; Phillips, D. L., *J. Am. Chem. Soc.*, (2004) **126**, 3119.
[15] Douvris, C.; Nagaraja, C. M.; Chen, C.-H.; Foxman, B. M.; Ozerov, O. V., *J. Am. Chem. Soc.*, (2010) **132**, 4946.
[16] Mandal, P. K.; Birtwistle, J. S.; McMurray, J. S., *J. Org. Chem.*, (2014) **79**, 8422.
[17] Wullschleger, C. W.; Gertsch, J.; Altmann, K.-H., *Chem.–Eur. J.*, (2013) **19**, 13 105.
[18] Chan, K.-P.; Ling, Y. H.; Chan, J. L.-T.; Loh, T.-P., *J. Org. Chem.*, (2007) **72**, 2127.
[19] Monguchi, Y.; Kume, A.; Hattori, K.; Maegawa, T.; Sajiki, H., *Tetrahedron*, (2006) **62**, 7926.
[20] Kannan, M.; Raichurkar, A. V.; Khan, F. R. N.; Iyer, P. S., *Bioorg. Med. Chem. Lett.*, (2015) **25**, 1100.
[21] Zhang, Y.-M.; Gu, M.; Ma, H.; Tang, J.; Lu, W.; Nan, F.-J., *Chin. J. Chem.*, (2008) **26**, 962.
[22] Bravo-Altamirano, K.; George, K. M.; Frantz, M.-C.; LaValle, C. R.; Tandon, M.; Leimgruber, S.; Sharlow, E. R.; Lazo, J. S.; Wang, Q. J.; Wipf, P., *ACS Med. Chem. Lett.*, (2010) **2**, 154.
[23] Ramanathan, A.; Jimenez, L. S., *Synthesis*, (2010), 217.
[24] Arcadi, A.; Cerichelli, G.; Chiarini, M.; Vico, R.; Zorzan, D., *Eur. J. Org. Chem.*, (2004), 3404.
[25] Sawama, Y.; Yabe, Y.; Shigetsura, M.; Yamada, T.; Nagata, S.; Fujiwara, Y.; Maegawa, T.; Monguchi, Y.; Sajiki, H., *Adv. Synth. Catal.*, (2012) **354**, 777.
[26] Nakao, R.; Rhee, H.; Uozumi, Y., *Org. Lett.*, (2005) **7**, 163.
[27] Uozumi, Y.; Yamada, Y., *Chem. Rec.*, (2009) **9**, 51.
[28] Logan, M. E.; Oinen, M. E., *Organometallics*, (2006) **25**, 1052.
[29] Redwan, I. N.; Dyrager, C.; Solano, C.; Fernández de Trocóniz, G.; Voisin, L.; Bliman, D.; Meloche, S.; Grøtli, M., *Eur. J. Med. Chem.*, (2014) **85**, 127.
[30] Bhattacharjya, A.; Klumphu, P.; Lipshutz, B. H., *Org. Lett.*, (2015) **17**, 1122.
[31] Maleczka, R. E., Jr.; Rahaim, R. J.; Teixeira, R. R., *Tetrahedron Lett.*, (2002) **43**, 7087.
[32] Rahaim, R. J.; Maleczka, R. E., Jr., *Tetrahedron Lett.*, (2002) **43**, 8823.
[33] Blum, J.; Bitan, G.; Marx, S.; Vollhardt, K. P. C., *J. Mol. Catal.*, (1991) **66**, 313.
[34] Li, J.; Zheng, T.; Sun, H.; Li, X., *Dalton Trans.*, (2013) **42**, 13 048.
[35] Janni, M.; Peruncheralathan, S., *Org. Biomol. Chem.*, (2016) **14**, 3091.
[36] Chen, J.; Zhang, Y.; Yang, L.; Zhang, X.; Liu, J.; Li, L.; Zhang, H., *Tetrahedron*, (2007) **63**, 4266.
[37] Rasolofonjatovo, E.; Provot, O.; Hamze, A.; Rodrigo, J.; Bignon, J.; Wdzieczak-Bakala, J.; Lenoir, C.; Desravines, D.; Dubois, J.; Brion, J.-D.; Alami, M., *Eur. J. Med. Chem.*, (2013) **62**, 28.
[38] Akzinnay, S.; Bisaro, F.; Cazin, C. S. J., *Chem. Commun. (Cambridge)*, (2009), 5752.
[39] Fujita, K.-i.; Owaki, M.; Yamaguchi, R., *Chem. Commun. (Cambridge)*, (2002), 2964.
[40] Cannon, K. A.; Geuther, M. E.; Kelly, C. K.; Lin, S.; MacArthur, A. H. R., *Organometallics*, (2011) **30**, 4067.
[41] Narayanam, J. M. R.; Tucker, J. W.; Stephenson, C. R. J., *J. Am. Chem. Soc.*, (2009) **131**, 8756.
[42] Senaweera, S. M.; Singh, A.; Weaver, J. D., *J. Am. Chem. Soc.*, (2014) **136**, 3002.
[43] McTiernan, C. D.; Pitre, S. P.; Ismaili, H.; Scaiano, J. C., *Adv. Synth. Catal.*, (2014) **356**, 2819.

Keyword Index

A

Acetamides, *N*-(1-arylethyl)-, from *N*-(1-arylvinyl)acetamides by rhodium-catalyzed asymmetric hydrogenation 20–23
Acetamides, *N*-(1-arylvinyl)-, rhodium-catalyzed asymmetric hydrogenation to give *N*-(1-arylethyl)acetamides 20–23
Acrylamide, *N*-phenyl-, bisperoxidation/cyclization/hydrogenation sequence to give 3-hydroxy-3-(hydroxymethyl)indolin-2-one 319
Acrylates, substituted, asymmetric homogeneous hydrogenation to give chiral carboxylates 47, 48
Acrylic acids, α-substituted, iridium-catalyzed asymmetric hydrogenation to give chiral α-methyl carboxylic acids 43, 44
Adipic acid, from glucaric acid by a one-pot deoxygenation procedure 232
Alcohols, aliphatic, catalytic hydrodeoxygenation to give alkanes 231–233
Alcohols, alkyl, from alkyl aryl ethers by nickel-catalyzed hydrogenolysis 244, 245
Alcohols, from alkyl 2-hydroxyethyl ethers by directed hydrogenolysis 253–255
Alcohols, from carbonates by heterogeneous catalytic reduction 279–281
Alcohols, β-nitro, from β-nitroalkyl peroxides by chemoselective hydrogenation 324
Alcohols, primary, from 2-substituted tetrahydrofurans by palladium-catalyzed selective hydrogenolysis in an ionic liquid 256, 257
Alcohols, use as reductants in the deoxygenation of sulfoxides to sulfides 339–344
Aldehydes, from alkenes by ozonolysis/reduction in methanol 332–334
Aldehydes, heterogeneous gold-catalyzed reaction with primary amines in the presence of hydrogen/carbon dioxide to give tertiary methylamines 308
Alkanes, fluoro-, chiral, from alkenyl fluorides by asymmetric homogeneous hydrogenation 58, 59
Alkanes, from aliphatic alcohols by catalytic hydrodeoxygenation 231–233
Alkanes, from alkenes by hydrogenation using commercial palladium nanoparticles 68
Alkanes, from alkenes by hydrogenation using core-shell iron/iron oxide nanoparticles 82
Alkanes, from alkenes by hydrogenation using graphene-supported iron nanoparticles 81
Alkanes, from alkenes by hydrogenation using iridium nanoparticles in an ionic liquid 84
Alkanes, from alkenes by hydrogenation using magnetic carbon-supported palladium nanoparticles 78
Alkanes, from alkenes by hydrogenation using palladium nanoparticles in biphasic media 73, 74
Alkanes, from alkenes by hydrogenation using palladium nanoparticles in mesocellular foam 69, 70
Alkanes, from alkenes by hydrogenation using palladium nanoparticles in poly(ethylene glycol) 69
Alkanes, from alkenes by hydrogenation using palladium nanoparticles in polystyrene 71, 72
Alkanes, from alkenes by hydrogenation using palladium nanoparticles supported on magnetic carbon-coated cobalt nanobeads 77
Alkanes, from alkenes by hydrogenation using palladium on amine-terminated ferrite nanoparticles 76
Alkanes, from alkenes by hydrogenation using phenanthroline-stabilized palladium nanoparticles in poly(ethylene glycol) 70, 71
Alkanes, from alkenes by hydrogenation using polymer-supported iron nanoparticles under flow conditions 82, 83
Alkanes, from alkenes by hydrogenation using unsupported iron nanoparticles 79, 80
Alkanes, from alkenes by transfer hydrogenation using hydrazine as hydrogen donor with supported nickel nanoparticles 85
Alkanes, from alkenes by transfer hydrogenation using isopropanol as hydrogen donor, nanoparticle catalyzed 86–88
Alkanes, from alkynes by hydrogenation 195, 196
Alkanes, from allenes by platinum- or palladium-catalyzed hydrogenation 221, 222
Alkanes, from cyclic ethers by total hydrodeoxygenation 257, 258
Alkanes, from furans by total hydrodeoxygenation using noble metal and acid catalysts 259, 260
Alkanes, from iodoalkanes by hydrodeiodination under base-mediated hydrogenation conditions 355, 356
Alkanes, from nonfunctionalized alkenes by iridium-catalyzed asymmetric hydrogenation 8–13
Alkanols, 3-fluoro, from 3-fluoro allylic alcohols by iridium-catalyzed asymmetric hydrogenation 58, 59

Alkanols, 3-methoxy-, from 3-methoxyallyl alcohols by iridium-catalyzed asymmetric hydrogenation 39, 40
Alkanols, secondary, chiral, from alkenyl silyl ethers by asymmetric homogeneous hydrogenation using a frustrated Lewis pair catalyst and deprotection 41, 42
E-Alkenes, aryl-, from arylalkynes by homogeneous catalyzed semihydrogenation 216, 217
Alkenes, asymmetric homogeneous hydrogenation 7–62
Alkenes, asymmetric hydrogenation using palladium nanoparticles 75
Alkenes, carbonyl substituted, asymmetric catalytic reduction 42–54
Alkenes, cyclic, from cyclic allenes by selective monohydrogenation 222
Alkenes, fluoro-, asymmetric homogeneous hydrogenation to give chiral alkyl fluorides 58, 59
Z-Alkenes, from alkynes by heterogeneous semihydrogenation 199–202
Z-Alkenes, from alkynes by heterogeneous transfer hydrogenation 202–204
E-Alkenes, from alkynes by homogeneous semihydrogenation 214–219
Z-Alkenes, from alkynes by homogeneous semihydrogenation 204–212
E-Alkenes, from alkynes by homogeneous transfer hydrogenation 219, 220
Z-Alkenes, from alkynes by homogeneous transfer hydrogenation 212–214
Alkenes, from dienes by partial reduction via a protection/hydrogenation/deprotection strategy 96
Alkenes, from diols by catalytic deoxydehydration 233–239
Alkenes, from *vic*-diols by reduction using formic acid 233–235
Alkenes, from iodoalkenes by heterogeneous light-mediated reductive deiodination 372
Alk-1-enes, from terminal alkynes by heterogeneous semihydrogenation 199, 200
Alkenes, hydrogenation using commercial palladium nanoparticles 68
Alkenes, hydrogenation using core-shell iron/iron oxide nanoparticles 82
Alkenes, hydrogenation using graphene-supported iron nanoparticles 81
Alkenes, hydrogenation using iridium nanoparticles in an ionic liquid 84
Alkenes, hydrogenation using magnetic carbon-supported palladium nanoparticles 78
Alkenes, hydrogenation using palladium nanoparticles in biphasic media 73, 74
Alkenes, hydrogenation using palladium nanoparticles in mesocellular foam 69, 70
Alkenes, hydrogenation using palladium nanoparticles in poly(ethylene glycol) 69
Alkenes, hydrogenation using palladium nanoparticles in polystyrene 71, 72
Alkenes, hydrogenation using palladium nanoparticles supported on magnetic carbon-coated cobalt nanobeads 77
Alkenes, hydrogenation using palladium on amine-terminated ferrite nanoparticles 76
Alkenes, hydrogenation using phenanthroline-stabilized palladium nanoparticles in poly(ethylene glycol) 70, 71
Alkenes, hydrogenation using polymer-supported iron nanoparticles under flow conditions 82, 83
Alkenes, hydrogenation using unsupported iron nanoparticles 79
Alkenes, nitrogen-substituted, asymmetric catalytic reduction 14–33
Alkenes, nonfunctionalized, iridium-catalyzed asymmetric hydrogenation 8–13
Alkenes, oxygen-substituted, asymmetric catalytic reduction 34–41
Alkenes, ozonolysis/reduction in methanol to give aldehydes or ketones 332–334
Alkenes, phosphorus-, boron-, or sulfur-substituted, asymmetric catalytic reduction 54–62
Alkenes, symmetrical 1,2-disubstituted, ozonolysis in methanol and hydrogenation to give methyl hemiacetals 331, 332
Alkenes, transfer hydrogenation using hydrazine as hydrogen donor with supported nickel nanoparticles 85
Alkenes, transfer hydrogenation using isopropanol as hydrogen donor, nanoparticle catalyzed 86–88
Alkenylboronates, asymmetric homogeneous hydrogenation to give chiral alkylboronates 57, 58
N-Alkenylcarboxamides, rhodium-catalyzed asymmetric hydrogenation to give *N*-alkylcarboxamides 24–26
Alkenylphosphine oxides, asymmetric homogeneous hydrogenation to give chiral alkylphosphine oxides 55, 56
Alkenylphosphonates, asymmetric homogeneous hydrogenation to give chiral alkylphosphonates 55, 56
Alkenylsilanes, ozonolysis/hydrogenation to give α-hydroxy ketones 321
Alkenyl silyl ethers, asymmetric homogeneous hydrogenation using a frustrated Lewis pair catalyst and deprotection to give chiral secondary alcohols 41, 42
Alkenyl sulfides, asymmetric homogeneous hydrogenation to give chiral alkyl sulfides 59–61
Alkenyl sulfones, asymmetric homogeneous hydrogenation to give chiral alkyl sulfones 59–61

Alkoxysilanes, chiral, from silyl enol ethers by asymmetric homogeneous hydrogenation 41, 42
Alkylbenzenes, from styrenes by hydrogenation using palladium nanoparticles on amphiphilic supports 72, 73
N-Alkylcarboxamides, from *N*-alkenylcarboxamides by rhodium-catalyzed asymmetric hydrogenation 24–26
Alkyl carboxylates, chiral, from enol esters by asymmetric homogeneous hydrogenation 35–37
Alkyl diphenylphosphinates, from vinyl diphenylphosphinates by iridium-catalyzed asymmetric hydrogenation 38, 39
Alkyl ethers, from enol ethers by iridium-catalyzed asymmetric hydrogenation 39, 40
Alkyl fluorides, chiral, from alkenyl fluorides by asymmetric homogeneous hydrogenation 58, 59
α-Alkylidene ketones, cyclic, iridium-catalyzed asymmetric hydrogenation to give α-chiral cyclic ketones 53
3-Alkylidenepiperidin-2-ones, iridium-catalyzed asymmetric hydrogenation to give chiral 3-alkylpiperidin-2-ones 50, 51
3-Alkylidenepyrrolidin-2-ones, iridium-catalyzed asymmetric hydrogenation to give chiral 3-alkylpyrrolidin-2-ones 50, 51
3-Alkylpiperidin-2-ones, chiral, from 3-alkylidenepiperidin-2-ones by iridium-catalyzed asymmetric hydrogenation 50, 51
3-Alkylpyrrolidin-2-ones, chiral, from 3-alkylidenepyrrolidin-2-ones by iridium-catalyzed asymmetric hydrogenation 50, 51
Alkynes, aryl-, homogeneous catalyzed semihydrogenation to give *E*-arylalkenes 216, 217
Alkynes, heterogeneous semihydrogenation to give *Z*-alkenes 199–202
Alkynes, heterogeneous transfer hydrogenation to give *Z*-alkenes 202–204
Alkynes, homogeneous semihydrogenation to give *E*-alkenes 214–219
Alkynes, homogeneous semihydrogenation to give *Z*-alkenes 204–212
Alkynes, homogeneous transfer hydrogenation to give *E*-alkenes 219, 220
Alkynes, homogeneous transfer hydrogenation to give *Z*-alkenes 212–214
Alkynes, terminal, heterogeneous semihydrogenation to give alk-1-enes 199, 200
Alkynes, total hydrogenation to give alkanes 195, 196
Allenes, cyclic, selective monohydrogenation to give cycloalkenes 222
Allenes, hydrogenation to give alkenes or alkanes 220–224
Allenyl ketones, palladium-catalyzed hydrogenation to give saturated ketones 223, 224
Allylic alcohols, alkenyl, selective hydrogenation to give alkenyl alcohols 99–102
Allylic alcohols, 3-fluoro, iridium-catalyzed asymmetric hydrogenation to give chiral 3-fluoro alkyl alcohols 58, 59
Allylic alcohols, 3-methoxy, iridium-catalyzed asymmetric hydrogenation to give chiral 3-methoxypropan-1-ols 39, 40
Amines, heterogeneous gold-catalyzed dimethylation using hydrogen and carbon dioxide to give dimethylamines 308
Amines, methylation to give methylamines using hydrosilanes/carbon dioxide 305
Amines, primary, heterogeneous gold-catalyzed disubstitution using hydrogen/carbon dioxide and aldehydes to give tertiary methylamines 308
Amines, reductive N-methylation using hydrogen and carbonates 282, 283
β-Amino acid esters, from α,β-unsaturated β-amino acid esters by rhodium- and ruthenium-catalyzed asymmetric hydrogenation 32, 33
α-Amino acid esters, from α,β-unsaturated α-amino acid esters by rhodium-catalyzed asymmetric hydrogenation 27–31
α-Amino acids, from α,β-unsaturated α-amino acids by rhodium-catalyzed asymmetric hydrogenation 27, 28
α-Amino acids, reduction/iodination/hydrodeiodination sequence to give ethylamines 355, 356
Anilines, *N*,*N*-dimethyl-, from nitrobenzenes by reductive dimethylation using hydrogen and carbon dioxide 308, 309
Anilines, dimethylation using dimethyl carbonate and molecular hydrogen to give *N*,*N*-dimethylanilines 282, 283
Anilines, heterogeneous palladium-catalyzed methylation using hydrogen and carbon dioxide to give *N*-methylanilines 307
Anilines, homogeneous ruthenium-catalyzed methylation using hydrogen and carbon dioxide to give *N*-methyl- or *N*,*N*-dimethylanilines 305–307
Anthracene, platinum-catalyzed hydrogenation to give 1,2,3,4-tetrahydro- or 1,2,3,4,5,6,7,8-octahydroanthracene 140
Anthracene, rhodium-catalyzed hydrogenation to give 1,2,3,4,5,6,7,8-octahydroanthracene 139
Anthracene, ruthenium-catalyzed hydrogenation to give 1,2,3,4-tetrahydro- or 1,2,3,4,5,6,7,8-octahydroanthracene 135, 136
Anthracenes, tris(pentafluorophenyl)borane/phosphine catalyzed hydrogenation to give 9,10-dihydroanthracenes 145
Arenes, deuterated, from haloarenes by catalytic deuterodehalogenation 364, 365

Arenes, from alkyl aryl ethers by nickel-catalyzed hydrogenolysis 244–246
Arenes, from aryl ethers by iron(III)- or cobalt(II)-catalyzed reductive cleavage with lithium aluminum hydride 248
Arenes, from bromoarenes by hydrodebromination under microwave-assisted copper-catalyzed concurrent tandem catalysis 367, 368
Arenes, from chloroarenes by hydrodechlorination using a polymer-supported nanopalladium catalyst 360, 361
Arenes, from chloroarenes by hydrodechlorination using homogeneous palladium(II) acetate/sodium formate 362
Arenes, from chloroarenes by palladium on charcoal catalyzed hydrodechlorination 357–359
Arenes, from chloroarenes by palladium on charcoal catalyzed transfer reduction using sodium formate 359
Arenes, from chloroarenes by rhodium-catalyzed transfer hydrodechlorination 367
Arenes, from diaryl ethers by nickel-catalyzed hydrogenolysis 244, 247
Arenes, from fluoroarenes by photocatalytic hydrodefluorination 370, 371
Arenes, from fluoroarenes by platinum on charcoal catalyzed hydrodefluorination 359, 360
Arenes, from haloarenes by hydrodehalogenation in water with in situ generated molecular hydrogen 362, 363
Arenes, from haloarenes by palladium(II) acetate catalyzed hydrodehalogenation 365, 366
Arenes, from iodoarenes by heterogeneous light-mediated reductive deiodination 372
Arenes, from phenols by catalytic hydrodeoxygenation 229, 230
Arenes, iridium-catalyzed hydrogenation to give cyclohexanes 133, 134
Arenes, monocyclic, reduction to cycloalkanes 127–135
Arenes, monocyclic, reduction to cycloalkenes 145–149
Arenes, palladium polyoxometalate catalyzed hydrogenation to give cyclohexanes 134, 135
Arenes, polycyclic, partial reduction 135–145
Arenes, reductive methylation to give methylarenes using hydrogen and carbon dioxide 310
Arenes, rhodium-catalyzed hydrogenation to give cyclohexanes 131, 132
Arenes, rhodium-catalyzed hydrogenation to give cyclohexenes 148
Arenes, ruthenium-catalyzed hydrogenation to give cyclohexanes 128–130
Arenes, ruthenium-catalyzed hydrogenation to give cyclohexenes 147
Azelaic acid, from oleic acid by oxidative cleavage 115
Azelaic acid and aldehyde, from oleic acid derivatives by ozonolysis and subsequent oxidative or reductive workup 334, 335

B

Benzaldehydes, one-carbon homologation involving hydrogenation of a sulfoxide catalyzed by ruthenium nanoparticles on titanium(IV) oxide to give phenylacetaldehydes 347
Benzene, evolution of catalysts for the reduction to cyclohexane 127, 128
Benzene, evolution of catalysts for the reduction to cyclohexene 145, 146
Benzene, from phenol by ruthenium-catalyzed hydrogenolysis 229, 230
Benzofurans, rhodium-catalyzed enantioselective hydrogenation to give chiral 2,3-dihydrobenzofurans 185, 186
Benzofurans, ruthenium/NHC-catalyzed enantioselective hydrogenation to give chiral 2,3-dihydrobenzofurans 182, 183
Benzonitriles, reductive dimethylation to give *N,N*-dimethylbenzylamines using hydrogen and carbon dioxide 308, 309
4*H*-1-Benzopyrans, 2-substituted, iridium-catalyzed asymmetric hydrogenation to give 2-substituted 3,4-dihydro-2*H*-1-benzopyrans 40
Benzotetraphenes, palladium-catalyzed hydrogenation to give di- or tetrahydrobenzotetraphenes 141–143
Benzothiophenes, ruthenium/NHC-catalyzed enantioselective hydrogenation to give chiral 2,3-dihydrobenzothiophenes 182, 183
Benzyl alcohols, chiral, from benzoates by asymmetric synthesis involving hydrogenation of a sulfoxide catalyzed by ruthenium nanoparticles on titanium(IV) oxide 347
Benzylamines, *N,N*-dimethyl-, from benzonitriles by reductive dimethylation using hydrogen and carbon dioxide 308, 309
Bicarbonate, sodium, hydrogenation to sodium formate using a homogeneous iron–PNP complex 271, 272
Bicarbonate, sodium, hydrogenation to sodium formate using homogeneous iron/tetradentate phosphine complexes 270, 271
Bicarbonate, sodium, hydrogenation to sodium formate using homogeneous phosphine-free iron complexes 272
Bicarbonates, electroreduction using polymer-coated palladium electrodes to give formates 284
Bicarbonates, heterogeneous catalytic reduction to give formates 275

Bicarbonates, homogeneous ruthenium-catalyzed transfer hydrogenation to give formates 273, 274
Bicarbonates, hydrogenation to formates using homogeneous metal–NHC complexes 273
Biodiesel, oxidative stability 118, 119
Biphenyl, from polychlorinated biphenyls by catalytic hydrodechlorination 366
Biphenyl-2-ol, from dibenzofuran by iron(III)- or cobalt(II)-catalyzed reductive cleavage with lithium aluminum hydride 248
Boronates, alkyl-, chiral, from alkenylboronates by asymmetric homogeneous hydrogenation 57, 58
α-Bromo amides, photocatalyzed reductive dehalogenation to give carboxamides 369
Bromoarenes, catalytic deuterodebromination to give deuterated arenes 364, 365
Bromoarenes, hydrodebromination in water with in situ generated molecular hydrogen to give arenes 362, 363
Bromoarenes, hydrodebromination under microwave-assisted copper-catalyzed concurrent tandem catalysis to give arenes 367, 368
Bromoarenes, palladium(II) acetate catalyzed hydrodebromination to give arenes 365
α-Bromo ketones, palladium(II) acetate catalyzed hydrodebromination to give ketones 365, 366
Bromotetrahydropyrans, chemoselective hydrodehalogenation to give tetrahydropyrans 356, 357
Buta-1,3-diene, catalytic 1,2- and 1,4-hydrogenation to give butenes 92
Buta-1,3-diene, from *meso*-erythritol by rhenium-catalyzed deoxydehydration 238
Butane-1,3-diol, from tetrahydrofuran-3-ol by directed hydrogenolysis 255
Butenes, from buta-1,3-diene by catalytic 1,2- or 1,4-hydrogenation 92
Butenes, 2-methyl-, from isoprene by semihydrogenation 92–95

C

Camelina oil, methyl esters, oxidizability and C18:3 content before and after hydrogenation 119, 120
Carbon dioxide, heterogeneous hydrogenation to give formic acid or formate salts 296–298
Carbon dioxide, heterogeneous hydrogenation to give methane 302–304
Carbon dioxide, heterogeneous hydrogenation to give methanol 301, 302
Carbon dioxide, homogeneous hydrogenation to give formate salts using nonprecious-metal-based catalysts 295, 296
Carbon dioxide, homogeneous hydrogenation to give formic acid or formate salts 290–296
Carbon dioxide, homogeneous hydrogenation to give methanol 299, 300
Carbon dioxide, homogeneous iridium-catalyzed hydrogenation to give formate salts 294, 295
Carbon dioxide, homogeneous rhodium-catalyzed hydrogenation to give formic acid or formate salts 291, 292
Carbon dioxide, homogeneous ruthenium-catalyzed hydrogenation to give formic acid or formate salts 292, 293
Carbon dioxide, hydrogenation 289–310
Carbonate, dimethyl, heterogeneous copper-catalyzed hydrogenation to give methanol 280, 281
Carbonate, dimethyl, homogeneous ruthenium-catalyzed hydrogenation to give methanol 275, 276
Carbonates, cyclic, hydrogenation to give 1,2-diols 276, 277
Carbonates, cyclic, ruthenium-catalyzed transfer hydrogenation to give 1,2-diols 278, 279
Carbonates, heterogeneous catalytic reduction to give alcohols 279–282
Carbonates, metal, catalytic reduction to methane 282
Carboxamides, chiral, from enamides by asymmetric homogeneous hydrogenation 20–27
Carboxamides, chiral, from α,β-unsaturated amides by asymmetric homogeneous hydrogenation 48–51
Carboxamides, *N*-cycloalkenyl-, rhodium-catalyzed asymmetric hydrogenation to give *N*-cycloalkylcarboxamides 24–26
Carboxamides, *N*-cycloalkyl-, from *N*-cycloalkenylcarboxamides by rhodium-catalyzed asymmetric hydrogenation 24–26
Carboxamides, from α-halo amides by photocatalyzed reductive dehalogenation 369
Carboxylates, α-acyloxy, from α,β-unsaturated α-acyloxy esters by rhodium- and ruthenium-catalyzed asymmetric hydrogenation 36, 37
Carboxylates, chiral, from α,β-unsaturated esters by asymmetric homogeneous hydrogenation 47, 48
Carboxylic acids, chiral, from α-substituted α,β-unsaturated acids by asymmetric hydrogenation using palladium nanoparticles/cinchonidine 75
Carboxylic acids, chiral, from α,β-unsaturated carboxylic acids by asymmetric homogeneous hydrogenation 43–46
Carboxylic esters, from α-chloro esters by photocatalyzed reductive dehalogenation 369, 370
Carboxylic esters, saturated, from α,β,γ,δ-unsaturated esters by palladium-catalyzed hydrogenation 224

Cardanol, homogeneous ruthenium-catalyzed partial hydrogenation to give monounsaturated alkenylphenols 113, 114
Cardoon oil, methyl esters, ruthenium-catalyzed partial hydrogenation to give methyl octadecenoates 112
Cardoon oil, noble metal based heterogeneous partial hydrogenation to give monounsaturated products 115
Cetane number, of hydrogenated vegetable oils 118, 119
α-Chloro amides, photocatalyzed reductive dehalogenation to give carboxamides 369
Chloroarenes, catalytic deuterodechlorination to give deuterated arenes 364, 365
Chloroarenes, hydrodechlorination in water with in situ generated molecular hydrogen to give arenes 362, 363
Chloroarenes, hydrodechlorination using a polymer-supported nanopalladium catalyst to give arenes 360, 361
Chloroarenes, hydrodechlorination using homogeneous palladium(II) acetate/sodium formate to give arenes 362
Chloroarenes, palladium on charcoal catalyzed transfer reduction using sodium formate to give arenes 359
Chloroarenes, palladium on charcoal catalyzed triethylamine-mediated hydrodechlorination to give arenes 357, 358
Chloroarenes, rhodium-catalyzed transfer hydrodechlorination to give arenes 367
Chlorobiphenyls, catalytic hydrodechlorination to give biphenyl 366
α-Chloro esters, photocatalyzed reductive dehalogenation to give carboxylic esters 369, 370
Chloropyrimidines, from dichloropyrimidines by palladium on charcoal catalyzed regioselective dechlorination 358, 359
Cholestanol, from cholesterol by hydrogenation using commercial palladium nanoparticles 68
Cholesterol, hydrogenation using commercial palladium nanoparticles to give cholestanol 68
Chrysene, palladium-catalyzed hydrogenation to give 5,6-dihydrochrysene 142
Citronellol, from geraniol by enantioselective semihydrogenation using a homogeneous catalyst 99, 100
Citronellol, from geraniol by semihydrogenation using a heterogeneous catalyst 101, 102
Cycloalkanes, from monocyclic arenes by reduction 127–135
Cycloalkenes, from monocyclic arenes by reduction 145–149
Cyclododeca-1,5,9-triene, catalytic semihydrogenation to give cyclododeca-1,5-dienes and cyclododecene 106–108
Cyclododecene, from cyclododeca-1,5,9-triene by catalytic semihydrogenation 106–108
Cyclohexane, from benzene by reduction, evolution of catalysts 127, 128
Cyclohexanes, from arenes by iridium-catalyzed hydrogenation 133, 134
Cyclohexanes, from arenes by palladium polyoxometalate catalyzed hydrogenation 134, 135
Cyclohexanes, from arenes by rhodium-catalyzed hydrogenation 131, 132
Cyclohexanes, from arenes by ruthenium-catalyzed hydrogenation 128–130
Cyclohexanes, from chloroarenes by palladium on charcoal catalyzed transfer hydrogenation using sodium formate 359
Cyclohexane-1,2,3,4-tetraols, from 2,3-dioxabicyclo[2.2.2]octane-5,6-diols by dihydroxylation/hydrogenation 329
Cyclohexanol, from methoxyphenols by hydrogenolysis using ruthenium-on-carbon and magnesium oxide 252
Cyclohexene, from benzene by reduction, evolution of catalysts 145, 146
Cyclohexenes, from arenes by rhodium-catalyzed hydrogenation 148
Cyclohexenes, from arenes by ruthenium-catalyzed hydrogenation 147
Cycloocta-1,5-diene, catalytic semihydrogenation to give cyclooctene 103–105
Cyclooctene, from cycloocta-1,5-diene by catalytic semihydrogenation 103–105
Cyclopentanol, from furan-2-carbaldehyde by catalytic reduction 260–262
Cyclopentanones, from furan-2-carbaldehydes by catalytic reduction 260–263

D

Decahydropyrene, from pyrene by rhodium-catalyzed hydrogenation 139
Decahydroquinolines, from quinolines by heterogeneous rhodium-catalyzed hydrogenation 154
cis-Decalin, from naphthalene by ruthenium-catalyzed hydrogenation 136
α,β-Dehydro-α-amino acids and esters, asymmetric homogeneous hydrogenation to give chiral α-amino acid derivatives 27–31
α,β-Dehydro-β-amino acids and esters, asymmetric homogeneous hydrogenation to give chiral β-amino acid derivatives 32, 33
Deoxygenation, of alcohols 229–239
2-Diazo-*N*-phenylacetamides, 2-cyano-, cyclization/peroxidation/hydrolysis/hydrogenation sequence to give 1-alkyl-3-hydroxy-2-oxoindoline-3-carboxamides 319, 320

Dibenzofuran, iron(III)- or cobalt(II)-catalyzed reductive cleavage with lithium aluminum hydride to give biphenyl-2-ol 248
Dibutyl muconate, from mucic acid dibutyl ester by rhenium-catalyzed reduction 237
Dienes, partial reduction via a protection/hydrogenation/deprotection strategy to give alkenes 96
Diethyl 2,3-dihydroxysuccinate, rhenium-catalyzed deoxydehydration to give diethyl fumarate 236
9,10-Dihydroanthracenes, from anthracenes by tris(pentafluorophenyl)borane/phosphine catalyzed hydrogenation 145
2,3-Dihydrobenzofurans, chiral, from benzofurans by rhodium-catalyzed enantioselective hydrogenation 185, 186
2,3-Dihydrobenzofurans, chiral, from benzofurans by ruthenium/NHC-catalyzed enantioselective hydrogenation 182, 183
3,4-Dihydro-2*H*-1-benzopyrans, 2-substituted, from 2-substituted 4*H*-1-benzopyrans by iridium-catalyzed asymmetric hydrogenation 40
Dihydrobenzotetraphenes, from benzotetraphenes by palladium-catalyzed hydrogenation 142, 143
2,3-Dihydrobenzothiophenes, chiral, from benzothiophenes by ruthenium/NHC-catalyzed enantioselective hydrogenation 182, 183
5,6-Dihydrochrysene, from chrysene by palladium-catalyzed hydrogenation 142
4,5-Dihydroimidazoles, chiral, from imidazoles by ruthenium-catalyzed enantioselective hydrogenation 181
2,3-Dihydroindoles, chiral, from indoles by iridium-catalyzed enantioselective hydrogenation 172, 174, 176
2,3-Dihydroindoles, chiral, from indoles by palladium-catalyzed enantioselective hydrogenation 186–189
2,3-Dihydroindoles, chiral, from indoles by rhodium-catalyzed enantioselective hydrogenation 185, 186
2,3-Dihydroindoles, chiral, from indoles by ruthenium-catalyzed enantioselective hydrogenation 179, 180
2,3-Dihydroindoles, from indoles by rhodium-catalyzed hydrogenation 166, 167
1,2-Dihydroisoquinolines, chiral, from isoquinolines by iridium-catalyzed enantioselective hydrogenation 172, 173, 176
1,2-Dihydrolimonene, from limonene by semihydrogenation 95–97
4,5-Dihydrooxazoles, chiral, from oxazoles by ruthenium-catalyzed enantioselective hydrogenation 182
9,10-Dihydrophenanthrene, from phenanthrene by palladium-catalyzed hydrogenation 142
9,10-Dihydrophenanthrene, from phenanthrene by ruthenium-catalyzed hydrogenation 136
4,5-Dihydropyrazoles, chiral, from pyrazoles by palladium-catalyzed enantioselective hydrogenation 187, 188
4,5-Dihydropyrene, from pyrene by palladium-catalyzed hydrogenation 143
4,5-Dihydropyrene, from pyrene by platinum-catalyzed hydrogenation 141
4,5-Dihydropyrene, from pyrene by ruthenium-catalyzed hydrogenation 137
3,4-Dihydropyrroles, chiral, from pyrroles by palladium-catalyzed enantioselective hydrogenation 187, 188
2,3-Dihydropyrroles, chiral, from pyrroles by ruthenium-catalyzed enantioselective hydrogenation 180, 181
3,4-Dihydroxydihydrofuran-2(3*H*)-ones, rhenium-catalyzed deoxydehydration to give furan-2(5*H*)-ones 237
1,4-Diketones, from furan-2-carbaldehydes by catalytic reduction 260–263
Dimethylamines, from primary amines by heterogeneous gold-catalyzed dimethylation using hydrogen and carbon dioxide 308
N,*N*-Dimethylanilines, from anilines by dimethylation using dimethyl carbonate and molecular hydrogen 282, 283
Dimethyl maleate, ozonolysis in methanol and subsequent hydrogenation to give methyl 2-hydroxy-2-methoxyacetate 332
Diols, catalytic deoxydehydration to give alkenes 233–239
Diols, from cyclic carbonates by heterogeneous catalytic reduction 279–281
Diols, from cyclic carbonates by homogeneous ruthenium-catalyzed hydrogenation 277, 278
1,2-Diols, from cyclic carbonates by ruthenium-catalyzed transfer hydrogenation 278, 279
1,2-Diols, from furan-2-carbaldehydes by catalytic reduction 260–262
1,ω-Diols, from α-hydroxymethyl cyclic ethers by directed hydrogenolysis 253–255
1,2-Diols, reduction to alkenes using formic acid 233–235
2,7-Dioxabicyclo[2.2.1]heptanes, from 2,3,5-trioxabicyclo[2.2.2]oct-7-enes by hydrogenation 330, 331
2,3-Dioxabicyclo[2.2.2]octane-5,6-diols, dihydroxylation/hydrogenation to give cyclohexane-1,2,3,4-tetraols 329
2,3-Dioxabicyclo[2.2.2]oct-5-enes, peroxide and alkene hydrogenation to give 4-hydroxycyclohexanone derivatives 329, 330
1,3-Dioxan-2-one, homogeneous ruthenium-catalyzed hydrogenation to give propane-1,3-diol 277, 278

1,2-Dioxins, dihydroxylation/hydrogenation to give 1,2,3,4-tetraols 327, 328
1,3-Dioxolan-2-ones, homogeneous ruthenium-catalyzed hydrogenation to give 1,2-diols 277
Diphenylacetylene, homogeneous copper-catalyzed semihydrogenation to give (*Z*)-stilbene 208, 209
Diphenylacetylene, homogeneous iridium-catalyzed transfer hydrogenation to give (*E*)-stilbene 219, 220
Diphenylacetylene, homogeneous iron-catalyzed semihydrogenation to give (*E*)-stilbene 218
Diphenylacetylene, homogeneous ruthenium-catalyzed semihydrogenation to give (*E*)-stilbene 217, 218
Diphenylacetylene, semihydrogenation using a rhodium–TangPhos catalyst to give (*Z*)-stilbene 204, 205
Disulfides, hydrogenation/hydrogenolysis using palladium on charcoal to give thiols 351, 352
Dodecahydrotriphenylene, from triphenylene by platinum-catalyzed hydrogenation 141
Dodecahydrotriphenylene, from triphenylene by rhodium-catalyzed hydrogenation 139
Dodecahydrotriphenylene, from triphenylene by ruthenium-catalyzed hydrogenation 137

E

Electrochemical reduction, of carbonates 283–285
Enamides, asymmetric homogeneous hydrogenation to give chiral amides 20–27
Endoperoxides, formed by [2 + 4] cycloaddition of singlet oxygen to cyclic 1,3-dienes, hydrogenation to effect 1,4-dihydroxylation 329–331
Ene reaction, of singlet oxygen with a C=C bond 325
Enol acetates, cyclic, rhodium-catalyzed asymmetric hydrogenation to give cycloalkyl acetates 37
Enol esters, asymmetric homogeneous hydrogenation to give chiral alkyl carboxylates 35–37
Enol ethers, asymmetric homogeneous hydrogenation to give chiral alkyl ethers 39–41
Enol phosphinates, asymmetric homogeneous hydrogenation to give chiral alkyl phosphinates 38, 39
Enones, asymmetric homogeneous hydrogenation to give α-chiral ketones 52–54
Enzymatic reduction, of carbonates 283–285
6-Epiplakortolide E 327
meso-Erythritol, rhenium-catalyzed deoxydehydration to give buta-1,3-diene 238
Ethers, alkyl aryl, nickel-catalyzed hydrogenolysis to give arenes and alkanols 244–246
Ethers, alkyl 2-hydroxyethyl, directed hydrogenolysis to give alcohols 253–255
Ethers, aryl, iron(III)- or cobalt(II)-catalyzed reductive cleavage with lithium aluminum hydride to give arenes 248
Ethers, cyclic, α-hydroxymethyl, directed hydrogenolysis to give 1,ω-diols 253–255
Ethers, cyclic, total hydrodeoxygenation to give alkanes 257, 258
Ethers, diaryl, nickel-catalyzed hydrogenolysis to give benzenes and phenols 244, 247
Ethers, methyl phenyl, iridium-catalyzed hydrogenolysis to give phenols 252
Ethylamines, from α-amino acids by a reduction/iodination/hydrodeiodination sequence 355, 356
Ethyl carboxylates, chiral, from vinyl carboxylates by rhodium-catalyzed asymmetric hydrogenation 35, 36

F

Farnesol, asymmetric transfer hydrogenation to give a diene 100
Fluoroarenes, from perfluoroarenes by homogeneous cobalt(0)-catalyzed selective defluorination 363, 364
Fluoroarenes, photocatalytic hydrodefluorination to give arenes 370, 371
Fluoroarenes, platinum on charcoal catalyzed hydrodefluorination to give arenes 359, 360
Formate, sodium, from sodium bicarbonate by homogeneous hydrogenation using iron/tetradentate phosphine complexes 270, 271
Formate, sodium, from sodium bicarbonate by hydrogenation using a homogeneous iron–PNP complex 271, 272
Formate, sodium, from sodium bicarbonate by hydrogenation using homogeneous phosphine-free iron complexes 272
Formates, from bicarbonates by electroreduction using polymer-coated palladium electrodes 284
Formates, from bicarbonates by heterogeneous catalytic reduction 275
Formates, from bicarbonates by homogeneous ruthenium-catalyzed transfer hydrogenation 273, 274
Formates, from bicarbonates by hydrogenation using homogeneous metal–NHC complexes 273
Formates, from carbon dioxide by heterogeneous hydrogenation 296–298
Formates, from carbon dioxide by homogeneous hydrogenation using nonprecious-metal-based catalysts 295, 296
Formates, from carbon dioxide by homogeneous iridium-catalyzed hydrogenation 294, 295

Formates, from carbon dioxide by homogeneous rhodium-catalyzed hydrogenation 291, 292
Formates, from carbon dioxide by homogeneous ruthenium-catalyzed hydrogenation 292, 293
Formic acid, from carbon dioxide by homogeneous rhodium-catalyzed hydrogenation 291, 292
Formic acid, from carbon dioxide by homogeneous ruthenium-catalyzed hydrogenation 292, 293
Formic acid, reversible synthesis from hydrogen and carbon dioxide and its potential use as a hydrogen-storage material 290
Fumarate, diethyl, from diethyl 2,3-dihydroxysuccinate by rhenium-catalyzed deoxydehydration 236
Furan-2-carbaldehyde, catalytic hydrogenation to give pentane-1,5-diol 258, 259
Furan-2(5*H*)-ones, from 3,4-dihydroxydihydrofuran-2(3*H*)-ones by rhenium-catalyzed deoxydehydration 237
Furans, catalytic reduction to give cyclopentanones, cyclopentanols, acyclic diols, or diketones 260–263
Furans, rhodium-catalyzed hydrogenation to give tetrahydrofurans 166, 167
Furans, ruthenium/NHC-catalyzed enantioselective hydrogenation to give chiral tetrahydrofurans 182, 183
Furans, total hydrodeoxygenation using noble metal and acid catalysts to give alkanes 259, 260

G

Geraniol, cyclization/hydrogenation using a heterogeneous catalyst to give isopulegol and menthol 101, 102
Geraniol, enantioselective semihydrogenation using a homogeneous catalyst to give citronellol 99, 100
Geraniol, partial reduction 99–102
Geraniol, semihydrogenation using a heterogeneous catalyst to give citronellol 101, 102
Glucaric acid, one-pot deoxygenation procedure to give adipic acid 232
Glycerol, homogeneous ruthenium-catalyzed deoxygenation to give propane 231, 232
Glycerol, selective copper-catalyzed deoxygenation to give propane-1,2-diol 231

H

Hemiacetals, methyl, from symmetrical 1,2-disubstituted alkenes by ozonolysis in methanol and hydrogenation 331, 332
Hexane-1,6-diol, from 2-(hydroxymethyl)tetrahydropyran by directed hydrogenolysis 253, 254
(*Z*)-Hex-3-en-1-ol, from hex-3-yn-1-ol by heterogeneous hydrogenation 201
Hex-3-yn-1-ol, heterogeneous hydrogenation to give (*Z*)-hex-3-en-1-ol 201
Hydrodehalogenation, metal catalyzed 355–369
Hydrodehalogenation, photoinduced 369–372
Hydrodeoxygenation, of aliphatic alcohols to give alkanes 231–233
Hydrodeoxygenation, of phenols to give arenes 229, 230
Hydrogenation, for deoxygenation of sulfoxides and sulfones to give sulfides 344–351
Hydrogenation, for reduction of disulfides to give thiols 351, 352
Hydrogenation, of alkenes, asymmetric, homogeneous catalysis 7–62
Hydrogenation, of alkenes, heterogeneous catalysis, mechanism 67, 68
Hydrogenation, of alkenes, using ionic-liquid-stabilized nanoparticles 83, 84
Hydrogenation, of alkenes, using iron nanoparticles 79–84
Hydrogenation, of alkenes, using palladium nanoparticles 68–78
Hydrogenation, of alkynes 195–220
Hydrogenation, of allenes 220–224
Hydrogenation, of arenes 127–149
Hydrogenolysis, of ethers 243–264
Hydrogenolysis, of furans 258–263
Hydrogenation, of hetarenes 153–189
Hydrogenolysis, of phenols to give arenes 229, 230
Hydroperoxides, from naphthalene derivatives by an ene reaction with singlet oxygen 325
Hydroperoxides, heterogeneous catalytic hydrogenation, mechanistic aspects 316–318
Hydroxenin, from oxenin by catalytic hydrogenation 197, 198
Hydroxybenzenes, from polyhydroxybenzenes by ruthenium-catalyzed mono-hydrodeoxygenation 230
α-Hydroxy carbonyl compounds, from α-peroxy carbonyl compounds by hydrogenation 321, 322
β-Hydroxy carbonyl compounds, from β-peroxy carbonyl compounds by hydrogenation 322
γ-Hydroxy carbonyl compounds, from γ-peroxy carbonyl compounds by hydrogenation 323
4-Hydroxycyclohexanones, from 2,3-dioxabicyclo[2.2.2]oct-5-enes by peroxide and alkene hydrogenation 329, 330
α-Hydroxy ketones, from alkenylsilanes by ozonolysis/hydrogenation 321
β-Hydroxy ketones, from homochiral β-peroxy ketones by hydrogenation 322
γ-Hydroxy ketones, from γ-peroxy ketones by hydrogenation 323

Hydroxyindolizinones, from peroxy-bridged indolizinones by hydrogenation 331

I

Imidazoles, ruthenium-catalyzed enantioselective hydrogenation to give chiral 4,5-dihydroimidazoles 181
Indoles, iridium-catalyzed enantioselective hydrogenation to give chiral 2,3-dihydroindoles 172, 174, 176
Indoles, methyl-, from indoles by methylation to give methylindoles using hydrogen and carbon dioxide 310
Indoles, methylation to give methylindoles using hydrogen and carbon dioxide 310
Indoles, palladium-catalyzed enantioselective hydrogenation to give chiral 2,3-dihydroindoles 186–189
Indoles, rhodium-catalyzed enantioselective hydrogenation to give chiral 2,3-dihydroindoles 185, 186
Indoles, rhodium-catalyzed hydrogenation to give 2,3-dihydroindoles 166, 167
Indoles, ruthenium-catalyzed enantioselective hydrogenation to give chiral 2,3-dihydroindoles 179, 180
Indoline-3-carboxamides, 1-alkyl-3-hydroxy-2-oxo-, from 2-cyano-2-diazo-*N*-phenylacetamides by a cyclization/peroxidation/hydrolysis/hydrogenation sequence 319, 320
Indolin-2-one, 3-hydroxy-3-(hydroxymethyl)-, from *N*-phenylacrylamide by a bisperoxidation/cyclization/hydrogenation sequence 319
Indolizines, ruthenium/NHC-catalyzed enantioselective hydrogenation to give chiral 5,6,7,8-tetrahydroindolizines 182, 183
Indolizinones, peroxy-bridged, hydrogenation to give hydroxyindolizinones 331
Iodoalkanes, hydrodeiodination under base-mediated hydrogenation conditions to give alkanes 355, 356
Iodoalkenes, heterogeneous light-mediated reductive deiodination to give alkenes 371, 372
Iodoarenes, heterogeneous light-mediated reductive deiodination to give arenes 371, 372
Iodoarenes, hydrodeiodination in water with in situ generated molecular hydrogen to give arenes 362, 363
Iodoarenes, in situ generation from bromoarenes and subsequent catalytic hydrodehalogenation to give arenes 367, 368
Iodoarenes, palladium(II) acetate catalyzed hydrodeiodination to give arenes 365
Iodotetrahydropyrans, chemoselective hydrodehalogenation to give tetrahydropyrans 356, 357
Isoprene, partial reduction 92–95
Isopulegol, from geraniol via cyclization using a heterogeneous catalyst 101, 102
Isoquinolines, heterogeneous platinum(IV)-catalyzed hydrogenation to give tetrahydroisoquinolines 156, 157
Isoquinolines, homogeneous rhodium-catalyzed transfer hydrogenation to give tetrahydroisoquinolines 165
Isoquinolines, iridium-catalyzed enantioselective hydrogenation to give chiral dihydro- or tetrahydroisoquinolines 172, 173
Isoquinolinium salts, iridium-catalyzed enantioselective hydrogenation to give tetrahydroisoquinolines 176

K

Ketones, α-chiral, from enones by asymmetric homogeneous hydrogenation 52–54
Ketones, cyclic, α-chiral, from α-alkylidene cyclic ketones by iridium-catalyzed asymmetric hydrogenation 52, 53
Ketones, from alkenes by ozonolysis/reduction in methanol 332–334
Ketones, from α-bromo ketones by palladium(II) acetate catalyzed hydrodebromination 365, 366
Ketones, γ-phenyl, from γ-peroxy γ-phenyl ketones by hydrogenolysis 323
Ketones, saturated, from allenyl ketones by palladium-catalyzed hydrogenation 223, 224

L

Lactams, α-chiral, from α,β-unsaturated cyclic amides by iridium-catalyzed asymmetric hydrogenation 50, 51
Lignin, model compounds 249, 250
Lignin, structure 249
Lignin model dimer, nickel-on-carbon catalyzed hydrogenolysis to give phenols 247
Lignin model dimer, Zn/Pd/C catalyzed hydrogenolysis of the β-O-4 linkage to give phenols 250, 251
Limonene, partial reduction 95–97
Lindlar catalyst, use in alkyne semihydrogenation reactions 199–201
Linolenic acid derivatives, selectivity in stepwise hydrogenation 116
Linseed oil, methyl esters, rhodium-catalyzed partial hydrogenation to give methyl octadecenoates 111

M

Maleic acid, from tartaric acid by rhenium-catalyzed deoxydehydration 237
p-Menthene, from limonene by semihydrogenation 95
Menthol, from geraniol by cyclization/hydrogenation using a heterogeneous catalyst 101, 102

Methanation, of carbon dioxide 302–304
Methane, from carbon dioxide by heterogeneous hydrogenation 302–304
Methane, from metal carbonates by catalytic reduction 282
Methanol, from carbon dioxide by heterogeneous hydrogenation 301, 302
Methanol, from carbon dioxide by homogeneous hydrogenation 299, 300
Methanol, from dimethyl carbonate by heterogeneous copper-catalyzed hydrogenation 280, 281
Methanol, from dimethyl carbonate by homogeneous ruthenium-catalyzed hydrogenation 275, 276
Methoxyphenols, hydrogenolysis using ruthenium-on-carbon and magnesium oxide to give cyclohexanol 252
Methylamines, from amines by methylation using hydrosilanes/carbon dioxide 305
Methylamines, from amines by reductive N-methylation using hydrogen and carbonates 282, 283
Methylamines, tertiary, from primary amines by heterogeneous gold-catalyzed disubstitution using hydrogen/carbon dioxide and aldehydes 308
N-Methylanilines, from anilines by heterogeneous palladium-catalyzed methylation using hydrogen and carbon dioxide 307
N-Methylanilines, from anilines by homogeneous ruthenium-catalyzed methylation using hydrogen and carbon dioxide 305–307
Methylation, catalytic reductive, using carbonates 282, 283
Methylation, of amines, using carbon dioxide 305–310
Methylation, of C—H bonds, using carbon dioxide 310
Methylation, reductive, using carbon dioxide 304–310
Mucic acid, rhenium-catalyzed reduction to give muconic acid 237
Mucic acid dibutyl ester, rhenium-catalyzed reduction to give dibutyl muconate 237
Muconic acid, from mucic acid by rhenium-catalyzed reduction 237
Myrcene, partial reduction 97, 98

N

Naltrexone derivatives, synthesis 329, 330
Naphthalene, ruthenium-catalyzed hydrogenation to give *cis*-Decalin 136
Naphthalene derivatives, ene reaction with singlet oxygen to give hydroperoxides 325
Naphthalenes, ruthenium-catalyzed hydrogenation to give tetrahydronaphthalenes 136–138
Naphthalenes, tetrapropyldiborane-catalyzed hydrogenation to give tetrahydronaphthalenes 143, 144
Naphthyridines, ruthenium-catalyzed enantioselective hydrogenation to give chiral tetrahydronaphthyridines 177, 178
Nitrobenzenes, reductive dimethylation to give *N,N*-dimethylanilines using hydrogen and carbon dioxide 308, 309

O

Octadecenoates, methyl, from polyunsaturated C_{18} fatty acid methyl esters by heterogeneous copper-catalyzed hydrogenation 117, 118
Octadecenoates, methyl, from vegetable oil methyl esters by rhodium-catalyzed partial hydrogenation 111, 112
Octahydroanthracene, from anthracene by platinum-catalyzed hydrogenation 140
Octahydroanthracene, from anthracene by rhodium-catalyzed hydrogenation 139
Octahydroanthracene, from anthracene by ruthenium-catalyzed hydrogenation 136
Octahydrophenanthrolines, chiral, from phenanthrolines by ruthenium-catalyzed enantioselective hydrogenation 177, 178
Octahydrotriphenylene, from triphenylene by platinum-catalyzed hydrogenation 140
Octahydrotriphenylene, from triphenylene by rhodium-catalyzed hydrogenation 139
Oleic acid, from rapeseed and soybean oils by selective hydrogenation using copper/silica 120, 121
Oleic acid, methyl ester, from methyl esters of polyunsaturated C_{18} fatty acids by partial hydrogenation 109
Oleic acid, oxidative cleavage to give azelaic acid and pelargonic acid 115
Oleic acid derivatives, ozonolysis and subsequent oxidative or reductive workup to give pelargonic and azelaic acids and aldehydes 334, 335
Oxazoles, ruthenium-catalyzed enantioselective hydrogenation to give chiral 4,5-dihydrooxazoles 182
Oxenin, catalytic hydrogenation to give hydroxenin 197, 198
Ozonides, reduction 331–337

P

Pelargonic acid, from oleic acid by oxidative cleavage 115
Pelargonic acid, from oleic acid derivatives by ozonolysis and subsequent oxidative or reductive workup 334, 335
Pelargonic aldehyde, from oleic acid derivatives by ozonolysis and subsequent oxidative or reductive workup 334, 335

Penta-1,3-diene, from xylitol by a two-step formic acid mediated deoxydehydration 235
Pentane-1,4-diol, from tetrahydropyran-3-ol by directed hydrogenolysis 255
Pentane-1,5-diol, from furan-2-carbaldehyde by catalytic hydrogenation 258, 259
Pentane-1,5-diol, from 2-(hydroxymethyl)tetrahydrofuran by directed hydrogenolysis 253–255
Pentan-2-ol, from 2-methyltetrahydrofuran by rhodium-catalyzed hydrogenolysis 256
Perfluoroarenes, homogeneous cobalt(0)-catalyzed selective defluorination to give fluoroarenes 363, 364
Peroxides, β-nitroalkyl, chemoselective hydrogenation to give β-nitro alcohols 324
Peroxo compounds, reduction 316–331
Peroxy acetal esters, unsaturated, catalyst-dependent selective hydrogenation 318, 319
α-Peroxy carbonyl compounds, hydrogenation to give α-hydroxy carbonyl compounds 321, 322
β-Peroxy carbonyl compounds, hydrogenation to give β-hydroxy carbonyl compounds 322
γ-Peroxy carbonyl compounds, hydrogenation to give γ-hydroxy carbonyl compounds 323
β-Peroxy ketones, homochiral, hydrogenation to give β-hydroxy ketones 322
γ-Peroxy ketones, hydrogenation to give γ-hydroxy ketones 323
γ-Peroxy γ-phenyl ketones, hydrogenolysis to give γ-phenyl ketones 323
Peruloside A 355
Phenanthrene, ruthenium-catalyzed hydrogenation to give dihydrophenanthrene 136
Phenanthrenes, palladium-catalyzed hydrogenation to give dihydro- or tetrahydrophenanthrenes 142, 143
Phenanthrolines, ruthenium-catalyzed enantioselective hydrogenation to give chiral tetrahydro- or octahydrophenanthrolines 177, 178
Phenol, from 2-phenoxy-1-phenylethanol by reductive cleavage of the β-O-4 linkage in the presence of base 250
Phenol, from phenyl 2-phenylethyl ether by reductive cleavage 251
Phenol, ruthenium-catalyzed hydrogenolysis to give benzene 229, 230
Phenol derivatives, peroxidation/hydrogenation to give pterocarpans 326
Phenols, catalytic hydrodeoxygenation to give arenes 229, 230
Phenols, from a lignin model dimer by nickel-on-carbon catalyzed hydrogenolysis 247
Phenols, from a lignin model dimer by Zn/Pd/C catalyzed hydrogenolysis of the β-O-4 linkage 250, 251
Phenols, from diaryl ethers by nickel-catalyzed hydrogenolysis 244, 247
Phenols, from methyl phenyl ethers by iridium-catalyzed hydrogenolysis 252
Phenylacetaldehydes, from benzaldehydes by a one-carbon homologation involving hydrogenation of a sulfoxide catalyzed by ruthenium nanoparticles on titanium(IV) oxide 347
Phenylacetylene, homogeneous iron-catalyzed hydrogenation to give styrene 210
Phosphinates, alkyl, chiral, from enol phosphinates by asymmetric homogeneous hydrogenation 38, 39
Phosphine oxides, alkyl-, chiral, from alkenylphosphine oxides by asymmetric homogeneous hydrogenation 55, 56
Phosphonates, alkyl-, chiral, from alkenylphosphonates by asymmetric homogeneous hydrogenation 55, 56
Photocatalytic reduction, of carbonates 283–285
Piperidinecarboxamides, from pyridinecarboxamides, heterogeneous palladium-catalyzed hydrogenation 158, 159
Piperidines, chiral, from 2-(2-oxooxazolidin-3-yl)pyridines by heterogeneous diastereoselective hydrogenation 159
Piperidines, chiral, from pyridines by iridium-catalyzed enantioselective hydrogenation 172–175
Piperidines, from pyridines by heterogeneous platinum(IV)-catalyzed hydrogenation 157, 158
Polyenes, cyclic, semihydrogenation 102–108
Polyenes, mechanism of catalytic C=C bond conjugation/isomerization via hydrogen addition 116
Polyenes, partial hydrogenation 91–122
Polyhydroxybenzenes, ruthenium-catalyzed mono-hydrodeoxygenation to give hydroxybenzenes 230
Polyunsaturated C_{18} fatty acids, methyl esters, heterogeneous copper-catalyzed hydrogenation to give methyl octadecenoates and other partially reduced products 117, 118
Polyunsaturated C_{18} fatty acids, methyl esters, partial hydrogenation to give methyl oleate 109
Polyunsaturated C_{18} fatty acids, oxidative stability and melting points 109, 110
Propane, from glycerol by homogeneous ruthenium-catalyzed deoxygenation 231, 232
Propane-1,3-diol, from 1,3-dioxan-2-one by homogeneous ruthenium-catalyzed hydrogenation 277, 278
Propane-1,2-diol, from glycerol by selective copper-catalyzed deoxygenation 231
Propane-1,3-diol, selective catalytic hydrogenolysis to give propan-1-ol 231

Propan-1-ol, from propane-1,3-diol by selective catalytic hydrogenolysis 231
Pterocarpans, from phenol derivatives by peroxidation/hydrogenation 326
Pyrazoles, palladium-catalyzed enantioselective hydrogenation to give chiral 4,5-dihydropyrazoles 187, 188
Pyrene, palladium-catalyzed hydrogenation to give 4,5-dihydropyrene 142
Pyrene, platinum-catalyzed hydrogenation to give 4,5-dihydropyrene 141
Pyrene, rhodium-catalyzed hydrogenation to give 4,5-dihydro-, 4,5,9,10-tetrahydro-, or 1,2,3,3a,4,5,5a,6,7,8-decahydropyrene 139
Pyrene, ruthenium-catalyzed hydrogenation to give 4,5-dihydropyrene 137
Pyridinecarboxamides, heterogeneous palladium-catalyzed hydrogenation to give piperidinecarboxamides 158, 159
Pyridines, heterogeneous platinum(IV)-catalyzed hydrogenation to give piperidines 157, 158
Pyridines, iridium-catalyzed enantioselective hydrogenation to give chiral piperidines 172–175
Pyridines, 2-(2-oxooxazolidin-3-yl)-, heterogeneous diastereoselective hydrogenation to give piperidines 159
Pyrimidines, iridium-catalyzed enantioselective hydrogenation to give chiral tetrahydropyrimidines 172–174
Pyrroles, palladium-catalyzed enantioselective hydrogenation to give chiral 3,4-dihydropyrroles 187, 188
Pyrroles, rhodium-catalyzed hydrogenation to give pyrrolidines 166, 167
Pyrroles, ruthenium-catalyzed enantioselective hydrogenation to give chiral pyrrolidines 180, 181
Pyrrolidine, heterogeneous palladium-catalyzed methylation using hydrogen and carbon dioxide to give 1-methylpyrrolidine 307, 308
Pyrrolidine, 1-methyl, from pyrrolidine by heterogeneous palladium-catalyzed methylation using hydrogen and carbon dioxide 307, 308
Pyrrolidines, chiral, from pyrroles by ruthenium-catalyzed enantioselective hydrogenation 180, 181
Pyrrolidines, from pyrroles by rhodium-catalyzed hydrogenation 166, 167
Pyrrolidines, from pyrroles by ruthenium-catalyzed enantioselective hydrogenation 180, 181

Q

Quinolines, heterogeneous gold-catalyzed hydrogenation to give tetrahydroquinolines 154, 155
Quinolines, heterogeneous iridium/BINAP-CMP catalyzed asymmetric hydrogenation to give chiral 2-substituted tetrahydroquinolines 155, 156
Quinolines, heterogeneous rhodium-catalyzed hydrogenation to give decahydroquinolines 154
Quinolines, heterogeneous rhodium-catalyzed hydrogenation to give tetrahydroquinolines 153, 154
Quinolines, homogeneous iridium/NHC catalyzed hydrogenation to give tetrahydroquinolines 160–162
Quinolines, homogeneous iridium-catalyzed transfer hydrogenation to give tetrahydroquinolines 162, 163
Quinolines, homogeneous rhodium-catalyzed transfer hydrogenation to give tetrahydroquinolines 163, 164
Quinolines, iridium-catalyzed enantioselective hydrogenation to give chiral tetrahydroquinolines 170, 171
Quinolines, palladium-catalyzed enantioselective hydrogenation to give chiral tetrahydroquinolines 187, 188
Quinolines, rhodium-catalyzed enantioselective hydrogenation to give chiral tetrahydroquinolines 184, 185
Quinolines, ruthenium-catalyzed enantioselective hydrogenation to give chiral tetrahydroquinolines 177–179
Quinoxalines, homogeneous iridium-catalyzed transfer hydrogenation to give tetrahydroquinoxalines 165, 166
Quinoxalines, homogeneous rhodium-catalyzed transfer hydrogenation to give tetrahydroquinoxalines 165
Quinoxalines, iridium-catalyzed enantioselective hydrogenation to give chiral tetrahydroquinoxalines 172–175
Quinoxalines, ruthenium-catalyzed enantioselective hydrogenation to give chiral tetrahydroquinoxalines 177, 178

R

Rapeseed oil, selective hydrogenation using copper/silica to give partially reduced products 120, 121

S

Semihydrogenation, of polyenes 91–122
Semihydrogenation, of terminal alkynes and *Z*-selective reduction of internal alkynes 199–214
Semihydrogenation, *E*-selective reduction of internal alkynes 214–220
Silyl enol ethers, asymmetric homogeneous hydrogenation to give chiral alkoxysilanes 41, 42

Soybean oil, selective hydrogenation using copper/silica to give partially reduced products 120, 121
(*Z*)-Stilbene, from diphenylacetylene by homogeneous copper-catalyzed semihydrogenation 208, 209
(*E*)-Stilbene, from diphenylacetylene by homogeneous iridium-catalyzed transfer hydrogenation 219, 220
(*E*)-Stilbene, from diphenylacetylene by homogeneous iron-catalyzed semihydrogenation 218
(*E*)-Stilbene, from diphenylacetylene by homogeneous ruthenium-catalyzed semihydrogenation 217, 218
(*Z*)-Stilbene, from diphenylacetylene by semihydrogenation using a rhodium–TangPhos catalyst 204, 205
Styrene, from phenylacetylene by homogeneous iron-catalyzed hydrogenation 210
Styrenes, hydrogenation using palladium nanoparticles on amphiphilic supports to give alkylbenzenes 72, 73
Sulfides, alkyl, chiral, from alkenyl sulfides by asymmetric homogeneous hydrogenation 59, 60
Sulfides, diaryl, from diaryl sulfoxides by deoxygenation using ruthenium nanoparticles on titanium(IV) oxide in a continuous flow reactor 346, 347
Sulfides, diaryl, from diaryl sulfoxides by deoxygenation using supported metal nanoparticle catalysts 341, 342, 345
Sulfides, from sulfones by deoxygenation using a titanium(IV) oxide supported platinum/ MoO_X catalyst with molecular hydrogen 349, 350
Sulfides, from sulfoxides by deoxygenation catalyzed by supported ruthenium nanoparticles using propan-2-ol as a reductant 342, 343
Sulfides, from sulfoxides by deoxygenation using a titanium(IV) oxide supported platinum/ MoO_x catalyst with molecular hydrogen 348, 349
Sulfides, from sulfoxides by deoxygenation using homogeneous molybdenum and rhenium catalysts with molecular hydrogen 344
Sulfides, from sulfoxides by deoxygenation using ruthenium nanoparticles on titanium(IV) oxide with molecular hydrogen 345–348
Sulfides, from sulfoxides by molybdenum-catalyzed deoxygenation using pinacol as a reductant 339–341
Sulfones, alkyl, chiral, from alkenyl sulfones by asymmetric homogeneous hydrogenation 59–61
Sulfones, deoxygenation to give sulfides using a titanium(IV) oxide supported platinum/ MoO_X catalyst with molecular hydrogen 349, 350
Sulfoxides, deoxygenation to give sulfides catalyzed by a titanium(IV) oxide supported platinum/ MoO_x catalyst with molecular hydrogen 348, 349
Sulfoxides, deoxygenation to give sulfides catalyzed by supported ruthenium nanoparticles using propan-2-ol as a reductant 341–343
Sulfoxides, deoxygenation to give sulfides using homogeneous molybdenum and rhenium catalysts with molecular hydrogen 344
Sulfoxides, deoxygenation to give sulfides using ruthenium nanoparticles on titanium(IV) oxide with molecular hydrogen 345–348
Sulfoxides, diaryl, deoxygenation to give diaryl sulfides using supported metal nanoparticle catalysts 341, 342, 345
Sulfoxides, diaryl, deoxygenation to give diaryl sulfides using ruthenium nanoparticles on titanium(IV) oxide in a continuous flow reactor 346, 347
Sulfoxides, molybdenum-catalyzed deoxygenation to give sulfides using pinacol as a reductant 339–341
Sunflower oil, methyl esters, rhodium-catalyzed partial hydrogenation to give methyl octadecenoates 111, 112
Sunflower oil, noble metal based heterogeneous partial hydrogenation to give monounsaturated products 114, 115

T

Tartaric acid, rhenium-catalyzed deoxydehydration to give maleic acid 237
Terpenes, polyunsaturated, selective hydrogenation 91–102
Tetracyclines, from 6-hydroperoxy-5a(11a)-dehydrotetracyclines by chemoselective hydrogenation 325
Tetradecane-1,2-diol, rhenium-catalyzed deoxydehydration to give tetradec-1-ene 236
Tetradec-1-ene, from tetradecane-1,2-diol by rhenium-catalyzed deoxydehydration 236
Tetrahydroanthracene, from anthracene by platinum-catalyzed hydrogenation 140
Tetrahydroanthracene, from anthracene by ruthenium-catalyzed hydrogenation 136
Tetrahydrobenzotetraphenes, from benzotetraphenes by palladium-catalyzed hydrogenation 141–143
Tetrahydrofuran, 2-(hydroxymethyl)-, directed hydrogenolysis to give pentane-1,5-diol 253, 254
Tetrahydofuran, 2-methyl-, rhodium-catalyzed hydrogenolysis to give pentan-2-ol 256
Tetrahydrofuran-3-ol, directed hydrogenolysis to give butane-1,3-diol 255

Tetrahydrofurans, chiral, from furans by ruthenium/NHC-catalyzed enantioselective hydrogenation 182, 183
Tetrahydrofurans, from furans by rhodium-catalyzed hydrogenation 166, 167
Tetrahydrofurans, 2-substituted, palladium-catalyzed selective hydrogenolysis in an ionic liquid to give a primary alcohol 256, 257
Tetrahydrofurans, total hydrodeoxygenation to give alkanes 257, 258
Tetrahydroindolizines, chiral, from indolizines by ruthenium/NHC-catalyzed enantioselective hydrogenation 182, 183
Tetrahydroisoquinolines, chiral, from isoquinolines by iridium-catalyzed enantioselective hydrogenation 172–174
Tetrahydroisoquinolines, chiral, from isoquinolinium salts by iridium-catalyzed enantioselective hydrogenation 176
Tetrahydroisoquinolines, from isoquinolines by heterogeneous platinum(IV)-catalyzed hydrogenation 156, 157
Tetrahydroisoquinolines, from isoquinolines by homogeneous rhodium-catalyzed transfer hydrogenation 165
Tetrahydronaphthalenes, from naphthalenes by ruthenium-catalyzed hydrogenation 136–138
Tetrahydronaphthalenes, from naphthalenes by tetrapropyldiborane-catalyzed hydrogenation 143, 144
Tetrahydronaphthyridines, chiral, from naphthyridines by ruthenium-catalyzed enantioselective hydrogenation 177, 178
Tetrahydrophenanthrenes, from phenanthrene by palladium-catalyzed hydrogenation 142, 143
Tetrahydrophenanthrolines, chiral, from phenanthrolines by ruthenium-catalyzed enantioselective hydrogenation 177, 178
Tetrahydropyran, 2-(hydroxymethyl)-, directed hydrogenolysis to give hexane-1,6-diol 253–255
Tetrahydropyran-3-ol, directed hydrogenolysis to give pentane-1,4-diol 255
Tetrahydropyrans, from 4-halotetrahydropyrans by chemoselective hydrodehalogenation 356, 357
Tetrahydropyrene, from pyrene by rhodium-catalyzed hydrogenation 139
Tetrahydropyrimidines, chiral, from pyrimidines by iridium-catalyzed enantioselective hydrogenation 172, 174
Tetrahydroquinolines, chiral, from quinolines by iridium-catalyzed enantioselective hydrogenation 170, 171
Tetrahydroquinolines, chiral, from quinolines by palladium-catalyzed enantioselective hydrogenation 187, 188
Tetrahydroquinolines, chiral, from quinolines by rhodium-catalyzed enantioselective hydrogenation 184, 185
Tetrahydroquinolines, chiral, from quinolines by ruthenium-catalyzed enantioselective hydrogenation 177–179
Tetrahydroquinolines, from quinolines by heterogeneous gold-catalyzed hydrogenation 154, 155
Tetrahydroquinolines, from quinolines by heterogeneous rhodium-catalyzed hydrogenation 153, 154
Tetrahydroquinolines, from quinolines by homogeneous iridium-catalyzed transfer hydrogenation 162, 163
Tetrahydroquinolines, from quinolines by homogeneous iridium/NHC catalyzed hydrogenation 160–162
Tetrahydroquinolines, from quinolines by homogeneous rhodium-catalyzed transfer hydrogenation 163, 164
Tetrahydroquinolines, 2-substituted, chiral, from quinolines by heterogeneous iridium/BINAP-CMP catalyzed hydrogenation 155, 156
Tetrahydroquinoxalines, chiral, from quinoxalines by iridium-catalyzed enantioselective hydrogenation 172–175
Tetrahydroquinoxalines, chiral, from quinoxalines by ruthenium-catalyzed enantioselective hydrogenation 177, 178
Tetrahydroquinoxalines, from quinoxalines by homogeneous iridium-catalyzed transfer hydrogenation 165, 166
Tetrahydroquinoxalines, from quinoxalines by homogeneous rhodium-catalyzed transfer hydrogenation 165
Tetrahydrothiophenes, chiral, from thiophenes by ruthenium/NHC-catalyzed enantioselective hydrogenation 182, 183
Tetrahydrotriphenylene, from triphenylene by ruthenium-catalyzed hydrogenation 136
1,2,3,4-Tetraols, from 1,2-dioxins by dihydroxylation/hydrogenation 327, 328
Tetraphenes, palladium-catalyzed hydrogenation to give 5,6-dihydrotetraphenes 142
Thiols, from disulfides by hydrogenation/hydrogenolysis using palladium on charcoal 351, 352
Thiophenes, ruthenium/NHC-catalyzed enantioselective hydrogenation to give chiral tetrahydrothiophenes 182, 183
Toluenes, from benzenes by methylation using hydrogen and carbon dioxide 310
Triglycerides, partial hydrogenation to give triolein 109
Triolein, from triglycerides by partial hydrogenation 109

1,2,4-Trioxane derivatives, hydrogenation to give 2,7-dioxabicyclo[2.2.1]heptanes 330, 331
Triphenylene, platinum-catalyzed hydrogenation to give octahydro- or dodecahydrotriphenylene 140, 141
Triphenylene, rhodium-catalyzed hydrogenation to give octahydro- or dodecahydrotriphenylene 139
Triphenylene, ruthenium-catalyzed hydrogenation to give tetrahydro- or dodecahydrotriphenylene 136, 137

U

α,β-Unsaturated acids, α-substituted, asymmetric hydrogenation using palladium nanoparticles/cinchonidine to give chiral α-substituted carboxylic acids 75
α,β-Unsaturated amides, asymmetric homogeneous hydrogenation to give chiral carboxamides 48–51
α,β-Unsaturated amides, cyclic, iridium-catalyzed asymmetric hydrogenation to give α-chiral lactams 50, 51
α,β-Unsaturated amides, α-substituted, acyclic, asymmetric hydrogenation to give α-chiral carboxamides 48–50
α,β-Unsaturated β-amino acid esters, rhodium- and ruthenium-catalyzed asymmetric hydrogenation to give chiral β-amino acid esters 32, 33
α,β-Unsaturated α-amino acid esters, rhodium-catalyzed asymmetric hydrogenation to give chiral α-amino acid esters 27–31
α,β-Unsaturated α-amino acids, rhodium-catalyzed asymmetric hydrogenation to give chiral α-amino acids 27,28
α,β-Unsaturated carboxylic acids, asymmetric homogeneous hydrogenation to give chiral carboxylic acids 43–46
α,β-Unsaturated esters, α-acyloxy, rhodium- and ruthenium-catalyzed asymmetric hydrogenation to give α-acyloxy carboxylates 36, 37
α,β-Unsaturated esters, asymmetric homogeneous hydrogenation to give chiral carboxylates 47, 48
α,β,γ,δ-Unsaturated esters, palladium-catalyzed hydrogenation to give saturated esters 223
α,β-Unsaturated heterocyclic acids, iridium-catalyzed asymmetric hydrogenation to give chiral saturated heterocyclic acids 45, 46
α,β-Unsaturated Weinreb amides, β,β-disubstituted, iridium-catalyzed asymmetric hydrogenation to give β-chiral Weinreb amides 48, 49

V

Vegetable oils, hydrogenated, industrial and health considerations 109, 110
Vegetable oils, selective hydrogenation 109–121
Vinylboronates, rhodium- and iridium-catalyzed asymmetric hydrogenation to give chiral alkylboronates 57, 58
Vinyl carboxylates, rhodium-catalyzed asymmetric hydrogenation to give chiral ethyl carboxylates 35, 36
Vinyl diphenylphosphinates, iridium-catalyzed asymmetric hydrogenation to give chiral alkyl diphenylphosphinates 38, 39
Vinyl fluorides, rhodium- and iridium-catalyzed asymmetric hydrogenation to give chiral alkyl fluorides 58, 59
Vinylphosphine oxides, iridium-catalyzed asymmetric hydrogenation to give chiral alkylphosphine oxides 55, 56
Vinylphosphonates, rhodium-catalyzed asymmetric hydrogenation to give chiral alkylphosphonates 55, 56
Vinyl sulfides, rhodium- and ruthenium-catalyzed asymmetric hydrogenation to give chiral alkyl sulfides 59, 60
Vinyl sulfones, rhodium- and iridium-catalyzed asymmetric reduction to give chiral alkyl sulfones 59–61
Vitamin A, synthetic approach involving oxenin hydrogenation 197, 198

W

Weinreb amides, β-chiral, from β,β-disubstituted α,β-unsaturated Weinreb amides by iridium-catalyzed asymmetric hydrogenation 48, 49

X

Xylitol, two-step formic acid mediated deoxydehydration to give penta-1,3-diene 235

Author Index

In this index the page number for that part of the text citing the reference number is given first. The number of the reference in the reference section is given in a superscript font following this.

A

Abbott, G. 93[18]
Abbott, M. 289[14]
Abboud, K. A. 27[87], 28[87], 29[87], 31[87]
Abdullina, R. M. 115[71]
Abraham, D. 285[48]
Abu-Omar, M. M. 250[15], 251[15]
Abura, T. 291[38]
Abu-Reziq, R. 76[21]
Adam, R. 282[37], 283[37]
Adams, D. R. 330[23]
Adams, R. 127[18]
Addis, D. 15[66], 20[66], 21[66], 23[66], 24[66]
Adkins, H. 196[6]
Agapie, T. 245[7]
Agbossou-Niedercorn, F. 170[40]
Ager, D. 7[11], 43[11]
Ager, D. J. 7[4]
Aguado, R. 339[34]
Aguilhon, J. 93[19]
Ahlquist, M. S. G. 270[6]
Ahmada, A. 302[91]
Ahmed, A. 153[9]
Aida, T. 339[19]
Aimino, D. 351[63]
Aizenshtat, Z. 134[34], 135[34]
Akashi, A. 157[19]
Aksoylu, A. 302[96], 303[96]
Akutagawa, S. 100[38]
Akzinnay, S. 366[38]
Alami, M. 157[22], 158[22], 365[37]
Alamillo, R. 253[37]
Albani, D. 201[47], 201[48]
Alba-Rubio, A. C. 253[40], 253[42]
Alberico, E. 85[47]
Albers, M. O. 209[84], 210[84], 210[85], 211[84]
Albers, P. W. 200[33]
Alberti, B. N. 285[50]
Alder, A. P. 224[145]
Alemán, J. 221[134]
Allen, P. 156[17], 157[17]
Allendorf, M. D. 251[19]
Allinger, N. L. 196[11], 196[12], 197[11], 197[12]
Almena Perea, J. 35[118], 36[118]
Almora-Barrios, N. 201[47]
Alonso, F. 86[50], 86[51], 87[51], 104[53], 105[53], 135[36], 355[10]
Alper, H. 76[21]
Al-Rawashdeh, M. 204[67]
Altmann, K.-H. 355[17]
Álvarez, E. 236[39]
Alvez, G. 96[28]
Amada, Y. 258[51], 258[52], 259[52]
Ameta, S. C. 284[41]
Amini, M. 339[35]
Amundson, L. M. 250[15], 251[15]
An, G. 69[9], 319[8], 320[8]
Anbarasan, P. 270[4], 291[36], 295[36], 296[36]
Anders, D. E. 335[36], 336[36]
Anderson, J. A. 198[25], 200[37], 202[37]
Anderson, J. R. 145[43]
Andersson, K. J. 282[35]
Andersson, P. G. 7[3], 8[3], 8[19], 8[23], 8[24], 8[27], 8[28], 8[29], 8[32], 8[34], 8[36], 10[19], 10[23], 10[24], 10[27], 10[28], 10[29], 10[32], 10[34], 10[36], 11[19], 11[23], 11[24], 11[27], 11[28], 11[29], 11[32], 11[34], 12[19], 12[23], 12[24], 12[27], 12[28], 12[29], 12[32], 12[36], 13[19], 13[23], 13[24], 13[27], 13[28], 13[29], 13[32], 14[28], 38[126], 38[127], 39[126], 39[127], 47[139], 48[139], 55[151], 56[151], 57[36], 57[153], 58[155], 59[155], 59[160], 59[161], 61[160], 62[160]
Andriollo, A. 210[89]
Ang, E. L. 232[24], 236[24], 237[24], 238[24], 239[24]
Ankudinova, T. V. 224[144]
Aoki, K. 221[125], 221[127]
Apesteguía, C. R. 120[87]
Aquila, W. 223[142]
Arai, M. 289[25]
Arakawa, H. 291[42]
Aramendía, M. A. 203[58]
Araujo, M. H. 98[36]
Arcadi, A. 359[24]
Arcelli, A. 203[56]
Arceo, E. 233[35], 234[35], 235[35], 236[35], 236[42], 238[42], 239[35], 239[42]
Archer, R. 232[28], 233[28]
Arena, F. 301[76], 301[81]
Aresta, M. 289[4], 289[6], 289[11]
Arif, A. M. 331[27]
Arisawa, M. 351[71]
Armbrüster, M. 200[35]
Armbruster, U. 229[7]
Arnáiz, F. J. 339[34], 339[49], 340[49], 341[49], 341[50], 343[49]
Arnold, H. R. 196[5]
Arnold, J. 209[82], 209[83]
Arpe, H.-J. 92[6], 102[44], 127[6]
Arterburn, J. B. 339[33]
Asada, D. 36[124], 37[124]
Asao, N. 204[59], 204[60]
Ashworth, T. V. 211[93]
Asiri, A. M. 239[50]
Assary, R. S. 257[47]
Astier, M. 146[45]
Atesin, A. C. 256[46], 257[46], 257[47]
Au, C.-T. 289[23]
Audemar, M. 260[73], 261[73]
Augustine, R. L. 91[1], 92[1], 96[28]
Augustynski, J. 284[43], 284[44]
Austin, R. N. 229[6], 229[10], 230[6], 230[10]
Au-Yeung, T. T.-L. 32[104], 33[104]
Avery, T. D. 327[19], 329[22]
Ayad, T. 15[77], 20[77], 24[77], 26[77], 174[56], 174[57], 175[56], 175[57]
Ayman, K. M. 229[9]
Aziz, M. A. A. 302[91]
Azua, A. 291[51], 291[52], 292[51], 293[51], 294[52], 295[52]

B

Baba, A. 231[14], 239[14]
Baba, N. 102[43]
Babu, S. 15[78], 20[78], 24[78], 26[78]
Bacchi, A. 206[73]
Bachir, R. 93[14]
Bachman, G. L. 7[5]
Badalo Branco, J. 93[15]
Baek, B. 229[10], 230[10]
Baeza, A. 40[129], 41[129], 176[60]
Bagherzadeh, M. 339[35]
Bahr, N. 223[141]
Bahrami, K. 339[29]
Bai, C. 15[62], 15[63], 20[62], 20[63], 21[62], 21[63], 22[62], 22[63], 23[62], 23[63], 24[62], 24[63], 25[63], 27[63], 28[63], 29[63], 30[63]
Bai, W.-J. 326[17]
Baik, J. H. 302[95], 302[98], 303[95], 303[98]
Baiker, A. 289[13]
Bailey, P. S. 332[29], 332[30]
Bailey, W. F. 355[9]
Baker, R. T. 339[15]
Bakermans, P. 200[43]
Balakrishna, B. 170[35]
Balakrishnan, M. 259[59]
Balaraman, E. 275[19], 276[19]
Balas, L. 199[29]
Balberan, N. T. 127[2]
Ball, Z. T. 214[101]
Bang-Andersen, B. 216[110]
Bansode, A. 301[79], 301[83], 302[79]
Bao, H. L. 308[117], 309[117]
Bao, J. 27[85], 28[85], 29[85], 31[85]
Bao, M. 204[59]
Bappert, E. 15[78], 20[78], 24[78], 26[78]
Barbaro, P. 103[49], 103[50], 104[49], 104[50], 104[52], 105[49], 105[52], 131[32], 132[32], 133[32]
Barbera, K. 301[76]
Barbier, J. 93[14]
Bargon, J. 214[105]
Baroudi, A. 276[22]
Bartók, M. 75[16]
Baruwati, B. 76[30]

Bashir, M. J. K. 289[15]
Bateman, M. 76[19], 76[20]
Bath, S. S. 332[30]
Baum, E. 272[10]
Baumann, W. 270[4], 270[5], 291[36], 295[36], 296[36]
Baumvol, I. J. R. 146[47]
Baxter, C. A. 157[23], 158[23]
Bayer, A. 8[27], 10[27], 11[27], 12[27], 13[27]
Beauchamps, M. G. 100[39], 101[39]
Beckerle, K. 236[40]
Bedford, C. T. 325[15]
Beiring, B. 182[70], 182[71], 182[73], 183[70], 183[71], 183[73], 184[70]
Beletskaya, I. P. 355[10]
Bell, A. T. 67[6], 259[59]
Beller, M. 15[66], 20[66], 21[66], 23[66], 24[66], 76[18], 80[37], 205[68], 270[3], 270[4], 270[5], 273[14], 274[14], 282[37], 283[37], 291[36], 295[36], 296[36], 304[104], 305[113], 305[114], 306[114], 307[114], 310[119]
Bellotti, D. 302[101], 303[101]
Beltramini, J. 301[78]
Beltran Toro, M. A. 201[48]
Ben-David, Y. 218[113], 271[8], 291[50], 295[50], 296[50]
Benedix, R. 205[70]
Bengali, A. A. 276[22]
Bennett, M. A. 128[24]
Berben, P. H. 200[43], 200[44], 201[44], 201[46], 202[44]
Berger, M. 333[31], 333[32]
Bergman, R. G. 209[82], 233[35], 234[35], 235[35], 236[35], 236[42], 238[42], 239[35], 239[42], 250[16]
Bergner, E. J. 91[4], 99[4], 100[4]
Bergquist, J. 8[32], 8[34], 10[32], 10[34], 11[32], 11[34], 12[32], 13[32]
Berhal, F. 174[57], 175[57]
Bernabeu, M. C. 156[16], 157[16]
Bernsmann, H. 15[58], 15[65], 20[58], 20[65], 21[58], 21[65], 22[58], 22[65], 27[58], 28[58], 29[58], 30[58]
Berque-Bestel, I. 157[22], 158[22]
Berthet, J.-C. 304[105]
Bertini, F. 271[7], 274[7], 291[47], 295[47], 296[47]
Bertóti, I. 75[16]
Beydoun, K. 304[110], 305[110], 306[110], 307[110]
Bhagwat, M. M. 222[138]
Bhattacharjya, A. 362[30], 363[30]
Bhattacharyya, A. 127[3]
Bi, H. 127[14], 128[14], 146[54]
Bianchini, C. 210[91], 210[92], 211[92]
Bignon, J. 365[37]
Birtwistle, J. S. 355[16], 356[16]
Bisaro, F. 366[38]
Bishop, A. 76[27]
Bishop, L. M. 250[16]
Bisson, L. 93[19]
Bitan, G. 362[33]
Blackmond, D. G. 8[13]
Blandy, C. 127[17], 128[17]
Blankenstein, J. 8[21], 10[21], 11[21], 13[21]
Blaser, H.-U. 100[40], 184[75]
Bleeker, B. F. 201[46]
Bliman, D. 362[29]
Blondiaux, E. 305[112]
Blum, J. 275[18], 362[33]
Boaz, N. W. 27[90], 28[90], 29[90]
Boccuzzi, F. 104[54], 117[54]
Boddien, A. 270[3]
Boerleider, R. 212[95]
Bogdał, D. 120[86]
Bogel-Łukasik, E. 96[26], 98[33]
Bogel-Łukasik, R. 96[26], 98[33]
Bohanna, C. 210[92], 211[92]
Bohnen, F. M. 8[12]
Boissière, C. 93[19]
Boland, S. 200[44], 201[44], 201[46], 202[44]
Boldrini, G. P. 85[48]
Bolm, C. 52[146], 53[146], 54[146]
Bomans, P. H. H. 204[65]
Bond, G. C. 92[11], 93[11], 197[16], 197[18], 197[19], 198[24], 198[27], 221[130], 221[133]
Bondioli, P. 117[82], 117[83], 118[82], 118[84], 119[83], 121[82]
Bonrath, W. 195[2], 196[9], 196[10], 197[15], 199[2], 199[15], 200[10], 200[39], 200[41], 202[54], 202[55], 221[15], 223[15], 223[140]
Bonura, G. 301[76]
Boogers, J. A. F. 79[35], 80[35]
Borah, B. J. 133[33], 134[33]
Borau, V. 203[58]
Borges, L. E. P. 229[3]
Börner, A. 32[101], 33[101]
Borodziński, A. 198[24], 198[27]
Borretto, E. 200[41]
Borsotti, G. 115[73], 115[74], 120[73], 120[74]
Boucher-Jacobs, C. 236[43], 239[43]
Boukherroub, R. 355[5]
Bouriazos, A. 111[65], 111[66], 111[67], 112[65], 112[66], 112[67], 112[68], 113[65]
Boussie, T. R. 232[28], 233[28]
Bouwens, T. 213[97]
Bovio, C. 15[64], 20[64], 21[64], 24[64], 25[64], 27[64], 28[64], 29[64]
Bowers, C. R. 214[106]
Boxwell, C. J. 128[25]
Boyce, N. 83[40]
Boymans, E. H. 200[44], 201[44], 202[44]
Bradbury, B. J. 156[18], 157[18]
Braden, D. J. 259[60]
Bradley, P. R. 156[17], 157[17]
Braidy, N. 76[19]
Brands, K. J. M. 157[23], 158[23]
Brandsma, L. 220[123]
Brandt, B. 67[7]
Brandt, P. 8[27], 8[28], 10[27], 10[28], 11[27], 11[28], 12[27], 12[28], 13[27], 13[28], 14[28]
Bravo-Altamirano, K. 358[22]
Bredig, G. 275[15]
Breesch, A. 115[70]
Breit, B. 81[38]
Bremeyer, N. 157[23], 158[23]
Bresó-Femenia, E. 135[37], 136[37], 137[37], 138[37]
Breuer, K. 223[142]
Brewer, S. E. 157[23], 158[23]
Breysse, M. 351[61], 351[62]
Bridier, B. 198[26], 201[49]
Brion, J.-D. 157[22], 158[22], 365[37]
Broggi, J. 103[47]
Brooks, A. 229[10], 230[10]
Brot, N. 339[8], 339[10]
Brown, H. C. 351[58]
Brückner, A. 76[18]
Brückner, R. 221[126]
Bruehwiler, A. 196[10], 200[10]
Bruijnincx, P. C. A. 243[1]
Bruneau, C. 15[76], 20[76], 24[76], 25[76], 32[100], 33[100]
Brunner, G. 96[26]
Buechter, D. D. 156[18], 157[18]
Buil, M. L. 131[31], 132[31], 133[31]
Bunnett, J. F. 355[13]
Buntara, T. 232[22], 253[33], 253[35], 253[36], 260[33], 261[33]
Burch, R. 351[68], 352[68]
Burgess, K. 8[25], 10[25], 11[25], 13[25], 14[25], 39[128], 40[128]
Burgoyne, E. E. 196[6]
Burhardt, M. N. 216[110]
Burk, M. J. 35[120], 36[120], 36[122], 37[122], 204[66]
Busca, G. 302[101], 303[101]
Buss, A. 200[39], 202[54]
Byers, J. A. 291[37], 295[37], 296[37]

C

Cabañas, A. 96[27]
Cabiac, A. 106[57], 107[57], 108[57]
Cabrero-Antonino, J. R. 282[37], 283[37]
Cabrita, I. 339[38]
Cadierno, V. 35[117], 35[121], 36[117], 36[121]
Cadu, A. 59[161]
Cai, C. 27[94], 28[94], 29[94], 30[94]
Cai, Q. 32[109], 33[109]
Cai, W. J. 301[82]
Cai, X.-F. 170[39], 172[42], 172[44], 173[42], 173[44], 187[86], 188[86]
Cai, Z. 15[70], 20[70], 24[70], 25[70], 27[70]
Calais, C. 351[61], 351[62]
Calderazzo, F. 289[8]
Calhorda, M. J. 344[53], 350[53]
Calvino, J. J. 105[56]
Calvo, B. 115[72]
Campeau, L.-C. 47[142], 48[142]
Campo, F. D. 260[73], 261[73]

Canaff, C. 301[87]
Cannon, K. A. 367[40], 368[40]
Cantat, T. 304[105], 304[106], 304[111], 305[106], 305[111], 305[112]
Cao, B. 15[54], 20[54], 21[54], 22[54], 23[54], 24[54], 26[54], 35[54], 36[54]
Cao, L.-L. 172[43], 173[43]
Cao, M. 15[74], 20[74], 24[74], 25[74]
Cao, P. 35[111], 37[111], 38[111]
Cao, Q. 253[38]
Cao, Y. 154[12], 155[12], 156[12], 281[34]
Capacci, A. G. 15[51], 20[51], 21[51], 22[51], 23[51]
Capozzoli, L. 207[76]
Capuzzi, L. 115[73], 115[74], 120[73], 120[74]
Cárdenas-Lizana, F. 199[28]
Carreño, M. C. 339[6]
Carroll, R. J. 330[23]
Carter, S. R. 275[15]
Cartigny, D. 174[56], 174[57], 175[56], 175[57]
Carvalho, W. A. 146[58]
Castillón, S. 135[37], 136[37], 137[37], 138[37]
Castonguay, A. 76[19]
Castrillón, J. P. A. 339[24]
Catalano, M. 103[49], 104[49], 105[49]
Catellani, M. 212[94], 214[94]
Caubere, P. 355[4]
Cavallo, L. 103[47], 304[108], 305[108]
Cavell, K. J. 206[74], 211[74], 212[96]
Cazin, C. S. J. 103[46], 103[47], 304[107], 304[108], 305[107], 305[108], 366[38]
Centi, G. 289[10], 289[18]
Cepeda, E. A. 115[72]
Cerichelli, G. 359[24]
Cha, J. S. 351[58]
Chakraborty, S. 299[72], 300[72]
Challenger, C. E. 156[17], 157[17]
Chan, A. S. C. 15[45], 15[57], 20[45], 20[57], 21[45], 21[57], 22[45], 22[57], 27[57], 27[96], 28[57], 28[96], 29[57], 29[96], 30[57], 30[96], 32[104], 33[104], 172[45], 173[45], 177[61], 178[66], 179[66]
Chan, J. L.-T. 356[18], 357[18]
Chan, K.-P. 356[18], 357[18]
Chan, L. K. M. 304[103]
Chan, S. 32[104], 33[104]
Chandrasekaran, K. 283[38]
Chang, F.-W. 302[92], 302[94], 303[92], 303[94]
Chang, H. 250[13]
Chang, J.-C. 93[13]
Chang, J.-R. 93[16]
Chang, M. 172[49], 173[49], 174[49]
Chang, T. 295[59], 296[59]
Chang, Y. 83[42], 83[45]
Chao, S. 275[16], 275[17], 284[17], 284[42]
Chapman, G. 236[37]
Chasar, D. W. 339[12]
Chatani, N. 245[8], 246[8]
Chatgilialoglu, C. 355[5]
Chatterjee, M. 259[55]
Chaudret, B. 80[37], 127[17], 128[17], 135[37], 136[37], 137[37], 138[37]
Chawla, H. P. S. 200[34]
Chazelle, V. 76[20]
Chen, B. H. 127[12], 128[12]
Chen, C. 44[136]
Chen, C.-H. 355[15]
Chen, C.-L. 231[17], 233[17]
Chen, E. Y.-X. 259[57]
Chen, F. 166[32], 167[32], 177[62], 177[63], 177[64], 177[65], 178[64], 178[65], 178[66], 179[66], 184[63]
Chen, H. 15[62], 15[63], 20[62], 20[63], 21[62], 21[63], 22[62], 22[63], 23[62], 23[63], 24[62], 24[63], 25[63], 27[63], 28[63], 29[63], 30[63], 146[52], 210[87], 302[93], 303[93], 304[93]
Chen, H.-B. 232[21]
Chen, J. 15[69], 20[69], 24[69], 27[96], 28[96], 29[96], 30[96], 187[84], 188[84], 260[81], 280[30], 281[30], 365[36], 366[36]
Chen, J. G. 289[17]
Chen, K. 253[28], 253[32], 254[28], 254[32], 255[32], 256[28]
Chen, L. 32[109], 33[109]
Chen, L.-Y. 204[59]
Chen, M.-W. 170[39], 172[42], 172[44], 172[48], 172[50], 173[42], 173[44], 173[48], 173[50], 174[50], 187[82], 187[83], 187[87], 188[82], 188[83], 188[87], 188[88], 189[88], 204[59]
Chen, P. 104[51]
Chen, Q.-A. 153[4], 172[48], 173[48], 186[81], 187[81], 187[83], 187[85], 188[83], 188[85], 189[81]
Chen, Q.-Y. 239[49]
Chen, T. 213[98], 220[98], 220[117]
Chen, W. 99[37], 101[37]
Chen, W.-S. 91[2]
Chen, X. 281[31], 281[32]
Chen, Y. 172[49], 173[49], 174[49], 289[24]
Chen, Y.-W. 8[17], 10[17], 11[17], 13[17], 146[65]
Chen, Y. Z. 293[55], 301[75]
Chen, Z.-P. 170[36], 172[50], 173[50], 174[50], 187[86], 187[87], 188[86], 188[87]
Cheng, C. 158[25], 159[25], 160[25]
Cheng, J. 156[18], 157[18]
Cheng, P. 127[17], 128[17]
Cheng, Y. 127[15], 128[15]
Cherkasov, N. 200[36], 200[37], 202[36], 202[37]
Cheruku, P. 38[126], 38[127], 39[126], 39[127], 55[151], 56[151], 57[153]
Chew, L. M. 104[51]
Chi, Y. 15[75], 20[75], 24[75], 25[75], 26[75], 27[75], 27[91], 28[91], 29[91], 30[91]
Chia, M. 232[19], 253[34], 253[37], 254[34]
Chiarini, M. 359[24]
Chiorino, A. 104[54], 117[54]
Choi, J. C. 289[3]
Choi, J. H. 15[41], 20[41], 21[41], 22[41], 23[41], 24[41], 25[41]
Choi, M. C. K. 27[96], 28[96], 29[96], 30[96]
Chorkendorff, I. 282[35]
Chou, L. 302[97], 303[97]
Chou, T.-C. 93[12], 93[13]
Choukroun, R. 127[17], 128[17]
Christensen, M. 47[142], 48[142]
Chueh, W. C. 289[14]
Chumachenko, V. A. 115[71]
Chung, B. 260[79], 261[79], 262[79], 263[79]
Chung, L. W. 44[136], 291[49]
Church, T. L. 8[32], 10[32], 11[32], 12[32], 13[32], 55[151], 56[151]
Cindric, B. 322[11]
Cirtiu, C. M. 82[39]
Civera, M. 15[64], 20[64], 21[64], 24[64], 25[64], 27[64], 28[64], 29[64]
Clacens, J. 260[73], 261[73]
Claridge, J. B. 260[71], 261[71]
Claus, P. 146[55], 200[38]
Claver, C. 20[80], 104[52], 105[52], 131[32], 132[32], 133[32]
Clement, N. D. 206[74], 211[74]
Clipson, A. 157[20], 158[20]
Closier, M. 156[17], 157[17]
Cobo, A. J. G. 146[58]
Cokoja, M. 273[13]
Coll, M. 8[30], 8[33], 10[30], 10[33], 11[30], 11[33], 13[30], 13[33]
Collière, V. 127[17], 128[17]
Comminges, C. 301[87]
Company, A. 339[41]
Concepción, P. 105[56]
Cong, Y. 253[39], 257[49], 259[58], 260[78]
Cooper, B. H. 135[35]
Copéret, C. 202[51]
Coq, B. 106[57], 107[57], 108[57]
Cordes, D. B. 304[107], 305[107]
Corma, A. 105[56]
Cornils, B. 291[33], 292[33]
Costa, N. J. S. 98[32]
Costa, P. J. 344[53], 350[53]
Coudurier, G. 146[44]
Coville, N. J. 210[85]
Cowan, J. C. 335[36], 336[36]
Cox, D. P. 330[23]
Cozzi, P. G. 8[16], 8[20], 10[16], 10[20], 12[20], 13[16], 13[20]
Crabbé, P. 220[120]
Crabtree, R. H. 7[2], 8[2], 160[29], 161[29], 162[29], 164[29], 294[58], 295[58]
Cravotto, G. 200[41]
Crawford, P. 351[68], 352[68]
Cremer, G. 115[70]
Crespo-Quesada, M. 199[28], 200[40], 202[53]
Crombie, L. 221[132], 222[139]

Cseri, T. 93[19]
Cuan, Q. 259[54]
Cui, F. 260[81]
Cui, M.-S. 259[53], 260[74], 261[74], 262[74]
Cui, X. 8[25], 10[25], 11[25], 13[25], 14[25], 307[115], 308[118], 309[115], 309[118], 310[118]
Cui, Y. 281[31], 281[32]
Cunill, F. 92[10]
Curtis, J. M. 335[35], 336[35]
Curtiss, L. A. 257[47]

D

Dagdagan, O. A. 196[12], 197[12]
Dai, C. 220[119]
Dai, W. 289[24]
Dai, W.-L. 127[4], 280[29], 281[31], 281[32], 289[23]
Dai, X. 308[118], 309[118], 310[118]
Dai, X.-J. 231[15], 239[15]
Dang, T.-P. 7[6], 14[6], 20[6]
Daniliuc, C. 59[157], 60[157], 62[157]
Daniliuc, C.-G. 319[9], 320[9], 321[9]
Das, P. 133[33], 134[33]
Das Neves Gomes, C. 304[106], 305[106]
Datye, A. K. 232[19], 253[34], 254[34]
Daud, W. M. A. W. 229[8]
Dauth, A. 271[9], 272[9]
David, W. 107[59], 108[60]
Davies, I. W. 172[49], 173[49], 174[49]
Davis, B. H. 229[3]
Davis, H. E. 196[11], 197[11]
Davis, R. J. 232[19], 253[34], 254[34]
Davis, R. W. 251[19]
Day, M. W. 245[7]
Debaun, J. R. 339[7]
Debenham, S. D. 27[90], 28[90], 29[90]
de Bruin, B. 213[97]
Decken, A. 339[15]
Deelman, B.-J. 204[65]
de Groen, M. 200[43]
de Jong, E. 243[3], 258[3]
de Julián Fernández, C. 79[36], 80[36]
Delahay, G. 106[57], 107[57], 108[57]
de la Piscina, P. R. 301[82], 301[89]
Delay, F. 355[3]
Delferro, M. 127[3]
Della Bella, L. 118[84]
Dell'Amico, D. B. 289[8]
Della Monica, C. 95[25]
Delmas, H. 108[61]
Deluzarche, A. 279[26], 279[27]
Demaerel, J. 323[12]
Demailly, G. 339[4]
Demonchaux, P. 32[100], 33[100]
Deng, F. 27[94], 28[94], 29[94], 30[94]
Deng, J. 15[46], 15[47], 20[46], 20[47], 21[46], 21[47], 22[46], 22[47], 24[47], 25[47], 55[149], 56[149], 259[53], 260[74], 261[74], 262[74]
Deng, J.-F. 127[4], 146[48], 146[50]
Deng, L. 322[11]
Deng, Y. 307[115], 308[118], 309[115], 309[118], 310[118]
Derakhshandeh, P. G. 339[35]
De Riccardis, F. 95[25]
de Rooij, R. M. 200[43]
Deshpande, M. 156[18], 157[18]
Desmond, R. 157[23], 158[23]
de Souza, P. M. 229[3]
Desravines, D. 365[37]
Dessimoz, A.-L. 199[28]
Dethlefsen, J. R. 236[41]
Dev, S. 200[34]
Devaprabhakara, D. 222[138]
de Vries, A. H. M. 7[4], 20[82]
de Vries, J. G. 7[4], 7[10], 7[11], 15[58], 15[65], 20[58], 20[65], 20[82], 21[58], 21[65], 22[58], 22[65], 27[58], 28[58], 29[58], 30[58], 32[106], 33[106], 43[11], 79[35], 79[36], 80[35], 80[36], 113[69], 114[69], 153[7], 172[46], 173[46], 185[79], 186[79], 204[61], 232[22], 243[3], 253[33], 253[35], 253[36], 258[3], 260[33], 261[33], 355[11]
de Winter, D. A. M. 201[46]
DeWitt, E. J. 67[2]
de Wolf, E. 204[65]
Dhakshinamoorthy, A. 85[49], 86[49]
Diao, Z.-F. 289[20], 299[20]
Dias, E. L. 232[28], 233[28]
Díaz, J. L. 156[16], 157[16]
Dibenedetto, A. 289[4]
Dickinson, J. G. 229[2]
Didillon, B. 93[14]
Diéguez, M. 7[3], 8[3], 8[23], 8[24], 8[29], 8[30], 8[33], 8[36], 10[23], 10[24], 10[29], 10[30], 10[33], 10[36], 11[23], 11[24], 11[29], 11[30], 11[33], 12[23], 12[24], 12[29], 12[36], 13[23], 13[24], 13[29], 13[30], 13[33], 57[36]
Diesen, J. 38[127], 39[127]
Diesen, J. S. 8[19], 10[19], 11[19], 12[19], 13[19], 58[155], 59[155]
Dietrich, P. J. 253[37]
Digioia, F. 115[73], 115[74], 120[73], 120[74]
Dijkstra, A. J. 116[78]
Di Mondo, D. 231[18], 232[18], 233[18]
Ding, D. 259[62]
Ding, G. 146[57], 260[82]
Ding, K. 15[60], 15[61], 20[60], 20[61], 21[60], 21[61], 22[60], 22[61], 23[60], 23[61], 27[60], 27[61], 28[60], 28[61], 29[60], 29[61], 30[60], 30[61], 35[116], 36[116], 36[123], 37[123], 38[123], 47[140], 48[140], 48[143], 49[143], 50[140], 51[140], 51[143], 52[140], 53[140], 54[140], 69[9], 277[23], 279[23], 280[30], 281[30]
Ding, R.-S. 154[12], 155[12], 156[12]
Ding, Z. 177[62], 178[66], 179[66]
Dinjus, E. 291[33], 292[33]
Dirat, O. 157[20], 158[20]
Dirnsteiner, T. 223[142]
Diskin-Posner, Y. 218[113], 271[8], 271[9], 272[9], 291[50], 295[50], 296[50]
Dixneuf, P. H. 15[76], 20[76], 24[76], 25[76]
Dixon, D. J. 214[104]
Do, D. T. 324[14]
Dobereiner, G. E. 160[29], 161[29], 162[29], 164[29], 294[58], 295[58]
Dobinson, F. 332[30]
Dobler, W. 223[141], 223[142]
Dobročka, E. 260[68], 261[68], 263[68]
Dobson, N. A. 197[17]
Dohle, W. 355[8]
Doi, J. T. 339[21]
Donato, A. 146[46]
Dong, W. 253[39]
Dong, X.-Q. 44[136], 48[145], 50[145]
Donkervoort, H. G. 200[43]
Donkervoort, J. G. 200[44], 201[44], 202[44]
Donohoe, T. J. 304[103]
Dormer, P. G. 58[156], 59[156]
dos Santos, E. N. 97[31], 98[31], 98[36]
Dou, R. 127[14], 128[14], 146[54]
Douvris, C. 355[15]
Doyle, A. M. 67[3], 67[4], 67[5]
Drabowicz, J. 339[13]
Drake, J. L. 291[37], 295[37], 296[37]
Drauz, K. 32[101], 33[101]
Drees, M. 273[13]
Driess, M. 339[41]
Drost, R. M. 213[97]
Drury, W. J., III 13[38]
Du, H. 41[131], 42[131], 210[87]
Du, R. J. 298[68]
Du, X. L. 308[117], 309[117]
Duan, X. 302[93], 303[93], 304[93]
Duan, Y. 172[48], 173[48], 187[82], 187[83], 188[82], 188[83], 188[88], 189[88]
Duan, Z.-C. 15[46], 15[47], 15[59], 20[46], 20[47], 20[59], 21[46], 21[47], 21[59], 22[46], 22[47], 22[59], 24[47], 24[59], 25[47], 25[59], 27[59], 28[59], 29[59], 30[59], 55[149], 56[149]
Dubois, J. 365[37]
Dubuis, R. 197[13], 198[13], 199[13]
Duin, M. A. 206[73]
Dumesic, J. A. 232[19], 253[34], 253[37], 253[40], 253[42], 254[34], 259[60]
Dunsford, J. J. 212[96]
Dupau, P. 15[76], 20[76], 24[76], 25[76]
Dupont, J. 83[43], 83[44], 84[46], 146[47]
Durand, R. 106[57], 107[57], 108[57]
Durand, T. 199[29]
Durupthy, O. 93[19]
Dussault, P. H. 318[7], 319[7], 320[7]
Dutta, D. K. 133[33], 134[33]
Dutton, H. J. 116[77]
Dworakowska, S. 120[86]
Dyrager, C. 362[29]

Dyson, P. J. 128[25], 128[26], 128[27], 148[68], 160[28], 270[3], 291[35], 293[35], 294[57]

E

Ebel, K. 91[4], 99[4], 100[4]
Edouard, G. 245[7]
Eggersdorfer, M. 197[15], 199[15], 221[15], 223[15]
Eglinton, G. 197[17]
Eichel, H. J. 232[20]
Eisenstein, O. 160[29], 161[29], 162[29], 164[29]
Ejiri, S.-I. 339[8]
Elageed, E. H. M. 278[24]
Elek, J. 291[34], 292[34], 293[34], 294[56]
El Gabaly, F. 251[19]
Elias, Y. 202[54]
Elliott, J. M. 157[20], 158[20]
Ellis, D. J. 128[25], 128[26]
Ellman, J. A. 233[35], 234[35], 235[35], 236[35], 236[42], 238[42], 239[35], 239[42], 250[16]
Elsevier, C. J. 7[10], 204[61], 205[69], 205[70], 205[71], 206[69], 206[72], 206[73], 206[74], 211[69], 211[74], 212[94], 212[95], 212[96], 213[97], 214[94], 355[11]
Emerson, K. M. 157[23], 158[23]
Engelhard, M. H. 229[9]
Engler, H. 323[12]
Englert, U. 291[46], 293[46], 294[46]
Engman, M. 8[32], 8[34], 10[32], 10[34], 11[32], 11[34], 12[32], 13[32], 57[153], 58[155], 59[155]
Enoki, T. 88[52]
Enthaler, S. 15[66], 20[66], 21[66], 23[66], 24[66], 339[39], 339[40], 339[41], 339[42], 339[44], 339[46], 339[47]
Erathodiyil, N. 69[10], 70[10]
Erlandsson, M. 294[57]
Erre, G. 15[66], 20[66], 21[66], 23[66], 24[66]
Esaki, A. 260[70], 261[70]
Escribano, J. 339[34]
Espenson, J. H. 339[32]
Esteruelas, M. A. 131[31], 132[31], 133[31], 204[64], 210[88], 210[89], 210[90], 210[92], 211[92]
Etayo, P. 35[114], 36[114]
Evans, D. G. 83[41], 302[93], 303[93], 304[93]
Evans, T. J. 156[17], 157[17]

F

Fache, F. 153[11], 154[11], 156[11]
Fachinetti, G. 298[65], 298[66]
Fajula, F. 103[50], 104[50]
Falter, C. 289[14]
Fan, B. 172[45], 173[45]
Fan, G.-Y. 146[52]
Fan, H. 146[62]
Fan, H.-J. 188[88], 189[88]
Fan, K. 146[51], 146[53], 146[54], 146[56], 146[64], 280[29]
Fan, K.-N. 146[59], 154[12], 155[12], 156[12]
Fan, Q.-H. 32[104], 33[104], 153[3], 166[32], 167[32], 172[45], 173[45], 177[61], 177[62], 177[63], 177[64], 177[65], 178[64], 178[65], 178[66], 179[66], 184[63]
Fan, W. 250[13]
Fan, Y. 93[17], 95[17]
Fang, M. 147[67], 153[9], 156[15]
Fang, R. 260[69], 261[69], 263[69]
Fang, W. 301[88]
Fang, X. 305[113]
Farès, C. 215[109]
Fecher, G. H. 83[43]
Federsel, C. 270[3], 270[4], 270[5], 291[36], 295[36], 296[36]
Federsel, H.-J. 100[40]
Fedorov, A. 202[51]
Fei, J. 296[61], 297[61], 298[61]
Feilchenfeld, H. 275[18]
Felkin, H. 7[2], 8[2]
Feng, Y. 76[23], 83[40]
Feng, Z.-G. 326[17]
Feringa, B. L. 15[58], 15[65], 20[58], 20[65], 20[82], 21[58], 21[65], 22[58], 22[65], 27[58], 28[58], 29[58], 30[58], 32[106], 33[106], 172[46], 173[46], 185[79], 186[79]
Fernandes, A. C. 233[33], 339[36], 339[37], 339[38], 339[45], 339[48]
Fernandes, J. A. 344[53], 350[53]
Fernandez, E. 20[80]
Fernandez, P. 157[23], 158[23]
Fernández de Trocóniz, G. 362[29]
Fernández-Pérez, H. 35[114], 36[114], 172[52], 173[52], 174[52]
Fernández-Rodríguez, M. A. 339[49], 340[49], 341[49], 341[50], 343[49]
Ferreira, A. 32[100], 33[100]
Fichtner, P. F. P. 83[44], 84[46], 146[47]
Fierro, J. L. G. 253[36], 301[84], 301[89]
Filonenko, G. A. 298[67]
Finke, R. G. 128[28]
Firouzabadi, H. 339[17], 339[27]
Fischer, C. 32[101], 33[101]
Fischer, J.-H. 67[7]
Fleming, I. 197[20], 214[100]
Flores-Alamo, M. 304[109], 305[109]
Floryan, L. 201[50]
Foley, J. 157[23], 158[23]
Foley, P. 335[35], 336[35]
Fonseca, G. S. 84[46]
Fonseca, I. 96[26]
Fout, A. R. 218[114]
Fox, M. E. 27[87], 28[87], 29[87], 31[87]
Foxman, B. M. 355[15]
Franciò, G. 35[115], 36[115], 38[115]
Frank, H. A. 221[125], 221[127], 221[128]
Frankel, E. N. 109[62], 110[62]
Frantz, M.-C. 358[22]
Fray, M. J. 156[17], 157[17]
Frederick, B. G. 229[6], 229[10], 230[6], 230[10]
Frederik, P. M. 204[65]
Frediani, M. 207[76]
Frediani, P. 210[91], 210[92], 211[92]
Fresco, Z. M. 232[28], 233[28]
Freund, H.-J. 67[3], 67[4], 67[5], 67[7], 94[20], 94[21]
Frihed, T. G. 214[103]
Fristrup, P. 236[41]
Frogneux, X. 304[106], 304[111], 305[106], 305[111], 305[112]
Fröhlich, R. 159[27]
Frost, C. D. 200[33]
Frost, J. W. 230[11]
Frusteri, F. 301[76], 301[81]
Fu, G. 187[84], 188[84]
Fu, G. C. 220[119]
Fu, P. P. 141[39], 142[39], 143[39]
Fu, Y. 15[56], 20[56], 21[56], 22[56], 23[56], 26[56], 27[97], 28[97], 30[97], 259[53], 260[74], 261[74], 262[74]
Fuchs, C. 91[3]
Fuchs, M. 215[108], 215[109]
Fujita, E. 270[2]
Fujita, K.-i. 162[30], 163[30], 367[39]
Fujita, S. 289[25]
Fujita, T. 204[59]
Fujiwara, Y. 359[25], 360[25], 361[25]
Fukuoka, Y. 146[66]
Fukutani, K. 67[7]
Fukuzawa, S.-i. 27[86], 28[86], 29[86], 55[150], 56[150]
Fukuzumi, S. 291[38], 291[39]
Fulajtarová, K. 260[65], 260[68], 261[65], 261[68], 263[65], 263[68]
Furikado, I. 253[27], 254[27], 256[27]
Furimsky, E. 229[1]
Furler, P. 289[14]
Fürstner, A. 157[24], 158[24], 214[102], 214[103], 214[107], 215[107], 215[108], 215[109], 216[107], 219[107]
Furukawa, N. 339[19]
Furukawa, S. 162[30], 163[30]
Fusi, A. 101[41], 102[41], 117[79], 117[80], 120[80], 121[80]

G

Gaillard, S. 272[12]
Galano, J.-M. 199[29]
Galarneau, A. 103[50], 104[50]
Galasso, I. 117[83], 119[83]
Galindo, A. 236[39]
Gallagher, J. R. 127[3]
Gallezot, P. 146[44]
Galvagno, S. 146[46]
Gancher, E. 333[31]
Ganić, A. 57[154], 58[154]
Gao, F. 15[49], 20[49], 21[49], 22[49], 247[10], 248[10]
Gao, G. 278[24]
Gao, H. 83[42], 83[45]
Gao, J. 301[88]
Gao, L. 128[30], 130[30]

Gao, M. 15[79], 20[79], 24[79], 26[79], 27[98], 28[98], 30[98], 31[98]
Gao, W. 59[158], 60[158], 62[158]
Gao, X. 231[16], 232[16], 339[16]
Garbarino, G. 302[101], 303[101]
García, D. 341[50]
García, J. J. 304[109], 305[109]
García, N. 339[49], 340[49], 341[49], 341[50], 343[49]
García, S. 105[56]
Garcia Fidalgo, E. 128[29], 129[29], 130[29]
García-García, P. 339[49], 340[49], 341[49], 341[50], 343[49]
García-Garrido, S. E. 35[117], 35[121], 36[117], 36[121]
García Ruano, J. L. 221[134]
Garcia-Sharp, F. J. 332[29], 332[30]
Gargano, M. 104[54], 117[54], 117[80], 120[80], 121[80]
Garrett, S. 157[20], 158[20]
Gassner, F. 291[33], 291[53], 292[33], 292[53]
Gates, B. C. 251[23]
Gaube, J. 107[59], 108[60]
Gaudino, J. 15[78], 20[78], 24[78], 26[78]
Gazdar, M. 339[18]
Geantet, C. 351[61], 351[62]
Genêt, J.-P. 174[55], 174[56], 174[57], 175[55], 175[56], 175[57]
Geng, H. 15[69], 20[69], 24[69]
Gennari, C. 15[64], 20[64], 21[64], 24[64], 25[64], 27[64], 28[64], 29[64]
George, K. M. 358[22]
Gerlach, A. 204[66]
Gertsch, J. 355[17]
Geus, J. W. 200[43], 200[44], 201[44], 201[46], 202[44]
Geuther, M. E. 367[40], 368[40]
Ghampson, T. I. 229[6], 230[6]
Giachi, G. 207[76]
Gianetti, T. L. 209[82]
Gibb, A. 27[99], 28[99], 31[99], 32[99]
Giersch, W. 221[135], 222[135]
Gildemeister, E. 201[45]
Gillon, A. 20[80]
Gin, M. E. 251[21]
Girgsdies, F. 201[50]
Giustra, Z. X. 153[6]
Gladiali, S. 85[47]
Glaisher, S. 76[27]
Gleeson, D. 104[55], 105[55], 108[55]
Glorius, F. 59[157], 60[157], 62[157], 159[26], 159[27], 160[26], 182[70], 182[71], 182[72], 182[73], 182[74], 183[70], 183[71], 183[72], 183[73], 183[74], 184[70]
Göbbel, H.-G. 223[142]
Godard, C. 104[52], 105[52], 131[32], 132[32], 133[32]
Goddard, R. 13[38], 159[26], 160[26]
Goeppert, A. 289[16]
Goesmann, H. 272[10]
Goess, B. C. 96[29], 97[29]
Gohil, S. 38[126], 39[126]
Gohke, K. 98[34]
Goia, D. V. 233[30]
Gokhale, A. A. 259[59]
Goldman, A. S. 276[22]
Gomes da Silva, M. 98[33]
Gommermann, N. 355[8]
Gong, J. 232[21], 289[7], 302[90]
Gong, X.-Q. 259[54], 259[63]
Gong, Y. 153[10]
Gonsalvi, L. 271[7], 274[7], 291[47], 294[57], 295[47], 296[47]
González, F. V. 324[13]
González, I. 204[64]
González-Carballo, J. M. 301[84]
González-Liste, P. J. 35[117], 35[121], 36[117], 36[121]
González-Sebastián, L. 304[109], 305[109]
Gopaladasu, T. V. 236[38]
Gordon, J. C. 260[76], 261[76], 262[76]
Gorgas, N. 291[47], 295[47], 296[47]
Gosavi, T. 174[56], 175[56]
Gostanian, T. M. 253[40], 253[42]
Goto, M. 213[98], 220[98]
Goulas, K. 259[59]
Goundie, B. 229[6], 229[10], 230[6], 230[10]
Govender, T. 59[160], 61[160], 62[160]
Goy, R. 202[55]
Grabow, L. C. 229[10], 230[10]
Grabowski, R. 301[80]
Graham, T. J. A. 96[29], 97[29]
Grasemann, M. 196[10], 200[10], 202[52], 202[53], 290[28]
Grass, R. N. 77[31]
Greck, C. 339[4]
Greene, D. J. 27[85], 28[85], 29[85], 31[85]
Gridnev, I. D. 15[52], 15[55], 15[67], 20[52], 20[55], 20[67], 21[52], 21[55], 21[67], 23[52], 23[55], 23[67], 32[102], 33[102]
Griffin, K. 351[68], 352[68]
Grigoropoulos, A. 260[71], 261[71]
Grin, Y. 200[35]
Grinberg, N. 15[51], 20[51], 21[51], 22[51], 23[51]
Grippo, A. 259[59]
Gross, E. 92[8]
Grøtli, M. 362[29]
Grunwaldt, J.-D. 239[48]
Gu, L. 177[61]
Gu, M. 229[9], 358[21], 359[21]
Gu, R. 318[5]
Gu, W. 127[3]
Gualandi, A. 153[5]
Guan, H. 299[72], 300[72]
Guan, J. 253[38]
Guan, X. 355[14]
Guerrero, M. 98[32]
Guillaumet, G. 355[4]
Guin, D. 76[30]
Guingouain, M. 15[76], 20[76], 24[76], 25[76]
Gülak, S. 273[14], 274[14]
Gunanathan, C. 275[19], 276[19]
Gunasekar, G. H. 289[19], 296[19], 296[63], 297[63], 298[63]
Gunbas, G. 259[59]
Guo, P. 146[51]
Guo, R. 27[96], 28[96], 29[96], 30[96], 32[104], 33[104]
Guo, R.-N. 170[39], 172[42], 172[43], 172[44], 173[42], 173[43], 173[44]
Guo, X. 231[13]
Gupta, P. 215[109]
Gusevskaya, E. V. 97[31], 98[31], 98[32], 98[36]

H

Haddad, N. 15[51], 20[51], 21[51], 22[51], 23[51]
Haenel, M. W. 291[46], 293[46], 294[46]
Hagen, C. M. 128[28]
Haghdoost, M. M. 339[35]
Hahn, T. 223[142]
Haile, S. M. 289[14]
Hakim, S. H. 253[40], 253[42]
Ham, J. 327[20]
Hamasaka, G. 79[33], 82[33], 83[33]
Hamze, A. 365[37]
Han, B. 69[9], 83[42], 83[45], 127[13], 128[13], 146[57], 146[62], 146[63], 250[14]
Han, H. 127[13], 128[13]
Han, J. 319[8], 320[8]
Han, L.-B. 213[98], 220[98], 220[117]
Han, X. 281[34]
Han, X.-W. 15[73], 20[73], 24[73], 25[73], 170[37], 172[41], 173[41], 174[37]
Han, Y. 69[10], 70[10]
Han, Z. 47[140], 48[140], 48[143], 49[143], 50[140], 51[140], 51[143], 52[140], 53[140], 54[140], 277[23], 279[23], 280[30], 281[30]
Hansen, C. A. 230[11]
Hansen, L. K. 8[28], 10[28], 11[28], 12[28], 13[28], 14[28]
Happer, D. A. R. 351[65]
Haque, D. M. 330[23]
Hardacre, C. 351[68], 352[68]
Harder, S. 113[69], 114[69]
Hardouin, C. 170[40]
Harley, R. A. 127[2]
Harrison, D. J. 339[15]
Hartwig, J. F. 244[6], 245[6], 246[6], 247[6], 247[9], 247[10], 248[10]
Harvey, R. G. 141[39], 142[39], 143[39]
Hasan, S. K. 339[25]
Hasanayn, F. 276[22]
Hasegawa, S. 221[128]
Hashiguchi, Y. 172[53], 173[53], 174[53]
Hashimoto, A. 156[18], 157[18]
Hashimoto, H. 221[128], 290[32]
Hashimoto, Y. 339[26]
Hattori, K. 357[19], 358[19], 361[19]
Haupert, L. J. 250[15], 251[15]

Hauwert, P. 212[94], 212[95], 212[96], 214[94]
Hay, A.-E. 15[76], 20[76], 24[76], 25[76]
Hayashi, H. 291[38], 291[39]
Hayashi, M. 13[38]
Hayashi, T. 38[125]
Hazari, N. 160[29], 161[29], 162[29], 164[29], 294[58], 295[58]
He, H.-Y. 154[12], 155[12], 156[12]
He, L. 154[12], 155[12], 156[12]
He, L.-N. 289[20], 299[20]
He, P. 281[34]
He, S. 302[93], 303[93], 304[93]
He, Y. 177[61], 178[66], 179[66]
He, Y.-B. 239[49]
He, Y.-M. 153[3], 177[62], 177[63], 177[64], 177[65], 178[64], 178[65], 184[63]
Heagy, M. D. 283[40]
Healy, M. P. 304[103]
Heck, R. F. 203[57], 204[57]
Hedberg, C. 8[27], 8[28], 8[34], 10[27], 10[28], 10[34], 11[27], 11[28], 11[34], 12[27], 12[28], 13[27], 13[28], 14[28]
Heeres, H. J. 113[69], 114[69], 232[22], 243[3], 253[33], 253[35], 253[36], 258[3], 260[33], 261[33]
Heitbaum, M. 159[27]
Heller, D. 205[68]
Henrick, C. A. 197[21]
Hensen, E. J. M. 298[67]
Hensley, A. J. R. 229[9]
Hermannsdörfer, J. 251[18]
Hermans, S. 104[55], 105[55], 108[55]
Hernández-Garrido, J. C. 105[56]
Herrero, J. 204[64]
Herrmann, A. T. 238[47]
Herrmann, W. A. 273[13], 291[33], 292[33]
Herzberg, D. 27[89], 28[89], 29[89]
Heslop, K. 20[80]
Hessel, V. 204[67]
Hesson, J. M. 127[2]
Hibbitts, D. 232[19], 253[34], 254[34]
Hibino, S. 339[14]
Hicks, J. C. 251[21], 296[62], 297[62], 298[62]
Hida, S. 174[58], 175[58]
Hietala, J. 290[27]
Higashi, N. 15[67], 20[67], 21[67], 23[67], 32[102], 33[102]
Higashida, K. 216[111], 217[111]
Hilgraf, R. 8[20], 10[20], 12[20], 13[20]
Hills, L. 236[39]
Himeda, Y. 270[2], 289[21], 290[21], 291[40], 291[41], 291[42], 291[43], 291[44], 291[45], 292[45], 293[45], 294[40], 294[43], 294[44], 294[45], 295[45]
Hindermann, J. P. 279[26], 279[27]
Hirano, S.-i. 295[60], 296[60]
Hirao, T. 174[55], 175[55]
Hirose, T. 291[41]
Hirsekorn, F. J. 128[22]
Hoang, D.-L. 229[7]
Hoang-Van, C. 146[45]
Hoarau, C. 326[18]
Hoen, R. 15[58], 15[65], 20[58], 20[65], 21[58], 21[65], 22[58], 22[65], 27[58], 28[58], 29[58], 30[58]
Hoffmann, F. 201[45]
Hofkens, J. 115[70]
Hoge, G. 15[55], 20[55], 21[55], 23[55], 27[85], 28[85], 29[85], 31[85], 32[103], 33[103]
Hoke, D. 355[6]
Holle, S. 159[26], 160[26]
Hölscher, M. 291[46], 293[46], 294[46]
Holz, J. 32[101], 33[101]
Homs, N. 301[82], 301[89]
Hong, C. H. 232[27]
Hong, S. H. 278[25], 279[25]
Hong, U. G. 302[95], 302[98], 303[95], 303[98]
Hong, Y. 229[9]
Hooper, D. S. 127[2]
Hoorn, J. 79[36], 80[36]
Höpner, T. 285[49]
Horáček, J. 229[5]
Horiuchi, Y. 15[52], 20[52], 21[52], 23[52]
Horiuti, I. 98[35]
Horiuti, J. 67[1], 68[1]
Hörmann, E. 8[35], 10[35], 12[35]
Hosaka, M. 27[86], 28[86], 29[86]
Hoshi, K. 153[8]
Hou, D.-R. 8[25], 10[25], 11[25], 13[25], 14[25]
Hou, G. 15[54], 15[69], 20[54], 20[69], 21[54], 22[54], 23[54], 24[54], 24[69], 26[54], 35[54], 36[54]
Hou, G.-H. 27[97], 28[97], 30[97]
Hou, M. 127[13], 128[13]
Hou, X.-L. 8[17], 10[17], 11[17], 13[17], 48[144], 49[144], 51[144]
Hough, J. J. 211[93]
Hoyos, L. J. 351[57]
Hronec, M. 260[65], 260[68], 261[65], 261[68], 263[65], 263[68]
Hsieh, M.-C. 302[94], 303[94]
Hu, A.-G. 15[56], 20[56], 21[56], 22[56], 23[56], 26[56]
Hu, L. 210[87]
Hu, P. 351[68], 352[68]
Hu, S.-C. 146[65]
Hu, T. 229[4]
Hu, W. 157[23], 158[23]
Hu, X. 15[62], 15[63], 20[62], 20[63], 21[62], 21[63], 22[62], 22[63], 23[62], 23[63], 24[62], 24[63], 25[63], 27[63], 28[63], 29[63], 30[63], 48[145], 50[145]
Hu, X.-P. 15[44], 15[46], 15[47], 15[59], 20[44], 20[46], 20[47], 20[59], 21[44], 21[46], 21[47], 21[59], 22[44], 22[46], 22[47], 22[59], 24[47], 24[59], 25[47], 25[59], 27[44], 27[59], 28[44], 28[59], 29[59], 30[44], 30[59], 32[105], 33[105], 55[149], 56[149]
Hu, Y. 158[25], 159[25], 160[25], 172[43], 173[43]
Hu, Z. 302[102], 303[102]
Hua, G. 339[28]
Hua, R. 213[99], 220[116]
Huang, F.-C. 91[3]
Huang, H. 15[62], 15[63], 20[62], 20[63], 21[62], 21[63], 22[62], 22[63], 23[62], 23[63], 24[62], 24[63], 25[63], 27[63], 28[63], 29[63], 30[63]
Huang, J. 83[42], 83[45]
Huang, J.-D. 15[46], 15[47], 20[46], 20[47], 21[46], 21[47], 22[46], 22[47], 24[47], 25[47]
Huang, K. 15[54], 20[54], 21[54], 22[54], 23[54], 24[54], 26[54], 35[54], 36[54]
Huang, T.-N. 128[24]
Huang, W. 32[109], 33[109]
Huang, W.-X. 172[51], 172[54], 173[51], 173[54], 174[51], 174[54], 187[86], 188[86]
Huang, Y. 127[15], 128[15], 172[49], 173[49], 174[49], 253[41], 254[41]
Huang, Z. 260[81], 280[30], 281[30]
Huber, G. W. 257[48], 259[61]
Huber, W. 197[23]
Huch, V. 52[148], 53[148]
Hudlický, M. 318[6]
Hudlicky, T. 330[23]
Hudson, R. 76[17], 76[19], 76[20], 76[23], 76[25], 76[26], 76[27], 79[33], 82[33], 82[39], 83[33]
Huff, C. A. 299[70], 300[70], 300[74]
Huheey, J. E. 195[4]
Huo, W. 250[13]
Hutchins, R. O. 355[6]
Huynh, T. M. 229[7]
Hwang, S. 302[95], 302[98], 303[95], 303[98]
Hyett, D. J. 20[80]

I

Ibhadon, A. O. 200[36], 200[37], 202[36], 202[37]
Ichikuni, N. 102[43]
Ienco, A. 271[7], 274[7]
Igawa, K. 321[10], 322[10]
Iimuro, A. 174[58], 175[58], 176[59], 177[59]
Ikariya, T. 295[60], 296[60]
Ikeda, R. 172[53], 173[53], 174[53]
Ikonomakou, E. 112[68]
Ikushima, Y. 259[55]
Ilyna, I. I. 316[2], 316[3], 317[2]
Imaizumi, S. 98[34]
Imamoto, T. 15[43], 15[52], 15[55], 15[67], 20[43], 20[52], 20[55], 20[67], 21[43], 21[52], 21[55], 21[67], 22[43], 23[52], 23[55], 23[67], 32[102], 33[102]
Imelik, B. 146[44]
Inoue, S. 100[38], 233[31], 339[46]
Inoue, Y. 290[32]
Inui, T. 299[69]
Iranpoor, N. 339[17], 339[27]
Ireland, T. 35[118], 36[118]
Iriarte-Velasco, U. 115[72]
Irran, E. 339[46]

Iseki, A. 219[115]
Iseki, K. 36[124], 37[124]
Ishibashi, J. S. A. 153[6]
Ishikawa, M. 252[24], 252[25]
Ishikawa, S. 76[17], 76[19]
Ishikawa, Y. 204[59]
Ishizuka, K. 172[53], 173[53], 174[53]
Isler, O. 197[22], 197[23], 198[22], 199[22]
Ismaili, H. 371[43], 372[43]
Italiano, G. 301[76]
Ito, T. 185[78], 186[78]
Ito, Y. 88[52], 166[33], 167[33], 185[77], 185[78], 186[77], 186[78]
Ivanov, A. 93[18]
Iwasaki, Y. 157[19]
Iwashita, T. 221[125], 221[128]
Iyer, P. S. 358[20]
Izquierdo, J. F. 92[10]
Izumida, H. 290[32]
Izzo, I. 95[25]

J

Jackson, M. 27[84], 27[87], 28[84], 28[87], 29[84], 29[87], 31[84], 31[87]
Jackson, S. D. 67[5]
Jackstell, R. 270[3], 270[4], 270[5], 273[14], 274[14], 291[36], 295[36], 296[36]
Jacobi von Wangelin, A. 79[34]
Jacobs, G. 229[3]
Jacobs, P. 115[70], 116[75]
Jacquet, O. 304[106], 304[111], 305[106], 305[111]
Jain, A. 324[13]
Jain, S. 284[41]
Jäkel, C. 91[5], 100[40]
Jalil, A. A. 302[91]
Jamalian, A. 339[27]
Jameel, H. 250[13]
James, B. R. 344[52]
Jang, Y. 339[31]
Janni, M. 364[35], 365[35], 369[35]
Jansen, M. 223[143], 224[143]
Jantke, D. 273[13]
Jardine, F. H. 128[20], 128[21]
Jarrell, T. M. 250[15], 251[15]
Jaworska, M. 157[21], 158[21]
Jefford, C. 355[3]
Jeger, O. 224[145]
Jenkins, P. A. 221[132], 222[139]
Jennerjahn, R. 270[3]
Jennings, M. 231[18], 232[18], 233[18]
Jensen, A. D. 239[48]
Jensen, P. A. 239[48]
Jeon, H. B. 339[31]
Jérôme, F. 260[73], 261[73]
Jerphagnon, T. 32[100], 33[100], 172[46], 173[46], 185[79], 186[79]
Jessop, P. G. 289[1], 289[26], 292[54], 293[54], 295[59], 296[59]
Ji, G. 296[64], 297[64], 298[64]
Ji, J. 44[136]
Ji, J.-X. 32[104], 33[104]
Ji, L. 278[24]
Ji, Y. 172[51], 173[51], 174[51], 278[24], 279[28]
Jia, L. 301[88]
Jia, S. 260[75], 260[79], 261[75], 261[79], 262[79], 263[75], 263[79]
Jia, W. 279[28]
Jia, X. 15[45], 15[57], 20[45], 20[57], 21[45], 21[57], 22[45], 22[57], 27[57], 28[57], 29[57], 30[57], 93[17], 95[17]
Jiang, G.-F. 187[82], 188[82]
Jiang, H. 232[28], 233[28]
Jiang, H.-L. 290[29]
Jiang, J. 15[71], 20[71], 24[71], 47[138], 48[138], 50[138], 51[138]
Jiang, Q. 35[111], 37[111], 38[111]
Jiang, T. 69[9], 83[42], 83[45], 127[13], 128[13], 146[57], 146[62], 146[63]
Jiang, Z. 166[32], 167[32], 308[117], 309[117]
Jiao, G. 15[70], 20[70], 24[70], 25[70], 27[70]
Jiao, H. 32[101], 33[101]
Jiménez, C. 203[58]
Jimenez, L. S. 358[23]
Jiménez, O. 156[16], 157[16]
Jin, M. 200[40]
Jin, S. 44[136]
Jin, T. 204[59]
Jing, X.-K. 232[21]
Jitsukawa, K. 260[77], 261[77], 262[77], 339[43], 341[51], 342[51], 343[51], 345[54], 346[54], 347[54], 350[54]
Johann, T. 91[4], 99[4], 100[4]
Johansson, L. E. 116[76]
Johnson, B. F. G. 102[45], 104[45], 104[55], 105[45], 105[55], 108[45], 108[55]
Johnson, C. D. 332[30]
Johnson, C. R. 196[7]
Johnson, J. S. 324[14]
Johnson, R. P. 222[136]
Johnsson, P. 290[27]
Johnston, S. K. 200[37], 202[37]
Jones, A. B. 157[20], 158[20]
Jongerius, A. L. 243[1]
Jonischkeit, T. 206[72]
Joó, F. 289[1], 291[34], 292[34], 293[34], 294[56]
Jorgensen, M. 289[22]
Jung, K.-D. 289[19], 296[19], 296[63], 297[63], 298[63]
Jung, M. 327[20]
Junge, K. 15[66], 20[66], 21[66], 23[66], 24[66], 80[37], 282[37], 283[37], 304[104], 305[113], 305[114], 306[114], 307[114], 310[119]
Junker, E. M. 27[99], 28[99], 31[99], 32[99]
Jurčík, V. 103[46], 103[47]

K

Kadyrov, R. 15[66], 20[66], 21[66], 23[66], 24[66]
Kagan, H. B. 7[6], 14[6], 20[6]
Kaibel, G. 223[142]
Kainz, Q. M. 77[31]
Kaiser, S. 8[16], 8[26], 10[16], 10[26], 11[26], 13[16], 13[26], 13[38]
Kajikawa, T. 221[125], 221[127], 221[128]
Kalberg, C. S. 36[122], 37[122]
Källström, K. 8[27], 8[28], 10[27], 10[28], 11[27], 11[28], 12[27], 12[28], 13[27], 13[28], 14[28]
Kameyama, N. 181[69], 182[69]
Kameyama, R. 208[80], 209[80], 211[80]
Kanao, R. 260[70], 260[72], 261[70], 261[72], 263[72]
Kandula, S. 176[59], 177[59]
Kaneda, K. 185[78], 186[78], 260[77], 261[77], 262[77], 339[43], 341[51], 342[51], 343[51], 345[54], 346[54], 347[54], 350[54]
Kanehira, K. 38[125]
Kang, K. H. 251[20]
Kang, K. Y. 68[8]
Kang, X. 127[13], 128[13]
Kannan, M. 358[20]
Kano, S. 339[14]
Karakhanov, E. 93[18]
Karamé, I. 195[2], 199[2], 289[21], 290[21]
Kardasheva, Y. 93[18]
Karimi, B. 339[23]
Karimipourfard, D. 251[23]
Karube, D. 185[77], 186[77]
Karunananda, M. K. 217[112], 218[112]
Kasahara, I. 100[38]
Kashiwabara, M. 179[67], 180[67], 180[68], 181[68], 184[68]
Kasuga, K. 291[42], 291[43], 291[44], 291[45], 292[45], 293[45], 294[43], 294[44], 294[45], 295[45]
Kataoka, Y. 174[55], 175[55]
Kato, K. 36[123], 37[123], 38[123]
Katsumura, S. 221[125], 221[127], 221[128]
Katz, J. L. 76[27]
Kaufmann, R. 201[48]
Kaukoranta, P. 8[34], 10[34], 11[34]
Kaushik, M. 83[40]
Kawai, M. 300[73]
Kawanami, H. 259[55]
Kawasaki, Y. 321[10], 322[10]
Kayaert, P. 115[70]
Kayaki, Y. 295[60], 296[60]
Kazmaier, U. 52[148], 53[148]
Kean, A. J. 127[2]
Keaveney, W. P. 333[31], 333[32]
Keen, S. P. 157[23], 158[23]
Keenan, M. 13[38]
Kelbichová, V. 229[5]
Kelley, P. 245[7]
Kelly, C. K. 367[40], 368[40]
Kelsen, V. 80[37]
Kempe, R. 243[5], 249[5], 251[18]
Kendall, G. R. 127[2]
Kennedy, J. W. J. 157[24], 158[24]

Kenttämaa, H. I. 250[15], 251[15]
Keogh, J. 355[6]
Kerdphon, S. 59[161]
Khai, B. T. 203[56]
Khairulina, L. N. 351[69], 351[70]
Khalil, L. B. 285[45]
Khan, F. R. N. 358[20]
Kharraz, E. 335[35], 336[35]
Khedri, M. 339[29]
Khodaei, M. M. 339[29]
Kieczka, H. 290[27]
Kieffer, R. 279[26], 279[27]
Kikuchi, S. 27[86], 28[86], 29[86], 339[26]
Kim, J. K. 251[20]
Kim, K. T. 339[31]
Kim, N. 68[8]
Kim, S. H. 278[25], 279[25]
Kim, Y. G. 232[27]
Kindler, A. 223[141]
Kinoshita, M. 339[1]
Kirby, F. 201[46]
Kircher, T. 223[140]
Kirchner, K. 291[47], 295[47], 296[47]
Kirchstetter, T. W. 127[2]
Kirkpatrick, D. 355[3]
Kischkewitz, M. 319[9], 320[9], 321[9]
Kissel, W. S. 27[85], 28[85], 29[85], 31[85]
Kistiakowsky, G. B. 221[129]
Kita, Y. 174[58], 175[58], 176[59], 177[59]
Kitamura, T. 355[12]
Kitanaka, T. 281[33]
Kitatsuji, C. 162[30], 163[30]
Kiwi-Minsker, L. 196[10], 199[28], 200[10], 200[40], 202[52], 202[53]
Klankermayer, J. 299[71], 300[71], 304[110], 305[110], 306[110], 307[110]
Klasovsky, F. 200[38]
Klauss, R. 272[10]
Klein, H.-F. 108[60]
Klein, I. 250[15], 251[15]
Klein Gebbink, R. J. M. 233[34], 236[34]
Kleman, P. 35[117], 35[121], 36[117], 36[121]
Klerk, C. 206[73]
Klibanov, A. M. 285[50]
Klimczyk, S. 216[110]
Klosin, J. 27[87], 28[87], 29[87], 31[87]
Klumphu, P. 362[30], 363[30]
Klussmann, M. 323[12]
Kluwer, A. M. 204[61], 206[72]
Kneisel, F. F. 355[8]
Knochel, P. 35[118], 35[119], 36[118], 36[119], 355[8]
Knölker, H.-J. 272[10]
Knowles, W. S. 7[1], 7[5], 196[8]
Kobayashi, T. 127[3]
Koblenz, T. S. 206[72]
Koch, G. 8[12]
Kofler, M. 197[23]
Kogan, V. 134[34], 135[34]
Koh, D. J. 302[95], 302[98], 303[95], 303[98]
Koharski, D. 355[6]
Kohno, K. 289[5]
Kohrt, C. 205[68]
Kojima, K. 128[30], 130[30]
Kok, S. H. L. 15[45], 20[45], 21[45], 22[45]
Komarov, I. V. 32[101], 33[101]
Kon, K. 307[116], 309[116], 348[56], 349[56], 350[56], 351[56]
Kong, H. 301[85]
Kong, L. L. 289[15]
Kong, X. 335[34], 336[34]
Konishi, H. 339[26]
Konishi, M. 146[66]
Konno, T. 55[150], 56[150]
Konrad, T. M. 35[115], 36[115], 38[115]
Konta, H. 339[3]
Kopp, F. 355[8]
Korach, M. 315[1]
Koritala, S. 116[77]
Korn, T. 355[8]
Koso, S. 253[27], 253[28], 253[29], 253[30], 253[31], 254[27], 254[28], 254[29], 254[30], 254[31], 255[29], 255[31], 256[27], 256[28], 256[29]
Köster, R. 143[40], 144[40]
Kosugi, H. 339[3], 339[5]
Kothe, J. 291[46], 293[46], 294[46]
Kou, Y. 83[41], 127[5], 128[5], 148[68]
Kouchi, M. 15[55], 20[55], 21[55], 23[55]
Kozłowska, A. 301[80]
Krabbe, S. W. 324[14]
Krackl, S. 339[41], 339[46]
Krause, J. A. 299[72], 300[72]
Krebs, F. C. 289[22]
Kreifels, S. 318[7], 319[7], 320[7]
Kretschmer, W. P. 251[18]
Krishnamurthy, D. 15[51], 20[51], 21[51], 22[51], 23[51]
Krishnamurthy, S. 351[63]
Krishnamurti, M. 197[17]
Krivorucho, O. P. 351[69]
Krska, S. W. 58[156], 59[156], 172[49], 173[49], 174[49]
Krüger, C. 8[12]
Kubička, D. 229[5]
Kubota, T. 253[28], 254[28], 256[28]
Kudo, A. 282[36]
Kuenzi, R. 223[140]
Kühn, F. E. 273[13]
Kuipers, B. W. M. 204[65]
Kuivila, H. G. 355[7]
Kumada, M. 38[125]
Kumar, A. 233[30]
Kume, A. 357[19], 358[19], 361[19]
Kumobayashi, H. 27[95], 28[95], 29[95], 100[38]
Kunieda, N. 339[1]
Kunimori, K. 253[27], 254[27], 256[27]
Kunishige, H. 347[55]
Kuo, E. 200[31]
Kuo, M.-S. 302[94], 303[94]
Kurahashi, T. 207[75], 211[75]
Kurokawa, T. 185[77], 185[78], 186[77], 186[78]
Kuroki, Y. 36[124], 37[124]
Kusano, H. 180[68], 181[68], 184[68]
Kusumoto, S. 252[26]
Kusumoto, T. 221[128]
Kuwano, R. 153[2], 166[33], 167[33], 172[53], 173[53], 174[53], 179[67], 180[67], 180[68], 181[68], 181[69], 182[69], 184[68], 185[77], 185[78], 186[77], 186[78]
Kuzuka, T. 253[30], 254[30]
Kwok, W. H. 27[96], 28[96], 29[96], 30[96]
Kwok, W. M. 355[14]
Kwon, M. S. 68[8], 71[12], 72[12]
Kwon, S. 232[27]

L

Labella, L. 289[8]
Lachowska, M. 301[80]
Lacroix, M. 351[61], 351[62]
Ladera, R. 301[84]
Lai, C.-C. 91[2]
Lam, K. 172[45], 173[45]
Lan, Y. 73[15], 74[15]
Landaeta, V. R. 294[57]
Landers, R. 146[58]
Landini, D. 339[20]
Lange, J.-P. 243[2], 258[2]
Langer, R. 271[8], 291[50], 295[50], 296[50]
Langler, R. F. 339[15]
La Pierre, H. S. 209[83]
Laquidara, J. M. 100[39], 101[39]
Large, S. E. 27[90], 28[90], 29[90]
Lau, C. P. 293[55]
Laudert, D. 197[15], 199[15], 221[15], 223[15]
Laurenczy, G. 128[28], 270[3], 290[28], 291[34], 291[35], 292[34], 293[34], 293[35], 294[56], 294[57]
LaValle, C. R. 358[22]
La-Venia, A. 330[25], 331[25], 331[26]
Lavilla, R. 156[16], 157[16]
Lawrence, R. 119[85], 120[85]
Lazier, W. A. 196[5]
Lazo, J. S. 358[22]
Lazreg, F. 304[108], 305[108]
Lee, H. J. 290[31]
Lee, H. M. 141[39], 142[39], 143[39]
Lee, I. Q. 318[7], 319[7], 320[7]
Lee, J. 47[142], 48[142], 290[31], 302[95], 302[98], 303[95], 303[98]
Lee, J. K. 15[41], 20[41], 21[41], 22[41], 23[41], 24[41], 25[41], 251[20]
Lee, J. S. 68[8]
Lee, J. W. 251[20]
Lee, S.-g. 15[41], 20[41], 21[41], 22[41], 23[41], 24[41], 25[41]
Lee, S. S. 69[10], 70[10]
Lefèvre, G. 304[105]
Leffmann, R. 236[40]
Lefort, L. 79[35], 79[36], 80[35], 80[36]
Lehmann, C. W. 159[26], 160[26]

Lehmann, H. 196[10], 200[10]
Lei, A. 15[69], 20[69], 24[69]
Leimgruber, S. 358[22]
Leisch, H. 330[23]
Leitner, W. 35[115], 36[115], 38[115], 291[33], 291[46], 291[53], 292[33], 292[53], 293[46], 294[46], 299[71], 300[71], 304[110], 305[110], 306[110], 307[110]
Leitus, G. 271[8], 271[9], 272[9], 291[50], 295[50], 296[50]
Leng, J. 239[50]
Lennon, I. C. 27[84], 27[87], 28[84], 28[87], 29[84], 29[87], 31[84], 31[87]
Lenoir, C. 365[37]
Leonard, D. P. 283[40]
Le Page, J.-F. 127[7]
Létinois, U. 197[15], 199[15], 221[15], 223[15]
Leung, A. W.-M. 8[32], 10[32], 11[32], 12[32], 13[32]
Leutzsch, M. 215[109]
Le Valant, A. 301[87]
Lew, W. 157[21], 158[21]
Lewis, M. L. 156[17], 157[17]
Li, B. 99[37], 101[37]
Li, C. 155[13], 155[14], 156[14], 184[76], 185[76], 186[76], 187[84], 188[84], 257[49], 302[93], 303[93], 304[93]
Li, C.-J. 76[19], 76[20], 79[33], 82[33], 83[33], 231[15], 239[15]
Li, F. 281[34], 289[24]
Li, G. 257[49], 259[58], 260[78]
Li, H. 146[48], 146[51], 146[59], 146[64], 153[10], 260[75], 261[75], 263[75], 276[21], 281[34]
Li, H.-X. 146[59]
Li, H.-Y. 301[85]
Li, J. 155[13], 155[14], 156[14], 213[99], 220[116], 363[34], 364[34], 368[34]
Li, J.-Q. 47[139], 48[139]
Li, L. 188[88], 189[88], 253[41], 254[41], 259[58], 365[36], 366[36]
Li, M. 253[39]
Li, M. M.-J. 259[63]
Li, N. 253[39], 253[41], 254[41], 257[49], 259[58], 259[61], 260[78]
Li, P. 48[145], 50[145]
Li, Q. 301[88]
Li, R.-X. 146[52]
Li, S. 43[132], 43[133], 43[134], 44[132], 44[133], 44[134], 45[133], 45[134], 46[132], 46[133], 260[78]
Li, W. 7[8], 7[9], 15[39], 15[40], 15[69], 15[72], 20[39], 20[40], 20[69], 20[72], 21[39], 21[40], 21[72], 22[40], 23[39], 23[72], 24[39], 24[69], 24[72], 25[39], 25[72], 27[40], 27[88], 28[40], 28[88], 29[40], 29[88], 30[40], 30[88], 32[108], 33[108], 59[157], 60[157], 62[157], 186[81], 187[81], 189[81], 250[13], 298[68]
Li, W.-D. 232[21]
Li, X. 15[45], 15[57], 20[45], 20[57], 21[45], 21[57], 22[45], 22[57], 27[57], 27[96], 28[57], 28[96], 29[57], 29[96], 30[57], 30[96], 32[104], 33[104], 127[15], 128[15], 232[24], 232[25], 232[26], 232[29], 235[36], 236[24], 236[45], 237[24], 237[45], 238[24], 238[45], 239[24], 239[36], 239[45], 253[43], 363[34], 364[34], 368[34]
Li, X.-J. 146[52]
Li, X.-L. 259[53], 260[74], 261[74], 262[74]
Li, Y. 48[143], 49[143], 51[143], 95[22], 95[24], 127[12], 128[12], 153[10], 231[16], 232[16], 260[69], 260[82], 261[69], 263[69], 279[28], 305[113], 305[114], 306[114], 307[114], 310[119]
Li, Y.-L. 91[2], 355[14]
Li, Y.-N. 289[20], 299[20]
Li, Y.-X. 170[38], 171[38]
Li, Z. 177[61], 178[66], 179[66], 257[47]
Li, Z.-C. 127[5], 128[5]
Lian, C. 279[28]
Liang, G. 15[49], 20[49], 21[49], 22[49]
Liang, S. 146[57], 146[63]
Liang, S.-P. 302[92], 303[92]
Liang, X.-L. 301[86]
Liang, X.-M. 15[59], 20[59], 21[59], 22[59], 24[59], 25[59], 27[59], 28[59], 29[59], 30[59]
Liaw, B. J. 301[75]
Lieberman, D. 27[99], 28[99], 31[99], 32[99]
Lightfoot, A. 8[13], 8[14], 10[14], 11[14], 13[14]
Liguori, F. 104[52], 105[52], 131[32], 132[32], 133[32]
Lim, H. 302[95], 302[98], 303[95], 303[98]
Lin, G.-D. 301[85]
Lin, M. 35[112], 37[112]
Lin, S. 245[7], 367[40], 368[40]
Lin, T.-B. 93[12]
Linares, N. 103[50], 104[50]
Lindhardt, A. T. 216[110]
Lindlar, H. 195[3], 197[3], 197[13], 197[14], 198[3], 198[13], 198[14], 199[3], 199[13], 199[14]
Linehan, J. C. 292[54], 293[54]
Ling, H. 318[5]
Ling, Y. H. 356[18], 357[18]
Linhardt, R. 77[31]
Lipshutz, B. H. 362[30], 363[30]
List, B. 52[147]
Liu, B. 146[50]
Liu, C. 95[23]
Liu, D. 8[15], 10[15], 11[15], 13[15], 15[53], 20[53], 21[53], 24[53], 27[53], 27[88], 28[53], 28[88], 29[53], 29[88], 30[88], 32[53], 33[53], 35[53], 35[113], 36[53], 36[113], 187[84], 188[84], 259[57]
Liu, F. 260[73], 261[73]
Liu, G. 15[70], 20[70], 24[70], 25[70], 27[70], 259[63], 335[34], 336[34]
Liu, H. 127[13], 128[13], 146[57], 146[62], 146[63], 260[69], 260[81], 261[69], 263[69], 280[30], 281[30]
Liu, H.-J. 202[51]
Liu, J. 146[53], 302[93], 303[93], 304[93], 365[36], 366[36]
Liu, J.-L. 146[59]
Liu, K. 260[75], 261[75], 263[75]
Liu, L.-J. 172[51], 173[51], 174[51]
Liu, N. 298[68]
Liu, Q. 95[24], 273[14], 274[14]
Liu, Q.-S. 233[32]
Liu, S. 146[49], 172[49], 173[49], 174[49], 257[50], 258[51], 258[52], 259[52]
Liu, S.-Y. 153[6]
Liu, T. 213[99]
Liu, T.-L. 15[68], 15[74], 20[68], 20[74], 21[68], 23[68], 24[74], 25[74]
Liu, W. 229[4]
Liu, X. 15[70], 20[70], 24[70], 25[70], 27[70], 47[140], 48[140], 50[140], 51[140], 52[140], 53[140], 54[140], 99[37], 101[37], 259[56], 259[62], 260[75], 260[79], 260[80], 260[81], 261[75], 261[79], 262[79], 263[75], 263[79]
Liu, X.-H. 259[54]
Liu, X.-M. 301[78]
Liu, X.-X. 259[53]
Liu, Y. 15[60], 20[60], 21[60], 22[60], 23[60], 27[60], 28[60], 29[60], 30[60], 35[116], 36[116], 36[123], 37[123], 38[123], 47[141], 48[141], 50[141], 51[141], 53[141], 155[13], 155[14], 156[14], 177[64], 178[64], 187[84], 188[84], 210[87], 279[28], 322[11]
Liu, Y.-M. 154[12], 155[12], 156[12]
Liu, Z. 83[45], 146[49], 146[50], 250[14], 296[64], 297[64], 298[64]
Liu, Z.-M. 301[86]
Lo, H.-K. 202[51]
Löber, O. 91[4], 99[4], 100[4]
Logan, M. E. 362[28], 368[28]
Loh, T.-P. 356[18], 357[18]
Lokteva, E. 355[1]
López, A. M. 210[88]
López, I. 324[13]
López, N. 198[26], 201[47]
Lotz, M. 35[118], 35[119], 36[118], 36[119]
Lou, X.-J. 232[21]
Lough, A. J. 210[86]
Louie, L. 259[59]
Lu, A.-H. 76[28]
Lu, G. 259[56], 259[62], 260[80]
Lu, G. Q. 301[78]
Lu, G.-Z. 259[54]
Lu, H. 279[28]
Lu, S. 155[13], 155[14], 156[14]
Lu, S.-M. 52[146], 53[146], 54[146], 153[4], 153[7], 170[37], 170[38], 171[38], 172[41], 173[41], 174[37]
Lu, T. 232[26]
Lu, T.-J. 325[16]
Lu, W. 15[71], 20[71], 24[71], 358[21], 359[21]
Lu, W.-J. 8[17], 10[17], 11[17], 13[17], 48[144], 49[144], 51[144]
Lu, X. 289[24], 322[11]
Luan, J. 339[16]

Luan, N. 145[42]
Lucas, M. 146[55]
Lucien, E. 156[18], 157[18]
Ludwig, W. 67[7]
Lunin, V. 355[1]
Lunow, T. 143[40], 144[40]
Luo, H. 15[62], 15[63], 20[62], 20[63], 21[62], 21[63], 22[62], 22[63], 23[62], 23[63], 24[62], 24[63], 25[63], 27[63], 28[63], 29[63], 30[63], 229[4]
Luo, S.-L. 289[23]
Luque, R. 260[69], 261[69], 263[69]
Luska, K. L. 82[39]
Lutz, M. 212[96]
Lv, H. 15[71], 15[74], 15[79], 20[71], 20[74], 20[79], 24[71], 24[74], 24[79], 25[74], 26[79], 27[98], 28[98], 30[98], 31[98], 32[109], 33[109], 59[158], 59[159], 60[158], 61[159], 62[158], 62[159]

M

Ma, H. 358[21], 359[21]
Ma, J. 289[12], 299[12]
Ma, L. 279[28]
Ma, R. 289[20], 299[20]
Ma, S. 15[51], 20[51], 21[51], 22[51], 23[51]
Ma, W. 177[64], 178[64]
Ma, X. 69[9], 289[7]
Ma, Y. 127[15], 128[15], 316[4], 317[4], 318[4]
McAllister, C. G. 107[58]
MacArthur, A. H. R. 367[40], 368[40]
McClain, J. M. 236[44], 239[44]
McClymont, T. 197[15], 199[15], 221[15], 223[15]
McCormick, R. L. 119[85], 120[85]
McCue, A. J. 198[25], 200[37], 202[37]
McDonald, R. 331[27]
McEwen, J.-S. 229[9]
McFarland, E. W. 302[99], 302[102], 303[99], 303[102]
Machado, G. 83[43], 83[44], 146[47]
Machalaba, N. 147[67], 156[15]
Machara, A. 330[23]
Machrouhi, F. 58[156], 59[156]
McIntyre, S. 8[35], 10[35], 12[35]
Mackenzie, E. B. 27[90], 28[90], 29[90]
McLaughlin, L. 351[68], 352[68]
McMurray, J. S. 355[16], 356[16]
McNamara, N. D. 296[62], 297[62], 298[62]
McTiernan, C. D. 371[43], 372[43]
Madesclaire, M. 339[11]
Maegawa, T. 157[19], 357[19], 358[19], 359[25], 360[25], 361[19], 361[25]
Maeno, Z. 260[77], 261[77], 262[77]
Maestri, G. 212[94], 214[94]
Magistri, L. 302[101], 303[101]
Mahon, J. J. 339[22]
Mahringer, A. 333[33], 334[33]
Maia, A. M. 339[20]
Maier, W. F. 200[31]
Main, A. D. 292[54], 293[54]
Maiti, S. N. 351[60], 351[67]
Maitlis, P. M. 128[23]
Maj, A. M. 170[40]
Malan, C. 27[89], 28[89], 29[89]
Malati, M. A. 285[46], 285[47]
Maldonado, M. F. 59[160], 61[160], 62[160]
Male, J. 251[22]
Maleczka, R. E., Jr. 362[31], 362[32]
Maligres, P. E. 47[142], 48[142]
Manbeck, G. F. 270[2]
Mandal, P. K. 355[16], 356[16]
Mandler, D. 283[39]
Mankad, N. P. 217[112], 218[112]
Mann, R. S. 221[131]
Manna, C. M. 291[37], 295[37], 296[37]
Manning, T. D. 260[71], 261[71]
Manorama, S. V. 76[30]
Manuel, G. 355[5]
Mapelli, S. 117[83], 119[83]
Marbet, R. 220[124], 223[124]
Marchetti, F. 236[39], 289[8]
Marcos, R. 270[6]
Marcos, V. 221[134]
Marcum, C. L. 250[15], 251[15]
Marecot, P. 93[14]
Marin, G. B. 127[9]
Marin, N. 96[28]
Marin, T. W. 285[48]
Marinas, J. M. 203[58]
Markley, K. S. 110[63]
Marks, T. J. 127[3], 256[46], 257[46], 257[47]
Marlor, C. W. 156[18], 157[18]
Marsden, P. 233[35], 234[35], 235[35], 236[35], 239[35]
Marshall, J. 127[18]
Marshall, J. A. 222[137]
Martin, A. 229[7]
Martin, G. A. 146[44]
Martin, N. J. A. 52[147]
Martorell, A. 20[80]
Marwani, H. M. 239[50]
Marx, S. 362[33]
Mashima, K. 174[55], 174[56], 174[57], 174[58], 175[55], 175[56], 175[57], 175[58], 176[59], 177[59], 216[111], 217[111]
Mashkina, A. V. 351[69], 351[70]
Masnadi, M. 76[19]
Mathias, J. P. 156[17], 157[17]
Matsubara, S. 207[75], 211[75]
Matsumoto, N. 295[60], 296[60]
Matsumoto, T. 233[31]
Matsumura, F. 339[9]
Matsushima, K. 259[55]
Matzinger, D. 351[59]
Maurer, F. 52[148], 53[148]
Maximov, A. 93[18]
Mazuela, J. 8[23], 8[24], 8[29], 8[36], 10[23], 10[24], 10[29], 10[36], 11[23], 11[24], 11[29], 12[23], 12[24], 12[29], 12[36], 13[23], 13[24], 13[29], 57[36]
Medevielle, A. 106[57], 107[57], 108[57]
Medlock, J. 195[2], 197[15], 199[2], 199[15], 200[30], 200[41], 202[54], 202[55], 221[15], 223[15]
Medlock, J. A. 200[39]
Meeldijk, J. D. 204[65]
Mehler, G. 15[58], 20[58], 20[81], 20[83], 21[58], 22[58], 27[58], 28[58], 29[58], 30[58]
Meiners, I. 236[40]
Meiswinkel, A. 20[83]
Mejorado, L. 326[18]
Meli, A. 210[91], 210[92], 211[92]
Melián-Cabrera, I. 232[22], 253[33], 253[35], 253[36], 260[33], 261[33]
Méliet, C. 170[40]
Mellone, I. 271[7], 274[7]
Meloche, S. 362[29]
Menapace, L. W. 355[7]
Ménard, A. 27[99], 28[99], 31[99], 32[99]
Meng, J. 15[79], 20[79], 24[79], 26[79]
Meng, J.-j. 27[98], 28[98], 30[98], 31[98]
Meng, J.-q. 127[5], 128[5]
Menges, F. 8[16], 8[22], 8[35], 10[16], 10[22], 10[35], 11[22], 12[35], 13[16], 13[22], 40[129], 41[129]
Menn, J. J. 339[7]
Mercadé, E. 104[52], 105[52], 131[32], 132[32], 133[32]
Mérel, D. S. 272[12]
Meriaudeau, P. 146[44]
Meriwether, H. T. 127[10]
Metiu, H. 302[102], 303[102]
Meulenberg, R. W. 229[6], 230[6]
Meyer, U. 210[89]
Mezzatesta, G. 301[81]
Miao, S. 69[9]
Micetich, R. G. 351[60], 351[67]
Michaelides, I. N. 214[104]
Mičušík, M. 260[68], 261[68], 263[68]
Midgley, P. A. 102[45], 104[45], 105[45], 108[45]
Mikami, Y. 339[43]
Mikkelsen, M. 289[22]
Mikołajczyk, M. 339[13]
Miller, C. 223[141], 223[142]
Miller, J. T. 127[3], 250[15], 251[15], 253[37], 253[40]
Miller, S. J. 160[29], 161[29], 162[29], 164[29]
Milstein, D. 218[113], 271[8], 271[9], 272[9], 275[19], 276[19], 291[50], 295[50], 296[50]
Minato, T. 204[59]
Minenkov, Y. 304[108], 305[108]
Minnaard, A. J. 15[58], 15[65], 20[58], 20[65], 20[82], 21[58], 21[65], 22[58], 22[65], 27[58], 28[58], 29[58], 30[58], 32[106], 33[106], 172[46], 173[46], 185[79], 186[79]
Mischne, M. P. 330[25], 331[25], 331[26]
Mitchard, D. A. 221[132]
Mitchell, J. W. 351[65]
Mitchell, S. 201[47], 201[48]
Mitra, J. 250[17]

Mitsudo, T.-a. 41[130]
Mitsudome, T. 260[77], 261[77], 262[77], 339[43], 341[51], 342[51], 343[51], 345[54], 346[54], 347[54], 350[54]
Mitsui, S. 98[34]
Mitten, J. V. 58[156], 59[156]
Miyai, T. 231[14], 239[14]
Miyamoto, M. 153[8]
Miyazawa, S. 291[41]
Miyazawa, T. 253[27], 254[27], 256[27]
Mizugaki, T. 260[77], 261[77], 262[77], 339[43], 341[51], 342[51], 343[51], 345[54], 346[54], 347[54], 350[54]
Moawad, M. M. 285[45]
Möbus, K. 200[33]
Moens, L. 119[85], 120[85]
Moglie, Y. 135[36]
Mohar, B. 15[50], 20[50], 21[50], 35[50], 36[50], 37[50]
Mohn, W. W. 355[2]
Molinaro, C. 27[99], 28[99], 31[99], 32[99]
Molnár, Á. 92[7], 106[7]
Monguchi, Y. 157[19], 357[19], 358[19], 359[25], 360[25], 361[19], 361[25]
Monsees, A. 32[101], 33[101]
Montilla, F. 236[39]
Moon, S. 232[27]
Moores, A. 76[17], 76[19], 76[20], 76[23], 79[33], 82[33], 82[39], 83[33], 83[40]
Mooter, G. 115[70]
Morais, J. 83[43], 146[47]
Moran, J. 220[118]
Moran, W. J. 57[152]
Mordenti, L. 355[4]
Moreno-Marrodan, C. 103[49], 104[49], 104[52], 105[49], 105[52], 131[32], 132[32], 133[32]
Morère, J. 96[27]
Moret, M.-E. 233[34], 236[34]
Moret, S. 291[35], 293[35]
Mori, K. 253[32], 254[32], 255[32]
Mori, R. 289[24]
Morioka, T. 245[8], 246[8]
Morken, J. P. 57[152]
Morokuma, K. 291[49]
Morris, G. E. 7[2], 8[2]
Morris, R. H. 210[86]
Mortensen, P. M. 239[48]
Motta, A. 127[3]
Motz, G. 251[18]
Moura, F. C. C. 98[36]
Mouratidis, K. 111[66], 112[66]
Mowrey, D. 58[156], 59[156]
Moyano, R. 236[39]
Mršić, N. 172[46], 173[46], 185[79], 186[79]
Mu, X. 253[38]
Mu, X.-d. 83[41], 127[5], 128[5], 148[68]
Muckerman, J. T. 270[2]
Mueller, T. 200[39]
Muetterties, E. L. 128[22]
Mülhaupt, R. 81[38]
Mullens, P. 157[23], 158[23]
Müller, C. 204[67]
Müller, U. 285[49]
Müller-Plathe, F. 107[59]
Munshi, P. 292[54], 293[54]
Murata, K. 301[77]
Murphy, V. J. 232[28], 233[28]
Musker, W. K. 339[21]
Muzzio, D. 157[23], 158[23]

N

Naccache, C. 146[44]
Nádasdi, L. 291[34], 292[34], 293[34], 294[56]
Nadeem, H. 289[15]
Nador, F. 135[36]
Nagahara, H. 146[66]
Nagano, T. 174[56], 174[57], 175[56], 175[57], 176[59], 177[59]
Nagao, K. 208[81], 209[81]
Nagaraja, C. M. 355[15]
Nagashima, H. 128[30], 130[30]
Nagata, M. 174[55], 175[55]
Nagata, S. 359[25], 360[25], 361[25]
Nagatsu, Y. 260[77], 261[77], 262[77]
Naglia, M. 206[73]
Nahra, F. 304[107], 305[107]
Nakagawa, Y. 252[24], 252[25], 253[28], 253[29], 253[31], 253[32], 254[28], 254[29], 254[31], 254[32], 254[44], 254[45], 255[29], 255[31], 255[32], 256[28], 256[29], 257[50], 258[51], 258[52], 259[52], 260[64], 281[33]
Nakao, R. 72[13], 73[13], 73[14], 360[26], 361[26]
Nakao, Y. 208[80], 209[80], 211[80]
Nan, F.-J. 358[21], 359[21]
Narayanam, J. M. R. 369[41], 370[41], 372[41]
Narayanan, B. 15[51], 20[51], 21[51], 22[51], 23[51]
Narender Reddy, A. V. 351[60], 351[67]
Narine, S. 335[34], 336[34]
Narula, A. P. S. 200[34]
Nasir Baig, R. 78[32]
Nassim, B. 220[120]
Nativi, C. 220[122]
Nazer, B. 351[58]
Negoro, N. 185[80], 186[80]
Nelson, D. M. 156[18], 157[18]
Nelson, R. C. 229[10], 230[10]
Nelson, T. D. 58[156], 59[156]
Neri, G. 146[46]
Nerlov, J. 282[35]
Netherton, M. R. 220[119]
Netscher, T. 195[2], 197[15], 199[2], 199[15], 221[15], 223[15]
Neumann, E. 8[37], 10[37], 13[37]
Neumann, G. T. 296[62], 297[62], 298[62]
Neumann, K. T. 216[110]
Neumann, R. 134[34], 135[34]
Neurock, M. 127[9], 232[19], 253[34], 253[36], 254[34]
Neuschütz, K. 220[119]
Newman, C. 229[6], 230[6]
Ng, F. T. T. 344[52]
Nguyen, D. A. 229[7]
Nicholas, C. P. 127[3]
Nicholas, K. M. 236[37], 236[38], 236[43], 236[44], 239[43], 239[44]
Nichols, C. L. 156[17], 157[17]
Nichols, J. M. 250[16]
Nie, H. 99[37], 101[37]
Niedzwiedzki, D. M. 221[125], 221[127], 221[128]
Nielsen, D. R. 315[1]
Niembro, S. 131[31], 132[31], 133[31]
Niessen, H. G. 214[105]
Nijhuis, T. A. 204[67]
Nimmanwudipong, T. 251[23]
Nishibayashi, R. 207[75], 211[75]
Nishide, K. 347[55]
Nishimura, S. 127[1], 195[1], 196[1], 197[1]
Nishitani, M. 101[42]
Niu, L. 260[66], 261[66]
Nkosi, B. S. 210[85]
Noack, K. 200[32]
Node, M. 347[55]
Noel, S. 232[22], 253[33], 253[35], 260[33], 261[33]
Nogueira, I. D. 98[33]
Nokami, J. 339[1]
Nolan, S. P. 103[46], 304[107], 305[107]
Nolting, A. 47[142], 48[142]
Noronha, F. B. 229[3]
Norrby, P.-O. 8[36], 10[36], 12[36], 57[36]
Noujima, A. 339[43]
Nova, A. 160[29], 161[29], 162[29], 164[29]
Novakova, E. K. 351[68], 352[68]
Noyori, R. 100[38]
Nozaki, K. 252[26], 291[48], 291[49], 294[48], 295[48]
Nunes da Ponte, M. 96[26], 98[33]
Núñez-Rico, J. L. 35[114], 36[114], 170[35], 172[52], 173[52], 174[52]

O

Oae, S. 339[19]
Obata, K. 347[55]
Oberhauser, W. 207[76]
O'Connor, E. M. 351[59]
Ogata, K. 55[150], 56[150]
Ogden, G. 67[1], 68[1]
Oger, C. 199[29]
Ogo, S. 291[38], 291[39]
Ohira, Y. 260[72], 261[72], 263[72]
Ohishi, M. 233[31]
Ohloff, G. 221[135], 222[135]
Ohmiya, H. 208[81], 209[81]
Ohno, M. 153[8]
Ohshima, T. 174[56], 174[57], 175[56], 175[57]
Ohsumi, M. 180[68], 181[68], 184[68]
Ohta, H. 88[52]

Ohta, T. 100[38]
Ohtake, A. 153[8]
Ohtsuki, A. 245[8], 246[8]
Ohyama, J. 260[70], 260[72], 261[70], 261[72], 263[72]
Oinen, M. E. 362[28], 368[28]
Ojeda, M. 301[84]
Ojima, I. 7[8]
Oki, H. 27[86], 28[86], 29[86]
Okuda, J. 236[40]
Okumura, K. 253[30], 253[31], 254[30], 254[31], 255[31]
Olah, G. A. 289[16]
Olcay, H. 257[48]
Oliván, M. 131[31], 132[31], 133[31]
Olpp, T. 221[126]
Olszewski, P. 301[80]
Omae, I. 289[2]
Omonov, T. S. 335[35], 336[35]
Ondruschka, B. 196[9]
O'Neill, B. J. 253[37], 253[40]
Onishi, Y. 231[14], 239[14]
Ono, M. 146[66]
Onodera, W. 307[116], 309[116], 348[56], 349[56], 350[56], 351[56]
Onozawa-Komatsuzaki, N. 291[42], 291[43], 291[44], 291[45], 292[45], 293[45], 294[43], 294[44], 294[45], 295[45]
Onsan, Z. 302[96], 303[96]
Oohara, N. 15[43], 20[43], 21[43], 22[43]
Ooi, S. 69[10], 70[10]
Ookawa, A. 351[64]
Oro, L. A. 204[64], 210[88], 210[89], 210[90], 210[92], 211[92]
Orpen, A. G. 20[80]
Orr, R. K. 47[142], 48[142]
Ortega, N. 182[70], 182[71], 182[72], 182[73], 182[74], 183[70], 183[71], 183[72], 183[73], 183[74], 184[70]
Orzechowski, L. 131[31], 132[31], 133[31]
Osako, T. 79[33], 82[33], 83[33]
Osantey, I. 104[53], 105[53]
Osborn, J. A. 128[20], 128[21], 204[62], 204[63]
Ou, Y. 156[18], 157[18]
Overman, J. D. 351[59]
Overman, L. E. 351[59], 351[66]
Owaki, M. 367[39]
Owen, B. C. 250[15], 251[15]
Ozerov, O. V. 355[15]
Özkar, S. 127[16], 128[16]

P

Paciello, R. 91[5], 100[40]
Pagán-Torres, Y. J. 232[19], 253[34], 254[34]
Pai, C.-C. 32[104], 33[104]
Pais, G. C. G. 156[18], 157[18]
Paiva, A. 96[26]
Pal, A. 233[30]
Palkovits, R. 231[12], 236[40], 243[4]
Pàmies, O. 7[3], 8[3], 8[23], 8[24], 8[29], 8[30], 8[33], 8[36], 10[23], 10[24], 10[29], 10[30], 10[33], 10[36], 11[23], 11[24], 11[29], 11[30], 11[33], 12[23], 12[24], 12[29], 12[36], 13[23], 13[24], 13[29], 13[30], 13[33], 57[36]
Pamingle, H. 221[135], 222[135]
Pampaloni, G. 289[8]
Pan, H. 283[40]
Pan, H.-B. 139[38], 140[38], 141[38]
Pan, J. 177[61]
Pan, Y. 319[8], 320[8]
Pandey, L. K. 339[30]
Pando, C. 96[27]
Pang, J. 235[36], 239[36], 253[43]
Pang, X. 253[43]
Panten, J. 97[30]
Papadogianakis, G. 111[65], 111[66], 111[67], 112[65], 112[66], 112[67], 112[68], 113[65]
Pape, F. 208[79]
Papp, G. 291[34], 292[34], 293[34]
Pappas, J. J. 333[31], 333[32]
Paptchikhine, A. 8[29], 8[32], 10[29], 10[32], 11[29], 11[32], 12[29], 12[32], 13[29], 13[32], 55[151], 56[151], 57[153], 58[155], 59[155]
Pardatscher, L. 273[13]
Park, C. M. 68[8], 71[12], 72[12]
Park, J. 68[8], 71[12], 72[12]
Park, J. N. 302[99], 303[99]
Park, K. 289[19], 296[19], 296[63], 297[63], 298[63]
Parker, D. G. 128[26]
Parker, S. F. 200[33]
Parsell, T. H. 250[15], 251[15]
Parthasarathi, R. 251[19]
Passaglia, E. 207[76]
Pasto, D. J. 214[100]
Pastor, A. 236[39]
Patel, B. A. 203[57], 204[57]
Pathak, U. 339[30]
Patricia, J. J. 355[9]
Paul, D. 182[73], 183[73]
Pecchia, P. 117[83], 119[83]
Pedersen, D. S. 327[21], 328[21]
Pedrosa, M. R. 339[34], 339[49], 340[49], 341[49], 341[50], 343[49]
Pei, Y. 127[14], 128[14], 145[42], 146[53], 146[54], 146[56], 146[59]
Pelagatti, P. 206[73]
Pelayo, C. 131[31], 132[31], 133[31]
Pelizzi, C. 206[73]
Pellegatta, J.-L. 127[17], 128[17]
Peña, D. 32[106], 33[106]
Pendse, H. 257[48]
Peng, G. 253[38]
Peng, Y.-K. 259[63]
Perathoner, S. 289[10], 289[18]
Perdriau, S. 113[69], 114[69]
Pereira Gonçalves, A. 93[15]
Pérez, A. 210[88]
Pérez-Alonso, F. J. 301[84]
Pérez-Ramírez, J. 77[31], 198[26], 201[47], 201[48], 201[49], 201[50]
Pericàs, M. A. 92[10]
Peris, E. 291[51], 291[52], 292[51], 293[51], 294[52], 295[52]
Perret, N. 260[71], 261[71]
Perry, M. C. 8[25], 10[25], 11[25], 13[25], 14[25], 339[33]
Peruncheralathan, S. 364[35], 365[35], 369[35]
Peruzzini, M. 210[91], 210[92], 211[92], 271[7], 274[7], 291[47], 294[57], 295[47], 296[47]
Peters, B. 59[160], 61[160], 62[160]
Peters, B. K. 59[161]
Pettinari, C. 236[39]
Pettman, A. 163[31], 164[31], 165[31], 167[31], 184[76], 185[76], 186[76]
Pettus, T. R. R. 326[17], 326[18]
Petty, S. T. 351[66]
Pfaltz, A. 8[12], 8[13], 8[14], 8[18], 8[20], 8[21], 8[22], 8[26], 8[31], 8[35], 8[37], 10[14], 10[18], 10[20], 10[21], 10[22], 10[26], 10[31], 10[35], 10[37], 11[14], 11[18], 11[21], 11[22], 11[26], 11[31], 12[18], 12[20], 12[35], 13[14], 13[18], 13[20], 13[21], 13[22], 13[26], 13[31], 13[37], 13[38], 14[31], 40[129], 41[129], 57[154], 58[154], 176[60]
Pflum, D. A. 27[85], 28[85], 29[85], 31[85]
Pham, H. N. 232[19], 253[34], 254[34]
Pham Minh, D. 93[19]
Phan, B. M. Q. 229[7]
Phansavath, P. 174[57], 175[57]
Philippaerts, A. 115[70], 116[75]
Philippot, K. 98[32], 127[17], 128[17]
Philipse, A. P. 204[65]
Phillips, A. D. 294[57]
Phillips, D. L. 355[14]
Phillips, J. G. 157[21], 158[21]
Phillipson, J. J. 92[9]
Pholjaroen, B. 253[39], 253[41], 254[41]
Phua, P. H. 232[22], 253[33], 253[35], 260[33], 261[33]
Phua, P.-H. 79[35], 79[36], 80[35], 80[36]
Piarulli, U. 15[64], 20[64], 21[64], 24[64], 25[64], 27[64], 28[64], 29[64]
Pidko, E. A. 298[67]
Pietropaolo, D. 146[46]
Pignataro, L. 15[64], 20[64], 21[64], 24[64], 25[64], 27[64], 28[64], 29[64]
Pilkington, C. J. 27[92], 28[92], 29[92]
Pillai, U. R. 70[11], 71[11]
Pinard, L. 301[87]
Pires de Mato, A. 93[15]
Pironti, V. 95[25]
Pitchumani, K. 85[49], 86[49]
Pitre, S. P. 371[43], 372[43]
Pizzano, A. 35[117], 35[121], 36[117], 36[121], 36[122], 37[122]
Plasseraud, L. 128[29], 129[29], 130[29]
Plée, D. 106[57], 107[57], 108[57]
Plummer, C. W. 47[142], 48[142]
Poater, A. 103[47], 272[12]
Po-Ba, Y. M. 156[17], 157[17]
Podos, S. D. 156[18], 157[18]

Pohl, M.-M. 76[18], 229[7]
Polanyi, M. 67[1], 68[1], 98[35]
Polborn, K. 35[119], 36[119]
Poli, N. 117[80], 120[80], 121[80]
Pollari, I. 290[27]
Pollhammer, S. 333[33], 334[33]
Pollock, R. A. 229[6], 230[6]
Polshettiwar, V. 76[22], 76[24]
Poole, D. L. 304[103]
Poole, T. H. 96[29], 97[29]
Porosoff, M. D. 289[17]
Porras, A. 203[58]
Post, J. A. 201[46]
Post, M. L. 76[21]
Pouilloux, Y. 301[87]
Pover, K. A. 224[146]
Powell, M. T. 8[25], 10[25], 11[25], 13[25], 14[25]
Prakash, N. 296[63], 297[63], 298[63]
Praliaud, H. 146[44], 351[57]
Prechtl, M. H. G. 83[40]
Preindl, J. 214[102]
Preti, D. 298[65], 298[66]
Prétôt, R. 8[12]
Pribbenow, C. 205[68]
Price, J. D. 222[136]
Price, R. 243[2], 258[2]
Primet, M. 351[57]
Pringle, P. G. 20[80]
Provot, O. 365[37]
Prunier, M. L. 118[84]
Pruski, M. 127[3]
Pryde, E. H. 335[36], 336[36]
Psaro, R. 101[41], 102[41], 117[79], 117[80], 117[81], 117[82], 118[82], 119[81], 120[80], 121[80], 121[81], 121[82]
Psaroudakis, N. 111[66], 112[66]
Pu, L.-Y. 43[135], 45[135], 46[135]
Pucci, M. J. 156[18], 157[18]
Pugin, B. 8[31], 10[31], 11[31], 13[31], 14[31]
Puskás, R. 75[16]

Q

Qi, W. 257[48]
Qiao, M. 127[14], 128[14], 145[42], 146[48], 146[51], 146[53], 146[54], 146[56], 146[64]
Qiao, M.-H. 127[4], 146[59]
Qin, H.-L. 239[50]
Qin, J. 177[62], 177[65], 178[65]
Qin, L.-Z. 231[17], 233[17]
Qiu, L. 32[104], 33[104]
Qiu, M. 15[47], 20[47], 21[47], 22[47], 24[47], 25[47]
Qiu, R. 213[98], 220[98]
Qu, B. 15[51], 20[51], 21[51], 22[51], 23[51]
Quan, X. 47[139], 48[139]

R

Rabelo-Neto, R. C. 229[3]
Rabten, W. 59[161]
Radivoy, G. 135[36]
Radkowski, K. 214[107], 215[107], 216[107], 219[107]
Radnik, J. 229[7]
Rageot, D. 8[31], 10[31], 11[31], 13[31], 14[31]
Rahaim, R. J. 362[31], 362[32]
Rahi, R. 153[9]
Rahimi, K. 236[40]
Rahimpour, M. R. 251[23]
Rai, R. S. 200[31]
Raichurkar, A. V. 358[20]
Raja, R. 102[45], 104[45], 104[55], 105[45], 105[55], 108[45], 108[55]
Rajaram, J. 200[34]
Raju, S. 233[34], 236[34]
Ramanathan, A. 358[23]
Ramp, F. L. 67[2]
Raney, M. 128[19]
Rangheard, C. 79[36], 80[36]
Rank, J. S. 197[18]
Rao, A. N. 339[30]
Raphael, M. W. 285[47]
Raphael, R. A. 197[17]
Rasolofonjatovo, E. 365[37]
Rasrendra, C. B. 243[3], 258[3]
Ratcliff, M. 119[85], 120[85]
Ratovelomanana-Vidal, V. 15[77], 20[77], 24[77], 26[77], 174[55], 174[56], 174[57], 175[55], 175[56], 175[57]
Rauchfuss, T. B. 250[17]
Ravasio, N. 101[41], 102[41], 104[54], 117[54], 117[79], 117[80], 117[81], 117[82], 117[83], 118[82], 118[84], 119[81], 119[83], 120[80], 120[86], 120[87], 121[80], 121[81], 121[82]
Ray, N. A. 256[46], 257[46]
Reader, M. 157[20], 158[20]
Rebrov, E. V. 200[36], 202[36]
Recchia, S. 117[80], 120[80], 121[80]
Redwan, I. N. 362[29]
Reese, C. N. 96[29], 97[29]
Reetz, M. T. 15[58], 20[58], 20[81], 20[83], 21[58], 22[58], 27[58], 28[58], 29[58], 30[58]
Rehren, C. 223[143], 224[143]
Reibenspies, J. H. 8[25], 10[25], 11[25], 13[25], 14[25]
Reis, P. M. 344[53], 350[53]
Reiser, O. 77[31]
Reller, A. 200[32]
Rempel, G. L. 344[52]
Ren, D. 154[12], 155[12], 156[12]
Ren, F. 279[28]
Ren, J. 260[80]
Ren, Y. 247[11], 247[12], 248[11], 248[12], 249[11], 250[11], 251[11]
Renaud, J. L. 15[76], 20[76], 24[76], 25[76]
Renaud, J.-L. 32[100], 33[100], 272[12]
Renken, A. 196[10], 200[10], 202[52], 202[53]
Rensel, D. J. 251[21]
Renuncio, J. A. R. 96[27]
Resasco, D. E. 229[3]
Resta, C. 298[65]
Reutemann, W. 290[27]
Reyniers, M.-F. 127[9]
Reynolds, M. 15[78], 20[78], 24[78], 26[78]
Rezaei, P. S. 229[8]
Rezayee, N. M. 300[74]
Rhee, H. 72[13], 73[13], 360[26], 361[26]
Riani, P. 302[101], 303[101]
Ribeiro, F. 250[15], 251[15]
Ribeiro, F. H. 253[37], 253[40]
Ricci, A. 220[122]
Ricci, M. 289[11]
Ricciuto, L. 210[86]
Richmond, E. 220[118]
Rideout, W. H. 315[1]
Riduan, S. N. 289[9]
Riente, P. 86[50], 86[51], 87[51]
Rinaldi, R. 116[75]
Rivada-Wheelaghan, O. 271[9], 272[9]
Riveira, M. J. 330[25], 331[25], 331[26]
Riviere, A. 82[39]
Robert-Lopes, M.-T. 220[120]
Robinson, T. V. 327[21], 328[21]
Robles-Dutenhefner, P. A. 97[31], 98[31], 98[36]
Roblin, J. 222[139]
Rockstroh, N. 273[14], 274[14]
Rodondi, M. 101[41], 102[41]
Rodrigo, J. 365[37]
Rodriguez, S. 15[51], 20[51], 21[51], 22[51], 23[51]
Rodríguez, S. 324[13]
Roeser, J. 339[44]
Roessler, F. 196[10], 200[10]
Rogers, D. W. 196[11], 196[12], 197[11], 197[12]
Rogolino, D. 206[73]
Rohmann, K. 291[46], 293[46], 294[46]
Roithner, E. 333[33], 334[33]
Rojas, S. 301[84]
Rolla, F. 339[20]
Roller, B. 295[59], 296[59]
Romanenko, A. V. 115[71]
Romão, C. C. 339[36], 339[45], 344[53], 350[53]
Ronco, A. 197[23]
Rong, H. 279[28]
Rong, L. 277[23], 279[23]
Rooney, D. W. 351[68], 352[68]
Rophael, M. W. 285[45], 285[46]
Rosenberg, E. 93[18]
Rosi, L. 207[76]
Rosner, T. 8[13]
Ross, Z. 229[6], 230[6]
Rosseinsky, M. J. 260[71], 261[71]
Rossi, L. M. 98[32], 146[47]
Rostami, A. 331[27]
Rothenberger, S. D. 222[137]
Rouvimov, S. 251[21]
Roy, R. 76[20]
Roy, S. 107[59]
Royo, B. 344[53], 350[53]
Rubio, R. 339[49], 340[49], 341[49], 343[49]

Ruck, R. T. 47[142], 48[142]
Rudolf von Rohr, P. 200[30], 202[54], 202[55], 301[83]
Ruhoff, J. R. 221[129]
Ruiz, P. 229[10], 230[10]
Rujirawanich, J. 59[161]
Rummelt, S. M. 214[102]
Ruppert, A. M. 231[12], 243[4]
Rupprechter, G. 94[20], 94[21]
Rusching, U. 285[49]
Rush, R. V. 333[32]
Russell, M. J. 128[23]
Russo, O. 157[22], 158[22]
Ryuhei, I. 181[69], 182[69]

S

Saaler, A. 196[10], 200[10]
Sabacky, M. J. 7[5], 196[8]
Sabatier, P. 127[11]
Sachse, A. 103[50], 104[50]
Sachtler, W. M. H. 197[18]
Saeys, M. 127[9]
Sahle-Demessie, E. 70[11], 71[11]
Saidi, M. 251[23]
Saikia, L. 133[33], 134[33]
Saito, M. 300[73], 301[77]
Saito, N. 282[36]
Saito, T. 27[95], 28[95], 29[95], 238[47]
Sajiki, H. 157[19], 357[19], 358[19], 359[25], 360[25], 361[19], 361[25]
Sajonz, P. 157[23], 158[23]
Sajtos, A. 332[28], 333[33], 334[33], 336[28]
Sakakura, T. 289[3], 289[5]
Sakamaki, Y. 36[124], 37[124]
Sakata, T. 282[36]
Salabas, E. L. 76[28]
Salden, A. 223[141]
Sale, K. L. 251[19]
Samimi, F. 251[23]
Sanchayan, R. 107[59], 108[60]
Sanchez, C. 93[19]
Sánchez-de-Armas, R. 270[6]
Sánchez-Delgado, R. A. 147[67], 153[9], 156[15], 210[89]
Sandbrink, L. 236[40]
Sandoval, C. A. 36[123], 37[123], 38[123]
Sanford, M. S. 299[70], 300[70], 300[74]
Sankar, G. 102[45], 104[45], 104[55], 105[45], 105[55], 108[45], 108[55]
Santoro, F. 120[87]
Santoro, O. 304[107], 304[108], 305[107], 305[108]
Sanz, R. 339[34], 339[49], 340[49], 341[49], 341[50], 343[49]
Sanz, S. 291[51], 291[52], 292[51], 293[51], 294[52], 295[52]
Sapountzis, I. 355[8]
Sárkány, A. 92[7], 106[7]
Sarmah, P. P. 133[33], 134[33]
Sasaki, Y. 290[32], 300[73]
Sasson, Y. 269[1], 275[18]
Sato, K. 166[33], 167[33], 185[77], 185[78], 186[77], 186[78]
Sato, M. 259[55]
Satsuma, A. 260[70], 260[72], 261[70], 261[72], 263[72]
Saucy, G. 220[124], 223[124]
Savage, P. E. 229[2]
Savoia, D. 85[48], 153[5]
Savourey, S. 304[105]
Sawama, Y. 359[25], 360[25], 361[25]
Sawamura, M. 208[81], 209[81]
Sayo, N. 100[38]
Scaiano, J. C. 371[43], 372[43]
Schaffner, S. 8[20], 10[20], 12[20], 13[20]
Schauermann, S. 67[7]
Scheeren, C. W. 83[44]
Scheinmann, F. 224[146]
Schlaf, M. 231[18], 232[18], 233[18], 260[76], 261[76], 262[76]
Schlägl, R. 200[32]
Schleis, T. 210[86]
Schlepphorst, C. 59[157], 60[157], 62[157]
Schley, N. D. 160[29], 161[29], 162[29], 164[29]
Schleyer, D. 214[105]
Schlögl, R. 200[35]
Schmeier, T. J. 294[58], 295[58]
Schmidt, A. 103[48], 104[48]
Schmitz, P. 35[115], 36[115], 38[115]
Schmoeger, C. 196[9]
Schneider, H. J. 196[6]
Schneider, M. 229[7]
Schneider, W. F. 296[62], 297[62], 298[62]
Schnider, P. 8[12], 8[13], 8[14], 10[14], 11[14], 13[14]
Scholten, J. J. F. 127[8]
Schomäcker, R. 103[48], 104[48]
Schouten, J. C. 204[67]
Schrems, M. G. 8[18], 8[37], 10[18], 10[37], 11[18], 12[18], 13[18], 13[37]
Schrock, R. R. 204[62], 204[63]
Schudde, E. P. 20[82]
Schuit, G. C. A. 197[18]
Schulte-Elte, K. H. 221[135], 222[135]
Schulz, M. 210[88]
Schumacher, N. 282[35]
Schünemann, V. 76[29]
Schuster, M. E. 201[50]
Schüth, F. 76[28]
Schütz, J. 195[2], 199[2]
Schwab, W. 91[3]
Schweitzer-Chaput, B. 323[12]
Scipio, D. 289[14]
Scopelliti, R. 270[3]
Scott, A. I. 325[15], 325[16]
Scott, J. P. 27[99], 28[99], 31[99], 32[99]
Scown, C. D. 259[59]
Seayad, A. M. 69[10], 70[10]
Segawa, Y. 145[41]
Segobia, D. J. 120[87]
Sell, T. 20[83]
Sels, B. 115[70], 116[75]
Semagina, N. 196[10], 200[10], 202[52], 202[53]
Semba, K. 208[80], 209[80], 211[80]
Semernina, V. 93[18]
Semikolenov, V. A. 316[2], 316[3], 317[2]
Semmeril, D. 204[66]
Senanayake, C. H. 15[51], 20[51], 21[51], 22[51], 23[51]
Senaweera, S. M. 370[42], 371[42]
Senda, Y. 98[34]
Sener, C. 253[40], 253[42]
Seo, J. G. 302[95], 303[95]
Sergeev, A. G. 244[6], 245[6], 246[6], 247[6], 247[9]
Sethupathi, S. 289[15]
Sevonkaev, I. V. 233[30]
Shafaghat, H. 229[8]
Shaikhutdinov, S. K. 67[3], 67[4], 67[5]
Shang, G. 7[8]
Shang, H. 247[12], 248[12]
Shang, J. 48[143], 49[143], 51[143]
Shao, Y. 259[56], 259[62]
Sharlow, E. R. 358[22]
Sharma, A. 284[41]
Sharma, S. 302[102], 303[102]
Shaterian, H. R. 339[17]
Shaw, D. 157[20], 158[20]
Shehzad, A. 289[15]
Shen, B.-x. 318[5]
Shen, D. 304[103]
Shen, R. 213[98], 220[98]
Sheridan, J. 221[133]
Shevlin, M. 27[99], 28[99], 31[99], 32[99], 47[142], 48[142]
Shi, F. 76[18], 307[115], 308[118], 309[115], 309[118], 310[118]
Shi, L. 59[159], 61[159], 62[159], 170[39], 172[42], 172[43], 172[44], 172[48], 172[54], 173[42], 173[43], 173[44], 173[48], 173[54], 174[54], 187[87], 188[87]
Shi, Q. 15[57], 20[57], 21[57], 22[57], 27[57], 28[57], 29[57], 30[57]
Shi, R. 95[22]
Shibasaki, M. 207[77], 207[78], 208[77], 208[78]
Shigeta, Y. 347[55]
Shigetsura, M. 157[19], 359[25], 360[25], 361[25]
Shimazu, S. 102[43]
Shimizu, H. 27[95], 28[95], 29[95]
Shimizu, K. 55[150], 56[150], 307[116], 309[116], 348[56], 349[56], 350[56], 351[56]
Shimon, L. J. W. 271[8], 275[19], 276[19], 291[50], 295[50], 296[50]
Shin, N. 232[27]
Shinmi, Y. 253[30], 254[30]
Shiraki, H. 347[55]
Shiramizu, M. 232[23], 236[23], 237[23], 237[46], 238[23], 238[46], 239[23], 239[46]
Shkrob, I. A. 285[48]
Shoemaker, J. 232[28], 233[28]
Shoinoya, M. 98[34]
Shu, Q. 127[12], 128[12]

Shylesh, S. 76[29]
Sicsic, S. 157[22], 158[22]
Siddiki, S. M. A. H. 307[116], 309[116], 348[56], 349[56], 350[56], 351[56]
Sidorov, A. I. 224[144]
Siegel, S. 197[20]
Sierra, I. 115[72]
Silks, L. A. P., III 260[76], 261[76], 262[76]
Silveira, E. T. 146[47]
Silvestre-Albero, J. 94[20], 94[21]
Simakova, I. L. 316[2], 316[3], 317[2]
Simmons, B. A. 251[19]
Simon, O. B. 355[11]
Singh, A. 370[42], 371[42]
Singh, M. P. 351[60], 351[67]
Singh, S. 251[19]
Singh, S. K. 290[29]
Singh, T. 59[161]
Singleton, E. 209[84], 210[84], 210[85], 211[84], 211[93]
Sisak, A. 355[11]
Skattebøl, L. 220[121]
Skrydstrup, T. 216[110]
Skrzypek, J. 301[80]
Slawin, A. M. Z. 103[47], 304[107], 305[107]
Słoczyński, J. 301[80]
Smeets, W. J. J. 205[70], 205[71]
Smidt, S. P. 8[26], 8[35], 10[26], 10[35], 11[26], 12[35], 13[26]
Smiles, S. 339[18]
Smith, H. A. 127[10], 221[129]
Snow, H. 156[17], 157[17]
Soai, K. 351[64]
Soares, G. V. 146[47]
Sodano, G. 95[25]
Sodeoka, M. 207[77], 207[78], 208[77], 208[78]
Sola, E. 210[89]
Solà, L. 92[10]
Solano, C. 362[29]
Solladié, G. 339[2], 339[4]
Sommer, H. 214[102]
Somorjai, G. A. 92[8]
Song, C. E. 15[41], 20[41], 21[41], 22[41], 23[41], 24[41], 25[41]
Song, H. 233[32], 302[97], 303[97]
Song, H.-J. 259[53]
Song, I. K. 251[20], 302[95], 302[98], 303[95], 303[98]
Song, J. 327[20]
Song, M.-J. 231[17], 233[17]
Song, S. 43[133], 43[134], 43[135], 44[133], 44[134], 45[133], 45[134], 45[135], 46[133], 46[135]
Song, Y. 156[18], 157[18]
Songis, O. 103[47]
Soo, Y.-L. 259[63]
Sooknoi, T. 229[3]
Sorribes, I. 304[104], 305[114], 306[114], 307[114]
Soták, T. 260[68], 261[68], 263[68]
Sotiriou, S. 111[67], 112[67]
Sousa, E. M. B. 97[31], 98[31]
Sousa, S. C. A. 233[33], 339[37], 339[38], 339[48]
Sowa, J. R., Jr. 100[39], 101[39]
Spadaro, L. 301[76], 301[81]
Spannenberg, A. 270[4], 291[36], 295[36], 296[36]
Spek, A. L. 204[65], 205[70], 205[71]
Spevak, P. 351[60], 351[67]
Speziali, M. G. 97[31], 98[31], 98[36]
Spichiger-Ulmann, M. 284[43], 284[44]
Spielkamp, N. 159[26], 160[26]
Spinacé, E. V. 146[60]
Spindler, F. 15[78], 20[78], 24[78], 26[78], 184[75]
Spod, H. 146[55]
Sprengers, J. W. 206[74], 211[74], 212[94], 214[94]
Squarcialupi, S. 298[65], 298[66]
Sreekumar, S. 259[59]
Srimani, D. 218[113]
Stair, P. C. 256[46], 257[46]
Staiti, P. 146[46]
Stalder, C. J. 275[16], 275[17], 284[17], 284[42]
Stalzer, M. M. 127[3]
Stanislaus, A. 135[35]
Stark, W. J. 77[31]
Stathis, P. 111[67], 112[67]
Stavila, V. 251[19]
Štávová, G. 229[5]
Stefaniak, M. H. 156[17], 157[17]
Stein, M. 81[38]
Steinfeld, A. 289[14]
Stephan, D. W. 145[41]
Stephan, M. 15[50], 20[50], 21[50], 35[50], 36[50], 37[50]
Stephenson, C. R. J. 369[41], 370[41], 372[41]
Steurer, P. 81[38]
Stevenson, E. D. 127[2]
Still, I. W. J. 339[25]
Stivala, C. E. 238[47]
Stoch, J. 301[80]
Stöger, B. 291[47], 295[47], 296[47]
Stolle, A. 196[9]
Strauch, P. 339[44]
Struijk, J. 127[8]
Stüber, F. 108[61]
Studer, A. 319[9], 320[9], 321[9]
Studer, M. 184[75]
Stumpf, A. 15[78], 20[78], 24[78], 26[78]
Su, D. S. 308[117], 309[117]
Su, H. 232[26]
Su, W.-B. 93[16]
Su, X. 232[24], 236[24], 237[24], 238[24], 239[24]
Subrahmanyam, A. V. 257[48]
Sugasawa, J. 27[86], 28[86], 29[86]
Sugata, C. 351[71]
Sugihara, H. 291[42], 291[43], 291[44], 291[45], 292[45], 293[45], 294[43], 294[44], 294[45], 295[45]
Sugino, E. 339[14]
Sugiya, M. 15[52], 20[52], 21[52], 23[52]
Suisse, I. 170[40]
Sul'mann, E. M. 224[144]
Summers, D. P. 275[16], 275[17], 284[17]
Sun, B. 127[14], 128[14], 146[54], 146[56], 322[11]
Sun, H. 319[8], 320[8], 363[34], 364[34], 368[34]
Sun, J. 229[9]
Sun, J. M. 289[25]
Sun, N. 289[12], 299[12]
Sun, R. 235[36], 239[36], 253[43]
Sun, W. 27[94], 28[94], 29[94], 30[94]
Sun, X. 127[13], 128[13]
Sun, Y. 58[156], 59[156], 166[32], 167[32], 289[12], 299[12]
Sundararaju, B. 214[107], 215[107], 216[107], 219[107]
Surburg, H. 97[30]
Surya Prakash, G. K. 289[16]
Süss-Fink, G. 128[28], 128[29], 129[29], 130[29]
Sutherlin, D. 15[78], 20[78], 24[78], 26[78]
Sutton, A. D. 260[76], 261[76], 262[76]
Suzuki, T. 259[55]
Szépvölgyi, J. 75[16]
Szmant, H. H. 339[24]
Szőllősi, G. 75[16]
Szőri, K. 75[16]

T

Tadaoka, H. 174[55], 174[56], 175[55], 175[56]
Taddei, M. 220[122]
Tagliavini, E. 85[48]
Taher, D. 231[18], 232[18], 233[18]
Tai, C.-C. 289[1], 292[54], 293[54], 295[59], 296[59]
Tait, B. D. 196[7]
Takahashi, H. 15[43], 15[55], 20[43], 20[55], 21[43], 21[55], 22[43], 23[55]
Takahashi, Y. 341[51], 342[51], 343[51], 345[54], 346[54], 347[54], 350[54]
Takaya, H. 100[38]
Takeda, T. 153[8]
Takegami, Y. 41[130]
Takeguchi, T. 299[69]
Tam, N. C. 339[15]
Tamura, K. 15[52], 20[52], 21[52], 23[52]
Tamura, M. 252[24], 252[25], 254[44], 254[45], 257[50], 258[51], 258[52], 259[52], 260[64], 281[33]
Tan, J. 166[32], 167[32]
Tan, L. 157[23], 158[23]
Tan, Q. 232[19], 253[34], 253[36], 254[34]
Tan, X. 127[14], 128[14], 146[53], 146[56]
Tanaka, M. 41[130]
Tanaka, R. 291[48], 291[49], 294[48], 295[48]
Tanaka, Y. 153[8], 339[14]
Tandon, M. 358[22]

Tang, D.-T. D. 182[74], 183[74]
Tang, G. 308[117], 309[117]
Tang, J. 187[82], 188[82], 358[21], 359[21]
Tang, M.-T. 93[16]
Tang, W. 7[7], 8[15], 10[15], 11[15], 13[15], 15[7], 15[42], 15[51], 15[70], 15[75], 20[42], 20[51], 20[70], 20[75], 21[42], 21[51], 22[42], 22[51], 23[42], 23[51], 24[42], 24[70], 24[75], 25[42], 25[70], 25[75], 26[42], 26[75], 27[42], 27[70], 27[75], 28[42], 29[42], 30[42], 32[107], 32[108], 32[110], 33[107], 33[108], 33[110], 35[112], 35[113], 36[113], 37[112], 47[137], 163[31], 164[31], 165[31], 166[32], 167[31], 167[32], 172[45], 173[45]
Tani, K. 219[115]
Tanielyan, S. K. 96[28]
Tao, F. 145[42]
Tarasenko, Y. A. 96[26]
Taurino, A. 103[49], 104[49], 105[49]
Taylor, D. K. 327[19], 327[21], 328[21], 329[22]
Teeter, H. M. 335[36], 336[36]
Teichert, J. F. 208[79]
Teichner, S. J. 146[45]
Teixeira, R. R. 362[31]
Teixeira, S. R. 84[46], 146[47]
Teschner, D. 201[50]
Texeira, S. R. 83[44]
Thai, T.-T. 272[12]
Thanassi, J. A. 156[18], 157[18]
Thibault, M. E. 231[18], 232[18], 233[18]
Thiel, K. 339[44]
Thiel, N. O. 208[79]
Thiel, W. 215[109]
Thiel, W. R. 76[29]
Thoma, C. L. 156[18], 157[18]
Thomas, A. 339[44]
Thomas, G. 200[31]
Thomas, J. K. 283[38]
Thomas, J. M. 102[45], 104[45], 104[55], 105[45], 105[55], 108[45], 108[55]
Thomazeau, C. 93[19]
Tian, F. 47[141], 48[141], 50[141], 51[141], 53[141]
Tian, M. 247[12], 248[12]
Tian, X. 247[12], 248[12]
Tidona, B. 301[83]
Tiedje, J. M. 355[2]
Tiekink, E. R. T. 327[19], 327[21], 328[21], 329[22]
Tisseraud, C. 301[87]
Tlili, A. 305[112]
To, D. E. 221[131]
Tobisu, M. 245[8], 246[8]
Tokmic, K. 218[114]
Tokonami, K. 254[44], 254[45]
Tölle, F. 81[38]
Tolman, W. B. 289[6]
Tolstoy, P. 8[32], 10[32], 11[32], 12[32], 13[32]
Tom, J. 238[47]
Tominaga, K.-I. 300[73]
Tomishige, K. 252[24], 252[25], 253[27], 253[28], 253[29], 253[30], 253[31], 253[32], 254[27], 254[28], 254[29], 254[30], 254[31], 254[32], 254[44], 254[45], 255[29], 255[31], 255[32], 256[27], 256[28], 256[29], 257[50], 258[51], 258[52], 259[52], 260[64], 281[33]
Tomooka, K. 321[10], 322[10]
Tompsett, G. A. 259[61]
Tomson, N. C. 209[82]
Tonbul, Y. 127[16], 128[16]
Torralvo, M. J. 96[27]
Toste, F. D. 209[83], 232[23], 236[23], 237[23], 237[46], 238[23], 238[46], 239[23], 239[46], 259[59]
Touchy, A. S. 348[56], 349[56], 350[56], 351[56]
Toyir, J. 301[82], 301[89]
Trapasso, L. E. 67[2]
Trasarti, A. F. 120[87]
Trens, P. 106[57], 107[57], 108[57]
Trifonova, A. 8[19], 10[19], 11[19], 12[19], 13[19]
Tristany, M. 79[35], 80[35]
Triwahyono, S. 302[91]
Trombini, C. 85[48]
Tromp, D. S. 212[96]
Trost, B. M. 197[20], 214[100], 214[101]
Trunfio, G. 301[81]
Tsang, S. C. E. 259[63]
Tsay, M.-T. 302[92], 302[94], 303[92], 303[94]
Tschumi, J. 223[140]
Tse, M. K. 76[18]
Tsichla, A. 111[65], 112[65], 113[65]
Tsuneto, A. 282[36]
Tsuyuki, T. 101[42]
Tucker, J. W. 369[41], 370[41], 372[41]
Tucker, M. H. 259[60]
Tudge, M. T. 47[142], 48[142]
Turnbull, K. 339[25]
Turney, T. W. 128[24]

U

Uda, H. 339[3], 339[5]
Ueba, M. 231[14], 239[14]
Ueda, N. 253[30], 254[30]
Uematsu, T. 102[43]
Uhm, S. 290[31]
Ulan, J. G. 200[31]
Ullrich, A. 52[148], 53[148]
Umani-Ronchi, A. 85[48]
Umpierre, A. P. 83[43], 84[46], 146[47]
Uozumi, Y. 72[13], 73[13], 73[14], 79[33], 82[33], 83[33], 88[52], 360[26], 360[27], 361[26]
Urakawa, A. 301[79], 301[83], 302[79]
Urban, S. 182[70], 182[71], 182[73], 182[74], 183[70], 183[71], 183[73], 183[74], 184[70]
Urbano, F. J. 203[58]

V

Valente, P. 329[22]
Valero, C. 210[89]
Valla, C. 40[129], 41[129]
Vallribera, A. 131[31], 132[31], 133[31]
van Asselt, R. 205[70], 205[71]
van Buijtenen, J. 243[2], 258[2]
van den Berg, M. 15[58], 15[65], 20[58], 20[65], 20[82], 21[58], 21[65], 22[58], 22[65], 27[58], 28[58], 29[58], 30[58]
van der Heide, E. 243[2], 258[2]
van der Waal, J. C. 243[3], 258[3]
van Esch, J. 20[82]
van Koten, G. 204[65]
van Laren, M. W. 205[69], 206[69], 206[73], 211[69]
van Leest, N. P. 213[97]
van Maanen, R. 201[46]
van Putten, R.-J. 243[3], 258[3]
van Walsum, G. P. 257[48]
Vardia, J. 284[41]
Varga, M. 92[7], 106[7]
Varma, R. S. 76[22], 76[23], 76[24], 78[32]
Vasiliou, C. 111[65], 112[65], 113[65]
Vaughan, W. E. 221[129]
Vávra, I. 260[68], 261[68], 263[68]
Vaz, J. M. 146[60]
Vedrine, J. C. 146[44]
Veiros, L. F. 291[47], 295[47], 296[47]
Verel, R. 201[47]
Verendel, J. J. 7[3], 8[3], 8[23], 8[24], 10[23], 10[24], 11[23], 11[24], 12[23], 12[24], 13[23], 13[24]
Vernuccio, S. 200[30], 202[55]
Vico, R. 359[24]
Vidal-Ferran, A. 35[114], 36[114], 170[35], 172[52], 173[52], 174[52]
Vieille-Petit, L. 128[28]
Vigier, K. D. O. 260[73], 261[73]
Vilé, G. 77[31], 201[47], 201[48], 201[49], 201[50]
Villar, L. 203[58]
Viney, M. M. 209[84], 210[84], 211[84]
Vineyard, B. D. 7[5]
Vinokurov, V. 93[18]
Vitale, C. 135[36]
Vizza, F. 210[91]
Vogels, C. M. 339[15]
Vogt, D. 200[44], 201[44], 202[44]
Voisin, L. 362[29]
Vollhardt, K. P. C. 362[33]
vom Stein, T. 299[71], 300[71], 304[110], 305[110], 306[110], 307[110]
Vono, L. L. R. 98[32]
Voropaev, I. V. 115[71]
Vrijburg, W. L. 298[67]
Vu, V. A. 355[8]
Vuong, H. V. 156[17], 157[17]
Vuori, A. 290[27]

W

Wada, Y. 355[12]
Wagh, Y. S. 204[60]
Wai, C. M. 139[38], 140[38], 141[38]

Wakamatsu, T. 208[81], 209[81]
Waldie, F. D. 260[76], 261[76], 262[76]
Waldkirch, J. P. 15[72], 20[72], 21[72], 23[72], 24[72], 25[72]
Wallace, D. J. 157[23], 158[23]
Wallace, T. J. 339[22]
Wan, H. 335[34], 336[34]
Wang, A. 235[36], 239[36], 253[39], 253[41], 253[43], 254[41], 257[49], 259[58], 260[78]
Wang, B. 281[31]
Wang, C. 163[31], 164[31], 165[31], 167[31], 184[76], 185[76], 186[76]
Wang, C.-J. 15[49], 15[68], 20[49], 20[68], 21[49], 21[68], 22[49], 23[68]
Wang, D. 76[21], 279[28], 355[14]
Wang, D.-S. 153[4], 170[38], 171[38], 186[81], 187[81], 187[82], 187[85], 188[82], 188[85], 189[81]
Wang, D.-W. 170[38], 171[38]
Wang, D.-Y. 15[46], 15[47], 20[46], 20[47], 21[46], 21[47], 22[46], 22[47], 24[47], 25[47], 55[149], 56[149]
Wang, F. 302[93], 303[93], 304[93]
Wang, G. 8[12]
Wang, H. 44[136], 251[22], 253[43], 260[80]
Wang, H.-z. 148[68]
Wang, J. 146[51], 146[64], 172[45], 173[45], 247[11], 247[12], 248[11], 248[12], 249[11], 250[11], 251[11]
Wang, J. Q. 308[117], 309[117]
Wang, L. 127[15], 128[15], 281[34]
Wang, L.-X. 15[56], 20[56], 21[56], 22[56], 23[56], 26[56], 27[97], 28[97], 30[97], 43[132], 44[132], 46[132]
Wang, P. 95[24]
Wang, Q. 32[109], 33[109], 99[37], 101[37], 156[18], 157[18], 247[12], 248[12]
Wang, Q. J. 358[22]
Wang, Q.-Y. 233[32]
Wang, R. 95[22]
Wang, S. 93[17], 95[17], 289[7]
Wang, S.-M. 239[50]
Wang, T. 177[61], 177[65], 178[65], 178[66], 179[66], 260[67], 261[67]
Wang, V. 251[22]
Wang, W. 35[112], 37[112], 47[137], 146[48], 146[57], 146[62], 146[63], 229[4], 289[7], 302[90]
Wang, W.-B. 32[108], 33[108], 170[37], 174[37]
Wang, W.-H. 270[2], 289[21], 290[21]
Wang, X. 155[13], 155[14], 156[14], 158[25], 159[25], 160[25], 213[98], 220[98], 235[36], 239[36], 253[39], 253[43], 257[49], 259[58], 260[78]
Wang, X.-B. 170[38], 171[38], 172[47], 173[47]
Wang, X.-L. 232[21]
Wang, X.-Y. 239[50]
Wang, Y. 146[64], 153[10], 229[9], 231[13], 259[56], 259[62], 259[63], 260[67], 260[80], 261[67], 331[27]
Wang, Y.-Q. 172[41], 173[41], 259[54]
Wang, Y.-Z. 232[21]
Wang, Z. 15[61], 20[61], 21[61], 22[61], 23[61], 27[61], 28[61], 29[61], 30[61], 35[116], 36[116], 36[123], 37[123], 38[123], 47[140], 48[140], 48[143], 49[143], 50[140], 51[140], 51[143], 52[140], 53[140], 54[140], 146[49], 146[61], 177[61], 253[39], 257[49], 277[23], 279[23]
Wang, Z.-X. 276[21]
Wang, Z.-Y. 330[24]
Warsink, S. 212[95]
Wassenaar, J. 206[74], 211[74]
Wasserman, H. H. 325[16]
Watanabe, F. 98[34]
Watanabe, H. 253[31], 253[32], 254[31], 254[32], 255[31], 255[32]
Watanabe, T. 300[73]
Watanabe, Y. 41[130], 339[5]
Watari, R. 295[60], 296[60]
Wdzieczak-Bakala, J. 365[37]
Weaver, J. D. 370[42], 371[42]
Webb, G. 221[130]
Webb, J. D. 247[9], 247[10], 248[10]
Wechsberg, M. 333[33], 334[33]
Weckhuysen, B. M. 243[1]
Wedemeyer-Exl, C. 184[75]
Wei, B. 59[159], 61[159], 62[159]
Wei, M. 302[93], 303[93], 304[93]
Wei, S. 41[131], 42[131]
Wei, W. 289[12], 299[12]
Wei, X. 15[51], 20[51], 21[51], 22[51], 23[51]
Weidauer, M. 339[42]
Weigand, J. J. 212[95], 212[96]
Weinberg, K. 231[12], 243[4]
Weinkauff, D. J. 7[5]
Weir, J. R. 203[57], 204[57]
Weisel, M. 47[142], 48[142]
Weissbach, H. 339[8], 339[10]
Weissermel, K. 92[6], 102[44], 127[6]
Weitekamp, D. P. 214[106]
Welch, M. 15[78], 20[78], 24[78], 26[78]
Wells, P. B. 92[9], 197[19], 221[130]
Welther, A. 79[34]
Welton, T. 128[25], 128[26]
Wen, C. 280[29], 281[31]
Wen, J. 47[138], 48[138], 50[138], 51[138]
Wen, M. 276[21]
Wen, S. 44[136]
Wenderski, T. A. 326[18]
Wendt, B. 80[37]
Werkmeister, S. 80[37]
Werner, H. 210[88], 210[89]
Werner, L. 330[23]
Wesselbaum, S. 299[71], 300[71]
West, F. G. 331[27]
West, R. M. 259[60]
Westcott, S. A. 339[15]
Wheeler, M. C. 229[6], 229[10], 230[6], 230[10]
White, C. 128[23]
Wichert, J. 201[49]
Wieland, J. 81[38]
Wiener, H. 269[1], 275[18]
Wienhöfer, G. 205[68]
Wilde, M. 67[7]
Wiles, J. A. 156[18], 157[18]
Wilkinson, G. 128[20], 128[21]
Willis, R. G. 197[17]
Willner, I. 283[39]
Willnow, P. 285[49]
Wilson, G. R. 92[9]
Wilson, R. D. 157[23], 158[23]
Winter, B. 221[135], 222[135]
Winterbottom, J. M. 221[130]
Wipf, P. 358[22]
Wise, C. 27[99], 28[99], 31[99], 32[99]
Witte, P. T. 200[42], 200[43], 200[44], 201[42], 201[44], 201[46], 201[47], 202[42], 202[44]
Woelk, K. 206[72]
Wolf, D. 200[38]
Wolf, H. R. 224[145]
Wolf, L. M. 215[109]
Wong, W.-T. 330[24]
Woodmansee, D. H. 8[31], 10[31], 11[31], 13[31], 14[31]
Woollins, J. D. 339[28]
Wrabetz, S. 201[50]
Wright, G. J. 351[65]
Wrighton, M. S. 275[16], 275[17], 284[17], 284[42]
Wu, B. 172[50], 173[50], 174[50]
Wu, D. 232[26]
Wu, H.-P. 27[85], 28[85], 29[85], 31[85], 32[103], 33[103]
Wu, J. 27[96], 28[96], 29[96], 30[96], 32[104], 33[104], 163[31], 164[31], 165[31], 167[31], 277[23], 279[23]
Wu, L. 273[14], 274[14]
Wu, R. 100[39], 101[39], 260[76], 261[76], 262[76]
Wu, S. 32[110], 33[110], 35[112], 37[112]
Wu, T. 146[57], 146[62], 302[100], 303[100]
Wu, T.-S. 259[63]
Wu, W. 83[42], 83[45], 91[2]
Wu, W.-P. 259[53], 260[74], 261[74], 262[74]
Wu, X. 184[76], 185[76], 186[76], 278[24]
Wu, Y. 146[49]
Wu, Z. 15[77], 20[77], 24[77], 26[77], 200[41]
Wuchter, N. 108[60]
Wullschleger, C. W. 355[17]
Wüstenberg, B. 195[2], 199[2]
Wysocki, J. 182[72], 183[72]

X

Xanthopoulos, N. 202[52]
Xi, J. 259[62]
Xia, C. 27[94], 28[94], 29[94], 30[94], 280[30], 281[30]
Xia, Q. 259[56], 259[62], 259[63]
Xia, Q.-N. 259[54]
Xia, W. 104[51]
Xia, X. 260[81]
Xia, Y. 200[40]

Xia, Z. 260[79], 261[79], 262[79], 263[79]
Xiang, J. 178[66], 179[66]
Xiao, B. 253[43]
Xiao, C.-x. 148[68]
Xiao, D. 15[40], 20[40], 21[40], 22[40], 27[40], 28[40], 29[40], 30[40], 35[111], 37[111], 38[111]
Xiao, F. 289[12], 299[12]
Xiao, G. 260[66], 260[67], 261[66], 261[67]
Xiao, J. 163[31], 164[31], 165[31], 166[32], 167[31], 167[32], 184[76], 185[76], 186[76], 220[117]
Xiao, R. 260[66], 261[66]
Xie, H. 302[100], 303[100]
Xie, J.-H. 7[9], 15[56], 20[56], 21[56], 22[56], 23[56], 26[56], 27[97], 28[97], 30[97], 43[134], 44[134], 45[134]
Xie, J.-R. 301[86]
Xie, S. 146[48], 146[50], 146[54], 146[64]
Xie, S.-H. 127[4]
Xie, Y. 47[141], 48[141], 50[141], 51[141], 53[141], 69[9]
Xing, L. 27[97], 28[97], 30[97], 158[25], 159[25], 160[25]
Xing, R. 257[48]
Xu, G. 15[70], 20[70], 24[70], 25[70], 27[70]
Xu, H. 250[14]
Xu, J. 158[25], 159[25], 160[25], 250[14]
Xu, L. 15[45], 15[57], 20[45], 20[57], 21[45], 21[57], 22[45], 22[57], 27[57], 28[57], 29[57], 30[57], 166[32], 167[32], 172[45], 173[45], 177[61], 177[62]
Xu, Q. 290[29]
Xu, W. 260[79], 260[80], 261[79], 262[79], 263[79]
Xu, X. 153[10]
Xu, Y.-J. 260[74], 261[74], 262[74]
Xu, Z. 260[75], 260[79], 261[75], 261[79], 262[79], 263[75], 263[79], 296[62], 297[62], 298[62]
Xue, L. 270[6]
Xue, P. 59[159], 61[159], 62[159]
Xue, Z. 15[74], 20[74], 24[74], 25[74]

Y

Yabe, Y. 359[25], 360[25], 361[25]
Yaguchi, K. 157[19]
Yalpani, M. 143[40], 144[40]
Yamada, N. 153[8]
Yamada, T. 359[25], 360[25], 361[25]
Yamada, Y. 360[27]
Yamada, Y. M. A. 79[33], 82[33], 83[33], 88[52]
Yamagata, T. 174[55], 175[55], 219[115]
Yamaguchi, M. 351[71]
Yamaguchi, R. 162[30], 163[30], 367[39]
Yamaguchi, Y. 36[123], 37[123], 38[123]
Yamaji, K. 176[59], 177[59]
Yamakawa, T. 260[77], 261[77], 262[77]
Yamamoto, Y. 204[59]
Yamano, T. 185[80], 186[80]
Yamashita, M. 185[80], 186[80], 291[48], 291[49], 294[48], 295[48]
Yan, B. 289[17]
Yan, J.-M. 290[29]
Yan, M. 204[59], 247[11], 248[11], 249[11], 250[11], 251[11]
Yan, N. 148[68]
Yan, P. 260[75], 260[79], 261[75], 261[79], 262[79], 263[75], 263[79]
Yan, S. 146[51]
Yan, T. 305[114], 306[114], 307[114], 310[119]
Yan, X. 146[61]
Yan, Y. 15[48], 20[48], 21[48], 22[48], 24[48], 25[48], 27[91], 28[91], 29[91], 30[91], 32[48], 33[48]
Yan, Z.-F. 301[78]
Yanagida, S. 355[12]
Yanagisawa, A. 15[52], 20[52], 21[52], 23[52]
Yang, D. 330[24]
Yang, G. 272[11]
Yang, J. 259[58], 302[97], 303[97]
Yang, K. 95[23]
Yang, L. 73[15], 74[15], 365[36], 366[36]
Yang, M. 95[22]
Yang, P. 259[62]
Yang, P.-Y. 15[73], 20[73], 24[73], 25[73], 170[37], 174[37]
Yang, X. 276[20]
Yang, Y. 229[4]
Yang, Z. 250[14], 296[64], 297[64], 298[64]
Yao, D. 47[141], 48[141], 50[141], 51[141], 53[141]
Yao, K. 247[11], 248[11], 249[11], 250[11], 251[11]
Yao, S. 233[32]
Yao, X. 15[57], 20[57], 21[57], 22[57], 27[57], 28[57], 29[57], 30[57]
Yarulin, A. 200[40]
Yasuda, H. 289[3]
Yasuda, M. 231[14], 239[14]
Yasuma, T. 185[80], 186[80]
Yasunori, Y. 41[130]
Yasutake, M. 15[67], 20[67], 21[67], 23[67], 32[102], 33[102]
Ye, Z.-S. 172[42], 172[43], 172[44], 172[48], 172[50], 173[42], 173[43], 173[44], 173[48], 173[50], 174[50], 187[85], 188[85]
Yee, N. K. 15[51], 20[51], 21[51], 22[51], 23[51]
Yi, G. 232[26]
Yin, A. 280[29]
Yin, H. 355[12]
Yin, S. 213[98], 220[98], 220[117]
Yin, S.-F. 289[23]
Yin, X. 59[159], 61[159], 62[159]
Ying, J. Y. 69[10], 70[10]
Yip, C. W. 15[45], 20[45], 21[45], 22[45]
Yokohama, S. 351[64]
Yokoyama, T. 259[55]
Yoneyama, K. 339[9]
Yoon, S. 289[19], 296[19], 296[63], 297[63], 298[63]
Yoshida, K. 15[52], 20[52], 21[52], 23[52]
You, J. 32[101], 33[101]
Younas, M. 289[15]
Young, J. F. 128[20], 128[21]
Youssef, N. S. 285[45]
Yu, B. 250[14], 296[64], 297[64], 298[64]
Yu, C. 127[12], 128[12]
Yu, C.-B. 172[51], 172[54], 173[51], 173[54], 174[51], 174[54], 186[81], 187[81], 187[82], 187[83], 187[87], 188[82], 188[83], 188[87], 188[88], 189[81], 189[88]
Yu, L. 154[12], 155[12], 156[12]
Yu, S.-B. 15[46], 15[47], 20[46], 20[47], 21[46], 21[47], 22[46], 22[47], 24[47], 25[47], 55[149], 56[149]
Yu, Y. 95[24], 296[61], 297[61], 298[61]
Yu, Y.-B. 43[132], 44[132], 46[132]
Yu, Z.-X. 178[66], 179[66]
Yuan, H. 32[109], 33[109]
Yus, M. 86[50], 86[51], 87[51], 104[53], 105[53], 135[36], 355[10]

Z

Zaccheria, F. 101[41], 102[41], 117[79], 117[80], 117[81], 117[82], 117[83], 118[82], 118[84], 119[81], 119[83], 120[80], 120[86], 120[87], 121[80], 121[81], 121[82]
Zafarana, G. 301[81]
Zaheer, M. 243[5], 249[5], 251[18]
Zahmakıran, M. 127[16], 128[16]
Zaidman, B. 269[1]
Zakarian, A. 238[47]
Zakzeski, J. 243[1]
Zale, S. E. 285[50]
Zalmanov, N. 275[18]
Zalucky, J. 204[67]
Zanella, M. 260[71], 261[71]
Zanobini, F. 210[91]
Zanotti-Gerosa, A. 27[89], 27[92], 28[89], 28[92], 29[89], 29[92]
Zanutelo, C. 146[58]
Zareyee, D. 339[23]
Zbinden, H. 200[32]
Zehbe, R. 339[44]
Zelinski, R. 232[20]
Zeng, H. 233[32]
Zeng, J.-B. 232[21]
Zeng, Q.-H. 15[46], 15[59], 20[46], 20[59], 21[46], 21[59], 22[46], 22[59], 24[59], 25[59], 27[59], 28[59], 29[59], 30[59]
Zeng, W. 172[47], 173[47]
Zhang, B. 260[82]
Zhang, C. 339[16]
Zhang, C.-M. 43[133], 43[134], 44[133], 44[134], 45[133], 45[134], 46[133]
Zhang, G. 127[3], 319[8], 320[8]
Zhang, H. 229[9], 232[24], 236[24], 237[24], 238[24], 239[24], 250[14], 296[64], 297[64], 298[64], 365[36], 366[36]
Zhang, H.-B. 301[85]

Zhang, J. 69[9], 177[63], 184[63], 275[19], 276[19], 299[72], 300[72], 339[16]
Zhang, L. 277[23], 279[23]
Zhang, M. 73[15], 74[15], 250[13]
Zhang, N. 127[12], 128[12]
Zhang, P. 146[57], 146[62], 153[10], 302[102], 303[102]
Zhang, Q. 146[61]
Zhang, S. 76[18], 99[37], 101[37], 289[24]
Zhang, T. 235[36], 239[36], 253[39], 253[41], 253[43], 254[41], 257[49], 259[58], 260[78]
Zhang, W. 15[69], 20[69], 24[69], 47[141], 48[141], 50[141], 51[141], 53[141], 73[15], 74[15], 187[84], 188[84]
Zhang, X. 7[7], 7[8], 7[9], 8[15], 10[15], 11[15], 13[15], 15[7], 15[39], 15[40], 15[42], 15[48], 15[53], 15[54], 15[68], 15[69], 15[71], 15[72], 15[74], 15[75], 15[79], 20[39], 20[40], 20[42], 20[48], 20[53], 20[54], 20[68], 20[69], 20[71], 20[72], 20[74], 20[75], 20[79], 21[39], 21[40], 21[42], 21[48], 21[53], 21[54], 21[68], 21[72], 22[40], 22[42], 22[48], 22[54], 23[39], 23[42], 23[54], 23[68], 23[72], 24[39], 24[42], 24[48], 24[53], 24[54], 24[69], 24[71], 24[72], 24[74], 24[75], 24[79], 25[39], 25[42], 25[48], 25[72], 25[74], 25[75], 26[42], 26[54], 26[75], 26[79], 27[40], 27[42], 27[53], 27[75], 27[88], 27[91], 27[93], 27[98], 28[40], 28[42], 28[53], 28[88], 28[91], 28[93], 28[98], 29[40], 29[42], 29[53], 29[88], 29[91], 29[93], 30[40], 30[42], 30[88], 30[91], 30[93], 30[98], 31[98], 32[48], 32[53], 32[107], 32[108], 32[109], 32[110], 33[48], 33[53], 33[107], 33[108], 33[109], 33[110], 35[53], 35[54], 35[111], 35[112], 35[113], 36[53], 36[54], 36[113], 36[123], 37[111], 37[112], 37[123], 38[111], 38[123], 44[136], 47[137], 47[138], 48[138], 48[145], 50[138], 50[145], 51[138], 59[158], 59[159], 60[158], 61[159], 62[158], 62[159], 172[49], 173[49], 174[49], 186[81], 187[81], 189[81], 289[12], 299[12], 365[36], 366[36]
Zhang, X.-B. 290[29]
Zhang, Y. 232[24], 232[25], 232[26], 232[29], 236[24], 236[45], 237[24], 237[45], 238[24], 238[45], 239[24], 239[45], 289[9], 296[61], 297[61], 298[61], 307[115], 308[118], 309[115], 309[118], 310[118], 365[36], 366[36]
Zhang, Y. J. 15[41], 20[41], 21[41], 22[41], 23[41], 24[41], 25[41]
Zhang, Y.-M. 358[21], 359[21]
Zhang, Z. 15[40], 15[52], 20[40], 20[52], 21[40], 21[52], 22[40], 23[52], 27[40], 28[40], 29[40], 30[40], 35[111], 37[111], 38[111], 44[136], 127[13], 128[13], 146[62]
Zhang, Z. C. 247[11], 247[12], 248[11], 248[12], 249[11], 250[11], 251[11], 260[75], 260[79], 261[75], 261[79], 262[79], 263[75], 263[79]
Zhao, B. 15[61], 20[61], 21[61], 22[61], 23[61], 27[61], 28[61], 29[61], 30[61]
Zhao, C. 148[68], 355[14]
Zhao, D. 182[74], 183[74], 339[16]
Zhao, F. 260[81]
Zhao, G. 83[42], 83[45], 279[28]
Zhao, H. 232[24], 236[24], 237[24], 238[24], 239[24]
Zhao, J. 302[97], 303[97]
Zhao, N. 289[12], 299[12]
Zhao, S. 146[49]
Zhao, Y. 213[98], 220[98], 296[64], 297[64], 298[64], 302[93], 303[93], 304[93]
Zheng, H. 231[16], 232[16], 260[82]
Zheng, J. 127[12], 128[12]
Zheng, M. 235[36], 239[36], 253[43], 260[78]
Zheng, T. 363[34], 364[34], 368[34]
Zheng, X. 296[61], 297[61], 298[61], 302[100], 303[100]
Zheng, Z. 15[44], 15[46], 15[47], 15[59], 15[62], 15[63], 20[44], 20[46], 20[47], 20[59], 20[62], 20[63], 21[44], 21[46], 21[47], 21[59], 21[62], 21[63], 22[44], 22[46], 22[47], 22[59], 22[62], 22[63], 23[62], 23[63], 24[47], 24[59], 24[62], 24[63], 25[47], 25[59], 25[63], 27[44], 27[59], 27[63], 28[44], 28[59], 28[63], 29[59], 29[63], 30[44], 30[59], 30[63], 32[105], 33[105], 55[149], 56[149]
Zho, Y. 250[14]
Zhong, X. H. 308[117], 309[117]
Zhou, G. 145[42], 146[53], 146[54], 146[56], 157[23], 158[23], 302[100], 303[100]
Zhou, H. 15[56], 20[56], 21[56], 22[56], 23[56], 26[56], 127[12], 128[12], 177[61]
Zhou, J. 231[13]
Zhou, M. 15[74], 20[74], 24[74], 25[74], 260[66], 260[67], 261[66], 261[67]
Zhou, Q.-L. 7[9], 15[56], 20[56], 21[56], 22[56], 23[56], 26[56], 27[97], 28[97], 30[97], 43[132], 43[133], 43[134], 43[135], 44[132], 44[133], 44[134], 45[133], 45[134], 45[135], 46[132], 46[133], 46[135], 168[34], 169[34]
Zhou, S. 272[11]
Zhou, T. 59[160], 59[161], 61[160], 62[160]
Zhou, W. 319[8], 320[8]
Zhou, X. 229[6], 230[6], 250[17]
Zhou, Y. 69[9], 213[98], 220[98], 220[117]
Zhou, Y.-G. 15[73], 20[73], 24[73], 25[73], 27[93], 28[93], 29[93], 30[93], 32[108], 33[108], 153[1], 153[4], 153[7], 170[36], 170[37], 170[38], 170[39], 171[38], 172[41], 172[42], 172[43], 172[44], 172[47], 172[48], 172[50], 172[51], 172[54], 173[41], 173[42], 173[43], 173[44], 173[47], 173[48], 173[50], 173[51], 173[54], 174[37], 174[50], 174[51], 174[54], 186[81], 187[81], 187[82], 187[83], 187[85], 187[86], 187[87], 188[82], 188[83], 188[85], 188[86], 188[87], 188[88], 189[81], 189[88]
Zhou, Z. 32[104], 33[104], 172[45], 173[45]
Zhu, A. 69[9]
Zhu, F. 272[11]
Zhu, H. 260[66], 261[66]
Zhu, L. 127[12], 128[12]
Zhu, L.-J. 146[59]
Zhu, M. 146[61]
Zhu, Q.-c. 318[5]
Zhu, Q.-C. 316[4], 317[4], 318[4]
Zhu, R. 158[25], 159[25], 160[25], 259[53], 260[74], 261[74], 262[74]
Zhu, S. 231[16], 232[16]
Zhu, S.-F. 43[132], 43[133], 43[134], 43[135], 44[132], 44[133], 44[134], 45[133], 45[134], 45[135], 46[132], 46[133], 46[135]
Zhu, Y. 39[128], 40[128], 231[16], 232[16], 260[82], 285[48]
Zhu, Z. 339[32]
Zhuang, J.-H. 146[59]
Zhuang, X. 259[63]
Zhu-Ge, L. 272[11]
Zhuo, L.-G. 178[66], 179[66]
Ziebart, C. 270[4], 270[5], 291[36], 295[36], 296[36]
Zimmermann, N. 8[13], 8[20], 10[20], 12[20], 13[20], 13[38]
Zolotukhina, A. 93[18]
Zong, B. 127[14], 128[14], 145[42], 146[53], 146[54], 146[56]
Zorzan, D. 359[24]
Zupančič, B. 15[50], 20[50], 21[50], 35[50], 36[50], 37[50]
Zvarec, O. 327[19]
Zwietering, P. 197[18]

Abbreviations

Chemical

Name Used in Text	Abbreviation Used in Tables and on Arrow in Schemes	Abbreviation Used in Experimental Procedures
(*R*)-1-amino-2-(methoxymethyl)pyrrolidine	RAMP	RAMP
(*S*)-1-amino-2-(methoxymethyl)pyrrolidine	SAMP	SAMP
ammonium cerium(IV) nitrate	CAN	CAN
2,2′-azobisisobutyronitrile	AIBN	AIBN
barbituric acid	BBA	BBA
benzyltriethylammonium bromide	TEBAB	TEBAB
benzyltriethylammonium chloride	TEBAC	TEBAC
N,*O*-bis(trimethylsilyl)acetamide	BSA	BSA
9-borabicyclo[3.3.1]nonane	9-BBNH	9-BBNH
borane–methyl sulfide complex	BMS	BMS
N-bromosuccinimide	NBS	NBS
tert-butyldimethylsilyl chloride	TBDMSCl	TBDMSCl
tert-butyl peroxybenzoate	TBPB	*tert*-butyl peroxybenzoate
10-camphorsulfonic acid	CSA	CSA
chlorosulfonyl isocyanate	CSI	chlorosulfonyl isocyanate
3-chloroperoxybenzoic acid	MCPBA	MCPBA
N-chlorosuccinimide	NCS	NCS
chlorotrimethylsilane	TMSCl	TMSCl
1,4-diazabicyclo[2.2.2]octane	DABCO	DABCO
1,5-diazabicyclo[4.3.0]non-5-ene	DBN	DBN
1,8-diazabicyclo[5.4.0]undec-7-ene	DBU	DBU
dibenzoyl peroxide	DBPO	dibenzoyl peroxide
dibenzylideneacetone	dba	dba
di-*tert*-butyl azodicarboxylate	DBAD	di-*tert*-butyl azo-dicarboxylate
di-*tert*-butyl peroxide	DTBP	DTBP
2,3-dichloro-5,6-dicyanobenzo-1,4-quinone	DDQ	DDQ
dichloromethyl methyl ether	DCME	DCME
dicyclohexylcarbodiimide	DCC	DCC
N,*N*-diethylaminosulfur trifluoride	DAST	DAST
diethyl azodicarboxylate	DEAD	DEAD
diethyl tartrate	DET	DET
2,2′-dihydroxy-1,1′-binaphthyllithium aluminum hydride	BINAL-H	BINAL-H
diisobutylaluminum hydride	DIBAL-H	DIBAL-H
diisopropyl tartrate	DIPT	DIPT

Chemical (cont.)

Name Used in Text	Abbreviation Used in Tables and on Arrow in Schemes	Abbreviation Used in Experimental Procedures
1,2-dimethoxyethane	DME	DME
dimethylacetamide	DMA	DMA
dimethyl acetylenedicarboxylate	DMAD	DMAD
2-(dimethylamino)ethanol	$Me_2N(CH_2)_2OH$	2-(dimethylamino)ethanol
4-(dimethylamino)pyridine	DMAP	DMAP
dimethylformamide	DMF	DMF
dimethyl sulfide	DMS	DMS
dimethyl sulfoxide	DMSO	DMSO
1,3-dimethyl-3,4,5,6-tetrahydro-pyrimidin-2(1*H*)-one	DMPU	DMPU
ethyl diazoacetate	EDA	EDA
ethylenediaminetetraacetic acid	edta	edta
hexamethylphosphoric triamide	HMPA	HMPA
hexamethylphosphorous triamide	HMPT	HMPT
iodomethane	MeI	MeI
N-iodosuccinimide	NIS	NIS
lithium diisopropylamide	LDA	LDA
lithium hexamethyldisilazanide	LiHMDS	LiHMDS
lithium isopropylcyclohexylamide	LICA	LICA
lithium 2,2,6,6-tetramethylpiperidide	LTMP	LTMP
lutidine	lut	lut
methylaluminum bis(2,6-di-*tert*-butyl-4-methyl-phenoxide)	MAD	MAD
methyl ethyl ketone	MEK	methyl ethyl ketone
methylmaleimide	NMM	NMM
4-methylmorpholine *N*-oxide	NMO	NMO
1-methylpyrrolidin-2-one	NMP	NMP
methyl vinyl ketone	MVK	methyl vinyl ketone
petroleum ether	PE[a]	petroleum ether
N-phenylmaleimide	NPM	NPM
polyphosphoric acid	PPA	PPA
polyphosphate ester	PPE	polyphosphate ester
potassium hexamethyldisilazanide	KHMDS	KHMDS
pyridine	pyridine[b]	pyridine
pyridinium chlorochromate	PCC	PCC
pyridinium dichromate	PDC	PDC
pyridinium 4-toluenesulfonate	PPTS	PPTS
sodium bis(2-methoxyethoxy)aluminum hydride	Red-Al	Red-Al
tetrabutylammonium bromide	TBAB	TBAB

[a] Used to save space; abbreviation must be defined in a footnote.
[b] py used on arrow in schemes.

Chemical (cont.)

Name Used in Text	Abbreviation Used in Tables and on Arrow in Schemes	Abbreviation Used in Experimental Procedures
tetrabutylammonium chloride	TBACl	TBACl
tetrabutylammonium fluoride	TBAF	TBAF
tetrabutylammonium iodide	TBAI	TBAI
tetracyanoethene	TCNE	tetracyanoethene
tetrahydrofuran	THF	THF
tetrahydropyran	THP	THP
2,2,6,6-tetramethylpiperidine	TMP	TMP
trimethylamine *N*-oxide	TMANO	trimethylamine *N*-oxide
N,*N*,*N*′,*N*′-tetramethylethylenediamine	TMEDA	TMEDA
tosylmethyl isocyanide	TosMIC	TosMIC
trifluoroacetic acid	TFA	TFA
trifluoroacetic anhydride	TFAA	TFAA
trimethylsilyl cyanide	TMSCN	TMSCN

Ligands

acetylacetonato	acac
2,2′-bipyridyl	bipy
1,2-bis(dimethylphosphino)ethane	DMPE
2,3-bis(diphenylphosphino)bicyclo[2.2.1]hept-5-ene	NORPHOS
2,2′-bis(diphenylphosphino)-1,1′-binaphthyl	BINAP
1,2-bis(diphenylphosphino)ethane	dppe (not diphos)
1,1′-bis(diphenylphosphino)ferrocene	dppf
bis(diphenylphosphino)methane	dppm
1,3-bis(diphenylphosphino)propane	dppp
1,4-bis(diphenylphosphino)butane	dppb
2,3-bis(diphenylphosphino)butane	Chiraphos
bis(salicylidene)ethylenediamine	salen
cyclooctadiene	cod
cyclooctatetraene	cot
cyclooctatriene	cte
η^5-cyclopentadienyl	Cp
dibenzylideneacetone	dba
6,6-dimethylcyclohexadienyl	dmch
2,4-dimethylpentadienyl	dmpd
ethylenediaminetetraacetic acid	edta
isopinocampheyl	Ipc
2,3-*O*-isopropylidene-2,3-dihydroxy-1,4-bis(diphenylphosphino)butane	Diop
norbornadiene (bicyclo[2.2.1]hepta-2,5-diene)	nbd
η^5-pentamethylcyclopentadienyl	Cp*

Radicals

acetyl	Ac
aryl	Ar
benzotriazol-1-yl	Bt
benzoyl	Bz
benzyl	Bn
benzyloxycarbonyl	Cbz
benzyloxymethyl	BOM
9-borabicyclo[3.3.1]nonyl	9-BBN
tert-butoxycarbonyl	Boc
butyl	Bu
sec-butyl	*s*-Bu
tert-butyl	*t*-Bu
tert-butyldimethylsilyl	TBDMS
tert-butyldiphenylsilyl	TBDPS
cyclohexyl	Cy
3,4-dimethoxybenzyl	DMB
ethyl	Et
ferrocenyl	Fc
9-fluorenylmethoxycarbonyl	Fmoc
isobutyl	iBu
mesityl	Mes
mesyl	Ms
4-methoxybenzyl	PMB
(2-methoxyethoxy)methyl	MEM
methoxymethyl	MOM
methyl	Me
4-nitrobenzyl	PNB
phenyl	Ph
phthaloyl	Phth
phthalimido	NPhth
propyl	Pr
isopropyl	iPr
tetrahydropyranyl	THP
tolyl	Tol
tosyl	Ts
triethylsilyl	TES
triflyl, trifluoromethanesulfonyl	Tf
triisopropylsilyl	TIPS
trimethylsilyl	TMS
2-(trimethylsilyl)ethoxymethyl	SEM
trityl [triphenylmethyl]	Tr

General

absolute	abs
anhydrous	anhyd
aqueous	aq
boiling point	bp
catalyst	no abbreviation
catalytic	cat.
chemical shift	δ
circular dichroism	CD
column chromatography	no abbreviation
concentrated	concd
configuration (in tables)	Config
coupling constant	*J*
day	d
density	*d*
decomposed	dec
degrees Celsius	°C
diastereomeric ratio	dr
dilute	dil
electron-donating group	EDG
electron-withdrawing group	EWG
electrophile	E^+
enantiomeric excess	ee
enantiomeric ratio	er
equation	eq
equivalent(s)	equiv
flash-vacuum pyrolysis	FVP
gas chromatography	GC
gas chromatography-mass spectrometry	GC/MS
gas–liquid chromatography	GLC
gram	g
highest occupied molecular orbital	HOMO
high-performance liquid chromatography	HPLC
hour(s)	h
infrared	IR
in situ	in situ
in vacuo	in vacuo
lethal dosage, e.g. to 50% of animals tested	LD_{50}
liquid	liq
liter	L
lowest unoccupied molecular orbital	LUMO
mass spectrometry	MS
medium-pressure liquid chromatography	MPLC
melting point	mp
milliliter	mL
millimole(s)	mmol
millimoles per liter	mM
minute(s)	min
mole(s)	mol
nuclear magnetic resonance	NMR
nucleophile	Nu^-
optical purity	op
phase-transfer catalysis	PTC
proton NMR	^1H NMR

General (cont.)

quantitative	quant
reference (in tables)	Ref
retention factor (for TLC)	R_f
retention time (chromatography)	t_R
room temperature	rt
saturated	sat.
solution	soln
temperature (in tables)	Temp (°C)
thin layer chromatography	TLC
ultraviolet	UV
volume (literature)	Vol.
via	via
vide infra	*vide infra*
vide supra	*vide supra*
yield (in tables)	Yield (%)

List of All Volumes

Science of Synthesis, Houben–Weyl Methods of Molecular Transformations

Category 1: Organometallics

- **1** Compounds with Transition Metal—Carbon π-Bonds and Compounds of Groups 10–8 (Ni, Pd, Pt, Co, Rh, Ir, Fe, Ru, Os)
- **2** Compounds of Groups 7–3 (Mn···, Cr···, V···, Ti···, Sc···, La···, Ac···)
- **3** Compounds of Groups 12 and 11 (Zn, Cd, Hg, Cu, Ag, Au)
- **4** Compounds of Group 15 (As, Sb, Bi) and Silicon Compounds
- **5** Compounds of Group 14 (Ge, Sn, Pb)
- **6** Boron Compounds
- **7** Compounds of Groups 13 and 2 (Al, Ga, In, Tl, Be ··· Ba)
- **8a** Compounds of Group 1 (Li ··· Cs)
- **8b** Compounds of Group 1 (Li ··· Cs)

Category 2: Hetarenes and Related Ring Systems

- **9** Fully Unsaturated Small-Ring Heterocycles and Monocyclic Five-Membered Hetarenes with One Heteroatom
- **10** Fused Five-Membered Hetarenes with One Heteroatom
- **11** Five-Membered Hetarenes with One Chalcogen and One Additional Heteroatom
- **12** Five-Membered Hetarenes with Two Nitrogen or Phosphorus Atoms
- **13** Five-Membered Hetarenes with Three or More Heteroatoms
- **14** Six-Membered Hetarenes with One Chalcogen
- **15** Six-Membered Hetarenes with One Nitrogen or Phosphorus Atom
- **16** Six-Membered Hetarenes with Two Identical Heteroatoms
- **17** Six-Membered Hetarenes with Two Unlike or More than Two Heteroatoms and Fully Unsaturated Larger-Ring Heterocycles

Category 3: Compounds with Four and Three Carbon—Heteroatom Bonds

- **18** Four Carbon—Heteroatom Bonds: X—C≡X, X=C=X, X_2C=X, CX_4
- **19** Three Carbon—Heteroatom Bonds: Nitriles, Isocyanides, and Derivatives
- **20a** Three Carbon—Heteroatom Bonds: Acid Halides; Carboxylic Acids and Acid Salts
- **20b** Three Carbon—Heteroatom Bonds: Esters and Lactones; Peroxy Acids and R(CO)OX Compounds; R(CO)X, X = S, Se, Te
- **21** Three Carbon—Heteroatom Bonds: Amides and Derivatives; Peptides; Lactams
- **22** Three Carbon—Heteroatom Bonds: Thio-, Seleno-, and Tellurocarboxylic Acids and Derivatives; Imidic Acids and Derivatives; Ortho Acid Derivatives
- **23** Three Carbon—Heteroatom Bonds: Ketenes and Derivatives
- **24** Three Carbon—Heteroatom Bonds: Ketene Acetals and Yne—X Compounds

Category 4: Compounds with Two Carbon—Heteroatom Bonds

25 Aldehydes

26 Ketones

27 Heteroatom Analogues of Aldehydes and Ketones

28 Quinones and Heteroatom Analogues

29 Acetals: Hal/X and O/O, S, Se, Te

30 Acetals: O/N, S/S, S/N, and N/N and Higher Heteroatom Analogues

31a Arene—X (X = Hal, O, S, Se, Te)

31b Arene—X (X = N, P)

32 X—Ene—X (X = F, Cl, Br, I, O, S, Se, Te, N, P), Ene—Hal, and Ene—O Compounds

33 Ene—X Compounds (X = S, Se, Te, N, P)

Category 5: Compounds with One Saturated Carbon—Heteroatom Bond

34 Fluorine

35 Chlorine, Bromine, and Iodine

36 Alcohols

37 Ethers

38 Peroxides

39 Sulfur, Selenium, and Tellurium

40a Amines and Ammonium Salts

40b Amine *N*-Oxides, Haloamines, Hydroxylamines and Sulfur Analogues, and Hydrazines

41 Nitro, Nitroso, Azo, Azoxy, and Diazonium Compounds, Azides, Triazenes, and Tetrazenes

42 Organophosphorus Compounds (incl. RO—P and RN—P)

Category 6: Compounds with All-Carbon Functions

43 Polyynes, Arynes, Enynes, and Alkynes

44 Cumulenes and Allenes

45a Monocyclic Arenes, Quasiarenes, and Annulenes

45b Aromatic Ring Assemblies, Polycyclic Aromatic Hydrocarbons, and Conjugated Polyenes

46 1,3-Dienes

47a Alkenes

47b Alkenes

48 Alkanes